Converting Units of Measure

To Convert from	To	Multiply by
Acre-feet	Gallons	3.26×10^5
Acre-foot	Cubic meters	1,233.5
Acres	Square feet	43,560
Barrels oil (bbl)	Cubic feet (ft^3)	5.61
Barrels oil	Gallons (gal)	42
Bars	Pounds/square inch (lb/in^2)	14.504
Centigrade (°C)	Fahrenheit (°F)	1.8, then add 32
Centimeters (cm)	Inches (in)	0.394
Centimeters per second (cm/s)	Feet per day (ft/day)	2,835
Centimeters per second	Gallons per day per square foot (gal/day/ft^2)	21,200
Centimeters per second	Meters per day (m/day)	864
Cubic centimeters (cm^3)	Cubic inches (in^3)	0.061
Cubic inches (in^3)	Cubic centimeters (cm^3)	16.387
Cubic feet (ft^3)	Barrels of oil (bbl)	0.18
Cubic feet	Cubic meters (m^3)	0.028
Cubic feet per second (ft^3/s)	Cubic meters per second (m^3/s)	0.003
Cubic meters (m^3)	Acre-feet	8.11×10^{-4}
Cubic meters	Cubic feet (ft^3)	35.249
Cubic meters per second (m^3/s)	Cubic feet per second (ft^3/s)	353.107
Cubic miles (mi^3)	Cubic kilometers (km^3)	4.167
Cubic kilometers (km^3)	Cubic miles (mi^3)	0.240
Fahrenheit (°F)	Centigrade (°C)	subtract 32, then divide by 1.8
Feet (ft)	Meters (m)	0.305
Feet per day (ft/day)	Centimeters per second (cm/s)	3.53×10^{-4}
Feet per mile (ft/mi)	Meters per kilometer (m/km)	.188
Gallons	Acre-feet	3.07×10^{-6}
Gallons per day per square foot (gal/day/ft^2)	Centimeters per second (cm/s)	4.72×10^{-5}
Grams (g)	Ounces (oz)	0.035
Hectares (ha)	Square feet (ft^2)	1.076×10^5
Inches (in)	Centimeters (cm)	2.540
Kilograms (kg)	Pounds (lb)	2.205
Kilometers (km)	Miles (mi)	.621
Liters (l)	Quarts (qt)	1.057
Meters (m)	Feet (ft)	3.281
Meters	Yards (yd)	1.094
Meters	Kilometers (km)	10^{-3}
Meters	Centimeters (cm)	10^2
Meters	Millimeters (mm)	10^3
Meters	Micrometers (microns; μm)	10^6
Meters per day (m/day)	Centimeters per second (cm/s)	0.00116
Meters per kilometer (m/km)	Feet per mile (ft/mi)	5.283
Miles (mi)	Kilometers (km)	1.609
Ounces (oz)	Grams (g)	28.350
Pounds (lb)	Kilograms (kg)	0.454
Quarts (qt)	Cubic centimeters (cm^3)	946.358
Quarts	Liters (l)	0.946
Square centimeters (cm^2)	Square inches (in^2)	0.155
Square feet (ft^2)	Square meters (m^2)	0.093
Square feet	Hectares (ha)	0.929×10^{-5}
Square inches (in^2)	Square centimeters (cm^2)	6.452
Square kilometers (km^2)	Square miles (mi^2)	0.386
Square meters (m^2)	Square yards (yd^2)	1.196
Square miles (mi^2)	Square kilometers (km^2)	2.589
Square yards (yd^2)	Square meters (m^2)	0.836
Yards (yd)	Meters (m)	0.914

Converting Units of Measure

How Much is a Part per Million, Billion, or Trillion?

Unit	1 part per million (ppm)	1 part per billion (ppb)	1 part per trillion (ppt)
Length	1 inch in 16 miles	1 inch in 16,000 miles	1 inch in 16,000,000 miles (A 6-inch leap on a journey to the sun)
Time	1 minute in 2 years	1 second in 32 years	1 second in 320 centuries
Money	1¢ in $10,000	1¢ in $10,000,000	1¢ in $10,000,000,000
Weight	1 oz salt on 32 tons potato chips	1 pinch salt on 10 tons potato chips	1 pinch salt on 10,000 tons potato chips
Volume	1 drop vermouth in 80 fifths of gin	1 drop vermouth in 500 barrels of gin	1 drop vermouth in 25,000 hogsheads of gin
Area	1 sq ft in 23 acres	1 sq ft in 36 sq miles	1 sq in. in 250 sq miles
Quality	1 bad apple in 2,000 barrels	1 bad apple in 2,000,000 barrels	1 bad apple in 2,000,000,000 barrels

Source: Crummet, W.B., *ChemEcology*, December 1976.

Our Geologic Environment

Harvey Blatt

Hebrew University of Jerusalem
(formerly of University of Oklahoma)

Prentice Hall
Upper Saddle River, New Jersey 07458

Library of Congress Cataloging-in-Publication Data

Blatt, Harvey.
    Our geologic environment / Harvey Blatt.
        p.   cm.
    Includes bibliographical references and index.
    ISBN 0-13-371022-X
    1. Environmental geology.   I. Title.
QE38.B57 1997
550--dc20

96-30973
CIP

Executive Editor: Robert McConnin
Editor-in-Chief: Paul Corey
Editorial Director: Tim Bozik
Assistant Vice President of Production and
    Manufacturing: David W. Riccardi
Executive Managing Editor: Kathleen Schiaparelli
Marketing Manager: Leslie Cavaliere
Manufacturing Manager: Trudy Pisciotti
Creative Director: Paula Maylahn
Art Director: Joseph Sengotta
Art Manager: Gus Vibal
Photo Research Director: Lori Morris-Nantz
Photo Research Administrator: Melinda Reo
Photo Researchers: Cindy-Lee Overton and Mira Schachne
Cover Designer: Kevin Kall
Cover photo: View down center of road in arid landscape
    to tornado on horizon. Tony Stone Images/Peter Rauter
Text Composition and Production Coordination:
    Custom Editorial Productions, Inc.

© 1997 by Prentice-Hall, Inc.
Simon & Schuster/A Viacom Company
Upper Saddle River, New Jersey 07458

Printed in the United States of America

10 9 8 7 6 5 4 3 2 1

ISBN 0-13-371022-X

Prentice-Hall International (UK) Limited, *London*
Prentice-Hall of Australia Pty. Limited, *Sydney*
Prentice-Hall Canada Inc., *Toronto*
Prentice-Hall Hispanoamericana, *S. A., Mexico*
Prentice-Hall of India Private Limited, *New Delhi*
Prentice-Hall of Japan, Inc., *Tokyo*
Simon & Schuster Asia Pte. Ltd., *Singapore*
Editora Prentice-Hall do Brasil, Ltda., *Rio de Janeiro*

Brief Contents

Civilization exists by geologic consent,
subject to change without notice.
Will Durant, Historian

We cannot protect what we do not understand.
Daniel Hillel, Soil Scientist

The future isn't what it used to be.
Unknown

Dedicated to my wonderful wife, Midge.

Contents

Chapter 4

Soil Problems—Erosion and Pollution 67

Chapter 5

Slope Stability and Landslides 91

Chapter 6

Clean Water 123

Chapter 7

Streams and Floods 161

Chapter 8

Water Quality and Pollution 193

Chapter 9

Solid Waste 229

Chapter 10

Shorelines, Erosion, Wetlands, and Pollution 257

Chapter 11

Earthquakes and Plate Tectonics 297

Chapter 12

Volcanoes and Eruptions 335

Chapter 13

Nonfuel Mineral Resources and the Environment 367

Chapter *14*

Energy from Fossil Fuels 401

Chapter *15*

Energy from Alternative Sources 431

Preface

It seems that each decade brings a new theme that galvanizes the psyche of scientists. In the 1950s, it was "nuclear." In the 1960s, "space" was in vogue. The 1970s saw "ecology" come into fashion, and the 1980s buzzed with "computers." For the present decade, there can be little doubt that the focus is on "the environment."

But what does a scientist mean when using the term *environment*? There are, of course, many possibilities. Social scientists may mean the psychological and emotional setting in which human interactions take place. Biologists may be referring to the chemical milieu in which cells perform their many functions. Meteorologists may use the term to indicate elevations above sea level, the nature of the topography of the ground, and the state of saturation of the atmosphere with water vapor. To the current generation of geologists, *environment* refers to the interaction between living organisms (particularly humans) and their nonliving surroundings. This is a relatively new interest for most geologists, a radical departure from our previous nearly total preoccupation with natural resources, particularly the search for oil and natural gas.

Those of us who have spent most or all of our scientific careers in colleges and universities are well aware of the incredibly rapid change in student interest within geology. Both majors and nonmajors have a strong and growing interest in our surroundings, particularly the aspects favored by print and visual media. A few years ago, concern focused on acid rain, its causes, effects, and possible cures. But in 1996, as this is written, that topic seems less exciting to the American public. Now, other atmospheric concerns enjoy the media spotlight, including air pollution and, most particularly, global warming. Every newspaper, scientific journal, and educational broadcast seems to offer words about global warming.

Because of the unusually rapid onset of this interest in environmental geology, few textbooks are available, either at the introductory or advanced level. That is the reason this book was written. Although this is not the only introductory environmental geology text available, it is different in useful ways.

Most important, this is not a physical geology book with some environmental material added to enhance its market appeal. Instead, its focus is on the interaction between humans and their physical environment. Two important chapters on soils and their relationship to crops and agriculture appear early in the book. Three chapters deal with the distribution and pollution of our water supply. And the final, and longest, chapter treats the atmosphere and its apparent rapid deterioration. In fact, these critical topics occupy more than one-third of the book.

Earth's raw materials—minerals and rocks—are the fundamental components on and in which environmental changes occur. It is essential that environmental geology students acquire a basic understanding of these Earth materials, so I present them straightforwardly in Chapter 2.

Volcanoes, earthquakes, and coastal processes also generate intense environmental interest, as recent eruptions, quakes, and losses of barrier islands clearly demonstrate. Fossil energy resources and their polluting effects also attract considerable interest. But soil, water, and air are most central to human concerns. We must eat, drink, and breathe before we can be concerned about where and when the next volcano or earthquake will occur.

I have written an "accessible" book aimed at the typical nonmajor student in the United States of the 1990s. Scientific jargon has been kept to the minimum consistent with accurate communication, and a glossary is provided.

The American system of units is used whenever possible, not because I prefer it (I don't), but because "2 miles" is more comprehensible to most students than "3.2 kilometers." And 68°F communicates temperature in a way that 20°C does not. Until metric or SI units are used consistently in the lower grades, I believe that using them to teach nonscience majors at the undergraduate level results only in loss of communication. This is sad but true. The abstraction that is mathematics is too often a turn-off for nonscience students, and unfamiliar units only exacerbate the problem.

Some types of data are seldom expressed in American units, so I have used metric units in these cases. For example, my presentation of the electromagnetic spectrum (introduced to discuss global warming and ozone depletion) employs metric units. An appendix offers conversions.

The following ancillaries are available with *Our Geologic Environment*:

- Instructor's Manual (0-13-525742-5)
- Transparencies (0-13-525767-0)

- Slides (0-13-471616-7)
  100 color slides of environmental features shown in the book. These depict specific environmental features such as slash-and-burn agriculture, smog, and so forth.
- Prentice Hall Custom Test
  Macintosh (0-13-525784-0);
  Windows (0-13-525776-X)
- Geodisc (0-13-304163-8)
- GEODe: Geologic Explorations on Disk
  (0-13-516545-8)
- *New York Times* "Themes of the Times"
  (0-13-738113-1), which features recent articles on dynamic geology applications.

I welcome your comments on this book: your experiences teaching with it, any material you feel is unclear, or omitted topics you would like to see included. It is through your helpful observations that improvements can be made. Please send your comments to: Executive Editor—Geology, Prentice Hall, Inc. 1633 Broadway, New York, NY 10019-6785.

Acknowledgments

No one can realistically claim expertise in all the topics that need to be considered in an environmental geology textbook. Without the help of many reviewers who are gracious enough to read drafts of chapters, there is no hope of a good textbook resulting from a single author's efforts. I am fortunate in having had an excellent group of reviewers to help me. They checked for accuracy as well as how my thoughts were expressed, and their efforts have been indispensable. Their efforts have greatly strengthened the final manuscript. I am grateful to them all. Some were chosen by professional editors and some by me. These reviewers were:

Herbert G. Adams, California State University, Northridge
Judson L. Ahern, University of Oklahoma
Jock Campbell, Oklahoma Geological Survey
Larry W. Canter, University of Oklahoma

Pascal de Caprariis, Indiana–Purdue University at Indianapolis
Edward Evenson, Lehigh University
Duncan Foley, Pacific Lutheran University
Dru Germanoski, Lafayette College
Gilbert Hanson, State University of New York at Stony Brook
Peter L. Kresan, University of Arizona
James Laguros, University of Oklahoma
Michael L. McKinney, University of Tennessee
Mark Meo, University of Oklahoma
David W. Miller, Jefferson Community College
Orrin H. Pilkey, Jr., Duke University
Nicholas Pinter, Southern Illinois University
Susan Postawko, University of Oklahoma
David Sabatini, University of Oklahoma
Neil Salisbury, University of Oklahoma
Paul A. Schroeder, University of Georgia
Robert D. Shuster, University of Nebraska, Omaha
Paul Skierkowski, University of Oklahoma
David Stearns, University of Oklahoma
Clifford Thurber, University of Wisconsin, Madison
Barry Weaver, University of Oklahoma

In addition to these colleagues, many individuals at Prentice Hall's editorial and production offices have worked long and hard to produce the quality of workmanship evident in this publication. Those of whom I am aware include Bob McConnin, Executive Editor; Fred Schroyer, Development Editor; and Lori Morris-Nantz, Photo Researcher. The attractive physical appearance of this book is the result of their efforts. I appreciate them all.

Harvey Blatt
Hebrew University of Jerusalem

Annotated Supplements List

Instructor's Manual (0-13-525742-5)

Transparencies (0-13-525767-0)

One hundred line drawings and tables shown in the text.

Slides (0-13-471616-7)

100 color slides of environmental features shown in the text. These depict specific environmental features such as slash-and-burn agriculture, smog, and so forth.

Prentice Hall Custom Test

Windows (0-13-525776-x); Mac (0-13-525784-0)

Based on the powerful testing technology developed by Engineering Software Associates, Inc., this supplement allows instructors to tailor exams to their own needs. With the on-line testing option, exams can be administered on-line, so data can be automatically transferred for evaluation. A comprehensive desk reference guide is included, along with on-line assistance.

Life in the Internet—Geosciences (0-13-266578-6)

Andrew Stull, California State University–Fullerton and Duane Griffin, University of Wisconsin–Madison

This unique resource gives clear, step-by-step instructions to access regularly updated geoscience resources, as well as an overview of the World Wide Web, general navigation strategies and brief student activities. *Available free when packaged with copies of the text.*

GEODe (0-13-516545-8)

Developed by Dennis Tasa in collaboration with Ed Tarbuck and Fred Lutgens. A fully interactive CD-ROM tutorial, designed to reinforce key geologic concepts through animation, interactive exercises, and expanded text art. Five fundamental topics are covered: the rock cycle, the hydrologic cycle, erosional forces, mountain building, and plate tectonics. Subjects that are often difficult for students to visualize, such as the crystal structure of minerals, mountain range formation, and convection currents, are clarified for students through various activities including labeling diagrams, locating epicenters, and plotting charts.

New York Times *Themes of the Times—***The Changing Earth** (0-13-738113-1)

This unique newspaper supplement featuring recent articles on dynamic geography applications from the pages of the New York Times encourages students to make connections between the classroom and the world around them. *Available free, in quantity, upon adoption of a Prentice Hall text.*

Geodisc (0-13-304163-8)

This full-color video disc is available to adopters of any Prentice Hall introductory geology or physical geography text. The **Geodisc** contains over 1200 photos, 200 drawings, 50 minutes of full-motion video, and 6 minutes of animations on various Earth Science topics to make classroom presentations more vivid. Short segments from television news coverage of earthquakes, floods, and volcanic eruptions bring Earth's dynamic processes into the classroom. Also included is The Movie Series (Earth, Mars, Miranda, L.A.) produced at the Jet Propulsion Laboratory and U.S. Geological Survey. These video segments are computer-generated virtual flights across the surface of the planets. Each item on the laser disc is described in an accompanying manual along with bar codes for easy access to the images.

Reprinted with special permission of King Features Syndicate.

"We literally move mountains to mine the earth's minerals, redirect rivers to build cities in the desert, torch forests to make room for crops and cattle, and alter the chemistry of the atmosphere in disposing of our wastes. At humanity's hand, the earth is undergoing a profound transformation—one with consequences we cannot fully grasp."

Sandra Postel, Environmentalist

The Environmental Problems We Face

1

This peaceful, relaxing coastal setting seems ideal for a get-away-from-it-all vacation: calm water, fantastic scenery, and—hopefully—reasonably priced accommodations. However, if you are concerned about our environment, some questions might cross your mind . . .

- Is the water always this calm? Do severe storms occur here, and high waves? I wonder which direction the waves come from. The beaches look great . . . but no one is swimming. Why? I certainly hope the community isn't dumping garbage here . . .

- Strange, how the low-lying fingers of land are so close to mountains. And the two fingers are at right angles to each other. I wonder how they formed. Curious, how nearly all the houses and boats are on one side of each land finger. Is something wrong with the other side?

- I wonder whether the lagoon is filling with debris from those mountains. Sure would ruin the place if it did . . . I wonder whether the people who live here investigate things like that. Could the mountains be volcanoes that might erupt? Have they ever had an earthquake here?

- The sky sure is clear. No pollution here. But how do the people earn their living? They can't *all* make a living from fishing. Are there industries nearby? Why don't they pollute this lovely scene?

- As you will learn in this course, environmental geologists are people who have learned how to answer these questions. By the time you finish this course, you will be able to answer these questions, too.

◄

The majestic fjords of northern Norway.
Photo: Charlie Crangle/JLM Visuals.

Your author is a geologist who focuses on environmental concerns. My aim is to show you how an understanding of Earth's natural processes can help us all live in harmony with our environment. We humans appear to be the only creatures who deliberately foul our own nest, despite the obvious fact that this Earth is the only nest we have! We cannot all relocate to some other friendly planet, so we'd best take care of the home we have. That is the theme of this book.

Before we begin our study of the environment and its relationship to human concerns, we must consider the nature of *geology* and its point of view. *Geologists* are scientists who study Earth, our home (Figure 1.1). Geologists examine how our planet originated, what its insides are like, how its surface has changed through time, and how the organisms that live on it have evolved.

The science of **geology** is divided into many specialties, including minerals and rocks, volcanoes, earthquakes, fossils, natural resources (such as petroleum and metallic mineral deposits), and the environment. In turn, the specialty of **environmental geology** is broken down into subspecialties. Some environmental geologists study water pollution. Others examine the effects of radioactivity on living organisms. Still others specialize in the causes and cures of pesticide-contaminated soil.

Environmental studies have not always been a major focus of geologists. Until recently, 80 to 90 percent of geologists spent their careers exploring for petro-

FIGURE 1.1 Scientists who study Earth.

Geologists in California examining soil development. The metal rods are braces to prevent collapse of the trench walls. *Photo: Y. Enzel.*

leum and natural gas, which occur below Earth's surface to depths perhaps as great as 5 miles. Only a small number of geologists paid attention to the effect of geologic phenomena in our surface environment. These phenomena include landslides, floods, soils, or human activities that relate directly to geology, such as garbage disposal and pollution.

Within the past ten to twenty years, however, a major reorientation has occurred among working geologists. The percentage of geologists searching for oil and gas has plummeted, and the percentage studying environmental issues has skyrocketed.

Here are some questions environmental geologists seek to answer:

- Can hazardous waste be buried safely below Earth's surface, or should it be incinerated?
- What effect does incineration of hazardous materials have on the air we breathe?
- Is our drinking water being polluted by rainwater that dissolves waste from mining of coal or metal ores? If so, can we afford to halt the mining?
- If we halt mining, where will we find copper to make wire for carrying electricity into our homes? And where will we get the coal that we burn in power plants to generate electricity in the first place?
- On the other hand, can we afford *not* to stop mining, given that it continues to pollute our water supply?
- Is there a happy medium between these extremes?

Environmental geologists seek solutions to these perplexing, expensive problems that affect every human being on this planet.

Most students enrolled in today's undergraduate geology programs plan environmental careers upon graduation. Whether you are planning a career in environmental geology, are taking this course to fulfill a science requirement, or are simply curious about geology and the environment, geologic knowledge is essential. Geologic variables determine where earthquakes occur, which areas are most susceptible to floods, the likely sites of rockfalls and avalanches, where volcanoes will erupt, where underground water supplies are likely to be found, and the best places to locate sanitary landfills (garbage pits) or storage facilities for hazardous waste.

Geology Versus People: Who Runs the Show?

Geology controls us all. We can't manufacture a valley with a clear stream for our use; instead, we find such a valley and move there. If you live in Hawaii, you are

TABLE 1.1 **Number Of Years Required to Add Billions of People to Earth's Population**

Billions of People	Date Achieved	Years Required	Number of People Added Per Year
First billion	1800 (A.D.)	All of human history	About 1000*
Second billion	1930	130	7.7 million
Third billion	1960	30	33.3 million
Fourth billion	1975	15	71.4 million
Fifth billion	1987	12	83.3 million
Sixth billion	1997	10	100.0 million

*Assuming about 1 million years of *Homo sapiens* reproduction.
Source: Population Reference Bureau estimates.

helpless to control the volcanoes on which you live. If you live in California, neither you nor the government can prevent the next earthquake. You can't stop a river or a hurricane from flooding your home. Geology always has run the show, and it always will.

But now, the relationship between geology and people is no longer a simple one-way street. Throughout the past few thousand years, we have created technology that enables us to modify our geologic surroundings and make them more to our liking. For example:

- Canals to funnel water from places of abundance to places of scarcity have been used for thousands of years.
- More recent attempts to control water supply have included the rerouting of large parts of river systems.
- People who live along coastlines want local or federal governments to erect barriers that will block the sea from advancing over their beachfront property.
- Hills are often flattened to form pads for home construction in hilly areas such as Los Angeles.
- Those who live in earthquake-prone areas want the federal government to find ways to reduce the potential for disaster.

In general, we seem determined to fight nature, rather than to harmonize with it. However, "Nature, to be commanded, must be obeyed," observed English philosopher Francis Bacon in 1620. That advice is even more meaningful today.

The human population is increasing rapidly, adding another 100 million people each year (Table 1.1). This growth can convert a noncritical situation into a serious scenario. For example:

- The number of people killed, injured, or displaced by natural disasters increased by 6 per-

TABLE 1.2 **Total Number of Events and Deaths in Natural Disasters, 1976–1991**

Type	Number of Events	Number Killed
Weather events		
Hurricanes, typhoons	894	896,063
Flood	1358	304,870
Storm	819	54,500
Cold and heat wave	133	4926
Drought	430	1,333,728
Total	3634	2,594,087
Associated with weather events		
Avalanche	29	1237
Landslide	238	41,992
Fire	729	81,970
Insect infestation	68	0
Famine	15	605,832
Food shortage	22	252
Epidemic	291	124,338
Total	1392	855,621
Geological		
Earthquake	758	646,307
Volcano	102	2764
Tsunami	20	6390
Total	880	655,461
Grand Total	5906	4,105,169

Source: World Meteorological Association

cent each year over the 25 year period between 1967 and 1991 (Table 1.2) and now totals more than 300 million. The 5906 natural disasters during the period killed more than 4 million people.

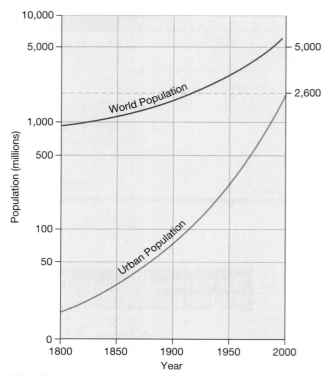

FIGURE 1.2   The growth of cities.

Urbanization compared to world population growth since 1800. *Modified from Davis, 1965, Scientic American.*

Water shortages (droughts) were responsible for about one-third of the deaths, and water surpluses (hurricanes, river floods, and storms) about 30 percent.

- *Garbage disposal.* There are now 6 billion people on Earth. Since 1950, the number of Earth's cities with populations exceeding 4 million has grown from 15 to 47. Twenty megacities now have populations exceeding ten million. This occurred as the number of people who live in urban areas increased from 740 million to 2.6 billion (Figure 1.2). In industrialized countries, 74 percent of the population now lives in cities; in the non-industrialized countries, it is 37 percent and increasing rapidly. A U. N.-sponsored study found that the population of the world's cities will almost double by 2025. By 2000, most of the world's people will live in cities. Consider the problem this growth creates for the collection and disposal of garbage. Obviously, it is more difficult to deal with garbage from a city of several million people than a town with 10,000.

- *Sea level rise.* A rise in sea level of a foot or two, due to possible global warming, is not an urgent matter along a sparsely populated shore area. But

it is significant to a city like New Orleans, most of which already lies below sea level. Many of the world's largest cities are along coastlines—New York, Rio de Janeiro, Singapore.

- *Water supply.* Low water supply in the United States is a more serious problem for desert cities like Phoenix today than in 1950, when its population was much smaller. The same is true of many non-desert cities, such as Los Angeles. An increasing number of urban areas have water-supply problems.

- *Air pollution.* The air in an increasing number of American cities is so bad that "pollution alerts" are commonplace. The intense burning of coal, oil, and gasoline have created an atmospheric disaster that has been recognized only recently. Chemicals and microscopic particles in the air now regularly affect the lungs of most Americans, as well as a large and growing number of people around the world.

The extent of some of these problems is shown in Figure 1.3. Such problems are more easily solved in wealthy countries such as the United States, England, or Germany. The high cost of addressing environmental troubles usually can be borne by well-off societies, despite much moaning and groaning from the citizenry. In poorer countries, however, resources are scarce. Money spent for environmental remedies is money that is not available for governmental assistance with food and shelter, and such assistance is needed by a large percentage of the population.

Earth, Air, Water, and Us

Our environment is controlled by three major variables: earth, air, and water. These are the essential resources for human existence; our survival as a species depends on the amount, distribution, and purity of each. In this context,

- *Earth* includes the (1) variety and distribution of minerals and rocks; (2) origin, variety, distribution, and pollution of soils; (3) causes, distribution, and prediction of earthquakes and volcanic activity; (4) cause and prevention of coastal erosion and slope failure; and (5) mineral and water resources.

- *Air* includes its (1) composition and circulation, (2) pollution by human activities, and (3) climate.

- *Water* includes its (1) distribution, (2) movement and flooding, (3) range in chemical composition, and (4) pollution by human activities.

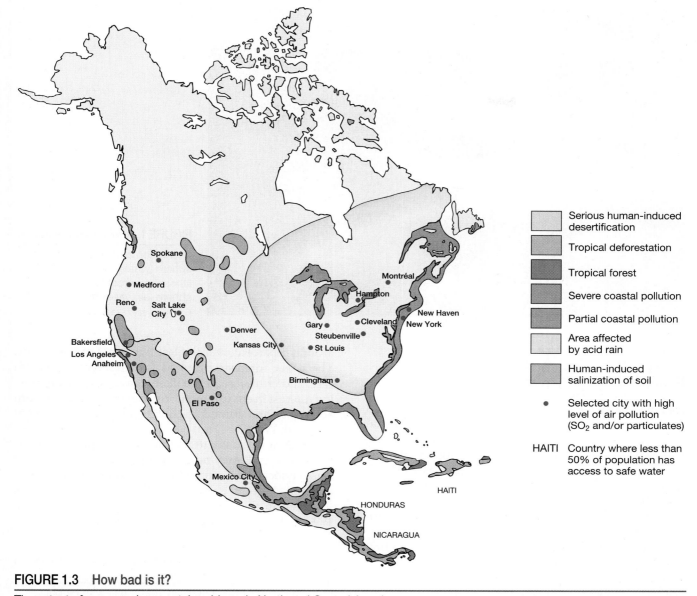

FIGURE 1.3 How bad is it?

The extent of some environmental problems in North and Central America. *Source: National Geographic Atlas of the World, 6th ed., 1992, p. 15.*

These three spheres—Earth, air, and water—are tied together, because water circulates freely among the atmosphere, ground surface, and subsurface (Figure 1.4). It does so in its three states, as a liquid, solid (ice), and gas (water vapor). Solid surface materials are transported to the sea by surface waters and are carried upward into the atmosphere as wind-borne dust. Materials dissolved from Earth's surface are carried by rivers to the oceans, and from there into the atmosphere by sea spray. The spray is carried landward by the wind and redeposited on the land surface as part of rainfall. You can see that no isolated compartments exist in the natural world; all things are interrelated.

In recent years, scientists have increasingly emphasized the role of living organisms in modifying their own natural environment. This thinking has been fueled by the **Gaia hypothesis,** which states that the whole Earth behaves like an organism, a self-regulating network of interdependent physical and biological systems. In this network, a disturbance in one part must result in adjustments in other parts.

Human civilization has proven to be a major disturbing influence in the natural world, as illustrated by water pollution, air pollution, and soil destruction caused by human excesses. As observed by the comic strip character named Pogo, "We have met the enemy and he is us." Hence, consideration of the environment *must include us.* We human beings and our living habits form an important part of this book.

FIGURE 1.4 Where water roams.

Interaction among the hydrosphere (water), biosphere (life), lithosphere (rocks), and atmosphere (air). *Photo: Robert Frerck/Odyssey Productions.*

How Interested Are We in Environmental Problems?

As noted earlier, the bulk of U.S. college students in a geology major are planning careers in environmental geology. This concern for the environment by young people, and the desire to do something about it, is growing in other industrialized countries as well. But college students are just one segment of the population. What about adults in general? Are those who are in controlling positions in industrial societies as progressive as current students?

Several surveys suggest some answers. The first counted the articles on four environmental topics that appeared in Canadian newspapers during 1977–1990: acid rain, global warming, ozone depletion, and tropical deforestation (devastation of rain forests). The results, graphed in Figure 1.5, show a strong and generally growing interest in environmental subjects.

Not surveyed was coverage of other environmental concerns, like smog, water supplies and pollution, floods, soil erosion, landslides, earthquakes, coastal erosion, and alternative sources of energy. Had these been included, the total number of articles published no doubt would have gone well over a thousand per year toward the end of the period. Nevertheless, over the 14-year period surveyed, Canadian press coverage of environmental topics increased at least 10 times.

A second survey in Great Britain during the six years from 1988 to 1994 asked 2000 adults about issues that were most important to them (Figure 1.6). Their interest in environmental concerns dropped from a high of 35 percent in 1989 to a fairly constant low

around 5 percent over the last two years of the survey. The other interests graphed (unemployment, health care, and crime) suggest a likely reason for the decreasing interest in environment. You can see that, as concern about jobs and crime grew, interest in the environment declined. Why?

As politicians are well aware, people vote their pocketbooks. It may be that people in Great Britain, at least, associate environmental remediation with a loss of jobs. A good example of this interpretation has occurred in the United States. Amendments made in 1990 to the Clean Air Act require electric utilities to

FIGURE 1.5 Number of articles in Canadian newspapers on four environmental topics.

More than half of the articles in 1990 dealt with global warming. *Source: Environment, October 1995, vol. 37, no. 8, p. 25.*

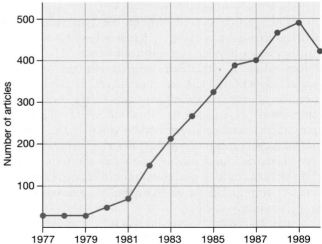

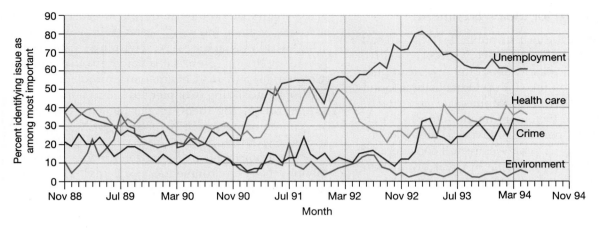

NOTE: The percentages shown are based on monthly surveys of 2000 adults in the United Kingdom.

FIGURE 1.6

Environmental attitudes in the United Kingdom, 1988–1984. *Source:* Environment, *October 1995, vol. 37, no. 8, p. 8.*

reduce air pollution from coal-burning power plants. One way to achieve this is to burn coal that contains less sulfur. Many miners of coal that has a higher sulfur content lost their jobs as a direct result of the new environmental laws.

Despite such occurrences, numerous surveys in the United States have found that the overwhelming cause of job displacement in industrialized countries is not environmental. It is the combination of business mergers and resulting downsizing. In California, a 1995 report by the State Senate concluded that most jobs in the state are lost because of leveraged buyouts (a technique of acquiring a large company with little cash), mergers, taxes, and similar changes in business practices. Job losses in the Los Angeles area in the few years before the report resulted more from defense cuts than from the state's implementation in the 1980s of the nation's strictest air-quality controls.

A national survey of employers taken in 1995 revealed that between 1987 and 1990, an average of only 0.1 percent of all large-scale layoffs were the result of environmental regulations. Falling product demand accounted for 21 percent of the job losses and reorganization and change in business ownership another 15 percent.

An economist at the U.S. Census Bureau's Center for Economic Studies has determined that plants with poor environmental records are no more profitable than cleaner plants in the same industry. And this is as true for pollution-intensive industries as it is for clean industries. He concluded that, "There is simply no evidence that superior environmental performance puts firms at a market disadvantage. Government regulations should not be weakened out of fear of damaging industries' market performance."

These data do not say that nobody has ever lost a job because of environmental laws, as the example of the coal miners shows. But it does demonstrate that it is unjustified to make these laws the scapegoat for other problems in our industrial economy. It is unfortunate that environmental laws and their short-term cost make a convenient whipping boy for the understandable concern of people about an apparent decrease in employment opportunity.

There is an alternative but related interpretation of the decline in environmental concern shown by the people surveyed in Great Britain. Faced with the immediate worries of unemployment and crime, these people considered environmental problems to be a long-term concern; therefore, finding solutions could be postponed until there is nothing more pressing to worry about. When you are out of work and bills are piling up, the problems of soil erosion, leaking oil storage tanks, or sea level rise seem less urgent.

But concerns that are postponed once often are postponed again and again. Somehow, the time never seems right to invest the time, energy, and money required to deal with the postponed problems. Proposed solutions seem to recede into the distance. Environmental problems, like the National Debt, can be foisted onto the next generation.

However, environmental problems will never go away, and they only will grow more severe the longer they are ignored. And, "more severe" always means more difficult to treat and more expensive to solve. Unlike the National Debt, which will, at worst, lead to financial collapse, failure to cleanse our water supply, or to stop polluting the soil with pesticides, or to stop injecting poisonous chemicals into the air may result in the illness or death of much of the American population. The time to act is now.

American Concern for the Environment

We have seen that the British public in 1994 did not view environmental concerns as very pressing relative to some other worries. How about the American public? A 1991 survey by the Gallup Organization found that 57 percent of us favor "immediate and drastic actions" to preserve the environment, but a 1995 survey found that the percentage favoring these actions had decreased to 35 percent. We are still considerably more concerned than the people in Great Britain, but our support for immediate action seems to have lessened in recent years, as did theirs.

As we will see later in this book, we spend billions of dollars each year trying to solve our environmental problems, but some areas of the country are apparently more concerned than others (Table 1.3). Alaskans are most concerned about the environment, with Vermonters (who would seem to have few environmental problems), Coloradans (landsliding, Chapter 5; pollution from metallic mining, Chapter 13), and New Jerseyites (water pollution, Chapter 8; mountains of municipal solid waste, Chapter 9) also showing serious concern. Among the less concerned citizens are Louisianans, whose $11.69 expenditure places them 29th on the list, despite the fact that their state has perhaps the worst water pollution problem in the nation (Chapter 8). West Virginia, a state with serious pollution problems related to coal mining (Chapter 14) and road maintenance problems resulting from landsliding (Chapter 15) seems almost uninterested in solving them. Its per-person expenditure on environmental problems is only $4.22, placing it last among the 50 states.

How concerned are the citizens of your state with solving their environmental problems? How concerned are you?

Your Role in Our Environment

What can you, as an individual, do to affect the ever-increasing degradation of our surroundings? Can you use less water? How? By taking fewer showers? By flushing the toilet less frequently? By drinking less liquid? By catching rainwater in a container and using it instead of tap water? By watering the lawn and shrubbery less often? Is anyone really going to do these things? What difference does one person make?

TABLE 1.3 Spending for Environmental Protection Varies from State to State

State	Per capita spending	State	Per capita spending	State	Per capita spending
Alabama	$ 8.22	Louisiana	$ 11.69	Ohio	$ 7.12
Alaska	101.55	Maine	18.89	Oklahoma	6.71
Arizona	6.20	Maryland	16.29	Oregon	12.70
Arkansas	6.61	Massachusetts	31.75	Pennsylvania	11.27
California	27.19	Michigan	12.15	Rhode Island	21.02
Colorado	41.02	Minnesota	11.55	South Carolina	10.36
Connecticut	15.72	Mississippi	11.55	South Dakota	10.80
Delaware	14.27	Missouri	30.70	Tennessee	9.92
Florida	13.74	Montana	12.26	Texas	11.26
Georgia	8.18	Nebraska	10.03	Utah	17.10
Hawaii	7.43	Nevada	11.70	Vermont	59.10
Idaho	21.55	New Hampshire	24.32	Virginia	12.84
Illinois	24.63	New Jersey	40.11	Washington	19.22
Indiana	6.29	New Mexico	11.38	West Virginia	4.22
Iowa	12.60	New York	25.29	Wisconsin	19.70
Kansas	9.19	North Carolina	6.98	Wyoming	14.11
Kentucky	8.08	North Dakota	5.21		

NOTE: Environmental expenditures comprise the following items: air quality, drinking water, hazardous waste, pesticides control, solid waste and water quality. Figures based on fiscal 1991 spending, 1990 population.

Source: National Academy of Public Administration calculation based on data from R. Steven Brown and K. Marshall, "Resource Guide to State Environmental Management," Third Edition, Lexington, Ky.: The Council of State Governments.

Similar questions can be asked about garbage disposal, industrial waste, maintenance of catalytic converters in cars, raising children, voting in elections, and many other activities.

The fact is, you as an individual *can* make a difference. Human behavior is such that good examples tend to be well regarded and imitated. This is the nature of leadership (Figure 1.7). Each of us has the opportunity to be a leader in some aspect of environmental protection or remediation.

This book exists to show you the essentials of how the Earth works, what it needs to continue to be a suitable home for us all, and what you as an individual can do to help keep this nice place habitable.

FIGURE 1.7 "I don't know about you, but I love 'em."

Most insects are 60–80 percent protein by weight, three to four times that of beef, fish, or eggs. Insects are widely available and do not pollute the land. In South Africa, "Mopani"—larvae of a species of emperor moth—is one of a number of insects that are eaten. In the Kalahari Desert, cockroaches are food; toasted butterflies are a favorite in Bali; french-fried ants are sold on the streets of Bogotá; and Malaysians love deep-fried grasshoppers. Will we follow their good example? *Photo: Anthony Bannister/Natural History Photographic Agency.*

How Much Time Before We Fill the Bottle?

Some prominent organizations, as well as some high-profile political and media personalities, insist that most, if not all, environmental problems are imaginary. They believe that Earth is so complex and self-regulating that it is essentially impossible to disturb it in any serious way. In other words, Earth's capacity to adapt to anything humans can do is limitless.

Well, maybe. On the other hand, this attitude brings to mind a story used by a professor in California to convince his students that it may be later than we think with respect to population growth and the resulting increase in capacity to pollute that it brings. He says:

> Think of yourselves as bacteria in a bottle. Suppose the bacteria in the bottle are growing at a rate that doubles their population every minute, and that, given its size, the bottle will be completely filled with bacteria in one hour. If that's true, and if it's eleven o'clock in the morning, and the bottle will be completely filled at noon, then when is the bottle going to be half-full?

Did you guess 11:30? Well, that's not quite right. The bottle will be half-full at 11:59, one minute before noon. One minute later, it will be completely full (see Figure 1.8). That's the nature of **exponential growth**—the kind of growth that is characteristic of human populations and many of our activities.

The political and governmental lesson from this is clear. If you were one of the bacteria in the bottle, you might not recognize how quickly things could change. At around 11:58—just two minutes before the bottle would be completely full—you would see that there was 75 percent open space in your bottle. Would you vote to control growth? Maybe not.

In the human world, our doubling time is measured in years, not in minutes, but the lesson is the same. It's definitely not too early to take growth and pollution issues seriously.

Environmental Problems We Will Examine

Television news, radio, newspapers, magazines, and books bombard us with a blizzard of scary stories about the environment. What is significant, and what is not? Which problems can we fix, and which ones are beyond our ability? To answer these questions, you first need a foundation in basic geology. That's the purpose of the next chapter. In it, we look at the materials of which the Earth's rocky crust is made. We answer such questions as: What is the nature of the ground we

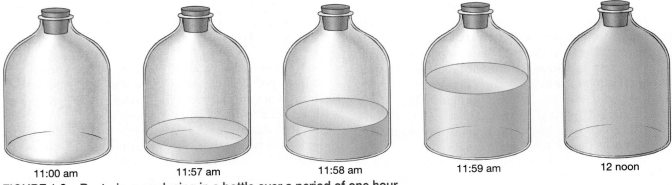

FIGURE 1.8 Bacteria reproducing in a bottle over a period of one hour.

| 11:00 am | 11:57 am | 11:58 am | 11:59 am | 12 noon |

For the first 57 minutes, population growth is deceptively small. In the last 3 minutes, disaster strikes. This is the nature of exponential growth, including that of human populations.

live on? In what ways do its characteristics contribute to many of society's problems?

Next, Chapters 3 and 4 consider how the atmosphere and water attack Earth's crust to create the soil in which our crops grow. This natural attack also creates surface instabilities, which lead to slope instabilities and landslides, which are the subject of Chapter 5.

Chapters 6 through 10 center around water existing at and near Earth's surface—in the ground, streams, lakes, rivers, and the ocean. These chapters examine water's benefits (drinking water and irrigation), harmful effects (floods), how we seriously damage this vital liquid (water pollution), and our interaction with the shoreline (coastal erosion).

Instabilities in Earth's interior cause earthquakes and volcanic eruptions, the topics of Chapters 11 and 12. Related to the locations of these phenomena are the locations of the world's mineral resources. Chapter 13 examines the material benefits and environmental problems created by mining activities.

Chapters 14, 15, and 16 are energetic! They look at the world's major energy resources. Chapter 14 covers crude oil, natural gas, and coal—and their polluting effects. These resources allow modern societies to be highly productive, yet cause major environmental problems. Because of their pollutant effects, intense efforts are underway to replace these conventional energy sources with alternative sources, such as wind power, the power of ocean waves and tides, and direct energy from the sun. These are the subject of Chapter 15. Another alternative energy source is nuclear energy, and its promises and problems are the subject of Chapter 16.

Last, we examine the atmosphere, the greenhouse effect, global warming, and the health problems caused by air pollution in Chapter 17.

The common thread that winds its way through this book's chapters and through all environmental problems is *human interaction with Earth's surface and the narrow zone within a few thousand feet above and below the surface.* This zone is part of the geologist's sphere of study. Hence, geologists are among the scientists most intimately involved with environmental problems.

In this book, you will see an intimidating collection of environmental problems and hazards. As noted, our environment not only changes on its own, in response to solar energy, gravity, and wear and tear; it also is challenged by human activities. This book will help you understand Earth's processes, see how human activities both mesh and clash with them, and give you the perspective needed to understand the continuing debate over environmental issues.

Summary

The range of environmental problems we face is large and varied. Many problems originate from irresponsible use of the land by human societies. Others arise from uncontrollable acts of nature, such as volcanoes and earthquakes. However, all Earth processes and organisms interact. Humans are not opponents of Earth in some "us versus them" system, but are an integral part of Earth's functioning.

The Earth's nonliving components—air, ocean, and land surfaces—form a complex system with the capacity to keep our planet a fit place for life. But this environment is not indestructible, nor are we. Humans have the capacity to render the Earth unsuitable for life through careless and preventable activities. The present six billion people are quite capable of destroying anything and we may already have begun to foul our habitat beyond repair. Only we are capable of doing this, and only we can stop it.

All things are interconnected . . . Whatever befalls the earth befalls the people of the earth. Man did not weave the web of life; he is merely a strand in it. Whatever he does to the web, he does to himself.

—attributed to Chief Seattle, 1854

Key Terms and Concepts

geology
environmental geology
exponential growth

Gaia hypothesis
natural disaster
pollution

population growth rate
public concern

Stop and Think!

1. List five environmental questions that you can ask your mayor or city manager at the next public meeting.

2. Give three examples of something you did during the past week that may have had a negative impact on the environment. Can you stop doing these things? If you can, will you?

3. Why do you believe humans frequently are in conflict with the natural world around them? Do you believe this is a solvable problem? How is it solvable?

4. List three ways in which the environment in your area would be adversely affected by a doubling of the population. How would you recommend the citizens deal with these problems?

5. How much water flows from your bathroom faucet each minute when it is turned wide open? Get a watch that displays seconds and a large container, and find out. Is it necessary for you to use as much water as you do each morning? Each evening? Will you change your water-using habits?

6. List why you think the per-person expenditure on environmental concerns is so different among the 50 states. One state spends only $4 per person, another more than $100. Where does your state place in the list of 50? Do you think it should do more? Less? About the same? Why?

References and Suggested Readings

Ausubel, J. H. 1986. Can technology spare the Earth? *American Scientist*, v. 84, p. 166–178.

Ausubel, J. H., Victor, D. G., and Wernick, I. K. 1995. The environment since 1970. *Consequences*, v. 1, no. 3, p. 3–15.

Caplan, Ruth. 1990. *Our Earth, Ourselves.* New York: Bantam Books, 340 pp.

Environmental Careers Organization, The. 1993. *The New Complete Guide to Environmental Careers.* Washington, D.C.: Island Press, 364 pp.

Engles, Nicholas. 1994. Environmental geology of urban areas. *Geoscience Canada*, v. 21, p. 158–162.

Goudie, Andrew. 1990. *The Human Impact on the Natural Environment, 3rd ed.* Cambridge, Massachusetts: The MIT Press, 388 pp.

Gough, L. P. 1993. Understanding our fragile environment. *U.S. Geological Survey Circular* 1105, 34 pp.

Homer-Dixon, T. F., Boutwell, J. H., and Rathjens, G. W. 1993. Environmental change and violent conflict. *Scientific American*, February, p. 38–45.

National Research Council. 1993. *Solid-Earth Sciences and Society.* Washington, D.C.: National Academy Press, 346 pp.

Perry, J. S. 1995. A sustainable future: catastrophe or cornucopia: crusade or challenge. *Climatic Change*, v. 29, p. 259–263.

Schneider, S. H., and Boston, P. J. (eds.). 1992. *Scientists on Gaia.* Cambridge, Massachusetts: The MIT Press, 433 pp.

Walker, Melissa. 1994. *Reading the Environment.* New York, New York: W.W. Norton, 597 pp.

Warner, D. J. 1992. *Environmental Careers: A Practical Guide to Opportunities in the '90s.* Boca Raton, Florida: Lewis Publishers, 288 pp.

Wilshire, H. G., Howard, K. A., Wentworth, C. M., and Gibbons, H. 1996. Geologic processes at the land surface. *U.S. Geological Survey Bulletin* 2149, 41 pp.

"It is somewhat difficult to realize the totality of our dependence on the Earth. The buildings in which we live are built of material manufactured from the Earth's crust. Heating, cooling, and lighting systems involve the use of energy derived from materials in the crust, energy stored from the sun by plants and animals million of years ago. Most of the clothing we wear, the food we eat, gasoline, cars, radios, TV—all these are triumphs of extraction of crude materials from the crust of the earth: dull, unpromising rocklike materials, discovered, extracted, concentrated, refined, and then manufactured into something of immediate value."

Frank Rhodes, Geologist and President, Cornell University

Minerals and Rocks

The photo shows fibers of a form of asbestos, greatly enlarged. Asbestos is a common mineral. Its removal from buildings has been a national obsession for more than 20 years, with concern about asbestos seeming to accelerate with time.

- What is asbestos?

- Why is it dangerous?

- What diseases can asbestos cause, and how common are they?

- Who is at risk and why?

- Are school children at risk from asbestos ceiling tiles in their classrooms?

- How effective is asbestos removal from schools?

At the conclusion of this chapter, we will examine the interesting story of asbestos.

Why should someone interested in the environment care about minerals and rocks? How are these "dead" materials relevant to us? Well, let us look at a few photos of our world, and discover how minerals and rocks underlie it all.

Understanding rocks helps us understand landslides (Figure 2.1). This landslide has made this house uninhabitable. But why did this disaster occur where it did and when it did? Could the landslide's cause have something to do with the internal structure of the underlying rock before it slid downward and broke into small pieces? What rock characteristics can lead to a catastrophe like the one shown here? Are all rocks equally susceptible to landsliding?

This slide happened after a heavy rain, which is typical of such disasters. Why did rainfall have this effect here, and not elsewhere? Why are some slopes less affected by rain than others? Only a careful examination of this event by a person knowledgeable about rocks and their interaction with water can answer these questions.

Understanding minerals helps us understand soil problems (Figure 2.2). This once-productive agricultural soil has been ruined by *salination* (a concentration of some sort of salt—that's the white crust you see). Formerly, this field yielded plenty of edible vegetation, but now it is useless. The farmer reported that the white crust started forming several years ago and his crops died as the crust grew. Also, the area covered by the crust continues to expand, even now that the crops are gone. On examining the crust, the farmer found

Chrysotile fibers (a form of asbestos) under high magnification. Width of individual fibers is 0.001 inches.
Photo: M. E. Gunter.

FIGURE 2.1 Landslide at Santa Cruz County, California.

This beachfront house will soon collapse. The slide happened after a heavy rain, which is typical of such disasters.
Photo: Alexander Lowry/Photo Researchers, Inc.

FIGURE 2.2 Salt crust on soil at a farm.

Dead cropland that has resulted from poor irrigation practices. Over a period of a few years, the soil developed a white coating and the growing crops became stunted and died. New seedlings were unable to sprout. Why does this happen? Can the farmer's remaining crops be saved? Can the area already affected be restored? Might a course in environmental geology have enabled the farmer to prevent this calamity? Should this farmer seek another line of work?
Photo: John D. Cunningham/Visuals Unlimited.

it apparently consists of minerals that have formed from evaporation of water in the soil over several years.

Exactly what are the minerals that have displaced his fruits and vegetables? Are all minerals equally harmful to crops? Do they all precipitate from soil water this easily? How did the minerals get into the soil water in the first place? Does it have anything to do with the source of his irrigation water? Why did this economic calamity happen to this farmer, but not to all farmers? Could this poisoning of his land have been prevented? Might some understanding of simple chemistry and mineral formation be able to answer these questions? Could this understanding prevent further damage to productive farmland?

Understanding rocks helps us understand water sources and water quality (Figure 2.3). This spring issues from a rock in a hillside. Residents and visitors alike have noted that the water pouring from the rock is always cold, winter or summer, and tastes better than the tap water in their homes. They also have observed that the water has flowed undiminished, even during dry years.

How did the water get into the rock? How did it travel from its point of entry to where we see it

FIGURE 2.3 This spring has produced pure, cool water day and night for thousands of years.

Springs of cool, clear water are common in hilly areas. Some springs form spectacular waterfalls as the water flows from otherwise ordinary-looking rock. Other springs ooze water slowly on gentle slopes. *Photo: John Colwell/Grant Heilman Photography, Inc.*

emerging? Can all rocks produce water like this? What types of rocks are most likely to yield this fresh water, and what characteristics must these rocks have? Where are they located? Why is its temperature so consistently cold year-round? Where might we look to find more springs of this good-tasting, free water? Is all spring water drinkable like this, or can it be salty and undrinkable—and why? Clearly, some understanding of rock distribution, rock characteristics, and rainfall and snowfall patterns is needed to answer such questions.

Understanding rocks and minerals helps us understand where metal ores can be found (Figure 2.4). This mine entrance has debris piles and nasty-looking polluted water draining from it. The mine produced lead-bearing rock for many years before it was abandoned.

Why did this particular rock contain valuable deposits of lead that are not found in neighboring areas? How can we find other such deposits that are needed by our industrial society? Does the presence of one metal indicate the possible presence of other metals? Do valuable metal deposits (in the form of metallic minerals) occur in all types of rocks? If not, which metals prefer which rocks?

When mining is finished, what should be done with the ugly piles of rock debris? Suppose the mining company has gone out of business or can't be located. Should the former owners, their descendants, the current landowners, or the government be liable for the clean-up? And what is causing that unsightly water that is draining from the debris? Is it poisonous, or simply a harmless attempt at abstract painting by Father Nature? How has rainwater interacted with the rock material to color the water? Again, a small amount of chemical knowledge about minerals and rocks may be the key to understanding potentially harmful processes in our environment.

The principal variables in the four examples we just visited are *minerals, rocks,* and *water.* These materials, and all others on Earth, are composed of the same fundamental particles: Earth's 90 natural elements. The names of some elements are familiar: silicon, oxygen, sulfur, uranium, carbon, and iron. To understand how Earth's materials interact and how they behave in the environment, you need to understand the nature of these elements.

Elements and Atoms

Elements are the fundamental ingredients in Nature's recipes for minerals, rocks, gases, liquids, and living things. About ninety natural elements exist. Each is a basic substance that has distinctive properties and behavior; if broken down any further, an element loses its identity. All matter on Earth—each of its natural elements—is formed of **atoms,** particles so small that 40 billion of them lying side by side would span only an inch. A human body contains 10^{28} atoms, the number 1 followed by 28 zeros. Atoms are composed of even tinier particles—protons, neutrons, and electrons.

A convenient way to visualize the arrangement of an atom is to imagine a tiny solar system. The "sun" at the center of an atom is its nucleus, which is a compact mass of protons and neutrons. The "planets" of an atom are its electrons, which orbit around the nucleus. Figure 2.5 shows a model of an atom. The electrons orbit the nucleus very rapidly, within fixed zones, so that, in effect, they create shell-like levels of electrical energy around the atom.

The electrons are extremely important. The entire universe literally runs on electricity, of the type we experience as "static cling," or as a shock of static electricity during dry weather, or as a lightning bolt during a thunderstorm. When we observe "static cling," we see objects clinging together because they have opposite electrical charges, positive and negative. In an atom, **protons** in the nucleus are positively charged and **electrons** are negatively charged, so they attract one another, holding the atom together. **Neutrons** are indeed neutral, having no charge at all.

In the common elements, the number of neutrons roughly equals the number of protons. If you add the protons and neutrons together, it gives you the **atomic**

FIGURE 2.4 Yak Tunnel Mine at Leadville, California Gulch, Colorado, has produced ore for many years.

This mine entrance has debris piles and nasty-looking polluted water draining from it. The mine produced lead-bearing rock for many years before it was abandoned. *Photo: Tim Haske/ProFiles West, Inc.*

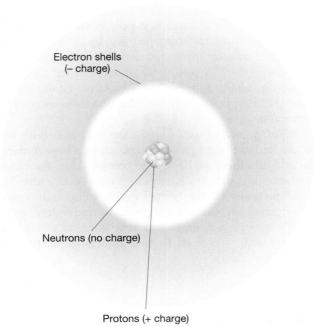

Electron shells
(– charge)

Neutrons (no charge)

Protons (+ charge)

FIGURE 2.5 Idealized model of an atom.

The atom includes a central nucleus composed of protons and neutrons, surrounded by rapidly orbiting electrons.

For example, the element carbon (atomic number six, for its six protons) has two stable isotopes. The commonest stable isotope has 6 neutrons, and this constitutes 98.893% of all natural carbon. This is called carbon-12 (because it has six protons and six neutrons). A rare isotope has seven neutrons (making it carbon-13), and constitutes only 1.107 percent of all carbon on Earth. Carbon-12 and carbon-13 total virtually 100 percent.

We say virtually because, if we extend the decimal places many times farther to the right, the total turns out to be slightly less than 100 percent. In very tiny quantities, carbon has several other isotopes. One of them is particularly important, the one with eight neutrons, carbon-14. This isotope is unstable, slowly disintegrating by giving off particles—a process called radioactive decay, or **radioactivity.** You may have heard of carbon-14, for it is widely used to determine the age of objects back to about 55,000 years before the present. (The method is called radioactive carbon-dating, or carbon-14 dating.) Unstable isotopes like carbon-14 are important in environmental concerns such as tracing groundwater movement (Chapter 6), nuclear power (Chapter 16), and radon contamination in our homes (Chapter 16).

The Importance of Electrons

Electrons are important because their electrical charge makes them interact with electrons that are orbiting around other atoms. This interaction allows some atoms to join into groups. These groupings may be gases, liquids, or solids, or may be dissolved in water. When we are concerned with natural chemical reactions on Earth (interactions between atoms), electrons are the active particles. Protons and neutrons are not involved. (Protons and neutrons participate only in nuclear reactions. These occur in nuclear power generation, nuclear weapons, and radioactive decay, such as in carbon-14, mentioned earlier.)

The electrons around each atomic nucleus are negatively charged. Their location is restricted to discrete zones around the nucleus termed **electron shells,** as shown in Figure 2.5. The innermost electron shell can hold only two electrons. The next shell outward can hold up to eight, the third shell can hold up to eighteen, the fourth up to thirty-six, and so on, with the number of electrons increasing as new shells are added for heavier elements.

Atoms always "desire" a full outer shell. Suppose an atom has three electron shells: two are complete (the innermost with two electrons, and the next shell with eight electrons), plus an incomplete third shell that has only one electron out of the 18 it could hold. Thus, it has a total of 2 + 8 + 1 = 11 electrons. Such an atom will try to unload this extra electron, giving it to a different

weight of the element. (Electrons have negligible weight compared to protons and neutrons.)

The number of protons in an atom gives each element its unique character. All atoms with eight protons form a unique gas that supports life on Earth (oxygen); all atoms with fourteen protons form a unique solid that makes great computer chips (silicon); all atoms with twenty-nine protons form a unique metal that makes beautiful jewelry and electrical wire (copper); and so on. The number of protons in the atom is the element's **atomic number.** The simplest element, hydrogen, contains only one proton (atomic number 1); the heaviest naturally occurring element, uranium, contains ninety-two protons (atomic number 92). These and the other elements are presented in the periodic table of the elements (Table 2.1).

Atoms are electrically balanced, so for each positive proton in an atom, a negative electron orbits the nucleus. Thus, a hydrogen atom, with its single proton, has a single orbiting electron. A uranium atom's nucleus, with its dense cluster of ninety-two protons, is surrounded by a busy swarm of ninety-two electrons, at various distances (shells) from the nucleus.

Isotopes

You might think that all atoms of any element are identical, and many of them are. However, the number of neutrons in the nucleus of an element can vary. This variation creates variants of an element, called **isotopes.**

TABLE 2.1 Periodic Table of the Elements

Legend:
- Group number — I
- Atomic number — 1
- Name — Hydrogen
- Chemical symbol — H

- The eight most abundant elements in Earth's surface rocks.
- Major economically important elements found in metallic ores.
- Major recognized pollutant elements.

Period	I	II											III	IV	V	VI	VII	VIII
1	1 Hydrogen H																	2 Helium He
2	3 Lithium Li	4 Beryllium B											5 Boron B	6 Carbon C	7 Nitrogen N	8 Oxygen O	9 Fluorine F	10 Neon Ne
3	11 Sodium Na	12 Magnesium M											13 Aluminum Al	14 Silicon Si	15 Phosphorus P	16 Sulfur S	17 Chlorine Cl	18 Argon Ar
4	19 Potassium K	20 Calcium C	21 Scandium Sc	22 Titanium Ti	23 Vanadium V	24 Chromium Cr	25 Manganese Mn	26 Iron Fe	27 Cobalt Co	28 Nickel Ni	29 Copper Cu	30 Zinc Zn	31 Gallium Ga	32 Germanium Ge	33 Arsenic As	34 Selenium S	35 Bromine Br	36 Krypton Kr
5	37 Rubidium Rb	38 Strontium Sr	39 Yttrium Y	40 Zirconium Zr	41 Niobium Nb	42 Molybdenum Mo	43 Technetium Tc	44 Ruthenium Ru	45 Rhodium Rh	46 Palladium Pd	47 Silver Ag	48 Cadmium Cd	49 Indium In	50 Tin Sn	51 Antimony S	52 Tellurium Te	53 Iodine I	54 Xenon Xe
6	55 Cesium Cs	56 Barium Ba	57–71 Series of Lanthanide Elements	72 Hafnium Hf	73 Tantalum Ta	74 Tungsten W	75 Rhenium Re	76 Osmium Os	77 Iridium Ir	78 Platinum Pt	79 Gold Au	80 Mercury Hg	81 Thallium Tl	82 Lead Pb	83 Bismuth Bi	84 Polonium Po	85 Astatine At	86 Radon Rn
7	87 Francium Fr	88 Radium Ra	89–103 Series of Actinide Elements	104 Rutherfordium Rf	105 Hahnium Ha	106 Seaborgium Sg												

Series of lanthanide elements:

57 Lanthanum La	58 Cerium Ce	59 Praseodymium Pr	60 Neodymium Nd	61 Promethium Pm	62 Samarium Sm	63 Europium Eu	64 Gadolinium Gd	65 Terbium Tb	66 Dysprosium Dy	67 Holmium Ho	68 Erbium Er	69 Thulium Tm	70 Ytterbium Yb	71 Lutetium Lu

Series of actinide elements:

89 Actinium Ac	90 Thorium Th	91 Protactinium Pa	92 Uranium U	93 Neptunium Np	94 Plutonium Pu	95 Americium Am	96 Curium Cm	97 Berkelium Bk	98 Californium Cf	99 Einsteinium Es	100 Fermium Fm	101 Mendelevium Md	102 Nobelium No	103 Lawrentium Lr

type of atom (element). Conversely, another atom (element) that needs just one more electron to complete its outer shell is happy to receive it. This interaction locks the two atoms together in electrical partnership, forming a **chemical bond** between the atoms. This is shown in Figure 2.6.

The element that released its "unwanted" electron now has one more proton than it does electrons, unbalancing the atom's electrical charge from neutral to +1. Conversely, the element that added the "unwanted" electron now has one more electron than it does protons, unbalancing this atom's electrical charge from neutral to −1. Such a charged atom is called an **ion.** The numerical values of charge are whole numbers and range from +6 to −2. Examples of ions are Ca^{+2} (calcium) and Cl^- (chlorine).

FIGURE 2.6 Ionic bonding.

Sodium and chlorine atoms in the process of transferring one electron from sodium to chlorine. Because of this transfer, sodium becomes a positive ion and chlorine a negative ion.

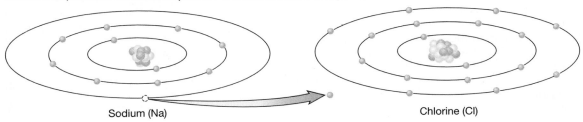

Sodium (Na)   Chlorine (Cl)

Chemical Bonds Hold It All Together

The interactions among electrons in elements cause elements to bond together. Four types of chemical bonds are important for environmental concerns. These are ionic bonds, covalent bonds, metallic bonds, and hydrogen bonds. (Don't be intimidated by these names—the concepts are simple, as you shall see.) These four types of bonds are important because each type has distinctive characteristics that affect the behavior of solid particles at the Earth's surface.

1. **Ionic bonds** occur in table salt, for example. Ionic bonds form when one element gives away an electron and another element accepts it. This bond is thought of as a total separation of an electron from its parent element and its legal adoption by an adopting element (Figure 2.6). Ionic bonds are best developed between two elements that need to gain or lose only a single electron to expose a full outer shell to the world. For example, sodium can accomplish this by giving away the lone electron in its outer shell, so that this old outer shell disappears and a full shell with eight electrons can be "seen." Chlorine exposes a full outer shell by adding a single electron to its outer shell, increasing the number of electrons from seven to eight, a full shell.

Elements with a single electron in their outer shells are those in group I in the periodic table (Table 2.1). Elements lacking a single electron in their outer shells are those in group VII. In the example shown in Figure 2.6, sodium (Na) in Group I bonds with chlorine (Cl) in Group VII, to form common sodium chloride (NaCl)—table salt.

Ionic bonds are easily disrupted by water, so that ionically bonded substances are very soluble. Sodium chloride, a fine example, is easily dissolved in water, as the world's salty oceans attest. (The most abundant elements in seawater are sodium, symbolized as Na^+, and chlorine, Cl^-.) Ionic bonds also occur between oxygen (which has space for two more electrons in its outer shell) and iron, calcium, sodium, magnesium, and potassium, all of which have a two-electron surplus in their outer shells.

So what, you may ask? Here is the environmental importance of this little bit of chemistry: A building should not be sited on a rock composed of ionically bonded minerals, *because they are water-soluble and therefore weak*. To illustrate with an extreme example, imagine building a 30-story office building on a foundation of salt, in a wet climate!

2. **Covalent bonds** are thought of as forming through a sharing of electrons (Figure 2.7). For example, if two atoms each have half of the electrons needed to fill their outer shell, they have equal urges to complete their outer shells. This results in a sharing of electrons, rather than a giving and accepting of electrons. Carbon-to-carbon bonds are a good example of covalency. The outer shell of each carbon atom has four of the eight electrons needed for a full second shell.

Organic compounds are rich in covalent carbon-carbon bonds, and are very insoluble in water. Examples include petroleum and natural gas. Among abundant elements of Earth's crust, covalent bonds occur between elements in the middle part of the periodic table—for example, between oxygen (negative charge) and aluminum or silicon (positive charge).

3. Metallic bonds occur in metals such as gold and copper. Metallic bonds form between elements whose outermost electron shells are held to the nucleus only weakly. All the electrons in this outer shell therefore can move rather freely, and it is this aggregate movement of large numbers of electrons that holds together the nuclei in a bonded state. When the electrons in a metal object are all forced to move in one direction because someone hooks a battery or an electrical generator to it, an electric current results. This is what happens along the copper wiring in your home when you

FIGURE 2.7 Covalent bonding.

A pair of electrons is shared between two carbon nuclei. Shown here is a network of these bonds in a carbon mineral such as graphite or diamond.

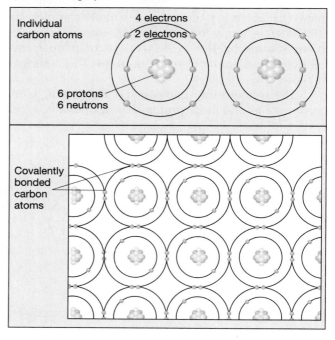

Individual carbon atoms

4 electrons

2 electrons

6 protons
6 neutrons

Covalently bonded carbon atoms

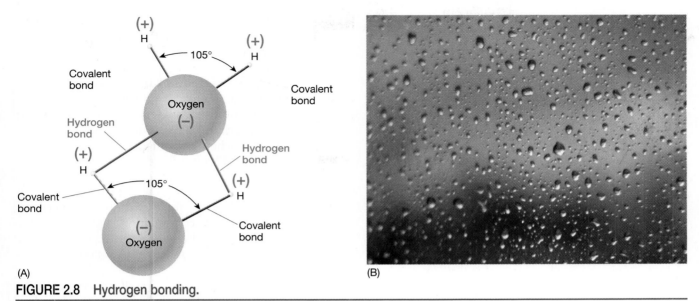

FIGURE 2.8   Hydrogen bonding.

Hydrogen bonding of water molecules, to each other and to molecules of other substances. (A) Idealized sketch of two water molecules loosely held together by hydrogen bonding. Note the size of the hydrogen ions, which are pinpoints compared to the oxygen ion. (B) Water clinging by hydrogen bonds to a glass surface. *Photo: Francis G. Sheehan/Photo Researchers, Inc.*

flip a light switch "on." Metallic elements are bonded by metallic bonds.

4. Hydrogen bonds are "secondary" bonds. A hydrogen bond forms between a hydrogen atom and a negative ion in another molecule. For example, water molecules (H_2O) loosely cling to one another by hydrogen bonds: the positive "hydrogen end" of one H_2O molecule is attracted to the negative "oxygen end" of another H_2O molecule. Hydrogen bonding of water is illustrated in Figure 2.8A. These hydrogen bonds make water stick to things, including to itself, holding together water droplets.

Hydrogen bonds are weak, but still require energy to be broken. This explains why heat energy must be added to ice to separate its hydrogen-bonded H_2O molecules (melting), and why even greater heat energy must be added to liquid water to separate its hydrogen-bonded H_2O molecules to cause evaporation.

Liquid water molecules form hydrogen bonds, not only among themselves, but with other negatively charged objects. This is why water coats surfaces that it cannot readily dissolve, like the walls of a glass shower stall or a urinal (Figure 2.8B).

Hydrogen bonds also are the reason that water is most dense at 39°F, and not at its freezing temperature of 32°, as you might expect. Because the structure of ice makes it less dense than liquid water, it floats. This is why ice cubes float on top of your drink, instead of sinking to the bottom. Were it not for this characteristic, lakes would freeze from the bottom up, instead of from the top down, and thus would be frozen solid during cold-climate winters, freezing plants and animals at the bottom of the lake (Figure 2.9).

To review the types of bonds, with examples: among the abundant elements in the Earth's crust, the only positive ions that *covalently* bond to oxygen are aluminum, silicon, and iron with a charge of +3. Forming *ionic* bonds with oxygen are iron with a +2 charge, calcium, sodium, magnesium, and potassium. *Metallic* elements are bonded by metallic bonds. *Hydrogen* bonds are very weak but extremely important in water and in clay minerals (Chapter 5).

FIGURE 2.9   Hydrogen bonds and freezing water.

Temperature layering in a lake because of hydrogen bonding.

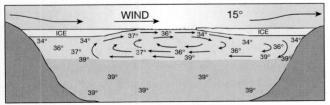

Elements in Earth's Surface Rocks

Our planet is composed of ninety natural chemical elements, plus about twenty more that are artificially generated in nuclear reactors. The basic ninety elements occur in very unequal amounts: just eight elements form 99.4% (by weight) of the upper twenty miles of Earth's **crust,** and the other eighty-two total only 0.6%.

TABLE 2.2 Abundant Elements in Earth's Crust

Element	Electrical Charge	Atomic Number (number of protons in nucleus)	Weight Percent
oxygen	O^{-2}	8	46.4
silicon	Si^{+4}	14	28.2
aluminum	Al^{+3}	13	8.3
iron	Fe^{+2}, Fe^{+3}	26	5.6
calcium	Ca^{+2}	20	4.1
sodium	Na^{+1}	11	2.4
magnesium	Mg^{+2}	12	2.3
potassium	K^{+1}	19	2.1
all others	--	--	0.6
			100.0

The eight most abundant elements are shown in Table 2.2. Note that oxygen is the only abundant negatively charged ion in the crust (O^{-2}); all the others are positively charged. This is why oxygen **oxidizes** (combines with) many substances. Note that the eight most abundant elements are all relatively light—they all have few protons, and therefore low atomic numbers. Only one (iron) has an atomic number above twenty.

Because these elements are the building blocks of Earth's minerals, their relative abundance in Earth's crust is reflected in their relative abundance in minerals. The most abundant minerals in the crust are composed largely of the top two elements—oxygen and silicon. These minerals are called **silicates** because of their silicon content.

Most of the economically important metallic elements needed for our industrial civilization are *not* among the abundant eight elements. For example, the abundance of titanium (used in spacecraft construction) is only 0.57%; manganese (used in steel alloys) is 0.09%; nickel, zinc, and chromium, 0.01% each; copper is 0.005%; and lead is 0.001%. Others, such as tin, molybdenum (used in specialty steels), and tungsten (used for light bulb filaments) are even less abundant.

Economically minable concentrations of these elements are scarce and are decreasing rapidly as our industrial consumption expands. Substitutes for most of them have yet to be found or synthesized (Chapter 13).

Minerals

When ions combine, the resulting change in electron distribution determines the characteristics (physical and chemical) of the resulting minerals. However, the properties of minerals cannot be predicted from the properties of the individual elements before they were combined.

For example, sodium (Na) is a metal that reacts with oxygen in air or water so ferociously that it is stored in kerosene (which contains only carbon and hydrogen). Chlorine (Cl) is a poisonous, greenish-yellow gas. But when combined as sodium chloride (NaCl, the mineral halite) the solid produced is ordinary table salt. The properties of the sodium-chlorine combination (very high solubility in water or steak juice) are determined by the distribution of electrons in the sodium chloride combination, just as the properties of the sodium and chlorine were determined by *their* electronic structures when they existed as uncombined elements.

Just as neutral atoms of sodium and chlorine have different properties as individuals than they have when combined into salt, sodium ions and chloride ions dissolved in water have yet a different set of properties. The "salty" taste created in water (or saliva) by sodium ions is well known to all and results from a biochemical interaction between the ions and your taste buds.

The bottom line of this discussion is this: the arrangement of electrons around atomic nuclei is the reason for the physical and chemical properties of all materials. These properties include size, weight, density, color, hardness, solubility, and so forth. But the two properties of greatest interest and importance to environmental scientists are hardness and solubility. For example:

- *Quartz (silicon dioxide, SiO_2) is very hard and very insoluble in water.* Water that has been in contact with quartz for an extended period is virtually unaffected.

- *Calcite (calcium carbonate, $CaCO_3$) is soft and moderately soluble in water.* Water that has been in contact with calcite for an extended period contains abundant calcium. It is called "hard water" or "limey water" because soap suds will not form well, and soapy hard water leaves a nasty scum in bathtubs.

- *Gypsum (calcium sulfate, $CaSO_4 \cdot 2H_2O$) is very soft and very soluble in water.* Water that has been in contact with gypsum even briefly is undrinkable because of its high sulfate content (the SO_4^{-2} ion is a diuretic, leading to frequent pit stops).

- *Halite (sodium chloride, NaCl) is very soft and extremely soluble in water.* Water that has been in contact with halite even briefly is undrinkable because of its high sodium content (salty taste).

(A) (B)

(C) (D)

FIGURE 2.10 The three most common minerals in Earth's crust.

(A) Quartz—a single quartz crystal shows well-developed crystal outlines and characteristic translucency. Quartz has no cleavage. (B) Feldspar—this variety is orthoclase feldspar, which has a characteristic salmon-pink color. Another variety is plagioclase feldspar (C), which has a characteristic white color and striations. (D) Mica—this variety is muscovite mica, which has a characteristic light color and silvery translucency. Note the sheetlike cleavage, which allows separation of thin sheets that feel roughly like thin plastic.

Important Minerals

Chemical bonds that form naturally often produce solids we call minerals. A **mineral** is carefully defined, and must meet all of these criteria:

1. A mineral *occurs naturally* (this excludes artificial materials).
2. A mineral is *inorganic* (meaning that it never was alive).
3. A mineral is a *solid at average temperatures on Earth's surface* (this excludes substances that are liquids and gases at average surface temperatures).
4. A mineral *has a regular, repeating internal structure* (it has a **crystalline** pattern, not a random arrangement of atoms).
5. A mineral has a *fairly definite chemical composition* (this excludes rocks like granite, which are mixtures of minerals).

It is interesting to test some common substances against these criteria to see whether they are minerals. Is a diamond a mineral? a brick? window glass? ice? Wood? How about a charcoal briquette?

The requirements of a fixed internal structure and definite chemical composition make constant the physical and chemical properties of a mineral. For example:

- Salt (crystalline structure, definite composition: NaCl) is always very soluble (through ionic bonding) and always tastes salty when dissolved in water (because of sodium ions).
- Quartz (crystalline structure, definite composition: SiO_2) always has a low solubility in water because the silicon and oxygen atoms are very tightly covalently bonded.
- Copper (crystalline structure, definite composition: Cu) always conducts electricity easily because of the metallic bonding among the copper atoms.

There are several ways to classify minerals. One is to group them according to the most significant ion groups they contain. Using this system, most of Earth's 3500 known minerals fall into the eight groups shown in Table 2.3):

1. **Silicates (SiO_4^{-4} ion).** The most abundant minerals in Earth's crust (about 65 percent) are silicates, minerals formed by groups of these silicon-oxygen structures. The structures are called tetrahedra for their shape, a four-sided pyramid. The most abundant silicate minerals are quartz, feldspar, and mica (Figure 2.10). Silicates are very insoluble and are used commercially in bricks, window glass, insulation, jewelry, and electronics.

TABLE 2.3 Minerals Important to Environmental Geologists

Mineral	Chemical Composition	Comments
Silicates (SiO_4^{-4} ion)		
Quartz	SiO_2	Most abundant mineral in sandstones
Orthoclase feldspar	$KAlSi_3O_8$	Most common minerals in Earth's crust
Plagioclase feldspar	$NaAlSi_3O_8$	
Hornblende	Ca, Na, Mg, Fe, Al silicate	Sources of iron for red soils and rocks
Augite	Ca, Mg, Fe, Al silicate	
Olivine	$(Mg, Fe)_2SiO_4$	
Chlorite	hydrous Mg, Fe, Al silicate	Commonly a factor in initiating landslides
Biotite mica	hydrous K, Fe, Mg, Al silicate	Commonly a factor in initiating landslides
Muscovite mica	$KAl_3AlSi_3O_{10}(OH)_2$	Commonly a factor in initiating landslides
Illite clay	hydrous K, Al, Fe, Mg silicate	Commonly a factor in initiating landslides
Montmorillonite clay	hydrous Na, Ca, Al, Fe, Mg silicate	Expands when wet; Commonly a factor in initiating landslides
Kaolinite clay	$Al_2Si_2O_5(OH)_4$	Commonly makes slopes unstable; Commonly a factor in initiating landslides
Carbonates (CO_3^{-2}) ion		
Calcite	$CaCO_3$	Host for commercial caves and sinkholes
Dolomite	$CaMg(CO_3)_2$	
Sulfates (SO_4^{-2} ion)		
Gypsum	$CaSO_4 \cdot 2H_2O$	Construction material
Phosphates (PO_4^{-3} ion)		
Apatite	$Ca_5(PO_4)_3F$	Agricultural fertilizer
Oxides (O^{-2} ion) (all are the major ores* of the metal shown)		
Hematite	Fe_2O_3	Iron ore
Rutile	TiO_2	Titanium ore
Cassiterite	SnO_2	Tin ore
Pyrolusite	MnO_2	Manganese ore

2. **Carbonates (CO_3^{-2}) ion.** The most abundant carbonate minerals are calcite (Figure 2.11) and dolomite. Calcite often forms **limestone,** a rock (rocks are aggregates of minerals). Because carbonate minerals are fairly soluble in water, most commercial caverns (like Mammoth Cave and Carlsbad Cavern) and sinkholes (like those in Florida and Kentucky) are formed in limestones. Calcite is the main mineral in cement and concrete. It also is used as a filler and whitener in plastic pipe, paper, and toothpaste.

3. **Sulfates (SO_4^{-2} ion).** Sulfate minerals are more soluble than the major carbonate minerals. The common sulfates are gypsum (Figure 2.12) and anhydrite. Gypsum is used to make the drywall plasterboard in houses, as a soil conditioner in agriculture, and as a hard cast to stabilize broken bones.

4. **Phosphates (PO_4^{-3} ion).** The only common phosphate mineral is apatite, used worldwide as an agricultural fertilizer. The use of phosphates as an agricultural fertilizer and in household laundry soaps is a source of pollution in lakes and streams (Chapter 8).

5. **Oxides (O^{-2} ion).** In oxide minerals, the negative oxygen ion is bonded to a positive *metallic* ion, unlike its bond to nonmetallic elements in the previous four groups. Examples include the bond with iron in hematite (Figure 2.13) and magnetite, and with titanium, tin, and manganese. Oxide

TABLE 2.3 *(Continued)*

Mineral	Chemical Composition	Comments
Sulfides (S^{-2} ion) (all but pyrite are the major ores* of the metal shown)		
Sphalerite (zinc)	ZnS	
Pentlandite (nickel)	$(Fe, Ni)_9S_8$	
Molybdenite (molybdenum)	MoS_2	
Cinnabar (mercury)	HgS	
Galena (lead)	PbS	
Chalcopyrite (copper)	$CuFeS_2$	
Cobaltite (cobalt)	$(Co, Fe)AsS$	
Orpiment, Realgar (arsenic)	As_2S_3, AsS	
Stibnite (antimony)	Sb_2S_3	
Pyrite (iron)	FeS_2	Major cause of acid mine waters
Halides (Cl^-, F^-, Br^-, I^- ions)		
Halite	$NaCl$	Table salt
Sylvite	KCl	Table salt substitute
Fluorite	CaF_2	Industrial processes
Native Elements		
Gold	Au	Jewelry
Silver	Ag	Photography
Platinum	Pt	Catalytic converters
Copper	Cu	Electrical wires
Sulfur	S	Insecticides
Diamond	C	Gem
Graphite	C	Pencils

*An ore is a rock from which an economically valuable metallic mineral can be extracted at a profit.

FIGURE 2.11 Calcite.

Large crystal of the mineral calcite, identified by its preferred cleavage at angles of 78° and 102° (due to weaknesses in its internal structure). Note the formation of double images viewed through clear crystals. Calcite is a mineral composed of calcium, carbon, and oxygen—$CaCO_3$.

FIGURE 2.12 Gypsum.

Gypsum is a sulfate mineral identified by its softness relative to a fingernail. Gypsum is made of calcium, sulfur, and oxygen, plus some water bound in—$CaSO_4 \cdot 2H_2O$.

FIGURE 2.13 Hematite.

This rock is formed of hematite spheres. The reddish color is very distinctive. Hematite is basically the same as rust found on steel, a compound of iron and oxygen—Fe_2O_3.

minerals are very important sources of these metals (ores) for our industrial civilization.

6. **Sulfides (S^{-2} ion).** In these minerals, sulfur is the major negative ion, rather than oxygen. Sulfides contain no oxygen. Instead, the sulfur is bonded to a metallic element. Consequently, sulfides are the main ores of zinc, nickel, molybdenum, mercury, copper, cobalt, arsenic, antimony, and lead (Figure 2.14A). Sulfide minerals are very unstable at Earth's surface and alter rapidly to sulfate ions, becoming an important source of water pollution (Figure 2.14B).

7. **Halides (Cl^-, F^-, Br^-, I^- ions).** Halides are minerals that contain either fluorine, chlorine, bromine, or iodine as an essential element. All are extremely soluble because of relatively weak ionic bonding between their atoms. The only important halide minerals are halite (NaCl) or ordinary table salt (Figure 2.15A); sylvite (KCl), used in fertilizers and as a "salt substitute" in medically prescribed "salt-free" diets; and fluorite (CaF_2), used in the chemical and steel industries. Today, salt is as important to us as it was to our ancestors (Figure 2.15B).

8. **Native elements.** Minerals that commonly occur simply as aggregates of atoms of a single element are gold, silver, platinum, copper, sulfur, and carbon in the forms of diamond and graphite (Figure 2.16). Gold, silver, platinum, and diamond are very valuable and are used in the jewelry industry; gold and platinum command about $400 per ounce of the metal (at 1996 prices). The native-element minerals have many other commercial uses. For example, the carbon mineral *graphite* is the metallic-looking writing material in "lead" pencils. (When graphite was discovered 400 years ago, it was thought to be a variety of lead.)

Identification of Minerals

Most common minerals, and those of particular importance in environmental concerns, can be identified by easily determined physical properties. The properties most generally useful in identification are hardness and cleavage.

Hardness is determined on the Mohs hardness scale, shown in Table 2.4. Mineral hardness ranges from 1 (the softest, like talc, from which baby powder is made) to 10 (the hardest, diamond). Harder materials like diamond (hardness 10) will scratch softer ones, like a glass table top (hardness 5) or marble surface (hardness 3). This is the key to determining mineral hardness: who will scratch whom? If someone hands you a sample of calcite and a sample of quartz, which look much alike, can you tell which is which by rubbing each against a knife blade? (Answer: see the table.) Few minerals have hardnesses greater than 7.

(A)

(B)

FIGURE 2.14 Sulfide minerals.

(A) Galena—the major ore of the metal lead. Galena is a mineral made of lead and sulfur—PbS. It has an outstanding cubic cleavage. Width of specimen is 3 inches. (B) Pyrite—large crystals of pyrite ("fool's gold") show its characteristic brassy yellow color. Pyrite, a mineral made of iron and sulfur, is the major cause of acid water that drains from coal mines (Chapter 13).

(A)   (B)

FIGURE 2.15   Halite.

(A) Halite is identified by its salty taste and cubic breakage (cleavage) pattern. The large fragment shows "rock" salt as it comes from the mine. To the left is a large cleavage fragment (not from the finer-grained large fragment). Next to the cleavage fragment is a pile of table salt from the author's kitchen. Large fragment is 3″ in length. Halite is a mineral made of sodium and chlorine—NaCl. (B) Guajiro Indians harvesting sun-dried salt at Manaure, Columbia, the last unmechanized flats of the national salt industry. *Photo: Loren McIntyre, National Geographic Society.*

Cleavage refers to the way that many minerals tend to break along particular planes, because of the weaker bonding across these planes. For example, calcite cleaves in two directions, with cleavage angles of 102° and 78° (Figure 2.11).

Color can help identify a mineral, because most minerals have a consistent color in freshly broken pieces. But color seldom is conclusive, because different specimens of the same mineral may vary in color due to impurities, and the color may change over time when a mineral surface is exposed to air and water. Those with variable color include quartz, calcite, and fluorite, although most quartz is colorless and translucent, and calcite is generally either clear or opaque white.

Color also can mislead. Real gold and pyrite ("fool's gold") have a similar shiny, golden color. Hardness is the key to telling them apart: real gold is soft, with a hardness of 3; pyrite is harder (6) than a knife blade (5).

FIGURE 2.16   Formation of a nugget.

Gold is shot through this 3.5-inch specimen of quartz from the Van Dyke gold mine in South Africa. *Photo: Bruce Cairncross/Rand Afrikaans University.*

TABLE 2.4   Mohs Hardness Scale

The scale was devised in 1824 by a mineralogist named Mohs.

Relative Hardness	Index Mineral	Common Objects
10	Diamond	
9	Corundum	
8	Topaz	
7	Quartz	Steel file—6.5
6	Orthoclase	Glass, knife, nail—5.5
5	Apatite	
4	Fluorite	
3	Calcite	Copper penny—3.0
2	Gypsum	Fingernail—2.5
1	Talc	

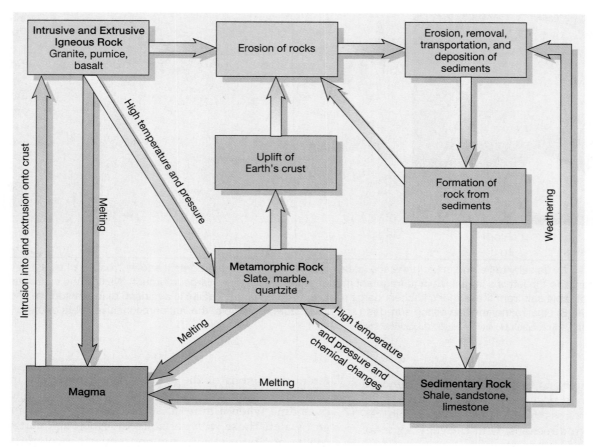

FIGURE 2.17 The rock cycle.

Various geologic processes change one rock type to another. *Source: Lutgens and Tarbuck,* Foundations of Earth Science, *New York: Macmillan, p. 31.*

Rocks

We must clearly distinguish between minerals and rocks. Recall that each *mineral* has a specific composition of one or more elements: quartz is SiO_2, calcite is $CaCO_3$, and so on. In contrast, a **rock** *is an aggregate of different mineral crystals.* For example, the rock we call *granite* is a mixture of three minerals: quartz, feldspar, and mica. The rock we call *shale* is a mixture of different clay minerals.

There are special cases, of course. Although most rocks are mixtures of two or more different minerals, some are composed of just one mineral. An example is very pure limestone that is all calcite ($CaCO_3$). Other special cases are volcanic glass and coal, both of which are composed of noncrystalline materials.

Geologists group rocks into three types, based on the process that formed them. These groups are *igneous rocks, sedimentary rocks,* and *metamorphic rocks.* These three groups are shown in the rock cycle diagram shown in Figure 2.17. The point of the diagram is the interrelationship that exists among the three types. Each can be converted into the other, as you shall see.

Igneous Rocks

Temperatures measured in underground mines and in deep holes drilled for petroleum and natural gas reveal that temperature increases with depth. The increase is referred to as the **geothermal gradient** and averages about 1°F per 60 feet of depth. In fact, miners in southern Africa's deep gold mines perspire heavily from the heat. Silicate minerals like quartz begin to melt at temperatures of 1200°F to 1300°F. So, at a depth of 16 miles (85,000 feet), where the temperature is around 1400°F, we would expect crustal rock to be molten. Indeed, some molten rock is generated at this depth.

But melting does not occur commonly until greater depths, because the increased pressure at depth compresses materials, forcing them to remain solid, even at temperatures that are above their melting point at surface pressure. Molten material may be produced at depths as shallow as perhaps 6 miles (because of an unusually high geothermal gradient) or as deep as 90 miles.

Molten silicate material is termed **magma.** Magma's most abundant elements are the eight listed in Table 2.2. In addition, most magmas contain a few percent of water and gases. Because of stresses that always exist in the crust, rocks above the magma are cracked. Through these fissures the magma can migrate upward to areas of lower pressure, sometimes all the way to Earth's surface, where it may emerge as a lava flow or form a volcano. These different forms of eruption are shown in Figure 2.18.

As the magma rises into cooler rocks near the surface, it chills and solidifies to form mineral crystals. Rocks made of these crystals are **igneous rocks** (the word *igneous* means "fire-formed"). If solidification (crystallization) occurs at great depth, the resulting rock is called **plutonic** for Pluto, god of the underworld

in Greek mythology. The crystals of the major minerals in plutonic rock are relatively large, typically 0.1 in. or more in diameter (Figure 2.19). This larger crystal size results from very slow cooling of the magma, giving the crystals time to grow.

Magmas that crystallize at shallow depths are fine-grained, often so fine-grained that the crystals are visible only through a high-powered microscope. Sometimes the magma rises all the way to the surface before it crystallizes and spills out onto the ground. Magma emitted at the surface is termed **lava.** Hardened lava is *volcanic rock.* The most common volcanic rocks are *basalt* (dark colored) and *rhyolite* (light colored) (Figure 2.20).

Much hardened lava is fully crystalline but exceedingly fine-grained, but other lava may have cooled so rapidly that the regular periodic crystal structure of minerals did not have time to form. This phenomenon occurs, for example, in Hawaii, where lava flows into the Pacific Ocean and is quickly

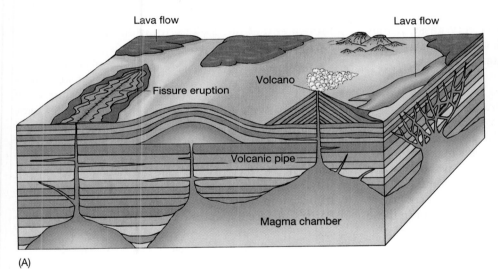

(A)

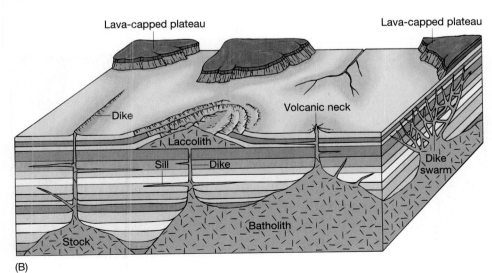

(B)

FIGURE 2.18 Different paths that magma may follow, and the results.

(A) Magma migrates toward the surface, with varying results.
(B) Long after the magma cools into rock, and the surface erodes, various igneous rock structures remain. *Source: E. J. Tarbuck and F. K. Lutgens, 1994,* The Earth, *4th ed.,* New York: Macmillan, p. 71.

FIGURE 2.19   Granite.

Granite is a rock—a mixture of several minerals, typically three or four: the pink mineral is orthoclase feldspar; the dark mineral is biotite mica; the clear grains are quartz; and the white grains are plagioclase feldspar. Note the relatively large crystal size, the product of slow cooling.

quenched. In this case, the resulting solid is *volcanic glass* (Figure 2.21A). If the lava is very gassy, it may quench into a frothy, glassy rock called *pumice* (Figure 2.21B). Some igneous rocks contain two sizes of crystals, reflecting two different rates of cooling.

Crystals that grow from the liquid magma interlock, so no empty cavities exist among the grains. Thus, unfractured igneous rocks are strong and contain scant water or gases, making these rocks good bases for sanitary landfills. The same quality also makes igneous rocks poor candidates for underground water supplies or oil and gas accumulations. However, some volcanic rocks become highly fractured during rapid cooling at Earth's surface and may contain enough water in these

fractures to serve as a water source for local communities (Chapter 6).

Sedimentary Rocks

Sedimentary rocks cover about two-thirds of Earth's land area, so they are the most common rock type. Some of them provide valuable resources upon which our industrial civilization depends, including ground water, oil, natural gas, and coal.

Sedimentary rocks do not form deep in Earth's interior, but form at Earth's surface, because they are made of **sediment** created by surface processes. These rocks have three origins: (1) they are bits of sediment cemented together, (2) they are precipitated directly from seawater, or (3) they are material left behind when water evaporates. Figure 2.22 shows sediment types and the Earth processes that act upon them to produce sedimentary rocks.

Sedimentary rocks are always layered, because that is how they are deposited. The breaks between layers commonly serve as surfaces of slippage, because rainwater that flows along these surfaces reduces the cohesion between them (Chapter 5).

Most sedimentary rocks are made of accumulated fragments (sediment) of existing rock that are transported and deposited by water or wind. Examples of such transport are avalanche deposits, river sand bars, tidal swamps, sand dunes, offshore ocean mud, and river deltas. Loose sediment is changed into hard rock by a combination of compaction under pressure and by cementing with minerals that precipitate from the water that flows through the sediment. This process of **lithification** (changing to rock) occurs from near Earth's surface to depths of tens of thousands of feet. It

FIGURE 2.20   Basalt and rhyolite.

(A) Basalt. A rock whose dark color suggests a large percentage of iron-rich minerals. The mineral crystals are too small to be visible to the unaided eye, the product of fast cooling. (B) Rhyolite. A rock whose pale pinkish color suggests the dominance of pink feldspar and quartz. The mineral crystals are too small to be visible to the unaided eye.

(A)

(B)

(A)   (B)

FIGURE 2.21 Examples of volcanic glass.

(A) Pumice, a glassy rock showing characteristic frothy texture. Pumice is used as an abrasive in Lava brand hand soap.
(B) Obsidian glass, a rock used in preindustrial societies for making cutting tools and weapons.

usually requires lots of time, from thousands to millions of years.

Sandstone, shale, and other sedimentary rocks are distinguished by particle size. Geologists define several sizes. The most significant are:

- "Sand size" clasts (particles of rock) are 0.08 inch down to 0.0002 inch in diameter. Sand grains cemented together form the rock called sandstone (Figure 2.23A).
- "Silt size" particles are smaller, 0.0002–00001 inch.
- "Clay size" particles are even smaller. This gives ordinary clay occurring in soil its fine, smooth texture. Cemented silt and clay grains form the rock called **shale** (Figure 2.23B).
- Coarser sediment is termed "gravel." When cemented, it forms the rock called **conglomerate** (Figure 2.23C).

The three most abundant types of sedimentary rocks are sandstone (20–25%), shale (65%), and limestone (10–15%). We will look briefly at each.

Sandstone

Sandstone (Figure 2.23A) is called a **clastic** rock, because it is formed of *clasts*, or fragments of existing rocks. These existing rocks were disintegrated by **weathering**. For example, rain, snow, and ice attack the surface of granite, slowly decomposing the crystals from which the rock is made. The resulting bits of rock (clasts) are washed into streams, which transport the clasts downstream until they are deposited. Many of these grains travel hundreds or thousands of miles to the ocean, to become sand on a beach or offshore. Other clasts become trapped in sand bars in the river and are

buried there. Some grains are blown by the wind and accumulate as sand dunes in desert areas. Many natural environments are the site of sand accumulations.

After deposition, the sand grains are covered and buried by new deposits. Most buried sand deposits remain in the subsurface for many thousands or millions of years. During this period, they are cemented into hard rocks by minerals such as calcite, quartz, or hematite, which precipitate from water into the spaces between the grains.

Commonly, these cementing materials do not completely fill the spaces between grains, leaving openings called **pores.** Thus, many sandstones eventually become filled with water, oil, or natural gas. Most porous rocks are also **permeable,** meaning their pores are interconnected, allowing fluids to migrate through the rock. Geologists who specialize in ground water and petroleum are vitally interested in how porous and permeable rocks are. In general, coarser-grained rocks contain larger pores and are more permeable.

Shale

In shale, about 60 percent of the rock is clay minerals. They are flexible and sheetlike, so they compact very tightly when buried (Figure 2.23B). This eliminates most pores and permeability. Thus, shales do not transmit fluids unless they are fractured after lithification. Precipitated cement is not required to turn mud into shale; compaction alone can do the job.

Shales are not the best base upon which to build a house. They are always highly fractured, and therefore tend to be unstable and move in response to gravity, freezing, thawing, and changes in water levels in the ground. Because they are composed predominantly of clay minerals, which have a sheeted structure

Sediment	Earth Process	Resulting Rock
Clay/silt deposits	Compaction cementation	Rock shale
Swamp	Compaction cementation	Coal
Gravel	Compaction cementation	Conglomerate
Sand	Compaction cementation	Sandstone

FIGURE 2.22 Where sedimentary rocks come from.

From left to right, you can see the type of sediment, the Earth processes that act upon it, and the result. *Photos: (Clay/silt) Peter L. Kresan Photography; (rock shale) Michael P. Gadomski/Photo Researchers, Inc.; (swamp) Mark C. Burnett/Photo Researchers, Inc.; (coal) Charles R. Belinky/Photo Researchers, Inc.; (gravel) Runk/Schoenberger/Grant Heilman Photography, Inc.; (conglomerate) Bill Bachman/Photo Researchers, Inc.; (sand) E. R. Degginger/Photo Researchers, Inc.; (sandstone) Peter L. Kresan Photography.*

(A)

(B)

(C)

FIGURE 2.23 Three common sedimentary rocks.

(A) Sandstone, coarse-grained in this sample. This sedimentary rock is composed of cemented-together grains of quartz and feldspar (white grains). The reddish color is a natural cement of hematite (iron oxide). *Photo: David McGeary.* (B) Shale, the most abundant type of sedimentary rock. It is made mostly of microscopic clay particles. Shale splits easily along thinly spaced planes because of the strongly parallel orientation of the clay minerals. The metal lens at center is 1.5 inches long.
(C) Conglomerate, a sedimentary rock that contains a large percentage of coarse grains. In this sample, both the pebbles and the sand are mostly quartz and chert. Width of specimen is 5 inches.

like micas, they weaken when wet. Also, some types of clay minerals swell and expand when wet, stressing and cracking building foundations and walls. In the United States, swelling clays are the most economically damaging geological hazard, outranking more spectacular catastrophic events such as earthquakes and volcanic eruptions. Before building a house, consult a geologist to learn what problems lie hidden beneath the harmless-looking surface.

Limestones

Limestones are sedimentary rocks composed almost entirely of calcite. Limestones form in two ways: they may be clastic like sandstones and shales—made of cemented fragments—or they may have precipitated from lake or sea water.

- In clastic limestones, the fragments most commonly are broken shells of clams, oysters, and other marine animals that live in shallow water near the beach (Figure 2.24). These creatures construct their shells of calcium carbonate (calcite). The cementing material in limestones also is calcite.

- In precipitated limestones, the texture is very fine-grained, and the grains have interlocking textures like an igneous rock—which also precipitates from a liquid (magma).

Evaporites

Sandstones, shales, and limestones form more than 95 percent of all sedimentary rocks. However, a few

FIGURE 2.24 Limestone, a sedimentary rock.

This specimen consists of fossil shells about 1 inch long, set in a matrix of calcite crystals about 0.02 inches in diameter. Specimen is 8 inches long.

less-common sedimentary rocks are of environmental significance. Foremost among these are **evaporite rocks,** which are composed of very soluble minerals you met earlier in this chapter: halite (NaCl), gypsum ($CaSO_4 \cdot 2H_2O$), and others. They are called evaporite rocks because at least 80 percent of a pool of seawater must be evaporated before they precipitate. The minerals in evaporite rocks are dominated by ionic bonds and are therefore very soluble. That is why most of the water must be evaporated for the minerals to precipitate.

Also important are **phosphorites,** dark-colored rocks composed of chemically precipitated apatite. Apatite is the same stuff your tooth enamel is made of: calcium phosphate. The phosphate in apatite is appetizing to plants, so these rocks are mined extensively for fertilizer. Unfortunately, the phosphate also can be a major water pollutant. An excess of phosphate in water stimulates algae to multiply uncontrollably and to cause the destruction of other life in a lake or stream, in a process called eutrophication (Chapter 8). Phosphorite deposits also contain commercially important uranium and vanadium (used in making specialty steels). In the United states, phosphorites are mined in Idaho, western Wyoming, and Florida.

Chert is another precipitated rock, composed of microcrystalline quartz (Figure 2.25). Chert is very hard and splintery when broken. It was used by Native Americans and later by European immigrants for arrowheads and cutting tools. Chert occurs in many colors, such as red (jasper), black (flint), and variegated (agate).

Metamorphic Rocks

Metamorphic means to change form, aptly describing **metamorphic rocks.** These rocks are the result of changes in the mineral composition and texture of existing rocks. The agents of change are heat and pressure.

Most metamorphic rocks form at temperatures between about 600°F and 1200°F, reached in the crust at depths of 6–30 miles. These temperatures are too low to melt the rock into liquid magma, but high enough to add a great deal of energy to the chemical elements that make up the minerals. This energy increase permits the ions (potassium, calcium, magnesium, etc.) to move along the crystal surfaces in thin films of water and recombine into other minerals—minerals in equilibrium with the new temperature conditions.

Pressure also increases with depth in the crust. If the pressure on a rock is equal in all directions, it will show a random orientation of minerals (as in the marble shown in Figure 2.26A). But rocks that form when the pressure is not equal in all directions show

FIGURE 2.25 Chert, a precipitated microcrystalline sedimentary rock composed of quartz.

Chert, also called flint, is very hard and can be sharpened to a blade, as was commonly done by North American Indians for use as arrowheads and spear points. Specimen is 6 inches long.

a preferred orientation of minerals, or a segregation of different minerals into distinct layers (as in the schist shown in Figure 2.26B). If the rock contains abundant sheet-like minerals such as micas, a small amount of water can cause slippage between the micas and slope instability, like the slippage between clay flakes in shales.

Rock Cycle. We have seen that one rock type can be changed into another by normal Earth processes, as shown in Figure 2.17. Sediments become sedimentary rocks and, with continued burial, the sedimentary rocks may turn into metamorphic rocks. If the temperature increases, these rocks melt and form magma. The magma may rise toward the surface, cool, and harden (crystallize) to form igneous rock. Igneous rocks that form at depth may subsequently be raised to the surface by stresses within Earth. The surface rocks are weathered by exposure to the atmosphere, separated into their constituent minerals, transported, and deposited as sediment to initiate a new cycle of rock formation and decay.

Minerals, Rocks, and People

Some environmental hazards result from direct human contact with minerals and rocks. High concentrations of asbestos, coal dust, and quartz dust in one's environment can cause cancer and lung disease.

Asbestos

The meaning of the term **asbestos** depends upon whom is speaking. To a mineral specialist, it is a group of silicate

(A)   (B)

FIGURE 2.26 Two examples showing how pressure influences the development of metamorphic rock.

(A) Marble is a metamorphic rock composed entirely of calcite and used commercially for coffee tables, window sills, and decorative walls in banks and public buildings. (B) Schist, also a metamorphic rock. The streaks and layering (called foliation) show how it was deformed. Schist is composed largely of micas with some fine-grained quartz.

minerals with similar internal structures and elongate shape (Figure 2.27). To an engineer, it is an industrial material with several useful properties. To a physician, it is an agent that might cause certain diseases. To a lawyer, asbestos signals a possible lawsuit; to a news reporter, a story; to an asbestos removal worker, a job; and to a public administrator or parent, a nightmare. From the viewpoint of governmental regulation, the key issue is the *character* of this mineral group. Exactly what may be dangerous and in what amounts, and what is the evidence regarding the possible danger?

The Occupational Safety and Health Administration (OSHA) defines asbestos as "any of the naturally occurring amphibole minerals (amosite, crocidolite, anthophyllite, tremolite, and actinolite) and the serpen-tine mineral chrysotile with dimensions greater than 5 micrometers [0.002 in.] long and less than 5 micrometers in diameter and an aspect ratio of 3:1." Translation: amphiboles and serpentines are two groups of minerals that are needle-shaped; aspect ratio means the ratio of length to width.

Thus, "asbestos" is not just one simple thing; the definition includes six different fibrous minerals. Approximately 95 percent of the asbestos mined and used in the United States is chrysotile, while crocidolite and amosite account for about 5 percent. Commercially, these three are known as "white" asbestos, "blue" asbestos, and "brown" asbestos, respectively.

Asbestosis is a chronic lung disease (technically, a pneumoconiosis) caused by foreign particles deposited

FIGURE 2.27 Asbestos.

(A) Natural asbestos as it occurs in metamorphic rocks. (B) Chrysotile fibers viewed through a special microscope. Width of individual fibers is about 0.04 mm (0.001 in.). *Photo: M. E. Gunter.*

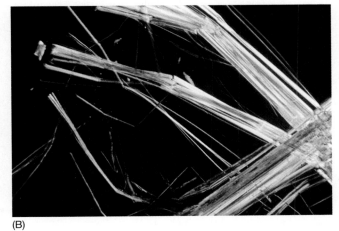

(A)   (B)

Minerals, Rocks, and People   **33**

in the lung through inhalation. It is the result of long-term inhalation of large amounts of asbestos. Lung tissue encapsulates the asbestos particles and hardens, thus decreasing lung efficiency in its fundamental job: the exchange of oxygen and carbon dioxide. The heart, in turn, must work harder, and death may result from heart failure. Approximately 200 deaths per year in the United States can be attributed to asbestosis (Table 2.5). *All deaths from asbestosis are directly linked to long-term occupational exposure in the pre-regulated work place.*

Note that asbestosis is not a cancer. There is a cancer related to asbestos, however, called mesothelioma, a rare disease of the lining of the lung and stomach that occurs mostly in men who worked in the construction industry. This disease usually results in death within one to two years of exposure but may not show up for several decades. Men born in the 1940s face the greatest risk because this group was working during the 1960s and 1970s when use of the most dangerous blue and brown asbestos was at its peak, and the hazards of asbestos to employees in the building trades were not yet recognized.

An extreme example of the dangers of asbestos is the Australian town of Wittenoom, nicknamed the "town of death." The Asbestos Diseases Society estimates that 2300 of the 20,000 people who lived or worked in the town between 1943 and 1966 have died of asbestos-related diseases. Blue asbestos was mined in the region during the 23-year period. The Australian government is dismantling Wittenoom piece-by-piece and burying the rubble in pits dug for the purpose at the old airport.

In 1972, OSHA set an acceptable upper limit for asbestos of 5 fibers/cubic centimeter in the work place, and this limit has been lowered (made more stringent) several times since (Table 2.6). From the table, it is clear that normal outdoor air contains a level of suspect fibers 60 percent higher than school air. In fact, the air in schools appears to be more free of asbestos fibers than indoor air in other types of structures.

Nevertheless, based on the public's perception that if large amounts of asbestos are harmful, then low levels—or even a single fiber—also must be harmful, "zero tolerance" is required. Responding to public

TABLE 2.5

Annual deaths in the United States from various causes (population approximately 255 million); Total deaths for 1994 = 2,286,000.

Rank	Cause of Death	Number	Percentage of Total Deaths
All causes		*2,286,000*	*100.0*
1.	Heart Disease	734,090	32.1
2.	Cancer	536,860	23.5
3.	Stroke	154,350	6.8
4.	Chronic obstructive lung disease and allied conditions	101,870	4.5
5.	Accidents and adverse effects	90,140	3.9
	Motor vehicle accidents	42,170	1.8
	All other accidents and adverse effects	47,980	2.1
6.	Pneumonia and influenza	82,090	3.6
7.	Diabetes melitus	55,390	2.4
8.	Human immunodeficiency virus (HIV) infection	41,930	1.8
9.	Suicide	32,410	1.4
10.	Chronic liver disease and cirrhosis	25,730	1.1
11.	Others		
	Mesothelioma	400	
	Asbestosis	213	
	Silicosis	135	
	Lightning	92	
	Bee stings	34	
	Spider bites	4	
	Snake bites	0	

Source: U.S. Public Health Service 1994 data.

TABLE 2.6

Fibers of asbestos per cubic centimeter (>5 micrometers with 3:1 ratio of length to width). Note the dramatic reduction of fibers with increasing regulation. (ACM = asbestos-containing material)

Fibers/cm³ in Work Place	Changing Regulations (1972–1992)	Ratio of Environment to Schools
>1640	Pre-regulated work place	>420,000
82	OSHA (1972)	21,000
33	OSHA (1976)	8,300
8.2	OSHA (1983)	2,100
3.3	OSHA (1986)	830
1.6	OSHA (1992)	420
0.006	Outdoor air	1.6
0.004	**Schools**	**1.0**
0.016	Indoor air, no ACM	4.1
0.009	Indoor air, nondamaged ACM	2.3
0.0012	Indoor air, damaged ACM	3.0

Source: Gunter, 1994, 20.

TABLE 2.7

All of these are considered acceptable risks by most people. Almost everyone will pet a dog, get married, and rest in bed. These "very risky" behaviors are not yet regulated by the federal goverment.

Event	The Chance It Will Happen to You This Year
Dying in airplane crash	1 in 40,000,000
Dying from a vaccination	1 in 25,000,000
Dying from an earthquake or volcano	1 in 11,000,000
Killed by a power mower	1 in 10,400,000
Dying from scholastic football	1 in 2,200,000
Dying from falling out of bed	1 in 2,000,000
Killed by a contraceptive	1 in 1,700,000
Killed by a home appliance	1 in 1,250,000
Injured during a shower fall	1 in 1,000,000
Struck by lightning	1 in 750,000
Poisoned in a fast-food restaurant	1 in 440,000
Electrocuted by turning on a lamp	1 in 350,000
Dying from long-term smoking or obesity	1 in 200,000
EPA action level	1 in 100,000
Drowning	1 in 50,000
Being killed by a meteor or comet impact	1 in 20,000
25-year-old woman having breast cancer	1 in 20,000
Dying during childbirth in the United States	1 in 12,000
Being murdered	1 in 11,000
Dying as a result of a motor vehicle accident	1 in 5800
Female being raped	1 in 2500
Infected with a flesh-eating bacteria	1 in 1500
Getting pregnant with faithful use of the pill	1 in 1000
Injured while resting in bed	1 in 400
Married couple getting divorced	1 in 96
Dying from a dog bite	1 in 12

pressure, Congress passed a law requiring that asbestos material that appears to be visibly deteriorating must be removed, irrespective of the number of fibers in the air. Hence, despite subsequent evidence that the removal process itself stirs up a higher level of asbestos fibers in the air than were present before the "clean-up," removal proceeds in all public buildings. A new industry, the asbestos removal industry, has been created. This industry is now quite large, and therefore has political clout, and has a vested interest in maintaining the status quo of federal regulations that support continuing asbestos removal.

There is no such thing as a risk-free environment. Experts at figuring odds say that the risk of a child dying from asbestos-related disease by attending a school containing asbestos is one-hundredth of the risk of that child being killed by lightning. The chance of being hit by lightning is one in 750,000 (Table 2.7). Ask yourself whether the money spent ($3 billion in 1992) and still being spent to remove asbestos from schools and other buildings (Figure 2.28) would be better applied to more immediate needs and more rational endeavors.

Considering the composition and distribution of materials on Earth's surface and in the atmosphere, the idea that even the minutest amount of carcinogen in the environment is unacceptable should be reevaluated. If not, billions more dollars will be wasted chasing trace amounts of suspect chemicals and minerals in an attempt to expunge them from our environment, where many exist naturally. This **no-threshold concept** being presented to the American public as part of a prudent approach to protecting the health of all U.S. citizens is forcing naturally occurring mineral materials to be classified as dangerous and possibly carcinogenic. The scientific information available does not support such a fearsome scenario, especially for the low levels of exposure encountered by the general population.

Black Lung and Silicosis

Similar diseases result from inhalation of particulate matter other than asbestos. *Black lung* is caused by inhalation of coal dust, *brown lung* is caused by inhalation of cotton fibers, and **silicosis** is caused by inhalation of silica dust.

FIGURE 2.28 Too cautious?

In late 1986, asbestos was discovered in the rubble produced by the demolition of the Circle Star Theater in Washington, D.C. The removal of the rubble changed from a bulldozer and open dumpsters to a high-technology, labor intensive, hand operation. First, a security and isolation fence was erected. Then, for several months, a 10- to 20-person crew dressed in environmental suits picked up and bagged every piece of debris into plastic bags. The bags were loaded into covered dumpsters and were removed by covered trucks. Where was the disposal site? *Photo: Bruce F. Molina, U.S. Geological Survey, Reston, Virginia.*

A less-common effect of rock particles on human health is the inhalation of coal dust by coal miners. Currently, active U.S. coal miners number about 120,000, and 40 percent of them work underground (the other 60 percent work above ground, either at surface jobs at underground mines or at surface mines). The underground mining environment is intrinsically unhealthy. Mining machines of various types rip into the coal, generating large amounts of fine coal dust. These particles are too small to see, but have disastrous effects upon human lungs, effects known collectively as black lung disease (Figure 2.29).

Black lung disease includes components of chronic bronchitis and emphysema, causing prolonged disability and suffering, and eventually death. The accumulation and retention of coal dust in miners' lungs is directly correlated with their years of underground exposure (Figure 2.30). Technological advances have automated mining to a large extent, sharply reducing the number of miners required. And the quality of ventilation in the mines has improved in recent years. However, coal mining remains a hazardous occupation.

Quartz dust is of concern to workers who perform sandblasting and rock drilling. Large particles of quartz that reach their noses are stopped by the tiny hair, mucous membranes, and other protective mechanisms of the respiratory tract and bronchi. But the smallest particles, those less than 20 micrometers in diameter, are carried to airways and the tiny air sacs in the lungs. Lung tissue reacts by developing fibrotic nodules and scarring around the trapped quartz particles. This *fibrotic* condition of the lungs is called silicosis. If the nodules grow too large, breathing becomes difficult and, in severe cases, death may result. Workers in the sandblasting and drilling trades should always wear respirator masks for protection. At present, many do not.

FIGURE 2.29 Drawing of the nose, mouth, throat, and lungs of the human body.

Natural filters keep out foreign particles such as coal dust. This figure shows the depth of penetration of different grain sizes of these particles. *Source:* Earth and Mineral Resources, *1990, vol. 59, no. 1, p. 7.*

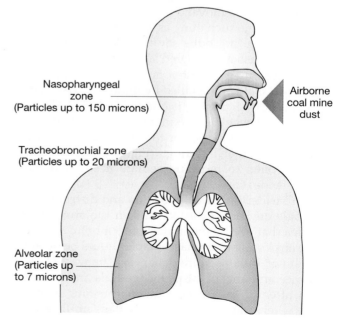

Nasopharyngeal zone
(Particles up to 150 microns)

Airborne coal mine dust

Tracheobronchial zone
(Particles up to 20 microns)

Alveolar zone
(Particles up to 7 microns)

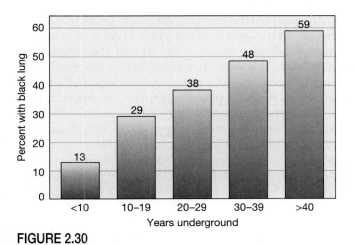

FIGURE 2.30

Relationship between years worked in underground coal mines and incidence of miners' pneumoconiosis. *Source: Helen Dwight Reed Educational Foundation.*

Government Regulation of Minerals Handling

Perhaps the strangest environmental action relating to minerals taken to date was made by the International Agency for Research on Cancer (IARC) which, in 1988, designated quartz as "probably carcinogenic to humans," *without reference to grain size.* This resulted in OSHA mandating that any U.S. product containing more than one-tenth of 1 percent quartz must display hazardous warning signs. The U.S. National Toxicology Program in 1991 also followed the IARC lead and listed quartz as "reasonably anticipated to be a carcinogen."

This is absurd, for at least 99 percent of all surface rocks and every beach in the world contains this much quartz. In fact, the average beach contains at least 50 percent quartz. Should we close all beaches as hazardous? Should children be prohibited from playing in sandboxes? Is it safe to drive on unpaved roads and stir up quartz dust? Most human exposure to quartz occurs from sources that cannot be controlled, and

never will be. Quartz is the most abundant mineral exposed at Earth's surface, so is it possible to avoid quartz during your lifetime? No. Clearly, living is hazardous to your health.

Beneficial Effects of Mineral Contacts

With the exception of asbestos, coal dust, and quartz dust, direct contacts between people and minerals are harmless. Holding minerals and rocks in your hand or walking barefoot on rock debris has no bad side effects. Some contacts may even be healthful.

An example is the reported beneficial effect of salt mines and salt caves for those suffering from asthma, bronchitis, and other respiratory ailments. In salt mines, there are tiny droplets of salt in the air, and as they are inhaled they relieve swelling of the mucous membranes in inflamed bronchial tubes by absorbing excess moisture. This shrinks the membrane, enlarges the tube's opening, and makes breathing easier. At present there are about 30 salt-cave clinics. One of them is above ground and was built with blocks of rock salt quarried from an underground mine in Russia and transported to Israel's arid climate.

The largest salt clinic is in western Ukraine and can treat more than 360 patients at a time, including children who eat and sleep at the clinic while they are being treated for asthma. One physician who uses the salt-air treatment says the children can be relieved of their asthma symptoms for six to twelve months after completing about a dozen treatments. For outpatients at the clinic, a treatment consists of sitting in a comfortable chair for an hour or two, reading magazines that are provided or listening to music through individual headphones, while breathing the salty air. In Israel, the treatment costs about $15 an hour.

Speleotherapy (speleo means cave) originated during World War II in Russia and has since spread to Ukraine, Eastern and Western Europe, and Israel. It has not yet been tried in North America.

Summary

An essential point of this chapter, and those that follow, is the *interconnectedness* of everything, from the smallest atoms to the largest bodies of rocks, from each individual person to the atmosphere, water, and soil.

Atoms in the presence of water react with other atoms to form larger groupings, which may be gases (carbon dioxide, CO_2), liquids (water, H_2O), or solids (quartz, SiO_2). Neither the physical state of the resulting compound (solid, liquid, or gas) nor its properties can be easily predicted from the properties of the original uncombined neutral atoms. The atomic aggregate has distinctive properties of its own, determined by the distribution of electrons in the aggregate.

Minerals are aggregates of atoms arranged in a regular periodic manner (crystalline) and that have formed naturally and inorganically. Each mineral has a distinctive chemical composition as well as crystal structure. About 3500 minerals occur in Earth, but fewer than twenty are abundant.

Minerals are the essential constituents of rocks, with a few significant exceptions (coal, volcanic glass). The three classes of rocks are igneous, sedimentary, and metamorphic; each can be changed into the other by natural processes. Each class has distinctive characteristics by which it can be recognized. Many of these characteristics are of major importance to environmentalists, such as solubility, hardness, strength, porosity, and permeability. Succeeding chapters will call on properties such as these to explain the origin of many environmental problems.

Many fears about pollution from naturally occurring mineral and rock particles are unfounded. Too many people who are responsible for protecting public health are ignorant of important aspects of science—aspects about which they should become better informed. This educational deficiency has resulted in many inaccurate and erroneous pronouncements by governmental and private agencies. Examples include the federal treatment of asbestos in public buildings and the issue of public exposure to radon gas (see Chapter 15). These pronouncements have caused the wasteful expenditure of billions of dollars that could have been used to counter legitimate environmental threats.

Key Terms and Concepts

asbestos	halide	permeable
asbestosis	hardness scale	phosphate
atom	hydrogen bond	phosphorite
atomic number	igneous rocks	plutonic
atomic weight	ion	pore
bonding	ionic bond	proton
carbonate	isotope	radioactivity
chemical bond	lava	rock
chert	limestone	rock cycle
clastic	lithification	sandstone
cleavage	magma	sediment
conglomerate	metallic bond	sedimentary rocks
covalent bond	metamorphic rocks	shale
crust	mineral	silicate
crystalline	native element	silicosis
electron	neutron	sulfate
electron shell	no-threshold concept	sulfide
element	oxide	weathering
evaporite rocks	oxidize	
geothermal gradient	periodic table	

Stop and Think!

1. Draw the structure of an atom with atomic number 12 that has eleven neutrons. When this atom occurs as an ion, will it have a positive or a negative charge? Why?

2. Describe the differences among an ionic bond, a covalent bond, and a metallic bond. What is the environmental significance of these differences?

3. What are the eight major groups of minerals? Name an environmentally significant use for a mineral in each group.

4. Describe the differences among igneous, sedimentary, and metamorphic rocks. Name two characteristics of each type that commonly serve to differentiate among samples when you find them.

5. Describe the operation of the rock cycle.

6. What are the differences among sandstone, shale, and limestone? Give an environmentally significant characteristic of each type of rock.

7. The asbestos-removal industry owes its size and resulting political clout to laws passed by the federal government. Can you name any other industries that have grown in recent years because of the public's growing environmental consciousness?

8. It is clear that when industrial workers inhale large amounts of very fine-grained asbestos, cotton, coal, or quartz, their health suffers. Can you think of some types of fine-grained particles in normal outdoor city air that you breathe daily which probably are bad for your health? What can we do about them?

References and Suggested Readings

Anonymous. 1995. Asbestos: Try not to panic. *Consumer Reports*, July, p. 468–469.

Chesterman, C. W. 1978. *The Audubon Society Field Guide to North American Rocks and Minerals*. New York: Alfred A. Knopf, 850 pp.

Desautels, P. E. 1974. *Rocks and Minerals*. New York: Grossett and Dunlap, 159 p.

Dietrich, R. V. 1989. *Stones: Their Collection, Identification, and Uses, 2d ed*. Prescott, Arizona: Geoscience Press, 191 pp.

Dorr, Ann. 1987. *Minerals—Foundations of Society, 2d ed*. Alexandria, Virginia: American Geological Institute, 96 pp.

Goldsmith, D. F. 1994. Health effects of silica dust exposure, *in* Silica: Physical Behavior, Geochemistry and Materials Applications. *Reviews in Mineralogy*, v. 29, p. 545–606.

Gunter, M. E. 1994. Asbestos as a metaphor for teaching risk perception. *Journal of Geological Education*, v. 42, p. 17–24.

Hardy, T. S. and Weill, H. 1995. Crystalline silica: Risks and Policy. *Environmental Health Perspectives*, v. 103 (2):152–55.

Lee, John and Manning, Lucy. 1995. Environmental lung disease. *New Scientist*, September 16, 4-page insert.

Mack, W. N. and Leistikow, E. A. 1996. Sands of the world. *Scientific American*, August, p. 44–49.

Mutmansky, J. M. 1990. The war on black lung. *Earth and Mineral Sciences*, v. 59, no. 1, p. 6–10.

Prinz, M., Harlow, G., and Peters, J. 1978. *Simon & Schuster's Guide to Rocks and Minerals*. New York: Simon and Schuster, 286 pp.

Shadmon, Asher. 1989. *Stone: An Introduction*. New York: Bootstrap Press, 140 pp.

Siever, Raymond. 1988. *Sand*. New York: Scientific American Books, 240 pp.

"Whoever could make two ears of corn . . . to grow upon a spot of ground where only one grew before, would deserve better of mankind, . . . than the whole race of politicians put together."
Jonathan Swift

Soils—Our Most Important Natural Resource

3

Most of the time, the ground we walk on is not hard rock but is a comparatively soft and loose material commonly termed "dirt." Typically, "dirt" is called **soil** by Earth scientists (if you have been digging in the ground, your dirty clothing truly may be "soiled"). Soil is the stuff in which crops grow. But of what is soil made? How does it form? Is soil the same everywhere? How can soil be protected against both natural and human damage?

From Rocks to Soils

If you look at a soil sample under a microscope, you can see that *soil is made of debris from crumbling rocks, decayed plants, and decayed animals* (Figure 3.1). Many soil varieties exist, varying in their proportion of rock fragments and organic bits. Let us examine how rocks decompose to create the small particles that make up a soil.

Igneous and metamorphic rocks typically form at great depth in Earth. Here, temperatures and pressures are much greater than at the surface. In addition, these rocks form where there is less water and less oxygen gas than are present in the atmosphere. As long as these rocks remain deep underground, they are "happy" to remain as they formed. However, forces within Earth, and relentless erosion of the surface, gradually expose these deep rocks to surface conditions. Surface conditions are dramatically different: temperatures are no longer hundreds or thousands of degrees; pressure from the air is only around 15 pounds per square inch, versus many tons per square inch underground; and water and oxygen are plentiful.

For these reasons, the rocks become "unhappy" (the scientific expression is "out of equilibrium") when raised to the surface by stresses within Earth. Accordingly, the rocks change their character to adjust to the new conditions. The mineral grains which had formed deep underground at high temperatures and pressures and under different chemical conditions react to form new substances in equilibrium with new conditions. These reactions occur at rock surfaces where they are in direct contact with the new conditions. This is the beginning of soil formation from "parent" rocks.

Broadly, two processes form soils from rock: chemical alteration and physical alteration. We will look at each.

Chemical Alteration of Rocks into Smaller Particles

Several types of chemical reactions occur in rocks. All are related to the differences between conditions deep in the crust and conditions at the surface. The new minerals produced contain much more water and more oxygen than the ones brought up from below, simply because water and oxygen are more available.

◀
Twelve row planter.
Photo: Arthur C. Smith III/Grant Heilman Photography, Inc.

FIGURE 3.1 The components in topsoil.

Major components in most soils include sand, silt, clay, organic matter from decaying plants and animals, and living roots, insects, and worms. *Photo: Edward A. Keller.*

Formation of Clay Minerals. Feldspars are the most abundant minerals in igneous and metamorphic rocks. When exposed to Earth's corrosive atmosphere, feldspars combine with rainwater and the hydrogen ions in the water to form clay minerals.

$$2KAlSi_3O_8 \; + \; 2H^+ \; + \; 9H_2O \; \rightarrow \; Al_2Si_2O_5(OH)_4$$
(orthoclase feldspar)   (acidic water)   (kaolinite clay)

$$+ \; 2K^+ \; + \; 4H_4SiO_4$$
(potassium ions)   (silica in solution)

There are many clay minerals; the equation shows a common example, kaolinite. Pure kaolinite is used to make fine porcelain. Note that the potassium (K) in the feldspar does not end up in the kaolinite clay. Instead, the potassium ions remain dissolved in the water. This makes the potassium ions available to plants when they take up the water through their roots (potassium is an essential plant nutrient).

More silica is present in the feldspar than is needed to form the kaolinite, so some silica also stays in solu-

tion, as dissolved silica. It is carried away with the potassium ions in the moving water and ends up in the ocean.

The most abundant minerals contain silicon and aluminum, and such minerals alter chemically ("rot" or decompose) to clay minerals. Thus, the *general* chemical reaction for the alteration at Earth's surface of minerals that contain silicon and aluminum is:

aluminum-silicon   +   hydrogen   +   water   →   clay
mineral   ion (+)   mineral

          +   metallic   +   soluble
             ion (+)   silica

With the exception of quartz, which has a very low solubility of 6 ppm, all the abundant minerals in igneous and metamorphic rocks react readily with rainwater to produce the results you see in the equation: (a) clay minerals; (b) one of the abundant metallic cations like potassium, calcium, or sodium; and (c) silica in solution. Which metallic cation is produced by this chemical alteration depends on which one is present in the mineral being altered; potassium happens to be the one in the example.

The only exception to this scheme occurs when the reacting mineral contains iron. Most of the iron in minerals occurs with a +2 electrical charge and, as it is released from the mineral during chemical alteration, the iron combines immediately with oxygen in the air and changes to a +3 electrical charge. Iron with a +3 charge is relatively insoluble, so it precipitates as the well-known iron oxide, red hematite:

$$4Fe^{+2} \; + \; 3O_2 \; \rightarrow \; 2Fe_2O_3$$
(iron in solution)   (oxygen gas)   (solid iron oxide: hematite)

Rust on iron and steel objects is composed of hematite. It is the rusty-red color of hematite that gives red sedimentary rocks and soils their ruddy color, although they usually contain only a few percent of hematite.

As noted, the most abundant minerals contain silicon and aluminum, and such minerals alter chemically ("rot" or decompose) to clay minerals. Therefore, we would expect that the dirt beneath our feet and in which our grass and crops grow is composed mostly of clay minerals and residual quartz grains from the parent rocks. This is correct: quartz grains (commonly called "sand") and clay *are* the major constituents of most soils, in addition to organic matter from the decayed plants that grew in the soil. Most soils also contain small percentages of the original minerals (feldspars, micas, and such) that have not yet decomposed completely.

Formation of Carbonate Minerals. Another important chemical process during soil formation is the combining of water with carbon dioxide gas, which produces carbonate ions:

FIGURE 3.2 Sheeting of a rock surface.

Expansion of the rock as erosion removes the overlying confining rock causes sheeting of a rock surface. The example here is granite in the Sierra Nevada mountains of California.

$$CO_2 + H_2O \rightarrow 2H^+ + CO_3^{-2}$$
(carbon   (water)   (hydrogen (carbonate ion)
dioxide)   ion)

The carbonate ion then is available to combine with calcium to form calcium carbonate in some soils (as described in Chapter 2).

Chelation. A final important chemical process during soil formation is **chelation** (*kee-LAY-shun*), in which positive ions released from decomposing mineral grains become incorporated within complex organic compounds released by plants. These organic complexes may then be carried away by moving water. Chelation increases the solubility of mineral grains, causing faster soil development from rock.

Physical Alteration of Rocks into Smaller Particles

The chemical changes we just examined are much more important than physical processes as agents of rock decay and soil formation. However, some physical processes can be significant.

One such process is **sheeting** (Figure 3.2). Sheeting is the separation of homogeneous and unlayered rocks, such as granite, into sheets that range in thickness from inches to feet. Sheeting results from the release of pressure as overlying rocks are removed by surface processes. Sheeting may extend to depths exceeding a hundred feet. This phenomenon is very important to businesses that quarry rock. If natural processes have already created geometrically broken surfaces, less labor and therefore less money are needed to obtain building stone from a rock outcrop.

Other physical weathering processes are frost wedging, salt wedging, and root wedging. Significant physical alteration at or near rock surfaces may result when water freezes, or when minerals crystallize from water, or when tree roots invade cracks in rocks. Water expands 9 percent when it freezes (crystallizes into ice), exerting a force great enough to split rock **(frost wedging)** (Figure 3.3). This effect can be seen on mountain slopes in temperate climates, as in the western United States , where the slopes are covered with slabs of surface rock that have resulted from repeated freeze-thaw cycles.

In more arid climates, such as the world's deserts, evaporating water leaves behind crystallized salts (like halite, NaCl) just below the rock surface. This creates pressure that causes flaking of the rock (**salt wedging,** shown in Figure 3.4).

Tree roots also are agents in the physical weathering of rocks by means of **root wedging,** as shown in

FIGURE 3.3 Frost wedging.

Frost wedging occurs when water seeps into fractures and freezes. The ice's volume is about 9 percent greater than that of liquid water, and this expansion forces the rock apart. This results in loose, angular fragments that move downslope by gravity. They accumulate at the base of the cliff as talus cones. *After: W. Kenneth Hamblin, Earth's Dynamic Systems, 6th ed., 1992, New York: Macmillan, p. 192.*

FIGURE 3.4 The Great Sphinx on the Giza Plateau near Cairo, Egypt.

Observe the deterioration of the limestone due mostly to salt crystallization. Air pollution and vibration from nearby rock quarrying are also factors in the deterioration. *Photo: F. El-Baz.*

FIGURE 3.5 Tree roots contribute to breaking up of rock.

This tree's roots are growing parallel to bedding, fracturing the rock. *Photo: Runk/Schoenberger/Grant Heilman Photography, Inc.*

Figure 3.5. You may have seen examples of root-heaving of sidewalks on campus.

The importance of physical weathering is that breaking a large rock into smaller pieces greatly increases the **surface area** of the sediment (Figure 3.6), better exposing it to chemical weathering. Chemical reactions occur only at surfaces, where water is in contact with the rock. The more surface exposed, the faster the rate of weathering.

Results of Chemical and Physical Alteration of Rocks

The results of chemical and physical alteration are visible not only on rocks in roadside outcrops but on building stones. Everyday examples are tombstones, bank building facades, statuary, and monuments (Figure 3.7). Some of this weathering is natural, but much of it results from human activities, particularly the burning of fossil fuels—gasoline and coal. Their combustion releases the gases sulfur dioxide (SO_2), nitrogen dioxide (NO_2), and carbon dioxide (CO_2). Each reacts with oxygen and water in the atmosphere to produce an acid:

$$2SO_2 + 2H_2O + O_2 \rightarrow 2H_2SO_4$$
(sulfur dioxide gas) (water) (oxygen) (sulfuric acid)

$$4NO_2 + 2H_2O + O_2 \rightarrow 4HNO_3$$
(nitrogen dioxide gas) (water) (oxygen) (nitric acid)

$$CO_2 + H_2O \rightarrow H_2CO_3$$
(carbon dioxide gas) (water) (carbonic acid)

Assaulted by these acids, stone decays rapidly (Chapter 15). Tombstones, especially those of calcium carbonate (limestone or marble), may become illegible within a century (Figure 3.8). In the case of sulfur dioxide, which generates sulfuric acid, weathering can be accelerated because this acid transforms moderately soluble calcite into gypsum:

$$CaCO_3 + 2H_2O + H_2SO_4 \rightarrow CaSO_4 \cdot 2H_2O$$
(calcite) (water) (sulfuric acid) (gypsum)

$$+ H_2O + CO_2$$
(water) (carbon dioxide)

Once gypsum is formed, it is easily washed away in rainy climates, because it is about 16,000 times more soluble than calcite. To the casual observer, the only

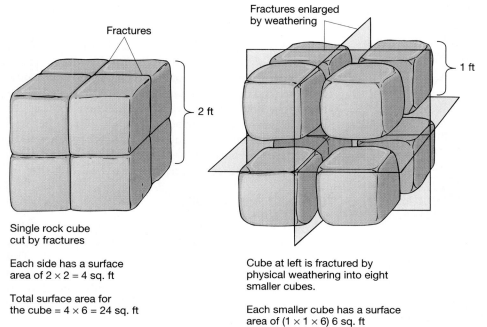

Fractures

Fractures enlarged
by weathering

2 ft

1 ft

Single rock cube
cut by fractures

Each side has a surface
area of 2 × 2 = 4 sq. ft

Total surface area for
the cube = 4 × 6 = 24 sq. ft

Cube at left is fractured by
physical weathering into eight
smaller cubes.

Each smaller cube has a surface
area of (1 × 1 × 6) 6 sq. ft

Total surface area for all eight
cubes = 6 × 8 = 48 sq. ft

FIGURE 3.6

Physical weathering increases the
surface area of rocks. *After: N. K.
Coch and A. Ludman,* Physical Geology,
1991, New York: Macmillan, p. 219.

sign of this process may be the altered surfaces of limestone (calcite).

The rate of deterioration (decomposition) of silicate building stones (like granite and basalt) is also increased by industrially produced acids in the atmosphere but the effects are less noticeable than with the much more soluble calcite. (This goes back to those stronger covalent bonds in silicates.)

FIGURE 3.7

This statue dramatically shows
weathering accelerated by acid
precipitation. *Photo: Simon
Fraser/Science Photo Library/Photo
Researchers, Inc.*

The next time you walk by a cemetery, bank, or monument, pause to look closely at the rock surface. You may be surprised to see chemical and physical weathering taking their toll, slowly generating the particles that will become new soil.

FIGURE 3.8

Tombstones are excellent indicators of weathering. Note the decreased legibility of the inscriptions. *Photos: Grunnitus/Monkmeyer Press and Breck P. Kent/Animals/Earth Scenes.*

Clay Minerals

Clay minerals continually form. This is due to chemical alteration of the aluminum-silicon minerals in abundant igneous and metamorphic rocks. Hence, clay minerals are the most abundant mineral type in sedimentary rocks, and sediments and sedimentary rocks cover most of Earth's land surface.

For the environmental geologist, clays are very important because they affect soil stability, plant growth, water-filtering, and choice of building sites.

Clay and Soil Stability

The shape of clay minerals is sheetlike. Consequently, a stack of clay minerals may, at great magnification, resemble a stack of playing cards (Figure 3.9). The weak bonds between adjacent clay flakes allow clays on a slope to slip past each other easily, resulting in slope failures (slumps, landslides).

If the clay flakes are stacked in an edge-to-face arrangement rather than face-to-face, space is created to hold water, which decreases the strength of the clay, also leading to slippage (Figure 3.9B).

FIGURE 3.9 Flakes of clay, with water in the pore spaces.

(A) When flakes of clay lie face-to-face, they are generally parallel, allowing little space between them to hold water. Such a clay is impermeable. (B) When flakes of clay lie more randomly face-to-edge, they form spaces for water, and water-filled pores occupy a higher percentage of the clay.

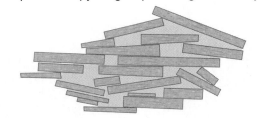

(A)

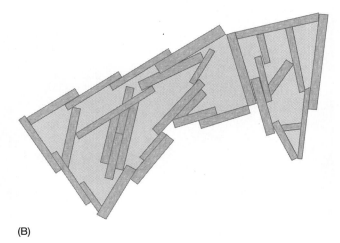

(B)

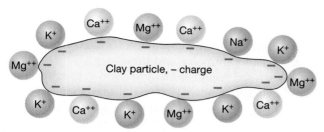

FIGURE 3.10 Positively charged ions surround a clay mineral grain in soil.

Large numbers of ions are held by each clay flake. The number of clay flakes in each cubic inch of soil is in the billions.

Clay and Plant Nutrition

The surfaces of clay mineral flakes attract positive ions that are important plant nutrients (Figure 3.10). The abundant ions present on clays are those most abundant in the waters surrounding the clay: potassium, sodium, and calcium. These three cations originated in silicate minerals like feldspar. Potassium and calcium are essential elements for plant growth, and the surfaces of countless clay mineral particles are a major storehouse for these nutrients. Clays are essential for good agricultural soils, residential lawns, and mountain forests.

Clay and Water Filtering

Some positive ions are attracted to clay surfaces more strongly than others. Larger positive ions with greater electrical charges will displace smaller ions with lower charges. Thus, ions of the toxic metals lead and cadmium, with their larger diameters and charges of +2, will displace smaller sodium ions with their charge of +1. Every atom of lead and cadmium held by a clay mineral is an atom of lead and cadmium that cannot enter our water supply.

Thus, clay minerals perform another important environmental function: they can prevent some harmful elements from contaminating our water. However, lead and cadmium that are not contaminating our water are instead contaminating the soil.

Other potential toxic elements that are relatively large and, therefore, would be preferentially held by clays include mercury (+2) and uranium (+3).

The bulk of organic pollutants are positively charged on one end, and thus will stick to clay mineral surfaces. Thus, it is not only single-atom inorganic ions that stick to clays but also complex organic materials that have a positive charge somewhere on their bodies.

Clay and Destructive Expansion of Soil

Because the bonding forces are so weak between adjacent clay flakes, water molecules worm their way between the parallel flakes, and can greatly increase the volume of the clay. Such soils are quite common in the United States (Figure 3.11). Soils that contain such clay are called expansive soils, and the amount of

FIGURE 3.11 Soil that expands.

Areas of swelling clays in the coterminous United States. *Source: U.S. Geological Survey.*

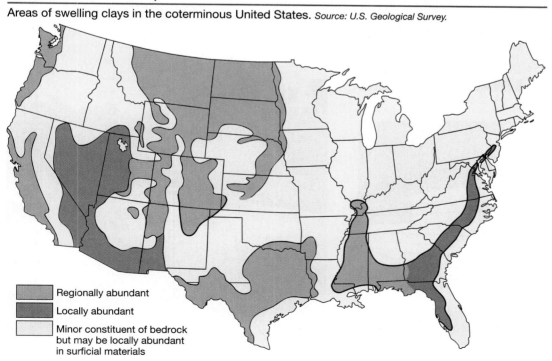

Regionally abundant

Locally abundant

Minor constituent of bedrock but may be locally abundant in surficial materials

swelling can be dramatic, up to 2000 percent (going from a bone-dry soil to very wet), although most expansions are much less. The force of this expansion is sufficient to lift and rotate small buildings and to crack foundations (Figure 3.12). Buried utility lines for water, gas, electricity, and telephone can be similarly disrupted. Such soils are termed **swelling soils** by geologists and engineers (Figure 3.12).

The Complexity of Soils

Considering the variety of Earth's rocks, you might expect different soils to develop from different rocks. Indeed, soils differ in thickness, composition, nutrient value to plants, permeability, susceptibility to erosion, and age—all important to environmental concerns.

Some soils remain in place above the underlying rocks from which they form **(residual soils).** Other soils have developed on rock debris transported and deposited by streams **(alluvial soils).** But whether residual or alluvial in origin, all soils pass through the same developmental stages and reflect the same variables: parent materials, climate, associated biotic activity (plants and animals), and time.

When soil scientists analyze a soil in its natural setting, they look at how its physical and chemical traits vary with depth. Their initial concern is with the **soil profile,** which is much like a highway road cut: a vertical slice extending from the surface down through layers of soil into the unaltered parent material (a sample soil profile is shown in Figure 3.13). A careful description of how the soil varies with depth provides valuable information about plant root environments, nutrient and moisture availability, clays, surface erosion, and other data necessary for wise land-use planning. Every well-developed, undisturbed soil has its own distinctive profile characteristics that determine how the soil can best be used.

As soil develops on either hard rock or on porous and permeable loose debris, a sorting of materials takes place. The soil becomes differentiated into layers with distinctive characteristics, called **soil horizons.**

(A)

(B)

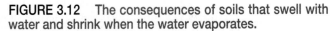

FIGURE 3.12   The consequences of soils that swell with water and shrink when the water evaporates.

(A) Popcorn-like texture of soils at the ground surface characterized by bulges and cracks is a characteristic of soils that contain swelling clays. Such clays have both positive uses and harmful effects. Swelling clays are used for applications as diverse as the manufacture of cosmetics and as a lubricant for drill bits when drilling for oil. (B) Stresses exerted by swelling soils have caused this block wall surrounding a condominium complex to tilt and break.
(C) Paved roads constructed on swelling soils without engineering precautions will exhibit heaving, undulation, and extensive cracking. *Photos courtesy E. Nuhfer.*

(C)

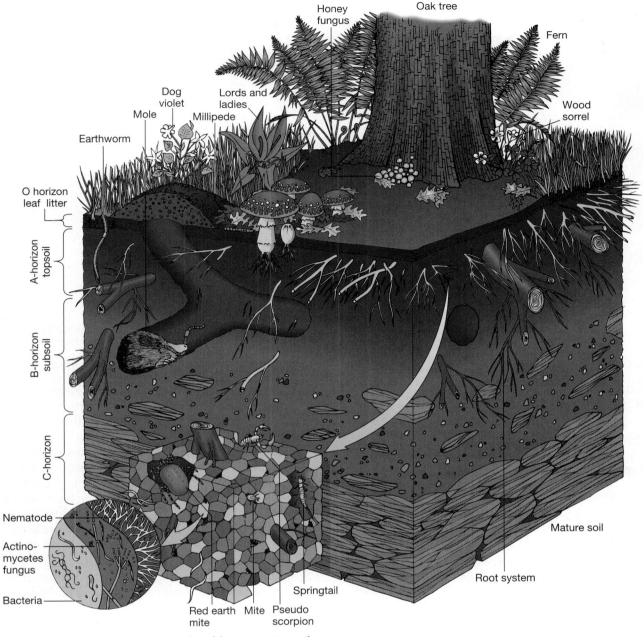

Honey fungus
Oak tree
Fern
Dog violet
Lords and ladies
Mole
Millipede
Wood sorrel
Earthworm
O horizon leaf litter
A-horizon topsoil
B-horizon subsoil
C-horizon
Nematode
Actino-mycetes fungus
Bacteria
Red earth mite
Mite
Pseudo scorpion
Springtail
Mature soil
Root system

FIGURE 3.13   Soil profile from a humid, temperate region.

The topsoil (A-horizon) contains a few pieces of organic matter. It is underlain by a subsoil (B-horizon), which is enriched in clay from the A-horizon. These horizons rest on broken rock (C-horizon), which is underlain by bedrock.

Soil horizons are roughly parallel to the ground surface (which is why they are called horizons).

Beneath the organic litter of decayed leaves, grass, and deceased bugs and worms (the O-horizon), is the **topsoil** (A-horizon; Figure 3.14). From the topsoil, soluble minerals and clay-size particles are transported downward by percolating water into a **subsoil**, or B-horizon. This B-horizon is in the process of developing into additional topsoil.

In general, the topsoil will be dark (gray or black) because of its content of carbon-rich decaying organic matter. The subsoil will be either reddish, due to hematite (iron oxide) accumulation, or white, due to calcite accumulation.

Below these two horizons lies the parent material of the soil. If the parent material is hard rock, such as granite, a C-horizon will clearly exist, composed of broken and partially altered bedrock immediately

FIGURE 3.14 The scene in healthy topsoil.

A-horizon as seen by plant roots.

above the granite. If the parent material is unconsolidated material such as old river sediment or glacial debris, the C-horizon may be vague and difficult to distinguish.

Soil Thickness

Soil thickness varies over Earth from none (bare rock or fresh sediment) to more than a hundred feet. Examples of this contrast are central Canada and the north-central United States.

Central Canada. This area contains flat areas lacking soil cover where erosion by streams or glaciers has been recent and intense. In central Canada, removal of soil cover by glacial scouring is so recent that new soil has not had time to develop on the impermeable granite bedrock. The scouring occurred perhaps 8,000 years ago, which is sufficient time for soil redevelopment in a warm climate on impermeable granite, but insufficient time in the cold climate of central Canada. Chemical reactions are very slow in regions of long, cold winter.

Northern Mississippi River Valley. In the more temperate climate of the Dakotas-Minnesota-Iowa-Nebraska-Kansas area, a thousand miles south of Central Canada, about three feet of new soil has formed on glacial debris during the 12,000–18,000 years since the ice melted. Both the warmer climate and the greater permeability of the glacial debris to water flow contributed to faster soil formation. Annual precipitation is about the same in the two areas, although more of it is liquid (not frozen) in the Mississippi Valley. The higher proportion of rain improves drainage, mineral decay, and the availability of nutrients to plants.

In most of the western United States, rainfall is less than in the eastern half of the country. Soils are, therefore, more poorly developed and plant cover is less extensive. Such land may be adequate for livestock grazing but not for crops, unless irrigation water is provided from more humid areas or from groundwater supplies below the surface (Chapter 6). Parts of the southern midcontinent in Oklahoma, Texas, New Mexico, and Arizona are made fertile by irrigation from in-the-ground water supplies formed hundreds or thousands of years ago.

How Fast Does Soil Form?

Rates of soil formation vary widely, depending on climate. The hotter and wetter the climate, the faster the soil develops. An average rate in a climate such as Iowa or Kansas is about one inch of soil per 250 years, which is instantaneous by geologic standards, but painfully slow when related to a rapidly growing population and its food consumption. Under unusual conditions, such as on unconsolidated and exceptionally porous and permeable sediment like sand dunes or modern river sand, rates of soil formation may increase greatly. In the humid, temperate climate of North Carolina, 5 inches of A- and B-horizon developed in only fifty years. Similar soil development on an impermeable rock such as granite would require thousands of years, which is "forever" in terms of human needs.

Existing soil needs to be protected from destruction by erosion and pollution. Lost or unusable soil takes too long to reform. Even fifty years is too long to wait for a usable soil to form, given the number of human mouths to feed and the rate at which our population is increasing. Under natural conditions, soil is practically a non-renewable resource.

The Role of Organic Matter in Soil

Organic matter is the stuff that makes topsoil dark and rich and good for growing vegetables, flowers, lawns, and crops. Specifically, organic matter is all the living and dead matter within and upon the soil. It includes **litter** (non-decomposed leaves, twigs, or stalks lying on the surface) and **humus** (plant and animal residues in the process of decomposition). This organic material, typically black or brown, exists in tiny particles less than 1/100,000 of an inch in size.

The capacity of humus to hold water and nutrient ions greatly exceeds that of clay, its inorganic counterpart. Humus is sticky and causes the soil to develop a loose "crumb structure," which greatly increases the ability of air and water to penetrate the soil (Figure 3.15). Small amounts of humus thus increase greatly the soil's capacity to promote plant growth. In an ideal agricultural soil, the amount of organic matter is about 5 percent, composed of both living and dead plants and animals.

Life in the Soil. A single cubic inch of fertile topsoil may contain over 16 billion bacteria. In a single ounce of soil, protozoa may number as many as 30 million under normal conditions. Earthworms may exceed 1.5 million per acre. The soil has numerous larger tenants as well. But microorganisms by far outnumber the larger plants and animals of the soil.

The soil provides microorganisms with a favorable habitat, including food and water. Unlike higher plants, which change solar energy into food through photosynthesis, microorganisms obtain their energy primarily from the tissue of these plants.

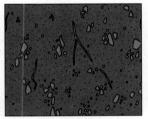 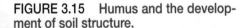 

Addition of organic matter and humus-forming process involving numerous organisms

Lack of humus causes compacted soil to have poor aeration and poor infiltration

Soil with humus has excellent aeration and excellent infiltration

FIGURE 3.15 Humus and the development of soil structure.

(A) Addition of humus improves compacted soil. (B) Crumb or granular structure around a plant root. Each crumb is composed of an aggregate of clay minerals, organic matter, and silt-size grains of other minerals.

(A)

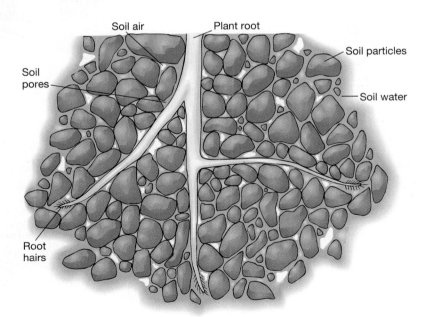

Soil air

Plant root

Soil particles

Soil pores

Soil water

Root hairs

(B)

As microbes decompose plant tissue, they not only obtain the energy they need, but also release back into the soil the many nutrients that exist as complex organic compounds within living plants. This return of nutrients to the soil is extremely important recycling. If plants absorbed most of the available nutrients in the soil and did not somehow return them, future plant growth would be limited by the diminished food supply. But microorganisms, through their feeding habits, release previously absorbed plant nutrients as soluble inorganic compounds. Thus, the flora and fauna of the topsoil are an essential part of a productive soil.

Soil Particle Size—Basis for Classifying Soils

The U.S. Department of Agriculture classifies soils by their proportions of three sizes of particles: sand (2.0–0.06 millimeter), silt (0.06–0.004 millimeter), and clay (smaller than 0.004 millimeter). (A millimeter equals 1/25 of an inch.) A convenient chart has been devised to show these proportions (Figure 3.16). As

shown, the ideal agricultural soil is a loam composed of roughly 40 percent sand, 40 percent silt, and 20 percent clay with a few percent of organic matter.

The predominance of sand and silt guarantees a reasonable porosity and permeability for water movement through the soil, and the clay and organic matter guarantee a readily accessible storehouse of nutrient elements for plant nourishment. Also, the sticky, adhesive character of most clay flakes causes the development of a granular structure of the fine particles so they do not clog the pores so essential for water flow. A soil that contains too much clay, however, will have its pores clogged with clay flakes.

Soil Water

Plants, like animals, need large amounts of water to survive. We can view a plant as a water "pump" that absorbs nutrient-laden water through its roots. The water rises through the stem or trunk and distributes nutrients to the plant's branches, leaves, and flowers. From the leaves, water transpires (evaporates) into the

FIGURE 3.16 Percentages of sand, silt, and clay in the major soil textural classes.

To use the diagram, locate the percentage of clay first and project inward as shown by the arrow. Do likewise for the percentage of silt or sand. The point at which the two projections cross identifies the class name. For example: A soil with the grain size distribution at **A** contains 60% clay, 40% silt, and 60% sand. The loam at **B** consists of 20% clay, 80% silt, and 50% sand. *Photos: Barry L. Runk/Grant Heilman Photography, Inc.*

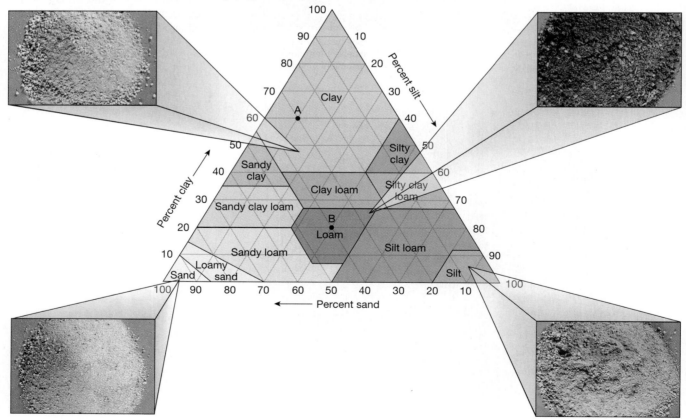

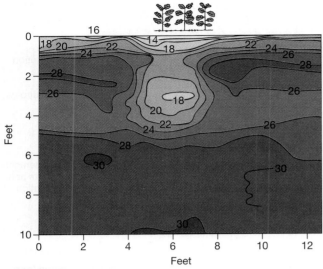

FIGURE 3.17 Soil moisture (%) sampled across a line of oak trees in a fallow field.

Note the depletion of moisture caused by the trees, especially in the top 5 feet of soil.

atmosphere. The net effect of this **transpiration** is that of a plant "pumping" water from the soil into the air. A large tree pumps many gallons of water a day in this manner (Figure 3.17).

In granular soils, water is either available or unavailable (Figure 3.18). Unavailable water is water that runs off on the soil surface or bypasses the plant in other ways. A more subtle type of unavailable water is water held to clay surfaces by hydrogen bonds. Consequently, this water cannot move through the soil.

Unavailable water may also be present in very small pores between the granules, where it is held by hydrogen bonding.

Available water is water held in large pores in the soil. The water in these pores is far distant from clay and other mineral surfaces, so it is free to move within the granular soil structure.

As Figure 3.19 shows, the relative amounts of unavailable and available water depend on the amount of clay minerals in the soil. A higher percentage of clay means more water is unavailable to the living plants rooted in the soil. Plants may wilt and die in soils containing more than 50 percent water because the water is unusable, being held to clay surfaces. Plants may also die of thirst in very sandy soils because the water drains too rapidly from the soil down into the underlying rocks.

A simple method for depicting an area's water supply-and-demand situation for soil moisture is shown for two locations in Figure 3.20. The blue lines graph precipitation over the year, and the red lines graph water demand (by plants and evaporation).

Brevard, North Carolina (A in Figure 3.20) has a low winter moisture demand from plants (less than 0.5 inch per month—note the red line) due to low temperatures and less daily sunlight. Plant demand increases rapidly through springtime as solar radiation intensifies and the duration of daylight increases. Peak water demand exceeds 5 inches during July, then drops rapidly in autumn. Because precipitation (blue line) exceeds moisture demand every month, a **moisture surplus** exists every month. As soil moisture is utilized, it is replenished, and the excess rainfall either drains through the soil or is lost as surface runoff.

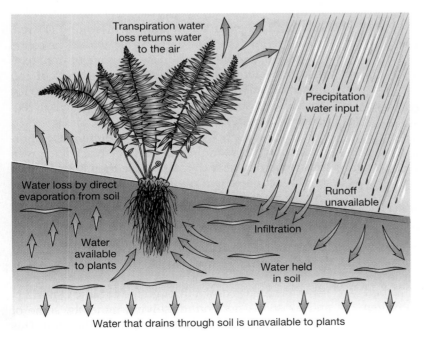

FIGURE 3.18 Water circulation in the vicinity of a plant.

The rainfall enters the soil, is used by the plant, and evaporates back into the atmosphere through the plant's leaves (a process called *transpiration*). Some water moves through the soil without encountering the plant's roots; it evaporates directly from the soil.

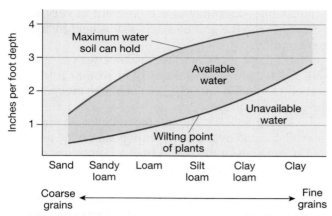

FIGURE 3.19 Water-holding capacities in soils of different texture.

The relationship between moisture in the soil and its availability to plants. Much of the water in a soil is held to clay mineral surfaces and cannot be removed by plant roots.

After: Donald Steila, and Thomas E. Pond, 1989, The Geography of Soils, *2nd ed., Savage, Maryland: Rowman & Littlefield Publishers, 239 pp.*

San Francisco (B in Figure 3.20) has a very different situation, with precipitation plummeting to essentially zero in June (blue line). The City on the Bay also has a greater moisture requirement than Brevard in winter and a smaller demand in summer. But, despite the smaller summer demand, the reduced ability of the atmosphere to supply rainfall creates a moisture deficiency from March to October. The loss of moisture from plant leaves and evaporation from the soil heavily taxes the soil reservoir until the moisture supply is nearly exhausted and a water deficiency exists. Irrigation is required for the survival of crops.

Soil Air

Intermixed with soil water is soil air (Figure 3.21). Soil air normally is more humid than the atmosphere above the ground, and it has a higher percentage of carbon dioxide. Carbon dioxide (CO_2) in soil air is often several hundred times greater than the 0.03% (three-hundredths of one percent) present in the atmosphere. Most of the CO_2 in soil air originates from the decay of plant and animal tissue:

organic matter + oxygen → carbon dioxide + water

The CO_2 cannot escape from the soil as rapidly as it is produced; hence the buildup. Soils with high CO_2 content can be very acidic and harmful to many crops.

Oxygen in soil air normally is less than in the atmosphere. Oxygen may be less than half the 21 percent present in the air above ground level. The oxygen is consumed by roots as they grow.

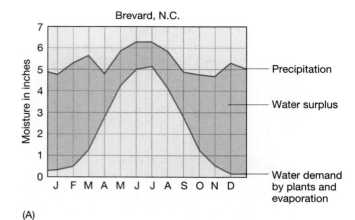

(A)

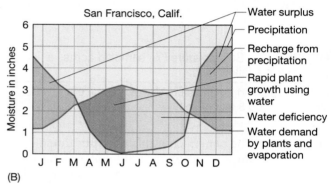

(B)

FIGURE 3.20 Water budgets of Brevard (in North Carolina's western mountains) and San Francisco.

Water deficiencies can be remediated by irrigation. *After Donald Steila, and Thomas E. Pond, 1989,* The Geography of Soils, *2nd ed., Savage, Maryland: Rowman & Littlefield Publishers, 239 pp.*

Why Is Soil Black, Brown, Red, Yellow, Green, or Gray?

Soils tend to exhibit two groups of colors: (1) red, brown, yellow, or green; and (2) gray-black. The colors red, brown, and yellow indicate the state of iron, oxygen, and water in the soil. Green indicates a lack of iron in the soil, and reflects the color of the dominant clay mineral. Red soil indicates oxidizing conditions, a requisite for the effective operation of septic tanks.

The gray-black color reflects a dominance of organic compounds in the soil horizon. Black and green soils tend to be associated because they both form in the absence of oxygen gas in soil water. The presence of organic matter masks the green clay color that would otherwise exist.

Classifying Soils

Water percolates down through the organic O-horizon and topsoil A-horizon, dissolving minerals, some of which become deposited in the subsoil B-horizon. This

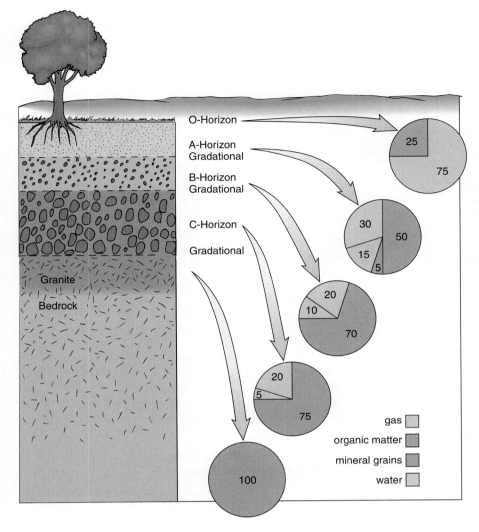

O-Horizon

A-Horizon
Gradational

B-Horizon
Gradational

C-Horizon

Gradational

Granite

Bedrock

gas
organic matter
mineral grains
water

FIGURE 3.21 Approximate compositions of a mature soil in a humid temperate region.

The O-horizon is composed of loosely packed organic matter in various stages of decay. The air in soil pores has the same proportion of gases (nitrogen, oxygen, carbon dioxide) as the atmosphere.

The A-horizon is dominated by mineral grains in various stages of decay, plus a small amount of organic matter. Soil pores contain water and air that is CO_2-enriched from decaying animals and plants.

The B-horizon is enriched in clay minerals and in hematite (Fe_2O_3) or calcite ($CaCO_3$), formed from iron and calcium from the A-horizon.

The C-horizon consists of broken but identifiable fragments of granite in very early decay.

is the basis for the broadest way to classify soils, which is by the minerals that accumulate in this B-horizon:

- If the subsoil is reddish because of the accumulation of iron oxide (hematite), the soils are called **pedalfers** (*ped-al-fer* = soil of *al*uminum and *fer*rum [iron]). These soils occur in wetter climates where annual precipitation exceeds about 30 inches. Note in Figure 3.22 that the wetter eastern United States generally has pedalfer-type soils. Figure 3.23A shows a typical iron oxide subsoil.

- If the subsoil is whitish because of calcite accumulation, the soils are called **pedocals** (*ped-o-cal* = soil of *cal*cium). These soils occur in dryer climates where annual precipitation is less than 30 inches. Note in Figure 3.22 that the dryer western United States generally has pedocal-type soils. Figure 3.23B shows a typical calcitic subsoil.

In wetter areas, ions of iron are leached from the mineral grains in the topsoil and carried downward

into the subsoil. The considerable organic matter in the topsoil makes it oxygen-deficient, so iron here remains in the soluble Fe^{+2} state.

Water transports these iron ions downward into the subsoil. The subsoil contains little organic matter, so more oxygen is available, and the Fe^{+2} iron is converted to insoluble Fe^{+3} iron, and it precipitates as red iron oxide (hematite, Fe_2O_3), shown in Figure 3.23A.

The red color of some soils may be inherited from the transported sediment of an earlier stage of soil development upstream or uphill, but the red color in that location originated in the same way.

In dryer areas, the iron atoms remain in the parent minerals and only potassium, sodium, magnesium, and calcium are leached into soil water. Of these, the calcium and magnesium are less soluble and tend to precipitate in some form in the subsoil. Also the subsoil contains a high concentration of carbonate ion, CO_3^{-2}, and it combines with the available calcium to form solid calcium carbonate: white calcite ($CaCO_3$). In the

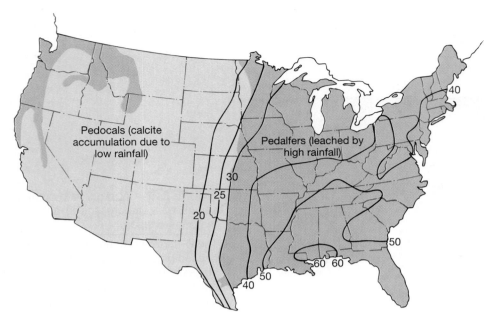

Pedocals (calcite accumulation due to low rainfall)

Pedalfers (leached by high rainfall)

FIGURE 3.22 Distribution of pedalfer and pedocal soils in the United States.

The contour lines show annual precipitation in inches. The boundary between pedalfers and pedocals approximately coincides with the contour line for 30 inches of annual precipitation.

Source: U.S. Department of Agriculture.

semiarid American Southwest, little or no A-horizon exists, so the B-horizon is at the surface (Figure 3.23B). The white calcite crust is called *caliche.*

Magnesium carbonate is more soluble than calcium carbonate and precipitates only in extremely dry climates. In desert areas, even potassium and sodium minerals may precipitate in the soil.

This simple division of soils is presented here to give you a general idea of soil chemistry. To soil scientists, the simple division of soils into pedocals and pedalfers is inadequate. The classification used by

the U.S. Department of Agriculture has ten major soil categories (orders), which are subdivided through five more levels of classification into about 12,000 soil series. Not even professional soil scientists can deal with so many terms and must keep reference books handy.

Laterite: An Interesting Soil

Laterite is a very widespread soil with extreme characteristics. It forms in very warm, very wet climates that

FIGURE 3.23 The B-horizon is a collecting level for various minerals.

(A) Accumulation of hematite in the subsoil (B-horizon) of a soil. *Photo: William E. Ferguson.* (B) Accumulation of calcite in the subsoil (B-horizon) of a soil. *Photo: Stephenie S. Ferguson/William E. Ferguson.*

(A)

(B)

FIGURE 3.24

Characteristic red color of a well-developed iron-rich lateritic soil. *Photo: GeoScience/PH.*

leached from the soil, but even covalently bonded silicon is largely or completely removed by chelation, leaving soil residues of aluminum and iron minerals. Aluminum-rich laterite, called *bauxite,* is the major ore of aluminum for the world's industrial economies (see Chapter 13).

Neither aluminum-rich laterite nor iron-rich laterite (Figure 3.24) contain the elements needed for plant nourishment. Farmers discover this, much to their chagrin, when they enter lateritic areas to practice slash-and-burn agriculture (Figure 3.25). In this type of farming, an area of tropical forest is felled and burned, and the land is cultivated for a year or two. The nutrients in laterite soils are located only in the upper few inches of soil (O-horizon), where they have been released by decaying plants. Below a few inches depth, the soil is barren of nutrients because laterite has been thoroughly leached of them.

Hence, farming cannot be continued for very long. When the shallow soil becomes exhausted, the farmers move on to another area to devastate the forest. Indigenous populations in areas of laterite soil were able to use slash-and-burn agriculture successfully because of their small numbers. They could deplete the soil in an area knowing that by the time they returned to it, it would be rejuvenated. Large-scale farming, however, is not possible.

The poor and temporary quality of lateritic tropical soils has been recognized by soil scientists and agriculturalists for a hundred years. Nevertheless, a few years ago, the World Bank (which is a branch of the United Nations) invested a large amount of mostly American money in an attempt at lateritic farming in

exist in equatorial regions, where many impoverished third-world countries are located. In most cases, the mean annual temperature exceeds 55°F and annual rainfall is at least 50 inches. Not only are ionically bonded potassium, sodium, calcium, and magnesium

FIGURE 3.25 Deforestation in Brazil's rain forest.

Here in the Amazon River basin, settlers and cattle ranchers are clearing the forest to open up new land and burning the trees as they go. During the 1980s, the rate of deforestation nearly doubled. Half of Earth's tropical forests already are gone. At the present rate, a few more decades could see the disappearance of these forests. *Photo: Asa C. Thoresen/Photo Researchers, Inc.*

(A)

(B)

FIGURE 3.26 Laterite as a building material.

(A) Quarry showing brick-grade laterite beneath about 32 inches of loose, granular overburden. The laterite is cut into slabs 6 feet thick, which are then split into bricks 8 by 16 inches in size. Near Toussana, Burkina Faso. *Photo courtesy G. J. J. Aleva.* (B) Buddhist temple at Angkor Wat, Cambodia, built of laterite bricks. *Photo: R. Ian Lloyd/The Stock Market.*

Brazil. The lush appearance of a tropical rainforest may imply good soil, but this is very misleading; in fact, the soil has limited usefulness as an agricultural resource.

Apparently, the World Bank neglected to note that in tropical, resource-poor countries, laterite soil is commonly used as a building material (Figure 3.26), much as adobe is used in the American Southwest. The word *laterite* is from the Latin term for *brick*, and bricks are impermeable to water and nutrients. Commercial fertilizers spread on laterite will not sink in.

Political efforts such as the laterite farming project may try to overcome scientific realities. But in the real world, realities triumph, and the expensive attempt at lateritic farming failed.

The Nutrients Soil Provides to Plants

Numerous chemical elements are present in the soil. Whether in solid minerals, organic matter, or soil pore water, these elements are all potential sources of nutrients for plants, which are rooted in the soil. But which of the ninety naturally occurring elements are essential for a plant to complete its life cycle?

This question was answered by comparing living plants in their natural setting (Figure 3.27A) and plants grown in the laboratory in controlled nutrient solutions (Figure 3.27B). By varying the chemical composition of laboratory solutions, scientists in the last century found that plants would only grow in solutions containing adequate water, nitrogen, potassium, calcium, magnesium, phosphorous, sulfur, and iron (Table 3.1). It was known that plants take up carbon from the air as carbon dioxide. So, adding the hydrogen and oxygen in water, ten elements were recognized as being essential plant nutrients (Table 3.1).

With the exception of iron, these elements are needed in relatively large amounts and are termed **macronutrient elements.** The elements needed in greatest quantity by most plants are nitrogen (N), phosphorous (P), and potassium (K). These are the key elements in most "NPK" fertilizers, such as those used to increase crop production, produce grassy lawns, and grow bountiful gardens. A common fertilizer composition for lawns is 10–20–10 (10% N, 20% P, 10% K).

In the 20th century, improved technology enabled the presence of more elements to be measured down to parts per million and parts per billion. Consequently, other elements were added to the nineteenth century list of plant nutrient essentials: manganese, zinc, copper, molybdenum, chlorine, and boron. For example, molybdenum present in nutrient solutions at a concentration of only 10 parts per billion (ppb) completely prevents the symptoms of molybdenum deficiency disease. These six elements (plus iron) are termed **micronutrient elements.**

Certain plants are known to require elements in addition to those listed in Table 3.1. Sodium is required for certain plants that live in saline (salty) soils. For

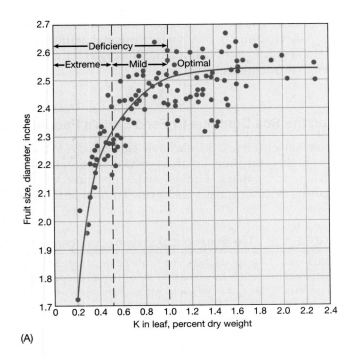

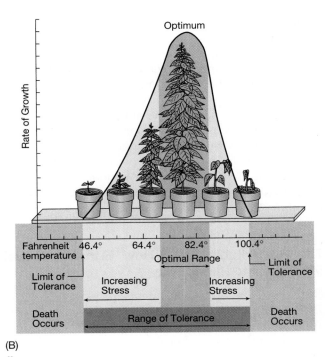

(A)   (B)

FIGURE 3.27 Determining what nutrients plants derive from soil.

(A) The relation between fruit size at harvest and the potassium concentration in leaves of 118 peach trees. The data reveal a critical potassium concentration of about 1.0 percent. An increase in potassium beyond this level does not increase fruit size. *After Lilleland et al., 1962,* Proceedings of the American Society of Horticultural Science *81:162-167.* (B) Relationship between plant growth and concentration in the soil solution of elements that are essential to the plant. Nutrients must be released (or added) to the solution in just the right amounts if normal plant growth is to occur. *After B. J. Nebel, and R. T. Wright, 1993,* Environmental Science, 4th ed., *Englewood Cliffs, NJ: Prentice-Hall, p. 33.*

TABLE 3.1	Essential Elements Needed by Higher Green Plants, and Sources of These Elements	

Element*	Direct Source	Dominant Ultimate Source
Macronutrients		
carbon	atmosphere, as CO_2	volcanic gas
nitrogen	atmosphere, as N_2	volcanic gas
oxygen	atmospheric O_2	algae
hydrogen (in water)	atmospheric H_2O and in soil pores	volcanic gas
potassium	surface of clay flakes	orthoclase feldspar
calcium	surface of clay flakes	plagioclase feldspar
magnesium	surface of clay flakes	ferromagnesian minerals
phosphorous	organic matter	apatite
sulfur	sulfate in pore waters	sulfide minerals
Micronutrients		
iron	surface of clay flakes	ferromagnesian minerals
manganese	surface of clay flakes	ferromagnesian minerals
zinc	surface of clay flakes	ferromagnesian minerals
copper	surface of clay flakes	ferromagnesian minerals
molybdenum	surface of clay flakes	ferromagnesian minerals
boron	surface of illite clay	granitic rocks
chlorine	pore waters of rocks	volcanic gas

*Greater or lesser amounts of these elements are also normally present in soil water and in pore waters of sedimentary rocks.

some of these plants, the requirement for sodium is so high it really is a macronutrient. Other unusual nutrients required for certain plants include selenium, cobalt, silicon, aluminum, gallium, and vanadium. As measurement techniques improve, other elements may be recognized as essential for plant metabolism.

Conversely, some of the ninety elements may be shown to be toxic to one or more varieties of plants. The definition of "pollutant" is not static; it changes as measurement technology improves and as we learn more about plant metabolism. An element thought to be a harmless contaminant today may be deemed a pollutant tomorrow.

The Remarkable Root: How Elements Travel from Soil into Plants

Of the essential nutrient elements for plants shown in Table 3.1, only carbon enters the plant directly from the atmosphere (in carbon dioxide). All the other nutrient elements enter plants when they take up water that surrounds their roots. Roots are a plant's agent in a key process: mining from the soil its store of nutrients, without which life as we know it could not exist. Roots are extremely efficient scavengers of the elements needed by plants. For example, the phosphorous taken up by roots typically is present in soil water at only one part phosphorous per million parts of water.

Because many nutrients exist in extremely low concentrations, higher plants have developed elaborate root systems. By progressively branching, they bring an astonishingly large surface area into contact with the soil (Figure 3.28). This improves their chance of finding all the nutrients they need. The area of contact is further increased by the growth of *root hairs,* very thin extensions of the root.

The dimensions attained by root systems are amazing. After growing only four months, one root system was carefully measured. Its total length (excluding root hairs) was 387 miles; the total surface area was 2554 square feet. When the root hairs were included, the total length increased to nearly 7000 miles and the area to 7000 square feet! *This single plant developed a surface area of 7000 square feet in direct contact with its nutrient supply.* (Seven thousand square feet is almost four times the floor area of an average American house.)

In addition to their incredible ability as "miners" of very dilute concentrations of nutrient elements from water, roots grow into small cracks in rocks and act like a wedge to pry the rock apart. The small pieces of rock thus created have more surface area than a single large rock and thus make available more nutrient elements to the plant. An additional aid to the plant is the soil-churning activity of burrowing organisms.

Earthworms, ants, termites, gophers, and many other animals continually mix surface organic matter with mineral grains from below, and this organic matter is rich in nutrient elements.

Can the Soil Continue to Feed Earth's People?

From ancient times, people have learned how to cultivate highly nutritional grains—wheat, corn, oats, rice, barley, and others. Grains make up most of the world's food. Countries that grow large quantities of

FIGURE 3.28 *The amazing root system of switchgrass.*

Actual size, 12 by 60 inches. The roots actually extend to a much greater depth. *From* Mineral Nutrition of Plants: Principles and Perspectives, © 1972 by John Wiley and Sons, Inc. From Weaver and Darland (1949). (J. E. Weaver, personal communication.)

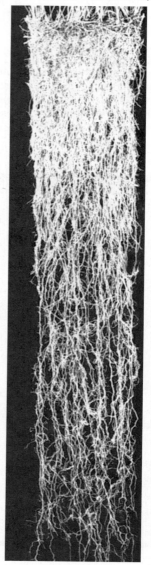

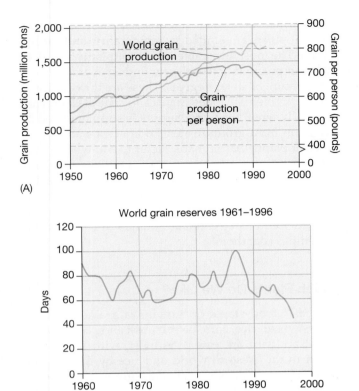

(A)

(B)

FIGURE 3.29 World food production falls behind population growth.

(A) Grains make up most of the world's food. Grains are significant even in the typical American college student's diet (cereal, chips, the wheat in pizza crusts and sandwich buns, and rice). From 1950 to about 1975, grain production grew faster than population. Since 1975, the amount of grain produced for each person on Earth has fallen as the population has grown faster than grain production. *Source: U.S. Department of Agriculture.* (B) World grain reserves are now at an all-time low.

grain are called the world's "breadbaskets." As agricultural technology has grown, so has the worldwide production of grains.

But now, world grain production is leveling off after forty years of sustained growth. The amount of grain available per person is decreasing significantly (Figure 3.29). Why?

Part of the reason is that each year we feed millions of additional people. There are simply more consumers to eat the annual harvest of grain. Africa now needs fourteen million tons more grain each year than it produces.

But another reason for the decrease in grain production per person is a decrease in Earth's soil resource, both in quantity and quality. More people take up more space, so more land is removed from crop production and converted into sprawling cities, suburbs, business districts, and highways. Every year, more than two million additional acres of American farmland stop sprouting grain and instead sprout housing developments, shopping malls, and other construction. There is about 15 percent less farmland in the United States today than in the 1960s (Figure 3.30). Along with that land, we lose not only food production but scenic open space and wildlife habitat.

From the beginning of agriculture around 8000 B.C. until 1950, nearly all growth in world food output resulted from increasing the number of acres being farmed. From 1950 to 1980, roughly four-fifths of the growth in world food output came from making those same acres more productive with new farming methods and chemicals. But, since 1980, *all* the increases in grain output have come from making the land work harder. Nonfarm uses of land are gobbling up more space and will continue to do so as Third World countries try to industrialize so they can become economically self-sufficient.

Further decreasing the land available for agriculture is **soil erosion.** Poor management of cropland has

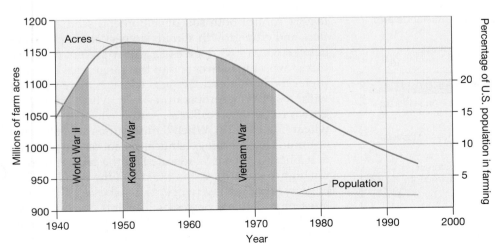

FIGURE 3.30 Shrinking farmland.

Trends in the number of acres farmed in the United States and the number of farm workers. Farmland started to decrease about 1950 and is now less than at any time since 1940. Only 2.5% of Americans now work on farms.

caused continual loss of topsoil (the A-horizon), washed away by water during heavy rainfall and irrigation. In recent years, the United States has made some progress in reducing soil loss, but developing countries filled with hungry people are hard-pressed to rescue rapidly eroding land and stop it from turning into wasteland.

Another reason for the decrease in productive cropland is diversion of irrigation water away from farmland for use in cities. For example, Arizona has purchased the water rights to large areas of previously irrigated cropland. The state has converted the cropland back to desert, so the water can be used in its cities.

Yet another reason for shrinking grain production is chemical pollution of farm land. This occurred in eastern Europe between 1945 and 1990 (see Chapter 8). Nuclear fallout on land in Ukraine and Belarus (formerly parts of the Soviet Union; see Chapter 16) have removed several million acres from cultivation, perhaps for hundreds of years.

Finally, there is the overall problem of mismanagement. Many of the world's nations have ineffective governments, criminal governments, or ineffective economic systems. In these impoverished countries, increases in grain production are possible only if the need is recognized and made a top government priority. Although wealthier countries can and should help, most of the burden for making this change must come from the affected nations themselves.

Why Do Plants Grow Where They Do?

Successful cultivation of crops and, indeed, any kind of plant, depends to a large extent on climate. Too little rain and plants wilt; too much rain and they drown. If temperatures grow too cold, plants freeze; too hot and their growth is stunted. Figure 3.31 shows how the types of plants are limited by temperature and precipitation. The most desireable region for plants has 20–40 inches of precipitation each year and an average yearly temperature of 50–65 degrees. These are the conditions in midcontinental United States and central California, especially when the water supply is supplemented by irrigation from groundwater. Edible crops are simply types of plants that humans prefer to eat: wheat, corn, oats, barley, rye, and rice, a very small percentage of Earth's vegetation. As was demonstrated by a health-food enthusiast a few decades ago, many weeds in vacant city lots are also edible, particularly when washed free of objectionable liquids and solids that sometimes coat them.

Which Plants Do We Like?

The basic food staple of humans is grain, which includes wheat, rice, corn, oats, barley, and rye. Of these, wheat and rice are eaten the most, with wheat preferred in the Western world and rice by East Asians. With the exception of rice, all of the grains are grown in abundance in the United States. We are a large country with enormous areas of flat ground suitable for crop farming, and are located in a temperate climatic belt, where summer temperatures average 70–75 degrees and rainfall is commonly adequate, although some farming areas require extra water from subsurface sources. In addition, and frequently overlooked as an important factor in our agricultural abundance, is our democratic form of government and capitalist economic system, which have been the foundation of our extraordinary national growth over the past 220 years. As a result of all these factors, the United States has been the world's breadbasket for many decades past and probably will continue to be for many decades into the future (Figure 3.32).

The areas suitable for crop farming are determined by temperature and precipitation. These two factors control both soil development, seed germination, and crop growth. Grains do best at temperatures between 65 and 85 degrees F. Wheat prefers temperatures toward the lower end of this range, rice and corn prefer the higher end. Because wheat is more versatile with respect to temperature, two wheat crops are grown in the United States, spring wheat and winter wheat (Figure 3.33). When temperatures are below the optimum for a grain, chemical reactions are slow, so that photosynthesis is retarded. If the temperature is too high, respiration is too rapid and the plant's ability to grow is harmed. At temperatures above 100 degrees, most physiological processes in plants decline due to the destruction of enzyme systems, chemical systems in the plant that make growth possible.

FIGURE 3.31 The control of vegetation types by temperature and precipitation.

Grains grow best in the area of overlap of grasslands and forests. *Source: National Science Foundation.*

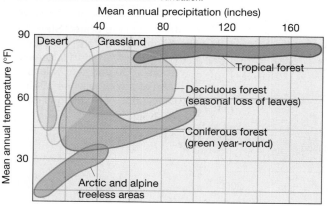

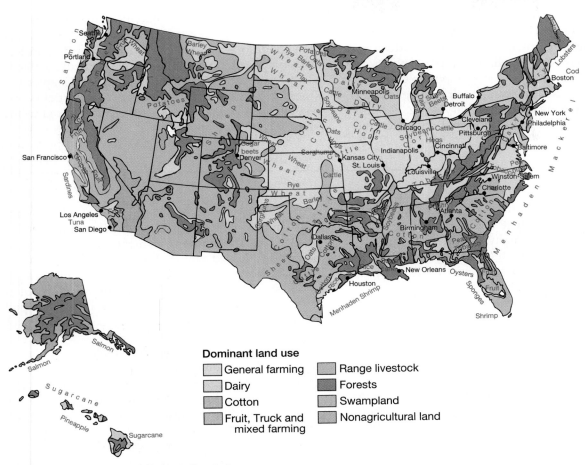

Dominant land use

- General farming
- Dairy
- Cotton
- Fruit, Truck and mixed farming
- Range livestock
- Forests
- Swampland
- Nonagricultural land

FIGURE 3.32 The World's breadbasket.

Areas of major farming in the United States and the products produced.

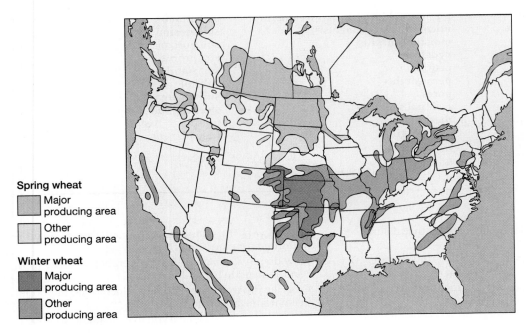

Spring wheat

- Major producing area
- Other producing area

Winter wheat

- Major producing area
- Other producing area

FIGURE 3.33 North America's wheat-producing areas.

Spring wheat is raised in the northern Great Plains states and in the Prairie Provinces. Severe winters prevent planting in these regions in the fall. The winter wheat belt extends from the southern Great Plains states through the eastern United States.

Rice is unique among the grains because it is grown under several inches of water. This requires that an area suitable for growing rice (a rice paddy) be located on a clayey rock or soil, so that permeability to water is low. Otherwise it will not be possible to keep the crop submerged. The water is drained either by pumps or by trenching just before harvesting. Because water in the field is stagnant (does not circulate), the dissolved oxygen in it is rapidly used up and methane (CH_4) is generated by the life forms that exist under such conditions. The methane gas released from rice fields (paddies) is a major source of the methane that is partly responsible for global warming (see Chapter 17). Rice is grown mostly in East Asia in small fields and very little is exported. Most of the crop is used in the country where it is grown.

Summary

Soil is our most important natural resource, but it has not received the attention it deserves and needs. A good agricultural soil normally requires several hundred years to form and is, therefore, an exhaustible resource in terms of immediate human needs. No matter how fat you are, you cannot survive more than a month or so without food, and without adequate soil there is no food.

The best soil for crops is a loam composed of about 40 percent sand, 40 percent silt, and 20 percent clay, supplemented by 5 percent organic matter. The nutrients for plant growth are stored almost entirely in the organic matter and clay minerals and are taken in by the plant through an elaborate root system. Carbon dioxide gas needed for photosynthesis is absorbed from the air by the leaves of the plant.

Bacteria, worms, and the myriad of other organisms that live in the soil are an essential part of the soil's functioning. The nutrient elements in humus (dead plants) are released again to the soil by bacterial attack, while the worms and other burrowing organisms keep the soil stirred. In this way, the nutrients are distributed in the soil and the soil structure is kept in the clumped condition (soil crumbs) that permits good drainage of water through the soil.

Explosive growth in world population and a decrease in the amount of good soil are a source of increasing concern. As the years pass, increasing numbers of people will be facing food shortages and starvation. Alleviating this growing problem will require considerable ingenuity by both the nations that have food surpluses and those who are needy.

Key Terms and Concepts

alluvial soil
bedrock
chelation
chemical weathering
clay minerals
fertilizer
frost wedging
grain
humus
laterite
litter
loam

macronutrient elements
mechanical (or physical) weathering
micronutrient elements
moisture surplus
pedalfer
pedocal
permeability
pores
residual soil
root wedging
salt wedging
sheeting

soil
soil crumb
soil erosion
soil horizons
soil profile
soil stability
subsoil
surface area
swelling soil
topsoil
transpiration

Stop and Think!

1. Sketch a soil profile and label the features that distinguish each horizon.
2. Explain why soils develop faster in hot rainy climates than in colder, dryer ones.
3. What are the two most abundant minerals in soil? Explain why.
4. Explain how chemical alteration of minerals makes nutrients available to plants.
5. Explain why the distribution of grain sizes in a soil is important for plant life.
6. How do physical and chemical alteration of rocks work hand-in-hand to break down rocks?
7. Describe the ways in which clay minerals are important in environmental concerns.
8. How thick is the soil where you live (or attend school)? Is the thickness different between hills and valleys? Why?

9. An organization such as the World Bank has many scientific, economic, and political advisors on its staff. Knowing this, what do you think can account for the wasteful expenditure of money on laterite farming?

10. Carefully remove a plant from its moorings in the soil. Is this difficult to do? Does the plant slip out easily or do you have to pull hard? Does "dirt" cling to the roots? What does this tell you about the relationship between the plant and its food supply?

11. A farmer friend gives you a handful of topsoil and asks your opinion of its quality. What features of the soil should you consider in answering?

References and Suggested Readings

Bongaarts, John. 1994. Can the growing human population feed itself? *Scientific American,* March, p. 18–24.

Brady, N. C. and Weil, R. R. 1996. *The Nature and Properties of Soils, 11th ed.* New York: Prentice-Hall, 740 pp.

Davidson, D. A. 1992. *The Evaluation of Land Resources.* New York: John Wiley & Sons, 198 pp.

Epstein, Emanuel. 1973. Roots. *Scientific American,* May, p. 48-58.

Gardner, Gary. 1996. Preserving agricultural resources. *State of the World,* Linda Starke, ed. New York: W. W. Norton, p. 78–94.

Hillel, Daniel. 1991. *Out of the Earth.* Berkeley, California: Univeristy of California Press, 321 pp.

Meyer, W. B. 1995. Past and present land in the USA. *Consequences,* v. 1, no. 1, p. 24–33.

Mount, T. D. 1994. Climate change and agriculture: a perspective on priorities for economic policy. *Climate Change,* v. 27, p. 121–138.

Ollier, Cliff. 1975. *Weathering.* London, England: Longman Group, 304 pp.

Pierzynski, G. M., Sims, J. T., and Vance, G. F. 1994. *Soils and Environmental Quality.* Boca Raton, Florida: Lewis Publishers, 313 pp.

Rozenzweig, Cynthia and Hillel, Daniel. 1995. Potential impacts of climate change on agriculture and food supply. *Consequences,* v. 1, no. 2, p. 22–32.

Tan, K. H. 1994. *Environmental Soil Science.* New York: Marcel Dekker, 304 pp.

Traeh, F. R. and Thompson, L. M. 1993. *Soils and Soil Fertility, 5th ed.* New York: Oxford University Press, 462 pp.

Wehrfritz, George. 1995. Grain drain. *Newsweek,* May 15, p. 22–28.

Wilken, Ellen. 1995. Assault of the earth. *Worldwatch,* March/April, p. 20–27.

"And the whole land thereof is brimstone, and salt, and a burning, that is not sown, nor beareth, nor any grass groweth therein."
Deuteronomy 29:22

Soil Problems—Erosion and Pollution

Agricultural production depends on soil quantity and quality. But soil ("dirt") is composed of loose sediment, so it is subject to removal by wind and water. How is soil naturally stabilized? Can stabilization be improved? How long does it take for a new soil to form after an existing soil has been removed? How do the specific minerals that make up a soil affect its productivity? Why must fertilizer be added to many soils to improve crop yields or to keep weeds from dominating the preferred grass on a residential lawn?

Soil Erosion

Soil erosion is natural and relentless. It is a process by which all land surfaces that protrude above sea level are worn down gradually by water and wind.

As erosion proceeds to wear down the land, great forces within Earth are uplifting and reforming Earth's crust. Unfortunately, the rate of this uplift is very slow. Human activities usually hasten the wearing-down process faster than uplifting processes can redress it.

Careful farming may build topsoil fast enough to compensate for the increased erosion that results from modern agriculture. Much more commonly, however, poor farming practices just accelerate erosion (Figure 4.1A). The U.S. Department of Agriculture estimates that 70 percent of U.S. soil erosion is caused by human activity. Loss of productivity can be concealed for a number of years by heavy use of fertilizer, but productivity must eventually decrease as more and more topsoil is lost. In the United States, the average soil loss is five to six tons per acre each year. We have lost about one-third of our topsoil since farming started about 350 years ago.

The U.S. Department of Agriculture's Soil Conservation Service estimates that about 44 percent of U.S. agricultural land is losing topsoil faster than it can be regenerated by natural soil-forming processes (Figure 4.1B). Many other countries are experiencing similar rates of soil loss. In tropical regions that have intense rainstorms and sloping ground, soil loss can be even greater. About 10 percent of the world's soil has been lost this century through deforestation, erosion, urban development, and other abuses of the land. Soil is as essential as air and water and should be given the same degree of protection.

Soil loss is a problem as serious as soil pollution. Media attention, however, is devoted almost entirely to soil pollution because the effect of chemical pollution on people is typically more immediate and striking. Sick people are more interesting to watch on the news than muddy water running off farmland in a Nebraska thunderstorm.

◄ *Has any of this toxic waste seeped into the ground?*
Photo: Tom Kelly/Phototake NYC.

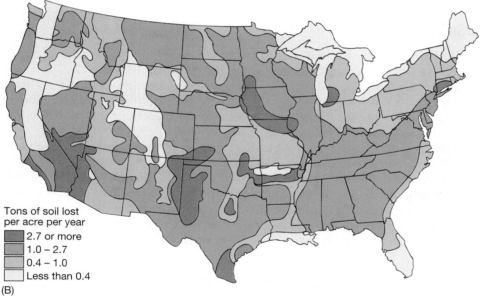

Tons of soil lost
per acre per year

- 2.7 or more
- 1.0 – 2.7
- 0.4 – 1.0
- Less than 0.4

(B)

FIGURE 4.1 Soil erosion in the United States.

(A) Erosion in this cultivated field in Oklahoma is carrying away vital topsoil. *Photo courtesy Soil Conservation Service.* (B) Average annual erosion of cropland soils in the United States. *Source: U.S. Department of Agriculture.*

Soil Erosion by Water

Most soil erosion is accomplished by water. Even soil that is carried away by wind is first detached from the ground by raindrops. A large raindrop falling in still air can reach a velocity of 20 miles per hour; in wind-driven rain, the velocity is greater, as everyone who has tried to walk into the teeth of a driving rain is well aware. Rainfall intensity is more important than the total amount of rain for the erosion of soil.

Each raindrop acts like a tiny hammer, dislodging and scattering soil particles (Figure 4.2). Considering that millions of raindrops strike the soil in a typical rainstorm, a large volume of soil can be loosened and transported downhill in a single rainfall.

The power of raindrops to erode bare soil is shown by the muddiness of the water draining from farmland. Mud is mostly clay and organic matter, and these are the two types of soil particles that hold nearly all the nutrient elements needed for plant growth. A rich and fertile

FIGURE 4.2 During a typical rainfall, millions of water drops strike the ground at nearly 20 miles per hour.

Each drop that strikes an exposed surface splashes soil particles, some up to three feet high. Some may land more than three feet away from the point of impact. Raindrop impacts thus break up soil so that it can be easily removed by sheet erosion. *Photo: Runk/Schoenberger/Grant Heilman Photography, Inc.*

loamy soil can be rapidly depleted of nutrients by sustained water erosion. In addition, the soil becomes sandier as mud is removed, so that its water-holding capacity is lost. This effect can be very serious for agricultural productivity in areas of marginal rainfall where annual precipitation is less than about 25 inches.

The channels carved by water in eroding farmland are divided into two types by soil scientists. **Rills** are channels small enough to be smoothed over during normal plowing. Typically, rills follow the lines formed by crop rows, because the crop roots bind the soil beneath them to resist erosion. The result of repeated rill erosion and later smoothing by plowing is similar to that of a sheet of water flowing over the field. A layer of soil is removed and the soil profile is made thinner.

Gullies are too large and steep-banked to be smoothed by normal plowing operations. They are at least one or two feet deep and often cross crop rows rather than following them. Gullies usually form where a concentration of water flows down a slope or over a bank, perhaps at the edge of the field where cultivation has been stopped because of an abrupt steep-

ening of the ground. Gullies tend to start at the edge of the field and erode headward into the field, toward the source of the water. A stream of water that is flowing harmlessly across a gently sloping field becomes highly erosive as it pours over a steep bank. A gully system can ravage a large area of farmland in only a few years, leaving islands of uneroded soil that are inaccessible and therefore unusable (Figure 4.1A).

Gullies remove entire soil profiles and often cut into unconsolidated material beneath the soil. Gully depth is limited either by erosion-resistant material such as bedrock or by the level of their outlet end.

Soil Erosion by Wind

Erosion by wind is a serious problem in many areas, particularly where rainfall is scant and wind velocities are high. In 1982, 37 percent of the soil moved by erosive forces in the United States was moved by wind erosion. Possibly 12 percent of the continental United States is somewhat affected by wind erosion, 8 percent moderately so, and perhaps 2–3 percent greatly. Soil loss from wind is greater than loss by water in Texas, Colorado, Montana, New Mexico, Oklahoma, and North Dakota. In the first four of these states, wind erosion on cropland is four to five times as great as water erosion.

The severity of wind erosion can be reduced significantly by proper land management. This is demonstrated by the location of a 1977 dust storm in north Texas. The dust storm did not cross over into New Mexico, due largely to a difference in state water laws:

- Texas allows landowners to use the land, and the groundwater under it, as they wish. Texas views any water restriction as an infringement on personal property rights. As a result, Texans heavily irrigate their farmland.
- In New Mexico, the state owns underground water. Permission to use it must be obtained from the State Engineer. In the area of the dust storm, the State Engineer had decided that the natural vegetation in eastern New Mexico is more suitable for livestock grazing than for farming. The drought in the late 1970s was accompanied by persistent strong winds which blew away topsoil in the affected part of Texas, while topsoil in eastern New Mexico remained in place.

The Dust Bowl. The devastation that can result from wind erosion was clearly shown to this nation by the disastrous **Dust Bowl** of the 1930s. Plowed land, combined with drought, left deep, rich topsoil at the mercy of high winds for several years. The affected area is shown in Figure 4.3.

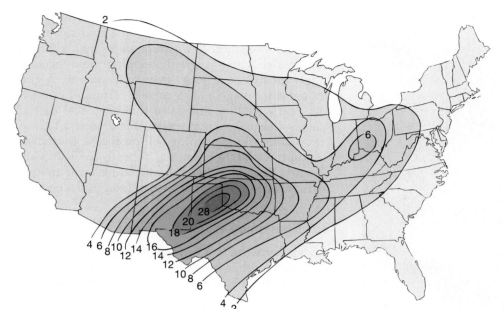

FIGURE 4.3 The contours indicate the number of dust storms or dusty conditions during March, 1936.

The center of the affected area varied somewhat from month to month during the 1930s, but this map is typical for the period.
Source: U.S. Weather Bureau.

Average annual precipitation in the region is about 20 inches, making it marginal for farming. In about one year out of five, the average drops to 15 inches, virtually a recipe for crop failure. The mid-1930s saw abnormally low rainfall. Native grasses that blanketed the region prior to farming withstood such drought; otherwise they would not have been there. But less-hardy crops like wheat perished. The soil was laid bare to wind erosion.

Homes, barns, tractors, and fields were buried under drifts of sand and silt up to 25 feet high. The sky could turn completely black in minutes (Figure 4.4). At

FIGURE 4.4 "Black blizzards" were common throughout the Dust Bowl during the 1930s.

Several minutes after the start of a storm, the sky would turn completely black, and the sun would be hidden from view for days at a time. *Photo: Library of Congress.*

times, dust obscured the sun for several days. Even wet towels stuffed in the cracks around windows could not keep the dust out. To breathe, people covered their faces with wet cloths. Each storm was followed by many cases of serious lung damage, some of which were fatal.

Numerous books chronicle the intensity of human and economic suffering in the Dust Bowl. The most famous is *The Grapes of Wrath* by John Steinbeck, which has been made into a classic film. Many present inhabitants of central California's agricultural region are descendants of the thousands of "Okies" (Oklahomans) who fled westward to escape the blowing dust.

In central California's San Joaquin Valley, the amount of rainfall is similar to that of the Dust Bowl area. But constant irrigation prevents drought from causing a dust bowl there. However, continual use of underground water to irrigate crops brings other environmental problems, including ground subsidence and a decrease in the water available for urban residents in California's ever-expanding cities (see Chapter 6).

Could another Dust Bowl occur in the Midwest? Yes, although it might not look as severe. The region is the nation's windiest flatland and is more susceptible than most areas to wind erosion. Today's farmers use special techniques to reduce topsoil loss, including planting of windbreaks in the form of rows of trees or bushes, but these cannot provide complete protection from strong winds. In fact, more land was damaged by wind between 1954 and 1957 than during the worst years of the Dust Bowl, although the dust storms were less spectacular.

Conservation Farming Can Reduce Soil Erosion

In conventional agriculture, soil is prepared for planting in several stages. First, remains of plants from the previous crop are plowed into the soil, which leaves bare soil exposed on the surface. The soil surface is then broken up to make a smooth bed for seeds. Once rows of crops are planted, a machine stirs the soil between the rows to rip out weeds. Thus, the soil is disturbed repeatedly, at great cost in time, equipment, and fuel. More important, however, the soil is left bare until later in the year when crops have grown enough that their roots can retain the soil. This means that the disturbed soil is left unprotected during springtime, when runoff of rainwater and erosion are greatest.

In recent years, conservation efforts known as **conservation farming** have reduced the amount of plowing on farmland (Figure 4.5A). Strategies vary from merely reducing plowing to no plowing at all, wherein seeds are planted directly through the residue of the previous crop (Figure 4.5B). For example, corn may be planted in wheat

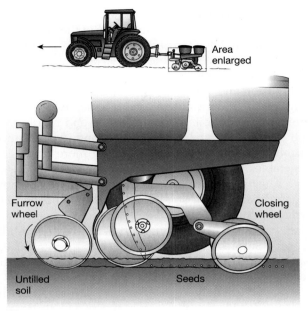

(A)

(B)

FIGURE 4.5 Conservation Farming.

(A) Conventional plowing leaves bare soil exposed to erosion. "No-till" planters cut only a narrow slit through sod and crop residues, drastically reducing erosion. (B) Crop residues left on the ground reduce the amount of soil carried away by wind and rain. *Photo: Anson Eaglin/USDA.*

stubble. Conventional planting leaves only 1–5 percent of the soil protected by crop residue. But the no-plowing method leaves 50–100 percent of the land protected and reduces soil erosion by more than 90 percent.

These differences have a marked effect on both soil erosion and water runoff. Plant-covered farmland contains abundant living plant roots, which retain moisture in the soil, reducing runoff, and holding the soil together. In well-drained farmland, conservation farming generally yields as much, or more, grain and vegetables than the harvest from conventional farming. This results from retaining more water, more organic matter in the soil, and more soil fauna, particularly earthworms. Farm equipment, with its sharp blades, is disastrous to long, soft-bodied creatures like worms. Studies indicate that about 85 percent of the earthworms in a field are turned into mincemeat by conventional mechanized farming.

Unfortunately, conservation farming has its down side. Plant stubble contains weed seeds and harbors insects. Therefore, as plowing is reduced, often the need for herbicides (to kill the weeds) and pesticides (to kill the bugs) increases. Herbicides and pesticides are chemical poisons. They are expensive and may contaminate streams, lakes, and groundwater. Also, on land that drains poorly, greater water content in the soil can encourage some plant diseases. There is no perfect solution to the dilemma of farming versus erosion.

Contour Plowing

Contour plowing slows the flow of water on sloping ground. This reduces water's ability to transport soil downhill. Even a small decrease in water velocity can result in a large decrease in erosion. Contour plowing produces ridges and furrows (channels) that run crosswise to water flow down a hill (Figure 4.6). This stops the formation of gullies.

Contour plowing is effective only on moderate slopes of up to about 8 degrees. Where land is scarce, forcing steeper slopes to be farmed, **terracing** can be effective. In terraced planting, a steep slope is cut into a series of broad steps that slant gently backward into the hill (Figure 4.7).

Soils, Forests, and Erosion

Growing crops and raising animals is not the only agricultural use for farmland. Growing trees and logging them is another. Using trees for building and burning them as a source of energy has been practiced since the earliest humans peopled Earth (see Chapter 14). The

FIGURE 4.6 A well-managed farm in the American Midwest employs both contour plowing and strip cropping.

Different crops are planted in adjacent strips. Some are legumes, a type of plant that harbors bacteria which convert nitrogen gas from the air into nitrate (NO_3^-) in soil water. Plants use the nitrate as a nutrient. *Photo: Grant Heilman/Grant Heilman Photography, Inc.*

harvesting of trees caused no environmental problems so long as it was not too intensive, due to Earth's formerly small human population.

However, in modern times, the intensity of tree cutting has become a worldwide epidemic, particularly in humid tropical areas where the luxuriant tree growth is often seen as a short pathway to economic prosperity. Most tropical countries are among the world's poorest, and, under such conditions, environmental concerns take a back seat to immediate financial need. As a result, tropical **rain forests** are disappearing (Figure 4.8). The area of forest lost each year is the size of the state of Missouri. More than half of Earth's rain forests have been toppled since 1945. At the present rate of destruction, they will all be gone in 30–50 years (Figure 4.9).

Poor tropical nations are not the only ones that are destroying rain forests. Hawaii's last remaining lowland rain forest is being cleared at a rate of about 10 acres per day to supply wood chips as fuel for an electrical power generating station. The resulting cleared area will be turned into pasture under tax incentives provided by the state and federal governments. Poverty is not the only reason for the destruction of rain forests.

Well, so what? What is the importance of rain forests?

FIGURE 4.7 Rice terraces in the Philippines.

The terraces sharply reduce soil erosion in this mountainous region and conserve water by reducing runoff. *Photo: D. Cavagnaro/Visuals Unlimited.*

The luxuriant tree growth we associate with tropical jungles serves several important functions in addition to its use as a firewood source:

- Perhaps most important is that the rain forests are the home for at least 50 percent of Earth's plant and animal species. Biologists estimate that half or more of them could be eliminated within 50 years if current rates of tropical forest destruction and degradation continue.

- Tropical rain forests supply hundreds of food products and industrial materials found nowhere else. Examples include coffee and rubber.

- So far, the main ingredients in 47 major medicinal drugs have come from tropical plants. Of the 3000 plants identified by the National Cancer Institute as having chemicals that fight cancer, almost three-fourths come from tropical rain forests.

In addition to these direct ecological effects, rain forest destruction increases erosion because the protective covering of trees and their leafy crowns is lost (Figure 4.9). Erosion caused by deforestation is not limited to cleared tropical hillsides. Its effect also is felt in temperate areas, such as the Pacific Northwest. Here, logging has increased soil erosion. Salmon reproduction has decreased due to sediment in the water. The shrinking salmon population in the Northwest's major waterways has terrible effects on the salmon industry and its employees, many of whom will lose their jobs. Coho salmon populations have fallen to less than 5 percent of their former numbers along the coast of Oregon and northern California.

These facts raise important social questions: Should logging be restricted or halted? Should the salmon industry be protected? How should society evaluate the conflicting interests of timber companies and the salmon industry?

All things in the natural world are interrelated. A change in one aspect of the environment is certain to affect other aspects. However, one can argue that some effects are more serious than others. In the example just cited, logging's effect on the salmon population is serious to us because salmon are commercially valuable fish. There are economic reasons for protecting salmon. But less clear is the value of a noneconomic species, like the spotted owl (Figure 4.10). This bird's habitat is first-growth forest in the Pacific Northwest. No one eats these owls, and few people have a "thing" for owls.

During 1990, timbering in the U.S. Northwest threatened the spotted owl with extinction. The question was raised: should tree harvesting be restricted to protect this owl, which has no commercial value? Is commercial value the best way to establish the value of a plant or animal? Do plants and animals have "rights"? In the United States they do, following passage of the federal Endangered Species Act in 1973. The Act prohibits the deliberate extinction of any species by humans, from cockroaches and silverfish to dogs and cats, because of their "aesthetic, ecological, educational, historical, recreational, and scientific value to the Nation and its people."

Soils, Forests, and Erosion **73**

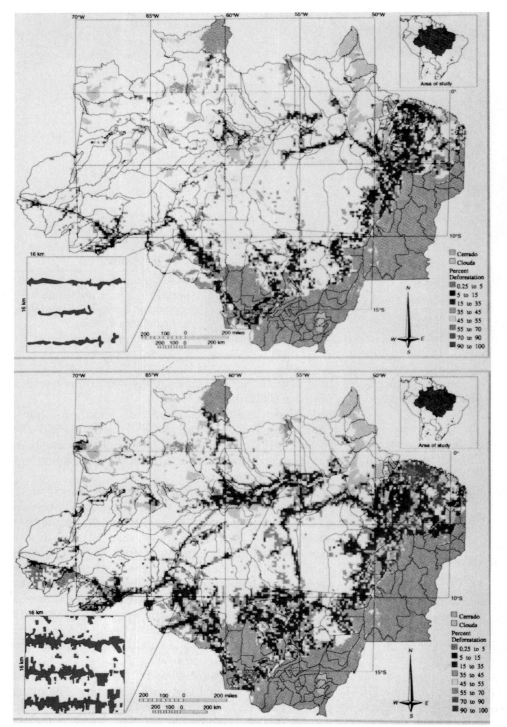

FIGURE 4.8

These two maps allow comparison of the extent of deforestation (percent) in the Brazilian Amazon rainforest over the decade between 1978 (A) and 1988 (B). Data are from the Landsat-4 and Landsat-5 satellites. *Source: NASA Headquarters.*

In the case of the spotted owl, the bird was protected and logging was curtailed, at the cost of some logging jobs. Other events have caused a greater loss of logging jobs. For example, many more jobs have been lost to automation in the industry, to decreased demand by the construction industry, and to the practice of shipping raw logs overseas. The spotted owl has cost relatively few jobs. However, placing an area off limits to protect certain species does cost some jobs. Suppose those who lost their jobs were members of your family. Would that color your view of the value of the spotted owl?

The challenge to societies is to decide when a human change in natural processes will be harmful enough in the long run to justify human suffering in

FIGURE 4.9 Deforestation in tropical Madagascar is increasing erosion and causing numerous deep gullies along the hillsides.

The stream is nearly choked with sediment. The ocean around Madagascar is colored blood-red by the lateritic mud that has been eroded from recently deforested areas.
Photo: © Frans Lanting/Minden Pictures.

FIGURE 4.10 A spotted owl.

Owls are cute, but how should their value be determined relative to human economic concerns? *Photo: Aaron Ferster/Photo Researchers, Inc.*

the short run. But who will bear the brunt of the suffering? Is possible extinction of the spotted owl a sufficiently important change in natural ecology to justify putting some loggers out of work? How should such issues be decided?

Sediment Pollution of Streams

Soil erosion contributes heavily to the sediment load of streams. This increased load can increase the frequency of floods, for two reasons. First, the mud increases the volume of the stream, which obviously increases the likelihood of overflow. Second, as excess mud is deposited along the sides of the stream channel, the channel is narrowed, encouraging the water to overflow more often.

The proportion of sediment load contributed by farmed soils is uncertain, but may be 30 to 35 percent (Figure 4.11). This means that 65 to 70 percent of stream sediment is of natural origin. When sediment is deposited, it is referred to as **silting.** Silting is considered a problem only because humans wish to interfere in some way in the natural cycle.

One of the most obvious problems is the silting of reservoirs. We think of dams as being permanent, trouble-free structures, but silting decreases their useful life by decades. Dams are constructed for hydroelectric power generation, or flood control, or to create a reserve of water for public supply.

An increase in stream sediment load resulting from agricultural practices is a water contaminant. But when a dam is built on the stream, the contaminant

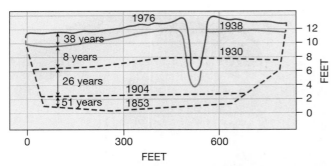

FIGURE 4.11 An example of sediment load contributed by farms.

Sediment accumulation in Coon Creek, Wisconsin between 1853 and 1976. From 1853 until 1904, the region had a very small population and little disturbance of the soil. Natural processes required 51 years to accumulate about 24 inches of sediment, an average of 0.5 inches/year. Between 1904 and 1930, as the population and number of farms grew, the rate increased to 2 inches/year (about 58 inches in 26 years). Soil erosion accelerated between 1930 and 1938 to 5 inches/year (40 inches in 8 years). Since 1938, soil conservation measures have decreased the rate to half inch per year that was typical of the pre-farming rate in the last century. *Source: Trimble, 1981, Science, v. 214. Permission of American Association of the Advancement of Science.*

FIGURE 4.12 Urbanization and erosion.

Urban construction turning firm ground into loose, easily eroded debris. *Photo: Mark E. Gibson/Visuals Unlimited.*

immediately becomes a pollutant. Few human-made objects are as useless as a sediment-filled former lake behind a dam.

Other sources of sediment pollution include:

- Urban construction projects that convert firm ground into loose piles of easily eroded debris (Figure 4.12).
- Surface mining ("strip mining") that converts hard rock into transportable sediment.
- Off-road vehicles, such as dune buggies, that destroy fragile, soil-preserving vegetation in dry areas. (This is a local, small-scale problem, not of the magnitude of the other sources of sediment.)

Soil Pollution

"Pollution" often seems to dominate the news media. Most pollution problems stem from industrial processes and the organic compounds they produce as byproducts. However, some organic compounds are manufactured deliberately because they have a beneficial effect when added to soils. Unfortunately, these soil additives have harmful effects as well.

Before we consider the pluses and minuses of these chemicals, we need to be clear about the meaning of the terms used by media people and scientists in talking and writing about pollution. Unless we agree on what the words mean, good communication is not possible. Consulting a dictionary yields descriptions of *pollution* such as impure, dirty, harmful, or contaminated. Although appropriate, such words do not provide a working definition that is useful when studying the environment.

Contamination Versus Pollution

Consider the words contamination and pollution. **Contamination** means *the concentration of a substance that is greater than would occur naturally, but the substance is not necessarily causing harm.* For example, your body may be able to handle a small amount of an artificial substance by decomposing it in your stomach or digestive system. You are able to tolerate small amounts of this contaminant, which causes you no ill effect. Many such chemicals are known. Of course, your body's tolerance for a contaminant varies with the nature of the substance and with your state of health. All living organisms tolerate variability in their surroundings, and without this tolerance they could not survive in the natural world. Some contamination is inevitable in our industrial society.

Pollution, on the other hand, means a *contaminant that is harmful to an organism.* For example, excessive zinc or copper in water is harmful to some types of fish, but not to humans. Therefore, an oversupply of zinc or copper is a *pollutant* to fish but only a *contaminant* to humans. The difficulty is that we generally do not know the specific level of each substance that each plant and animal can tolerate, including us. Typically, the public interprets this—incorrectly—to mean that *no* amount of contaminant is safe (see the asbestos discussion in Chapter 2).

You can see that the distinction between contamination and pollution is often uncertain biochemically. However, under the doctrine of "better safe than sorry," the distinction between the two words often is ignored completely in the media. The distinction seems often to be drawn politically rather than scientifically. No doubt this confusion will continue.

At present, 70,000 chemicals are in use. Hundreds, and possibly thousands, of new ones are created each year. Many are designed for use in agriculture to control bugs, weeds, or rodents. Obviously, it is impossible to test all of these substances to determine which may be harmful to a desirable plant or animal, and in what amounts. As a result, no health data exist on 70 percent of the chemicals in use.

Further, most tests conducted to determine safe levels for humans are not performed on people over long trial periods, but are done over brief periods using small animals that are plentiful and inexpensive, commonly mice (Figure 4.13). Typically, the mice are given doses of suspect chemicals that are unrealistically large in proportion to amounts likely to be ingested during a human lifetime. The rationale is that the high doses bring rapid reactions and that only the rate of onset of a harmful result is affected. The same result would occur with smaller doses, it is argued, but would simply take longer.

For many substances, however, this assumption is known to be false. For many chemicals, there exists a threshold below which no harm occurs. How frequently this is the case is unknown. So, once again, based on the "better safe than sorry" doctrine, the possibility of the existence of a threshold value for some or all plants and animals is neglected in the interpretation of test results. If *any* dose given to the mice causes cancer, the chemical has been required by law to be labeled as probably cancer-causing in humans (Delaney clause of the Food and Drug Administration, 1958; repealed in 1996). No way has been found to overcome this problem.

Types of Soil Pollution

The most common type of soil pollution occurs because of the deliberate application of pesticides. These are artificial organic (carbon-rich) compounds designed to

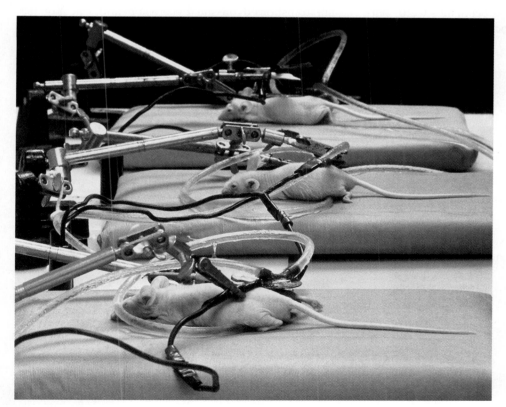

FIGURE 4.13   Is this necessary?

Laboratory testing of chemicals on mice. *Photo: Ben Edwards/Tony Stone Images.*

control "pests." Pesticides are manufactured to control any of the three types of organisms that are harmful to crops: herbicides (to kill weeds), insecticides, and rodenticides.

Here we must make an important distinction between organic and inorganic. Generally, *organic* refers to the chemistry of carbon compounds, which make up living things. *Inorganic* refers to the chemistry of noncarbon compounds. Loosely speaking, *organic* substances contain carbon; *inorganic* ones do not.

Although organic pesticides are the most harmful soil pollutants to humans, there also are inorganic pollutants. For example, crusts of harmful salts are produced by repeated irrigation of crops in areas with less than adequate rainfall, such as in the San Joaquin Valley of central California. And, in some agricultural areas, unusually high concentrations of elements occur in the soil. These can harm particular kinds of plants. We will now consider each of these pollution problems.

Pesticide Pollution of Soil

Every plant and animal has "enemies"—usually something that eats them. Several thousand life forms are known to cause problems: more than 10,000 insect species, 600 weed species, 1500 plant diseases (bacteria, viruses, and fungi), 1500 parasitic worm species (a group called nematodes, which includes hookworms and pinworms), and assorted rodents. Chemical **pesticides** have been used to control these pests for 2500 years. For most of this time, these pesticides were "natural" chemicals, not some concoction from a laboratory.

But in the middle of our own twentieth century, use of **synthetic organic pesticides** became widespread (Figure 4.14). Some 600 chemical poisons in about 50,000 combinations now are used worldwide to control pests. The use of agricultural pesticides in the United States occurs on approximately 75 percent of all cropland. About 70 percent of all livestock are treated with these substances.

The beneficial effects of pesticides cannot be doubted. They have controlled or eradicated numerous human diseases, for example, by killing the mosquitoes that carry yellow fever and malaria. They have protected crops and livestock against insects and diseases. Without the assistance of chemical weed control, conservation farming would be difficult. Also, pesticides reduce the spoilage of food as it moves from the farm, through food processors, through the food distribution system, to your dinner table (Figure 4.15).

However, these chemicals also enter the soil. Most are **biodegradable,** which means they are attacked by bacteria and other microorganisms that use the chemicals as food. In biodegradation, a chemical that is harmful to humans is converted into generally harmless inorganic products such as carbon dioxide, water, and soluble ions. Biodegradation is most effective in well-drained, aerated soils. Modern technology has created some genetically engineered microbes that thrive on specific organic pollutants.

A few chemicals are not readily biodegradable and persist in the soil or water for many years. Further, they are transported off-site by erosion and leaching (see Chapter 8). Some chemicals inadvertently kill nontarget

FIGURE 4.14 Airborne spraying of pesticide on a large farm in California.

This is a common method of pest control. Aircraft are used to apply 60 percent of the pesticides used on cropland in the United States. *Photo: Jack Fields/Photo Researchers, Inc.*

FIGURE 4.15

Calibrated spraying equipment minimizes the amount of pesticide used. *Photo: Spraying Systems Company.*

organisms, such as beneficial insects (yes, there are "good" ones that prey on harmful organisms or aid pollination) and soil organisms (like earthworms).

Moreover, as plant debris that contains these chemicals is consumed by earthworms, the chemicals tend to concentrate in their bodies. When birds and fish eat the worms, the pesticides concentrate further, to lethal levels. Damage to these larger creatures sounded the warning that pesticides can have devastating environmental complications. This awareness became widespread as people read Rachel Carson's classic 1962 book, *Silent Spring*.

Who could fail to be moved by its evocative introduction, in which she foretold of a town in the heart of America where all life seemed to live in harmony with its agricultural surroundings, until . . .

> . . . a strange blight crept over the area and everything began to change. . . . Everywhere was the shadow of death. The farmers spoke of much illness among their families. In the town the doctors had become more and more puzzled by new kinds of sickness appearing among their patients. There had been several sudden and unexplained deaths, not only among adults but even among children, who would be stricken suddenly while at play within a few hours . . . (p. 2)

Carson's "blight" was synthetic chemical pesticides. Encompassed within her bleak vision of 35 years ago are controversial issues that are still with us today. Are pesticides necessary? Are suitable substitutes available? Is there cause for concern about pesticide residues in our food supply?

Organic Pollution of Soil

The most-used pesticides are **herbicides.** They are formulated to kill "herbs," or weeds. (A **weed** is defined as any plant not wanted by the planter.) Herbicides are generally biodegradable and most have relatively low toxicity to mammals, including humans. However, some are quite toxic to fish and other wildlife. This emphasizes the need to consider the indirect effects of pesticide use.

Insecticides are the second most commonly used pesticide. Many are harmful to birds, fish, and humans. Three types are widely used:

- Chlorinated hydrocarbons are chemicals whose main elements are chlorine, hydrogen, and carbon. The most famous example is DDT. These chemicals were the most extensively used in the United States until the 1970s. It then was recognized that they persist in the environment and are toxic to birds and fish. Chlorinated hydrocarbons are used little in the United States today, although they are still common in some other parts of the world.

- Organophosphates (organic compounds rich in phosphorous) are a group of insecticides that biodegrade naturally in the environment but are quite toxic to humans.
- Carbamates (nitrogen-based organic compounds) are popular insecticides among environmentalists because of their biodegradability and relatively low toxicity to mammals.

Less-used pesticides include **fungicides** (used mostly to kill fungus on fruits and vegetables) and **nematocides** (used to kill nematode worms). Some of these are known to pollute water supplies because of their high solubility.

Clearly, the whole topic of pesticides is contentious. They are believed by most argiculturalists to be necessary to the large-scale agriculture that feeds modern civilization, but their use brings numerous problems. Each year, thousands of people, most of them not farmers, are admitted to U.S. hospitals for treatment of pesticide poisoning. Pesticides should not be used indiscriminately.

Because of their harmful side effects, the use of pesticides in U.S. agriculture decreased 15 percent from 1982 to 1992. New methods of pest control are emerging; the buzzword for these techniques is **integrated pest management.** This refers to a combination of using hardy plants, rotating crops from field to field (not planting the same crop in the same field year after year—this disrupts the pests' happy home, limiting their numbers), improved plowing practices, biological controls (such as release of predatory organisms to eat or disease the pests), minimal use of chemical pesticides, and even vacuum cleaners.

Organic Farming

One possible solution to the problem of pesticide pollution is **organic farming.** In this type of agriculture, no synthetic pesticides are used on the crops. Pest control is achieved by using crop varieties most resistant to pests, spraying soap instead of insecticides when pests appear, introducing ladybugs into the fields to prey on insects, and by alternating the types of crops grown in a field. Animal manures and crop residues are used as fertilizer rather than artificial compounds. At present, only two percent of American crops are grown in this way, and those who will eat only organically-grown produce are generally considered members of a nutritional cult near the lunatic fringe of American society.

Is it possible to conduct large-scale agriculture by organic methods and grow the same amounts and varieties of crops? And what about the cost? Those who patronize organic food stores are well aware that they have higher food bills than their pesticide-consuming neighbors. The data on these issues are not consistent. Some studies appear to show little or no loss of crop amounts and varieties and almost no increase in cost to the consumer. Other studies indicate somewhat decreased production and a significant cost difference.

The following is a testimonial by K. C. Livermore, a Nebraska farmer, who grows alfalfa, oats, soybeans, and corn on 260 acres. He says:

> We've done much better without chemicals. We hurt some at first when we switched over because we had to get the soil back in balance, get the poisons worked out of it. But in our fourth year, there was a big turnaround and now we're outyielding our "chemical neighbors" by far.

Mr. Livermore says about insect and weed problems:

> We don't have an insect problem like our chemical neighbors do. We don't have an altered plant. Our plants are natural and healthy. They pick up antibiotics from the soil, which turns insects away as nature intended. And we have insects, like ladybugs, which fight off the enemy insects. Ladybugs thrive on our farm.
>
> Also, as soon as you get natural, healthy soil, there isn't any weed problem. Nature puts in weeds to protect the soil. Weeds grow down in the soil and pick up trace minerals, and as they die they deposit these minerals on the soil's surface.
>
> And when you have your soil in balance, weeds just don't grow as fast and you don't grow as many of them. Another thing is that when we used chemicals we had a clotty soil. Now it will run through your hands just like flour at times. Earthworms and other life in the soil are alive and can loosen it. It's easy to push the weed right over when we cultivate.

Dow Chemical, the major manufacturer of pesticides, on the other hand, has published a book, *Silent Autumn*, that describes the plagues of Biblical times, as well as a variety of famines caused by uncontrolled pests in recent years. The book emphasizes the idea that careful use of pesticides is necessary to feed people in an ever more crowded world.

For organic farmers, though, K. C. Livermore sums it up:

> We'd like to see this thing get turned around. . . . We'd like to see the wildlife and the birds back here like it was in the 1940s and 50s. Is that a profitable way to farm? You bet it is. We use one-fourth less input and get as much or more back than anybody else. That should be real easy to calculate in your mind. . . .

EPA Journal, March 1978, pp. 24–25.

Does It Cost More to Farm Organically? The economics of organic farming are difficult to compare with the costs of conventional agriculture because conventional agricultural products do not carry their full cost. They are heavily subsidized by the federal government while the much smaller organic farms are not. In addition, the prices of organically-grown products are increased by factors such as small-scale production, widely dispersed farms, separate packing facilities, more expensive recycled packaging, and the pricing policies of the stores that carry the products. Most or all of these factors would disappear if organic farming were a larger-scale operation. A large part of the higher cost of organic produce is generated after the product has left the farm.

Inorganic Pollution of Soil

Inorganic pollutants are chemicals that contain little or no carbon and have not been produced by living organisms. Humans have been contaminating the soil with metals as long ago as 3200 B.C. when southern Europeans began smelting copper ore to make weapons, tools, and jewelry (see Chapter 13). Major inorganic soil pollutants are listed in Table 4.1.

Some inorganic pollutants are micronutrients, discussed earlier in Chapter 3 (see "The Nutrients Soil Provides to Plants"). High doses of these micronutrients can injure plants and animals, including humans, waterfowl, and other aquatic organisms. Fortunately, inorganic contamination of agricultural soils is not widespread. This fact helps protect the human food chain, because most individuals in wealthy nations consume food grown across a wide geographic area. For example, in the United States, we eat wheat from Oklahoma, oranges from Florida, and vegetables from California.

Soil pollution from inorganic sources is a greater problem in some nonfarm areas, particularly around old mines. Of greatest concern in the United States are lead-zinc mines, because the mine debris is high in lead, and sometimes in another heavy metal, cadmium. Both lead and cadmium are large atoms with a $^{+2}$ charge. Thus, they are adsorbed readily by clay minerals in the soil. Both metals can be toxic under the right circumstances. The world's largest lead-zinc deposits occur where three states adjoin: southwestern Missouri, southeastern Kansas, and northeastern Oklahoma.

Nevertheless, dietary studies of people living downstream from old lead and zinc mines and processing plants have shown that persons could ingest

TABLE 4.1 Sources of Selected Inorganic Soil Pollutants

Element	Major Uses	Sources of Contamination
Arsenic (As)	pesticides, plant desiccants (drying agents), animal feed additives, detergents	coal, petroleum, mining debris, metal processing, atmospheric deposition
Cadmium (Cd)	electroplating of nuts and bolts, paint pigments, plastic manufacturing, batteries	atmospheric deposition, metal processing, production of chemicals
Chromium (Cr)	stainless steel, chrome-plated metals, paint pigment, brick manufacture	manufacturing of metals, domestic wastewater, sewage sludge
Copper (Cu)	electric wires, plumbing, brass electroplating, antifouling paint, bronze	metal manufacturing, sewage sludge, atmospheric deposition, metal processing
Lead (Pb)	batteries, metal products, pigments, chemicals, combustion of oil, coal, and gasoline	atmospheric fallout, metal manufacturing, sewage sludge
Mercury (Hg)	manufacturing of plastics and electrical equipment, production of chemicals, pesticides	coal-burning power plants, atmospheric fallout
Nickel (Ni)	metal alloys, chemical production, batteries, combustion of oil, gas, and coal	domestic wastewater, metal processing, sewage sludge, atmospheric fallout
Zinc (Zn)	coatings on metals, alloys brass, batteries, rubber manufacturing	metal manufacturing, domestic wastewater, atmospheric fallout

excessive amounts of lead and cadmium by consuming some of their home-grown vegetables, milk, and meat. The lead can cause mental impairment, particularly in children.

Cadmium toxicity has been a major problem in Japan, where its effects are known as *itai-itai* disease. *Itai* is Japanese for "ouch," and the disease is characterized by bone loss which produces severe localized pain. Kidney function also is affected.

More than one billion people worldwide have been affected by lead poisoning and about half that number by cadmium. Arsenic and mercury poisoning affect about 200,000 individuals.

Of recent interest is the possibility of subclinical effects from exposure to pollutants in small amounts that are only slightly greater than normal. Subclinical effects are symptoms or diseases not normally diagnosed by physicians, such as ADD (attention deficit disorder) and hyperactivity in children. Proving cause-and-effect relationships for such behavioral problems is extremely difficult. Our understanding of human biochemistry and the social and psychological influences on behavior is inadequate.

Human Exposure to Soil Pollution

Most humans do not purchase soil at the supermarket or order soils as appetizers in restaurants. So how does soil pollution affect us? Soil pollutants enter our bodies at various points along the food chain. For example, plants rooted in the soil take in the pollutants; animals eat the plants; and then humans eat the animals.

As the pollutant is transferred from soil, to plant, to animal, to human, various processes influence the amount of pollutant that eventually reaches our bodies. For example, the laws of the U.S. Environmental Protection Agency concerning the maximum permissible amounts of lead and cadmium in soils allow much higher concentrations of lead than cadmium, even though lead is more toxic. This apparent error by the EPA is, in fact, not an error. Plants absorb cadmium with greater relish than lead. Consequently, at equal concentrations of lead and cadmium in the soil, lead concentrations in plants will be much lower than cadmium concentrations. Consequently, transfer of lead through the food chain will be less. In addition, often different parts of a plant accumulate different amounts of contaminant, and humans may eat one part of a plant but not another, perhaps the leaves but not the stems.

The transfer of pollutants from plants to us has an added complexity. As in plants, pollutants can become partitioned into different body parts within an animal. Some of your tissues may accumulate a particular pollutant element or pesticide component more than other tissues. Fat, organs, and muscle tissue normally accumulate different amounts of a particular pollutant. The reason is not always known, although it is recognized that many chemicals are fat-soluble and strongly concentrate in fatty tissue. Each type of pollutant can be expected to have different degrees of absorption from the soil and different amounts of partitioning into animal tissues.

Clearly, the transfer of pollutants from the soil to you is not a simple matter. Our present understanding of the mechanisms is quite primitive. Therefore, it makes sense to avoid polluting our soil and water.

Salinization of Soil

When most people talk about soil pollution, they are referring to toxic chemicals. They may not recognize that 20–25 percent of America's irrigated farmland is affected by salinization and that large acreages are rendered useless by salt. *Salinity* is saltiness. Salinity includes not only sodium chloride (halite, or table salt), but other salts as well (such as the calcium and magnesium carbonates, calcite and magnesite). **Salinization** refers to the accumulation of these salts at and immediately beneath the soil surface (Figure 4.16).

Soil salinization is a problem mostly in arid areas where heavy irrigation is needed to make the soil productive. The reason is that irrigation water, like all water, contains dissolved salts. When the irrigation water evaporates, the salts are left behind in the soil. Their high concentration stunts crop growth, decreases yields, eventually kills plants, and makes the land unproductive.

Soil salinization is a serious problem on about 25 percent of the world's cropland (Figure 4.17). The percentage is increasing yearly as farmers in poor nations coax productivity from marginally suitable soil. However, the problem also exists in the western United States, where low annual precipitation requires extensive irrigation. In the San Joaquin Valley of California, for example, reports from more than a century ago tell of abandoned farmland that became too salty for crops. In Utah, a highway outside of Salt Lake City offers a view of barren fields crusted with white carbonate salts.

Irrigation does not automatically mean soil salinization. Careful irrigation and drainage can prevent and even reverse salinization. This is accomplished by burying perforated drainage pipes and using them to flush the soil regularly with large volumes of low-salt water. This process is very expensive, particularly where most salinization occurs, because the low-salt water must be obtained from somewhere. After all, if abundant low-salt water were available, it would have been used for irrigation in the first place!

Also, flushing only slows the buildup of soil salinity; it cannot stop the process. Further, flushing salts from an upstream farm increases the salinity of

FIGURE 4.16 The white crust reveals a salinized agricultural soil.

The soluble salt is running off the soil into an irrigation ditch spreading its harmful effects to other parts of the farm. *Photo: Science VU/Visuals Unlimited.*

water downstream, which is used by other farmers. Flushing simply moves one farmer's problem downstream to a neighbor's farm.

In some cases, soil-flushing waterlogs (saturates) the soil, because soil permeability is low and drainage through it is poor. Water that cannot drain away as fast as it is applied accumulates underground and saturates the soil. The salty water envelopes plant roots and kills them. In addition, waterlogged soil becomes deficient in oxygen, because oxygen-rich surface water is not replenished from the atmosphere by water circulation through the soil.

Within the last few years, a new method has been devised to help farmers detect saltwater below ground before it ruins the farmland above. Armed with this information, farmers can plant trees wherever the salt level is rising. Trees keep the salt at bay because they suck up water as they grow and lower the water table (Figure 3.17).

The new salt detection system, designed in Australia, is adapted from the metal detectors used by geologists to search for ore deposits (Chapter 13) and is carried by an airplane as it flies over the farm at a low altitude. An electrical current is directed from the plane

FIGURE 4.17

Worldwide distribution of salt-affected soils. *Source: Szabolcs, István, 1989,* Salt-Affected Soils. *Boca Raton, Florida: CRR Press.*

into the ground and causes electrical currents in the ground wherever there is a good electrical conductor present. Salt water is an excellent conductor of electricity. The current induced in the ground is detected by equipment in the aircraft and the strength of the current indicates the salinity of the water beneath the farm. The system can detect salty water at depths between 15 feet and 90 feet underground. The technique works best in relatively flat areas, exactly the kind of ground surface on which large-scale farming occurs.

Cleaning the Soil

What can be done to cleanse a soil once it has been contaminated? There is no simple answer, because soil is a complex mixture of gas, liquid, and solid materials, each of which responds in different ways to different contaminants. In addition, the solid part of the mixture contains minerals that have very different physical and chemical properties. These range from quartz, which does not interact with contaminants, to various clay minerals, which are quite reactive.

Five main methods are used to remediate contaminated soil: dilution with water, soil-washing with water, vapor removal, use of contaminant-eating bacteria, and use of contaminant-eating plants. The choice depends on how much soil is being treated, the soil's composition, and the contaminant type:

1. *Dilution with water* is feasible only if the volume of soil to be treated is small. A large amount of "wash water" is required, and it must be disposed of. Dilution is best for soils contaminated by salt due to excessive irrigation in dry climates.

2. *Washing with water* involves running the soil through screens to separate metal-bearing clay

FIGURE 4.18 High-technology restaurant for bacteria.

Over time, bacteria will decontaminate this soil. *Source:* In situ Bioremediation: When Does It Work? *1993, National Academy of Sciences. Courtesy of the National Academy Press, Washington, DC.*

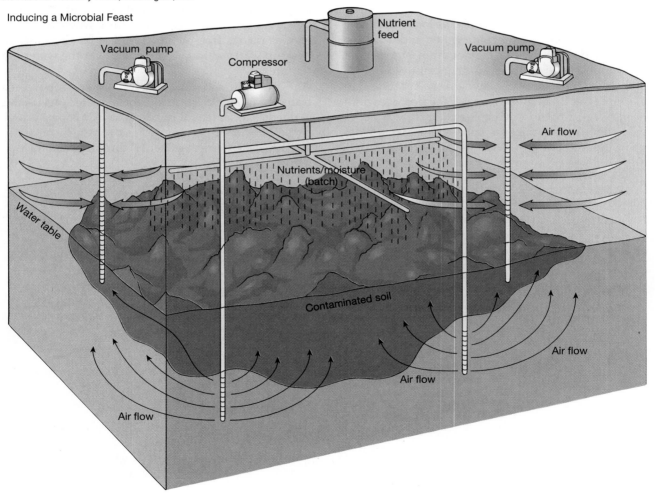

minerals from coarser soil particles (mostly quartz, which does not react with toxic metals). Washing has the same problems as dilution: a large amount of "wash water" is required, and it must be disposed of. Further, washing generates large volumes of wet, fine-grained muck to be disposed of off-site. Disposal of muck that is contaminated with potentially harmful chemical elements is difficult and subject to restrictions.

3. *Vapor extraction* can be used to remove volatile organic contaminants, which evaporate rapidly from soils. Air is pumped through the soil to flush it of contaminant gases in the soil's tiny pore spaces. This usually requires the drilling of shallow wells to place plentiful air in contact with the soil, and the use of pumps to move the air. This method is particularly useful for spills of hydrocarbons (kerosene, gasoline, or other petroleum products) that contaminate inaccessible places such as soils under buildings and highways. Remediation typically can be achieved in one to two years, while the building or highway remains in use.

4. *Contaminant-eating bacteria* are abundant in most soils and will "eat" certain organic contaminants. Different organisms attack oil, raw sewage, sewage sludge after treatment, and other wastes. Within one or two years, the amount of contaminant can be reduced to safe levels if the soil is kept aerated (to provide plenty of oxygen to the microorganisms) and nutrients are supplied to increase the population of the desired organisms (Figure 4.18). Because of advances in genetic engineering, research is underway to create new organisms that find particular contaminants ex-ceptionally tasty.

5. *Contaminant-eating plants* actually cleanse the soil. This vegetation accelerates the breakdown of organic chemical residues in soils. Also, some plants can extract certain heavy metals from soil and concentrate them in their leaves and stems (Figure 4.19). Unfortunately, this method may require up to ten years to reduce contamination to safe levels. If a plant exists that can accumulate the undesired element from the soil, and time is not a concern, such **accumulator plants** can be quite useful in remediation.

Case Study: Soil Remediation at Hill Air Force Base, Utah

In 1988, approximately 25,000 gallons of jet fuel were spilled at Hill Air Force Base in Ogden, Utah. Contamination was limited to the upper 65 feet of soil, a very

FIGURE 4.19 Nature's kidneys.

Some plants, such as parrot feather, can absorb metals, solvents and pesticides from soil and water. *Photo: Steven C. McCutcheon/U.S. Environmental Protection Agency.*

porous soil composed of mixed sand and gravel with occasional thin clay layers. Mean annual precipitation is only 16 inches, so no permanent water existed within the contaminated zone.

Shallow wells were drilled in the affected area to determine its extent below the surface and for remediation. Concentrations of jet fuel as high as 20,000 parts per million (2 percent by volume) were found, but the average concentration was only 400 parts per million. Concentration varied through the soil. Starting in December 1988, air was pumped into injection holes. This forced jet fuel vapors to emerge from discharge holes that had been drilled; the vapors were captured and burned. During the next 23 months, more than 35 million cubic feet of soil gas were extracted, laden with jet fuel vapor.

Simultaneously with injection of air into the soil, its resident organisms were "eating" the jet fuel, breaking it down into carbon dioxide and water. Calculations indicated that almost as much fuel was destroyed by the organisms as was vented by the air injection. Within two years, nearly all the hydrocarbons were cleaned from the soil.

Soil Surveys and Land-Use Planning

Most of the U.S. land surface has developed some soil. But soils vary widely in character. What is the soil in a particular area good for? How much can be farmed? Where is it safe to construct homes? Where is drainage too poor to permit road construction?

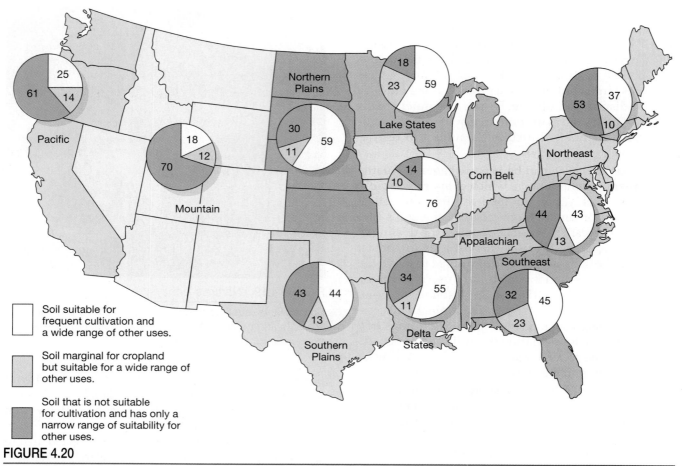

FIGURE 4.20

Percentage of land suitable for farming in the lower forty-eight states. *Source: U.S. Department of Agriculture.*

Legend for Figure 4.20:

- Soil suitable for frequent cultivation and a wide range of other uses.
- Soil marginal for cropland but suitable for a wide range of other uses.
- Soil that is not suitable for cultivation and has only a narrow range of suitability for other uses.

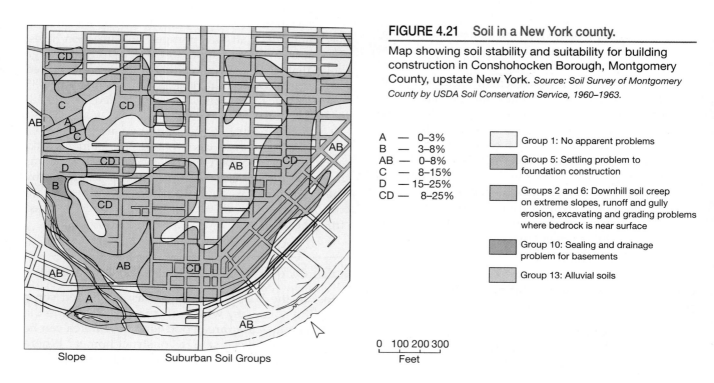

FIGURE 4.21 Soil in a New York county.

Map showing soil stability and suitability for building construction in Conshohocken Borough, Montgomery County, upstate New York. *Source: Soil Survey of Montgomery County by USDA Soil Conservation Service, 1960–1963.*

A — 0–3%
B — 3–8%
AB — 0–8%
C — 8–15%
D — 15–25%
CD — 8–25%

- Group 1: No apparent problems
- Group 5: Settling problem to foundation construction
- Groups 2 and 6: Downhill soil creep on extreme slopes, runoff and gully erosion, excavating and grading problems where bedrock is near surface
- Group 10: Sealing and drainage problem for basements
- Group 13: Alluvial soils

Slope   Suburban Soil Groups

0   100 200 300
Feet

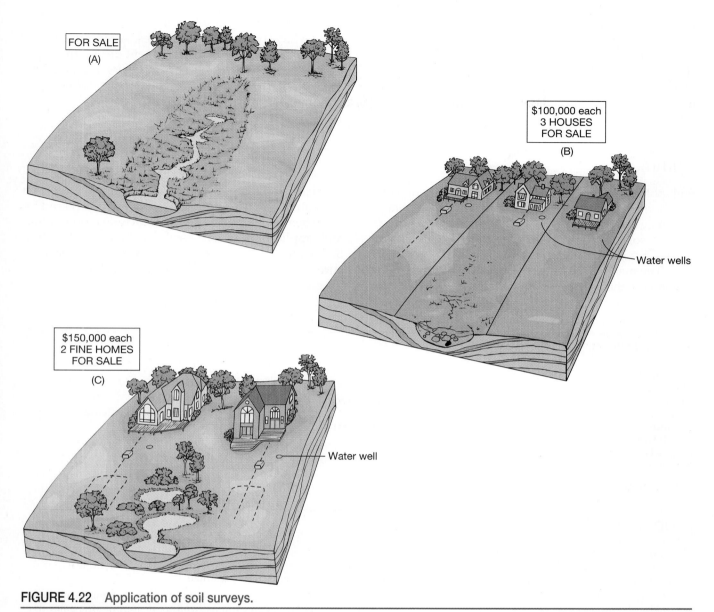

FOR SALE
(A)

$100,000 each
3 HOUSES
FOR SALE
(B)

Water wells

$150,000 each
2 FINE HOMES
FOR SALE
(C)

Water well

FIGURE 4.22 Application of soil surveys.

Two alternatives for a rural land developer. *After C. C. Mathewson, 1981,* Engineering Geology, *Columbus, Ohio: Merrill Publishing Co.*

Intelligent **land-use planning** requires answers to these questions.

To help answer such questions, a general soil survey was made by the U.S. Department of Agriculture in 1981 (Figure 4.20). Many states also have published detailed soil maps. About 44 percent of U.S. land is suitable for normal farming and another 13 percent is marginal for growing crops. The remainder is suited, and used, primarily for grasslands (grazing of cattle and sheep) and forests.

Soil surveys are important in urban areas, as well as in agricultural ones. Soil characteristics affect the stability of building foundations and the occurrence of landslides (Chapter 5), and the likelihood of water leaking into the basements of houses. Modern

soil maps are designed to consider these as well as other problems that are traceable to soil characteristics, such as waste disposal (Chapter 9) and highway construction.

Figure 4.21 shows the suitability of soils for building in a New York county. Soils were grouped on the basis of their susceptibility to erosion and landslides, their permeability to rainwater (drainage), and their suitability as foundations for buildings. More soil surveys are needed to make efficient use of soil, which is our most important natural resource.

A small-scale example of the use of such information about soils to a rural land developer is shown in Figure 4.22. Two types of soils are present. On the sides of the plot, the soil is dry and well-drained; the soil in the middle is wet and marshy. The developer might build three lower priced homes (smaller plots) as shown but, if he does, the middle home will have serious septic tank problems because of the poor drainage evident on the homesite. In time, the septic tank problems might affect the water supply in the wells of the two adjoining homes. Alternatively, the developer might use a development design that uses the wet area for vegetation, perhaps making less profit in the short run but enhancing his or her reputation for the long run.

Summary

Soil pollution and soil erosion are very significant worldwide. Both problems are largely controllable if adequate care is taken by users of the soil—us. Without healthy soils, humans will suffer and die. The fate of our food is in our hands.

The major pollutants are pesticides and salts precipitated in the soil as irrigation water evaporates. U.S. pesticide use is decreasing as its harmful effects are becoming widely recognized, and is being replaced by more benign methods. Examples include crop rotation and application of genetically engineered bacteria. Salinization is difficult to prevent and expensive to cure. A good solution to this problem has yet to be found.

Key Terms and Concepts

accumulator plant
biodegradable
conservation farming
contamination
contour plowing
dust bowl
fungicide
gully

herbicide
insecticide
integrated pest management
land-use planning
nematocide organic farming
pesticide
pollution
rain forest

rill
salinization
silting
synthetic organic pesticide
terracing
weed

Stop and Think!

1. Name three things people do that increase the rate of erosion. How might we do less of each of these things?
2. List some characteristics of a soil that help increase its resistance to erosion. Explain how each characteristic is beneficial.
3. What is conservation farming? What are its benefits and drawbacks?
4. Commercial tree-harvesting companies that supply wood for home construction say they plant at least one tree for every tree they cut, so that the forests always will be replenished. However, they generally plant only fast-growing trees, so harvesting can be done frequently. Most commonly, this means they plant fast-growing pine trees, not slower-growing hardwoods like maple and oak. Can you think of any drawbacks to this policy? Are there any alternatives?
5. Contact the Soil Conservation Service (consult your phone book) or state geological survey (in your state's capital) and ask for information about soil erosion problems in your state. What are the state government and individual farmers doing to combat soil erosion?
6. What is meant by *biodegradable?*
7. What is the difference between a contaminant and a pollutant? Are there major contaminants in your area? Are there also serious pollution problems in your area? Ask the state geological survey what is being done about any problems they recognize.
8. We usually are quick to deal with a chemical pollutant that threatens human health. But suppose the amount of a chemical is too small to threaten humans, but sufficient to threaten dogs? Your author likes dogs, so he votes to reduce the amount of the chemical so dogs are protected. But how about protecting cats, striped bass, pigeons, mosquitoes, leeches, rattlesnakes, or cockroaches? How much are you willing to spend to protect which creatures? Or, trees, rosebushes, poison ivy, or broccoli? Analyze your feelings and justify them.
9. Do you believe organic (chemical-free, "natural") farming should receive the same government benefits as conventional farming, which uses pesticides? Why? Contact your state agricultural office and ask about organic farming in your area. Contact your state or federal representative for his or her view of organic farming.
10. What types of minerals do you think contribute most heavily to soil salinization? What type of chemical bonding do these minerals have?
11. What methods are available to clean polluted soil? In what situations can each be used?

References and Suggested Readings

Alexander, Martin. 1995. How toxic are toxic chemicals in soil? *Environmental Science and Technology,* v. 29, p. 2713–2717.

Board, Paul. 1996. Contaminated lands. *New Scientist,* (insert), October 12, 4 pp.

Calvet, R. 1989. Adsorption of organic chemicals in soils. *Environmental Health Perspectives,* v. 83, p. 145.

Cheng, H. H., ed. 1990. *Pesticides in the Soil Environment: Processes, Impacts, and Modeling.* Madison, Wisconsin: Soil Science Society of America.

Goldman, Marvin. 1996. Cancer risk of low-level exposure. *Science,* v. 271, March 29, p. 1221–1222.

Klinkenborg, Verlyn. 1995. A farming revolution. *National Geographic,* v. 188, no. 6, p. 60–89.

Lang, Leslie. 1993. Are pesticides a problem? *Environmental Health Perspectives,* v. 101, no. 7, p. 578–583.

Norris, R. D. et al. 1994. *Handbook of Bioremediation.* Boca Raton, Florida: Lewis Publishers, 257 pp.

Oldeman, L. R., Hakkeling, R. T. A., and Sombroek, W. G. 1991. *World map of the status of human-induced soil degradation.* Nairobi, Kenya, International Soil Reference and Information Centre: United Nations Environment Program.

Pimental, David, and nine others. 1992. Environmental and economic costs of pesticide use. *Bioscience,* v. 42, p. 750–760.

Pimental, David and ten coauthors. 1995. Environmental and economic costs of soil erosion and conservation benefits. *Science,* v. 267, Feb. 24, p. 1117–1123.

Ross, S. M., ed. 1994. *Toxic Metals in Soil–Plant Systems.* New York: John Wiley, 469 pp.

Sawhney, B. L. and Brown, K. W. (eds.). 1989. *Reactions and Movement of Organic Chemicals in Soils.* Madison, Wisconsin: Soil Science Society of America Special Publication no. 22, 494 pp.

Smith, M. A., ed. 1985. *Contaminated Land: Reclamation and Treatment.* New York: Plenum Press, 433 pp.

Stigliani, William, and Salomons, Wim. 1993. Our fathers' toxic sins. *New Scientist,* Dec. 11, p. 38–42.

Szabolcs, István. 1989. *Salt-Affected Soils.* Boca Raton, Florida: CRR Press, 274 pp.

Let every valley be lifted up,
And every mountain and hill be made low;
And let the rough ground become a plain,
And the rugged terrain a broad valley.

Isaiah 40:4

Slope Stability and Landslides

Slide movements of large masses of rock or sediment, called *landslides*, commonly are triggered by human action. An example is home or highway construction in which rock or sediment is removed from the base of a slope to create a pad for a house or a bed for a new highway. Another example is the weakening of sediment by saturating it with water, through irrigation, like daily watering of a lawn. Yet another example is increasing the area of paved ground, which reduces the opportunity for water to soak in, thus increasing nearby streamflow and therefore erosion of stream banks.

- Are these occurrences an inevitable part of expanding civilization, or can they be controlled?

- How does the type of rock or sediment on a slope affect the slope's stability?

- Should municipalities spend tax dollars to create maps showing where susceptible slopes are located?

All things on Earth succumb, sooner or later, to the invisible, relentless pull of gravity. Soil moves downhill, water flows lower and lower until it reaches the sea, humans shorten with age and their flesh sags, and the gas molecules in the atmosphere are concentrated near the ground surface. These gravity-driven processes have been occurring since Earth was formed 4.5 billion years ago, with the only variation being in the types of natural features or organisms that sag.

Over time, gravity is irresistible; it always wins the tug-of-war. But for human civilization, the key variable is *how long* gravity is resisted. At one extreme is the slow creep of soil downhill on a gentle slope; the effect of gravity is slowed considerably by the cohesive properties of the soil and plant roots that help retain it. At the other extreme is a rockfall or avalanche in which there is negligible resistance to gravity, dooming people and houses in its path.

Landslides or mass-wasting always occur on slopes, and consist of downslope movement of a mass of soil and bedrock. Some of history's worst disasters have been caused by these **mass movements.** At Kansu, China in 1920, 200,000 people were killed in an earthquake-triggered mass movement of dry material. In the Huascaran area of Peru in 1970, an entire city was obliterated by an earthquake-triggered mass, killing 21,000 people. As the debris rushed down the mountainside, part of it became airborne as it flew over a ridge, launching boulders as large as 60 tons. These impacted over a broad downslope area as much as 2.5 miles distant from the launch point! The deadly

◀
Landslide at Yosemite.
Photo: Robert W. Cameron.

91

barrage of boulders wreaked havoc among people, livestock, and vegetation in the central part of the avalanche's path. The impact velocity of the larger boulders exceeded 500 miles an hour, clearly sufficient to crush anything in their path.

The incidence of landslides has been increasing throughout the twentieth century. Why? The answer goes back to human activity. Expanding urbanization (Figure 5.1) and more construction create more opportunities for landslides. Also, restrictions on the use of floodplains adjacent to rivers have forced many building sites to occupy sloping terrain that would have been avoided in earlier times. At present, landslide damage in the United States, coupled with the cost of landslide-control measures, totals several billion dollars annually.

Through both brute force and indirect influence, people move about 40 billion tons of soil and rock each year, a volume that equals or exceeds material transported by any other single agent, such as water, wind, or ice. After 4.5 billion years of sculpting the land surface, natural processes sometimes take a back seat to a scrawny, squawking ape that has enough intelligence to magnify its own meager muscle power through machinery.

What are the types of landslides or mass movements, and what factors determine which type is most likely to occur in various localities?

The Role of Gravity and Slope in Landslides

The most important factor that determines the stability of loose grains on a sloping ground surface is the force of gravity. **Gravity** tries to pull all objects toward the center of Earth. Its pull is constant, as shown by the fact that you cannot stand upright for very long. After awhile, your body grows weary of fighting the pull of gravity, and you sit or lie down.

If a sand or gravel particle is resting on a flat surface (Figure 5.2A), gravity will hold it in place. However, if the particle is lying on a sloping surface, a part of the gravitational force is oriented parallel to the ground slope (Figure 5.2B). So long as the part of the force oriented perpendicular to the surface (g_p) is greater than the part parallel to the surface (g_t), the particle will not move. Friction between the particle and the ground surface holds the particle in place. When the slope is steeper, g_t will exceed g_p, the fric-

FIGURE 5.1 Aerial photograph showing the result of the most destructive landslide in the history of Hong Kong, June 18, 1972.

Almost 26 inches of rain fell during the preceding two days, saturating and undermining the steep slope, which is composed of soil and underlying unlithified (unconsolidated) coarse sediment. The slope failure was 220 feet wide and destroyed a 4-story building and a 13-story apartment building; sixty-seven people were killed. Hong Kong's population density and topographic relief make it impossible to completely avoid construction in unsafe areas. *Courtesy Hong Kong Government.*

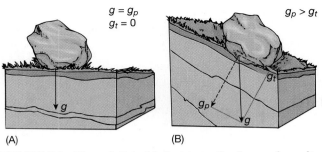

$$g = g_p$$
$$g_t = 0$$

$$g_p > g_t$$

FIGURE 5.2 The relationship between the force of gravity and the slope of the ground.

On a flat surface, all of the gravitational force (g) is directed perpendicular to the surface (g_p). On a sloping surface, there is a component of the force that acts tangentially to the surface (g_t). So long as g_p is larger than g_t, the particle will not move. But if g_t exceeds g_p—in other words, if the slope is steeper—the particle moves downhill.

tional force will be overcome, and the particle will move down the slope.

Normally, a large number of particles rests on the solid ground surface, rather than only one. On a gently sloping surface, the **friction** of particle against particle may be sufficient to keep them from moving. But if the slope increases because of erosion or because of an excavation at its base, the friction will be overcome and the mass will slide downslope. The mass of grains may slide on even very gentle slopes if the particles are wet. The presence of water between particles decreases frictional resistance to movement.

The Role of Water in Landslides

The presence of water in loose sediment is an important factor in causing landslides. Picture a mass of loose sand on a slope. The mass is stable, at least temporarily, because the frictional effect of grain against grain is sufficient to keep the grains from moving downslope. As the amount of water in the sediment mass increases, however, the pressure exerted by the water **(pore pressure)** increases and forces the grains apart so the amount of contact they have with each other decreases. As this happens, the sediment-water mass loses its frictional resistance to movement and moves downslope—a landslide. This is why landslides are more common during the rainy season.

A second effect of increasing the amount of water in a sediment occurs if the sediment contains clay minerals. Most clay minerals are thin and flat, like tiny flakes of paper. The clay minerals are held together in a muddy, cohesive mass by the ions on their surfaces. Surfaces of tiny clay particles have a negative charge that must be balanced by a positively charged ion in contact with the clay surface. This ion

is shared between adjacent clay flakes. As increasing amounts of water percolate between adjacent flakes, the ions are washed out, with the result that the clays slide past each other like a tilted deck of cards, and a landslide results.

A third way in which water can instigate landsliding is by dissolving the mineral cements that hold sand grains together. For example, the most common cement that binds grains together in sandstones is calcite. Calcite is very soluble compared to the common minerals that compose the "sand" in sandstones (grains of quartz and feldspar). Thus, over time, the calcite will dissolve first, releasing the grains, turning the sedimentary rock back into a sediment.

If the rock/sediment is on a sloping surface, a landslide may result. The much greater solubility of calcite compared to silica-rich minerals such as quartz is clearly seen by comparing tombstones in graveyards. Grandfather's tombstone made of soluble calcite (either the sedimentary rock limestone or the metamorphic rock marble) may be illegible, but grandmother's headstone made of the relatively insoluble igneous rock, granite, appears almost as new as it was the day she died.

Role of Earthquakes in Landslides

As those who live in hilly terrains in the western United States are well aware, earthquakes commonly trigger landslides. The reason earthquakes cause landslides in loose sediment is similar to the reason described for water. With water, the grains are forced apart by increased water pressure. In an earthquake, the grains are separated by shaking. The earthquake jostles the grains so that they lose cohesion and separate for a moment, sufficient time for gravity to cause a slope failure.

Earthquakes also can cause landsliding for another reason. Earthquake waves passing through solid rock can cause fractures, and the fracturing of the rock, combined with the jostling caused by the quake, allows gravity to have an instantaneous effect. Catastrophic downslope movement occurs. As noted earlier, some of the worst landslide disasters in recorded history have been caused by slides initiated by earthquakes.

Slides, Flows, and Creeps

Are all gravitationally controlled movements of loose sediment, soil, and massive rocks the same? No. You may be surprised to discover the variety of ways in which earth materials move. Many classifications and numerous terms have been used to describe them, such

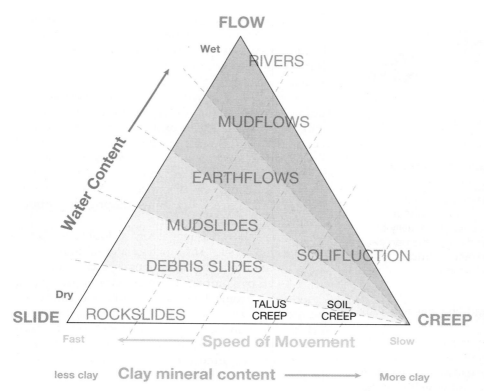

FLOW

Wet

RIVERS

MUDFLOWS

EARTHFLOWS

Water Content

MUDSLIDES

DEBRIS SLIDES

SOLIFLUCTION

Dry

SLIDE   ROCKSLIDES   TALUS   SOIL   CREEP
CREEP   CREEP

Fast   ← Speed of Movement →   Slow

less clay   Clay mineral content ——→   More clay

FIGURE 5.3 Types of slope failures differ.

They differ in speed of movement (ranging from creep to slides moving over 200 miles per hour) and type of movement (slide vs. flow). The type of movement depends upon how much water is in the slide debris. With increasing water content, the movement changes from sliding to flowing. The destruction caused by a slope failure depends on its speed and especially its size.
After Carson and Kirkby, 1972, Hillslope Form and Process, *Cambridge University Press.*

as rockslide, mudflow, slump, creep, avalanche, and others. Distinctions between these categories are not always clear, and they may vary with the person doing the naming. One system of naming is shown in Figure 5.3. It is based on the relative amounts of water, clay minerals, and nonclay sediment in the earth movement or landslide.

Slides

The corner of the triangle in Figure 5.3 labeled SLIDE represents movements that involve only small amounts of water or clay minerals, or perhaps none at all. Exemplifying this type of mass movement are rockfalls (Figure 5.4), rockslides, and debris slides.

Slides are slope failures that are initiated by slippage along a well-defined plane surface. The sliding mass is essentially undeformed and tends to move as an integrated mass. However, it may partially disintegrate during the sliding motion, giving rise to a flow movement late in the event. The plane of sliding may penetrate to depths of tens of feet as a concave surface separating undisturbed rock below from disturbed rock above.

Flows

At the corner labeled FLOW, the ratio of water to solid material is very high. The sediment volume is normally one percent or less of the water-sediment mass.

Creep

At the corner of the triangle labeled CREEP, the critical material is clay minerals. Because of cohesion between the clay minerals, sediment rich in clay typically moves slowly (Figure 5.5). In clay-rich sediments, water is always present, adhering to clay mineral surfaces. Water also commonly occurs in larger amounts in pores among the clay particles.

Recognition and identification of the types of mass movements began in antiquity, at least as early as 186 B.C. in China. But systematic identification and mapping of all landslide types in specific areas developed mainly during the past few decades. This research was spurred by increasing urbanization, which has intensified human interaction with unstable surface materials.

For example, geologic maps published one hundred years ago did not show a single landslide in the San Francisco Bay region of northern California, although they certainly were present. By 1970, 1200 landslides had been identified in that area. By 1980, 70,000 landslides had been recognized. In 1982 alone, high-intensity rainfall triggered 18,000 new mass movements, most of them in areas where landslides had not previously been recognized. The search for probable future landslide sites is more intense in California than elsewhere because of the rapid population growth and the natural instability of Earth and its surface materials there.

FIGURE 5.4 Rockfalls are a common hazard in mountainous regions.

They may injure motorists whose vehicles strike rocks that have fallen onto highways. Even trains can be derailed by rocks that fall onto railroad tracks. This boulder on Highway 6 in Colorado took the life of the driver of a gasoline tanker. *Photo: R. Van Horn/USGS.*

The next five sections of this chapter are in a specific order, matching the labels on the triangle in Figure 5.3, growing increasingly watery *from slide to flow:* rockslides, debris slides, mudslides, earthflows, and mudflows. You may find it helpful to refer to this figure as you study.

Rockslides and Snowslides

The most damaging type of slope failure is a rockslide. A **rockslide** is a free-fall or rapid movement of a large mass of material, usually composed of large rock fragments. Fall velocities may exceed 500 miles per hour. Cold and mountainous climates experience snowslides, or **avalanches,** which can have effects just as disastrous as some rockslides.

The annual property damage from rockslides approaches $2 billion in the United States alone, and no doubt is many times that amount when urban areas in other countries are included. The cost in human lives has been considerable throughout history (Table 5.1). As world population continues to increase, the human cost will increase correspondingly. The current worldwide death toll from rockslides averages about 600 per year. Snow avalanches kill about 20 people per year in the United States, and many more in alpine areas of Europe and Asia.

Rockslides produce both primary and secondary effects. The damage caused by the direct impact of the fragments on people and property is a primary effect. Secondary effects may occur when the mass of rock and sediment plummets into a large body of water or a river. For example, a 1964 rockslide at Lituya Bay, Alaska, sent

FIGURE 5.5 Soil creep.

Headstones tilting in a hillside cemetery due to creep. *Photo: John D. Cunningham/Visuals Unlimited.*

TABLE 5.1 Historical Earth Movements Known to Have Killed at Least 500 People

Most catastrophic earth movements are rockslides and debris slides and result in fewer than 100 deaths.

Year	Location	Approximate Deaths
1920	Kansu, China[1]	200,000
1970	Yungay, Peru (avalanche)[2]	25,000
218 B.C.	Alps (avalanches—Hannibal's army)	18,000
1949	Khait—near Garm, Tadzhikistan[2]	12,000
1916–1917	Alps, Italy (avalanches)[3]	10,000
1941	Huaraz, Peru (avalanche and mudflow)	7,000
1248	Alps, France (avalanche)	5,000
1962	Yungay, Peru	4,500
1991	Ormoc, Philippines	4,081[5]
1934	Bihar, Nepal[2]	3,500
1963	Vaiont Reservoir, Italy[4]	3,000
1967	Rio de Janeiro, Brazil	2,700
1618	Mount Conto, Switzerland	2,430
1889	South Fork Dam, Johnstown, PA[4]	2,200
1991	Uttarkashi, India	2,000
1945	Kure City, Japan	1,154
1958	Shizuoka, Japan	1,100
1971	Hindu Kush Mountains, Afghanistan[4]	1,000
1970	Alaknanda River, India[4]	600
1512	Brenno Valley, Switzerland	600
1928	St. Francis Dam, southern California[4]	500

[1]Loess cliffs containing homes collapsed and flowed (dry) during an earthquake, accounting for about half of the cited fatalities.
[2]Landslides generated during earthquakes.
[3]Italian and Austrian armies during World War I.
[4]Slide into river, lake, or reservoir produces flood; victims drowned.
[5]Total for mudflows, mudslides, and flash floods associated with typhoon.

a massive wave across the bay that reached heights of 1700 feet on opposite hillslopes. For perspective, that is higher than the world's tallest building, the Sears Tower in Chicago (1454 feet). Other rockslides into lakes behind dams have caused collapse of the dams and severe flood damage and loss of life downstream.

More common than dam collapse and overtopping is the formation of temporary dams across rivers and streams by rockslides and debris slides. Depending on the amount of sediment in the slide, the width of the stream channel, and velocity of the streamflow, a large lake can form behind the river blockage. Eventually, water leakage through pores in the debris and the pressure of lake water against the sediment dam will cause its collapse and a massive flood downstream. Because we cannot accurately predict the date on which such a catastrophe will occur, the best way to minimize flooding may be to dynamite the debris dam soon after it forms, before a great volume of water can accumulate behind it.

The character of earth materials is a crucial control on the occurrence of rockfalls and debris falls. Broken rock is more easily displaced than solid rock. So, not surprisingly, mountain slopes covered with large fragments of rock formed by frost wedging (described in Chapter 3) are prime locations for these falls.

Also clearly important is the steepness of the slope on which the fragments rest. The steeper the slope, the easier it is for the rock-against-rock friction to be overcome and the mass to slide rapidly downslope. A heavy rain may be sufficient to initiate a slide. When considering "solid rock" on a slope, factors such as the number and type of fractures, layering, rock composition, and degree of cementation are important.

Although a hillslope may be stable under usual conditions, a failure can be produced by one of the four most common trigger events: earthquakes, excessive precipitation, wave activity along shorelines, and the activities of modern civilization. The vertical and lateral ground motion caused by earthquakes is well known

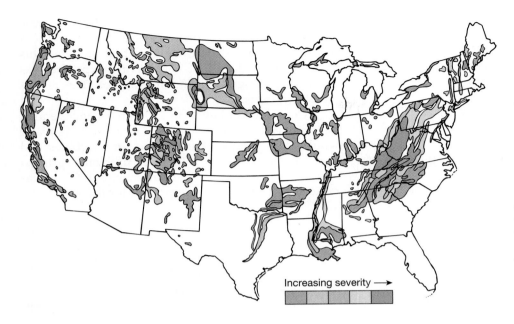

FIGURE 5.6 Areas of mass movement problems in the contiguous United States.

Areas too small to show clearly at this scale are not colored. *After American Institute of Professional Geologists, 1993, Citizens' Guide to Geologic Hazards, Arvada, CO. Used with permission.*

Increasing severity ⟶

and, thanks to television, has been seen by most Americans. In the United States, the effects are most clearly seen in California, where the combination of frequent tremors, steep slopes, and abundant poorly cemented sediment are a prescription for disaster. Other susceptible areas are the Rocky Mountains and the Appalachians (Figure 5.6).

The Gros Ventre Rockslide, Wyoming

In the Gros Ventre Valley, near Jackson Hole, Wyoming, the largest historical rockslide in the United States occurred in 1925 (Figure 5.7). Within a three-minute period, about 1.4 billion cubic feet of weathered sandstone, limestone, and water-saturated sediments roared across the valley at a speed of 100 miles per hour to heights of more than 260 feet on the opposite side of the valley.

A dam 230 feet high was created across the Gros Ventre River, composed of large blocks of rock broken during the fall, fine-grained debris of sandstone, shale, and soil, and broken and torn remains of the dense pine forest that had covered the hillside. The river was completely blocked, creating a lake. It filled so quickly that a house which had been 60 feet above the river was floated off its foundation 18 hours after the slide. Two years later, the lake overtopped the dam, partially draining itself and causing a devastating flood in the town of Kelly, 6.5 miles downstream. Six lives were lost.

The trigger for the Gros Ventre rockslide was a small earthquake. Its vibrations overcame the frictional resistance in the mass of water-saturated sedimentary materials. An added factor in the rockslide was the orientation of the rocks: they were parallel to the hillslope. Water-saturated shale is easily disturbed because of its high content of sheet-like clay minerals. Further, support for the sloping rocks had been undermined by river erosion at its base.

The slide scar on the hillside has begun to revegetate in the 70 years since the slide, the lake level has

FIGURE 5.7 In Wyoming, the Gros Ventre rockslide occurred in 1925, nearly three-quarters of a century ago.

Yet, its scar on the side of Sheep Mountain is still prominent. Trees are well-established in the rockslide debris in the foreground. *Photo: Stephen Trimble.*

lowered, and sometime soon the remaining part of the dam will be washed away naturally and the pre-slide river flow will resume. In another 50 to 100 years, no evidence may remain of the infamous flood of 1925. Given time, the environment heals itself.

Vaiont Dam Rockslide, Italy

A more damaging landslide event was the Vaiont Dam disaster of 1963 in the mountains of northern Italy (Figure 5.8). The region needed additional elec-

tric power, so it was decided to build a hydroelectric station. This required constructing a high dam across the Vaiont River. From the events that followed this decision, it is clear that engineering expertise was abundant, but geologic input was not. The results were disastrous.

Construction began in 1956 on a concrete arch dam across the valley to create a large reservoir. Water would flow through the dam to turn turbines and generate electric power. However, problems surfaced almost immediately. In the summer of 1957, it was noticed that concrete being poured for the dam foun-

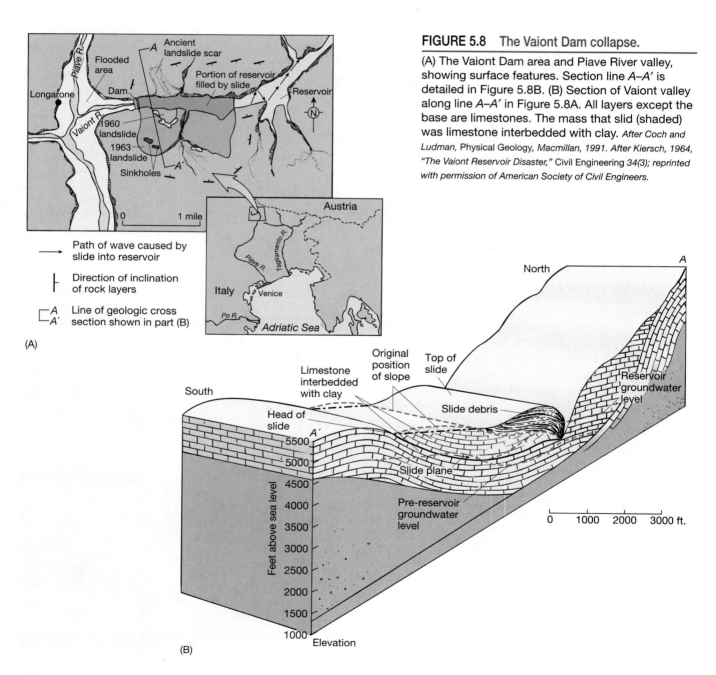

FIGURE 5.8 The Vaiont Dam collapse.

(A) The Vaiont Dam area and Piave River valley, showing surface features. Section line *A–A'* is detailed in Figure 5.8B. (B) Section of Vaiont valley along line *A–A'* in Figure 5.8A. All layers except the base are limestones. The mass that slid (shaded) was limestone interbedded with clay. *After Coch and Ludman, Physical Geology, Macmillan, 1991. After Kiersch, 1964, "The Vaiont Reservoir Disaster,"* Civil Engineering *34(3); reprinted with permission of American Society of Civil Engineers.*

dations was disappearing. It was filling fractures and solution cavities in the bedrock that was to support the dam, planned to be 856 feet high. The supporting rock is a clayey limestone, pockmarked with large holes. Adding to this problem was the orientation of the limestone, which dips toward the river on both sides (Figure 5.8B). Nevertheless, the dam was completed in 1959.

As engineers allowed the reservoir to fill with water, the level of water in the lake rose. The rate of surface creep on the hillside increased to eight times its normal rate. Although not a serious hazard itself, an increased rate of creep can be an important warning of potential rapid slope failure. The pressure of the increased volume of water in the reservoir forced the water into the limestone under the base and sides of the dam. It also saturated the clays in the rock. The excess pressure and the water in the rock pores had the effect of making the limestone "buoyant," and the water-soaked clay decreased the frictional resistance to sliding.

The result was predictable. In 1960, a mass of rock and soil with a volume of 25 million cubic feet slid off the south slope into the reservoir, generating a wave 6 feet high that moved across the lake. But this still was just a warning. At 10:39 P.M. on October 9, 1963, following heavy rains, more than 400 times the volume of the earlier slide—10.5 *billion* cubic feet of limestone—tore loose from the top of the south slope and roared as one intact block downward into the reservoir. The base of the slide was a sequence of slick, clayey limestone beds. The limestone slid into the reservoir, abruptly elevating the river by 1000 feet. This displaced about 1.8 billion cubic feet of water, forming a colossal water wave 650 feet high.

The wave wiped out lakeside villages, overtopped the dam, and plunged into the narrow gorge below. Sweeping westward out of the gorge (Figure 5.8A), the surging torrent of water swept across the Piave River valley, destroying the village of Langarone and its 1450 residents, then swept upriver and downriver to destroy other villages along the banks. In all, 1899 people perished. The officials involved were prosecuted, and the chief engineer committed suicide.

The Vaiont Dam rockslide could have been avoided by consulting a geologist, who would have told them not to build the dam there because of the unstable nature of the surrounding rocks. The Vaiont Dam still stands today, generating no electricity, slowly filling with sediment, a silent monument to inadequate planning on a grand scale.

Predicting Rockslides. Because of the catastrophic nature of rockslides, attempts have been made to determine the likelihood of their occurrence at a particular locality, based on past occurrences nearby. One rating system, recently devised but not yet adequately tested, is shown in Table 5.2. Thirteen important features of the landscape are listed, with the number of points awarded for each feature, depending on variations in the feature. (For example, the higher the slope and the steeper the slope, the more points these features are given.) The larger the number, the more dangerous the site.

Snow Avalanches

The United States has about 100,000 avalanches of snow and ice per year. Fully developed avalanches may have masses of one million tons (2 billion pounds) and travel at speeds of 200 miles per hour, twice that of a free-falling sky diver (Figure 5.9). Their impact pressure may reach at least 13.5 tons per square foot. To give you an idea of the significance of this force, only

FIGURE 5.9

Snow avalanche on Mt. Rainier. *Photo: Peter Dunwiddie/Visuals Unlimited.*

TABLE 5.2 | Colorado Rockfall Hazard Rating System

Factor	Rank			
	Slope Profile			
	3 Points	**9 Points**	**27 Points**	**81 Points**
Slope Height	25 to 50 ft	50 to 75 ft	75 to 100 ft	100ft
Segment Length	0 to 250 ft	250 to 500 ft	500 to 750 ft	750 ft
Slope Inclination	15 to 25 degrees	25 to 35 degrees	35 to 50 degrees	50 degrees
Slope Continuity	Possible launching features	Some minor launching features	Many launching features	Major rock launching features
	Geologic Characteristics			
Average Block or Clast Size	6 to 12 in.	1 to 2 ft	2 to 5 ft	5 ft
Quantity of Rockfall Event	1 cu ft to 1 cu yd	1 to 3 cu yds	3 to 10 cu yds	10 cu yds
	Case 1			
Structural Condition	Discontinuous fractures, favorable orientation	Discontinuous fractures, random orientation	Discontinuous fractures, adverse orientation	Continuous fractures, adverse orientation
Rock Friction	Rough, irregular	Undulating, smooth	Planar	Clay, gouge infilling, or slickensided
	Case 2			
Structural Condition	Few different erosion features	Occasional erosion features	Many erosion features	Major erosion features
Difference in Erosion	Small difference	Moderate difference	Large difference	Extreme difference
Climate and Presence of Water on Slope	Low to moderate precipitation; no freezing periods; no water on slope	Moderate precipitation or short freezing periods, or intermittent water on slope	High precipitation or long freezing periods of continual water on slope	High precipitation and long freezing periods, or continual water on slope and long freezing periods
Rockfall History (From Ride Through)	Few falls	Occasional falls	Many falls	Constant falls
Number of Accidents Reported in Mile	0 to 5	5 to 10	10 to 15	15 and over

Source: Colorado Geological Survey

0.3 ton/square foot pressure will destroy a wood-frame house; 1 ton/square foot pressure will uproot a mature tree; and 10 tons/square foot pressure will move a reinforced concrete building.

Consequently, the chance of survival and rescue for anyone caught in an avalanche is poor. Twenty percent of avalanche victims are killed immediately, and the rest must be extracted from the snow and ice within an hour to have any chance of survival. If found within 30 minutes, a victim has a 50 percent chance of surviving. Generally, however, rescue teams do not arrive in time; they find only 5 percent of victims alive.

In the United States, property damage due to avalanches is less than $1 million per year, and annual loss of life is around two dozen people. Of course, if you or a loved one is one of these victims, it is for you a major disaster indeed—for example, see Figure 5.10. But on a national scale, the loss of only a few people per year from a recurring natural event is negligible. (Compare this figure to the thousands killed in alcohol-related traffic accidents each year.)

The worst recorded U.S. avalanche disaster occurred in 1910 in Washington State, when 96 people died as their snowbound train was swept off the track in a steep canyon. In 1926, the mining community of Bingham Canyon, Utah was hit by a snow avalanche, killing 40. Outside the United States, many ice and snow avalanche disasters have occurred, the worst in Peru in 1962, when the front of a hanging glacier on Nevado Huascaran collapsed. The ice mass of 100 million cubic feet cascaded 2.5 miles downslope, destroying nine small towns and killing 21,000 people.

Avalanches are a growing problem as more people become involved in winter sports and recreation. The most effective tool against avalanche disasters is planning (Figure 5.11). Roads and buildings should not be sited where avalanches have occurred in recent years. Potential losses can be reduced by erecting walls or baffles to protect property and divert the flow of snow (Figure 5.12). Snow fences and wind baffles at strategic locations may prevent snow buildup in the first place. A common method in some recreational areas is the use of explosives to start small snow slides, thus defusing the possibility of major avalanches.

Debris Slides

Debris slides occur when loose, granular material on a slope is caused to move rapidly downhill (Figure 5.13). Loose sediment that does not contain clay will rest happily on a slope whose angle does not exceed about 30 degrees. At slope angles less than 30 degrees, the friction of grain against grain is sufficient to hold the grains in place. But at steeper angles, the friction is inadequate. The critical angle of about 30 degrees is called the **angle of repose** (Figure 5.14). Other things being equal, coarser fragments can maintain a steeper angle than finer fragments because of increased frictional contact. Clay particles in the sediment add cohesion and allow a steeper angle of repose.

Debris slides are triggered by earthquakes, precipitation, and human interference with natural processes, just like rockslides. Let us look at each "trigger."

Earthquake Vibrations Trigger Debris Slides

The effect of earthquake vibrations on debris resting on a slope is clear. The debris is destabilized as shaking separates the particles. This effectively disables the friction that restrained particle movement.

FIGURE 5.10 "The Iceman," oldest virtually intact human being yet discovered.

Radiocarbon dating indicates he is about 5200 years old. Medical investigators have determined that he had osteoarthritis, arteriosclerosis, and had broken some ribs during his fall. Melting ice and snow uncovered his body in 1991 in the Alps on the border of Italy and Austria. Did he have a paid-up life insurance policy? How much interest might his descendants be owed on the unpaid claim filed by their ancestors? *Photo: Sygma.*

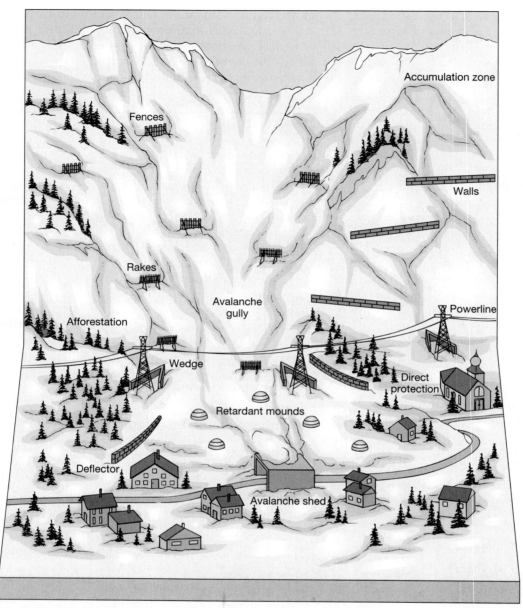

Labels in figure: Accumulation zone, Fences, Walls, Rakes, Avalanche gully, Afforestation, Powerline, Wedge, Direct protection, Retardant mounds, Deflector, Avalanche shed

FIGURE 5.11 Planning to avert disaster.

Various avalanche defenses reduce the likelihood of avalanching, and diminish the impact should one occur. Of course, not all of these are used in a single area. *After A. Robinson,* Earth Shock, *p. 230.*

Precipitation Triggers Debris Slides

The effect of precipitation is similar to that of earthquakes, in that the frictional restraining effect is lost. But the destabilizing by precipitation results because the debris contains significant clay minerals. First, water adheres to them and this decreases frictional resistance to particle movement. Second, some types of clay swell when wet, further reducing frictional contact between larger particles.

But third, and probably most important, clay hinders the free movement of water through the coarser debris. In effect, a column of water is formed from the clay concentration upward toward the ground surface. It's a crude dam. This column of water exerts pressure that forces grains apart. As the grains separate, the loss of grain-to-grain contact friction results in slippage of particles downslope. This may occur fairly rapidly during a heavy rainstorm, or it may build slowly during several lighter rainfalls. In either case, the exact time when the debris will move is difficult to predict.

During the 1994 Labor Day weekend, Interstate 70 in western Colorado was blocked for two days fol-

FIGURE 5.12

This snowshed protects a railway from avalanches. *Photo: Corbis/Bettman.*

lowing a debris flow close to the city limits of Glenwood Springs. Mountain slopes in this area had suffered a devasting wildfire in July, denuding the slopes of vegetation.

Consequently, heavy rains caused debris flows along many of the mountain drainage channels, pouring mud, burnt ash, timber and boulders onto Interstate 70 and almost damming the Colorado River, which runs parallel to the highway. Since Interstate 70 provides the only access through the area, trucking, tourist,

and emergency vehicle traffic had to be diverted through the northern part of the state. Thirty cars were swept into the Colorado River, but miraculously, because the debris flows began late at night, traffic was light and only one person was injured.

Human Activities Trigger Debris Slides

Human activities can cause debris slides in several ways. Debris on a slope can be set into motion by

FIGURE 5.13 Debris slide.

A debris slide has undercut the foundation of this California home, removing the sidewalk and threatening the road in the foreground. Obviously, slides affect buildings and transportation. Special engineering design, zoning, and land-use management can greatly reduce the frequency of scenes like this. *Photo: J. T. McGill/USGS.*

FIGURE 5.14 Critical angle.

Angle of repose is roughly 30 degrees for most granular material—sand, gravel, sawdust, salt, sugar, and so on. *Photo: George Leavens/Photo Researchers, Inc.*

removal of support at the toe of the slope. This situation often occurs during construction on a slope or at its base. Roads built at the base of slopes commonly undermine the debris in this way.

Movement can also result from loading the slope from above with significant weight. Loading from above occurs when a structure—a building, storage tank, or highway—is built on materials that cannot support the additional heavy load (Figure 5.15).

In the Los Angeles Basin, many people live where debris flows and slides are common. The affected residents have built their houses either on the lower part of unstable slopes or at their base. One resident described his location as "living in a drainage ditch." Many of

FIGURE 5.15

Failure of fill on Interstate Highway 80 east of San Francisco. *Photo: USGS.*

these homeowners knew of the debris slide problem before building their homes, but land shortage made it impossible to avoid unstable areas. Noted environmental writer John McPhee (1988) provides an excellent description of the problems faced by these residents.

Several factors combine to make major earth movements in southern California almost an annual event. The sedimentary rocks surrounding the Los Angeles Basin are mostly shales, highly fractured and thinly bedded. The steep hillsides need the binding of plant roots to remain stable. Hence, forest fires become an important precursor for several types of mass movements, including debris slides and mudslides. When a forest fire occurs, ignited either by lightning or human negligence, the loss of vegetation adds to the natural instability of the shale on the slopes. When spring rains come, slopes are rapidly destabilized as the clay flakes in shale lose cohesion. Mass movements of one type or another become inevitable, as many homeowners have learned to their sorrow.

Because of the need to build despite the danger, the governing board of Los Angeles County decided to use public money to help alleviate the problem. It was noticed that an area had been protected from a debris slide by a gravel quarry located on the slope above some homes. Recognizing the effectiveness of such excavations, more than 120 bowl-shaped "debris basins" were excavated by the county, using bulldozers on slopes above endangered houses in the San Gabriel Mountains near Los Angeles (Figure 5.16). At the lower end of each basin, a concrete or earth-fill dam was constructed. The debris basins, some as large as a football stadium, serve as giant colanders or sieves. They stop downhill movement of the bouldery debris, while letting the water that was mixed with the debris drain slowly downslope.

When a basin is filled nearly to capacity with debris, it is removed by large trucks. Clearing the basins is difficult during years in which storms are frequent. Measurements reveal that an average of 7 tons of debris moves downslope from each acre of hillslope each year in this area of the San Gabriel Mountains. Because of the variable volume of debris in each slide, a basin is occasionally overtopped in a single slide event, but even in these instances damage to homes below is reduced.

What about insurance? Insurance companies are in business to make money, but they lose each time they pay a claim. Not surprisingly, therefore, private insurance against destruction of homes by mass movements is very expensive in such unstable areas.

Talus and Talus Creep

Talus is the loose debris that accumulates at the base of a cliff or high area (Figure 5.17). Talus piles may originate quickly as the large deposit of a debris slide. However, the small accumulation of weathered debris that slowly builds at the base of any rock outcrop is also a talus pile. Obviously, this slowly accumulating type of talus pile is unrelated to mass movement.

A debris slide may come to rest at the base of a slope in an apparently stable configuration. However, the talus pile is composed of loose sediment, and thus has little resistance to earth tremors and heavy rainfalls. The result is slow, perhaps imperceptible movement outward into the valley. This phenomenon is called *talus creep*. Its only impact on humans occurs

FIGURE 5.16 Debris basin.

One of a series of small debris basins designed to contain and direct debris flows. This structure is located in an unnamed canyon near Pasadena in Los Angeles County, California. It could contain a flow of approximately 100 cubic yards in volume. *Photo: John Tinsley/USGS.*

FIGURE 5.17 Talus at the base of steep cliffs in the Chugach Mountains, Alaska.

The rockfall debris was channelled down a gully and produced a well-defined talus cone at the base of the cliff.

Photo: Steve McCutcheon/Visuals Unlimited.

when the talus migrates across roadways and creates a hazard, particularly in mountain environments. Talus creep poses no danger to human life.

Mudslides

The most common type of **mudslide** is a **slump,** a downslope movement of a coherent slab of muddy sediment along a rupture surface on its base (Figure 5.18). Usually the slumped material moves slowly and does not travel very far from its origin. Although the slump may remain a single mass, secondary ruptures commonly develop, having orientations that are parallel to the major initial break.

A slump mudslide usually occurs because the supporting sediment at its base has been removed. In this respect, the mudslide is similar to a debris slide. The principal difference between the two types of mass movement is the *coherence* of the mudslide as it moves downslope, which results from its higher content of cohesive clay minerals.

Just as debris slides are common in southern California, so are mudslides. Here, they have caused untold millions of dollars in property damage, as well as significant loss of life. As with debris slides, an increase in water pressure between sediment particles is often the precipitating cause of the mass movement. The sediment in a mudslide is richer in clay than the sediment in a debris slide and is, therefore, less permeable to water (see Chapter 6). Thus, the increase in water pressure may happen more quickly in sediment that will become a mudslide mass.

Earthflows

Earthflows contain sufficient water so that they move as coherent masses, often on rather gentle slopes. They may not originate from a sharp upslope rupture, as does a slump, but start as a stretching of the soil and upper rock layers that gradually increases in amount downslope. The slowly flowing mass of sediment oozes irregularly across the landscape (Figure 5.19). Active earthflow surfaces are lumpy and have wavy surfaces.

Seasonal soaking of muddy soil on slopes causes movement of the soggy mass, which becomes the earthflow. Studies indicate that after initial wetting during the first two or three rains, major episodes of movement coincide with periods of heavy rainfall. If the toe of the earthflow protrudes into a stream, the rate of earthflow movement can be greatly increased as the toe is eroded and its sediment carried away downstream.

Although earthflows move very slowly and are never life-threatening, they can cause significant property damage, as shown in Figure 5.19. At several locations in this photo, the earthflow globs are about to overtop the major highway. The highway engineer would have been well advised to locate the road farther out into the valley to avoid the now-imminent—and perhaps perpetual—expense of clearing the highway of debris.

Soil Creep

Earthflows grade in aspect into **soil creep,** which is the movement of clayey soil slowly downslope under the influence of gravity. The creeping movement occurs mostly in the upper few feet of soil and decreases rapidly with depth. Movement is slow because of the very high content of sticky clay minerals in most soils, but it is faster on steeper slopes and with greater percentages of soil moisture. Most rates are less than 0.5 inch per year on vegetated slopes.

Despite the slow rate of movement, the effects of soil creep are numerous and easy to see: downslope curvature of rock layers (Figure 5.20), tilted headstones

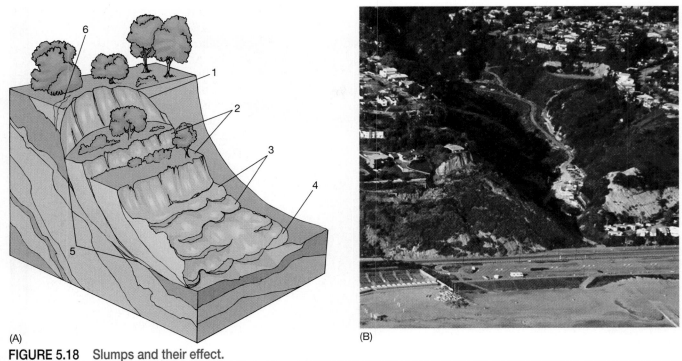

(A)

(B)

FIGURE 5.18 Slumps and their effect.

(A) Slumps often show features of sliding near their tops and flowing failure near their bases. The slide scarp (1) usually displays concave crescent-shape. Slump blocks (2) show rotational movement. Hummocks (3) are uneven mounds and ridges produced during soil flowage. Toe of slide (4) displays fluid movement. Failure plane (5) bounds the base. Fissure (6) indicates that slump is still active and that more movement is likely. Removal of toe at the base of a slope often triggers a slump failure. (B) This large slump is over a highway near the Bel Aire Bay Club, California. *Photo: Frank M. Hanna/Visuals Unlimited.*

FIGURE 5.19 Earthflows in the Horse Heaven Hills, Washington.

Lobes are encroaching on the highway, which will soon need extensive and expensive maintenance. *Photo: D. A. Rahm, courtesy Rahm Memorial Collection, Western Washington University.*

(A)

(B)

FIGURE 5.20 Creep and its effects.

(A) Soil creep seen along the side of a hillslope. The thinly-layered rock and overlying soil move slowly downslope in response to the pull of gravity. *Photo: John D. Cunningham/Visuals Unlimited.* (B) Diagram showing creep and its effects (1) Moved joint blocks; (2) trees with curved trunks concave upslope; (3) downslope bending and drag of bedded rock or weathered veins, also present beneath soil elsewhere on the slope; (4) displaced posts, poles, and monuments; (5) broken or displaced retaining walls and foundations; (6) roads and railroads moved out of alignment; (7) turf rolled downslope from creeping boulders; (8) stone-line at approximate base of creeping soil. (1) and (3) represent rock-creep; all other features shown are due to soil-creep. Similar effects may be produced by some types of landslides.

(Figure 5.5), and trees growing with their trunks bent upslope as they try to remain upright. Prevention of creep requires the construction of retaining walls, but even these may give way over time under the continuing and increasing pressure of accumulating creepy soil.

Permafrost and Solifluction

Permafrost is the permanently frozen ground in extremely frigid areas of northern Canada, Alaska, and Siberia (Figure 5.21A). Permafrost underlies 26 percent of Earth's surface, so it is not rare, but it is seldom seen

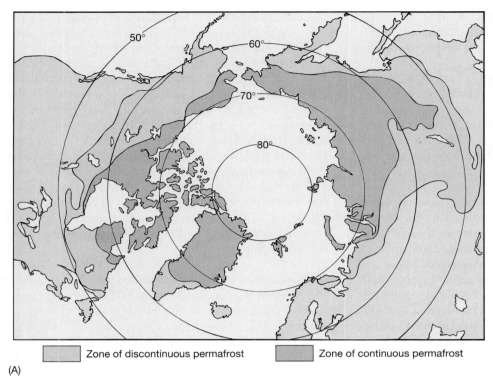

Zone of discontinuous permafrost   Zone of continuous permafrost

(A)

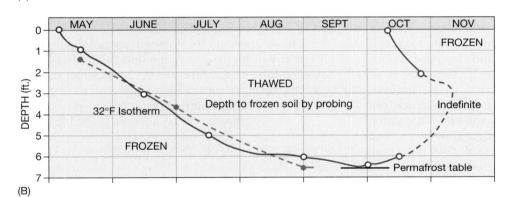

(B)

(C)

FIGURE 5.21 Permafrost.

(A) Extent of continuous and discontinuous permafrost in the Northern Hemisphere. (B) Depth of summer thawing of frozen soil at Fairbanks, Alaska.
(C) Solifluction lobes northeast of Fairbanks, Alaska. Solifluction occurs when the active layer thaws in summer. *Photo: James E. Patterson.*

because few people live in permafrost areas. Most permafrost is a few hundred feet thick, although thicknesses of several thousand feet have been recorded in Siberia.

The soil directly above permafrost is subject to **solifluction** (literally, *flowing soil*). The cause of solifluction is repeated freezing and thawing of the uppermost few feet of soil on a sloping ground surface. During a thaw, the water released in this active layer cannot penetrate the solid permafrost. Thus, soils in the summer often become saturated, and the resulting loss of friction and cohesion causes them to flow like viscous (thick) fluids (Figure 5.21B). Such materials can flow on slopes as gentle as 1°, but maximum displacement occurs between 5° and 20°. Soils on slopes steeper than 20° tend to drain easily, so the water escapes as surface runoff.

Solifluction is intermediate between earthflow and soil creep. It is identified by a water content greater than that for soil creep and a similarly high clay content. It generally involves both the soil and broken rock and sediment beneath it. Movement is less than about 12 inches per year, generally so slow as to be detectable only by measurements made over several seasons. But the result is the distinctive pattern you see in Figure 5.21C.

The construction of buildings, roads, and utility lines in permafrost areas is very difficult. Bedrock or coarse-grained sediment is reasonably stable because it drains easily, but soils rich in silt and clay pose serious problems for engineers. The active freeze-thaw layer near the surface becomes a quagmire in the summer and cannot support the weight of a building, truck, car, or train without deforming. Consequently, differential settling occurs in buildings, bridge abutments, and other structures. Also, because the soil does not drain,

the water in it refreezes when the temperature drops. The 9 percent expansion when water freezes causes irregular heaving of the soil during each freeze.

To avoid cracking of foundations, most buildings are built atop pilings driven deep into the permafrost layer. The base of the building is 3 to 6 feet above the ground surface so that air can circulate freely in the open space between the floor and the ground (Figure 5.22). This prevents thawing of the soil due to heat from the building during the winter, and minimizes thawing during the summer because the ground surface is shaded by the building. The pilings that support the building are driven to a depth at least twice the thickness of the active freeze-thaw surface layer so they will not be loosened by the annual heaving of the soil each winter.

Trans-Alaska Pipeline. Much of the crude oil produced from the prolific Alaskan oil fields comes from the northern coast of the state, in the vicinity of Prudhoe Bay on the Arctic Ocean. The oil travels by overland pipeline south to the port of Valdez on the Pacific Ocean. The Alaska Pipeline is 760 miles long, roughly the distance from Chicago to the Gulf of Mexico. It cost $7.7 billion and was built between 1974 and 1977 (Figure 5.23).

The soil beneath the northern part of the pipeline is permafrost. As the pipeline enters southern Alaska, however, the temperature during the summer months becomes warm enough to melt the soil ice to depths of several feet, causing solifluction.

Thus, engineers had to address three big problems in designing the pipeline. The pipeline had to (1) avoid damaging the fragile permafrost, and at the

FIGURE 5.22 These buildings in Alaska are raised above ground so air can circulate beneath them.

This prevents melting of the permafrost. *Photo: David Matherly/Visuals Unlimited.*

FIGURE 5.23 Alaska Pipeline zigzags so the pipe can flex but not break when the soil shifts.

The pipeline also contains sections that have special joints that allow the pipe to shift as much as 20 feet without rupturing during earthquakes. The pipeline also is high enough off the ground that caribou (reindeer) can migrate beneath it. *Photo: Arco; (inset) John Swedberg/Bruce Coleman, Inc.*

same time, (2) avoid damage from frost heaving and earthquakes, which in turn (3) could rupture the pipeline, spilling oil into the environment. The more you read about the Alaska Pipeline, the more you will appreciate it as a marvel of environmental engineering.

The pipeline is 4 feet in diameter. When full of oil, it weighs more than 6000 pounds *per foot* of length, so it is prone to sink. Further, the oil is hot. It is heated to make it flow easier and it enters the pipeline at 136°F. Friction between the oil and the pipe during flow increases the temperature considerably. Even with cooling devices along the route, the temperature is reduced only to 145°F, so it obviously could melt the permafrost layer beneath the pipe.

How did engineers protect the permafrost and protect the pipeline from differential heaving and settling of the soil over its 760-mile length, so that it will not bend or crack and spill millions of gallons of crude oil into the environment? Depending on local conditions, the pipe either is buried or is suspended above the ground:

- Burying the pipe works well in coarse-grained soils that are well drained and contain little ground ice. In other areas, a refrigerant is run along the pipe to keep the ground frozen.
- Suspending the pipe above the ground is necessary where the soil is particularly susceptible to thawing. Suspension also is needed where the route crosses the path of migratory animals such as caribou, so the animals can pass underneath. The above-ground suspension requires a dense network of pilings to provide the needed strength, and these are inserted to a depth of many feet. In addi-

tion, some of the supporting piles are refrigerated so they won't melt the permafrost. Others are distributed in a zigzag fashion to absorb the movement of heaving and earthquakes (Figure 5.23).

Since 1977, when oil started flowing through the pipeline, some minor breaks have occurred. For example, in 1981, 200,000 gallons of oil spilled when a valve ruptured. Other leaks are inevitable on a run of pipe this long. Overall, however, the design of the pipeline has effectively preserved both the permafrost and the structural integrity of the pipeline.

Mudflows

Mudflows, the most watery of mass movements, contain up to 30% water. They are common on slopes in semiarid areas. Their specific cause is infrequent but intense, short-lived rainstorms that quickly convert surface debris into a mass of viscous mud and rock that moves downslope at high velocities (Figure 5.24). Mudflows are very likely when heavy rains follow a drought, or when forest fires destroy slope vegetation and increase erodibility.

A common trigger for mudflows in elevated regions is the rapid melting of deep winter snows. This can cause surging flows of muddy debris that last for weeks. The mud may travel many miles from the edge of the mountains on slopes less than 1°. Velocities of several feet per second are common, especially on steeper slopes. The advancing fronts of the more fluid flows may splash about like a rapidly advancing tongue of water, although the amount of water is no

FIGURE 5.24 Major mudflow.

An earthquake in January 1989 in Tadzhikistan produced mudflows 50 feet high on a slope weakened by rain. *Photo: Vlastimir Shone/Gamma-Liaison, Inc.*

more than 30 percent of the mass. Parts of the mudflow that contain less water can be viscous enough to carry boulders on their surface.

Mudflows have become a serious environmental problem in recent years in southern California, as the expanding urban areas reach deeper into formerly uninhabited desert areas. In addition, numerous isolated resort communities have sprung up and grown outward toward the bases of the various mountain ranges of southern California. As a result of such growth, new environmental concerns have arisen. Mudflows, like most natural phenomena, are only problems when humans object to their occurrence or get in the way.

Liquefaction

Liquefaction is the transformation of soil from a solid to a liquid condition. How can this occur? It usually happens when an earthquake vibrates loose, water-saturated granular sediment until it loses its cohesiveness (Figure 5.25). The ability of the sediment to bear heavy loads such as buildings is lost almost instantaneously when liquefaction occurs.

The sediment that gives way is commonly a sand located beneath an impermeable soil (which essentially is a barrier to water flow). The soil's weight is partially supported by the water in the pores between sand

FIGURE 5.25 These buildings were an elementary school in Anchorage, Alaska.

Their clayey foundation experienced liquefaction during the Alaskan earthquake of March 27, 1964. However, note that these framed wooden buildings, although destroyed, are, for the most part, still integral units. Loss of life is minimized by such construction. The ground in the center of the photo has settled about 12 feet from its original elevation, seen on the left. *Photo: USGS.*

grains. When the sand is shaken by the earthquake, the grains are pushed apart, grain-to-grain contact is lost, and the sand-water mixture suddenly behaves as a liquid. Such sand often is called **quicksand.** It spreads laterally or moves downslope, and any overlying structures sink into the "liquid." Liquefaction during a minor earthquake can cause damage that seems out of proportion to the intensity of the quake.

Some clay sediment can also liquefy. When clay minerals are arranged in an edge-to-face arrangement (Figure 3.9B), the mass is stabilized by the abundant positive ions in the sediment pore water. When the sediment is raised above sea level, either by lowering of sea level or uplift of the land, rainwater may wash out these stabilizing ions, decreasing the strength of the clay. This is a serious problem in the St. Lawrence River valley in eastern Canada and in Alaska. Clay aggregates with the potential to liquefy are termed **quickclays.**

How Bad Is the Damage?

Let's say that you have been either unlucky or imprudent and, as a result, you have been victimized by slope instability. Your dwelling is damaged. How bad is the damage? You want to know for reasons of safety. Can you continue to live there? The insurance adjuster wants to know to determine whether you are covered,

and for what amount. The state or federal government may want to know to determine whether you are eligible for disaster assistance. Of course, any organization that might give you money to help with repairs is likely to send a representative for an on-site evaluation.

Scales have been devised to classify degrees of damage, one of which is shown as Table 5.3. The objective of the classification is to create a series of sequential categories to express the damage caused by landslides to buildings. Hence, the categories of damage relate to degrees of cracking, tilting, collapse, extension, or compression in affected houses. The use of complementary scales for roads, utility lines, buildings, bridges, and so on, enables an intensity map to be drawn for overall patterns of damage, which can be related to the physical characteristics of the movement, including its speed and what caused it.

How to Recognize Areas Prone to Slope Instability

Mass movements can be anticipated, based on the criteria that follow. Keep in mind, however, that these criteria only can suggest the places most likely to have problems. There are no guarantees for any location, only probabilities. Further, even when a

TABLE 5.3	Intensity Scale for Landslide Damage
Grade	**Description of Damage**
0	*None:* Building is intact.
1	*Negligible:* Hairline cracks in walls or structural members; no distortion of structure or detachment of external architectural details.
2	*Light:* Building continues to be habitable: repair not urgent. Settlement of foundations, distortion of structure, and inclination of walls are not sufficient to compromise overall stability.
3	*Moderate:* Walls out of perpendicular by 1–2 degrees, or substantial cracking has occurred to structural members, or foundations have settled during differential subsidence of at least 15 cm; building requires evacuation and rapid attention to ensure its continued life.
4	*Serious:* Walls out of perpendicular by several degrees; open cracks in walls; fracture of structural members; fragmentation of masonry; differential settlement of at least 25 cm compromises foundations; floors may be inclined by 1–2 degrees, or ruined by soil heave; internal partition walls will need to be replaced; door and window frames too distorted to use; occupants must be evacuated and major repairs carried out.
5	*Very serious:* Walls out of plumb by 5–6 degrees; structure grossly distorted and differential settlement will have seriously cracked floors and walls or caused major rotation or slewing of the building (wooden buildings may have detached completely from their foundations). Partition walls and brick infill will have at least partly collapsed; roof may have partially collapsed; outhouses, porches, and patios may have been damaged more seriously than the principal structure itself. Occupants will need to be rehoused on a long-term basis, and rehabilitation of the building will probably not be feasible.
6	*Partial collapse:* Requires immediate evacuation of the occupants and cordoning off the site to prevent accidents involving falling masonry.
7	*Total collapse:* Requires clearance of the site.

After B. E. Alexander, 1986, *Environmental Geology and Water Science*, v. 8, p. 147–151.

mass movement locality is identified, the exact timing of the next movement is uncertain. And, as noted before, timing is everything.

Under favorable situations, an area is susceptible to mass movements if:

1. *The area has a history of past landslides nearby, in the same rock units as the site in question.* Landslides of relatively recent vintage, say during the past one hundred years, can be recognized by the remaining scar and relatively unvegetated appearance (Figure 5.7). Wherever geologic mapping has been done in the world, the maps can be used to spot areas having known slide-prone rocks.

2. *The area has soil types rich in clay minerals, particularly those of the expanding type.* Abundant clay, particularly on a sloping surface, is an almost certain prescription for a landslide of some type. For example, the Pierre Shale of the northern mid-continental United States and adjacent Canada is rich in expandable clay minerals. Consequently, it is the site of many and repeated slope failures. In some places, the clay is so moist after rains that it is nearly impossible to drive on the slippery stuff, even up a gentle slope. A geologic map that shows where the Pierre Shale is exposed at the ground surface reveals these potentially dangerous areas.

3. *The area has downslope orientation of planes of weakness in bedrock.* Sliding is promoted by water acting to decrease cohesion between rock and mineral surfaces. You can anticipate sliding wherever weak places in rocks are parallel to the slope of the ground. Common situations where this occurs are (1) surfaces between sedimentary rock beds, called *bedding planes* (a good example is illustrated by the Vaiont Dam slide, Figure 5.8B); (2) planes of mica flakes in the metamorphic rock we call schist; or (3) rock fractures (joints).

4. *The area has slope undercutting.* Landslides are particularly common along stream banks, highway road cuts, and seacoasts during storms, all places where the natural support for a slope has been removed.

5. *The area has experienced removal of vegetation.* The removal of vegetation from a slope, whether by grading, logging, overgrazing, or fire, often decreases slope stability. Plant roots stabilize slopes by binding soil together or by binding soil to bedrock. Vegetation also reduces surface runoff of rainwater by using the water, decreasing the amount of water in the soil by taking it in through roots and evaporating it from leaf surfaces. The certainty of mudflows in southern California following the near-annual forest fires in the mountains provides a good example of the beneficial effect of vegetation on soil stability.

6. *The area experiences earthquake tremors.* Marginally stable slopes that have seemed safe for decades can be quickly rendered unstable by earthquake vibrations. All types of landslides may be generated, from rockslides to mudflows to liquefaction. There probably is no completely safe area in a place such as western California, where high topographic relief, soft and clayey sedimentary materials, repeated seismic activity, and intense urbanization are everywhere.

Landslide Maps

In recent years, more landslide maps have been published. Some are based on remote sensing from aircraft and satellites. Others are prepared from on-the-ground field studies (Figure 5.26). Most of the remote sensing investigations use aerial photography at a scale of at least 1:48,000 (1 inch on the photo = 48,000 inches, or 0.75 miles, on the ground). Although aerial photos are a powerful tool, there is no substitute for "going there"—actually visiting the site. Moreover, many important features of slope failure may be partially or completely obscured in areas of dense forest, or the photos may not show critical conditions of ground water that change radically with the seasons.

Where no landslide maps are available, or at least geological maps, ground studies are necessary to determine what variations in rock properties, orientation of rock layers, streams, and ground water influence slope stability, so that assessment of potential instability may be made for slopes that have not yet failed. It may be necessary to prepare a specialized engineering geological map for this purpose. The environmental geologist must be a keen observer and interpreter of the landscape.

Can Landslides Be Prevented?

Although human civilization creates many new opportunities for mass movement of earth materials, such movements have been going on since Earth was formed about 4.5 billion years ago. Movement in response to the pull of gravity cannot be stopped. What can be done, however, is to anticipate mass movements and to prepare for them. Landslides are one of the most easily predicted of geological hazards (the likelihood of their occurrence, if not the timing). The cost of preventing landslides is less than the cost of correcting them, except for small slides that can be handled by normal maintenance procedures.

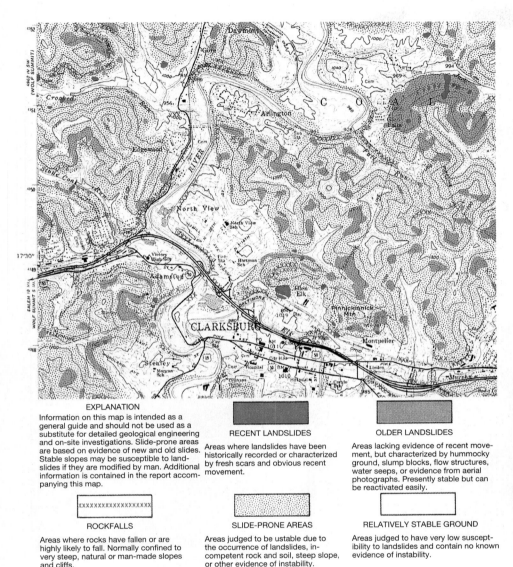

EXPLANATION

Information on this map is intended as a general guide and should not be used as a substitute for detailed geological engineering and on-site investigations. Slide-prone areas are based on evidence of new and old slides. Stable slopes may be susceptible to landslides if they are modified by man. Additional information is contained in the report accompanying this map.

RECENT LANDSLIDES

Areas where landslides have been historically recorded or characterized by fresh scars and obvious recent movement.

OLDER LANDSLIDES

Areas lacking evidence of recent movement, but characterized by hummocky ground, slump blocks, flow structures, water seeps, or evidence from aerial photographs. Presently stable but can be reactivated easily.

XXXXXXXXXXXXXXXXXXXX

ROCKFALLS

Areas where rocks have fallen or are highly likely to fall. Normally confined to very steep, natural or man-made slopes and cliffs.

SLIDE-PRONE AREAS

Areas judged to be ustable due to the occurrence of landslides, incompetent rock and soil, steep slope, or other evidence of instability.

RELATIVELY STABLE GROUND

Areas judged to have very low susceptibility to landslides and contain no known evidence of instability.

FIGURE 5.26 Portion of a mass-movement potential map of Clarksburg, West Virginia.

This map was created by geologists at that state's Geological Survey. Various government agencies publish similar maps for other areas. People seeking homesites should study such maps prior to building. *Source: West Virginia Geological and Economic Survey.*

In the Los Angeles area, the potential dollars to be lost from landslides will exceed dollars spent on preventive planning by between 50:1 and 110:1, it has been estimated. At least a 95 percent reduction in damaging land failures is technically attainable today with proper use of regional, local, and site investigations.

Several mechanical and electronic devices have been devised to warn of mass movement. The Denver and Rio Grande Railroad through the southern Rocky Mountains has developed electric circuits that are broken by a rock slide, triggering a stop signal for oncoming trains. Similarly, fences and barriers can be wired to identify pressures against them caused by mass movements and to transmit warning signals. Strain meters can detect increases in stress that precede large rockfalls. Of course, such equipment is expensive and cannot be set up everywhere it might be useful.

Ways to Reduce Incidence and Severity of Landslides

Several relatively inexpensive methods can reduce damage from landslides and perhaps even prevent them at a particular locality.

1. *Site investigation by both a geologist and an engineer.* Engineers are experts at construction but are not always aware of rock and drainage problems that a geologist would notice immediately. The Vaiont Dam disaster is a good example of this.

2. *Slope reduction.* Construction on slopes cannot always be avoided, but chosen slopes should be as gentle as possible (Figure 5.27). Grading may be necessary to reduce the slope before construction begins (Figure 5.28).

3. *Control of water movement on and within the slope debris.* Water can be rerouted at the top of the

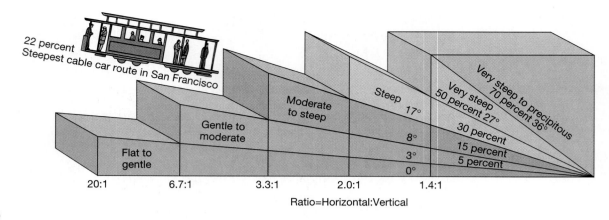

(A)

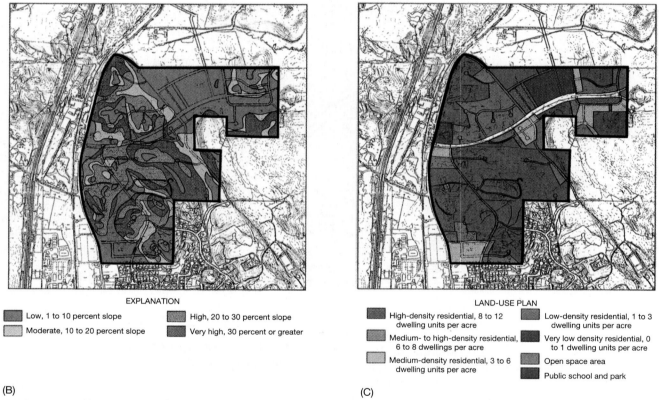

EXPLANATION

▮	Low, 1 to 10 percent slope	▮	High, 20 to 30 percent slope
▮	Moderate, 10 to 20 percent slope	▮	Very high, 30 percent or greater

LAND-USE PLAN

▮	High-density residential, 8 to 12 dwelling units per acre	▮	Low-density residential, 1 to 3 dwelling units per acre
▮	Medium- to high-density residential, 6 to 8 dwellings per acre	▮	Very low density residential, 0 to 1 dwelling units per acre
▮	Medium-density residential, 3 to 6 dwelling units per acre	▮	Open space area
		▮	Public school and park

(B)   (C)

FIGURE 5.27   Construction on slopes.

Maps that are part of a specific land-use plan required by the city of Healdsburg, California, before annexation of a 233-acre area north of its present boundary. (A) The slope is an important criterion for stability of the land surface. Some municipalities require soil and geologic studies of all proposed construction sites with slopes greater than a certain amount. (B) Slope map of area to be annexed. (C) Land-use plan for area to be annexed. Low-density residential uses are proposed for most of the areas with steep slopes.

slope. Also, drainage pipes can be set on or within the slope debris to channel the water away from areas that are particularly susceptible to mass movement.

4. *Retaining walls.* A concrete barrier or buttress called a **retaining wall** can be constructed on or at the base of the slope to prevent landslides (Figure 5.29). Commonly, this is done together with improved water drainage from the sediment behind the wall. Drainage pipes are inserted into the sediment to decrease water in it and improve stability. The pipes are perforated

FIGURE 5.28   Hillside terracing for a residential subdivision in southern California, a common practice in hilly areas.

Without significant buttressing at the base of each vertical cut, there is considerable potential for catastrophic mass movements involving the houses that will be built on the flat pads. *Photo courtesy County of Los Angeles.*

with penny-sized holes so that gravel cannot enter the pipe but water can leave, to drain away.

5. *Bolting.* Another, less commonly used method for slope stabilization is to drive giant steel bolts to anchor an unstable "sandwich" of rock layers. Each long bolt is driven through an upper, stable rock layer, through an unstable layer (such as shale) below, and into a stable layer beneath the shale, where the bolts are anchored (Figure 5.30).

6. *Fences, gabions, or covering with chain-link fencing.* Where a hillslope is too high for a cement buttress, fencing can be used (Figure 5.31A). Or, rock-filled wire enclosures called **gabions** may work best (Figure 5.31B). Alternatively, chain-link fencing can be laid on the slope to stabilize it and decrease landslide intensity.

7. *Planting vegetation.* Vegetation leaves deflect raindrops from direct, erosive impact on the soil (recall Figure 4.2), and their roots help to anchor soil particles. A revegetation technique often used today is called *hydroseeding*, shown in Figure 5.32.

Figure 5.33 shows a fine example of a total failure to apply methods 1 through 6 that you just read. No geologic input was sought before or during construction of the home for the unfortunate inhabitants. The hill behind the house is unstable because of its slope and the character of the rocks that form it. The concrete slab on which the house sits is unstable, unconsolidated debris that may give way at any moment, with disastrous results for the inhabitants. (Will it occur at night while they are sleeping?) Further, their water drainage problem can be solved only at great cost. The house is in serious danger of collapsing into the growing solution cavities beneath it.

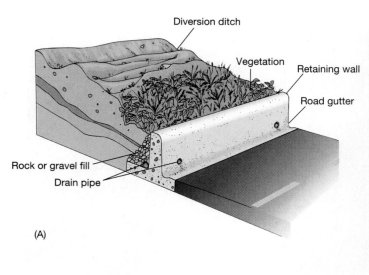

FIGURE 5.29 Slope-protection measures.

(A) Cross-section of retaining wall. A cut slope, if properly planted, supported, and drained, can be completely stabilized. Such precautions are not always taken, however. (B) The effectiveness of slope-protection measures is demonstrated along this west-facing graded slope above Calle Familia, San Clemente, California. *Photo: Siang Tan, courtesy of California Division of Mines and Geology.*

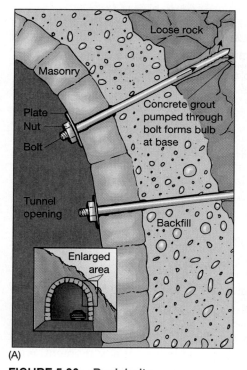

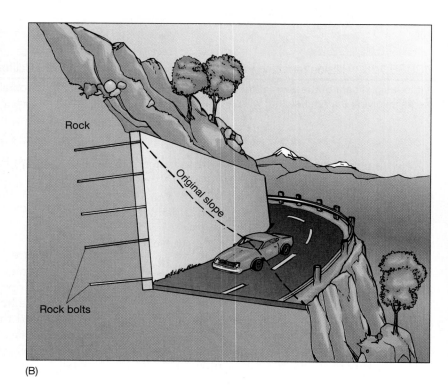

FIGURE 5.30 Rock bolts.

Rock bolts are used to retain unstable rock in tunnels, road cut slopes, and in the roofs of underground mines.

(A)

(B)

FIGURE 5.31 Measures to reduce the consequences of slope instability.

Fence (A) and gabions (B) protect roads from rockfall. *Photo: Dru Germanoski.*

FIGURE 5.32 Hydroseeding.

Hydroseeding sprays a one-shot mixture of plant seeds, fertilizer, water, and protective mulch on this bare ground. Planting vegetation like this helps stabilize the surface and reduces the risk of landslides. *Photo: Granitsas/The Image Works.*

The only question these homeowners might have is whether the fill or the permeable sandstone under their house will collapse first! If you were an insurance agent, would you insure this house? Would you want to live in this house? After completing this course and reading this textbook, would you buy the house? (See, you *are* getting something useful from your college education.) When the inevitable happens, who will be legally responsible?

The Uniform Building Code

The construction industry is well aware of the difficulties engineers encounter when building on slopes, on fractured rock, or on soils that contain expanding clays. A section of their **Uniform Building Code** was written as a guide for construction engineers when building in such areas, and these recommendations have been adopted by many municipalities. The Code considers such geologic features as layering, vertical fractures (joints), and slope angle. Experience shows that honoring even the minimum Code recommendations can cut losses from ground problems by more than 95 percent. Does construction in your community follow the Uniform Building Code?

Can Landslides Be Prevented? **119**

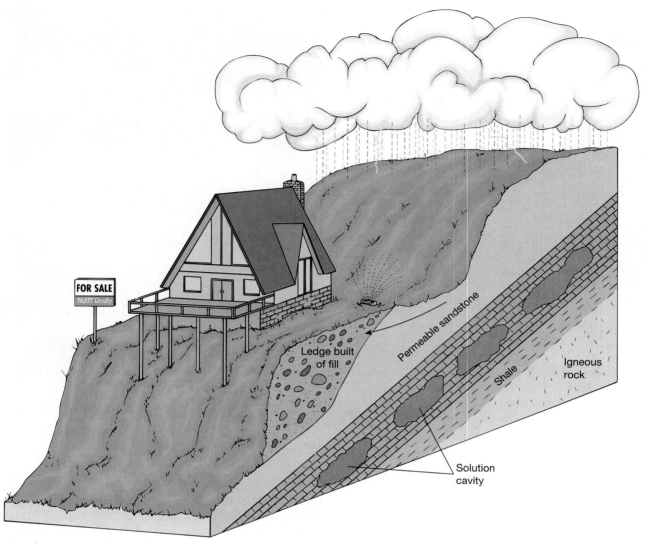

FIGURE 5.33 A realtor's challenge.

This recipe for an unhappy home includes solution cavities, a house built on unstable fill, and porous sandstone and limestone supplying water to a swelling shale. Is this house a good buy or a good-bye? Some day, when you seek your dream chalet in the country, remember this illustration.

Summary

Numerous types of mass movements or landslides are common. Most occur on slopes, in response to gravity's relentless tug. The specific type of movement is determined by the amount of water involved in the slide and by the percentage of clay minerals in the unstable rock or sediment. Water content of sediment in a mass movement ranges from zero up to about 30 percent and the percentage of clay minerals ranges from zero to 100 percent. Landslide velocities range from barely noticeable to rocks falling freely at several hundred miles per hour.

Landslide damage that is costly to humans can be largely prevented if existing scientific knowledge is applied. Reducing slope gradients, increasing the amount of vegetation on the slopes, and preventing buildup of water in the sloping soil or rock are all useful as preventive measures. Several types of specialized construction to stabilize slopes also may be helpful. Of course, no one can guarantee the safety or permanence of any house, bridge, or other structure that is built where the geology is prone to mass movements—especially in earthquake country.

Key Terms and Concepts

angle of repose
avalanche
creep
debris
earthflow
flow friction
gabion
gravity
landslide

liquefaction
mass movement
mudflow
mudslide
permafrost
pore pressure
quickclay
quicksand
retaining wall

rockslide
slides
slump
soil creep
solifluction
talus
uniform building code

Stop and Think!

1. What features of rock and sediment make slope failure likely? Explain why slope failure is likely for each feature.
2. What human activities near your home or near your school might increase the possibility of landslides?
3. How does water promote landslides?
4. How does the presence of clay minerals affect landsliding?
5. What are the differences among a slide, flow, and creep? Which type of movement is more common where you live? Why?
6. What environmental problems are created by each of the three types of movements mentioned in the previous question? Which are more life-threatening? Which do you think are more costly over a ten-year period? Ask your state geological survey about landsliding in your area.
7. In the geology section of your college library, seek maps that consider landslide potential as a factor in land-use planning. Does your state geological survey publish such maps (some states do)? If not, should they?
8. Ask your city planning commission whether they subscribe to the Uniform Building Code in city construction. If not, ask what city laws govern construction on slopes, on clayey soils, or on fractured rock.

References and Suggested Readings

Armstrong, Betsy and Williams, Knox. 1992. *The Avalanche Book, 2nd ed.* Golden, Colorado: Fulcrum, Inc.

Baum, R. L. and Johnson, A. 1996. Overview of landslide problems, research, and mitigation, Cincinnati, Ohio, area. *U.S. Geological Survey Bulletin* 2059-A, 33 pp.

Costa, J. E. and Wieczorek, G. F. (eds.). 1987. Debris flows/avalanches: Process, recognition, and mitigation. *Geological Society of America, Reviews in Engineering Geology,* volume VII, 239 pp.

Crozier, M. J. 1986. *Landslides: Causes, Consequences and Environment.* New York: Chapman and Hall.

Ellen, S. D. and Wentworth, C. M. 1995. Hillside materials and slopes of the San Francisco Bay region, California. *U.S. Geological Survey Professional Paper* 1357, 215 pp.

Jochim, C. L., Rogers, W. P., Truby, J. O., Wold, R. L., Jr., Weber, G., and Brown, S. P. 1988. Colorado landslide hazard mitigation plan. *Colorado Geological Survey Bulletin* 48, 149 pp.

Kiersch, G. A. 1964. The Vaiont Reservoir disaster. *Civil Engineering,* v. 34, p. 32–39.

McPhee, John. 1988. The control of nature: Los Angeles against the mountains (2 parts). *The New Yorker,* Sept. 26, p. 45–78; Oct. 3, p. 72–90.

Schuster, R. L., Varnes, D. J., and Fleming, R. W. 1981. Hazards from ground failures: Facing geologic and hydrologic hazards, earth-science considerations. *U.S. Geological Survey Professional Paper* 1240-B, p. 54–85.

Scott, K. M., Vallance, J. W., and Pringle, P. T. 1995. Sedimentology, behavior, and hazards of debris flows at Mt. Rainier, Washington. *U.S. Geological Survey Professional Paper* 1547, 56 p.

Tufnell, Lance. 1983. *Glacier Hazards.* New York: Longman, 97 pp.

Turney, J. E. 1985. Subsidence Above Inactive Coal Mines. *Colorado Geological Survey Special Publication* 26. Denver, Colorado: Colorado Geological Survey Department of Natural Resources, 32 pp.

Tyler, M. B. 1995. Look before you build. *U.S. Geological Survey Circular* 1130, 54 pp.

Wold, R. L., and Jochim, C. L. 1989. *Landslide Loss Reduction: A Guide for State and Local Planning.* Denver, Colorado: Colorado Geological Survey Department of Natural Resources, 50 pp.

". . . a good land, a land of brooks, of water, of fountains and depths that spring out of valleys and hills . . ."

Deuteronomy 8:7

Clean Water 6

Travel around America and ask people, "What is our most serious wa[ter] problem?" Responses will vary widely: In northern New York State, w[ith] plentiful precipitation (50 inches a year) and low population density, th[e] greatest concern might be the increasing acidity of lake water and the resulting decrease in fish population. In Mississippi River towns, the commonest complaint would be repeated flooding. Near Houston and New Orleans, chemical pollution of drinking water by petroleum refineries might be on everybody's lips. In southern Arizona, the most likely concern would be loss of irrigation water for agriculture, because the water is being piped overland to Los Angeles for household use—including lawn-watering. (Are lawns more important than food?)

The common thread in these concerns is the *location* of water, its *movement* over the land surface, and how we adversely affect its *purity.* Natural variation in water supply and quality is normal. But most land planning and development ignores these variations and fails to prepare for them.

What controls natural variations in water supply and quality? Are these variations made more extreme by our own carelessness? How rapidly is subsurface water replaced by natural processes? As water supplies dwindle, can seawater be desalinated cheaply to provide drinking water?

◄

Springs issuing from caves in the Redwall Limestone at Vasey's Paradise in the Grand Canyon, Arizona. The luxuriant vegetation could not exist here without watering by numerous springs. In relation to the springs shown here, where is the water table? Do springs flow near where you live?
Photo courtesy J. E. Warme.

We Are What We Drink

As scientist Vladimir Vernadsky wrote a hundred years ago, "Life is animated water." Indeed, water is the most essential compound for life on Earth. Although we appear to be solid, we are really bags of water, held together by a small amount of carbon-rich organic matter, much like gelatin, which also seems to be solid but is largely water. Living plants are about 90 percent water by weight, and so are we at birth.

Although we tend to dry up a bit as we grow older, we remain mostly water; men are about 60 percent water in adulthood, women 50 percent (Figure 6.1). The water is enclosed in the trillions of cells that form our bodies. Our brains are, incidentally, among the more mushy of our organs, composed of 85 percent water; most of the rest is fat. (Take it as a simple statement of fact if someone calls you mushy-headed or a fathead.)

Humans can survive more than a month without food but only a week without water. Water is lost continuously from our bodies by perspiration, and it must be replaced. In temperate climates, we must drink about 2 quarts a day, but people working in a desert climate perspire as much as 10 quarts per day and must replace it by drinking heavily. Table 6.1

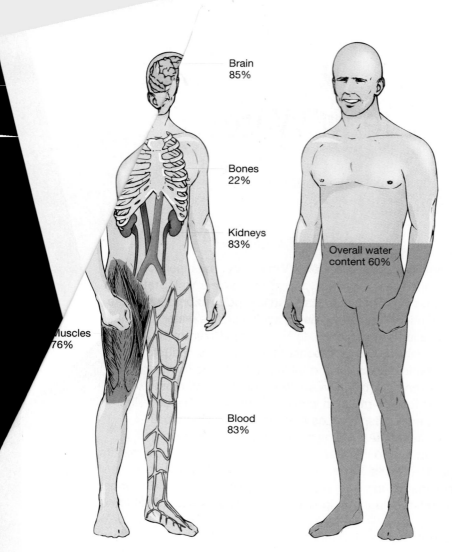

Brain
85%

Bones
22%

Kidneys
83%

Overall water
content 60%

Muscles
76%

Blood
83%

FIGURE 6.1 Water content of the human body.

Note that the bones are relatively dry (only 22 percent water); "dry bones" are mentioned in the Bible (Ezekiel 37:1–14) and in a popular song of the 1940s.

shows human input and output of water. Without dependable access to a stream, spring, well, pipeline, or reservoir, human activity in an arid region is impossible, and life itself is in danger.

The Hydrologic Cycle (Water Cycle)

The total amount of water at and near Earth's surface is immense. It is estimated to be the volume of a cube 148 miles on a side. However, most of this water is not

TABLE 6.1 Input and Output of Water by Humans

**Loss of 10 percent of body-water renders body immobile.
Loss of 20 percent of body-water causes death.**

50% drunk as liquid	15% exhaled as vapor
40% from solid food	25% perspiration
10% from body processes	60% urine and feces

directly available for human use. The salty oceans and saline lakes hold 97.2 percent, so most of Earth's water is not **potable,** or drinkable. Another 2.1 percent lies frozen in polar glaciers. That means less than 0.7 percent of all Earth's water that is thawed and sufficiently salt-free is usable by humans (Figure 6.2A).

To put this in perspective, if Earth's water totaled 50 gallons, our usable freshwater supply would be only a teaspoonful! This tiny amount includes freshwater in every stream, river, lake, pond, puddle, and stored in the soil and rocks. Consumption by Earth's people of this few tenths of a percent has increased dramatically during this century, from about 141 cubic miles per year in 1900 to around 1000 cubic miles per year today (Table 6.2).

At present growth rates, water demand is doubling every 21 years. Already facing chronic water shortage are 80 countries, with 40 percent of the planet's population. Water will be the dominant international issue for water-short countries for the coming decades. Almost 150 of the world's 214 largest river

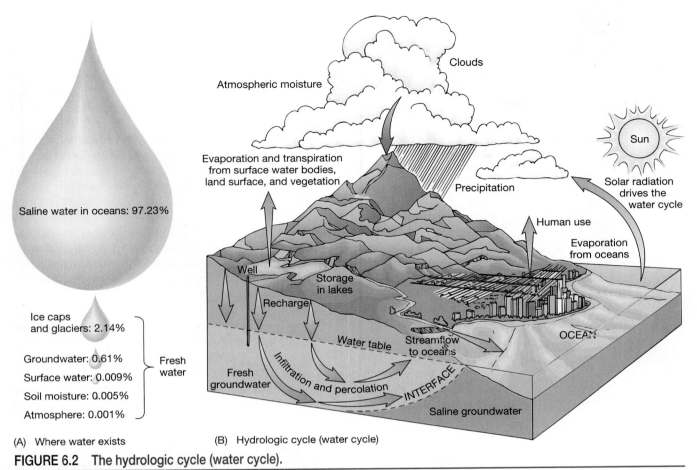

Saline water in oceans: 97.23%

Ice caps and glaciers: 2.14%

Groundwater: 0.61%
Surface water: 0.009%   } Fresh water
Soil moisture: 0.005%
Atmosphere: 0.001%

(A) Where water exists

Atmospheric moisture

Clouds

Evaporation and transpiration from surface water bodies, land surface, and vegetation

Precipitation

Sun

Solar radiation drives the water cycle

Human use

Evaporation from oceans

Well

Storage in lakes

Recharge

Water table

Streamflow to oceans

OCEAN

Fresh groundwater

Infiltration and percolation

INTERFACE

Saline groundwater

(B) Hydrologic cycle (water cycle)

FIGURE 6.2   The hydrologic cycle (water cycle).

(A) Freshwater comprises less than 3 percent of all water. More than 98 percent of freshwater is underground and must be pumped out for use. (B) How water travels through the hydrologic cycle. Human use for households and irrigating crops is the most important component to us, but is a trivial percentage of all water.

TABLE 6.2   Human Water Usage in the Twentieth Century

Overall usage will be about nine times greater in 2000 than in 1900. During the same 100-year period, world population has increased less than four times.

Annual Consumption (cubic miles)					Number of times greater from 1900 to 2000
Region	1900	1950	1990	2000	
Asia	99	206	586	800	8 times
N. America	19	69	174	191	10
Europe	9	23	133	162	18
Africa	10	13	56	76	8
S. America	3.6	14	36	52	14
Australia/ Oceania	0.5	2.4	9.1	11	22
Total cubic miles	141.1	327.4	994.1	1292	9 times (average)

systems are shared by two countries, and 50 are shared by three to ten countries. The main issue in future wars will be water rights.

The percentage of Earth's water that is in different environments (ocean, rivers, soil, and so on) varies little with time. But the water molecules themselves are continually shifting locations, as shown in Figure 6.2B. Water molecules evaporate from the surface of the vast ocean, rivers, lakes, and land. These molecules, now in the form of water vapor, are precipitated back to the surface (of both land and sea) as rain, snow, hail, sleet, frost, and dew.

Some of this water evaporates directly back into the atmosphere. Some soaks into the soil and is absorbed by plant roots; it eventually evaporates from leaf surfaces back into the atmosphere. Some water infiltrates into the rocks beneath the soil. Some runs off into streams and back to the ocean.

The name for this continual movement is the **hydrologic cycle** or **water cycle** (Figure 6.2B). It operates year-round, 24 hours a day, and has for several billion

FIGURE 6.3 "Waterfall," a lithograph by Dutch artist Maurits Escher, 1961.

In this artistic view of the water cycle, where does it start? Where does it end? *Source: Maurits Cornelius Escher. "Waterfall" 1961, Lithograph. © Cordon Art/Art Resource, NY.*

years. This cycle illustrates the interconnection among components of Earth's dynamic systems of air, soil, rock, and living things. The hydrologic cycle is as crucial to life's existence on Earth as is blood circulation in animals or sap circulation in trees. Another view of the water cycle is in Figure 6.3.

Water on Earth's Surface

Each year, more water precipitates onto the land than evaporates from it. About 22,600 cubic miles of water precipitates onto the land, but only 14,000 cubic miles of water evaporates, leaving 8600 cubic miles to drain back into the oceans. Our focus in this chapter is upon this 8600 cubic miles of water, for it is most of the water available to us. Some of this water keeps streams flowing or is stored in lakes. This is **surface water.** Some water sinks into the soil and underlying rocks, becoming stored in tiny spaces between rock grains or in fractures. This is **groundwater.**

Uneven Distribution of Surface Water

How much precipitation falls in each region of the United States? What problems are created by its uneven distribution? How does American society use—and abuse—the water? Let us examine the 470 cubic miles (1.4 trillion gallons) that runs off the continental United States.

The average U.S. precipitation is 30 inches per year, but its distribution is very uneven (Figure 6.4). The consequence of this is shown in Figure 6.5, which highlights water surplus and deficiency. The southeastern part of the country and part of the Pacific northwest have an overabundance compared to their needs, whereas most of the western U.S. has a shortage. The shortage may result from rapid evaporation or from rapid **runoff,** which is the downhill surface flow away from the site of the rainfall. Availability of surface water is summarized in this simple equation:

availability of surface water = precipitation – evaporation – surface runoff – infiltration into soil and rock

Areas with less than 20 inches of precipitation are marginal for agriculture and require supplemental water from irrigation. Areas with less than 10 inches of annual precipitation are barely suitable for human habitation, much less agriculture.

Water that runs off the land and into streams (runoff) reflects the amount of precipitation. High-volume precipitation is followed by high-volume runoff. (Some variations occur, reflecting differences in the rock exposed at the surface. For example, the Appalachian Mountains that trend northward from Alabama to New England contain many igneous and metamorphic rocks. In most cases, water cannot flow through these rocks, so runoff is high.)

The amounts of precipitation and stream runoff are yearly averages from government records that span many decades. But such averages can deceive. Everyone knows that variation exists around an average, and such variations often cause trouble when preparations are not made. Even areas where average precipitation exceeds 40 inches per year can suffer drought. This occurred during the summer of 1988 in the normally wet U.S. Southeast, when unusual weather patterns brought unusually scant rainfall. Conversely, rainfall that vastly exceeded the average brought devastating floods to the U.S. Midwest in 1993 (Chapter 7).

Overall, however, problems are more severe where precipitation is low. In a humid area, a 25 percent annual decrease (from 40 inches to 30 inches) is worrisome but unlikely to affect agriculture or human activities very seriously. But in a dry area, the same 25 percent decrease (but from 20 inches to 15 inches) will seriously affect agriculture. And a 25 percent decrease

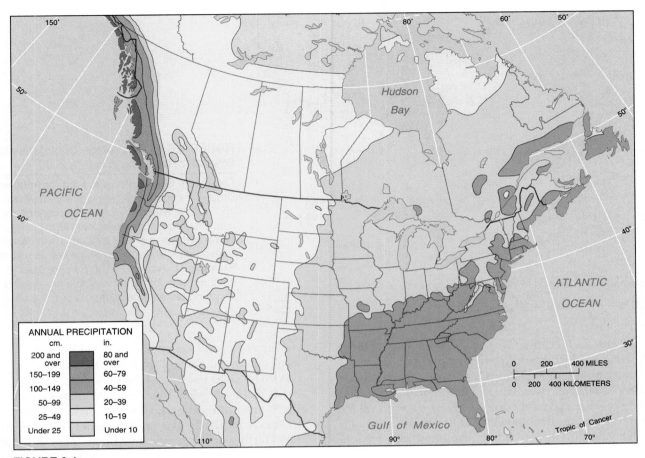

FIGURE 6.4

Annual precipitation in the United States. *Source: Robert W. Christopherson,* Geosystems, *2nd ed.,* Prentice Hall, 1994.

from 8 inches to 6 inches might be a calamity, requiring massive importation of water from a high-rainfall area some distance away, at great cost (Figure 6.6).

A prolonged drought affected most of California from 1987 to early 1993. Reservoirs dropped to their lowest levels in 15 years. Evaporation continued as always, but rainfall did not. The state legislature instituted water-use restrictions as the drought continued, despite strong resistance by the state's powerful agricultural interests.

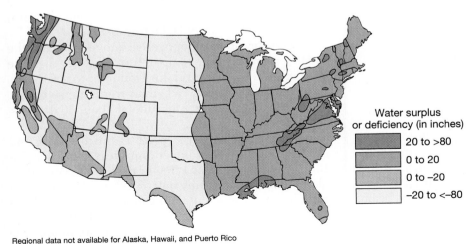

Regional data not available for Alaska, Hawaii, and Puerto Rico

FIGURE 6.5 Water surplus and deficiency in the United States.

Areas with less than 25–30 inches of precipitation per year have inadequate surface water. *Source: U.S. Geological Survey.*

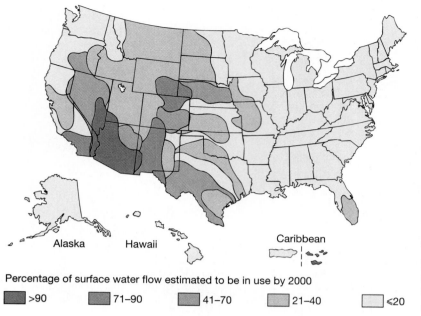

Percentage of surface water flow estimated to be in use by 2000

| >90 | 71–90 | 41–70 | 21–40 | ≤20 |

Alaska   Hawaii   Caribbean

FIGURE 6.6   Percentage of surface-water flow estimated to be in use by 2000.

Clearly the Southwest is headed for trouble. Droughts occur about once every 20 years and reduce water flows to 30 percent of average. Therefore, each year, only about one-third of the average runoff is available with certainty. Within a few years, most of the western United States will be above the 30 percent level of consumption, making inevitable severe and recurring water shortages. *Source: U.S. Water Resources Council.*

Nationwide, agriculture uses about 43 percent of all water withdrawals (from streams and from the ground), but neither California nor any other state has a comprehensive, coordinated plan to deal with water shortages that inevitably occur from time to time in the West. It is an unfortunate fact of life that governments at all levels tend toward screaming ambulance responses rather than practicing the preventive medicine of informed legislation. Nothing is done until the patient is near death, when panic sets in. Of course, this is not just a failure of elected officials to act; it reflects public opinion. How many people go to the doctor when they are feeling well?

What Controls the Distribution of Precipitation?

Three factors determine the pattern of precipitation on Earth: (1) the sun-driven circulation of the atmosphere; (2) distance from the ocean; and (3) changes in elevation of the land surface. Let us examine briefly each of these precipitation controls.

Circulation of the Atmosphere

Solar energy drives the motions in our atmosphere. However, the sun's radiation is received unevenly, imparting energy to the atmosphere unevenly. At the equator, the sun's rays strike Earth nearly perpendicularly, and thus most intensely, like a flashlight beam shining straight-ahead on a wall. But in higher latitudes toward the poles, the sun's rays arrive at a smaller angle, and are spread out, just like a flashlight beam when you shine it on a wall at an angle (Figure 6.7A). The solar radiation is spread over a wider area and is diffused. Also, there is a secondary effect that results from the difference in the angle at which the sun's rays hit Earth. The rays at higher latitudes must travel farther through the atmosphere to reach the surface, and during passage through the atmosphere some of the sun's heat is absorbed by the air. As a result, the amount of solar radiation reaching the ground at the equator is greater than elsewhere.

The daily input of solar energy drives the vertical atmospheric circulation, shown in Figure 6.7B. Warm air along the equator, being less dense, rises. It cools to the temperature of the air at higher altitude and moves laterally toward higher latitudes (poleward). The height to which the equatorial air rises is roughly 10 miles (50,000 feet). This altitude marks the boundary between the lower atmosphere in which we live (**troposphere**) and the upper atmosphere (**stratosphere**). This boundary is the **tropopause.**

The air at the tropopause moves toward the polar regions until its temperature is colder than the temperature of the air on the ground at that latitude. Cold air is denser (heavier) than warm air, so the air sinks. This happens at a latitude of about 30 degrees, as shown in the figure.

The air mass at the poles is much colder and denser than the air at lower latitudes. Thus, the altitude of the tropopause is lower than at the equator—about 5 miles lower, only half of its height at the equator. The cold, dense polar air cannot rise and so slides laterally toward lower latitudes (toward the equator). The boundary between the polar air mass and the equatorial air mass is called the **polar front.** Lobes of it move southward during U.S. winters, generating cold waves, storms, and even blizzards. An example is the blizzard that struck the northeastern United States in January, 1996.

Now, what does all this have to do with the global precipitation pattern? First, it causes rain along the equator: the rising air at the equator cools as it rises, and cold air cannot hold as much moisture as warm air. Air at 85°F can hold three times the amount of moisture of air at 52°F and six times the moisture of air at 34°F. So the air dumps most of its moisture within 5 degrees of the equator. This rain, combined with equatorial heat, permits the luxuriant vegetation of the rain forests (jungles). (These are the rain forests now under attack by slash-and-burn agriculture; see Chapter 4.)

Second, atmospheric circulation causes the world's deserts: at about 30 degrees latitude (Sahara desert, Australian desert), the cold air, having dumped most of its moisture, descends and warms as it sinks toward the warm ground surface. This air is generally dry, and warming increases its capacity to hold moisture. It spreads horizontally on the ground. The part that moves toward the equator grows warmer and dryer. Thus, the world's major desert belts occur between 30 and 20 degrees latitude. You can see these **latitudinal deserts** on the map of arid regions (Figure 6.8).

Distance from the Ocean

Not all deserts are caused by latitude. Permanent deserts exist in the interiors of continents because this land simply is too far from the ocean, the major source of water vapor. The farther from the ocean, the less moisture is available to fall on the land surface. The Gobi Desert in central Asia is a fine example of a desert resulting from being far from the ocean.

FIGURE 6.7   Solar radiation drives Earth's atmospheric circulation.

(A) Winter in the Northern Hemisphere (note tilt of Earth). Solar radiation is spread over a large area in higher latitudes, so that a unit of Earth's surface receives less radiation than at low latitudes. (B) Circulation of the troposphere (the lower atmosphere, where "weather" occurs). Note the average temperatures at the ground surface and at the top of the troposphere, the tropopause. The polar front forces its way farther southward during winter months than during summer months. The jet streams are rivers of air that meander around Earth in a west-to-east direction just below the tropopause. Their speeds can exceed 300 mph. *Source: World Climate News, World Meteorological Organization.*

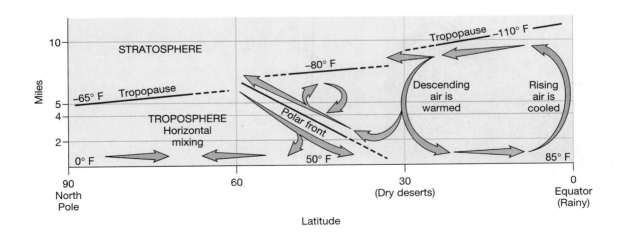

(A)

(B)

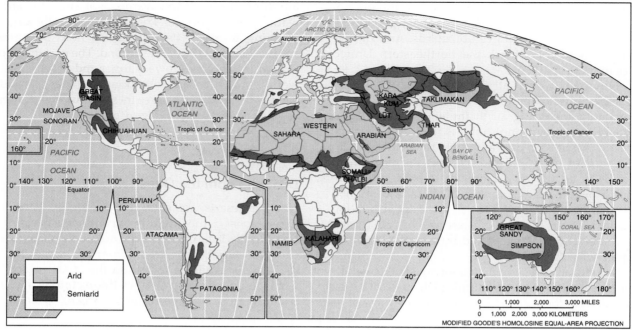

FIGURE 6.8 Earth's arid and semiarid areas.

Major deserts are named. *Source: Robert W. Christopherson*, Geosystems, 2nd ed., *Prentice Hall, 1994.*

Changes in Elevation of the Land Surface

Of particular importance in the western United States is aridity on the sheltered side (lee side) of mountain ranges. Figure 6.9 shows how this happens. As air moves landward from the ocean and encounters a mountain range, it is forced to rise. This cools the air, its moisture condenses, and precipitation falls. Hence, the windward side of coastal mountains is generally wet. Once the chilled air is pushed over the crest of the mountains, it sinks down the other side, warming as it goes. Having lost most of its moisture earlier, it has none to fall upon the sheltered side of the mountains, creating a **rain shadow,** or dry region.

The southwestern United States is a series of north–south mountain ranges, an area called the *Basin and Range Province,* shown in Figure 6.10. To the east, each basin is dryer than those in the west, because the

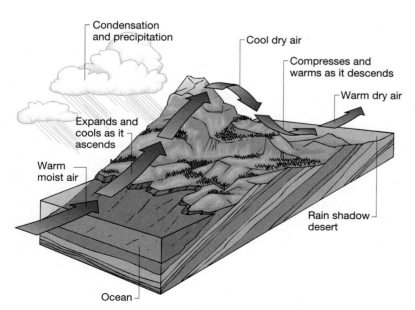

FIGURE 6.9   The orographic effect.

A mountain range can be wet on one side and dry on the other, creating a rain shadow.

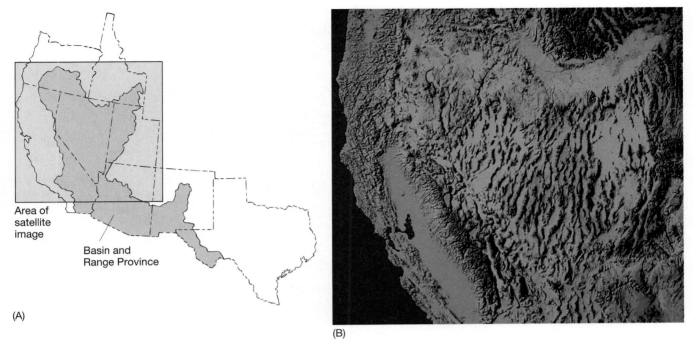

FIGURE 6.10 Northern part of the Basin and Range Province as seen by satellite.

The alternating basins and mountain ranges are the surface expression of the underlying faulted blocks. Moist air from the Pacific grows progressively dryer from repeated rise-and-fall cycles as it passes over the basins and ranges. *Satellite and computer image prepared for* Scientific American *by Thomas H. Jordan and J. Bernard Minster. Used by permission.*

repeated rise and fall of eastbound air makes it dryer with each up-and-down cycle. This has created the desert area in the American West shown in Figure 6.8.

Surface Water: The Problem of Sharing

People need freshwater (not salty seawater) for households (drinking, cooking, bathing), for irrigating agricultural crops, and for industry. We withdraw some freshwater from wells by drilling into the groundwater supply, but we also withdraw plenty of freshwater from surface-water sources: rivers, streams, and lakes.

This raises the question of ownership: who owns the water in a river, stream, or lake? If a few towns withdraw household water from a river, no one cares. But when large cities, irrigated farms, several factories, and an electrical power plant all compete for the same river water, ownership of that water suddenly becomes very important. How much water can farms upstream withdraw without infringing on the rights of private and industrial users downstream? How much water can industry and towns demand, when they depend on irrigated farms upstream to provide food? And what about the desert town 100 miles distant that wants to pipe some of the river's water to slake its thirst?

Such questions are the subject of continuing court battles, because the very survival of a city or an industry may depend on the outcome. Perhaps the most publicized example of such a controversy is the allocation of water from the Colorado River in the western United States (Figure 6.11A).

The Colorado is among the most heavily plumbed rivers in the world. Figure 6.11B shows how the Colorado's flow is regulated by 10 major dams. Colorado River water now irrigates about 2 million acres of farmland, provides water for the households of at least 21 million people, and generates nearly 12 billion kilowatt-hours of electricity annually.

The Colorado serves far more consumers than those who live along its course from Colorado and Wyoming into the Gulf of California in Mexico. Its waters fill swimming pools and sprinkle green lawns 250 miles away in Los Angeles, power neon lights in Las Vegas casinos, and irrigate thirsty crops in semiarid southern California, southern Arizona, and northern Mexico. The dams along the Colorado River store the river's water so that users can depend on a steady supply. Total storage capacity is vast, more than 2.6 quadrillion gallons.

Of course, the Colorado's flow varies with the seasons and each year's precipitation peculiarities. These irregularities are smoothed out by the dams, which create reservoirs to store water. During periods

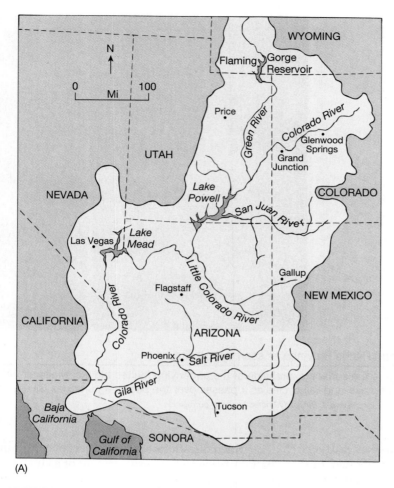

(A)

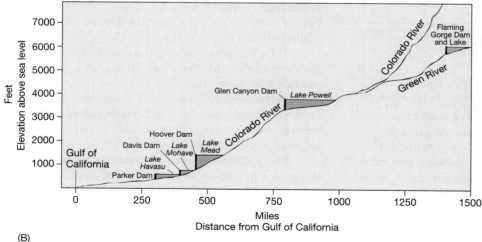

(B)

FIGURE 6.11   Two views of the Colorado River.

(A) Drainage area of the Colorado River. The water originates as precipitation and snowmelt in the highlands of western Colorado, western Wyoming, and northern Arizona (see Figure 6.4). It originally flowed into the Gulf of California, but so much of its water now is withdrawn that it no longer does so. (B) Profiles of the Green and Colorado Rivers show how they drop in elevation from their headwaters (at right) to sea level at the Gulf of California. Note locations of major dams and reservoirs. *Source: U.S. Bureau of Reclamation.*

of plentiful precipitation, most of the water is stored; during dry periods, this water is released as needed.

As Figure 6.12 shows, extensive aqueducts now channel the Colorado's water to the water-deficient areas of southern Arizona and California. Southern California also obtains water from wetter northern California through the California Water Project. Water from both the Colorado River and from northern California is used for agriculture and human consumption.

All of this sounds fine, except the Colorado does not always have enough water to satisfy both California and Arizona. Both southern California (Los Angeles, San Diego) and southern Arizona (Phoenix, Tucson) are fast-growing areas with fast-growing water demand. The

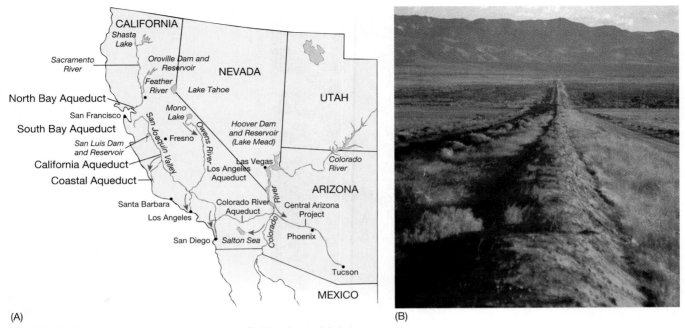

(A)   (B)

FIGURE 6.12   Overland piping of water in California and Arizona.

(A) The California Water Project and the Central Arizona Project provide large-scale transfers of water from water-rich drainage basins to water-poor ones. Arrows show general direction of water flow. (B) The Los Angeles Aqueduct pipeline, built in 1913, carries water hundreds of miles south from Mono Lake and the Owens River Valley. *Photo: W. J. Andrews.*

inevitable legal battles have reached the U.S. Supreme Court as both states claim rights to the limited water supply.

Adding to the water-shortage problem in the Los Angeles area is the resistance of people in northern California to provide more water from their region. They fear damage to fishing in the Sacramento River and a reduction in the natural flushing action that helps purge San Francisco Bay of industrial pollutants. (Which raises a question: Why not stop the pollution?)

Apparently, no permanent solution will satisfy everyone. The pie is small and cannot grow bigger, but the appetites are large and continually growing. This problem exists anywhere that too many people demand too much of a limited resource, and conflicts can arise between countries. Expect to hear about international "water wars" in your lifetime.

Groundwater: Stored Rainwater and Snowmelt

The ground beneath your feet contains water to some degree, stored in countless tiny pore spaces between grains of soil and rock. Drilling a well into this underground water supply is commonplace. In fact, groundwater from wells is used by 96 percent of U.S. rural

dwellers, and by numerous towns that do not have a large river handy.

But few users of groundwater think of how the water gets there. Where does it come from? How can wells continue to produce for centuries, despite continuing withdrawal of water? Why do some wells produce more than others, and why do some run dry?

The Nature of Groundwater

We use the term "solid as a rock." For example, a truck advertisement uses the phrase "like a rock" to promote the vehicle's durability. But in reality, the sedimentary rocks that dominate Earth's surface are full of tiny holes, called **pores.** Most pores have been in these rocks since they were deposited as small grains, produced by erosion of other rock.

All sandstones and many limestones originally were individual sand-size bits of quartz, feldspar, basalt, clam shells, and so on (Chapter 2, Sedimentary Rocks). Such grains are rounded or irregular in shape, and do not fit together neatly, like jigsaw puzzle pieces. Instead, they contact one another at random points around each grain. Plentiful spaces exist between them, and the amount of space is called the **porosity** of a rock (Figure 6.13).

Freshly deposited river or beach sand has porosities of 30 to 40 percent—in other words, the sand is

FIGURE 6.13 Fractured surface of an ancient quartz sandstone at high magnification.

Note the generally equal size of grains, about 0.01 inch or 0.2 millimeter long. Also note the abundant spacious pores among the grains. A small amount of clay is attached to the surface of many grains. *Photo: Chevron Oil Field Research Company.*

really 30 to 40 percent empty space. Although the grains later may become cemented, which reduces this percentage, porosities of sandstones are commonly 10 percent or greater. Most cementation does not completely fill the pores. Rainwater and water from melting snow enter these pores, accumulating to create nearly all of our groundwater.

Groundwater is not static, but constantly migrates through rocks, from pore to pore. It may move several feet per hour, or only inches per year, depending on the rock. It is replenished from surface rainfall and snowmelt, and, ultimately, it drains away underground, into streams, and into the sea. This replenishment of an aquifer is called **recharge.**

We cannot overstate the importance of groundwater. In the United States, the volume of water stored in rock pores underground, down to a few thousand feet, is 20 to 30 times greater than all the water in lakes, rivers, and streams combined. In addition, groundwater is free of sediment, and so needs no treatment before using. Unless polluted by human activities, groundwater generally is of good quality and safe to drink as is. It usually is cold and maintains a constant temperature in any local area. Approximately 800,000 water wells are drilled in the United States each year for domestic, commercial, and industrial use (Figure 6.14A).

Do you know the source of your drinking water, both at home and at school? There is a one-in-four chance that it is groundwater (Figure 6.14B). In 1985, the United States used nearly 338 billion gallons of water every day; about 22 percent of it was groundwater. In 1990, it increased to 25 percent. Groundwater wells supply about 37 percent of all "city water," about 96 percent of rural domestic supplies, and about 34 percent of water used in agriculture.

Were it not for this huge underground storage, many irrigation systems in arid and semi-arid regions of the United States could not continue. Agriculture and industry alike use groundwater because it usually is available at the point-of-use and does not require transportation over long distances.

How Do We Find Groundwater?

As we have seen, groundwater is present in pores of buried rocks, and the rocks most likely to have pores are sandstones and limestones, because they were deposited as irregularly shaped fragments. These two types of rocks, or their uncemented parent sediment, are present in the upper few hundred feet of Earth's crust over most of the world's land area. Hence, groundwater is present in most places. As a result, it is

(A)

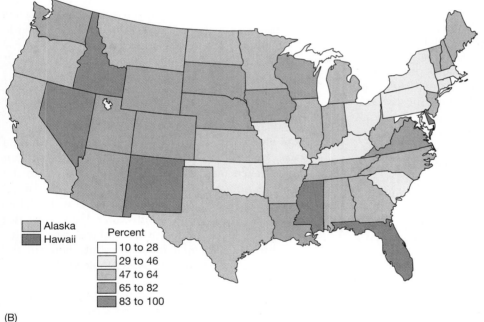
(B)

FIGURE 6.14   Two views of groundwater.

(A) This well gushes more than 200 gallons per minute, enough for about eighty people to take showers simultaneously. It was drilled as part of a water supply development project in south-eastern New York State. This area is known for low-yield wells, so geologists traced rock fractures to find water-bearing ones, with obvious success. This is an artesian well which flows without pumping. *Photo: Allison Kozak.*
(B) Percentage of the American people who depend on groundwater for household use. *Source: Environmental Protection Agency.*

Alaska
Hawaii
Percent
10 to 28
29 to 46
47 to 64
65 to 82
83 to 100

easier to find groundwater than not to find it, if you know a few basic geologic principles.

Two different methods are used, scientific and unscientific. They are totally unrelated, and their practitioners seldom interact. The two methods are those of **hydrogeologists** (geologists who are concerned with groundwater) and those of **dowsers,** or water-witchers.

A hydrogeologist studies rocks and maps, and uses the principles of hydrology to find a good place to drill a well. A dowser seeks groundwater by holding a forked stick (or wire) called a **divining rod** while walking over an area. The stick is supposed to deflect or twist downward of its own accord when the dowser passes over water.

A study of thousands of Australian wells drilled in the early 1900s revealed that sites suggested by a dowser produced water 85 percent of the time, and 70 percent of these wells produced more than 100 gallons per hour. Not bad for a technique most of us might laugh at. Many public-spirited dowsers do not charge for their services, which, combined with their success rate, explains the continued popularity of dowsing over the centuries.

But what is the success rate of hydrogeologists, based on the Australian records? They reached water 93 percent of the time and 83 percent of the wells produced at least 100 gallons per hour. Undeniably better, but would you choose a free dowser with an 85 percent success rate or a not-so-free consulting hydrogeologist with a 93 percent success rate? If the dowser and hydrogeologist charged equally, you might choose the hydrogeologist. But price makes the decision less clear-cut.

A third alternative is to combine the information in this chapter with the experience of your neighbors and pick a well site yourself. Your chances of hitting a winner are pretty good if you drill in an area underlain by sands or sandstones or limestones that are known for good wells.

Aquifers: Rock and Sediment Layers that Store Water

A layer of rock or sediment that yields water in amounts large enough to be useful is an **aquifer** (Latin for "water carrier"). The most common aquifers are porous sand, porous sandstone, and porous limestone. However, igneous and metamorphic rocks, which are mostly nonporous, often are fractured. The fractures provide an alternative to pores for groundwater movement and storage, creating good aquifers in some areas. Poor aquifers lack pore space, such as clay, shale, and unfractured igneous rock. To qualify as an aquifer, a layer of rock or sediment must contain a useful volume of water and be able to transmit it to a well.

Figure 6.15 shows how pore space can vary in rocks. The greater the percentage of pore space, and the more the pores are interconnected, the better water can permeate an aquifer (**permeability**), and the faster it will migrate through.

Another factor in the performance of an aquifer is the nature of rock layers above and below it. Figure 6.16 shows aquifers in two situations, unconfined and confined:

- In an **unconfined aquifer,** the water-bearing layer extends all the way to the ground surface. The water is under no pressure and flows wherever gravity leads it.
- In a **confined aquifer,** the water-bearing rock is capped by an impermeable confining layer. This impermeable layer is called an **aquiclude** if it completely blocks (occludes) the flow of water upward, or an **aquitard** if poor permeability retards water

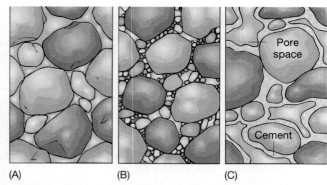

(A)   (B)   (C)

FIGURE 6.15   Porosity and permeability in different sandstones.

(A) has 30 percent pore space and excellent permeability (large pores). (B) shows small grains clogging the pores, reducing porosity to about 15 percent and significantly reducing permeability (due to smaller pores). (C) shows decrease in porosity resulting from partial cementation.

flow rather than occluding it completely. (A leaky aquifer is capped by an aquitard.) Commonly, an aquiclude is a mud or unfractured shale, a rock composed of tightly compacted flat clay flakes. It is impermeable to water. However, the aquiclude also may be a tightly cemented sandstone or limestone, or a layer of unfractured igneous rock such as a buried volcanic layer.

Water Table

Look at Figure 6.16. If you start digging anywhere at the surface, you will unearth damp soil, but it will not become soaked (saturated) until you get down to the water table. Above the water table, pore spaces are filled with air or part-air/part-water. Below the water table, soil and rock pores are filled with water. Thus, the **water table** is the upper boundary of soil and rock that is saturated with water. A well must penetrate beneath the water table. (Compare the two water-table wells; the dry well does not reach the water table.)

As Figure 6.16 shows, water in the unconfined aquifer does not rise to the ground surface; it can rise only to the water table. A pump is needed to push the water the rest of the way to the surface. This is typical of a well in an unconfined aquifer.

Artesian Water

Some wells flow at the surface, all by themselves, without pumping. Is this because the water table is higher than the well? Usually not. Wells that flow by themselves are under natural pressure to do so. The pressure exists because the well is drilled into a confined aquifer. In Figure 6.16, imagine the confined aquifer as

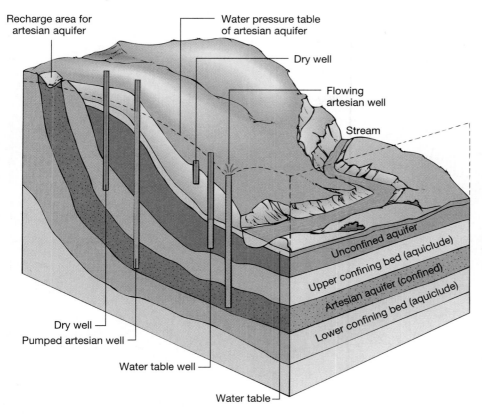

Recharge area for artesian aquifer

Water pressure table of artesian aquifer

Dry well

Flowing artesian well

Stream

Unconfined aquifer

Upper confining bed (aquiclude)

Artesian aquifer (confined)

Lower confining bed (aquiclude)

Dry well

Pumped artesian well

Water table well

Water table

FIGURE 6.16 Cross-section showing an unconfined and a confined aquifer.

Artesian well on the hill must be pumped to raise the water to the surface but the well in the valley flows without help. The unconfined aquifer must be pumped. Of course, water is available from the stream as well. *After* Groundwater: Issues and Answers, *American Institute of Professional Geologists, 1983.*

a huge pipe, with its open end at the upper left (the recharge area). Rainwater and snowmelt enter the recharge area and fill the confined aquifer. Because the recharge area is at a high elevation, water lower down in the aquifer is under pressure from the weight of the water above.

Any well that penetrates lower down on this confined aquifer will be under considerable pressure and will shoot up the well *to the height of the recharge area* (almost). The height to which water in such a well will rise is called the **potentiometric surface,** because it is the potential height to which water can rise in this situation. (Water in an unconfined aquifer does not rise because it is not under such pressure. An unconfined aquifer is not like a pipe; it is more like a wet sponge.)

Water from a confined aquifer is called **artesian water,** named for the French province (Artois) where this phenomenon is common and was first recognized. In Figure 6.16, artesian well number 2 flows at the surface without pumping, because the ground surface is lower than the potentiometric surface. (Such a well may even fountain—see Figure 6.14A.) Artesian well number 1 extends above the potentiometric surface, so, although water rises to this potential level inside the well, a pump must be used to push the water to the top of the well pipe.

The physical principle you just saw—that water will rise nearly to the height of its recharge site (artesian

flow)—is used in many communities to supply water at good pressure to homes. The "recharge site" is the familiar water tower, shown in Figure 6.17. The higher the tower, the greater the water pressure in city lines. In times of drought, the level of water in the tower decreases, and water pressure decreases in the homes being serviced. The city then may ask people to conserve water until the reservoir is replenished.

How Deep to Drill the Well?

How deep must a well be drilled to reach an aquifer? The answer is different for each well, because it depends on how deep the aquifer happens to be (Figure 6.16). In a valley, where the water table of an unconfined aquifer is near the surface—or even *at* the surface as a swamp, wetland, or stream—a 10-foot deep well may suffice. But on a high hilltop, a thousand-foot well may not be deep enough to reach an aquifer. The depth to the aquifer is important, as the cost of drilling a water well averages about $15 per foot in the United States. Thus, a common 100-foot well would cost around $1500 to drill, whereas a thousand-footer might exceed one's ability to pay the driller.

All sedimentary rocks begin as horizontal layers of sediment, deposited on the seafloor, or on a river's floodplain, or in the bottom of a lake. Even sediments deposited on slopes slide downward to

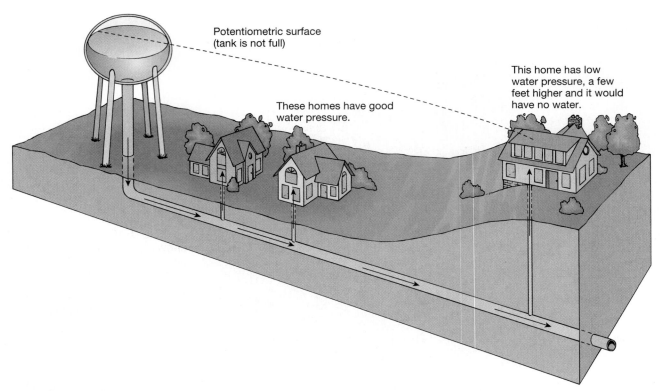

Potentiometric surface
(tank is not full)

This home has low
water pressure, a few
feet higher and it would
have no water.

These homes have good
water pressure.

FIGURE 6.17 An artificial artesian well system found in many American communities: a water tower.

The water will rise in the houses to the level in the tower. Water is pumped into the tower from a source such as a well, stream, or river.

rest on flat surfaces, where they eventually become cemented to form rock. However, when seen in outcrop, many rocks no longer are horizontal, but are tilted. They have been deformed after burial.

Figure 6.18 shows sedimentary rocks that have been tilted and now **dip** rather steeply. When drilling a water well in such an area, the driller needs to know how the rocks lie underground. You can see that a well drilled near the aquifer's outcrop at upper left will reach water at a shallower depth than a well drilled downdip. If the angle at which an aquifer dips downward into the ground can be measured, then the depth to the aquifer at any point on the ground surface can be determined easily.

Wells should be drilled into the lower part of the aquifer, so the well will not immediately go dry. Few drilled wells actually go dry; rather, they stop yielding water because the pump intake was not placed deep enough to allow for even a slight decline in water level.

The maximum depth to which a water well can be drilled depends on more than cost. As one drills deeper, groundwater generally becomes saltier because temperature increases with depth and minerals are more soluble at higher temperatures. The dissolved minerals are the "salt" in salty water. At depths greater than a few hundred feet, water is likely to be too salty for most uses,

such as drinking or irrigation. An expensive solution to the saltiness is desalination, but this is seldom done (see "Desalination of Seawater" later in this chapter).

So, how deep to drill the well? Study the rocks. Contact a hydrogeologist (try a local college geology or civil engineering department, or the *Yellow Pages* under Geologists).

Cone of Depression

The rate of withdrawal from a well normally is greater than the rate at which water can flow through the aquifer. As a result, the water table drops rapidly immediately around the well bore. This creates a circular depression in the water table called a **cone of depression** (Figure 6.19).

If multiple wells tap the aquifer, a cone of depression will form around each well. As withdrawal of water continues, each cone of depression enlarges until they merge, causing an overall lowering of the water table. If a well is not drilled deep enough below the water table, the well soon may go dry as the cone of depression around it deepens. If this happens, the driller must return to the well and drill deeper.

Thus, if an aquifer sandstone is 100 feet thick, the well should be drilled 90 feet into it rather than 10 feet.

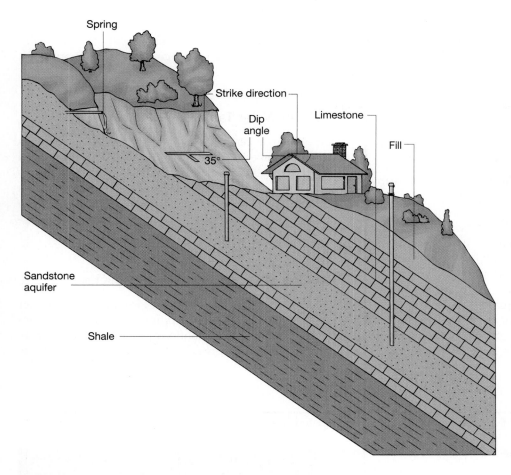

Spring
Strike direction
Dip angle
Limestone
Fill
35°
Sandstone aquifer
Shale

FIGURE 6.18 The strike and dip of rock layers are important to understanding underground water movement.

Strike is the compass trend of a rock layer: it is the compass direction (for example, north–south) of the intersection formed by the rock layer and a horizontal plane. *Dip* is the downward slope of the rock layer, measured perpendicular to the strike. The standard **T** symbol for strike and dip is shown on the ground surface. The long line is the strike; the short line is the dip, and the dip angle is shown. The strike and dip of the house roof shown are the same as the rock layers.

Don't congratulate yourself on your shrewd "saving" of 80 feet of drilling cost for the shallower well. You may pay for it later in waterless times and additional drilling cost for a second visit from a smug driller who knew better.

Springs, Oases, and Swamps

The drilling of wells to obtain groundwater has been practiced for thousands of years. The evidence is in ancient writings by tribes in the Middle East, the cradle of

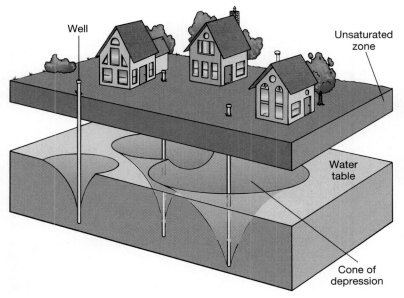

FIGURE 6.19 Cones of depression in the water table under three homes, caused by pumping the water wells.

The rate of pumping has been so high that the cones have partly merged. *Source: U.S. Geological Society.*

Well
Unsaturated zone
Water table
Cone of depression

Western civilization. These peoples no doubt knew the fundamentals of hydrogeology, for without knowledge of how to find water, life would have been even more precarious than it was. Two essential pieces of information they possessed were the locations of springs and oases.

A **spring** is a natural flow of groundwater from a rock surface. The most obvious springs emerge on a hillside (Figure 6.20). Their uninterrupted flow indicates they are emerging below the local water table, where rock pores are saturated. Were this not the case, the flow would be intermittent, flowing only during seasonal precipitation.

An **oasis** is a desert depression where the water table intersects the desert floor, so that water oozes out. Commonly, the water table may be 10 feet or deeper below the general level of the sandy desert floor, but it becomes exposed when wind scours a depression in the sand.

In humid climates, the equivalent of an oasis is a **swamp.** In both arid and humid climates, the intersection of the water table with the ground surface permits the growth of vegetation. In the desert, this vegetation

is a few palm or date trees. In the rainy climate, luxuriant vegetation adorns the swamp, which eventually may be transformed into a coal deposit (Chapter 14, Fossil Fuels).

Perched Water Table

Sometimes a secondary aquifer and water table is perched above the main one (Figure 6.21). A **perched aquifer** usually is smaller and harder to locate, either by a dowser or a hydrogeologist. Perched aquifers commonly do not reach the ground surface and are unconfined. When one does intersect the ground surface on a hillside, a spring appears.

Relation of Groundwater and Streams

Water that leaves an aquifer at the ground surface—either on its own (artesian) or by pumping—must be replenished if the aquifer is to continue producing useful water. In humid areas, recharge occurs through the ground sur-

FIGURE 6.20

Springs fed by groundwater emerging along bedding planes have created spectacular frozen waterfalls along this road cut.
Source: John Gerlach/Visuals Unlimited.

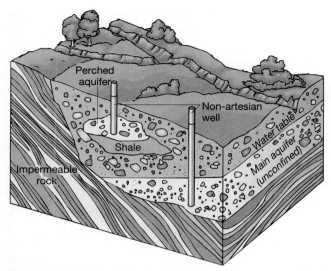

FIGURE 6.21 **A perched aquifer overlying an unconfined aquifer.**

Both aquifers are underlain by impermeable aquicludes and overlain by permeable sand.

face that lies between streams, wherever rainfall and runoff can reach the aquifer rock. In these humid regions, water tends to flow into streams because the recharge area is above the stream, as shown in Figure 6.22A.

In dry regions, such as much of western United States, recharge conditions are more complex. Most recharge occurs in mountain ranges, where rainfall and snowmelt are more available. In dry areas, the water table of an unconfined aquifer is far below the ground surface. As a result, streamwater itself helps recharge aquifers by draining downward to raise the level of the water table (Figure 6.22B).

Aquifer Characteristics

Environmental geologists are concerned with how groundwater moves, because it is so easily contaminated. For example, oil, gasoline, industrial chemicals, sewage, and fertilizer all enter the soil and rock with ease, and can contaminate groundwater locally.

For any location, environmental geologists need to understand the entire groundwater system. They need to know:

1. *Aquifer recharge sites.* For an *unconfined* aquifer, this may be the entire overlying surface, through which rainwater, snowmelt, and pollutants infiltrate and percolate down to the water table. For a *confined* (artesian) aquifer, the recharge area(s) may be in distant highlands or mountains.
2. *Water table.* How far below the surface does it lie?
3. *Water direction and velocity.* Which way does water flow in the aquifer, and how fast? To an investigator who is tracing a pollutant spill, this indicates how fast the contaminant might move, and in which direction.
4. *Discharge (water volume).* How much water are we dealing with? **Discharge** is the volume of water or saline passing a point per unit of time.

We consider these aspects below. (The specific problems of water pollution are examined in Chapters 8 and 9.)

How Fast Does Groundwater Flow?

How fast does groundwater flow? Rates varies widely, from feet per minute to feet per decade. Flow is faster in rocks with greater permeability and greater slope (dip).

FIGURE 6.22 **The relationship between surface water and groundwater is different in wet and dry climates.**

The wet-climate stream is fed continuously by a water-table aquifer. In contrast, the dry-climate stream works the opposite: it feeds water to adjacent areas. This is characteristic of streams that flow through desert areas (like the Colorado and Nile Rivers).

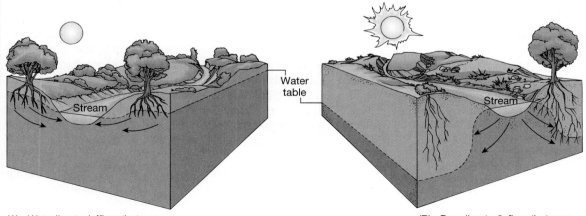

(A) Wet climate (effluent) stream

(B) Dry climate (influent) stream

The term used by hydrogeologists and environmental scientists to describe how easily water will flow through a rock is **hydraulic conductivity,** which simply means water conduction by rocks. Hydraulic conductivity is greater in coarser-grained rocks. For example, it is dramatically greater in a gravel or conglomerate than in a sand or sandstone. This is true because, as grain size increases, so does the size of the pore spaces among the grains.

Also, hydraulic conductivity is greater in deposits of clastics (rock fragments) that have a narrow range of grain sizes than in deposits with a wide range. For example, hydraulic conductivity is greater in a beach sand (Figure 6.13) than in a mixture of sand and clay mud. The reason is that the clay flakes are very small relative to the sand grains, so they settle among the sand grains and clog the pores. The slope of the water table is important because water flows faster on a steeper slope than on a gentler one, as is true of water at the surface.

Calculating Aquifer Discharge (Darcy's Law)

In 1856, French engineer Henri Darcy first recognized the relationship among the factors that control an aquifer's discharge—permeability, slope, and aquifer area. He wrote this relation, known as **Darcy's Law:**

Aquifer discharge (Q)	=	hydraulic conductivity (permeability of aquifer) (K)	×	slope of water table (dh/dl)	×	aquifer cross-sectional area (A)

The slope of the water table is determined by dividing dh, the difference in height of the water table at two points along the direction of flow, by dl, the horizontal distance between the two points.

So, written as a formula, Darcy's Law is:

$$Q = K \frac{dh}{dl} A$$

Using this formula, we can calculate the volume of water in a water-bearing rock unit. For example, suppose a confined aquifer is 60 feet thick and 5 miles wide. Wells have been drilled 1 mile apart in the direction of flow. The level of water in one well is 150 feet and the other is 140 feet, which defines the slope of the water table. The hydraulic conductivity of the aquifer is 4.5 feet/day. How much water will flow through the aquifer each day?

$$Q = K \frac{dh}{dl} A \text{ (which is aquifer width} \times \text{aquifer thickness)}$$

Plugging in the numbers,

$$Q = (4.5 \text{ ft/day}) \frac{150 \text{ ft} - 140 \text{ ft}}{5280 \text{ ft}} (60 \text{ ft}) (26,400 \text{ ft})$$

$$Q = 13,600 \text{ cubic feet/day.}$$

Such a discharge would cover a 300-foot-by-100-foot football field standing about 6 inches deep in water. Or, it would fill a 1700-square-foot house with a standard 8-foot-high ceiling.

How Fast Can an Aquifer Recharge?

Rates of aquifer recharge vary widely, from essentially zero in desert areas to more than 1 million gallons per square mile each day in humid rural areas that are underlain by very permeable soils and rocks. If replenishment is consistently slower than discharge, the aquifer will dry up.

One important difference between an aquifer's recharge area and discharge area is that the recharge area always is much larger than the discharge area, which normally is restricted to water wells. However, too many wells pumping too fast can deplete even the largest and most heavily recharged areas. This is, in fact, occurring in some heavily urbanized and heavily irrigated parts of the United States—Houston, New Orleans, Phoenix and Tucson in Arizona, the San Joaquin Valley and Long Beach in California.

How Much Water Can an Aquifer Provide?

An aquifer's **storage capacity** is simply the volume of pore space in it. Normally, pore spaces are smaller than the diameters of the grains that make up the rock (Figures 6.13 and 6.15). Despite their small size, pores in sedimentary rocks can be so numerous as to hold a surprising volume of fluid.

Consider, for example, a sandstone layer 10 feet thick that extends over an area of only 1 square mile. Simple arithmetic tells us that the volume of this rock is 278,784,000 cubic feet (1 square mile = 5280 feet × 5280 feet × 10 feet thick = 278,784,000 cubic feet). If it contains 10 percent pore space, the volume of pore space is one-tenth, or 27,878,400 cubic feet. One cubic foot of water is 7.5 gallons, so the pores in this layer of sandstone can hold more than 200 million gallons of water (27,878,400 cubic feet × 7.5 gallons/cubic foot). The average American family uses about 500 gallons per day for bathing, cooking, and drinking, so 200 million gallons of water would supply the household needs of a city of 400,000 families for one day.

This calculation is somewhat deceiving, because it assumes that all the water in the sandstone can be withdrawn, and quickly. In fact, this "bank" cannot be completely robbed of its deposits. Some of the water adheres tightly to the surfaces of the grains (remember the hydrogen bonds described in Chapter 2?) and cannot be pried loose by pumping. This amount of unavailable water is not trivial; it may be 10 percent to 40 percent of the total water in the aquifer, depending on grain size, clay mineral content, and other factors.

Case Study: The Ogallala Aquifer

The largest and best-studied aquifer in the United States is the Ogallala Formation. Since early this century, it has been a major water supplier to the midcontinental United States. Today, it irrigates more than 14 million acres of farmland. The Ogallala extends from South Dakota to Texas (Figure 6.23). It averages 200 feet thick and has a maximum thickness of about 1400 feet. The Ogallala is overlain only by modern sands from rivers and sand in wind-blown dunes.

The Ogallala Formation is composed of sandstone and gravel. These are partly cemented channel deposits of sediment-charged streams that flowed eastward from the Rocky Mountains 10 to 20 million years ago.

The sediments were deposited in east-west–trending valleys, carved by streams into pre-Ogallala rocks.

The Ogallala aquifer's area exceeds 170,000 square miles. It holds more than 70 quadrillion gallons of water (70,000,000,000,000,000 gallons). The aquifer is recharged from an annual rainfall of only 16 to 20 inches, but it had millions of years to fill, undisturbed. Today, this area has developed into one of the most important agricultural regions in the United States—enabled purely by irrigation with Ogallala water. The Ogallala serves an area that produces about 25 percent of U.S. feed-grain exports and 40 percent of wheat, flour, and cotton exports.

The aquifer can yield as much as 1000 gallons per minute, 24 hours a day. But recharge is very slow because

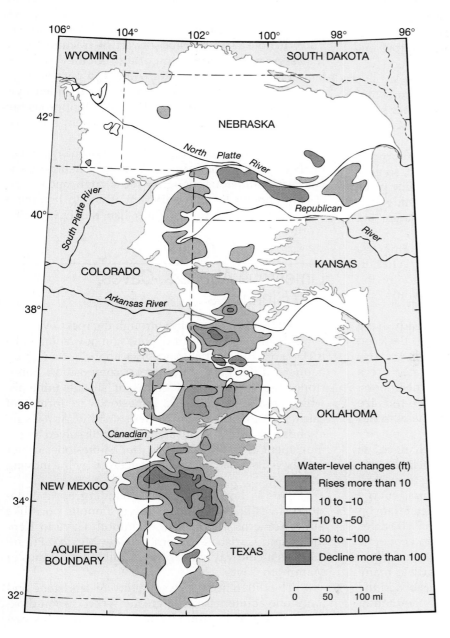

FIGURE 6.23 The Ogallala or High Plains aquifer, largest in the U.S.

Shown are water-level changes from its condition prior to development until 1980. Lowering of the water table has continued since then. *Source: U.S. Geological Survey.*

Water-level changes (ft)
- Rises more than 10
- 10 to −10
- −10 to −50
- −50 to −100
- Decline more than 100

of the low annual rainfall. Without Ogallala water, agricultural production would drop to a third of its present volume.

The bad news, of course, is that withdrawals of Ogallala water from the thousands of wells that tap it greatly exceed the amount of recharge, by up to twenty times in some areas. As a result, the water table is dropping rapidly (Figure 6.23). To date, only about 5 percent of the total groundwater resource has been used up, but water levels have declined 30 to 60 feet in large areas of Texas. As the water table lowers, well yields decrease and energy cost to pump the water goes up.

If present usage continues, the Ogallala will be effectively dry within a few decades, with disastrous effects on the economy of a large area of the United States. Some areas of Texas and New Mexico already have reached the point at which irrigation from the Ogallala aquifer no longer is practical. Our present ability to irrigate at low cost will not last forever.

Governmental Response. Existing governmental policies give farmers no incentive to use water sparingly. An example is corn, a very "thirsty" crop. Current federal policy provides price support to corn growers. This means that, should the market price drop, farmers will be paid the difference by the government, so they will avoid financial loss. Also, federal tax policy gives tax breaks to farmers who use pumped groundwater, and those who use more groundwater get larger tax breaks.

If you were President or in the U.S. Congress, what would you propose to do about this? What do you think the thousands of farmers served by the Ogallala and other overdrawn aquifers would say about your proposals?

Farmers in the Texas panhandle already have tasted the economic effect of overpumping. Between 1940 and 1959, before the Ogallala water table was significantly lowered by overpumping, irrigated areas flourished compared to their dryland neighbors. Irrigators had large increases in the number and size of farms and had higher incomes than those who did not irrigate.

But in the late 1970s, the situation changed, as groundwater levels in the irrigated areas fell sharply. Between 1978 and 1984, the cost of pumping from ever-greater depths drove up the average annual energy cost $7800 more on farms using Ogallala water than on dryland farms. Irrigation was halted on 367,000 acres in the irrigated area, and the area slid into economic decline. Farm income fell 18 percent, population 3 percent, and total employment 7 percent. Nearly as many providers of agricultural services lost their jobs in the period as farmworkers did, a ripple effect common in an integrated farm economy. Dryland farms adjacent to the irrigated ones, however, showed healthy growth during the same period.

Fractured Aquifers

Nearly all groundwater is produced from the pore spaces among rock grains. But some important underground reservoirs exist in rocks that lack pores large enough to permit the passage of water. These rocks may be sandstones or limestones in which the original pores have become completely plugged by cements, or they may be rocks that never had any pores, like granite. Yet, such rocks can become aquifers if they are fractured.

Rocks are brittle, and when they are squeezed by forces in Earth's crust, they break. If you look at rock where it is exposed in a hillside or highway road cut, you will note a large number of vertical fractures. They may seem narrow and unimportant, but they can transmit a large amount of water. Water can pass through openings too small to be seen by the unaided eye. Rocks such as granite that are normally nonporous can be important unconfined aquifers when fractured (Figure 6.24).

Flows of lava commonly develop shrinkage cracks on their upper surfaces during cooling. Sometimes these cracks are numerous enough, wide enough, and extend deeply enough to hold sufficient water to become a water source in rainy climates like the U.S. Pacific Northwest.

Limestone Aquifers, Caves, and Sinkholes

Does all this water running through the rocks ever dissolve them? Yes and no; it depends on how soluble the rocks are. Sandstones and limestones, often good aquifers, differ greatly in mineral composition. Sandstones are composed of grains of silicate minerals, which are not very soluble. Limestones are formed of calcite, which has a much greater solubility. As a result, a sandstone aquifer is not significantly dissolved by the movement of water through it, but a limestone aquifer may have its innards largely eaten away by the moving water.

This is how thousands of underground limestone caves are formed, including famous ones like Luray Caverns in Virginia, Mammoth Cave in Kentucky, and Carlsbad Cavern in New Mexico. Figure 6.25A is a cutaway view of the soluble limestones of Mammoth Cave.

The difference in the solubility of sandstone and limestone aquifers is made greater when carbon dioxide is dissolved in groundwater. This forms weak car-

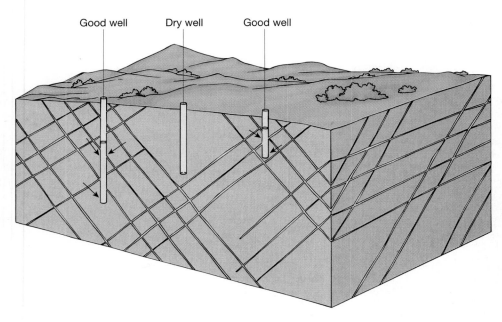

Good well   Dry well   Good well

FIGURE 6.24 Aquifers can occur in nonporous rocks such as granite and basalt, if they are fractured.

Wells usually must intersect a large number of fractures to provide an adequate water supply for a community.

bonic acid. Calcite solubility is greatly increased by even a small addition of acid, but the solubility of quartz remains unchanged. The carbon dioxide is acquired when rainwater passes through CO_2- enriched air in the soil:

$$CaCO_3 + H_2O + CO_2 \rightarrow Ca^{++} + 2HCO_3^-$$

| calcite (limestone) | from rain | carbon dioxide from air in soil | calcium ion in water | bicarbonate ion in water |

As water passes through the soil, it dissolves carbon dioxide gas that is released by dead and decaying plants. This water is able to dissolve more limestone (calcite) than otherwise would be possible. Almost all commercial caverns have formed, and continue to be formed, by this phenomenon.

The process just described occurs when a cave is filled with water. But what happens when CO_2 charged water enters an open cave whose air exchanges freely with the air above ground? Carbon dioxide in the groundwater dripping from the cave roof releases its dissolved CO_2 to the air. In other words, the chemical reaction shown above is reversed: the arrow that shows the direction of action is flipped, carbon dioxide is released into the air, and a tiny bit of calcium carbonate is deposited.

Over many years, those tiny deposits of calcium carbonate build up, forming the beautiful stalactites and stalagmites that fascinate visitors to caves. All result from the release of carbon dioxide from dripping water as it enters the cave from above.

In areas where caves develop, much of the stream drainage may be underground. Streams may flow for some distance at the ground surface, then plunge downward into hidden passageways and continue out of view, perhaps to reappear elsewhere as a spring. Such *disappearing streams* are common in areas underlain by highly soluble limestone.

One hears tales of "underground lakes or streams." But the only true underground lakes and streams are the "disappearing streams" just described in cave country. Remember that when we refer to the "water reservoir" underground, we almost always mean an aquifer formed of rock or sediment with saturated pores, not an underground lake or stream.

Sinkholes

In areas underlain by limestone—for example, most of Florida—the continual dissolution of the limestone by groundwater eventually leads to collapse of the surface. This forms **sinkholes** (Figure 6.26). The collapsed area may contain only soil, or soil and a thin layer of the bedrock beneath it. Sinkholes occur in many areas of the United States, especially in high-rainfall areas where limestone bedrock is particularly abundant (Figure 6.27). For example, southern Indiana may have up to 300,000 sinkholes. The lumpy, irregular ground surface that results over time is called **karst topography,** after a region in Slovenia where it was first described and is very widely developed.

Dangers exist for builders in areas underlain by limestone. An example is the region near Easton, Pennsylvania, on the eastern side of the Appalachian Mountains. The Pennsylvania Geological Survey estimates that more than 5200 sinkholes exist between Easton and the Maryland state line, only 75 miles to the south. That is an average of about 70 per mile, or one sinkhole each 75 feet. Sinkholes have claimed sections of Interstate

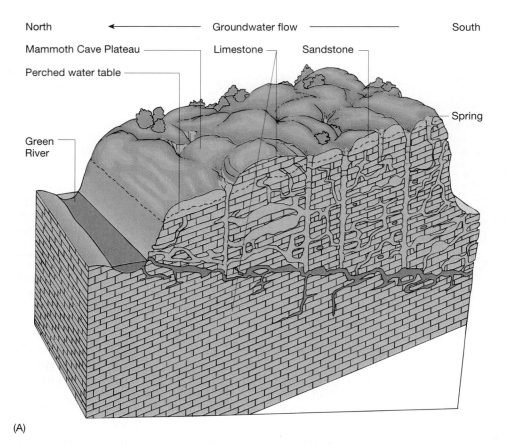

North ← Groundwater flow South

Mammoth Cave Plateau — Limestone Sandstone
Perched water table
Spring
Green River

(A)

(B)

FIGURE 6.25 Hydrology of Mammoth Cave in Kentucky.

(A) Cross-section through the Mammoth Cave Plateau of Kentucky. This huge limestone aquifer is composed of several limestone layers. Several hundred million years ago, the layers were deposited on the seafloor. Then they became elevated and tilted, so they now dip northward. Groundwater dissolved some of the limestone, forming the cavernous underground drainage shown. Groundwater flow is from south to north. (B) Tubular passage forming at or near the water table in the lowest level of Kentucky's Mammoth Cave. Periodic floods fill the entire passage with water. Note the banks of mud deposited after the high flows subsided. The commercial part of the cave used by visitors lies above the water table shown in the photo.
Photo courtesy A. N. Palmer.

Highway 78, numerous homes, and even a few lives in the past few years.

The problem is so severe that many municipalities are adopting "limestone ordinances" that require developers to search for potential sinkholes before development can begin. These ordinances may increase the cost of home construction by $5000 to $7000 and are not popular with builders. The up-front

FIGURE 6.26 A Florida sinkhole.

Removal of ground support by dissolution of underground limestone causes the sudden development of a sinkhole, such as this one in 1991, which consumed a home in the Florida town of Frostproof. *Photo: St. Petersburg Times/Gamma Liaison, Inc.*

cost to developers can be large enough to make profitable construction impossible. An added problem is that locating underground caves (potential sinkholes) is a difficult task when no visible surface evidence exists. Quoting a member of a regional planning commission, "You can do everything right and still end up with a sinkhole, or you can do everything wrong and have no sinkholes. It's just a matter of shifting the odds."

Within the last few years, development of **ground-penetrating radar** has shifted the odds toward detecting concealed sinkholes. Like conventional radar,

FIGURE 6.27

Areas susceptible to sinkholes and other types of subsidence. *Source: U.S. Geological Survey.*

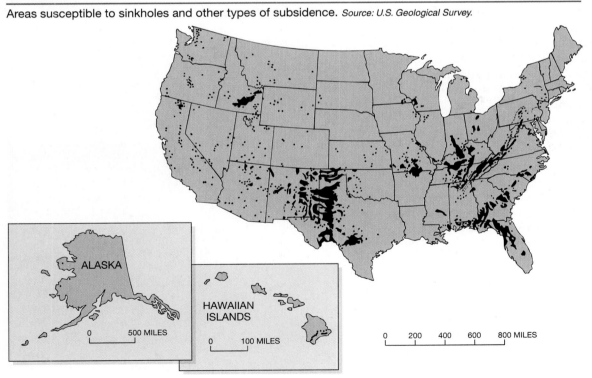

ground-penetrating radar transmits pulses of high-frequency radio energy and then watches for reflections. However, the signal is directed down into the ground to spot sinkholes, rather than into the air to spot aircraft or tornadoes, as in conventional radar. The cart-mounted radar equipment is towed slowly across the study area, continuously recording a strip chart of reflections from depths down to 30 feet. Because radar reflections differ among materials, the chart reveals different layers of rock and sediment underground. Analysis of the patterns can identify locations of potential sinkholes.

Although ground-penetrating radar is too expensive for individual home construction, it is justifiable for large structures such as schools, power plants, large residential developments, or hospitals. It also is valuable when landfills are being sited, because sinkholes could lead to pollution problems if toxic materials were exposed by the collapse of a cave roof (see Chapter 8).

Ground Subsidence due to Withdrawing Groundwater

Where limestone underlies the soil, we expect ground subsidence and sinkholes to appear as groundwater dissolves the limestone. In sandstone aquifers, such dissolution does not occur, so we would not expect ground subsidence. But, when water is pumped from the aquifer faster than it can be recharged, the water level in the aquifer declines. Then we discover that water, which does not compress, was strengthening the sandstone by filling its pores, helping support the weight of overlying rocks. Without this pore water, the sandstone partially collapses, and some compaction of the aquifer occurs. The ground subsides to create a broad, shallow, bowl-shaped depression that may extend over thousands of square miles.

Subsidence due to withdrawal of groundwater is a very serious problem in several areas of the United States, particularly California's San Joaquin Valley. Since 1925, the ground has subsided as much as 7 inches each year (Figure 6.28). Annual rainfall averages only 8–12 inches, much too little to support extensive agriculture. Nevertheless, with irrigation from underground aquifers, the area produces two-thirds of the nation's vegetables. But the cost is great: 85 percent of California's entire water use.

So, what should be done about the subsidence problem? California's agricultural industry and its tens of thousands of employees depend on continued pumping of groundwater. This industry is central to California's economy and indispensable to the nation's food production. Eventually, the majority of the aquifer will have collapsed, and the rate of subsidence will decrease.

But in the interim, citizens of the region must contend with bent and ruptured water and sewer pipes, cracked building foundations, ruptures in the ground surface because of unequal settling, and nearshore flooding in coastal areas.

Other densely populated areas that are seriously affected by groundwater-withdrawal subsidence include Houston-Galveston, Las Vegas, Tucson, New Orleans, Baton Rouge, and Savannah. For example,

FIGURE 6.28 How far can land subside when water is pumped from it?

This example is in California's San Joaquin Valley. Signs on the utility pole show the position of land surface in 1925, 1955, and 1977. Land surface was lowered 30 feet during this period; the original vegetation was rooted 30 feet above present ground level. *Source: U.S. Geological Survey.*

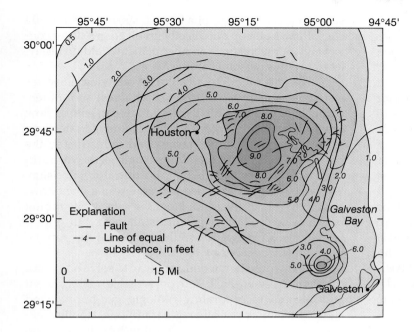

FIGURE 6.29

Contours show Texas-scale ground subsidence around Houston and Galveston. *Source: U.S. Geological Society.*

between Houston and Galveston, subsidence ranges from 5 feet to 9 feet (Figure 6.29). As a result, about 30 square miles of this low-lying coastal land have been permanently flooded. The risk of flooding from frequent hurricanes now extends further inland. Subsidence also has increased shoreline erosion.

Problems When Pumping Stops

With water shortages and lowering of aquifers worldwide, it is hard to imagine that a *rising* level of groundwater could be a problem. But it is in some areas, both in the United States (for example, Dayton, Ohio) and in other countries. Rising water tables occur where pumpage has been curtailed after years of intensive groundwater withdrawal that kept the water table below its natural level. Curtailment of pumping allows the water table to rise to its previous natural level, which may flood underground structures that were built after the water table was lowered. Foundation failures also occur, especially where the natural water level is within 10 feet of the land surface. Many basements built in dry, unconsolidated material that had remained dry for decades now have become permanently wet.

The public's first reaction may be that recent heavy precipitation has raised the water table or created a temporary perched-water system, when in fact the problem will continue unless pumping is resumed to again lower the water table.

Water-level records, where they exist, can indicate whether the water table is responding to weather events or whether it is recovering after pumping has

ceased. More localities are being dewatered temporarily for mining or industrial use, but such uses eventually will end. Serious problems may begin as the water table rises to its former height.

Managing Water Supplies

> Now we are entering a new era—an era of water scarcity. Already 30 countries have more people than their water supplies can sustainably support. We hear rumblings of potential war over water in the Middle East. And competition for supplies is brewing between city-dwellers and farmers around Beijing, New Delhi, Phoenix, and other water-short areas.
>
> Sandra Postel, *Last Oasis*, 1992

As noted, much of Earth's land area is naturally arid or semiarid. In these areas, large-scale agriculture is possible only through intensive use of groundwater. This water entered the aquifers long ago, when they were untapped and could be filled to capacity, or when populations were smaller and the technology for producing water from aquifers was less-developed.

In the late twentieth century, however, human populations are exploding (recall the bacteria in the bottle illustration from Chapter 1). Per capita use of water has increased. The result is a nine-fold increase in water usage worldwide between 1900 and 2000 (Table 6.2).

Streams and groundwater reserves are being depleted rapidly. Ten countries in the Middle East or North Africa currently are overdrawing their surface water account and pumping groundwater to make up the difference. Six other countries in the region are consuming more than half of their renewable water resources. Most also have high rates of population growth. Their water use probably soon will surpass their renewable supply. In Egypt, this could happen during the present decade.

Seawater has invaded Israel's overpumped coastal aquifer, one of the country's three principal freshwater sources. Wells are growing too salty for agriculture—10 percent now, expected to jump to 50 percent within 25 years. In Libya and Saudi Arabia, groundwater is being mined to support agricultural expansion. These water shortages will make lasting peace in the Middle East even harder.

Lack of water is limiting the acreage planted in the central Asian republics of the former Soviet Union. So much water has been diverted from streams into the Aral Sea that it has shrunk 30 percent in the last 30 years and the fishing industry has collapsed. In China, with 20 percent of Earth's people, critical water shortages have developed in recent years. The North Plain of China, where one-fourth of that country's crops are grown, now is restricting planted acreage due to scarce water.

Areas likely to face increased water shortages in the 1990s and beyond also include parts of India (the second most populous country), Mexico, North Africa, the Middle East, and parts of the western United States. Lowering of the water table in the Ogallala aquifer is only one of the groundwater overdraft problems we face (Figure 6.30).

Solutions, Ideal and Real

How can we tackle the water crises that have been present in the United States for many years? First, let us con-
sider an ideal way: use of a logical, fair-minded outside consultant, a person not directly involved in a water controversy. Someone from another planet would do nicely; let us call this person HeorShe.

We give HeorShe a U.S. map showing all water resources: streams (with their discharges) and all recognized aquifers (with their storage capacities, flow rates, and recharge rates). We ask Heorshe to devise a plan for moving people, farms, and industries to the best locations to meet their long-range needs. We Earth citizens all agree to abide by this arbitrator's decision. We each draw our assigned location from a giant bowl and go to live in the place HeorShe selects. Swapping locations with another person is allowed, but only if both agree. Each location could not exceed the population number decided by HeorShe.

This would be a wonderful way to solve our water problems, a solution based on sensible management and fairness, incorporating geologic, geographic, and climatic factors, rather than the traditional troublemakers: economic, social, and political factors. It would do away with decisions based on big money and power politics, to the lasting benefit of all of Earth's citizens.

Unfortunately, this is improbable in the extreme. In the real world, we must deal not only with the existing distribution of population, farming, and industry, but with human nature, which generally opposes the common interest. "Me first" has been the operating motto of nearly all humans since the dawn of history and is not going to change, short of a spiritual miracle that has yet to occur.

So now we must get real. The alternative methods of managing the finite supply of water near the Earth's surface are (1) to conserve water, (2) to redistribute

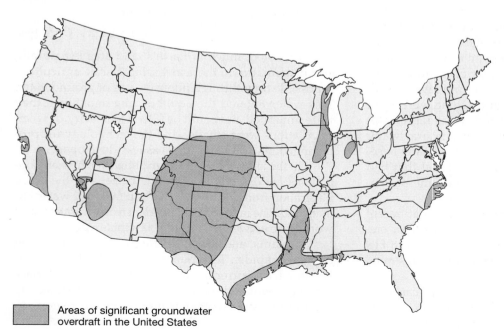

FIGURE 6.30   Groundwater overdraft in the United States.

Overdraft occurs when water is withdrawn from sources that cannot be renewed, or is withdrawn more quickly than sources can be recharged.

Source: Council on Environmental Quality.

Areas of significant groundwater overdraft in the United States

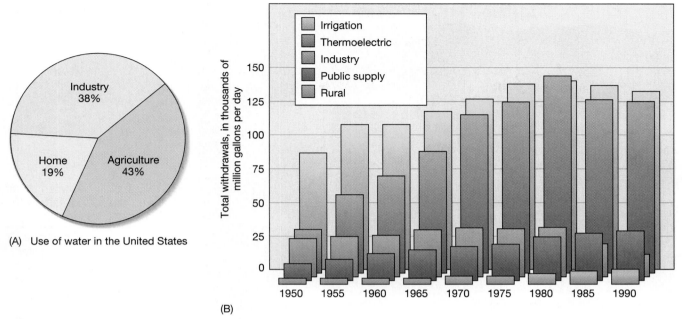

(A) Use of water in the United States

(B)

FIGURE 6.31 Use of water in the United States.

As in nearly all countries, agriculture is the biggest user. *Source: U.S. Geological Survey.*

---

water, or (3) to develop nontraditional water sources. Let us look at each option.

Conserving Water

> When the well's dry we know the worth of water.
>
> Benjamin Franklin

Figure 6.30 shows where groundwater aquifers are being overpumped. By comparing the map of annual precipitation (Figure 6.4) with Figure 6.30, you can see areas deficient in both surface water and groundwater. The three worst are southern Arizona (6–16 inches annual rainfall), the central and southern mid-continent (16–20 inches), and the San Joaquin Valley of central California (8–12 inches). Arizona's problem comes from irrigation and rapid population growth in Phoenix and Tucson. (In Tucson, the water table has dropped 100 feet in recent years.) The other two areas are in trouble from extensive crop irrigation rather than direct human consumption, so this is where the biggest savings can be made.

Conserving on the Farm. Agriculture is the biggest water user, worldwide and in the United States (Figure 6.31). Much U.S. cropland is irrigated with large sprinklers that spray into the air about 10 feet above the ground (Figure 6.32). This height and a little wind combine to distribute the water over a wide area. However, much of the water evaporates directly into the air before reaching the ground. Another common, evaporation-prone method is to run water in open concrete channels into the fields. Hydrologists estimate that

only 45 percent of the water reaches plants via air spraying or channel irrigation.

Computerized drip irrigation is far more efficient. Water drips through tiny holes in pipes on or just under the ground surface, just at the root level of plants (Figure 6.33). Almost all of the water reaches its intended target; efficiencies can reach 95 percent. This method now is used on nearly half of America's irrigated cropland. It has reduced water usage per acre by an average of one-third, while increasing crop yields.

Computerized drip irrigation was pioneered in Israel in the 1960s. It has made this tiny desert country a major supplier of fruit and vegetables to Europe. Agriculturists from countries as distant as China and India visit Israel to study the technique; some even come from Moslem countries. Plastic piping stretches through Israeli fields for many miles, watering each plant individually. Sensors measure wind velocity and air humidity and feed these data to computers, which determine the amount of water and dissolved fertilizer each plant requires. The Israeli agricultural areas have the same scant rainfall as the three most heavily irrigated U.S. areas.

Should the U.S. government offer tax incentives to farmers to encourage conversion from inefficient sprinkler and channel irrigation to the drip method? If farmers fail to conserve more water soon, what will be the future of our major agricultural areas?

Conserving at Home. Figure 6.34 shows how we use water daily. Table 6.3 is interesting, for it shows ways to reduce your water consumption with little effort.

FIGURE 6.32   Above-ground irrigation pipes on a large farm in Nebraska.

The pipes move in circles as a well taps the underlying aquifer rock and supplies the sprinklers along each pipe. The pipe makes one complete circle each day.
Photo: Bill Kamin/Visuals Unlimited.

Another way to conserve water is to become aware of "silent losses." These are leaks in municipal water pipes and in your dwelling, whether home, apartment, or residence hall. Many water pipes in American cities are old (a century is not unusual) and leaky. New York City loses 15 percent of its municipally piped water to leaks. Buffalo, New York, loses 40 percent. In Great Britain, about *one-third* of the water pumped by British water companies is estimated as lost to pipe leaks. In Manila, the Philippines' capital city, 58 percent is lost. It is estimated that 20 percent of all toilets in American homes leak. A severe toilet leak can silently waste hundreds of gallons per month.

To check for toilet leaks, put some food coloring in the tank and wait a few minutes. If you see color in the toilet bowl, you probably have a leak. More obvious leaks are those of dripping or dribbling faucets, inexpensively stopped by replacing the rubber washer inside.

Other water-savers include:

- Wash cars with a bucket of water rather than a running hose (use the hose for rinsing).
- Water lawns at night, when it is cool, rather than during the daytime heat (which promotes rapid evaporation). This lets the water fully soak into the soil pores.
- Don't let the water run continuously when washing dishes by hand.

A serious impediment to water conservation is the unrealistically low cost of our public water. Price does not reflect limited and diminishing supplies. For example, a typical urban family of four uses about 15,000 gallons of water per month, costing only about $20. At such low prices, there is no financial incentive to conserve. If utilities doubled or tripled their charges to reflect national water scarcity, conservation might become more popular. Given human nature, conservation will not become widespread until water shortages become more widely understood and felt in the wallet.

FIGURE 6.33   Computerized drip irrigation.

Delivers a precise amount of water directly to the roots of each plant; this method eliminates nearly all waste—evaporation and runoff—from traditional irrigation methods. *Photo: Sylvan Wittwer/Visuals Unlimited.*

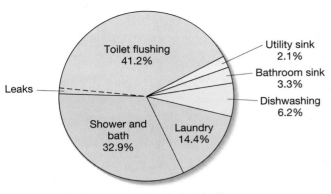

Daily water use for a typical family
of four, using 243 gallons/day

FIGURE 6.34 Use of water in an average American home.

Average indoor home use is approximately 75 gallons per person per day. About 75 percent of this is used in the bathroom (flushing toilets, showers, etc.).

Redistributing Water

Moving water to places of shortage has been practiced since biblical times in Egypt, Iraq (Babylonia, Mesopotamia), and Greece. In the United States, moving water is very costly and legally complex, but the demand for water transfer is great (Figure 6.35). Several interstate plans have been suggested:

- Pipe Mississippi River water from below New Orleans to parched west Texas. This plan foundered on the rocks of shared cost and responsibility.

- Pipe water from southeastern Oklahoma (50+ inches of annual rainfall) to western Oklahoma (less than 25 inches). The water would be pumped uphill, up to 300 miles, and stored in reservoirs. However, evaporation powered by 90–100°F summer temperatures may make the proposed project not cost-effective.

- Expand New York City's water supply with a third underground aqueduct to transfer water from the Catskill Mountains, 100 miles north of the city. New York now is served by two aqueducts, built in 1917 and 1936, but population and commercial growth soon will render them inadequate. The downhill route is almost entirely

TABLE 6.3 **Household Use of Water in the United States**

The rate of flow of water from the faucet, hose, or shower head is assumed to be 5 gallons per minute.

Use category	Water used*		Potential savings (in qts)	Water-saving suggestions
	Amount in qts.	Assumptions		
Drinking	3	Daily requirement	—	
Toilet	20	Per flush	5	Tank displacement.
Brushing teeth	40	Leave water on for 2 minutes	35	Turn off water while brushing.
Washing hands	20	Leave water on for 1 minute	15	Turn off water while soaping hands.
Shower	100	5-minute shower	40	Take 3-minute shower.
Washing clothes	120	1 load	20	Washing full loads could save as much as 17%.
Washing dishes	100	1 load, automatic dishwasher	17	Washing full loads could save as much as 17%.
Washing car	100	5 minutes to complete	60	Turn off water when not washing.
Lawn watering	250	Apply 1 inch to 100 sq. ft.	150	Use native plants or plants that thrive on little water. Save as much as 60%.

*Source: Denver Water Department

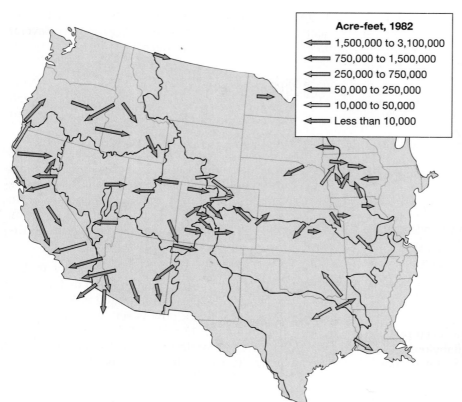

Acre-feet, 1982

⟵ 1,500,000 to 3,100,000
⟵ 750,000 to 1,500,000
⟵ 250,000 to 750,000
⟵ 50,000 to 250,000
⟵ 10,000 to 50,000
⟵ Less than 10,000

FIGURE 6.35   Transfers of surface water in western United States.

An acre-foot of water is the amount that covers one acre to a depth of one foot, 326,000 gallons.

gravity-powered (no pumping is required). But experience suggests that the $5 billion construction estimate will multiply by the time of completion in 2025.

Developing Nontraditional Water Sources

Three non-traditional water sources have received serious attention. Two were pioneered by Saudi Arabia, a mostly desert nation: desalination of ocean water and the towing of icebergs from Antarctica. The third is recycling and purification of wastewater. Here is a look at each.

Desalination of Seawater.   Desalination is practiced extensively by the Saudis and more sparingly elsewhere. About 8000 desalination plants operate worldwide, producing more than 150 million gallons per hour of drinking water. Two-thirds of the plants desalinate seawater; the others process less saline (brackish) water. Nearly 30 percent is produced in the Middle East and 12 percent in the United States. Globally, however, desalinated water totals only 0.1 percent of human usage.

Average ocean water contains 3.5 percent salt (ions in solution). (This also is expressed as 35,000 parts per million or ppm.) This doesn't sound like much, but if you have tasted seawater, you know that it is undrinkable. Farm animals can tolerate up to 7000 ppm depending on their size; irrigated crops, up to 700 ppm depending on the type of crop; but humans should not drink water containing more than about 1000 ppm. Average streamwater contains 120 ppm. Hence, to desalinate seawater for irrigation requires removal of at least 98 percent of the dissolved materials. For human drinking water, at least 99.7 percent of the ions must be removed. This is approaching the purity of distilled water.

There are six different methods for removing the ions from seawater. All are expensive, the major cost being the large amount of energy that is required to remove 99.7 percent of the salt. To this must be added the cost of transporting the purified water from the desalinating plant at sea level to the consumer. Saudi Arabia is short of freshwater, but has numerous desalination plants because the country has virtually unlimited energy in the forms of petroleum, natural gas, and intense sunlight.

Today, desalinated drinking water from the ocean in the Middle East costs about $2 for 250 gallons. Compare this to $0.01 to $0.05 per 250 gallons paid by farmers in the western United States for irrigation water, and about $0.30 paid by urban users for drinking water. The U.S. cost of desalinated water is about five times the cost of public water from surface or underground sources. Therefore, if all Americans today had to drink desalinated water from the ocean, an average water bill might be $100 per month, instead of the $20 we now pay for the family of four. This still is less than the average gas and electric bill, and still is a bargain, considering that water is fundamental to life.

The cost of desalinated ocean water will drop as technology improves. As the combination of dwindling groundwater supplies and continuing pollution make drinkable water less available, the sea will increase in importance as a resource. It is possible that, within one hundred years, most Americans will be drinking desalinated ocean water.

Towing Icebergs. The only other nontraditional source of water that has received serious consideration is towing of icebergs to Saudi Arabia from the Antarctic ice sheet. Because inexpensive energy is abundant for many countries in the arid Middle East, the idea of towing icebergs 10,000 miles is not as ridiculous as it might seem.

Cost analyses, however, are not promising. A large initial investment would be required in design and construction of huge, fast tugboats. A major problem would be melting as the ice moves northward into warm equatorial water. Ice melts rapidly at room temperatures, to say nothing of the higher temperatures in the Middle East. To demonstrate the effect, place an ice cube in a glass filled with warm tap water. How long would even a massive iceberg last in warm water?

Australia, largely desert because of the tropospheric circulation discussed earlier, also lacks abundant surface water and groundwater, so Australians have considered harvesting icebergs. Their southern coast is only 3000 miles from the Antarctic continent, and the route consists of cool-to-cold ocean water. Two university professors have suggested wrapping Antarctic icebergs in strong, lightweight fabric to retard melting during transport. Glacial ice might be cost-competitive with other freshwater sources in the country, even considering the expense of harnessing, wrapping, and towing the icebergs.

Purification of Wastewater. Americans are accustomed to the concept of recycling as it applies to metals, glass, plastic, and paper. But have you thought about recycling wastewater, such as sewage water? For some reason, the thought of drinking water that has passed through other people and a sewage treatment plant has a negative image.

Yet, hundreds of American cities already are recycling wastewater for *non*drinking purposes, such as crop irrigation in semiarid western states and landscaping at Florida's Disney World. Other cities like El Paso and Los Angeles are using wastewater to recharge aquifers used for drinking water, because the aquifers purify water. Many U.S. communities are studying the safety, economics, and feasibility of directing treated sewer water into the ground to replenish dwindling aquifers, even those tapped for public drinking water.

This practice, known as **artificial groundwater recharge,** typically injects treated city wastewater directly into confined aquifers, or spreads it on the ground to infiltrate the soil into an unconfined aquifer below. Letting wastewater infiltrate through soil removes chemicals, harmful bacteria, and viruses that survived the city's treatment plants (see Chapter 8). Clay minerals in the soil adsorb many harmful chemicals and immobilize harmful microorganisms so they do not reach the aquifer.

Currently, thirty-six states regulate water recycling, but only ten list permissible other-than-agriculture uses for reclaimed wastewater (Figure 6.36). No federal regulations directly govern recharge practices, such as pretreatment and post-treatment monitoring to assure water quality. However, all water consumed directly (drinking) or indirectly (irrigation of crops) must meet purity standards in the Environmental Protection Agency's Safe Drinking Water Act, passed about twenty-five years ago. In other words, wastewater *input* to aquifers is not the federal government's concern, but the water *withdrawn* from the aquifer for your use is.

Try your hand at devising an advertising campaign that would sell the public on wastewater recycling:

Need some fresh water?

Want it real pure?

That's why we're recycling

The stuff from your sewer!

Israel projects that one-third of its water needs in 2010 will be met by reclaimed and recycled sewage water.

FIGURE 6.36 Control of water recycling.

Water recycling is controlled in thirty-six states (light gray) but other-than-agricultural use for such water is regulated in only nine (dark green). *Source: Environmental Protection Agency.*

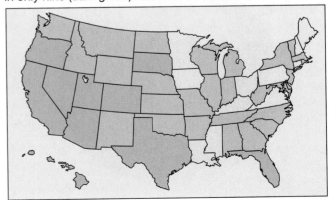

Water Quality: Hard and Soft

In Figure 6.37, a water pipe has been sawed open so you can see how older homes get clogged arteries, like people. If you live in an older building, its pipes might look like this. How do they get this way? A noticeable characteristic of water is its "hardness." A quick test for hardness is to wash your hands with a small amount of soap. If the soap lathers easily, the water is relatively "soft." If it is difficult to raise suds, the water is relatively "hard."

Chemists define **water hardness** as the quantity of ions in the water that have a +2 charge. Usually, this means calcium (Ca^{2+}) and magnesium (Mg^{2+}). The average amount of these ions in stream water is just under 60 parts per million (ppm). However, the amount varies widely with geographic location. Areas underlain by limestone ($CaCO_3$) may contain as much as 600 ppm of calcium, which is extremely hard water. The U.S. Geological Survey classifies degrees of water hardness into four categories:

0–60 ppm = soft

60–120 ppm = moderately hard

120–180 ppm = hard

>180 ppm = very hard

When soap is used with hard water, a reaction occurs between the calcium or magnesium ions in the water and the chemicals in the soap (which are sodium stearate, palmitate, and oleate). This forms calcium or magnesium compounds that are poorly soluble in water, and hence incapable of taking part in the cleansing action for which the soap is intended. These calcium and magnesium compounds are light and float on the water, forming the "soap scum" that we associate with hard water. This is undesirable, for it demands greater quantities of soap, more rinsing, and results in deposits of insoluble calcium compounds on fabrics, producing a dingy grayness instead of the desired whiteness.

Special hard-water soaps incorporate a water-softening agent (like sodium carbonate) to prevent formation of the troublesome calcium or magnesium compounds. More commonly, however, people who live in areas with hard water (Figure 6.38) install a water-softening device on the water supply pipe to their dwelling. This removes the calcium and magnesium ions before the water is used.

The water softener swaps sodium ions for the calcium and magnesium. This solves the hard-water problem, but may create a new problem for those on a low-sodium diet. The cartridges in these devices eventually become saturated with calcium and magnesium ions and must be replaced or recharged. Also, water softeners can waste from 15 to 120 gallons of water for every 1000 gallons of water processed. And water softening is expensive. Hence, many people live with hard water.

Another problem arises when calcium-rich waters are heated. The calcium ions react with the bicarbonate ions in the water to form calcium carbonate ($CaCO_3$):

$$Ca^{2+} + 2HCO_3^- + heat \rightarrow CaCO_3 + CO_2 + H_2O$$

The carbon dioxide gas travels away with the water. But the solid $CaCO_3$ (our old friend calcite) coats the inside of hot-water pipes (Figure 6.37) and the insides of tea kettles, thereby reducing heating efficiency. These deposits of *boiler scale* or *lime scale*, as they often are called, can be especially serious in boilers in which the water is heated under pressure in pipes running through a furnace. Formation of scale reduces the efficiency of heat transfer severely enough to result in a possible melting of the pipes.

Drinking hard water may have a positive side. Recent evidence suggests a relationship between higher water hardness and a lower incidence of heart disease. More than two dozen studies in several countries have found such a relationship, but the cause is unknown.

FIGURE 6.37   Talk about clogged arteries!

This metal water pipe's useful interior diameter has been reduced by 85 percent. Calcium carbonate (calcite) has precipitated from the hard water that flowed through the pipe over many years. Pure calcite is white; the brownish color of the precipitate near the pipe wall results from iron oxide (hematite, Fe_2O_3) in the calcite. What is the cure for a clogged plumbing system like this? Extensive bypass surgery: new plumbing for the home. *Photo courtesy G. Atkinson.*

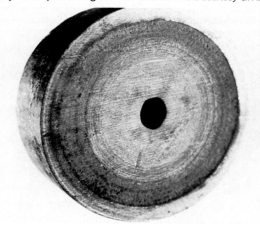

Medicinal Waters

Bottled water sold for medical purposes has a long tradition in Europe but is little used in the United States today. Most **medicinal water** has a clear medicinal effect, such

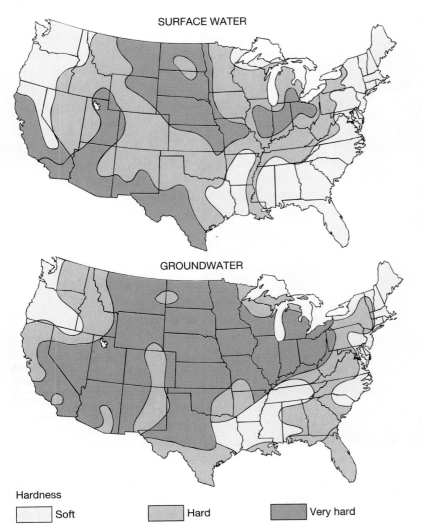

SURFACE WATER

GROUNDWATER

Hardness

☐ Soft   ▨ Hard   ▨ Very hard

FIGURE 6.38 Approximate hardness of surface water and groundwater in the United States.

Only small sections of the northeast, northwest, and south-central U.S. have soft surface water *and* soft groundwater. Hard groundwater occurs more widely than hard surface water. *Source: U.S. Geological Survey.*

as one sold in Germany that is rich in sodium bicarbonate, which reduces the acidity of an "acid stomach." In Great Britain, a water containing more than 310 ppm of iron (Fe^{+2}) is licensed and sold for preventing iron deficiency. Americans are not into "waters," instead taking Tums, Rolaids, or Maalox for an acid stomach and iron-rich "mineral" supplements for iron-deficiency. Your author knows of no studies that compare the effectiveness of European waters and American tablets.

Another health use of water that has been popular in Europe for 2500 years, and in the United States until early this century, is soaking in a **spa** under traditional medical supervision. In continental western Europe, a thousand spa resorts employ more than 300,000 skilled workers, technicians, engineers, and doctors. These spas are part of the social health-care systems of these countries (Figure 6.39). Germany defines acceptable spa waters as containing at least 1000 ppm dissolved solids, among which must be at least 20 ppm iron, 1 ppm sulfur, 1000 ppm carbon dioxide, 1 ppm fluorine, and radium at 18 nanocuries per liter. (The

nanocurie, one-billionth of a curie, is a measure of radioactivity.) It is claimed that bathing in spas is an excellent treatment for arthritis, rheumatism, skin disease, and stress reduction.

Case Study: Dead Sea Therapy

The Israeli government qualifies injured war veterans for two weeks of spa therapy a year, and the bulk of those taking advantage of this benefit go to the Dead Sea. Government studies indicate measurable medical benefits from the Dead Sea's water and its products, particularly salts and muds. The Dead Sea is the world's saltiest natural water body, with a salinity almost ten times that of normal sea water.

Muds from the Dead Sea contain high levels of potassium (which relaxes), calcium (which soothes), and magnesium (which heals)—a combination that appears to help relieve pain, especially the stiff and swollen joints of ailments like rheumatoid arthritis. Tests on forty

FIGURE 6.39 A modern health spa, Leuze in Stuttgart, Germany.

Such spas offer all the traditional cures, plus long-term fitness programs. *Photo: Werner H. Muller/Peter Arnold, Inc.*

arthritis sufferers in 1990 at an Israeli medical center indicated that treatment with Dead Sea mudpacks and sulfur baths from springs around the Sea produced "significant and lasting improvements in disease severity."

The Dead Sea also is well known for its symptomatic relief of skin ailments. In a 1983 experiment on psoriasis patients, 85 percent experienced "complete clearing or excellent improvement" of their condition. Perhaps the medical profession in the United States could learn something from these results, as they have from the effectiveness of Chinese acupuncture.

Summary

Water is the most important compound for life on Earth, but its supply is limited. Although atmospheric circulation and the hydrologic cycle cause water movement from place to place, they do not increase the supply available for human consumption.

Freshwater is not distributed evenly over Earth or the United States. The eastern U.S. has a surplus for its needs, and the western half has a shortage. Despite this imbalance, population growth is faster in the West. The two most extensive crop production areas are in the West, the San Joaquin Valley of California and the mid-continent belt from Texas north to the Dakotas.

Water needed for drinking or crop irrigation can be obtained and used without purification from either surface water or groundwater. Most water used for these purposes is groundwater. Because surface streams are totally allocated in many water-short areas, groundwater is increasingly tapped, so water tables are dropping nationwide. New water sources must be found or we will be in very serious trouble.

Additional water can be made available by several methods. We can conserve, and there are many ways to do so. Farmers can use proven newer irrigation techniques that are much less wasteful. Our personal use of water is extremely wasteful, in part because of the unrealistically low cost of public water. We can desalinate seawater, which would be costly, but not impossibly so. We can easily recycle wastewater so it can be used for irrigation, flushing toilets, and other uses that do not require the high level of purity required of drinking water by the American public.

Key Terms and Concepts

aquiclude
aquifer
aquitard
artesian water
artificial groundwater recharge
atmospheric circulation
computerized drip irrigation
cone of depression
confined aquifer
Darcy's Law
desalination
dip
discharge
divining rod
dowsers
ground-penetrating radar

groundwater
hydraulic conductivity
hydrogeologist
hydrologic cycle
karst topography
latitudinal desert
medicinal water
oasis
perched aquifer
permeability
polar front
pores
porosity
potable water
potentiometric surface
rain shadow

recharge
runoff
sinkhole
solar energy
spa
spring
storage capacity
stratosphere
surface water
swamp
tropopause
troposphere
unconfined aquifer
water cycle
water hardness
water table

Stop and Think!

1. Describe the hydrologic cycle. In what ways do human activities disturb it?

2. Describe the circulation of air in the troposphere.

3. Explain why permanent deserts exist in low latitudes.

4. How do the geographic orientation and height of mountain ranges affect the distribution of rainfall near coastlines?

5. What factors control the amount of pore space in a rock? Why are sandstones and limestones better aquifers than igneous or metamorphic rocks?

6. What is the difference between an unconfined aquifer and a confined aquifer? In which type do you think recharge might be more rapid? Why?

7. Draw a cross section showing a confined aquifer. Explain what determines whether it will flow at the surface without help, or whether it will need to be pumped.

8. What is permeability? Explain how a sandstone can be porous but not permeable to water.

9. What is the water table? How far below the surface is it where you live? Has its depth changed in recent years? Who would you contact to find out?

10. Does your community get its water from surface sources (streams, lakes, reservoirs) or subsurface sources (unconfined aquifer, confined aquifer)? Phone the city water department to find out.

11. What are sinkholes and where are they most likely to form?

12. The federal government already plays a role in agricultural water usage. What do you believe it should do to promote water conservation?

13. The federal government subsidizes (partially pays the cost of) many activities in our society, including the use of water for irrigation. Do you believe it should subsidize the construction of desalination plants? Remember, nothing is free. The only question is: Who pays?

References and Suggested Readings

American Institute of Professional Geologists. 1985. *Ground Water Issues and Answers.* Arvada, Colorado, 24 pp.

Back, W., Rosenshein, J. S., and Seaber, P. R. (eds.) 1988. Hydrogeology. *The Geology of North America*, v. 0–2, 524 pp., Boulder, Colorado: Geological Society of America.

Bertoldi, G. L., Johnston, R. H., and Evenson, K. D. 1991. Ground water in the Central Valley, California—a summary report. *U.S. Geological Survey Professional Paper 1401-A*, 44 pp.

Carrier, Jim. 1991. The Colorado: A river drained dry. *National Geographic*, June, p. 4–35.

Clarke, Robin. 1991. *Water: The International Crisis.* Cambridge, Massachusetts: The MIT Press, 193 pp.

Environmental Geology, 1996, v. 27, no. 2. *Special Issue: Springs and bottled waters.* pp. 74–142.

Fetter, C. W. 1994. *Applied Hydrogeology, 3rd ed.* New York: Macmillan College Publishing Company, 691 pp.

Gleick, P. H., (ed.) 1993. *Water in Crisis.* New York: Oxford University Press, 473 pp.

Gutentag, E. D., Heimes, F. J., Krothe, N. C., Luckey, R. R., and Weeks, J. B. 1984. Geohydrology of the High Plains Aquifer in Parts of Colorado, Kansas, Nebraska, New Mexico, Oklahoma, South Dakota, Texas and Wyoming. *U.S. Geological Survey Professional Paper* 1400-B, 63 pp.

Heath, R. C. 1989. Basic ground-water hydrology. *U.S. Geological Survey Water-Supply Paper* 2220, 84 pp.

Maybank, J. and six others. 1995. Drought as a natural disaster. *Atmosphere–Ocean*, v. 33, no. 2, p. 195–222.

Postel, Sandra. 1992. *Last Oasis: Facing Water Scarcity.* New York: W. W. Norton & Company, 239 pp.

Shiklomanov, I. A. 1990. *Global Water Resources: Nature and Resources,* v. 26, no.3, p. 34–43.

Smith, Z. A. 1989. *Groundwater in the West.* New York: Academic Press, 308 pp.

Van der Leeden, F., Troise, F. L., and Todd, D. K. 1990. *The Water Encyclopedia, 2nd ed.* Chelsea, Michigan: Lewis Publishers, 808 pp.

White, W. B., and four others. 1995. Karst lands. *American Scientist*, v. 83, p. 450–459.

"And the rain fell upon the earth for forty days and forty nights. . . .Then the flood came upon the earth for forty days . . ."
Genesis 7

Streams and Floods

7

It is common knowledge that flooding afflicts high-rainfall areas, like New Orleans. But serious flooding also occurs in desert areas of Arizona and New Mexico. What determines the location, timing, and frequency of these floods? How far from a river, or how high above it, should a house be for safety from floodwaters? Do flood waters carve the ground surface? Can they excavate buried fuel storage tanks, old "sanitary landfills," or coffins from cemeteries?

Streams

Of all geological processes, the movement of water on the land surface probably has the greatest direct effect on human civilization. We depend on streams and rivers for energy, travel, drinking water, and irrigation. Because of humankind's fundamental need for water, civilization began on **floodplains,** the flat plains that flank a stream or river. Worldwide, populations generally are more dense along waterways. Most Egyptians still live on the floodplain of the Nile River (Figure 7.1). In the United States, the New Orleans area long has been a commercial hub because of its location on the lower floodplain of the Mississippi River.

Stream characteristics and their development have been intensely studied by scientists and engineers for as long as science has existed because of the obvious impact flowing water has on people's lives. Interestingly, it was not until the seventeenth century that measurements of the volume of rain and snow convinced investigators that precipitation was adequate to account for the amount of water we see in streams.

How Streams Get Started

Streams do not exist at random. They have definite patterns, the most common being like the roots of a tree, where many tiny streams flow into larger ones, converging to form a single large river (the trunk). Thus, streams generally form a network. This is useful knowledge if you are lost in the woods: Walk downslope until you reach a stream, then follow it downstream until it flows into another, and so on. Eventually you will reach civilization, because people build homes near streams.

The origin and evolution of the stream pattern in any environment can be studied simply by looking at the ground around your feet when it rains. As the raindrops hit the ground on a sloping surface, the water initially runs off as **sheetflow,** a thin layer of water flowing downhill. Within a short distance, however, surface irregularities focus the water into small channels.

The location of these channels may seem to be random, but closer examination often reveals a pattern. Perhaps the channels follow depressions along a path that animals have

◀

Sandbaggers hastily build a levee on the Raccoon River in Van Meter (just west of Des Moines) on July 10, as the river rose faster than expected.

Photo by Gary Fandel; courtesy of *The Des Moines Register.*

FIGURE 7.1 Life began on a floodplain.

Today most Egyptians still live on the floodplain of the Nile River, as they have for millenia.
Photo: Corbis-Bettman.

taken for years to reach a sheltered area. Or, perhaps they flow straight to the lowest area. Perhaps the rocks at the surface or under the soil are layered and tilted, so that the water finds an easy path along their surface. Or, possibly the rocks have been broken into parallel fractures (joints) that channel the water. Perhaps the tilted rocks are a combination of strong and weak rock, such that the weaker rock erodes faster, creating channels for the water. The possibilities are many.

Whatever the cause, tiny water channels called rills form quickly (Figure 7.2). Even at this early stage, the water is picking up a load of sediment to carry. As they merge downslope, the channels become larger. The main channel that develops is said to have **tributaries,** which are smaller feeder streams that supply water to the main stream. The region from which the stream complex draws water is its catchment area for precipitation, called its **drainage basin.** Every bit of water that reaches the ground as rain, snow, sleet, hail, dew, or frost—except the portion that evaporates into the air—becomes part of the water in the drainage basin. Drainage basins may have areas as small as a tabletop in a rill system or as large as the vast catchment of the Mississippi River (Figure 7.3).

Drainage basins differ, not only in size, but in the type of rocks exposed, soil thickness, amount of rainfall and snowfall received, and the slope of tributary streams within them. Each of these characteristics affects the flooding tendency in the basin. The drainage basin is a key concept in determining flood potential (Figure 7.4).

The boundaries between the drainage basins are called **drainage divides.** They are easily traced on a map by finding the upper ends of tributaries (Figure 7.4). Larger streams have larger drainage basins. Drainage basins enlarge over time as the upper end of each tribu-

tary stream erodes the rock and soil, nibbling its way, slowly lengthening the stream headward. This is **headward erosion.** If the upper end of a stream erodes its

FIGURE 7.2 Rill system forming on new slope.

Photo: P. Carrara/U.S. Geological Survey.

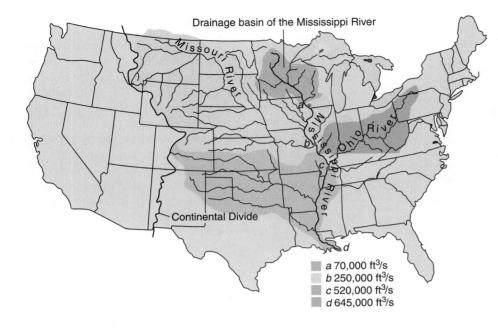

Drainage basin of the Mississippi River

Missouri River

Mississippi River

Ohio River

Continental Divide

d

■ a 70,000 ft³/s
■ b 250,000 ft³/s
■ c 520,000 ft³/s
■ d 645,000 ft³/s

FIGURE 7.3 The Mississippi River's drainage basin.

Combining its own basin with those of the Missouri, Ohio, and other rivers, the vast Mississippi River flows 2350 miles from its source in central Minnesota to its rendezvous with the Gulf of Mexico. At *a*, low discharge reflects the small drainage basin upstream of that point. At *b*, the dramatic discharge increase (357 percent) reflects addition of the far-reaching Missouri River basin discharge. At *c*, another large increase—743 percent over point *a*—results from addition of the Ohio River basin drainage, a high rainfall area. Heavy rain in any part of the basin can cause flooding at any point lower in the basin. This was demonstrated by the disastrous flooding of 1993, caused by heavy rainfall over Iowa and adjacent states.

way long enough, it eventually will meet another stream. Whichever of the two streams flows more steeply will *capture* the drainage of the other stream.

We think of mountain streams as being steep and tumbling, and lowland streams as being more level and slower. This is generally correct, as Figure 7.5 shows. Any

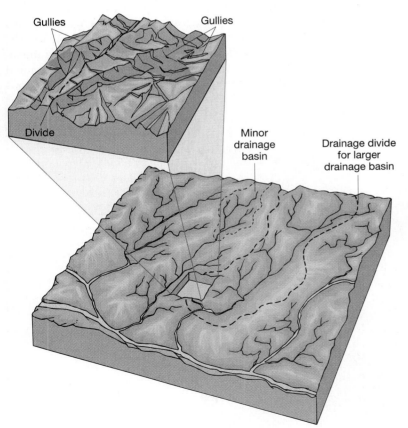

Gullies

Gullies

Divide

Minor drainage basin

Drainage divide for larger drainage basin

FIGURE 7.4 What is a drainage basin?

A drainage basin is the land drained by a main stream and its tributaries. The boundary between any two drainage basins is a ridge called a drainage divide. *After Robert W. Christopherson, Geosystems, Second Ed.,* Englewood Cliffs, New Jersey: *Prentice Hall, 1994.*

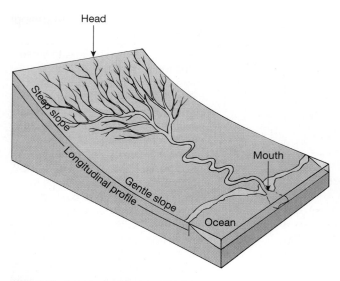

FIGURE 7.5 Longitudinal profile.

Any established stream has this longitudinal profile, with a steeper slope upstream toward its head and a much gentler slope downstream toward its mouth.

established stream has the **longitudinal profile** shown, with a steeper slope upstream toward its head and a much gentler slope downstream toward its mouth.

Given millions of years, a few large streams come to dominate a continent's drainage. Usually, only one or two such *master streams* develop on each continent, such as the Amazon River in South America, the Nile and Congo Rivers in Africa, the Mississippi River in North America, and the Chang Jiang River (formerly Yangtze) in eastern Asia (China).

Stages of Stream Development

Glance at aerial photos of streams, and you will see clear differences among them. Some stream valleys are deep and V-shaped, with the stream occupying the entire lower portion of the *V* (Figure 7.6). Other stream valleys are very wide and flat, with the stream width being only a very small percentage of the valley width (Figure 7.7A). What causes such differences? Of what significance are they to environmental concerns?

V-shaped valleys are formed during the early stage of a stream's evolution. This is evident in mountainous regions, where the stream's headwaters lie in a notch that the stream has carved into a ridge. Stream level may be many hundreds of feet below the level of the ridge peaks at its sides. The rocks of the ridge are strong enough to resist major landsliding into the valley, at least temporarily. However, the common presence of boulders in such a stream indicates that some landsliding (rockfall) is indeed occurring. Few people live in mountain highlands, so this mass movement is not an important environmental concern, except to the snowbirds who enjoy frigid weather.

As a stream descends from the mountains, its steep slope sharply levels out where the mountains end and join a plain. Velocity decreases, so the larger sediment particles—mostly gravel and some sand—are deposited in the stream bed. This increases the proportion of fine mud particles to sand in the water. As this happens, the stream begins to weave from side to side, forming snakelike **meanders** (Figure 7.7B). The meander shapes migrate downstream as their development proceeds and the weaving widens the valley. Previous

FIGURE 7.6 Yellowstone River.

The waterfall and rapids on the Yellowstone River in Wyoming indicate that the river is actively downcutting through the rock. Note the V-shaped profile and lack of flood plain, typical of any fast-cutting stream. *Photo: Gilda Schiff/Photo Researchers, Inc.*

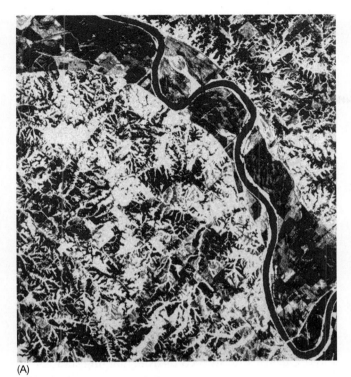
(A)

(B)

FIGURE 7.7   America's great meandering rivers.

(A) The Missouri River meanders its way across central Missouri. The wide, dark band along the river is its floodplain, formed through countless floods over thousands of years. Note the intricate pattern of tributary streams (bright lines). The different vegetation types and soil types show clearly, as do breaks in the levees (shoulders) along the Missouri's banks. The breaks (black spots, upper center) resulted from the 1993 flood. Radar image near Glasgow, Missouri, 1994, from Center for Remote Sensing, Boston University. (B) Over the centuries, the muddy lower Mississippi River has repeatedly shifted course, like a slowly writhing snake. This has left a legacy of old meander loops and bends, visible in this infrared satellite image. Old meanders are particularly clear at upper left. The red color indicates healthy vegetation. *Source: NASA/Science Photo Library/Photo Researchers, Inc.*

meander paths are commonly evident in aerial photos (Figure 7.7B).

The rocks in lowland areas are generally sedimentary rocks, with shale being most abundant. Shale is very thinly layered and always fractured so it is not difficult for the river to widen the valley. If the lowland materials are loose sediment, deposited by earlier episodes of stream development, then valley widening will occur very rapidly, perhaps within a few tens of years. The wide, flat valley created by the weaving meanders is called the floodplain (Figure 7.7A), the zone to which floods are usually confined when the river overflows its banks. When the river is in flood, the floodplain is called the **floodway.**

Stream Discharge

The most common stream characteristic determined by hydrologists is **stream discharge.** Discharge is the volume of water flowing past a point each second (or minute, or hour, or day). Thus, discharge involves both the size of a stream and how fast it flows. Discharge is calculated like this:

$$\underset{\substack{(\text{cubic feet}/\\ \text{second})}}{\text{discharge}} = \underset{(\text{feet})}{\text{stream width}} \times \underset{(\text{feet})}{\text{stream depth}} \times \underset{(\text{feet}/\text{second})}{\text{stream velocity}}$$

You can see that a stream's width, length, and velocity each control its discharge volume. This also is shown in Figure 7.8.

Figure 7.9 shows a stream at four different times—prior to a flood and at three stages during a flood. Note what happens to its discharge and depth.

Stream channels differ in the amount of frictional resistance to water flow offered by the stream bed and banks. Resistance to flow is large in a broad, shallow channel. It is much less in a deep, narrow channel. Also, resistance to flow caused by a stream's banks slows water near the banks, creating the drag shown in Figure 7.10. Consequently, stream velocity is fastest at the center of a stream, and near the surface.

Stream discharges increase downstream as tributary streams add their water to the amount in the **trunk stream.** In very large streams like the Mississippi River, the average discharge is almost beyond comprehension—553,000 cubic feet per second—a volume of more

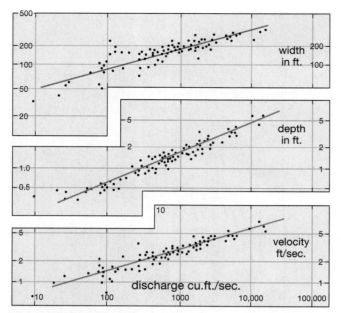

FIGURE 7.8 Stream discharge.

Discharge of a stream increases with its width, depth, and velocity. This example is the Powder River at Locate, Montana. *Source: L. B. Leopold and Thomas Maddock, Jr., U.S. Geological Survey Professional Paper 252, 1953.*

than 300 billion gallons of water per day, measured at the river's mouth, where it discharges into the Gulf of Mexico.

Because discharge varies with the season and precipitation throughout a stream's basin, a single measurement of a stream's discharge means little. Discharge must be measured repeatedly at the same locality to provide a useful scientific picture of a stream's discharge, power, and potential destructiveness. This gives hydrol-

ogists a better understanding of how flow varies with time.

In the United States, such sampling localities, called **gauging stations,** have been in operation for more than 100 years (Figure 7.11). The U.S. Geological Survey (USGS) maintains more than 11,000 gauging stations on principal streams and their tributaries. Therefore, a wealth of flow data for a large number of streams is available from the USGS and many state geological surveys.

For any stream, a graph can show its discharge versus water height above a selected level (called a *datum*). The best-fit line through the data points is called a **rating curve** (Figure 7.12). Once this curve is constructed, the only variable observed directly is the height of the river above the datum, called the **stage.** Stage is used to predict discharge, essential in flood management.

Historical Trends in Discharge. The United States, at about 220 years of age, is a young country compared to many others, some of which are ten times that age, or more. Our streamflow records from gauging stations are less than 200 years old. Consequently, no direct measurements of stream discharge are available with which to identify periods of drought or water abundance for earlier times. Such information is quite important, however, now that we are concerned with the possible existence of long-term global cycles, cycles that last much longer than 200 years. Without long-term data, going back hundreds or even thousands of years, it is difficult to accurately determine long-term trends.

How can we tell whether we are interfering in natural processes, unless we really know how they work? There is a strong suspicion among scientists that

FIGURE 7.9 Stream channel changes.

Four views of the San Juan River near Bluff, Utah, show how a stream channel changes shape and depth during a flood. At (A), the stream is shallow and quiet, flowing normally. At (B), its discharge has increased about ten times, its level rising. At (C), the peak of the flood, discharge has increased ninety-five times above normal, and the great force of rushing water is scouring the stream channel several feet deeper. At (D), flood waters are subsiding, with discharge down to twenty-eight times normal. As the water slows, sediment drops out, refilling the stream channel.

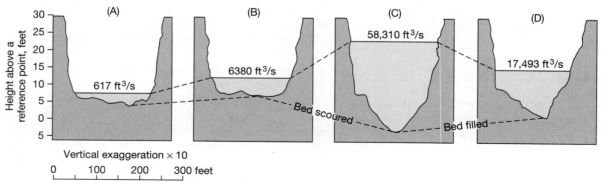

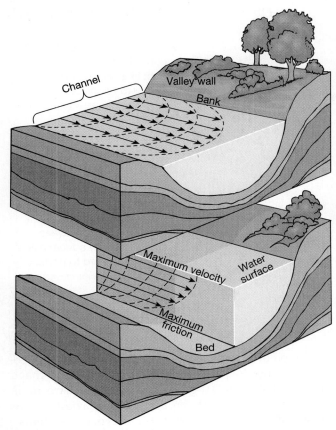

FIGURE 7.10 Stream flow velocity.

Stream flow is not uniform in velocity. It is most rapid near the center of a stream, and just below the surface.

many natural processes are cyclic. For example, because atmospheric temperature affects evaporation and precipitation, the amount of rainfall may be cyclic. Another example, to be considered in Chapter 17, is global warming (the global heat balance). Whether or not such long-term cycles exist, short-term ones have practical consequences, so it is important to determine whether they exist. Discharge measurements extending back many years, if they existed, would provide information useful in evaluating cyclic precipitation.

Despite our lack of stream-discharge measurements for the United States beyond 200 years ago, there is another way to approach the problem. Stream discharge is closely related to surface runoff: the greater the runoff, the more water in the streams and the higher their discharge. Using this concept, long-term stream flow records for the Colorado River in the West have been constructed by examining the growth rings of old trees. Annual runoff has been correlated with the width of these annual rings.

Gauging stations have been present on the Colorado for about 100 years, so the period from 1899 to 1963 was used as a calibration period. If rings for times before gauging are thicker than during the 1899–1963 period, then runoff, discharge, and precipitation were greater than today. If thinner, then runoff, discharge, and precipitation were less than today.

The results of this study are shown in Figure 7.13, from which several facts are evident. First, the period 1907–1930 contained the longest series of high-flow years in the entire 450-year record. Only one other period in the early 1600s is comparable. Second, droughts between 1868 and 1892 and between 1564 and 1600 are of longer duration and of greater magnitude than for any other period during the gauged record. Unfortunately, it was during the 1907–1930 high-flow period that the Colorado River water-allocation pact was completed, apportioning this scarce water resource to Arizona ranchers, the city of Los Angeles, and others.

What can be the meaning of a water pact whose allocations of water were based on years of unusually great supply? And how are we to handle droughts that can last for at least thirty years? The continual court battles of recent years over the distribution of Colorado River water suggest that the answer to these questions is perpetual litigation.

Stream Velocity

Recall the longitudinal profile of streams in Figure 7.5. Although streams all have similar profiles, the steepness, or gradient, varies from stream to stream. This is important, because the main control of **stream velocity** is **stream gradient,** which is the vertical drop of a stream per unit distance.

For example, if a stream drops 20 feet in elevation over the distance of a mile, its gradient is 20 feet per mile. The gradient of some mountain streams may be as great as 200 feet per mile. In contrast, portions of the lower Mississippi have a very gentle gradient of only 0.5 foot per mile.

As a stream leaves a highland area and heads across lowlands toward the ocean, its gradient decreases. Does this mean that the stream's velocity will decrease, too? Not necessarily. There is more water in the channel downstream, and this may nullify the effect of decreasing stream gradient. Stream velocity actually may increase downstream despite the lower gradient.

As the stream moves over the bottom it causes or creates friction with the channel base and sides. This drag creates turbulence along the water-sediment contact. Water cannot maintain a linear (laminar) flow pattern and is disrupted into a chaotic pattern (Figure 7.14). Spinning **eddies** rise from the stream bottom, carrying with them sediment that was deposited at an earlier time. Because turbulence is present in all streams,

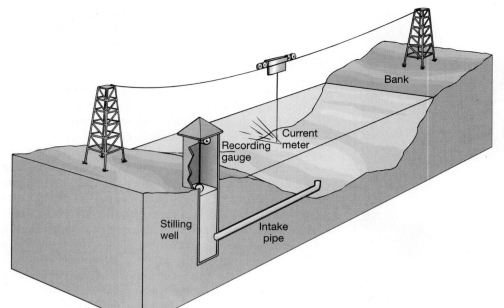

(A)

FIGURE 7.11 Gauging floods.
(A) A stream-gauging station. It measures stream velocity with a current meter and stream height in a stilling well.
(B) Record floods of the Mississippi River, marked on a flood wall at Vicksburg, Mississippi, 170 miles from the river's mouth at the Gulf of Mexico. Note the top mark: "1927, if levees had held, 62.2 [feet]." But they didn't, so the river rose to 56.2 feet (the mark below). The water poured through the broken levees, inundating many square miles of land they failed to protect.
Photo: U.S. Army Corps of Engineers.

(B)

the shape of the channel is continually changing, perhaps getting deeper, perhaps getting wider, perhaps getting shallower as sediment eroded upstream is deposited at the site. Even the slowest, laziest stream experiences turbulence. Streams are dynamic systems.

Stream Load

The material carried by a stream is its **stream load.** It commonly is divided into three parts:

- **Bed load** is the gravel-to-sand-sized coarser sediment that is too heavy to be lifted from the stream bottom. It still is moved by the water, transported by rolling or sliding. Sand-sized grains can be lifted from the bottom, but are too heavy to be carried downstream suspended in the water, so they move in a hopping motion called **saltation.**
- **Suspended load** is the sediment carried more or less continuously within the body of the water. This sediment is so fine-grained (silt-sized and clay-sized particles) that, once suspended by turbulence, it is kept suspended almost indefinitely by ever-present eddies in the water.
- **Dissolved load** is the dissolved material in the water, the ions that were removed from minerals and rocks during weathering.

In most large streams, 80–90 percent of the load is suspended load because clay is produced in such great abundance by weathering processes. The bulk of the suspended sediment in streams is eroded soil on its

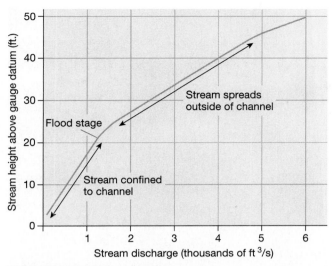

FIGURE 7.12 Rating curve.

Rating curve for a location on Black Bear Creek, Garfield County, Oklahoma. *Source: Oklahoma Geological Survey.*

way to the ocean. The amount of suspended sediment that a stream can carry increases very rapidly as water discharge increases and is, for practical purposes, unlimited. In some streams, the amount of mud is so great that the water looks like tomato soup because of the red color of the clay. Figure 7.15 shows suspended mud being discharged by the Mississippi River into the Gulf of Mexico.

Why Does It Precipitate?

Streams are fed by precipitation: rain, snow, sleet, and hail. But why does it precipitate? Why does the moisture in the air fall to the ground, sometimes gently and briefly, and other times in a deluge for hours?

The answer lies in the ability of warm air to house more moisture (water molecules) than cold air (Figure 7.16). Recall that this is why it rains so frequently and heavily in the tropics (Chapter 6). Because of the more direct impact of nearly perpendicular solar radiation at the equator compared to the poles (Figure 6.7), the ground at the equator is continually warmed and thus is a continual source of heat. Consequently, air at the ground always is quite warm compared to the air higher up. This warm air is less dense than cold air, and thus is buoyant, so it rises. As it rises, it expands and cools. As the air cools, it can hold less moisture, so the moisture is "squeezed out" (condenses). Gravity then pulls the moisture back to Earth's surface—in other words, it rains. This is why the equatorial region is rainy and steamy.

Hot, buoyant air that rises and cools is only one way to generate precipitation. A second way to cool warm, moist air is to force it to rise over a mass of colder air. This is what happens when warm, moist air moving north from the Gulf of Mexico meets cold air moving southward. This collision along the polar front is shown in Figure 7.17. The cold polar air is denser and hugs the ground, whereas the warm air full of Gulf moisture is less dense, and must rise over the cold air. What happens as the warm, wet air rises? It cools, producing rain, just like over the equatorial region.

This phenomenon, occurring repeatedly, was the cause of the great Midwest flood of 1993. That year, the northward-bound warm air was stalled by the stationary jet stream. So it rained, and it rained, and it rained, as warm, moist air kept pushing northward from the Gulf, a virtual river of water vapor pushed up over colder air, condensing and falling. This resulted in one of America's worst floods.

A third way of forcing warm, moist air to rise is to push it up over a mountain range. This occurs on the west side of the Sierra Nevada in California, where that north–south string of mountains gets in the way of Pacific moist air that is pushing eastward. The air must rise to go over the mountains and, as it does, it cools. In rising air, the temperature drops about one degree Fahrenheit for every 300 foot increase in elevation. The rising air expands, cools, and its moisture condenses and falls. As a result, the west side of the Sierra Nevada is well-watered and has abundant vegetation. The eastern side, however, gets what is left, which is little. As

FIGURE 7.13 Tree ring clues.

Long-term reconstruction of the annual runoff at Lees Ferry, Arizona, using tree ring data. *Source: C. W. Stockton, Long-Term Streamflow Records Reconstructed from Tree Rings, Papers of the Laboratory of Tree-Ring Research, No. 5, Tucson: University of Arizona Press, 1975.*

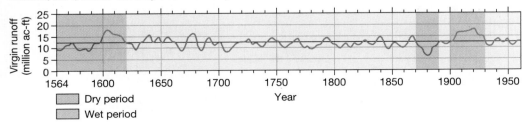

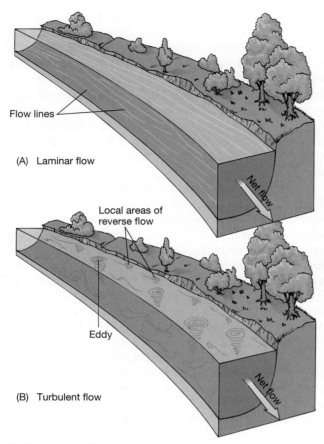

(A) Laminar flow

Flow lines

Net flow

Local areas of
reverse flow

Eddy

(B) Turbulent flow

Net flow

FIGURE 7.14   Types of streamflow.

At low velocity, (A), *laminar* flow lines are relatively straight
and parallel. At high velocity, (B), *turbulent* flow lines are
erratic, forming swirling eddies, and water is mixed top-to-
bottom. Despite the swirling, the net flow is downstream.

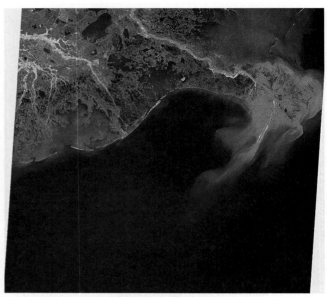

FIGURE 7.15   Mississippi River Delta.

False color image of the Mississippi River Delta, from
Landsat. The cloudy, light blue area is suspended mud
being discharged by the river into the Gulf of Mexico. *Photo:
NASA Headquarters.*

one moves progressively eastward across the Basin and
Range Province of Nevada and Utah, each valley grows
dryer than the one before (Figure 6.10).

The world's greatest annual rainfalls occur where
mountain barriers lie across the paths of moisture-bear-
ing winds. A famous example is Cherrapunji, India, on
the southern edge of the Himalaya, Earth's highest
mountain range. Warm, moist air moving northward

**FIGURE 7.16   How much water
can the air hold?**

Maximum amount of water that
can be held in air at different temperatures.

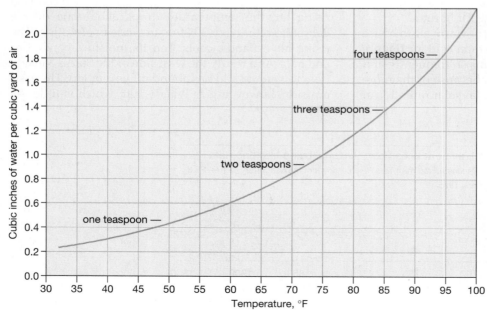

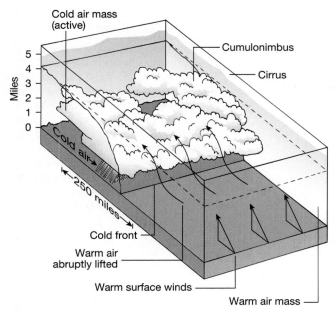

Cold air mass (active)

Cumulonimbus

Cirrus

Miles
5
4
3
2
1
0

Cold air

250 miles

Cold front

Warm air abruptly lifted

Warm surface winds

Warm air mass

FIGURE 7.17 Cold front coming through.

Warm, moist air is forced upward abruptly by denser advancing cold air. As the air rises, it is cooled by expansion to the dewpoint temperature. The condensed moisture forms clouds that may produce large raindrops, heavy showers, lightning and thunder, and hail. *After Robert W. Christopherson,* Geosystems, Second Ed., *Englewood Cliffs, New Jersey: Prentice Hall, 1994, p. 220.*

from the Indian Ocean is forced to rise, expand, cool, and drop its bountiful load of moisture, with the expected result. The recording station at Cherrapunji averages 450 inches of rain a year (37½ feet). In 1873, the total was a relatively low 283 inches; in 1861, the annual total was an incredible 905 inches (75 feet!), with 366 inches (30 feet) falling in the month of July alone. For comparison, the average annual precipitation in the United States is about 30 inches, and the highest rainfall areas in the central Gulf Coast and extreme northwestern coast receive perhaps 80 inches (Figure 6.4).

On our continent, the flash flood on the Big Thompson River that killed 145 people in 1976 was caused by cool air rising up a mountain slope, but it was a rise caused by the daily temperature change. In the late afternoon, cool afternoon breezes flowed up the side of the Big Thompson Canyon, cooled the warmer moist air sitting atop the canyon slopes, and triggered the catastrophic deluge. Such strong thunderstorms are quite common in this part of the Rocky Mountains and in Florida (Figure 7.18). See the section "Case Study: Big Thompson Canyon, Colorado, 1976" later in this chapter for more detail about these storm prone areas. Florida, however, has no mountains and the rains there are caused in another way.

Florida demonstrates the final way to make warm air rise, expand, cool, and rain. This occurs almost every afternoon in Florida and other southeastern states. Warm air near the ground is confined by a blanket of cooler air above. But intense sunshine through the day heats the air, building pressure. By late in the day, the air finally breaks through the blanket of cool air to rise, condense to form clouds (Figure 7.19), and create thunderstorms. These rains can cause flooding because Florida's landscape is flat as a pancake and poor drainage in some areas cannot remove the water as fast as it accumulates.

The formula for precipitation is always the same: warm, moist air is forced to rise by some mechanism; it expands, cools, and rain, snow, sleet, or hail falls. The only difference from place to place is the mechanism that causes the warm air to rise.

Lightning

About 2000 thunderstorms are booming over Earth at any moment. Meteorologists estimate that these generate about 8 million lightning strikes per day, averaging almost 100 strikes a second (Figure 7.20). Lightning, of course, is not distributed uniformly. You may see no lightning for long periods, or you may see many flashes within minutes.

These 8 million daily flashes worldwide generate a million-million watts of electrical power, more than the combined output of all the electric power generators in the United States. About 100 people are killed by lightning in the United States each year, about two-thirds of them in June, July, and August, the months when thunderstorms are most common. Three-quarters of fatal lightning strikes hit men because they spend more time outdoors than women. Golfers are particularly prone to being hit.

Your chances of getting hit by lightning are very slight: only 100 people out of the U.S. population of about 255 million get nailed each year. But one very unlucky park ranger in Virginia was struck by lightning eight times before his natural death (not from lightning). Lightning caused him to lose his big toe in 1942, his eyebrows in 1969, and his hair was set afire twice. Only slight burns resulted from his other unplanned electrical connections.

Benjamin Franklin was first to establish that lightning is electricity. In 1752, he flew history's most famous kite into a thunderstorm and watched the sparks jumping from a key hanging on the kite string to the knuckles of his hand. Luckily, his kite did not receive a direct hit by a lightning bolt; if it had, history would remember fried Franklin. Although two-thirds of those involved with lightning make a full recovery, it is probably because they were not directly hit.

Lightning occurs because of positive and negative charges that build at the tops and bases of clouds (Figure 7.21). The charges build by friction between

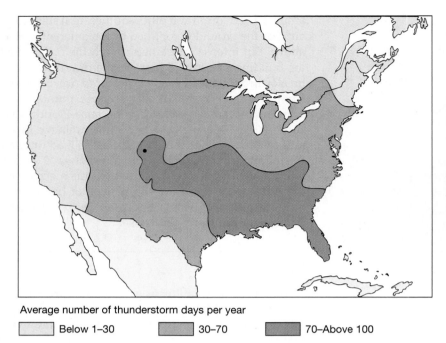

(A) Map shows the average number of days with thunderstorms per year. Thunderstorms are most frequent in the warm, humid southeast and in the central Great Plains. The dot shows the location of Big Thompson Canyon. (B) Diagram of the thunderstorm that caused the flood at Big Thompson Canyon. Moist air rose and cooled dramatically, its water vapor condensing into too many raindrops for Big Thompson River to carry away.

Average number of thunderstorm days per year

Below 1–30   30–70   70–Above 100

(A)

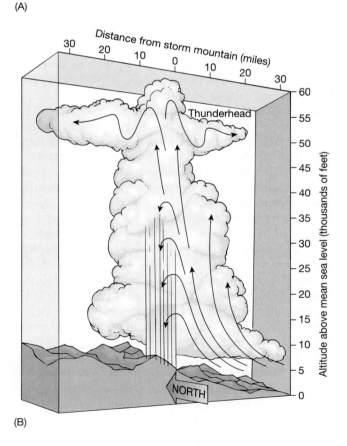

(B)

power the resistance of the atmosphere, a lightning bolt is formed.

The atmosphere is a poor conductor of electricity, so charges must build to at least 100 million volts before a lightning bolt can flash over. Compare this voltage with that used in portable electronics (3–9 volts), a car (12 volts), a table lamp (115–120 volts), or electric range (220–240 volts).

A lightning flash typically lasts a brief fraction of a second but generates a stunning 40 million kilowatts of power in a channel that measures only 0.1 inch to 4 inches in diameter. Concentrating this extreme amount of energy into such a tiny area superheats the air surrounding a lightning bolt to more than 55,000 degrees Fahrenheit.

The number-one rule of lightning safety is to stay indoors during a storm, away from metal plumbing and wiring. Stay out of the shower and off the phone (unless it's cordless or cellular). If you are caught outside, don't seek shelter beneath a tree, for trees are wonderful electrical conductors and favorite targets for lightning, as is demonstrated by the many farm animals that die beneath the oak they thought was a good place to wait out the rain. Your best protection if trapped outdoors is to lie in a low place, for lightning seeks high points like trees, antennas, utility poles, and church steeples.

rising and falling air drafts. It is similar to the static charge you can generate by rubbing your feet on a rug. The upper part of the cloud develops positive charges, the base negative charges. Opposite charges attract, so when the charges become strong enough to over-

Thunder, Son of Lightning

Thunderstorms are so called for the obvious reason that they produce thunder. But they really should be called lightning storms, for without lightning, there would be

FIGURE 7.19 Clouds over Florida.

Notice how the state's land area is well defined by the moisture condensing above it. *Photo: NASA/Johnson Space Center.*

no noise. The enormous heat generated by a lightning bolt causes the air near it to expand explosively, producing the intense sound we call thunder. No lightning, no thunder. Lightning and thunder occur at the same time. But light energy travels at 186,000 miles per second, whereas sound energy (which is the mechanical shoving of air molecules) travels only 1100 feet per second, so you see the lightning long before you hear the thunder.

To get a rough estimate of the distance between you and the lightning bolt, count the seconds from the time you see the flash until you hear the thunder. Every five seconds is a distance of about 5500 feet, or a little over a mile. Thunder normally can be heard at least 10 miles from the lightning strike, and occasionally as far as 15 miles.

Tornadoes

Thunderstorms produce not only rain, hail, lightning, and thunder, but sometimes tornadoes (Figure 7.22A). About 800 tornadoes are spotted in an average year in the United States (Figure 7.22B). Canadians observe about twenty. About one percent of thunderstorms give birth to tornadoes, usually out of the backside of the storm.

Even when using the most sophisticated equipment, tornado occurrence is unpredictable. They pop out of the dark clouds, extending Earthward in a dark mass of rotating wind from 5 yards to 300 yards wide, moving along the ground a few miles or tens of miles at 30–40 mph, and then retreat and disappear back into the clouds. Some tornadoes have been clocked at 60–65 mph. Some contact the ground and then lift, only to strike again at a distance of a mile or two. The devastating force of a tornado comes from the increasing speed with which the air rotates as it tightens into a funnel, in the same way a figure skater accelerates her spin by pulling her arms to her chest.

The word *tornado* is a Spanish verb that means "to turn." In shape, a tornado resembles an elephant's trunk, formed by winds rotating at 300 mph or more (Table 7.1). Inside, the tornado acts as a vacuum cleaner, sucking up anything loose it encounters. Black soil vacuumed by a tornado gives its funnel its usually dark color. Multiple funnels may descend from the same cloud.

Tornadic winds are strong enough to raise the roof and collapse a substantial house, suck up a railroad car, or open the fibers of a telephone pole and drive a straw through it. Paradoxically, the winds can carry a jar of pickles unbroken or pluck the feathers from a chicken. According to calculations by structural

(A)

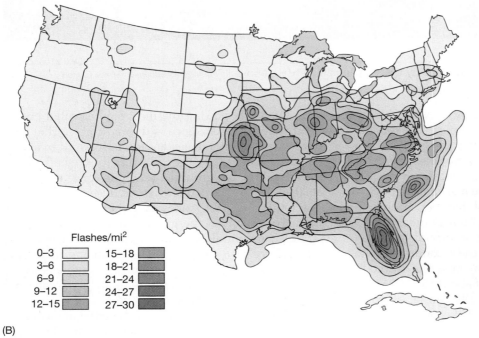

Flashes/mi²

0–3	15–18
3–6	18–21
6–9	21–24
9–12	24–27
12–15	27–30

(B)

FIGURE 7.20   Lightning.

(A) Dramatic lightning streaks through the early evening sky over Tampa, Florida. *Photo: Nelson Medina/Photo Researchers, Inc.*
(B) Annual lightning flash density contours (cloud-to-ground flashes per square kilometer) for 1989. *Source:* Monthly Weather Review, *American Meteorological Society.*

engineers, a 160-mph rotating wind will produce a lifting force of over 30 tons on a typical house. If the wind speed doubles to more than 300 mph, the lifting force is 100 tons.

In the most deadly recorded tornado, 689 people died and more than 2000 were injured in three midwestern states in 1925. In April 1991, a four-state area was devastated when more than seventy tornadoes touched down from Texas to Nebraska, killing at least thirty people and causing millions in property dam-

age. In April 1974, the Great Plains experienced 127 tornadoes, killing 315 and injuring over 6000 in eleven states, with damage exceeding $600 million. The Red Cross estimated that 27,590 families suffered some kind of loss.

Almost three-quarters of Earth's tornadoes occur in the continental United States, and about a third of these happen in Texas, Oklahoma, and Kansas, a swath known as "tornado alley" (Figure 7.22B). Half of these tornadoes occur in May and June.

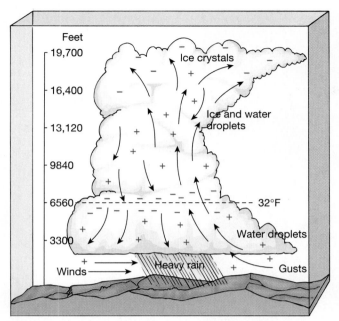

FIGURE 7.21   Anatomy of a thunderstorm cloud.

Structure and form of a cumulonimbus thunderstorm cloud. Violent updrafts and downdrafts mark the circulation within the cloud. Blustery wind gusts occur along the ground.

From Kansas in June 1928 comes an extraordinary account from a farmer named Will Keller. He saw a tornado coming and, calling to his family, ran to his tornado cellar. But just before slamming the door, for a few seconds he took a good look . . .

> At last the great shaggy end of the funnel hung directly overhead. Everything was as death. There was a strong gassy odor and it seemed I could not breathe. There was a screaming, hissing sound coming directly from the end of the funnel. I looked up and to my astonishment I saw right up into the heart of the tornado.

> There was a circular opening in the center of the funnel, about 50 to 100 feet in diameter, and extending upward for at least one-half mile, as best I could judge under the circumstances. The walls of this opening were of rotating clouds and the whole was made brilliantly visible by constant flashes of lightning which zigzagged from side to side.

> Around the lower rim of the great vortex small tornadoes were constantly forming and breaking away. These looked like tails as they writhed their way around the end of the funnel. It was these that made the hissing noise. I noticed that the direction and rotation of the great whirl was anticlockwise [counterclockwise], but the small twisters rotated both ways— some one way and some another.

It is not possible to design structures to withstand tornadoes; the winds are simply too strong, too freakish, and too uncommon for such design to be economically feasible. The National Weather Service in the United States forecasts probable tornado conditions and locally tracks the formation and progress of the storms. Most tornadoes move from southwest to northeast.

A *tornado watch* is announced when conditions make severe thunderstorms and tornadoes possible. However, 50–70 percent of the warnings issued are not followed by tornadoes. Tornado watches can cover thousands of square miles. A *tornado warning* is announced when a tornado actually is sighted, and people need to take cover in a tornado cellar or basement. If you don't have a tornado cellar or basement in your home, stay in a windowless room in the center of the house, such as a closet. Most tornado deaths result from head injuries caused by flying and falling debris produced by the high winds. If you protect your head and neck, you have a much better chance of surviving. If you have time, turn off the main gas and electric service to reduce the likelihood of fire.

If you are outside, get inside. Any structure is better than being out in the open. If you are unable to get inside, run or drive at right angles to the tornado's path. If there is no time to escape, lie flat in the nearest depression or ditch. Don't stay in your vehicle. Most tornado deaths occur in mobile homes and vehicles.

TABLE 7.1   Rating the Power of Tornadoes

Meteorologists use a system known as the Fujita-Pearson Tornado Intensity Scale to rate tornadoes.

Scale	Category	Force (mph)	Path Length (miles)	Path Width (yd/miles)	Expected Damage
F0	Weak	40–72	0–1	0–17 yds.	Light
F1	Weak	73–112	1–3.1	18–55 yds.	Moderate
F2	Strong	113–157	3.2–9.9	56–175 yds.	Considerable
F3	Strong	158–206	10–31	76–556 yds.	Severe
F4	Violent	207–260	32–99	.34–.9 mi.	Devastating
F5	Violent	260–318	100–315	1–3.1 mi.	Incredible

(A)

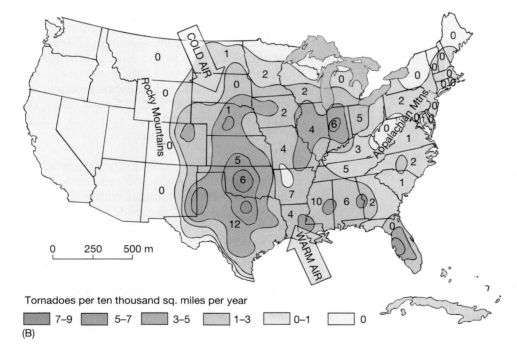

Tornadoes per ten thousand sq. miles per year

7–9   5–7   3–5   1–3   0–1   0

(B)

FIGURE 7.22   A spectacular twister.

(A) The awesome power of a tornado can shatter our illusion of security. *Photo: Howard Bluestein/Photo Researchers, Inc.*
(B) Map shows where tornadoes occurred in the U.S. from 1950 to 1985. Numbers on the map are average annual tornado-related death tolls in each state. *Source: NOAA.*

Floods

Precipitation—caused by the mechanisms we explored in the last section—can lead to a rapid increase in a stream's discharge (Figure 7.24). When a stream's discharge becomes so great that it exceeds the capacity of its channel, it overflows its banks in a **flood.** Floods are the most commonly experienced natural hazards (Figure 7.23). In the United States, rainstorms and their resulting floods and debris flows accounted for 61 percent (337 of 531) of federally declared disasters during 1965–1985.

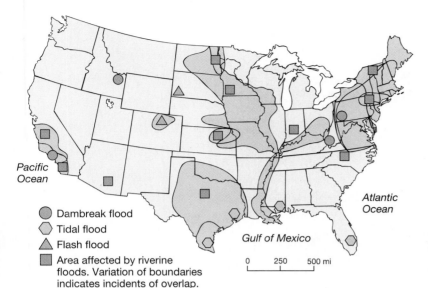

Dambreak flood
Tidal flood
Flash flood
Area affected by riverine
floods. Variation of boundaries
indicates incidents of overlap.

Pacific
Ocean

Atlantic
Ocean

Gulf of Mexico

0   250   500 mi

FIGURE 7.23   Great floods.

Map shows major floods in the conterminous
United States in the years 1889–1993. *Source:
U.S. Geological Survey.*

Case Study: "Slow-Rise" Flooding, U.S. Midwest, 1993

Flooding was widespread in the Midwest during the summer of 1993 (Figure 7.25). This was the costliest, most devastating flood in U.S. history. Forty-eight people died and damage totaled $15 to $20 billion in Illinois, Iowa, Kansas, Minnesota, Missouri, Nebraska, Wisconsin, and the Dakotas. About 100,000 housing units (homes and apartment buildings) were flooded. The failure of key infrastructure, including 388 wastewater facilities, spread the flood's effects far beyond the actual flooded areas. Hazardous waste was released into floodwaters from fifty-four sites on the federal government's most-polluted list (see Superfund Sites in Chapter 9).

The statistics from this event are impressive:

- During the seven months from January through July, more than an average year's rainfall fell in the upper Mississippi River drainage basin.
- Ten representative weather stations recorded greater than normal precipitation for the period, and eight received more than twice their normal rainfall. In July, three stations received more than four times that month's normal precipitation.
- Streamwater discharges of a level that occurs only once in ten years were recorded at 154 locations. Volumes of streamwater than normally occur only once in a hundred years were recorded at 45 locations. At 41 stations, new records for streamflow were set, to the dismay of the new record holders.

What caused this calamity? Obviously, it was the weather, but the specifics are important to know. The jet stream, a west-to-east-flowing river of air that normally flows at high altitude over Canada during the summer months, stalled over Iowa and the Midwest. It formed a barrier to moisture-laden air that was moving northward from the Gulf of Mexico. That air bumped into the jet stream over Iowa and dumped its moisture almost without letup.

It rained for fifty out of fifty-five days in Iowa, and rivers rose as high as 18.5 feet above their banks (Figure 7.26). Average rainfall in July is normally between 2.8 inches and 3.8 inches over the state, but, in 1993, rainfall exceeded 10 inches, the highest July total in 121 years of state record-keeping. The Boyer River at Logan in southwestern Iowa rose 15.7 feet in 24 hours on July 9.

In Van Meter, just west of Des Moines, a sandbag wall built by hundreds of volunteers and National Guard troops was simply no match for a river that was rising nearly one foot per hour (see chapter opening photo). The floodwater swept under the sandbags and pushed them aside on its way eastward into Des Moines. The USGS estimated that a flood of this severity occurs only once in every 500 years, on average. Damage exceeded $1 billion in the city (Figure 7.27).

The viewpoint of many Iowans in the summer of 1993 was well-expressed by Mark Twain in his classic, *Life on the Mississippi*: "You can plan or not plan and it doesn't make a hell of a lot of difference. What makes a difference is how much it decides to rain." Was he correct? Can large floods be predicted? After all, we know a lot more about these things now than when Twain wrote those words in 1883.

Some floods can be predicted. Iowa's 1993 flooding was an example of a "slow-rise" flood. Such flooding occurs where a drainage basin is very large. Here, the rate and timing of downstream flooding can be predicted days in advance from upstream information because one can see it coming.

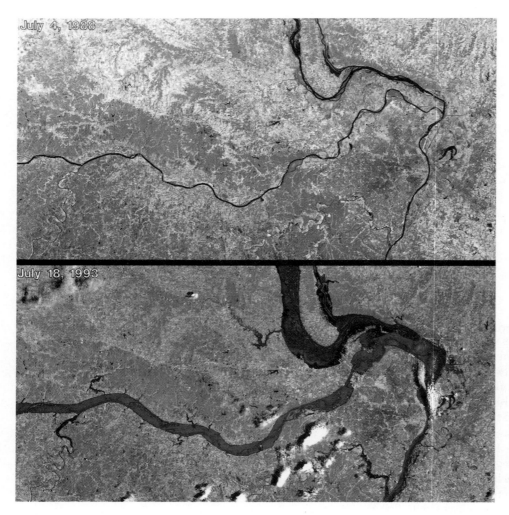

July 4, 1988

July 18, 1993

FIGURE 7.24 A visible difference.

Satellite views of the Missouri River (center) flowing into the Mississippi River. St. Louis is just south of their confluence. Upper image shows rivers at low level during the drought and record heat wave of summer 1988. Lower image depicts the peak of the 1993 flood. *Photo: Earth Observation Satellite Company.*

FIGURE 7.25 Courage.

Fearless flood victim spells it out in sandbags during the 1993 flooding in St. Charles County, Missouri. *Photo: Lee Stone/Sygma.*

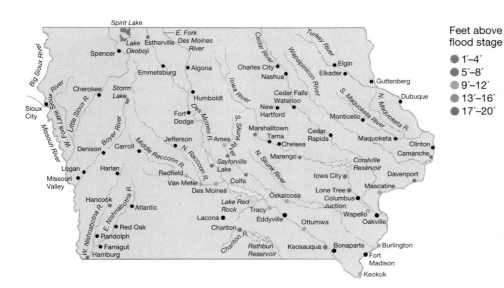

FIGURE 7.26 Above flood stage.

Flood levels and peaks of Iowa's major rivers and creeks during the 1993 flooding. *Wegner, Boone, and Cochran, eds., 1993,* Iowa's Lost Summer. The Flood of 1993. *Des Moines, Iowa: Des Moines Register and Tribune Company, p. 2.*

Flash Floods

In contrast, a **flash flood** occurs very rapidly in a small drainage basin. The flood comes roaring from a canyon in a breaking wave of water often 10–15 feet high, churning together mud, rocks, and debris, moving boulders and trees, and destroying buildings in its path. Flash floods happen in semiarid mountainous regions where precipitation occurs as intense thunderstorms and stream gradients are high.

The Rocky Mountain region of the western United States is a classic example of such an area. Because of the high velocity of the water, people downstream may have less than an hour's warning before the deluge hits. Consequently, there is limited opportunity for an organized, informed response, and emphasis turns toward saving lives, rather than reducing property damage.

Based on the experience of many flash floods, a relationship has been determined between the number of hours of warning before a flash flood and loss

FIGURE 7.27 Taylor Stadium in Des Moines on July 11, 1993.

The stadium is home to the Iowa Cubs professional baseball team. *Photo by Bob Nandell; courtesy of* The Des Moines Register.

prevention (Figure 7.28). No warning, of course, results in the maximum casualties and damage. If any warning is given, the reduction in loss of life and property damage is significant. However, the usefulness of a warning decreases for times greater than 12–15 hours. Warning times greater than one or two days are of no real use. Further, the volume of water in a flash flood is so great that most structures in the path of the flood cannot be protected, regardless of warning time. As the graph shows, it is essentially impossible to protect more than about one-third of them.

Case Study: Big Thompson Canyon, Colorado, 1976. The best-studied recent example of a devasting flash flood occurred in Big Thompson Canyon, Colorado, on July 31/August 1, 1976. The flood and associated debris flows claimed 139 lives and caused $50 million damage to highways, roads, bridges, homes, and small businesses (Figure 7.29).

The brunt of the storm occurred over the Big Thompson River basin between Drake and Estes Park. Rainfalls up to 12 inches were reported. At Glen Comfort, 7.5 inches fell during 70 minutes. Peak discharge at the canyon mouth near Drake was 31,200 cubic feet per second, more than four times the highest previous discharge recorded in eighty-eight years of flood records there. The flood crest moved through the 7.7 mile stretch between Drake and the Canyon mouth in about 30 minutes, an average rate of 15 miles per hour, or 23 feet per second.

It is easy to understand why hundreds of residents, campers, and tourists were caught with little or no warning. Carbon-14 dating of previous flood deposits in Big Thompson Canyon indicate that a flood as large as the one in 1976 had not occurred for at least 1000 years and perhaps not for 7000 years or more.

FIGURE 7.28 The value of a warning.

Relation between damage reduction and warning lead time. No more than about one-third of the damage can be prevented, even with two days of advance warning. *Source: C. Barrett, 1990,* Handbook on Automated Local Flood Warning and Response Systems, *Sunnyvale, California: Handar.*

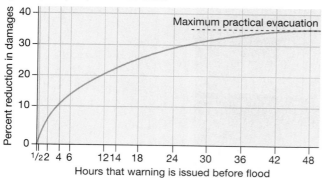

Why Do Floods Occur?

Stream channels develop during times of normal rainfall, so they are adjusted to normal streamflow. They can contain higher flows, up to a point. For example, in the humid northeastern United States, studies reveal that stream banks can contain the highest discharge that occurs within a 28-month period, on average. Most floods result when rainfalls are so great that temporary storage in soil pores is insufficient to keep the stream from rising above bank level.

Flooding is a normal, inevitable part of any stream's life over the years. Whether or not the flooding becomes an environmental problem depends on how close the stream is to population centers. It also depends on the elevation difference between a population center and the top of the stream bank.

Floods also can be caused by snow, ice, and landslides. In regions of heavy snowfall, such as the western United States, unusually high spring temperatures can rapidly melt snow, causing flooding. Snowmelt is produced faster than it can be absorbed by saturated or frozen ground. Flooding also is common in northern climates, where river ice essentially dams the water, forming a lake behind the ice. When the ice is breached it collapses, releasing the water in a flood. Floods from ice dams are common in mountainous western Canada. In warmer climates, temporary sediment dams formed by landslides replace ice dams as a cause of floods. A classic example is the 1925 slide in the Gros Ventre Valley of Wyoming and the subsequent flood in 1927 (see Chapter 5).

How Often Do Floods Occur?

Floods are natural and recurrent events. Over the years, repeated floods build the floodplain of a river. The function of a floodplain is to provide a pathway for excess water—water that won't fit within the stream channel. Unfortunately, human societies all too often ignore this floodplain function and colonize the floodplain for its economic advantages: level ground for construction, fertile soils for crops, ease of access, and ready supply of water. Too often, flooding is viewed like an earthquake—it always happens to someone else, somewhere else.

Living on a floodplain is a gamble like playing dice in Las Vegas. The same rules of chance apply: the stakes are high, but the long-run odds are against winning. The stakes are high because you win only if your flood losses are less than the value gained from living in the floodplain. The long-run odds are poor because floods are bound to occur. The biggest losses in built-up areas come from catastrophic floods like the 1993 Iowa disaster or the 1976 Big Thompson Canyon flood. Floods of this magnitude are rare, but even a small chance for such

(A)

(B)

FIGURE 7.29 Flood damage in Big Thompson Canyon, Colorado.

Village of Waltonia on the Big Thompson River before and after the flood, which claimed 140 lives. Two motels and other buildings were washed away, whereas structures on higher ground were safe. *Photos: U.S. Geological Survey.*

an occurrence is a matter of concern. The more severe a flood, the less its likelihood of occurrence.

Calculating Flood Frequency. How large a flood can be expected, and how often? The data needed to calculate a **flood frequency curve** are obtained from stream-gauging stations where streamflow data are continuously recorded. The data recorded include stream discharge, from which hydrologists calculate flood frequencies. The method for calculating flood frequency of a stream or river is simple and can be used by anyone who contacts their state's geological survey to obtain the necessary data:

1. Obtain the highest discharge for the stream for each year for as long as records have been kept. If there is more than one gauging station on the stream, choose the station with the longest record. The more data points you have, the more accurate will be your estimate of flood frequency.

2. Rank the discharges in decreasing order, from highest to lowest.

3. To determine how often you can expect each magnitude of discharge to occur, apply this formula:

$$\text{average recurrence interval} = \frac{\text{number of years of records} + 1}{\text{rank of discharge on list (3d highest, etc.)}}$$

For example:

$$\text{average recurrence interval} = \frac{10 \text{ years of records} + 1}{\text{5th highest discharge}} = \frac{11}{5} = \begin{array}{l}2.2 \text{ (one flood this size every } 2.2 \text{ years}\end{array}$$

Table 7.2 shows the recurrence interval of floods on the Danger River based on ten years of data.

A graph can be constructed showing how frequently any level of discharge is likely to occur. For this example, we used data from Table 7.2 in the equation above to generate the curve in Figure 7.30. There are only ten years of records. The highest data point is at a recurrence interval of 11 years (number of years of records + 1). Note that the estimate of discharge at 11 years is difficult to determine because the data point (1700 cubic feet per second) is an extreme value, much higher than and, therefore, far from the other data points on the graph.

As noted earlier, stream banks are high enough to hold in the stream channel the highest discharge that occurs within a 28-month period, or 2.33 years. In our Danger River example, this is a discharge of 520 cubic feet per second. The projected fifty-year flood discharge of 2500 cubic feet per second is 4.8 times the river's normal "full" discharge.

Determining the highest discharge to expect every 50 or 100 years is much more problematic. It depends very heavily on the highest discharge recorded during the ten-year period. For example, if the highest discharge recorded was 20 percent higher than the 1700 cubic feet per second in our example, the projected discharge of the fifty-year flood would be 3800 cubic feet per second. This dramatic change might mean the difference between minor property damage with small loss of life and a major economic and personal disaster for a community.

Note that all values other than the measured ten data points are *expectations*. Based on the ten data

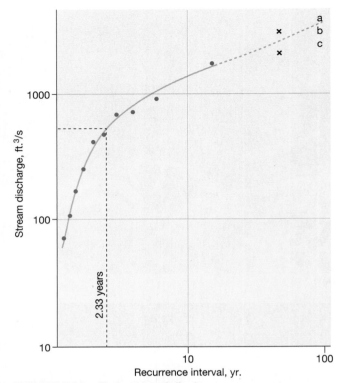

FIGURE 7.30 Estimates of discharges.

Plot of maximum stream discharge of the Danger River, 1985–1994, versus recurrence interval for the data in Table 7.2. Point a is the extrapolation if the highest discharge were 20 percent higher than recorded; b is based on the measured value; c is based on a 20 percent lower value than recorded.

points, we *expect* a discharge of 3500 cubic feet per second every 100 years—but, it is quite possible that the highest discharge between 1994 and 2094 will be only 3000 cubic feet per second; or perhaps it will actually be 4000 cubic feet per second. The small number of data points on which our extrapolation is based ensures that there is a large degree of uncertainty in our expectation.

In natural situations, we are always dealing in probabilities, not certainties. We use real data to make our estimates of future stream behavior, but must not forget that our estimates are only that—estimates. The larger our data set of past stream behavior, the better we will be our estimate of future stream behavior.

Flood hazard (or discharge hazard) estimates sometimes are expressed as **flood probabilities** rather than as recurrence intervals. One value is the reciprocal of the other. (A reciprocal is 1 divided by the number; for example, $1/100$ is the reciprocal of 100.) A 100-year flood, for example, has a $1/100$ (or 1%) chance of occurring in any given year. A 50-year flood has a $1/50$ (or 2%) chance. The possibility of two 50-year floods occurring in one year is small, only $1/50 \times 1/50 = 1/2500$, or 1 chance in 2500 (or 0.04%). But it can happen: in Houston, three

TABLE 7.2 Data for the Maximum Discharge on the Danger River, 1985–1994

The last column shows how often each size of flood is likely to occur.

Year	Maximum Discharge (ft³/s)	Discharges Arranged in Rank Order	Recurrence Interval (Years)
1985	419	1700	11.00
1986	687	900	5.50
1987	728	728	3.66
1988	1700	687	2.75
1989	254	468	2.20
1990	109	419	1.83
1991	71	254	1.57
1992	168	168	1.38
1993	900	109	1.22
1994	468	71	1.10

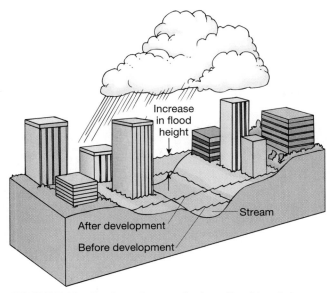

Increase in flood height

After development
Before development
Stream

FIGURE 7.31 Paving of ground raises flood levels.

The cross-section shows flood stages and the resulting property damage.

(column 5), a 65 percent chance within ten years (column 4), a 94 percent chance within twenty-five years (column 3), and a 99.9 percent chance of occurring within fifty years. A 100-year flood has a 63 percent chance of happening within a 100-year period, but only a 1 percent chance in a particular year.

A good analogy is a roulette wheel. The number of black and red numbers is very close to equal so that ten spins of the wheel should result in five black numbers and five red numbers. Nevertheless, there will be numerous ten-spin sequences in which the result will be six black and four red, or seven red and three black. Occasionally, the result may be nine and one, or rarely, ten and zero. Flood probability estimates are similar. In fifty spins of the calendar, there should occur only one fifty-year flood, but it is possible for two or three of them to occur in a single year. Worse yet, which year it will be cannot be determined in advance.

Urbanization and Flooding

The relationship among rainfall, stream discharge, and time is of great significance when we think about floods. Time is more important than many people realize. For example, during a rainfall, one very important factor that slows the rush of sheetflow runoff into a stream channel is vegetation. First, the plant leaves shield the soil from the full impact of raindrops, so that the water is more likely to infiltrate than to erode the

100-year floods occurred in a single year, 1979. The chance of that happening is one in one million! The real world is never risk-free.

Table 7.3 summarizes flood probabilities. For example, a flood with a ten-year return period (column 6) has a ten percent chance of occurring in any one year

TABLE 7.3 Likelihood of Floods of Different Magnitudes

Chance (%) of at Least One Flood of at Least this Size in a Certain Number of Years					
One Hundred years	*Fifty Years*	*Twenty-five years*	*Ten years*	*Any one year*	*Return Period, Years*
				50	2
				40	
				30	
				25	
				20	5
		99	80	15	
	99.9	94	65	10	10
	90.5	71	40	5	20
86	63	40	18	2	50
63	39	22	9.6	1	100
39	22	12	5	0.5	200
18	9.5	5	2	0.2	500
9.5	4.8	2.5	1	0.1	1000
5	2.3	1.2	0.5	.05	2000
2	1.0	0.5	0.2	.02	5000
	0.5	.25	0.1	.01	10,000

Source: B. M. Rich, *Water Resources Bulletin*, 9:187, 1973. Copyright © 1973 American Water Resources Association, Bethesda, Maryland. Reprinted by permission.

soil. Second, plant roots and soil pores can hold a large amount of water—at least temporarily—and thus reduce the maximum stream discharge. This creates a lag between the time of rainfall and the time of peak discharge. Put simply: if the Appalachian Mountains suddenly were stripped of all plants, residents would experience catastrophic flash floods.

Thus, it is no surprise that **urbanization**—paving over land, building construction, stripping vegetation and soil—both increases the highest discharge that local streams attain and decreases the lag time between the rainfall event and flooding (Figure 7.31). This increase clearly shows on a **hydrograph,** a plot of stream discharge over time.

Two hydrographs appear in Figure 7.32. The upper curve shows streamflow under pre-urban conditions, where plenty of vegetation and few impermeable surfaces exist. The lower curve shows the greater stream discharge and briefer lag time between rainfall and peak discharge that result from urbanization.

Smaller floods are more affected by urbanization than are larger, less frequent floods. A 50-year or 100-year flood is hardly affected at all by an increase in the amount of impermeable area caused by urbanization, because such extreme floods overwhelm the storage capacity of soil cover, no matter how extensive.

Anticipating and Controlling Floods

Advance planning for floods or other calamities can save both lives and property. But planning costs money, and the more comprehensive the plan, the higher the cost. How much are people who live on a river's floodplain willing to pay, and for what degree of protection? Most communities that engage in flood protection planning use as a guide the *100-year floodway.* This is the area along the river (floodplain and the zone surrounding it) that is likely to be flooded once each 100 years.

As you have seen, the 100-year discharge can be estimated. By using topographic maps (maps that use contour lines to show the shape and elevation of the land surface) it is easy to see the outline of the area that would be under water during an average 100-year flood. However, because of increasing urbanization, both the 100-year stream discharge and the contours of the land surface around a stream may change over time, so the 100-year floodway may change as well.

Figure 7.33 shows the San Lorenzo River flowing through the town of Felton, California, south of San Francisco. It shows the flood-prone zone along the river. At normal streamflow, the water surface is at an elevation of 240±5 feet, but **floodplain zoning** now prevents most construction below an elevation of 260–265 feet. Unfortunately, a large number of homes and other structures already exist on the east side of the

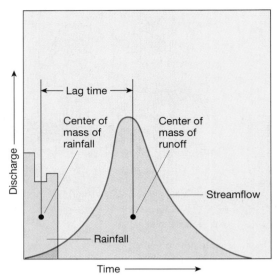

(A) Typical lag time between rainfall and runoff

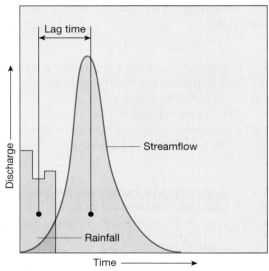

(B) Lag time between rainfall and runoff after urbanization

FIGURE 7.32 Urbanization and flooding.

When a rural area becomes urban, vegetation is removed, and more surface is paved, both of which shorten the lag time between rainfall and flood peak. Note that the flood peak also is higher following urbanization. *After L. B. Leopold, U.S. Geological Survey.*

river (center of map; note elevations). This is within the designated danger zone. Felton's citizens probably are unhappy about decreased property values and higher home insurance premiums that have resulted from the zoning ordinances.

Floodplain Zoning

For the reasons just described, floodplain zoning has not received an enthusiastic response from local govern-

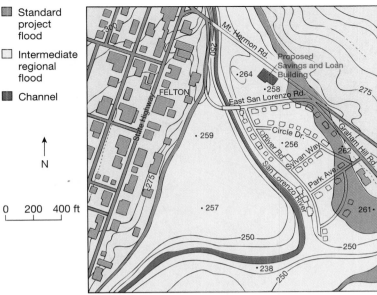

Standard project flood

Intermediate regional flood

Channel

N

0   200   400 ft

Ground elevation in feet above sea level.
Minimum contour interval 5 feet.

FIGURE 7.33   You can't build it here!

Site of a proposed savings and loan building on a floodplain, Santa Cruz County, California. The site of the proposed structure was inundated by up to 3 feet of water during flooding in 1955 and again in 1982. Consequently, the structure was not approved, nor was another proposed commercial complex, and the site was rezoned to a floodplain designation. Now only uses that are compatible with the flood hazard are permitted. *Source: U.S. Army Corps of Engineers,* Flood Plain Information—San Lorenzo River, Santa Cruz County, California, *1973.*

ments. Lack of scientific personnel, lack of money, local political pressure against unpopular restrictions on development, and the higher cost of floodproofing or of elevating structures on stilts combine to make difficult the passage of zoning ordinances. Real estate developers, in particular, almost always are opposed to restricting the use (residential, commercial, industrial) to which they may put their property.

Should communities pass laws restricting the use of private property? Would it be satisfactory simply to inform prospective builders or buyers of the danger, and allow them to do as they wish? Should the government subsidize the increased cost of insurance for those who build in dangerous areas if the town favors the development and no other location is suitable?

Flood Control

Several methods are used to control floods, none of which is without drawbacks. They include channelization, levees, and dams.

Channelization of Streams.   **Channelization** is the name for various methods of improving the stream's channel to increase stream discharge. The purpose is to help the stream carry away water faster from a threatened area. Channelization may increase discharge by dredging to straighten, widen, or deepen the channel.

Within cities, stream channels often are lined with cement to straighten them and keep them from meandering, as streams commonly do (Figure 7.34). Unfortunately, such major modifications in a natural stream course not only are unsightly but have ripple effects both upstream and downstream. Increasing the stream

gradient at one location causes it to increase upstream as well, which accelerates erosion. Down-stream, the incidence of flooding is greatly increased because of the increased volume of water funneled there more hastily

FIGURE 7.34   Stream channelization.

Photo: Stephenie S. Ferguson/William E. Ferguson.

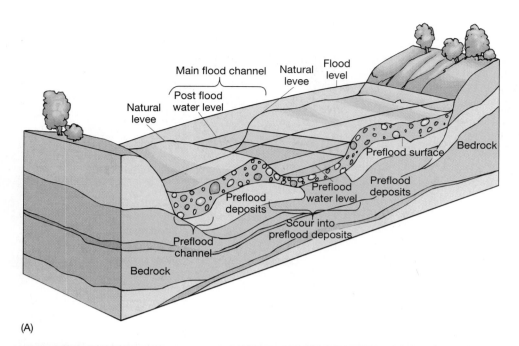

(A)

(B)

FIGURE 7.35 Natural levees and artificial levees.

(A) Cross-section of a natural levee after a flood. As the flood-water overflows its banks, the coarsest (heaviest) sediment is deposited first, closest to the channel. Sediment becomes progressively finer-grained and less abundant away from the banks. (B) Artificial levee keeps floodwaters in foreground from inundating the developed area behind the levee in the back-ground. *Photo: Steve McCutcheon/Visuals Unlimited.*

by the increase in stream velocity upstream. In addition, any change in the natural characteristics of a stream affects the ecosystem of which it is an integral part, usually negatively.

Building Levees along Streams. A second method of flood control is to build artificial levees (also known as dikes), which are raised banks that run along the top edge of the stream channel on each side. Streams build natural levees (Figure 7.35A) when they overflow their banks, because the water slows and sediment is deposited. Natural levees usually are low, capable of con-

taining only the greatest discharge that occurs during an average 28-month period. They afford no protection against major floods. By constructing higher levees along the channel margins, the channel can hold more water, reducing the occurrence of floods (Figure 7.35B).

Artificial levee construction has been advocated by the federal government for more than a century, but they now are reassessing the practice. Over the years, the U.S. Army Corps of Engineers has built numerous levees along the Mississippi that now total more than 2500 miles in length. Some of these levees are 30 feet high.

These levees have contained numerous floods. But the great flood of 1993 *breached* (broke through) some 800 levees, creating serious flooding on land that long had been protected by them. Further, fending off a flood upstream simply creates more severe flooding downstream.

Calculations indicate that, had the agricultural levees along the Mississippi been made high enough to contain the 1993 floodwaters, the river would have risen about 6 feet higher at St. Louis, putting much of that city under water. Also, the projected cost of elevating the levees would be several billion dollars. So, although the 1993 floods in the midcontinent were a disaster for towns and farms along the Mississippi River, the Corps of Engineers does not expect to build levees high enough to prevent similar floods. Instead, they are recommending government buyouts of property in flood-prone areas and improved flood insurance for buildings and crops.

Building Dams across Rivers.   Another widely used flood-control method is a **dam;** 75,000 exist in the United States. Hundreds of thousands more exist throughout the world. If all the water stored in dams were released to the ocean, sea level would rise about 3 inches. These huge concrete or earthen structures span a river and impound water that continually pours in from upstream (Figure 7.36). Some rivers need only a single dam, whereas along others, two or three may be necessary to manage the water volume. The Mississippi has twenty-eight locks and dams, and forty-five major dams have been built along its tributaries.

The pinnacle of dam construction in the United States was reached with the completion in 1976 of a 34-dam complex along the Tennessee River and its tributaries (Figure 7.37). At the time they were built, the sole purpose of these dams was the generation of hydroelectric power for an economically depressed region. Since then, however, the lakes created by these dams and the water resources they make available have led to extensive reforestation, tourism, and development of mineral resources. The dams were constructed by the Tennessee Valley Authority (TVA).

Dam Problems.   Most dams built since the TVA project have been multipurpose. They generate electric power, irrigate crops, and provide recreational activities such as swimming, camping, fishing, and boating. Until about thirty years ago, few questioned the good brought to an area by a dam. Since then, problems have been recognized, resulting in near-total abandonment of dam building.

Among the worst problems are catastrophic dam failures. During the last few decades, *several hundred thousand people worldwide* have been killed by dam failures. These resulted either from inadequate geologic investigation or poor engineering during construction.

FIGURE 7.36   Glen Canyon Dam.

This famous dam spans the Colorado River in Arizona.
Photo: Lowell Heaton/Arizona Department of Transportation.

(Recall the infamous Vaiont Dam disaster in Italy, discussed in Chapter 5.)

A 1992 national dam survey classified almost one-third of this country's 75,000 dams as hazardous, 10,000 as having high hazard potential, and another 13,500 as having significant hazard potential.

Another problem associated with dams is inundation of large amounts of property by the lake that forms behind the dam. This drives out homeowners and wildlife, displaces farms, kills vegetation, and alters ecosystems. The dams also create mud deposits in the reservoir, because the entering stream stops and hence drops its sediment load. In 1941, a USGS study reported that 39 percent of existing reservoirs would be largely filled with sediment before 2000. (Figure 7.38 shows an example from France.) More recent data suggest that only 54 percent of American reservoirs will function for more than a century, and 21 percent will be in use less than fifty years. About one-third of the sediment in the reservoirs may come from eroded cropland (see Chapter 3).

Just how much sediment accumulates annually in an "average" reservoir? Could we dredge the sediment

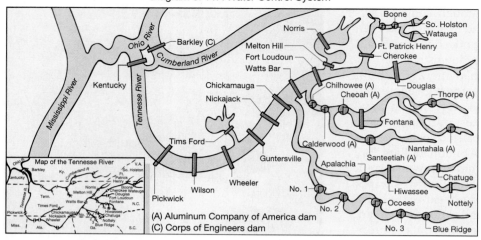

Diagram of TVA Water Control System

(A) Aluminum Company of America dam
(C) Corps of Engineers dam

FIGURE 7.37 The TVA Project.

Essential features of the Tennessee Valley Authority Project. *Source: Tennessee Valley Authority.*

from the reservoir and transport it to a useful place, like a construction site? What would this cost?

Sediment yield from southeastern U.S. drainage basins has been measured, in pounds per square mile per year:

- 110,000 from forested areas (low erosion rates)
- 1,700,000 from rangeland
- 850,000–60,000,000 from construction sites, where land is temporarily exposed in a highly disturbed and erodible form.

Let us assume a conservative amount of sediment entering a reservoir: only 1 million pounds per square mile per year. Dry mud weighs about 150 pounds per cubic foot. Hence, the 1 million pounds of sediment would occupy about 6700 cubic feet, roughly the volume of a 30 × 30-foot bungalow. Each year, the reservoir would accumulate 6700 cubic feet of mud from each square mile of drainage basin. The cost of dredging the mud from a reservoir is about 10 to 15 cents per cubic foot, so the yearly cost of dredging would be at least $670 per square mile of drainage basin.

When this per-square mile cost is multiplied by the 75,000 reservoirs in the nation, it appears that annual dredging cost would be $50 million per square mile of drainage basin. Drainage basins typically encompass many hundreds or thousands of square miles. So, the total dredging cost for American reservoirs would be in the billions annually! Further, our calculation does not include the cost of transporting mud, which might equal the cost of dredging, depending on

FIGURE 7.38 Drought at Sombernon Reservoir, France.

Dried and sun-cracked accumulation of mud exposed by drought and shrinking of the reservoir. *Photo: Didier Dorval/Explorer/Photo Researchers, Inc.*

distance. We conclude that the cost of removing the mud from the nation's reservoirs is prohibitive.

Below the dam, the water that is released after passing through hydroelectric turbines in the dam is free of sediment, all of which was trapped behind the dam. As the water plunges down into the stream in front of the dam, it picks up new sediment, and the increased erosion may extend for many miles downstream. The increased depth of the master stream that results from the increased erosion has effects on its tributaries as well. They also increase their downcutting because the elevation at which they join the master stream has been lowered.

Case Study: Egypt's Aswan Dam. Of even greater importance in some areas downstream from dam sites is the ecological damage and loss of nutrients needed for agriculture. A good example of this effect is the delta of the Nile River in Egypt. From before recorded history, flooding of the delta region at the mouth of the Nile was an annual event, important because it replenished agricultural soils with fresh sediment and water. But in 1968, the Aswan Dam across the Nile was completed. Its purpose was to generate hydroelectric power and control flooding, and both objectives were accomplished.

Unfortunately, unforseen problems soon became evident. After about thirty years, the reservoir behind the dam still is not full, nor is it expected to rise higher. The reasons are simple: evaporation into the hot, dry desert air and infiltration of the reservoir water into the permeable sandstone on which the dam was built. Evaporation is 50 percent greater than pre-construction calculations indicated. And the mud settling in the reservoir has not sealed the permeable sandstone, as had been expected. The reduced reservoir depth has reduced power generation considerably.

But downstream effects of the dam are even worse. Because of the loss of sediment that used to be an annual deposit in the delta and an increasing number of irrigation canals that trap sediment, the delta front is being eroded by currents along the shoreline of the Mediterranean Sea, reducing the land area available for agriculture. Parts of the delta coastline are receding at a rate of more than 300 feet per year. In addition, the drifting desert sand that encroaches on the fringes of the fertile corridor of the Nile valley during the dry season was stabilized by the river-borne mud deposited during the annual flood. Now that the flood has been eliminated, sand encroachment is much more difficult to control because it overwhelms the irrigated lands at the western edge of the Nile River valley. Further, reduced Nile flow is now inadequate to wash away the salts from the soil that are harmful to plant life.

Even worse, the floods that formerly deposited a new layer of nutrient-rich mud have ended. Artificial fertilizers now are needed, an expense not easily borne in a poor country. And finally, the freshwater snails that carry parasitic Bilharzia larvae have become much more abundant in the Nile water used for irrigation, causing a dramatic increase in the incidence of this debilitating and often fatal intestinal disease. The Aswan Dam and reservoir have proven to be much less than the salvation event that was prophesied in the 1960s.

Case Study: Valmeyer—A Town with Good Sense. Humans seem slow to recognize that we are an integral part of the natural world, not some special group apart from it. The natural environment, such as a river system, contains innumerable and often invisible interactions that have developed over billions of years to stabilize Earth as a place fit for the life that inhabits it. Despite our technological capabilities, we humans are a numerically trivial group compared to ants, termites, bacteria, and so on. However, we have larger, versatile brains. We are fully capable of "thinking" our way to extinction, despite our good intentions. It is not possible to improve on Father Nature.

One small group that has recognized the futility of fighting catastrophes that happen repeatedly in the same place is the people of Valmeyer, Illinois. Valmeyer is on the Mississippi River's east bank about 25 miles south of St. Louis. The river's floodwaters inundated the town twice in the summer of 1993, and 900 citizens saw 90 percent of their homes, offices, and public buildings destroyed (Figure 7.39). So they voted to construct a new Valmeyer about two miles east, away from and higher above the river. The Illinois State Geological Survey was asked to investigate and evaluate natural hazards at the proposed new site of about 500 acres.

At a town meeting in December 1993, Valmeyer's citizens voted to ask the federal government for funds to defray the cost of the relocation and new construction. Approval was granted in 1994 and 95 percent of Valmeyer's buildings and homes are being rebuilt at the new site. Only 5 percent of the citizens chose to remain at the old townsite. The federal government is moving or rebuilding about 6600 structures damaged in the floods. The estimated cost of the move is $9.6 million, of which the federal government (the nation's taxpayers) will pay $7.2 million. The remaining $2.4 million, divided among 900 residents, comes to less than $3000 per person. The cost for a family of four is about $11,000.

As a condition of the grant, the town must adopt drainage, sediment, and erosion controls; a storm-water management plan; construction guidelines; and other mitigation ordinances to protect against activating the geologic hazards found by the scientists from the Illinois State Geological Survey. The citizens of Valmeyer believe it is in their best interest to adapt to their surroundings rather than continue to fight what is surely a losing battle.

FIGURE 7.39 A town with good sense.

The 900 townspeople of Valmeyer, Illinois, voted to move themselves to higher ground, rather than face another flood like that of 1993. *Photo: Mary Butkus/AP/Wide World Photos.*

Summary

Of every 100 water molecules on and near Earth's surface, 97 are in the oceans. Most of the remaining three lie frozen in glaciers. Although this has been true for all of human history, it does not mean that the water molecules do not change location. They are always moving in the hydrologic cycle among the oceans, atmosphere, soil, and subsurface rocks. These movements are crucial for human existence.

The size of streams varies with geographic location, local topography, nature of the underlying bedrock, stage of stream development, and the amount of human intervention in the natural cycle. All of these factors influence the possibility of flooding. Although the probability of future floods can be calculated, the estimate is based on past occurrences and is a statistical average.

Human civilization always interferes with the natural development of streams and their water-carrying capacity. Urbanization, artificial channeling, and dam construction alleviate human problems in the short run (the term in office of elected politicians) but are harmful in the long run. And the long run commonly is not really that long, only a few tens of years. It is advisable not to live on a floodplain near an active river. As long as people do such inadvisable things, human disasters will continue to be more frequent than would otherwise occur.

Key Terms

bed load
channelization
dam
dissolved load
drainage basin
drainage divide

eddies
flash flood
flood
flood frequency curve
flood probabilities
floodplain zoning

floodway
gauging stations
headward erosion
hydrograph
levee
longitudinal profile

meander
rating curve
saltation
sheetflow
stage

stream discharge
stream gradient
stream load
stream velocity
suspended load

tributary
trunk stream
urbanization

Stop and Think!

1. What is a drainage basin? Examine a map of the area where you live (preferably a topographic map) and outline the drainage basin for the main stream in the area. How large is the basin? What is its shape?

2. Are the size and shape of the drainage basin related to the geology? How? What kinds of rocks are exposed in the basin?

3. What are the three types of stream load? What geologic or climatic factors determine the relative proportions of each in a stream?

4. Explain why meanders develop, and how they move. Can they pose a threat to communities located along the stream banks? If so, what should the community do about it?

5. What is the purpose of a guaging station? Are there any on the main stream in your area? Phone the local office of the U. S. Geological Survey to find out.

6. How do soil development, vegetation, and ground slope affect the development of floods?

7. What is the meaning of the expression "50-year flood"? If two of them happen in the same year, does it shake your confidence in the usefulness of such expressions? Explain your answer.

8. Keep in mind that the federal government is you (directing federal policy when you vote) and your tax money (paying for what gets done). With that in mind, what is the appropriate role of the federal government in alleviating the hardship and suffering caused by flooding?

9. List three ways in which human activities increase flood risk. Can any of these activities be stopped?

10. What dam is nearest to where you live? When was it built? For what purpose? On what types of rocks does it rest? Phone the local office of the U. S. Geological Survey to find out.

11. List three ways in which urbanization and associated construction activities affect the likelihood of floods. Can anything be done about these three influences?

References and Suggested Readings

Collier, M., Webb, R. H., and Schmidt, J. C. 1996. Dams and rivers. *U.S. Geological Survey Circular* 1126, 94 pp.

Davies-Jones, Robert. 1995. Tornadoes. *Scientific American*, August, p. 34–41.

Leopold, L. B. 1968. Hydrology for urban land planning—a guidebook on the hydrologic effects of urban land use. *U.S. Geological Survey Circular* 559, 18 pp.

Leopold, L. B. 1994. *A View of the River*. Cambridge, Massachusetts: Harvard University Press, 298 pp.

Lopez, R. E. and Holle, R. L. 1996. Fluctuations of lightning casualities in the United States: 1959–1990. *Journal of Climate*, v. 9, p. 608–615.

McCain, J. F., and many others. 1979. Storm and flood of July 31-August 1, 1976, in the Big Thompson River and Cache La Poudre River basins, Larimer and Weld Counties, Colorado. *U.S. Geological Survey Professional Paper* 1115, 152 pp.

Muir, Hazel. 1996. Striking back at lightning. *New Scientist*, October 7, p. 26–30.

Pearce, Fred. 1994. High and dry in Aswan. *New Scientist*, May 7, p. 28–32.

Pendick, Daniel. 1995. Tornado troopers. *Earth*, October, p. 40–49.

Platt, R. H. 1995. Sharing the challenge: Floodplain management into the 21st Century (book review). *Environment*, v. 37, no. 1, p. 25–28.

Russell, C. T. 1993. Planetary lightning. *Annual Review of Earth and Planetary Science*, v. 21, p. 43–87.

Watson, I. and Burnett, A. D. 1993. *Hydrology: An Environmental Approach*. Ft. Lauderdale, Florida: Buchanan Books, 702 pp.

Wegner, Michael, Boone, Lyle, and Cochran, Tim, (eds.). 1993. *Iowa's Lost Summer. The Flood of 1993*. Des Moines, Iowa: Des Moines Register and Tribune Company, 108 pp.

Williams, G. P. and Wolman, M. G. 1984. Downstream effects of dams on alluvial rivers. *U.S. Geological Survey Professional Paper* 1286, 83 pp.

Wolman, M. G. and Riggs, H. C. (eds.). 1990. Surface water hydrology. *The Geology of North America*, v. 0-1, Boulder, Colorado, Geological Society of America, 374 pp.

"And the fish that are in the Nile will die, and the Nile will become foul; and the Egyptians will find difficulty in drinking water from the Nile."

Exodus 7

Water Quality and Pollution

In nature, water has dissolved within it a wide variety of chemical elements and compounds. The inorganic ones come from weathering of rocks and probably include all ninety natural elements in Earth's crust, although some are too scarce to detect. The organic substances come from the decomposition of living things.

The quantity of these substances dissolved in water varies widely. The inorganic substances range from nearly zero in rainwater, to 120 ppm (parts per million) in average river water, to 35,000 ppm in the ocean, to 270,000 ppm in Utah's Great Salt Lake, to 320,000 ppm in the Dead Sea between Israel and Jordan. Organic substances in rivers generally are present at only 10–30 ppm, but in some slow-moving streams that flow through peat bogs, concentration can reach 500,000 ppm (50 percent), a fairly thick soup of many complex molecules.

Added to this natural variation are substances generated by people: inorganic and organic materials from fertilizers, pesticides, feedlots for farm animals, sewage, chemical manufacturing plants, and oil refining facilities. All of these materials are *contaminants.* But which ones are *pollutants*? How do we determine when a contaminant becomes a pollutant? A large number are at pollutant levels in many places. How do we deal with them?

The ongoing pollution of our water resource is one of our top-three most serious environmental problems, which we will address in this chapter. The others are soil pollution (Chapter 4) and air pollution (Chapter 17).

◀

Toxic effluent discharged into the Hudson River near New York City triggers cancer in 90 percent of adult tomcod fish.
Photo: David Woodfall / Tony Stone Images.

Defining "Uncontaminated Water"

Long ago, when people lived off the land, long before plumbing was invented, drinking water usually came from rivers and streams. Over ten thousands of years, we adapted to this water's typical combination of dissolved constituents. Hence, a reasonable definition for "pure" or **uncontaminated water** might be "average stream or river water that is unaffected by human activities" (Figure 8.1). This "pure" average stream water contains about 120 ppm of dissolved substances in the form of ions, shown in Table 8.1.

Examine this table and you will see that only the first nine substances are present in amounts greater than 1 ppm. About two-thirds of the ninety elements present in Earth's crust have been detected in stream water, most in exceedingly small amounts, and not all of them in every stream. No doubt the other third will be detected as measuring techniques improve.

FIGURE 8.1 Appearances deceive.

This sparkling stream water comes with a product warning.
Photo: Sygma.

TABLE 8.1 The Composition of Average River Water Worldwide

These substances exist as ions (electronically charged molecules) dissolved in water. They are listed from most common to least.

Ion name	Ion	ppm
Bicarbonate and Carbonate	HCO_3^-, CO_3^{2-}	58.8
Calcium	Ca^{2+}	15.0
Silica	H_4SiO_4	13.1
Sulfate	SO_4^{2-}	11.2
Chloride	Cl^-	7.8
Sodium	Na^+	6.3
Magnesium	Mg^{2+}	4.1
Potassium	K^+	2.3
Nitrate	NO_3^-	1.0
Iron	Fe^{2+}, Fe^{3+}	0.7
Aluminum	$Al(OH)_4^-$	0.2
Fluorine	F^-	0.1
Strontium	Sr^{2+}	0.1
	Total	120.7

Source: U.S. Geological Survey.

Elements in Water and Human Health

Water we actually drink often deviates from the average water in streams. This deviation can be good, indifferent, or bad. For example, the element fluorine (F) in small quantities is essential for sound teeth and bones. Silicon (Si), abundant in river water, is indifferent, with no known use in the human body or most plants. The approximately sixty elements that normally exist in small amounts in streamwater are harmless. However, they can become harmful in larger amounts; examples are lead and uranium. The relation between human health and the elements in uncontaminated stream water is not at all clear.

Major Ions in Drinking Water: Good and Bad

Table 8.1 shows average streamwater, but does not show the wide range of variation that occurs. This variation depends on the rocks with which the water associates.

Here are some examples of ions that are troublesome when present in excess.

Calcium ion is enriched in streams that drain large areas of limestone, usually several hundred parts per million rather than the 15 ppm shown in the table. Is this good or bad for your health? As noted in Chapter 6, hard water ("hardness" is essentially the amount of calcium ion in the water) seems to delay the onset of heart disease, although the reason is unknown.

Sulfate ion averages only 11 ppm in river water, but is extremely high (perhaps thousands of ppm) in streams that drain gypsum-bearing rocks. Gypsum is calcium sulfate that has water bound into it ($CaSO_4 \cdot H_2O$). Thus, it dissolves rapidly in water, releasing calcium ions and sulfate ions. Water containing more than 500 ppm sulfate ion tastes bitter, and if there is more than 1000 ppm, the water becomes a diarrhetic. This is a bit uncomfortable, but neither fatal nor long-lasting, curable with Kaopectate® or similar over-the-counter drugstore product. (By the way, the active ingredient in Kaopectate® is the clay mineral, kaolin.)

Sulfate ion makes bottled water sales brisk in some parts of west Texas where sulfate flavors the "drinking water" to intolerable levels for most visitors and many residents. The cause is abundant, thick gypsum deposits. The calcium ions and sulfate ions in the water may recombine in water pipes and precipitate a heat-retarding calcium sulfate scale, much like the "boiler scale" of calcium carbonate.

Sodium ion in drinking water, if excessive, causes a "salty" taste, as it does in seawater. In addition, ingesting high levels of sodium raises your blood pressure

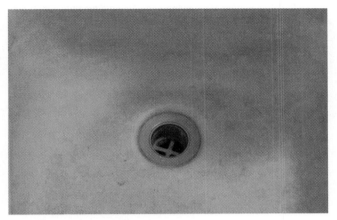

FIGURE 8.2 Iron damage.

This sink probably is coated with rust (iron oxide, Fe_2O_3, the mineral hematite). *Photo: Page Poore/Simon & Schuster/PH College.*

Fluoride ion at 0.1 ppm in average streamwater is harmless, and necessary for strong tooth enamel and bones. Fluoride gets into your teeth and bones because they are made of the mineral apatite. Apatite is calcium, phosphate, and a mixed bag of fluoride (F), chlorine (Cl), and hydroxyl ion (OH). The formula for apatite is written like this: $Ca_5(PO_4)_3(F,Cl,OH)$. The (F,Cl,OH) are shown this way to indicate that they substitute for one another freely in the apatite crystal structure. If they do so in the mineral apatite, they also do it in natural teeth and bones. A large share of these ions in teeth and bones is the hydroxyl ion, OH. Apatite with hydroxyl ion is more soluble than apatite with fluoride ion, making teeth weaker. But if apatite—and teeth—are exposed to more fluoride ion, some of it will replace the OH, creating a stronger crystal structure—and therefore, stronger teeth and less tooth decay.

Consequently, some claim that a daily intake about ten to forty times stronger (1–4 ppm) than average streamwater levels (0.1 ppm) helps prevent tooth decay. However, some evidence exists that this increased level of fluoride is harmful. This led to 1977 and 1993 National Research Council reports, which concluded that an intake of 1–4 ppm of fluoride is beneficial and not harmful, except for some particularly susceptible individuals (Figure 8.3).

If your drinking water is naturally low in fluoride ion, the water company probably is adding soluble fluoride compounds to your water supply to reduce tooth cavities. This is **fluoridation.** Most of these compounds are byproducts of fertilizer production. Recall from Chapter 2 that natural fertilizer comes from deposits of apatite, and all of this apatite is rich in fluoride ion.

and is risky for those with circulatory problems. Drinking sodium-rich water makes you thirstier than you were before you drank the water. (No matter how thirsty you are when trapped in a lifeboat after your yacht hits a reef, don't drink the seawater!)

Iron ion at the 0.7 ppm level in average streamwater is not objectionable in any way. However, when the amount increases to where the water is brownish (a few ppm), your plumbing needs help because its insides probably are coated with rust (iron oxide, Fe_2O_3, the mineral hematite). So will be your bathtub, laundry, and cooking utensils (Figure 8.2). The water may have a metallic taste as well.

FIGURE 8.3 Fluoride controversy.

Fluoride in drinking water affects teeth, particularly during childhood, depending on one's level of exposure and sensitivity. Small amounts strengthen teeth, but excessive fluoride causes *fluorosis*. The level of fluoride that is excessive is believed by most scientists to be more than 4 ppm. Several of the teeth in (B) show evidence of fluorosis. *Source:* Chemical and Engineering News, *August 1, 1988, v. 66, p. 26–42. Photos: David Kennedy, DDS.*

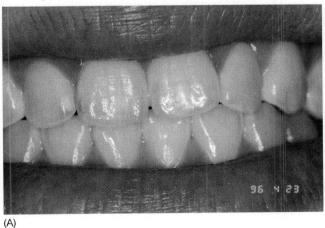

(A)

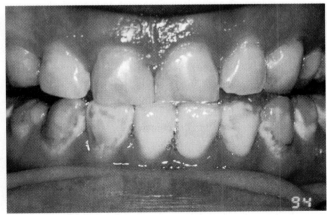

(B)

Despite the reassuring National Research Council reports, the controversy over public water fluoridation continues. Opponents believe that the NRC should admit that federal and state governments have been harming the public for the past forty years. They cite numerous studies indicating that fluoride levels which exceed those in average streamwater (0.1 ppm) increase the incidence of arthritis, cancer, kidney disease, infertility, and harm to the body's DNA repair system. They note that fluoride compounds are used in rat poisons and claim that 10,000 excess cancer deaths per year are linked to water fluoridation in the United States. But again, the NRC says that fluoridation is safe for most people.

Few Americans are aware of this controversy (you are now). At present, 132 million of us (about 50 percent) drink water that is either naturally or unnaturally high in fluoride.

Small But Mighty: Trace Ions in Drinking Water

Table 8.1 lists only about 10 percent of the natural elements as being present in average river water at levels of 0.1 ppm or greater. The preceding section showed that these can be troublemakers in excessive quantities. But the most serious water-pollution problems (other than those from organic chemicals) actually result from trace levels of some of the other 90 percent of natural elements. Examples of these **trace ions** are arsenic, lead, nitrate, chromium, cadmium, mercury, and selenium.

Arsenic

The toxicity of arsenic once made it popular as a means of murder and suicide. In mystery novels and plays, arsenic often was an agent of death (for example, the movie *Arsenic and Old Lace*). Today, we know that smaller doses of arsenic can be just as fatal if absorbed over long periods. Arsenic attacks the nervous system and causes cancer. In the hefty doses used in fiction, death resulted quickly from nervous system damage. Cancers result from lower doses over longer periods.

Problems with arsenic in drinking water have been noticed only recently. A study of 40,000 Taiwanese people exposed to high levels of arsenic in drinking water revealed a clear correlation between the arsenic and more than 400 cases of skin cancer. Similar correlations have been found in other countries. Farm and garden soils around Cornwall, England, contain concentrations up to 900 ppm, a residue from tin mining over many centuries. This is twenty-five times the British government's recommended limit for arsenic in

soil. The children of Cornwall, like all children, play in the dirt, swallowing several times the normal dietary intake of arsenic each day.

In the United States, some water supplies in California and Utah contain up to 500 ppb (0.5 ppm) of arsenic, but no nationwide study of arsenic in drinking water has been conducted. The EPA is considering a sharp reduction in the amount permissible in U.S. drinking water.

Lead

Lead attacks the brain and central nervous system, causes anemia, retards brain development in children, and lowers sperm count. Perhaps the greatest sufferers from lead poisoning were the citizens of ancient Rome some 2000 years ago. That the Romans swallowed a great deal of lead is evident from analyses of their bones, and it has been suggested that lead poisoning was an important factor in the decline of the Roman Empire, an event that plunged Europe into the Dark Ages.

In what ways did the Romans ingest lead, and why? Lead is a soft metal, easily melted and formed. It makes good water piping. The ruling class used lead pipes for plumbing, contaminating their water supply. They drank from leaded cups, also not a good idea. They used medicines containing lead and cosmetics containing lead pigments. The Romans also had a sweet tooth, and used a substance called *sapra*, made by boiling sour wine in lead pans, to sweeten their dishes. The lead compound thus formed was lead acetate, primarily responsible for the sweet taste. In fact, lead acetate is called "sugar of lead." A best-selling cookbook of the Roman Empire mentions sapra in eighty-five recipes.

Historians argue that gradual lead pollution and poisoning among the hereditary upper class led to widespread stillbirths, deformities, brain damage, and a decline in leadership that eventually caused the collapse of Roman civilization. Does this theory sound a bit far-fetched? Ice cores from Greenland dated radiometrically (see Chapter 16) show that for the 800-year interval from 500 B.C. to A.D. 300, the core period of Roman civilization, concentrations of lead in the ice are about four times normal. This has been interpreted as coming from atmospheric pollution caused by the large amounts of lead mined and smelted by the Romans. After Rome declined, so did lead concentrations in the atmosphere.

For the next 1500 years, lead pollution was not a serious problem. Then came the Industrial Revolution of the early 1800s and a sharp increase in lead use. Today, almost 200 years later, lead concentrations in humans worldwide are 500 to 1000 times the levels that existed before the Industrial Revolution. How serious a problem is lead in today's environment?

Commercial testing laboratories analyze water for lead content for between $20 and $100. If your home's lead level is above 15 ppb (the EPA "action level") in first-draw water (water that has stood in the pipes overnight) or 5 ppb in water that has run for a minute or more, consider doing something about it. The cheapest solution is to buy a water-treatment filter that fits under the sink. A pricier solution is to buy bottled water (see below).

Cities with higher-than-average lead in their drinking water include Chicago, Boston, San Francisco, New York, and Washington. Of Washington's water pipes, 16 percent are made of lead, according to the city's Water and Sewer Administration. (It is unknown whether the lead in Washington's drinking water has affected legislative judgment over the years.)

The seriousness of lead pollution in drinking water in the United States at present is uncertain. In 1993, EPA sampled water from houses with plumbing thought to have high lead levels. (Interestingly, Camp Lejeune, the Marine base in North Carolina, was found to have lead at 0.5 ppm, a level known to cause brain damage in children.) Based on this biased sample, EPA reported that 819 water systems serving 30 million people had excessive levels of lead. A critic pointed out that even in this sample, biased toward houses thought to have lead problems, high lead levels were found in only a few dwellings connected to the water systems serving 30 million people. It was, therefore, highly inaccurate to report lead problems for 30 million Americans! (One purpose of the environmental geology course you now are taking is to help you detect untrustworthy studies and news reports.)

The publicity about lead pollution in children has come from studies of minority children in older housing. Older dwellings are more likely to contain the technology of their time, which meant lead water pipes (until 1960), lead-based solder to seal the pipes (until 1988) and lead-containing water faucets. (Modern homes use plastic or copper pipe.) Older homes also have older paint. Lead-bearing paints (pigments) were common in houses built before 1960 but are no longer used. Thus, older buildings may have lead-bearing paint peeling from window frames and walls. Children explore their world by putting things in their mouths, so they may consume lead-based, sweet tasting peeling paint, resulting in some cases of retardation, mild to severe (Figure 8.4).

Lead and Delinquency

A four-year study involving 301 public school boys has shown that exposure to lead makes youths more aggressive. None of the children examined suffered from clinical lead poisoning, so the researchers measured the amount of metal accumulated in leg bones. Consistently, boys having higher lead levels were deemed more violent by parents and teachers.

Nitrate

Nitrates (compounds containing NO_3^-) occur normally in water and food, but also are a contaminant or pollutant in our water supply. Fertilizer is the most common source, but manure from concentrations of farm animals (feedlots) can be significant locally. When nitrates enter the body, they are converted to nitrites (NO_2^-). These reduce the blood's ability to transport oxygen. This is one of the causes of *cyanosis* (the "blue baby" syndrome), a serious disorder in very young children.

FIGURE 8.4 Lead-based paint peeling from tenement walls.

Children living in older buildings may ingest lead-based paint, particularly from windowsills, a common chewing surface. Lead-bearing paint dust also is ingested. *Photo: K. McGlynn/The Image Works.*

Nitrates in municipal drinking water is monitored and hence is not a cause for concern. However, if you live in a rural area, especially near fertilized fields or areas of manure concentration, and drink well water, a periodic check of nitrate content would be wise.

Heavy Elements

Although arsenic and lead have received the most publicity, many other heavy elements with high atomic weights are known to cause serious health problems when they occur at elevated levels in the environment. Chromium and cadmium attack your liver and kidneys. Cadmium replaces the calcium in the apatite of your bones, making them brittle. Mercury prefers your stomach, intestines, and kidneys (Figure 8.5). Selenium causes cancers and nervous system problems. Clearly, it is not a good idea to swallow pollutant levels of heavy elements. The problem is determining for each contaminant how much makes it a pollutant.

Organic Compounds

Ten thousands of organic compounds are used in the United States and worldwide, and many have found their way into our drinking water. Most are produced by the chemical industry, for use in manufacturing and agriculture. They are part of the technological society that provides our high living standard. The main problem lies in disposal of these chemicals after use.

In the 1960s, along the banks of rivers polluted with flammable hydrocarbons, children would toss matches into the water, watching gleefully as the surface ignited. But in 1968, Americans found it less amusing when the Cuyahoga River near Cleveland burst into flames and burned for a half hour because it contained so much oily waste and pollution! Industrial facilities often were built on river banks and lake shores so their waste chemicals could be dumped into the water. The Cuyahoga was a flaming example.

Today's problems are less blatant, but perhaps more serious because they are less visible. What is the most widespread problem in the beautiful Great Lakes? Toxic pollution (Figure 8.6). The eight states bordering the lakes restrict consumption of fish from the lakes because concentrations of pesticides, PCBs, dioxins, and mercury in fish tissue exceed human health standards. Some adverse health effects linked to these pollutants are birth defects, cancer, neurological disorders, and kidney ailments. The problems in the Great Lakes are symptomatic of America's water-pollution problems. The EPA reported in 1994 that, of America's rivers and lakes, *40 percent are no longer suitable for fishing and swimming.*

(A)

(B)

FIGURE 8.5 Mercury in the environment.

(A) Mercury pollution is dangerously high in portions of the Everglades. *Photo: John Shaw/Tom Stack and Associates.* (B) Liquid mercury scooped from a stream bottom with bare hands (not a good idea) by a miner in a gold mining area of the Amazon region, Brazil. *Photo: Sue Cunningham Photographic.*

Waterborne Diseases

Only 200 years ago, microbe-caused disease was poorly understood. Few people associated contaminated drinking water with disease. The first modern study demonstrating that polluted water causes disease was

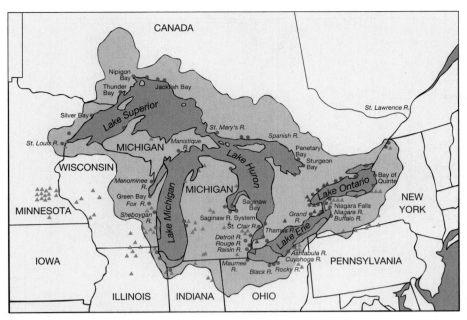

Great Lakes drainage basin

● Most polluted areas, according to the Great Lakes Water Quality Board

● Other "hot spots" of toxic concentrations in water and sediments

▲ U.S. Superfund sites (not including those added after October 1983)

▲ Canadian industrial waste sites identified by the Ontario Ministry of the Environment as "needing monitoring"

Eutrophic (oxygen-starved) areas

FIGURE 8.6   The Great Lakes basin and its major pollution sites.

Source: U.S. Environmental Protection Agency.

conducted in 1854 in London, during a cholera outbreak. Those who died lived near a public hand pump where residents obtained water. John Snow, a physician, discovered that the well beneath the pump was contaminated with sewage. When the pump handle was removed, the cholera epidemic ended (Figure 8.7).

Today, waterborne disease is uncommon in the United States, but it is the single greatest environmental health hazard in poor countries, where most of Earth's 5.8 billion people live. The vice-president of the World Bank says that "most rivers in and around cities and towns in these countries are little more than open stinking sewers that not only degrade the aesthetic life of the city but also constitute a reservoir for cholera and other water-related diseases." He also reported that about 95 percent of the world's sewage and a growing amount of industrial waste are now being dumped directly into rivers and streams. Eighty percent of the disease in these countries results from impure water and poor sanitation, killing 10 million people annually. Figure 8.8 shows access to safe water and the prevalence of waterborne diseases worldwide; note particularly the situation in Africa.

Waterborne diseases are caused by bacteria, viruses, or tiny parasites that are transmitted through polluted water, and food prepared with such water. Examples are typhoid, cholera, polio, hepatitis, dysentery, malaria,

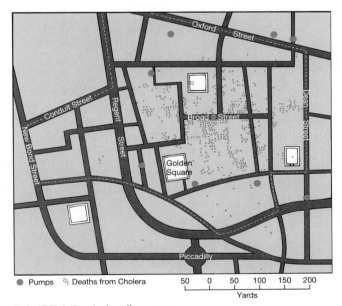

● Pumps   ⁖ Deaths from Cholera

FIGURE 8.7   A deadly pump.

The first modern study demonstrating that sewage-polluted water causes disease was conducted in 1854 in London, during a cholera outbreak. Based on the pattern of cholera deaths shown, which pump do you think was the likely culprit? When the pump handle was removed, the cholera epidemic ended.

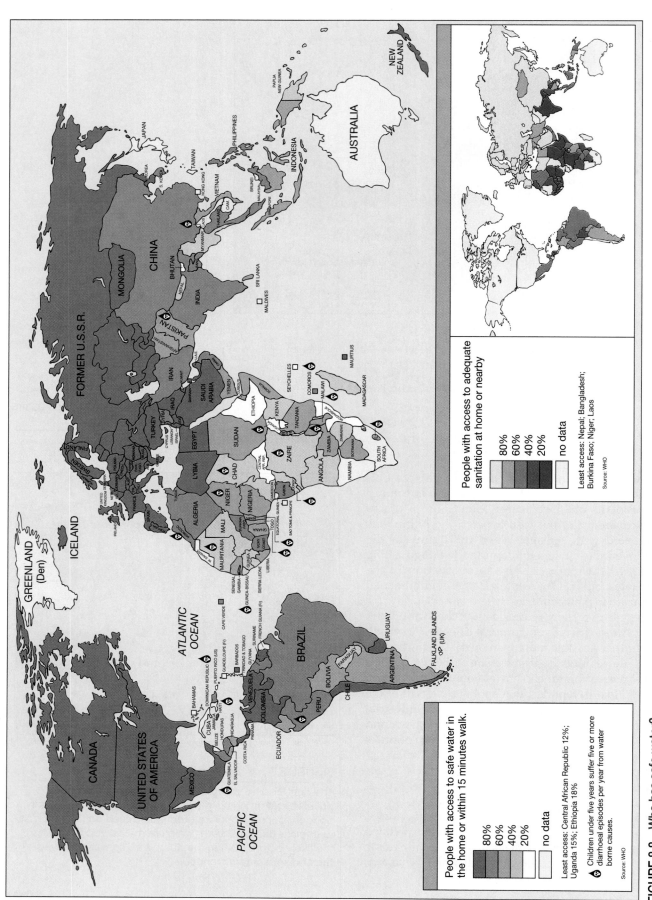

Figure legend (main map):

People with access to safe water in the home or within 15 minutes walk.

- 80%
- 60%
- 40%
- 20%
- no data

Least access: Central African Republic 12%; Uganda 15%; Ethiopia 18%

⚉ Children under five years suffer five or more diarrhoeal episodes per year from water borne causes.

Source: WHO

Figure legend (inset map):

People with access to adequate sanitation at home or nearby

- 80%
- 60%
- 40%
- 20%
- no data

Least access: Nepal; Bangladesh; Burkina Faso; Niger; Laos

Source: WHO

FIGURE 8.8   Who has safe water?

The percentage of Earth's people who have access to safe water in the home or within a 15-minute walk. Inset shows the percentage who have access to adequate sanitation at home or nearby. Data are for the 1980s.

FIGURE 8.9   Would you drink this water?

A combination water supply and garbage dump in Port-au-Prince, Haiti, one of the poorest nations in the Western Hemisphere. *Photo: Armando Waak, PAHO.*

yellow fever, and a host of parasites unheard of by most Americans. Some cause "only" debilitation; others can be fatal if untreated. In the world's impoverished nations, more than 400 children die *every hour* from these easily preventable diseases.

In 1990, a typical year, 4.2 million people died of bacterial diarrhea contracted from polluted water, 1–2 million people died of malaria spread by mosquitoes living in surface ponds and puddles, and 1-2 million succumbed to infectious hepatitis. Hundred thousands more died of schistosomiasis (contracted from a parasitic worm in water) and a host of other afflictions from polluted water.

To stem this threat, physicists in California have built a device that uses ultraviolet light to rid water of disease-producing organisms. The tabletop system takes in water, bathes it with ultraviolet radiation from a mercury-vapor lamp, and sends it out free of germs. The current model of the device can disinfect 15 gallons of water per minute at a cost of two cents per 265 gallons of water. It weighs only 15 pounds, costs $300, draws only 40 watts of power supplied by solar cells, and can operate unsupervised in remote locations. The device has a life expectancy of 15 years.

Testing for Waterborne Disease Organisms

Many disease-producing organisms thrive in sewage, and get into our water through sewage contamination. The common intestinal bacterium *Escherichia coli,* or *E. Coli,* typically is used as an indicator of how much sewage is present in water. E. coli is perfect for monitoring sewage because it exists only in human and animal feces, where it abounds. EPA recommends that safe drinking water contain no more than one E. coli

bacterium per half cup of water (0.2 pint), and that swimming water (some of which is swallowed during a swim) contain no more than 200. The bacteria themselves may cause disease and indicate the likely presence of other disease-causing agents in the water.

American Drinking Water

In the United States in 1993, the public water supply in Milwaukee was contaminated with *cryptosporidium* bacteria that multiplied in overloaded sewage treatment plants. About 400,000 people fell ill, and the bacteria may have hastened the deaths of forty-two patients with AIDS and one apparently healthy senior citizen. As tragic as these deaths are, their number is insignificant compared to the millions who die each year in less-developed countries. Waterborne diseases in these lands could be almost completely prevented if sanitation methods taken for granted in most industrialized nations were used (Figure 8.9).

Water Safety Guidelines

We generally do not know how much of most substances can be tolerated by plants, animals, or people. Even highly trained and experienced scientists disagree on how much of a substance is safe. The public understandably interprets this with extreme conservatism, usually viewing even a tiny amount of contaminant as unsafe. With this attitude of "better safe than sorry," public funds sometimes are spent to remedy problems that later turn out to be pseudo-problems. Journalists often amplify the problem, for few have training in science and they unwittingly spread misinformation. Also, some writers stretch the truth to present a sensational story.

TABLE 8.2 EPA Standards for Substances in Drinking Water

Primary drinking water standards, effects of pollutant levels on human health, and typical sources of the pollution.

Contaminants	Health Effects	Maximum Permissible Amount	Sources
Microbiological			
Coliform bacteria	Not necessarily disease-producing themselves, but coliforms can be indicators of other organisms that cause gastric infections	1 per 100 milliliters	Human and animal fecal matter.
Turbidity	Interferes with disinfection	1–5 turbidity units	Erosion, runoff, and sediment discharges
Inorganic chemicals			
Arsenic	Skin and nervous system toxicity, possible cancer risk	0.05	Pesticide residues, industrial waste and smelter operations, rocks.
Barium	Cardiac, gastrointestinal, and neuromuscular effects	1	Coal-fired power plants, filler for automative paints, specialty compounds used in bricks, tiles and jet fuels, rocks.
Cadmium	Kidney effects, hypertension, anemia, and altered liver activity	0.01	Mining, smelting, fossil fuel use, fertilizer application, sewage sludge, rocks.
Chromium	Liver, kidney effects	0.05	Abandoned mining operations and electroplating operations, rocks.
Lead	Nervous system damage. Kidney effects. Highly toxic to infants and pregnant women. Impaired mental performance in children	0.015	Leaches from lead pipe and lead-based solder pipe joints. Airbone lead from leaded gasoline combustion.
Mercury	Nervous system disorders; kidney effects	0.002	Used in manufacture of paint, paper, vinyl chloride. Used in fungicides, rocks.
Nitrate (as nitrogen, N)	"Blue baby syndrome" which results in asphyxia, cancer risk	10	Fertilizer, sewage, feedlots, rocks.
Selenium	Gastrointestinal effects	0.01	Coal burning, mining, smelting, selenium refining, glass manufacture, fuel oil, combustion, rocks.
Silver	Skin discoloration	0.05	Mining and processing, rocks.
Fluoride	Skeletal damage	4	Additive to drinking water, toothpaste, foods processed fluoridated water, rocks.
Organic chemicals			
Endrin	Nervous system/kidney effects	0.0002	Insecticide used on cotton, small grains and orchards (now prohibited in U.S.).
Lindane	Nervous system, kidney and liver effects; possible cancer risk	0.0004	Insecticide used on seed and soil treatments, foliage applications, wood protection.
Methoxychlor	Nervous system/kidney effects	0.1	Insecticide used on fruit trees, vegetables.

(Continued on next page)

TABLE 8.2 *(continued)*

Primary drinking water standards, effects of pollutant levels on human health, and typical sources of the pollution.

Contaminants	Health Effects	Maximum Permissible Amount	Sources
Organic chemicals			
2, 4-D	Liver/kidney effects	0.1	Herbicide used to control broadleaf weeds in agriculture, used on forests, range, pastures, and aquatic environments.
2, 4, 5-TP Silvex	Liver/kidney effects	0.01	Herbicide, canceled in 1984.
Toxaphene	Cancer risks	0.0005	Insecticide used on cotton, corn, grain.
*Benzene	Cancer-causing	0.005	Fuel (leaking tanks), solvents commonly used in manufacture of industrial chemicals, pharmaceuticals, pesticides, paints, and plastics.
*Carbon tetrachloride	Possible cancer risk	0.005	Common in cleaning agents and tetrachloride industrial wastes from manufacture of coolants.
*p-Dichloro-benzene	Possible cancer risk	0.075	Used in insecticides, moth balls, and air deodorizers
*1, 2-Dichloro-ethane	Possible cancer risk	0.005	Used in manufacture of insecticides.
*1, 1-Dichloro-ethylene	Liver/kidney effects	0.007	Used in manufacture of plastic, dyes, perfumes, paints and SOCs.
*1, 1, 1-Trichloro-ethane	Nervous system effects	0.2	Used in manufacture of food wrappings, synthetic fibers.
*Trichloroethane	Possible cancer risk	0.005	Waste from disposal of dry cleaning materials. From manufacture of pesticides, paints, waxes and varnishes, paint stripper, and metal degreaser.
*Vinyl Chloride	Cancer risk	0.002	Polyvinychloride (PVC) pipes and solvents used to join them. Industrial waste from manufacture of plastics and synthetic rubber.
Total trihalomethanes			
Chloroform and others	Cancer risk	0.1	Primarily formed when surface water containing organic matter is treated with chlorine.
Radioactive materials			
Gross alpha particle and others	Cancer-causing	15	Radioactive waste, uranium deposits (in billionths of a curie/liter).
Gross beta particle and others	Cancer-causing	4 mrem/yr	Radioactive waste, uranium deposits (in millirems of radiation/year).
Radium 226 & 228 (total)	Cancer-causing	5	Radioactive waste, rocks (in billionths of a curie/liter).

*Effective January 9, 1989.

Source: "You and Your Drinking Water," *EPA Journal* vol. 12, no. 7 (September 1986).

So how are we to decide how much of each constituent in water is safe? For many substances, the answer is that we cannot. The general principles to follow when determining water safety are:

1. If water has a composition pretty close to average river water as described in Table 8.1, it probably is all right. But if you suspect pollution, analysis for trace elements may be necessary.
2. If any of the elements shown in the table is present in extraordinary quantity are present, the water may not be safe to consume.
3. If large amounts of heavy metals or synthetic substances like pesticides and effluent from chemical plants and refineries are present, the water is likely to be polluted.

EPA Water Standards

To protect the public, the U.S. **Environmental Protection Agency** sets **water standards** for contaminant levels in water. Its authority to police the nation's water supply comes from the Clean Water Act, passed by Congress in 1972. Current EPA standards for substances in drinking water are shown in Table 8.2.

The acceptable limits for these compounds change—most commonly decreasing—as new data about their effects are produced. For example, EPA is considering reducing acceptable arsenic levels from 0.05 ppm to only 0.002 ppm. Barium may be raised from 1 ppm to 5 ppm, and chromium from 0.05 ppm to 0.1 ppm. A new asbestos standard also is being established.

Based on current EPA standards, more than 25 percent of all U.S. drinking-water supplies violated federal water-purity standards at least once during 1993–1994. Common contaminants were fecal matter (as indicated by E. Coli), lead, radioactive material, and pesticides. Funding and bureaucratic delay have slowed federal and state correction of these problems.

EPA also sets water standards for livestock and irrigation, generally tolerating higher contaminant levels than are acceptable for humans.

Which Pollution Problems Are Most Serious?

Consider the following pollution problems, and rank them in order of urgency to you:

a. Builders are erecting low-income housing on an abandoned factory site in your hometown when they discover dozens of rusted barrels leaking a green, foul-smelling liquid, perhaps looking like the scene in Figure 8.10.

b. Your local newspaper reports that the sewage treatment plant is operating beyond capacity

FIGURE 8.10 **High visibility pollution.**

Hazardous waste leaking from rusty drums in a storage facility in Baja California. *Photo: Diane Perry/University of California at Los Angeles.*

and runs the risk of being inundated by a 100-year flood.

c. Government scientists warn that fertilizer runoff from farmland in your state may result in fish kills in many areas.

If you are like most people, you would rank these in urgency in the same order in which they are listed: (a), (b), (c). And this illustrates the problem that the EPA has in deciding where to spend its money. Local pollution that can be seen, such as old dump sites, is considered a high-risk threat by the public. Sewage treatment plants, although not viewed as dangerous, command strong local attention, for none of us wants a basement knee-deep in raw sewage. Consequently, more than half of EPA's current spending goes toward cleaning up old dump sites and helping municipalities build sewage-treatment plants.

More frightening large-scale contamination receives relatively small amounts of money. The EPA spends just 20 percent of its budget on what it considers to be the most serious pollution problems (Table 8.3), whereas the rest goes toward what the public considers more urgent. Local pollution obviously is important to each of us and must be addressed. Unfortunately, money for remediation is limited, and no one

TABLE 8.3 EPA Versus Public Concerns

The EPA focuses upon global and long-range problems, while the public worries about local problems.

The EPA's Top 11 Concerns (not in rank order)	Public Concerns, in rank order (Bold items also are on EPA's list)
Ecological Risks	1. Active hazardous waste sites
Global climate change	2. Abandoned hazardous waste sites
Stratospheric ozone depletion	3. Water pollution from industrial wastes
Habitat alteration	4. **Occupational exposure to toxic chemicals**
Species extinction and biodiversity loss	5. Oil spills
	6. **Destruction of the ozone layer**
Health Risks	7. Nuclear power plant accidents
Criteria air pollutants (e.g., smog)	8. Industrial accidents releasing pollutants
Toxic air pollutants (e.g., benzene)	9. Radiation from radioactive wastes
Radon	10. **Air pollution from factories**
Indoor air pollution	11. Leaking underground storage tanks
Drinking water contamination	12. Coastal water contamination
Occupational exposure to chemicals	13. Solid waste and litter
Application of pesticides	14. **Pesticide risks to farm workers**
Stratospheric ozone depletion	15. Water pollution from agricultural runoff
	16. Water pollution from sewage plants
	17. **Air pollution from vehicles**
	18. Pesticide residues in foods
	19. **Greenhouse effect**
	20. **Drinking water contamination**
	21. Destruction of wetlands
	22. Acid rain
	23. Water pollution from city runoff
	24. Nonhazardous waste sites
	25. Biotechnology
	26. **Indoor air pollution**
	27. Radiation from X-rays
	28. **Radon in homes**

Source: "Counting on Science at EPA." *Science* 249 [1990], 616. L. Roberts. Copyright 1990 by the AAAS.

wants to pay more taxes, so local problems get attention while global problems simmer on the back burner.

Point Versus Nonpoint Pollution Sources

The sources of contamination/pollution of surface and ground waters usually are divided into "point" and "nonpoint" sources. A **point source** has smaller dimensions, such as a waste disposal pond, a small dump or waste burial, a waste-disposal well, or a mine. A **nonpoint source** is much larger in size, such as fertilizer runoff from a large farm or salt runoff from all of a county's roads that were sprinkled with salt to melt winter ice. Examples of both are shown in Figure 8.11.

Distinguishing between point and nonpoint sources is tricky because sources grade from a spilled can of gasoline to vast fertilizer runoff from thousand-acre farms in a large county. A pipe dripping toxics may have a diameter of 2 feet, a pond 20 feet, a small landfill 200 feet, a stockyard filled with cattle 2000 feet, and an agricultural field may cover several square miles. Defining "point" and "nonpoint" depends on your purpose.

Since passage of the Clean Water Act in 1972, more than $540 billion has been spent on water pollution controls. Nearly 90 percent of it was spent on point sources, because these are the easiest to isolate and fix.

What Is Hazardous Waste?

Each year, American industry produces more than 300 million tons of material officially designated as hazardous waste. One-third of it comes from Texas. Ninety percent of it is liquid. This is enough to fill the New Orleans Superdome 1500 times over. What is this stuff? According to the EPA, **hazardous waste** is any discarded chemical that can cause harm because it has any of the following properties (Figure 8.12):

Flammable—includes waste petroleum products, such as the used motor oil that many Americans spread on the ground or pour down a sewer; organic solvents that remove the food stains from your carpet; and some toxic compounds such as PCBs (polychlorinated biphenyls).

Unstable—likely to explode or release toxic fumes. Examples include many chemical mixtures (recall experiments from high school chemistry) and cyanide solvents.

Corrosive—liquids such as strong acids or bases that attack metals or human tissue. Common examples are household cleaners that are strongly acidic or alkaline.

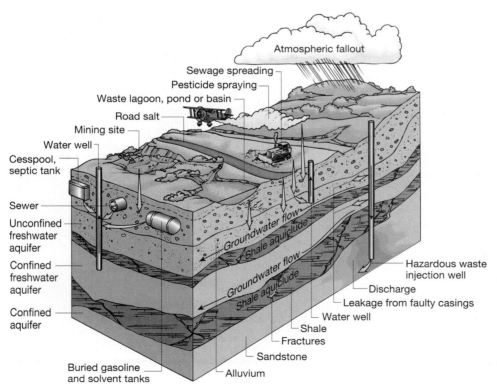

FIGURE 8.11 Principal types of groundwater contamination.

Shown are both point sources (tanks, injection well, pond, mine) and nonpoint sources (road salt, atmospheric fallout) in the United States.

Corrosivity

Reactivity

Ignitability

Toxicity

FIGURE 8.12 Properties of hazardous waste.

Any discarded chemical that has any of the following characteristics is hazardous waste: flammable, unstable, corrosive, toxic.

Toxic—if released into the environment. This includes pesticides that contain DDT (production of which is now banned in the U.S.), dioxins, PCBs, and various compounds of arsenic, mercury, and lead.

You can see that almost any facility can produce hazardous waste: chemical plants, manufacturing facilities, sanitary landfills (garbage dumps—see Chapter 9), and so on. However, some waste is excluded by law from being labeled hazardous:

- Items discarded by households and hotels.
- Oil-well drilling waste.
- Municipal incinerator ash.
- Most hazardous waste generated by the military.
- Radioactive waste.
- Waste from small businesses and factories that produce less than 220 pounds of stuff per month.

Some of these exclusions appear to have political motivation, not environmental. But even using EPA's definition, the agency reports 2.3 billion pounds of hazardous waste were produced by American industry in 1994. This was a decrease of 9 percent from 1993 and a 44 percent decrease since 1988. The American Chemical Society estimates toxic waste discharges to be two to ten times greater than EPA estimates. The federal government released 10 million pounds of toxic waste, most of it by the Department of Defense. The United States leads the world in producing hazardous waste. By 1995, EPA had approved 1,276 hazardous waste storage sites, an average of 25 per state, with at least one actually in each state. There are many more thousands of waste sites in

the United States that have not received the EPA's seal of approval. About 20 percent of the American public lives within four miles of a hazardous waste site. Recent epidemiological studies have shown an association between living near one of these hazardous waste sites and increased risks of several types of cancers and increased risks of birth defects. Groundwater is thought to be the main source of contamination.

Types of Pollution Sources

Pollution sometimes is classified by the type of contaminant—heavy metals, complex organic compounds, or inorganic light elements, as in Table 8.2. But it also is useful to look at the types of pollution sources. The following list of pollution sources is in generally decreasing order of importance. Some might differ with the order given here, but environmentalists agree these are the most serious offenders.

1. *The federal government.* This secretive organization refuses to allow regulation or supervision of its polluting activities, under the shield of "national security." The U.S. military produces more than 1 billion pounds of hazardous waste per year, which is more than the top five chemical companies combined.

2. *Chemical plants/oil refineries.* After our own government, these are the nation's worst polluters, generating a vast brew of organic contaminants. The witches in *Macbeth* never cooked up anything to equal this.

3. *Sanitary landfills for solid waste* (Chapter 9). What goes into these? Almost anything you can imagine. Where can we put it all? Certainly not near my house—but this is what every homeowner says, so we obviously have a problem.

4. *Pesticides (insecticides, herbicides, and othercides).* Includes roach motels, bug sprays, and fluoride compounds similar to those in your toothpaste.

5. *Sewage and septic systems* (Chapter 9). What happens to your residues after you flush them? Can they creep back to haunt you?

6. *Radioactive waste* (Chapter 16). To many people, "radwaste" is the most difficult and terrifying of all human creations. Where do we get rid of something that stays "hot" for thousands of years?

7. *Seawater incursion.* If you live near a coastline, is saltwater seeping into your groundwater? Is something fishy about your iced tea?

8. *Gasoline, motor oil, and other petroleum products.* After you change your car's oil, do you also change your groundwater by pouring the dirty stuff on the ground? What do garages do? Do

CAA	= Clean Air Act	HMTA	= Hazardous Materials Transportation Act
CPSA	= Consumer Product Safety Act	OSHA	= Occupational Safety & Health Act
FFDCA	= Fed. Food, Drug & Cosmetic Act	PPPA	= Poison Prevention Packaging Act
FFA	= Flammable Fabrics Act	RCRA	= Resource Conservation & Recovery Act
FHSA	= Fed. Hazardous Substances Act	SDWA	= Safe Drinking Water Act
FIFRA	= Fed. Insecticide, Fungicide, & Rodenticide Act	TSCA	= Toxic Substances Control Act
FWPCA	= Fed. Water Pollution Control Act		

FIGURE 8.13 Major federal laws controlling pollution.

This figure illustrates important legislation designed to lessen pollution of air, land, water, and workplaces.

rusted and leaky gasoline storage tanks lurk beneath old gasoline stations in your town?

9. *Acid mine drainage* (Chapter 13). Coal brings us electricity, but at what cost to streams, groundwater, and the landscape? Is the soil too rich near a gold mine?

10. *Farm animal waste.* The smell of the great outdoors, but does it leak into your water supply?

11. *Salt applied to roadways in winter.* Where does it go when it melts the snow? Do all those sodium and chloride ions do something to groundwater?

12. *Cemeteries.* The mummy's revenge: pollution after death?

13. *Thermal pollution.* Hot water is great in your shower, but do fish like it pouring from a power plant into their cool stream?

A variety of federal, state, and local laws now control the use and disposal of waste. Some control the generation of waste components. Some control manufacturing and processing of products. Others regulate transportation, and still others control consumer and industrial use and disposal.

The more important federal laws aimed at reducing air, land, and water pollution are shown in Figure 8.13.

Chemical Plants/Oil Refineries

Aside from the military, most U.S. chemical hazardous waste is generated by industry. In 1991, approximately 60,000 industrial plants legally discharged 29 million gallons of wastewater containing potentially toxic chemicals. This was an increase of 24 percent over 1990, and an average of nearly 500 gallons per plant. Two-thirds of this wastewater came from chemical manufacturers (Figure 8.14).

These discharges have high concentrations of heavy metals (Table 8.4) and synthetic organic chemicals (Table 8.5). They went directly into surface waters—one of several disposal methods for such waste. As you read these words, chemical manufacturing facilities are piping these dangerous wastes into streams adjacent to the plants. In fact, many plants were built next to large streams so they could be used as sewers for the chemical waste.

Figure 8.15 shows locations of disposal sites for hazardous waste. Let us look at the different methods used—underground injection, incineration, open pits, and export.

Out of Sight, Out of Mind, But... Another disposal method for hazardous waste is to pump it into deep disposal wells. This is called **underground injection.** Although regulated by the federal government, individual state laws largely govern the siting of injection wells. Underground injection of hazardous liquid waste is concentrated in fewer parts of the country than most other types of hazardous waste disposal. This is for geologic and economic reasons; manufacturing clusters near resources and transportation, such as around Houston and

FIGURE 8.14   The Cancer Corridor.

Aerial photo of chemical plants and/or oil refineries in Louisiana between Baton Rouge and New Orleans, the "Cancer Corridor."

Photo: Eastcott/M./The Image Works.

TABLE 8.4   Principal Trace Metals in Wastewater

These trace metals contaminate industrial wastewater. Their major sources appear in the left column.

Industry	Metals
mining and ore processing	arsenic, beryllium, cadmium, lead, mercury, manganese, uranium, zinc
metallurgy, alloys	arsenic, beryllium, bismuth, cadmium, chromium, copper, lead mercury, nickel, vanadium, zinc
chemical industry	arsenic, barium, cadmium, chromium, copper, lead, mercury, tin, uranium, vanadium, zinc
glass	arsenic, barium, lead, nickel
pulp and paper mills	chromium, copper, lead, mercury, nickel
textiles	arsenic, barium, cadmium, copper, lead, mercury, nickel
fertilizers	arsenic, cadmium, chromium, copper, lead, mercury, manganese, nickel, zinc
petroleum refining	arsenic, cadmium, chromium, copper, lead, nickel, vanadium, zinc

Source: From J. E. Fergusson, *Inorganic Chemistry and the Earth*, p. 268. Copyright © 1982 Oxford, England, Pergamon Press. Reprinted by permission.

in the Baton Rouge—New Orleans corridor of Louisiana (Figure 8.15, blue).

Nearly three times the liquid waste is injected compared to the volume that is discharged into surface waters. Some 85 million gallons were legally pumped down 187 deep wells in 1991, mostly by the industries that produced the waste. The injection site normally is chosen by a geologic study, with public input, but uncertainties always remain about the subsurface geology:

- Rock layers thought to be aquicludes may have unseen fractures or changes in clay content that turn them into leaky aquitards (Figure 8.11).
- Small earthquakes may occur, shifting the plumbing of permeable rock layers.
- The steel well casing (pipe) that lines a well to prevent leakage from one underground aquifer into another may develop leaks as it ages.

From this, it is clear that liquid hazardous waste has seeped into groundwater at various sites. Aquifer contamination from underground injection has been reported in about half the states, from Massachusetts to California. Who is legally (financially) responsible: the federal government, which has regulatory authority, or the state, which authorized injection of the waste and supervised site location, or the chemical company, which generated the waste? The answer normally depends on the agility and quantity of lawyers who argue these cases in court, and the decision never satisfies everyone.

Up in Smoke.   **Incineration** is another method of hazardous waste disposal. It destroys waste by thermally

TABLE 8.5 | Common Synthetic Organic Chemical Wastes

Toxic organic compounds frequently found in chemical wastes and their effect on people when present at pollutant levels.

Chemical	Mutations	Carcinogenic	Birth Defects	Still Births	Nervous Disorders	Liver Disease	Kidney Disease	Lung Disease
Benzene	X	X	X	X				
Dichlorobenzene	X			X	X	X		
Hexachlorobenzene	X	X	X	X	X			
Chloroform	X	X	X		X			
Carbon tetrachloride	X		X	X	X	X		
Chloroethylene (vinyl chloride)	X	X			X	X		X
Dichloroethylene	X	X		X	X	X	X	
Tetrachloroethylene		X			X	X	X	
Heptachlor	X	X		X	X	X		
Polychlorinated biphenyls (PCBs)	X	X	X	X	X	X		
Tetachlorodibenzo-dioxin	X	X	X	X	X	X		
Toluene	X			X	X			
Chlorotoluene	X	X						
Xylene			X	X	X			

Known Health Effects*

*Determined from tests on experimental animals.

Source: Adapted from S. Esptein, L. Brown, and C. Pope. *Hazardous Waste in America.* Copyright © 1982 by Samuel S. Epstein, M. D., Lester O. Brown, and Carl Pope. Reprinted with permission of Sierra Club Books.

decomposing organic compounds. The Netherlands incinerates about half its hazardous waste. The EPA estimates that 60 percent of U.S. hazardous waste could be incinerated, but only 20 percent is burned at present. Assuming proper air-pollution controls so that the more volatile heavy metals are not spewed into the air, incineration is a potentially safe method of disposal for most hazardous waste. But it is also the most expensive. We will pay more for any products made using the hazardous chemicals.

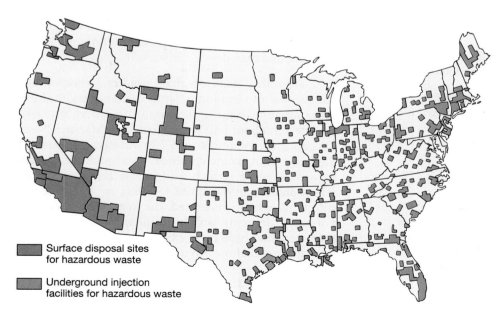

Surface disposal sites for hazardous waste

Underground injection facilities for hazardous waste

FIGURE 8.15 Disposal locations of hazardous waste, 1991.

Surface disposal sites are shown in red and underground injection sites are in blue. Note the small number of underground sites compared to surface sites.

Ash that remains after incineration must be disposed of, but it often contains toxic heavy metals. Such ash is not classified as hazardous by the EPA and can be disposed of in a secured sanitary landfill (see Chapter 9) or underground vault, much as radioactive waste may be handled (see Chapter 16). However, according to the Office of Technology Assessment, eventually even the best-designed landfill or vault will leak and can threaten groundwater supplies. As has been said, "there is no away, only delay."

Open Pits Are the Pits.   Much of the country's hazardous waste is deposited in ponds, pits, or lagoons. Their bottoms are supposed to be sealed with a plastic liner, although most experts consider it only a matter of time until a liner leaks. But, for most of these pits, it makes no difference: according to EPA, 70 percent of the pits, ponds, and lagoons that now store hazardous waste *have no liners!* Up to 90 percent of these pits containments may threaten groundwater. Three-quarters of our fifty states now have groundwater contamination caused by these surface impoundments.

The federal government's Office of Technology Assessment estimates that about 2 trillion gallons of liquid waste are dripping into the soil each year from hazardous waste pits. Also, as the hazardous liquid sits in the pond for months, volatile organic compounds evaporate into the air, move downwind, and eventually fall to the surface in rainfall, contaminating waters far from the original disposal site. This is certainly an undesirable form of interstate commerce.

America's Nasty Export.   As the cost of hazardous waste disposal has risen, we and other industrialized nations have been exporting our waste to other countries for disposal. Most legal exports go to Canada and Mexico, but there is evidence of a growing illegal trade in hazardous waste, with shipments to less-developed countries in Asia, Africa, and Latin America. These poor nations find it hard to resist the cash that comes with accepting the legal shipments, and officials there are easily bribed to facilitate entry of the illegal shipments.

Louisiana's "Cancer Corridor."   One of the best-studied areas of water pollution from the chemical industry is in Louisiana. Its so-called **Cancer Corridor** stretches 150 miles along the Mississippi River, from the state capitol in Baton Rouge to the nation's twenty-fourth largest city, New Orleans. This is the nation's "toxic hotspot," with 500 hazardous waste sites and the highest concentration of manufacturers, users, and disposers of toxic chemicals in the entire country. At least 136 major industries line the river's banks. The majority are petrochemical plants, which manufacture products from petroleum: organic chemicals, pesticides, gasoline, plastics, and synthetic fibers. They have located in the Corridor to take advantage of low taxes and the area's vast resources of petroleum, natural gas, and surface water.

Most of Louisiana's 500 hazardous waste sites are located in the Corridor. Numerous old dump sites dot areas near the Corridor's chemical facilities. Millions of pounds of unidentified chemicals buried at these sites threaten groundwater and surface supplies (Figure 8.16). Of seventy-five Louisiana landfills and waste ponds tested in 1986, at least thirty showed signs of leaking (Figure 8.17). As an additional threat to water supplies, at least seventy-two injection wells were recently operative in the state, most owned and operated by the waste-generating industries. Chemicals discharged by Louisiana industries are regularly found in public drinking water.

The most plentiful contaminants in the Mississippi are manufactured in large quantities in the Cancer Corridor. The organics with the highest concentrations are all major Cancer Corridor products. Other chemicals produced in the Corridor also are present, although in lesser amounts. The thirteen Louisiana parishes (counties) that depend on the Mississippi for drinking water have among the highest U.S. mortality rates for several forms of cancer—including rectal cancer, a disease often linked to drinking water. Among the thirteen parishes, rectal cancer rates are highest among those living downstream from or within the Cancer Corridor. It is all very suspicious, to say the least.

Pesticides

Pesticides are chemicals that kill organisms that humans consider undesirable (pests); they also are called **biocides** (Chapter 4). About 25,000 different pesticide products are sold in the United States. Their use has many short-term benefits, including better yields of grain, vegetables, and fruit, plus control of mosquitoes that transmit the disease malaria. Most widely used are insecticides (insect killers), herbicides (weed killers), fungicides (fungus killers), and rodenticides (rodent killers).

In the United States, approximately one billion pounds of pesticides are used each year at a cost of $4.1 billion. More than 90 percent of pesticides are synthetic organic compounds. Forty-six of the compounds have been detected in groundwater in thirty states and approximately 35 percent of the foods purchased by U.S. consumers contain detectable amounts of pesticide residues. The National Academy of Sciences estimates that pesticide pollution is responsible for 20,000 cases of cancer in the United States each year.

Despite the widespread use of pesticides in the U.S., animal and plant pests destroy 37 percent of all potential food and fiber crops. Although the use of pesticides is generally profitable, their use does not always

FIGURE 8.16 A polluted lake in downtown Baton Rouge.

Business and local governments pay more than 80 percent of the cleanup bills.

FIGURE 8.17 Oops!

When certain amounts of hazardous waste are spilled or released, a company must notify appropriate governmental agencies. An emergency response plan should be in place.
Photo: David M. Doody.

decrease crop losses. Even with a tenfold increase in insecticide use in the U.S. between 1945 and 1989, losses from insect damage have nearly doubled, from 7 percent to 13 percent.

Insecticides. As noted in Chapter 4, the most serious contamination/pollution problems with pesticides involve insecticides. Insecticides are widely used, can persist in the environment for many years, and commonly are very harmful to a wide variety of living organisms. The best-studied example of an insecticide hazard is DDT, widely used in the United States until its ban in 1972. (DDT abbreviates di-chloro-di-phenyl-tri-chloro-ethane). It is a *chlorinated hydrocarbon*, or a complex compound of chlorine, hydrogen, and carbon. Although banned in the United States twenty-five years ago, traces of it still are found in mammal's milk, sperm, and body fat; in the sediments of lakes, river, and harbors; and in fish.

As a group, chlorinated hydrocarbons are long-lasting in the environment, with a half-life of two to five years. (The half-life is the time it takes for half of the amount present to be broken down.) Even worse, this group of chemicals is fat-soluble and therefore becomes stored in fatty tissue. This results in a strong **biological magnification**, in which amounts of the contaminant become progressively concentrated with each step up the food chain (Figure 8.18). Unfortunately, it took several decades for biologists to fully recognize this magnification. By then, DDT was widespread in the environment. Considerable evidence exists that widely used pesticides suppress the immune system, making people more susceptible to infectious diseases and certain cancers. The

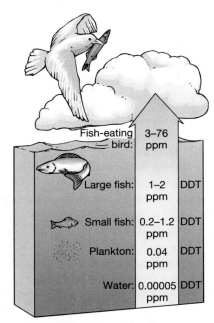

FIGURE 8.18 Biological magnification.

In 1967, fish-eating birds from a salt-marsh estuary along New York's Long Island contained almost a million times more DDT than could be found in the water. (Compare the levels, from 0.00005 ppm in the water to 3–76 ppm in the osprey.) At each step of the food chain, DDT was being concentrated as organisms consumed and absorbed more DDT than they were able to excrete. *Source: W. Keeton*, Biological Science, 3rd ed., *New York: Norton, 1980.*

problem is most acute in less-developed countries, where infectious diseases cause nearly half of all deaths and where pesticide use is increasing.

DDT and other pesticides now have contaminated the remotest areas of Earth. Traces of twenty-two insecticides have been found in tree bark from ninety locations around the world, including areas as remote as the virtually uninhabited rain forests in South America. The compounds probably became airborne through crop spraying and evaporation from plant surfaces and the ground and were transported by wind. They generally are concentrated at higher latitudes because of evaporation in hotter low latitudes and condensation in cooler high latitudes. Thus, polar regions contain high concentrations of these contaminants, even though none of the chemicals was released there.

A further complication is the increasing immunity of insects to insecticides. The number of insect species resistant to insecticides is on the rise:

1940: 10 (approximately)
1950: 15
1960: 135

1970: 250
1980: 450 (nearly)
1990: 500

The use of pesticides causes natural selection to occur: only the resistant individuals survive (survival of the fittest). Less-resistant individuals die off, so that only those who can tolerate the applied chemical survive and produce offspring that also are resistant. Given the rapid breeding rate of insects, it is not long before all members of a species become resistant to the insecticide. It may take only weeks for an insect population to evolve **pesticide resistance.** It takes longer to develop a new pesticide in the laboratory than it takes for insect species to develop resistance.

Other Pesticides. Herbicides decompose fairly rapidly in soil and water and are generally low in toxicity to animals. However, there are prominent exceptions, like Agent Orange, a "defoliant" that sickened a number of soldiers during the Vietnam War in the 1960s and 1970s and is believed responsible for birth defects of 300,000 children in Vietnam. Weeds, like insects, are becoming increasingly resistant to chemicals designed to kill them. In 1980, twelve weed species were resistant to herbicides. In 1990, more than 100 species were resistant and the number is still growing. Fungicides used on fruits and vegetables are sometimes a problem to humans but are used much less than herbicides. Rodenticides are the least used of the pesticide group.

Other pesticide types have half-lives of weeks or a few months. They lack the intense biological magnification typical of chlorinated hydrocarbon insecticides. The continual invention of new pesticides and the increasing variety of synthetic organic compounds they contain makes adequate testing extremely difficult.

Alternative Pest Controls. Over the long run, the chemical approach to pest control is bound to fail. We cannot know the effects of all chemicals on the large variety of living organisms. The use of synthetic chemicals assumes that one type of organism, deemed a pest to humans, can be eliminated without harmful effects to the ecosystem as a whole. But ecosystems are dynamic, operating through intricate interactions in which every single organism has a role. It seems clear that a chemical assault on one species will inevitably unbalance the system and produce other undesirable effects. Pesticide use, particularly insecticides, should decrease, and be carefully controlled.

However, the United States is a major breadbasket for the world. Several billion people depend on American foodstuffs to stave off widespread starvation. As the third-world population swells uncontrollably, American agriculture will be asked to produce in even greater abundance. Can we accomplish this without pesticides? Can we minimize our interference with nature? Certainly we can do better than our current spraying of one billion pounds of pesticides each year.

Several alternative methods of pest control have been introduced, but none has yet gained widespread use or acceptance (see also Chapter 4). These methods include:

1. *Control by natural enemies.* Various natural predators and parasites can be introduced (Figure 8.19A). In the United States, natural enemies have been used to control about seventy types of insect pests. This type of biological control nor-

mally affects only the target species and is non-toxic to other species, including people. Once a population of natural predators or parasites is established, control of pest species is often self-perpetuating and the development of genetic resistance is minimized.

2. *Genetic engineering.* Recent technological developments have made possible the development of plants that are resistant to attacks by pests (Figure 8.19B). It is now possible in some cases to actually manipulate the genetic code of organisms to make them less palatable to their

FIGURE 8.19 Alternative pest-control methods.

(A) An adult ladybug sucking dry an aphid, an activity that illustrates biological control of pests. *Photo: Peter J. Bryant/Biological Photo Service.* (B) Genetic engineering can reduce pest damage and maximize yields. An extra chromosome was added to the tomato on the left, producing a larger fruit. *Photo: Leonard Lessin/Peter Arnold, Inc.* (C) Care for some lemonade? Probably not from this fruit, infested with red scale mites. Pheromones are now being used to help control mite populations. *Photo: William E. Ferguson.* (D) A natural insecticide, pyrethrum, is harvested from the heads of these flowers. *Photo: Nigel Cattlin/Holt Studios International/Photo Researchers, Inc.*

(A)

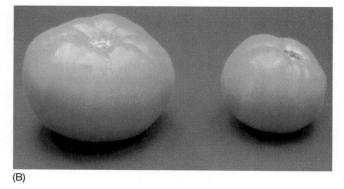

(B)

(C)

(D)

enemies. As might be expected, however, the pests and plant diseases respond by developing new strains that are able to attack the new varieties of plants. However, sometimes the pests are foiled. In 1996, genetic engineers created tropical rice plants that produce their own supply of a potent insecticide. The toxin is very effective against several major insect pests but is harmless to birds and mammals, including humans. The batttle continues.

3. *Sex attractants.* In one of the sneakier attacks, scientists have isolated the sexual pheromone of some insects (Figure 8.19C). Pheromones are chemicals released by the female members of many insect species that inform the male that the female is ready to mate. It doesn't take much—only about one ten-millionth of an ounce—and the male gets the message. So far, sex-attractant pheromones have been isolated for about 450 insect species and have been used either to attract predators of the undesirable species or to induce the male to mate with females who have been sterilized. These chemicals work on only one species, do not cause genetic resistance, and are harmless to nontarget species.

4. *Natural insecticides.* The chrysanthemum flowers in Figure 8.19D contain *pyrethrum,* a natural insecticide, in their heads. Harvested flower heads are ground into a powder and used as a commercial insecticide.

Purifying Pesticide-Contaminated Water. When a pesticide contaminates surface water or groundwater, there is no adequate way to purify it. Surface water eventually will be flushed away by rainfall, so it is just a matter of waiting a few months or a few years, depending on the level of contamination, the flow through the water body, and climate. But a pesticide in groundwater is another matter. In an aquifer, contaminants do not spread evenly or at constant rates. Commonly, the contamination is not recognized until long after it occurs. Given the complexity of most aquifers and the different properties of contaminants, serious contamination of an aquifer may be permanent. The aquifer may become closed as a water resource.

The effects of pesticide contamination may be calamitous for a sizable portion of a state, and perhaps for the nation as a whole. Estimates by the EPA indicate that up to 25 percent of usable groundwater is contaminated with synthetic organic chemicals, pesticides, or fertilizer. In some areas, up to 75 percent is contaminated. In New Jersey, every major aquifer is contaminated. In California, pesticides contaminate the drinking water of more than 1 million people. Imagine what

widespread pollution of the Ogallala aquifer (Chapter 6) would mean to U.S. agricultural production.

Seawater Incursion

Most of the world's largest cities are located along coastlines. This is where immigrants landed and remained, or where people long have worked in ocean transport or fisheries—Boston, New York, Miami, New Orleans, Houston, Los Angeles, San Francisco, Seattle. The need for freshwater to supplement surface supplies soon caused the drilling of water wells in increasing numbers as populations grew. Many aquifers beneath the land extend to the sea. The aquifer water under the land is fresh, but as the sea is approached, water in the aquifer grows saltier (Figure 8.20A). If too much water is withdrawn from the aquifer beneath the land, salty water migrates in to replace it. The aquifer becomes polluted with saltwater, unfit to drink (Figure 8.20B). This is called **seawater incursion.**

An example of seawater incursion can be seen in water wells at Union Beach, New Jersey (Figure 8.21). The concentration of chloride ion is commonly used to measure the presence of seawater. Until the early 1960s, chloride ion concentration remained static at about 2 ppm, despite heavy pumping from wells, which lowered the water table by 15 feet. But as water use increased and the water table continued to drop

FIGURE 8.20 Coastal aquifers and the ocean.

(A) Normal relation of a freshwater aquifer and salty seawater. Freshwater from the aquifer discharges into the ocean, and a natural boundary exists between saltwater and freshwater. (B) Excessive pumping of wells causes saltwater to encroach into a freshwater aquifer. This is an increasingly serious problem in many fast-growing coastal areas like Florida and southern California. *Source: U.S. Geological Survey.*

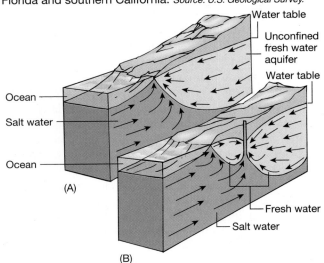

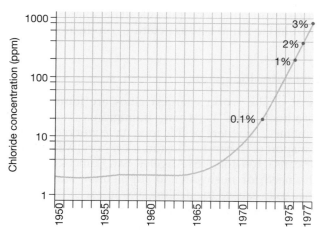

FIGURE 8.21 An example of seawater incursion.

Chloride concentrations in water samples from Union Beach Borough well field, 1950–1977. *Source: U.S. Geological Survey.*

below sea level, the pressure of the freshwater became too weak to keep seawater from infiltrating the aquifer. By 1977, the chloride concentration reached 660 ppm (about 3.4 percent seawater; seawater contains 19,000 ppm chloride ion). Chloride concentrations greater than 250 ppm exceed the EPA's standard for drinking water.

The problems caused by seawater incursion will become more severe as sea level continues to rise because of global warming (Chapter 17). The freshwater–seawater interface will naturally move inland, much to the distress of the nation's coastal residents.

Gasoline, Motor Oil, and Other Petroleum Products

The United States has some 200 oil refineries, 2300 bulk petroleum terminals, 10,000 wholesale distribution sites, 115,000 gasoline stations, 170,000 miles of petroleum pipelines, 281,000 miles of natural gas pipelines, perhaps 3 million underground storage tanks, and vast areas that produce oil and gas from drilled wells. This mammoth industry has mammoth environmental problems. In the past decade, new environmental regulations have been passed that concern all of these facilities.

The EPA has written extensive new rules for the existing underground gasoline storage tanks that lie buried a few feet below the surface. Almost all are located at present or former gasoline stations. Studies of groundwater beneath the tanks reveal that many tanks leak. They are made of steel, have been buried for decades, and the metal has been attacked continuously by shallow groundwater, with its mildly corrosive load of hydrogen ions. Recall that rainwater is naturally acidic, even without acidic additions from industrial smoke stacks (see Chapter 17).

Consequently, most tanks that leak do so from having rusted through. An entire new industry has arisen to test the soil and groundwater around these buried gasoline tanks to determine whether leakage has occurred. This industry, like the asbestos removal industry, did not exist only a short time ago.

Leaky tanks would not be so bad if they were in solid rock. But a high percentage of underground storage tanks are installed in unconsolidated material, because cities usually are built near water. Many inland cities are built on the unconsolidated sediment of floodplains or river terraces. Coastal cities are built on the unconsolidated sediment of old beaches or glacial outwash.

Because these materials are permeable, releases from underground storage tanks drain readily onto the water table (Figure 8.22). Groundwater flow in uncemented sediment is rapid, and plumes of contamination can spread quickly over wide areas. It is easy for the content of a leaky gasoline tank to reach unconfined aquifers in sandy unconsolidated materials, which commonly are used for domestic water supply.

Near the leak, the gasoline occurs as both a liquid and a vapor. Vapor exists only in the unsaturated zone above the water table. Liquid gasoline exists as a coating on sediment grains above the water table, as a dissolved fluid in the groundwater, and as a separate liquid. If the liquid is lighter than water, as is gasoline, it will float atop the water table. If it is heavier, it will be present along the base of the aquifer, immediately above the aquiclude. However, very few petroleum products are heavier than water. The common ones, like gasoline and motor oil, are 20–30 percent lighter than the freshwater in an aquifer, and so will float atop the water table.

Despite the old saying that "water and gasoline do not mix," in fact they do somewhat. Small amounts of gasoline can dissolve in water, certainly enough to cause sickness. Solubilities of the different kinds of molecules in gasoline range from about 0.1 ppm to 600 ppm in water. For comparison, the solubility of quartz in groundwater is 6 ppm, calcite perhaps 500 ppm.

A common indicator of gasoline contamination in groundwater is the benzene molecule. Benzene is abundant in gasoline, but absent from clean water. In a water-gasoline mixture that has had time to stabilize, about 50–60 parts per million of benzene will dissolve in the water. This is 50,000–60,000 parts per billion. Under current EPA guidelines, the maximum benzene allowed in public drinking water is only 5 parts per billion. Thus, a saturated water-gasoline mixture contains ten thousand times the benzene deemed safe for drinking. Therefore, one gallon of gasoline can make 10,000 gallons of groundwater unfit to drink.

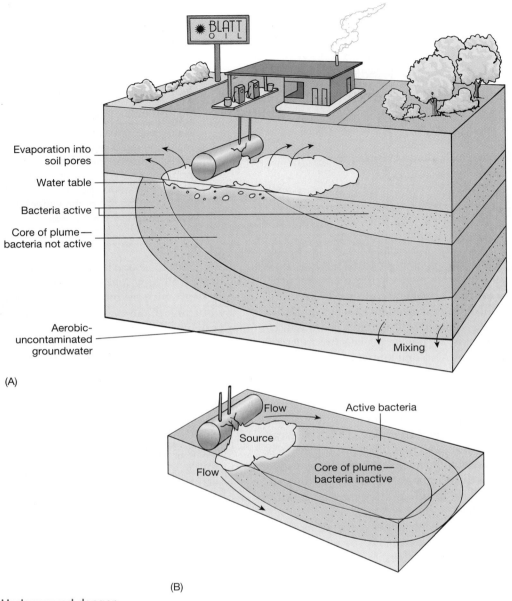

Evaporation into soil pores

Water table

Bacteria active

Core of plume— bacteria not active

Aerobic- uncontaminated groundwater

Mixing

(A)

Flow

Source

Flow

Active bacteria

Core of plume— bacteria inactive

(B)

FIGURE 8.22 Underground danger.

(A) Cross section and (B) plan view of a leaking buried tank.

Case Study: USA. In 1988, an underground gasoline seep was detected at a storage facility near Indianapolis (Figure 8.23). A nearby creek flowed from the storage area through an urban recreational area frequented by large crowds. Further downstream, the creek also flowed through a glacial aquifer that was part of a municipal water supply. A boom (surface water skimmer) was placed in the stream to catch floating gasoline and a "recovery well" was drilled to stop further seepage into the creek.

The concentration of dissolved benzene in the affected groundwater attained 12 ppm (12,000 parts per billion), or 2400 times the 5 ppb EPA standard. To remediate this pollution, 11 recovery wells were drilled (Figure 8.23) to recover gasoline floating atop the water table and gasoline dissolved in the water. The wells were shallow and inexpensive because of the shallow depth of the water table, only a few tens of feet. Fluids pumped from the wells were delivered to an oil/water separator to remove floating gasoline and then treated to remove dissolved gasoline.

In the process of pumping, the water table was lowered. This allowed sediment particles that were coated with gasoline to become exposed to the air, so

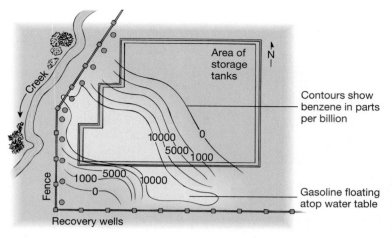

Area of storage tanks

N

Creek

Fence

Recovery wells

Contours show benzene in parts per billion

Gasoline floating atop water table

FIGURE 8.23 A lingering problem.

Gasoline-contaminated area near Indianapolis impacting regional water supply. Contours show dissolved benzene concentration in parts per billion (ppb) one year after the leak. Benzene contamination exceeded 12,000 ppb only in wells that had gasoline floating on the top of the water table. *Source: American Association of Petroleum Geologists.*

some organic molecules became oxidized. Also, some were evaporated by solar heat during warm months. As shown in Figure 8.23, benzene concentrations in the groundwater still greatly exceeded EPA's standard a year after the leak occurred. How many years will be required to reduce existing benzene levels (up to 10,000 ppb) to less than 5 ppb? No one knows.

Case Study: Russia. Corrosion and shoddy construction have created countless leaks in the ten thousands of miles of pipelines in the former Soviet Union. Large spills of crude oil from defective pipes are common. A member of the Russian Academy of Sciences admits that no one knows the number of spills, where they are flowing at the surface, or where they may be sinking into the groundwater.

Some of Russia's oil pipelines have an average life of only five or six years before corrosion makes its first hole, yet almost no devices exist for detecting leaks. Some experts claim that 5–10 percent of Russia's oil now leaks from pipelines. Spills are usually dealt with by plowing the oil into the ground or by burning it. No system is in place to cleanse land or remove oil.

An event in 1994 awakened the world to this oil-contamination disaster that had existed for years, but was concealed by the former Soviet Union. In a major rupture, about 50 million gallons of oil spilled from a corroded pipeline onto the Arctic tundra (Figure 8.24). This was nearly six times the oil spilled from the *Exxon Valdez* in the Gulf of Alaska in 1989. The ensuing interaction between politics and the environment is a frightening example of what happens in impoverished nations, which Russia clearly has become. Komineft, the nearly bankrupt regional oil company that operated the pipeline, decided it couldn't afford to halt production, despite the leak of its lifeblood onto the tundra. So, they just kept pumping crude oil through the leaking pipe! They built dams to capture the leaking oil.

FIGURE 8.24 The tundra despoiled.

Cleaning up the 1994 oil spill in the former Soviet Union. *Photo: Robert Wallis/SABA Press Photos, Inc.*

However, the dams soon broke, so Komineft finally stopped pumping. With the help of American and Australian cleanup experts, a $99 million "loan" from the World Bank, $25 million from the European Bank for Reconstruction and Development, and $16 million from Komineft, the Russian government tried to burn off or skim the gelatinous muck from Arctic streams before it flowed from the Kolva River into the Arctic Ocean. However, success was limited.

Ecology experts say that Russia averages two oil spills, mostly minor, *per day*. The cumulative effect of this on the always-stressed Arctic flora and fauna is likely to be disastrous. As a result of this oil spill and earlier ones in the same area, the fish catch has declined from 68 tons in 1976 to 8.4 tons in 1994, a decline of almost 90 percent.

Imagine such a situation occurring in the United States and the resulting media coverage, public outrage, government involvement, environmental activists on the march, political battles, damage-control ads from the oil company, legal warfare, billions of dollars flowing freely for the cleanup, and new taxes to fund it all. Different countries certainly handle environmental problems in different ways.

Used Motor Oil from Your Car.   When the oil is changed in your car, where does it go? In most cities and towns, gas stations and garages save waste oil from oil changes for recycling. But what if you change the oil yourself, as 60 percent of Americans do? EPA estimates that every year about 180 million gallons of used motor oil—the equivalent of more than 16 *Exxon Valdez* oil spills—are poured down storm drains and end up in our rivers and lakes and probably our groundwater as well. Does anyone want to drink used motor oil?

If you change your own oil, disposal is easy. Collect the oil in old plastic jugs, such as gallon-size milk or detergent containers. Find a local gas station that will let you pour it into their oil-recycling tank.

Farm Animal Waste

The general trend in agriculture is toward concentrating animals. The purpose is to reduce cost: they occupy less real estate, require less travel to care for, and can be raised in a more efficient "factory" arrangement. This is done with animals that become meat for our dinner tables: cattle, hogs, and chickens. But, concentration creates problems.

When animals graze on pasture and rangeland, their manure is dispersed across a large area and decomposes quickly. When animals are concentrated in a small farm feedlot, a chicken coop, or an ark, problems arise. The manure cannot decompose as fast as it is produced, and **feedlot waste** piles up (Figure 8.25).

Manure from these animal concentrations is a valuable source of major and minor plant nutrients. Major nutrients include nitrogen, phosphorous, and potassium, the main ingredients in commercial fertilizer. Minor nutrients include zinc, magnesium, sulfur, and copper. Thus, manure is an excellent source of nutrients for adding to soil, as any gardener knows.

But manure contains excess nitrates. Cattle and hogs digest their feed and excrete 75 percent of its nitrogen (in the form of nitrates and other nitrogen compounds). The EPA's 10 ppm nitrate limit for drinking water (Table 8.2) easily can be exceeded in water contaminated by feedlot manure. So, manure also is a contaminant and, in concentration, can become a pollutant. Water becomes polluted when rainfall drains through the manure, leaching nitrates into surface water and groundwater. Regulations governing livestock waste differ from state to state but most are not enforced.

Manure also pollutes the air with odorless and odorful gases. Odorless methane, generated as manure rots, is a greenhouse gas that retains heat in the atmosphere, possibly contributing to global warming, considering the volume of manure involved. The odorful gases are pollutants only in the sense that they offend some people. According to North Carolina's Swine Odor Task Force, living downwind of hog operations makes people tense, angry, and depressed, as well as tired and confused. A member of the task force concluded that "the amount of anger people have over odor pollution is tremendous. The smell gets into bedding, carpets, and drapes. People can't sell their homes because no one wants to live near a hog farm." Present research in manure management includes genetically engineering microbes that control odor.

Five Midwestern states (Nebraska, Texas, Kansas, Iowa, and Colorado) collectively feed about two-thirds of our cattle. Of these, 84 percent are fed in feedlots that hold at least a thousand animals. Beef cattle feedlots in Texas can hold up to 100,000 animals. A feedlot with 100,000 cattle produces as much waste as a city of 1.1 million people. Most of the pies accumulate in the five midwestern states that feed the cattle.

But cattle are not the only problem in animal waste disposal. The 60 million hogs in the United States produce an estimated 100 million tons of waste each year and, during the summer of 1995, a series of hog waste spills fouled streams and rivers in Iowa and North Carolina, the top two hog-producing states.

The nation's 7.5 billion chickens and turkeys are also warehoused for efficiency. Broiler chicken farms in Arkansas, the major producing state, raise as many as 400,000 chickens at at time. The nation's poultry generate 300 million tons of waste annually. All together, livestock and poultry generate a billion tons of waste each year.

(A)

(B)

FIGURE 8.25 Concentrating animals for efficient production.
(A) Feedlot for cattle showing normal concentrations of waste-water, urine, and manure. *Photo courtesy Soil Conservation Service.*
(B) Mass-production chicken coop, a new high (or low) in food production. Buildings like this hold 100,000 chickens and generate enormous amounts of waste. *Photo: Garry D. McMichael/Photo Researchers, Inc.*

Experts in nutrient management agree: no runoff from animal feedlots or livestock manure should flow into any stream. The nitrates feed algae, which consume the stream's oxygen. The microorganisms in animal feces can cause cholera, typhoid, and hepatitis. Heavy concentrations of farm animals should be located away from known aquifers and preferably in areas of impermeable bedrock, such as shale. However, present locations of feedlots were established long before water pollution became a serious problem. Feedlots are where they are for easy access by farmers and ranchers. Who is to bear the financial cost of relocating feedlots now thought to be dangerous to human health?

Salt on Roadways

During the winter months in the U.S.–Canadian snow belt, road crews spray pellets of "rock salt" (NaCl, halite) onto highways and parking lots to melt ice and snow. The United States alone uses about 8 million tons each winter. Why salt? A mixture of water and salt has a lower freezing point than does water alone (Figure 8.26). Pure water freezes at 32 degrees Fahrenheit, but the presence of positively charged ions such as sodium interferes with hydrogen bonding, which links water molecules together to form ice crystals. Seven ounces of salt will melt one pound of ice.

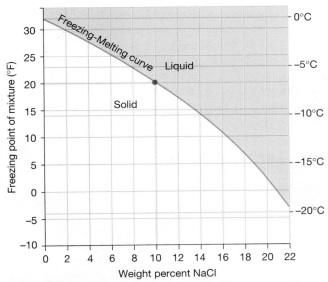

FIGURE 8.26 Effect of adding salt (NaCl) to road surfaces.

For example, when 10 percent by weight salt is added, the mixture freezes at 20 degrees Fahrenheit rather than at 32 degrees. It takes half a pound of salt to melt a pound of ice at 32 degrees.

FIGURE 8.27 Why we salt our roads.

Sprinkling salt on wintry roadways can help prevent this.
Photo: Roberto Borea/AP/Wide World Photos.

Salting icy highways makes travel much safer (Figure 8.27), but it also causes rapid rusting of metal on cars, kills vegetation alongside roads, and causes disintegration of concrete, the material many roads are made of.

What happens to the salty water after it melts the ice? Like all water on the land surface, some becomes runoff into streams and some soaks into the soil and may enter the local groundwater supply. In metropolitan Toronto, roadways receive more than 100,000 tons of salt each year, an amount equal to 7 ounces of salt for each 10 square feet of land. Only 45 percent of the salt is flushed annually from the area in surface runoff. The rest enters temporary storage in shallow subsurface water. If present rates of salt application continue, sodium concentration in the area's groundwater will increase.

One result will be that freshwater springs near Toronto will discharge saltier water, reaching 425 ppm in about 20 years. This is more than fifty times the sodium content of average stream water (6.3 ppm, Table 8.1). It is more than twice the acceptable limit of 200 ppm for drinking water set by the Canadian government (the EPA recommendation is 250 ppm). The values of 200 ppm and 250 ppm reflect the sodium taste threshold for most people (between 200 and 300 ppm).

However, more than taste is involved in determining an acceptable sodium level in water. Elevated sodium is harmful to people with hypertension, high blood pressure, and heart disease. Physicians advise these people to pursue a diet that is salt-free (which really means sodium-free).

This creates an environmental dilemma: should Torontonians continue to use road salt, allowing people to travel to work and school faster and more safely, even though it will make local spring water and groundwater unfit for human consumption unless treated? Discontinuing the use of road salt unquestionably would increase automobile accidents during winter months.

Cemeteries

Environmentally Friendly Casket—Swiss engineered, recycled cardboard, no trees must die when you do. Mahogany type finish. No tool assembly. Use for storage or Halloween while alive. $199, while supplies last. 1-800-555-5555.

Washington Post classified advertisement, 1993

The disposal of deceased people is closely regulated by state governments. Bodies may be either cremated or buried, usually in wooden caskets. Wood, like bodies, decomposes. Even musicians decompose. Can subsurface water be polluted with organic chemicals through decomposition? Is it safe to use groundwater near cemeteries? Does the character of the sediment that

FIGURE 8.28   Resting place for the terminally weary.

Have liquids from old and decayed coffins contaminated the peaceful lake in the background? Is someone supposed to investigate the quality of this water? *Photo courtesy Myrle Farley.*

surrounds the casket affect the mobility of the potential pollutants? Are there any regulations concerning such things? Are water wells located adjacent to cemeteries in *your* community? Will our final act on Earth be to pollute groundwater, whether we like it or not?

It is distasteful to think of cemeteries as sources of water pollution, but they can be. In 1994, there were 1.8 million burials in the United States, and all of these bodies are decomposing as you read this. Decomposing bodies and their decomposing coffins leak fluids into groundwater (Figure 8.28). These can contaminate groundwater, particularly in areas of high rainfall and high water tables. Historical cases include a reported higher incidence of typhoid among people living near cemeteries in Berlin between 1863 and 1867. Typhoid is caused by a bacterium in drinking water. Authorities in Paris reported a "sweetish taste and infected odor" in water from wells close to cemeteries in the area, especially in hot summer periods.

Where are cemeteries usually located? In areas with hills, they usually are placed on a hill so that the deceased

loved ones will be at least physically closer to God. Hilltops also provide a pleasant view for visitors. The permanent residents, of course, are unable to appreciate the view. The town's living inhabitants likely reside in the stream valley below, because towns are more easily built on flat land near a ready supply of water. Because sedimentary rocks cover about two-thirds of Earth's land surface, and because sedimentary rocks are easier to excavate to a depth of 6 feet than are igneous and metamorphic rocks, cemeteries nearly always are located in sedimentary rocks. Unfortunately, sedimentary materials also have the highest permeabilities to water.

As a result of our tendency to locate cemeteries in higher places where we can dig easily, the fluids emanating from decomposing bodies in decomposing coffins are able to move through the sediment and bedrock around most cemeteries and into the surface and subsurface water supply. Further, corpses routinely are preserved with formaldehyde, a known carcinogen.

An average human corpse contains about 22 pounds of protein, 11 pounds of fat (35 percent of it

TABLE 8.6	Groundwater Quality in Relation to Distance from a Row of Graves in Holland					
Distance from graves, feet	1.6	4.9	8.2	11.5	14.8	18.0
Bacterial count per cubic inch	350	500	500	220	75	10

saturated fat), and 1.1 pounds of carbohydrate. Estimates by a Dutch necrobiologist indicate that complete decay of this material requires about ten years at a burial depth of around 8 feet in sandy soil. Water flow through such a grave was estimated at 106 gallons/year (1.2 quarts/day). Table 8.6 shows the bacterial count in the groundwater at various distances from a row of graves. Because of studies such as this, many European countries have passed laws requiring a minimum distance between cemeteries and wells used for drinking water. In England, it is 300 feet; in France, 325 feet; in Holland, 165 feet. For aesthetic reasons alone, separation distances should always be as large as possible. In Connecticut, the required distance depends on the rate of flow of the water from the well.

<10 gallons/minute	>75 ft
10–50 gallons/minute	>150 ft
>50 gallons/minute	>200 ft

Does your state have a law regulating the location of water wells in relation to cemeteries? Should it?

Thermal Pollution

Electrical generating plants and some factories use water to remove heat produced by their operations. Cool water is borrowed from a nearby stream and circulated through the plant, where it absorbs heat energy from pipes, motors, or whatever needs cooling. It then is returned to the stream, at an elevated temperature due to the heat it has absorbed. The amount of water returned to the stream is the same as that removed, and its chemical composition usually is unchanged. Thus, this use of water is described as *nonconsumptive*.

The only change in property of the returned water is that its temperature is higher than normal stream temperature (Figure 8.29). Thus, the water is contaminated with heat energy, and if it is hot enough to injure aquatic life in the stream, it is **thermal pollution.** The temperature increase varies, but commonly is around 15 Fahrenheit degrees, not enough to kill outright the plants or fish near the return point for the heated water, but enough to affect their life cycles.

FIGURE 8.29 Thermal pollution.

Thermal pollution emission from an electric power plant affects the dissolved oxygen level as well as the temperature of the water downstream.

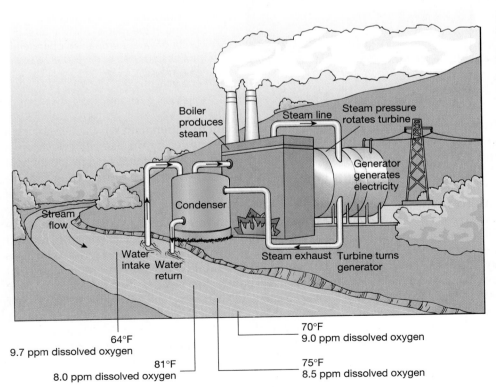

Boiler produces steam

Steam line

Steam pressure rotates turbine

Generator generates electricity

Condenser

Steam exhaust

Turbine turns generator

Stream flow

Water intake Water return

64°F
9.7 ppm dissolved oxygen

81°F
8.0 ppm dissolved oxygen

70°F
9.0 ppm dissolved oxygen

75°F
8.5 ppm dissolved oxygen

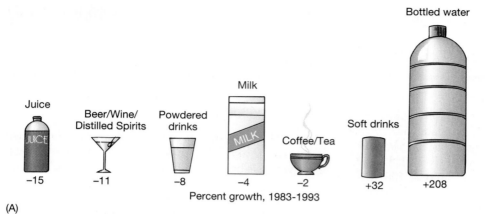

Bottled water

Juice

Beer/Wine/
Distilled Spirits

Powdered
drinks

Milk

Coffee/Tea

Soft drinks

−15   −11   −8   −4   −2   +32   +208

Percent growth, 1983-1993

(A)

(B)

FIGURE 8.30 American drinking habits.

(A) A clear trend: Americans are drinking more bottled water and less coffee, tea, milk, and alcohol.
(B) Bottled water is heavily promoted in the media. More than 225 bottled water companies operate in the United States. *Photo: Dion Ogust/The Image Works.*

Temperature influences reproductive cycles, digestion rate, and respiration rate. At higher temperatures, fish need more food to maintain their body weight, and they have shorter life spans. Also, warmer water contains less oxygen gas, which all animals need.

Seaward of Turkey Point on Biscayne Bay, Florida, in the early 1970s, biologists discovered a 75-acre area that was barren of aquatic life. Surrounding this, another 100 acres were only sparsely populated. This was disturbing, for Biscayne Bay is a breeding ground for fish and shellfish and an abundant source of lobster and crab. Investigation revealed that the barren area was centered around an effluent canal from Florida Power and Light's Turkey Point Power Plant. The heated water was causing this biological desert in the Bay.

It is important to understand why water is such a great coolant. The answer lies in its *hydrogen bonding.* Hydrogen bonding allows water to absorb lots of heat energy without its own temperature rising significantly. The heat energy absorbed from equipment in a plant goes toward breaking hydrogen bonds in the water, rather than in causing the molecules to jump around more (become "hotter"). This jumping around of molecules or atoms creates what we sense as hotness in a material. This ability of a substance to absorb heat without becoming hot itself is its **heat capacity.** The heat capacity of water is greater than that of any other common substance except liquid ammonia (another substance that has hydrogen bonds). Water is used because it is more available and cheaper than liquid ammonia.

Bottled Water

Concern over pollution in the public water systems that serve 97 percent of all Americans has sent sales of bottled water soaring over the past decade (Figure 8.30A). Today, most U.S. grocery stores have a bottled water display. Hundreds of bottled water companies worldwide participate in this multibillion-dollar business.

In 1979, 0.5 billion gallons of bottled water were sold in the United States. By 1991, this quadrupled to 2.1 billion gallons. In 1993, sales reached 2.3 billion gallons, and consumption of bottled water now exceeds that of tea, wine, liquor, powdered drinks, and fruit juice. You may be among the 43 percent of Americans who drink bottled water at least some of the time, or among the 8 percent who use it exclusively.

But bottled water costs 700 times more than tap water. Is it worth the cost? Are you buying water from "pure mountain springs"? Or are you buying intensively advertised hogwash with a scenic label on a designer bottle? A picture on the label may represent the water source as a glacier, mountain lake, or waterfall, when in fact the water comes from the public water system. Also, there are no accepted definitions for terms commonly used by bottlers such as "pure," "crystal pure" "premium," or "purified."

As we have seen, EPA monitors the quality of the nation's tap water. **Bottled water,** however, is considered a food and is regulated by the U.S. Food and Drug Administration. Although the FDA in 1994 raised their standards for water purity to those of the EPA, their jurisdiction is limited to interstate commerce. On products made and sold within a state (intrastate), regulation is up to the individual state. About two-thirds of bottled water is shipped intrastate. As is true of almost all state rules governing commerce, laws differ from one state to another. Some states are stricter than others. What are the standards for bottled water purity in your state?

The ion content of bottled springwater varies tremendously among brands (Figure 8.30B), from a low of 13 ppm to a high of 3000 ppm, twenty-five times that of average streamwater. Some bottled waters are quite high in sodium, and can be harmful to those on low-sodium diets. Most bottled waters are several times saltier than streamwater because they come from springs where water has moved slowly, percolating through the pores of buried rocks, dissolving sodium. The longer water is in contact with rocks, the more dissolved ions it is likely to contain. Streamwater contains fewer ions because it is mostly rapid surface runoff of precipitation.

In addition to higher ion content, springwater can become contaminated with synthetic chemicals. In 1990, the cancer-causing chemical benzene was detected in bottles of Perrier at four times the EPA limit for tap water. Perrier recalled more than 170 million bottles.

In summary, there is no guarantee that bottled water is better for you than municipal tap water. The quality of different brands can differ as much as that of tap water in different cities. If you do not investigate the purity of the bottled water you are buying, you may be increasing the cost of your drinking water by 700 times, with no benefit.

Summary

The increasing contamination of our nation's water resources is one of the three most important environmental problems in this country. The other two are soil pollution (Chapter 4) and air pollution (Chapter 17). Immense societal effort and vast amounts of money are needed to correct existing contamination/pollution, which has built up over two centuries in the United States. Unless we begin now, the problems may become unsolvable. The old adage "out of sight, out of mind" is true, but is disastrous when applied to polluted groundwater, which cannot be ignored. Once surface or groundwater becomes polluted, it is very difficult—often impossible—to purify it. An ounce of prevention is worth a ton of cure.

The largest non-governmental polluters are the chemical industries. They unavoidably produce large amounts of synthetic and nonsynthetic compounds that become effluent from their plants. State governments have controlled these effluents inadequately as they strive to offer a "friendly" environment for industrial growth. The result of this policy is clearly evident in the lower Mississippi River area, particularly the delta region of southern Louisiana, where the river is a soup of toxic chemicals, pesticides, fertilizers, and misplaced nutrients.

Pesticides, poorly designed sanitary landfills, and inadequate sewage treatment are other major sources of water pollution. The use of pesticides should be decreased and eventually abandoned. Communities should closely regulate landfill construction and sewage treatment. Protection of a community's drinking water must receive high priority.

Key Terms

benzene	fluoridation	seawater incursion
biocide	hazardous waste	thermal pollution
biological magnification	heat element	trace ion
bottled water	incineration	uncontaminated water
Cancer Corridor	nonpoint source	underground injection
Environmental Protection Agency	pesticide resistance	water standards
feedlot waste	point source	waterborne diseases

Stop and Think!

1. The chapter assumes that a good standard for "pure" water is *uncontaminated stream water*. But chemical composition of streamwater varies considerably from different areas. Suppose we take local rainwater as our standard. Would the EPA need to change its standards for purity? Explain.

2. What properties cause a waste to be classified as hazardous? What products in your house, apartment, or dormitory are hazardous according to EPA criteria?

3. The federal government is the nation's biggest polluter. However, it has exempted itself from regulation on the basis of national security. Is this valid? Should this be changed? If you believe the government's polluting activities should be regulated—like yours are—who should do the regulating? How can the law be changed to make this happen?

4. We have not been paying the true cost of producing chemicals we need, or think we need. Part of this true cost is disposal of chemical waste, which has been done cheaply but not effectively, resulting in very serious pollution of surface and subsurface waters. Are you prepared to pay double the price you are now paying for almost everything so that hazardous waste can be incinerated and the ash safely buried? If you are not, what alternative do you suggest?

5. As you read these words, poisons are drip, drip, dripping into our soil and water supply. The dripping continues twenty-four hours a day, year-round. Is this happening near you? Are there unlined containment pits near your town? Are industrial plants pouring unregulated effluent into a stream near you? Should you contact your congressional representatives for information?

6. If some of our hazardous waste is not exported, our disposal problem worsens rapidly. Should we ship some of our waste to the "underpolluted" areas of less-developed countries? Or is this taking unfair advantage of less fortunate people and cultures?

7. Suppose you receive a single job offer upon graduation, to work and live along the Mississippi River between Baton Rouge and New Orleans. If you turn it down, it means flipping burgers at your hometown fast-food outlet, keeping the 1979 Subaru, and accumulating more interest on your unpaid student loans. Just how seriously do you view water pollution as a personal danger?

8. What are the types of pesticides? Which type is generally more dangerous to humans?

9. Pesticide residues are present on a large number of the fruits and vegetables we eat. Are you going to switch to organic produce because of it? Where would you buy it? How would you know that it is "organically grown?" Could you afford it? Do you want your state or federal government to give tax breaks to organic farmers to make them more competitive with farmers who use pesticides? (Remember, this will raise your tax burden; nothing is free.)

10. Explain why salt water has been creeping into the groundwater supply in many areas. Name three factors that have made seawater incursion an increasingly important pollution problem in recent years. Is salt water incursion a looming problem for your community? (You don't have to live near the coast for this to be a problem.) Call the local office of the U.S. Geological Survey and ask.

11. Why are gasoline storage tanks so commonly buried where leaks can imperil the water supply? Are there leaky gasoline storage tanks in your community? Phone the city manager's office and request information, and get an opinion from the nearest U.S. Geological Survey or state geological survey.

12. Visit the bottled water section of your local food market and examine the labels. Are they informative? How does the content of dissolved substances compare with the much less expensive water you are drinking from the faucet at home? Does your community have a printed analysis of its water purity they can send you? Phone the city water department and ask for one.

References and Suggested Readings

Breslin, Karen, 1993. In our own backyards; the continuing threat of hazardous waste. *Environmental Health Perspectives*, v. 101, p. 484–489.

Bruce, L. G. 1993. Refined gasoline in the subsurface. *American Association of Petroleum Geologists Bulletin*, v. 77, p. 212–224.

Eisinger, Josef. 1996. Sweet poison. *Natural History*, v. 106, no. 7, p. 48–53.

Fricke, M. 1993. Natural mineral waters, curative-medical waters and their protection. *Environmental Geology*, v. 22, p. 153–161.

Goldman, B. A. 1991. *The Truth About Where You Live*. New York: Random House, 416 pp.

Green, J. and M. W. Trett. 1989. *The Fate and Effects of Oil in Freshwater*. New York: Elsevier, 338 pp.

Hamilton, P. A. and R. J. Shedlock. 1992. Are fertilizers and pesticides in the ground water? *U.S. Geological Survey Circular* 1080, 15 pp.

Hazardous Waste: Issues and Answers. 1985. American Institute of Professional Geologists. Arvada, Colorado: AIPG, 24 pp.

Howard, K. W. F., and J. Haynes. 1993. Groundwater contamination due to road deicing chemicals—salt balance implications. *Geoscience Canada*, v. 20, p. 1–8.

Ingram, Collin. 1991. *The Drinking Water Book*. Berkeley, California: Ten Speed Press, 195 pp.

Meadows, Robin. 1995. Livestock legacy. *Environmental Health Perspectives*, v. 103, p. 1096–1100.

Moore, J. W. 1991. *Inorganic Contaminants of Surface Water*. New York: Springer-Verlag, 334 pp.

Pimental, D., and others. 1992. Environmental and economic costs of pesticide use. *Bioscience*, v. 42, p. 750–760.

Water Quality: Agriculture's Role. 1992. Ames, Iowa: Council for Agricultural Science and Technology Task Force Report No. 120, 103 pp.

"And willful waste, depend upon 't,
Brings, almost always, woeful want!"
Ann Taylor, poet

"Out of sight, out of mind."
*Thomas à Kempis, 1420
(popularly adapted)*

Solid Waste

One of the certain byproducts of human life is garbage. It may be buffalo remains left by American Indians, turkey bones tossed by the Pilgrims, rifles left to rust on the battlefield of Gettysburg during the Civil War, Bulgarian dictionaries lost by twentieth century immigrants, or plastic packaging and cans of hair spray discarded by recent suburbanites. But whatever it is, there is a lot of it. We call it **municipal solid waste,** because its disposal is a service we expect from municipal governments, and it is solid as opposed to liquid sewage or wastewater. U.S. solid waste has increased 135 percent between 1960 and 1993, a reflection of our increasingly wasteful society—from 88 million tons (1960) to 207 million tons (1993). During the same period, the population increased by only 40 percent.

Solid Waste

Ask Americans to name the most abundant sources of solid waste, and many would respond "household garbage" (municipal solid waste) and "old cars." But in reality, neither of these is a big contributor. More than three-quarters of solid waste is produced by activities related to mining and farming (Figure 9.1). The amount of rock debris generated by the removal of coal and minerals from near Earth's surface is enormous (see Chapters 13 and 14), matched only by the bodily excretions of cattle, swine, and chickens. Neither mining waste nor farm manure is collected for systematic disposal but both can be a problem locally.

Third in abundance as a solid waste are crop residues, the unusable parts of cultivated plants. Fortunately, plants are biodegradable and their decomposition products do not pollute, so they do not pose a serious disposal problem.

Most media reports concerned with solid waste talk about the 5 percent that comes from urban population centers. It gets people's attention because it is located where people are, it is very visible, it possibly is odiferous, it must be transported for disposal, and it can be a source of disease. It also costs a lot of money to deal with. A significant and increasing part of our city tax burden goes to disposing of our trash. In addition, most municipal garbage contains some **hazardous solid waste,** such as paint solvents, batteries, insecticides, and other materials.

◀

Class visit to landfill.
Photo: Mark A. Stallings.

Municipal Solid Waste

The first known municipal dump in the Western world was established in Athens, Greece, in 400 B.C. By A.D. 1400, waste from Paris was piled so high outside the city gates that it interfered with the city's defenses against its enemies, and this presaged one of the more

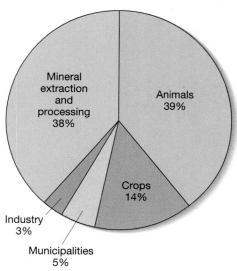

FIGURE 9.1   Source of solid wastes.

Mining and farming generate the lion's share of solid waste, but most of it stays at the mine or on the farm. The 3 percent generated by industry and the 5 percent generated at home are much smaller in volume, but a much bigger problem.

interesting military errors in medieval warfare. In 1415, a dump made history when it was "captured" during a Portuguese attack on a Moroccan city. The attackers thought it was a strategic hill.

Today, another hill is growing steadily in New York City. It is the largest landfill on our planet, located at Fresh Kills, on Staten Island (Figure 9.2). Fresh Kills (*kil* is Dutch for *stream*) was originally a marshland, a tidal swamp. When the landfill opened in 1948, New York planned to fill the marshland with waste and then develop the site for residential use and light industry. Now, fifty years later, the landfill rises more than 150 feet above low-lying Staten Island, occupies an area of nearly 3000 acres (4.7 mi²)—nearly 8 percent of the island's area—and has a volume estimated at 2.9 billion cubic feet (Figure 9.2), a volume larger than the Great Wall of China. The **Fresh Kills landfill** now is projected to rise about 400 feet above sea level when it is closed in 2001, making it the highest topographic feature along a 1500-mile stretch of the Atlantic seaboard from Boston to Florida.

Each working day, Fresh Kills increases 75 feet in length, receiving 14,000 tons of municipal and commercial waste generated in New York City, its suburbs, and parts of New Jersey. The waste travels from dumpsters behind city buildings to garbage trucks, and then to barges that cross the waterways among the various parts of the city. When the barges dock near the landfill, they are met by a fleet of vehicles that take the waste the rest of the way to the Fresh Kills dump. Five hundred people work there in shifts 24 hours a day, six days a week, at a cost to the city of $100 million a year.

Fresh Kills is now so famous (or infamous) that it has become a tourist spot. New York City's Department of Sanitation has produced a color brochure and video to help visitors find their way around. About 300 people a week visit the facility to take a 90-minute guided tour. According to the department's assistant commissioner of public affairs, "It is a very sensory experience visually. As far as the eye can see is garbage; aurally, yes, with all that heavy equipment, and in terms of your olfactory senses, you'll certainly know where you're standing. It is one of those things in life you have to experience to truly appreciate."

Fresh Kills sanitary landfill is not a place to be lost, particularly after sunset. Staten Islanders have long been unhappy with the dump on their doorstep, despite landscaping to screen it from nearby communities and scenting the rubbish with pine oil. In recent years, there has been a grassroots campaign to secede from New York City and close Fresh Kills. The garbage collected in American cities each year totals about 150 million tons. New York City and its environs have 8 percent of the nation's urban population and correspondingly produce 8 percent of all urban garbage. Thus, New York City is no more full of it than other American cities. California, on the other hand, has about the same urban population as New York City but produces twice as much garbage.

Nevertheless, the Fresh Kills landfill is projected to be full by the turn of the century. What will happen when this occurs? Can the 380,000 people who live on the island be compacted like the garbage is each day, to make more space? Should they be relocated? Or might a fifty–fifty division of the island between garbage and people be acceptable? Might this be a plank in the platform of New York's next mayoral candidates?

In 1993, each American generated an average of 4.4 pounds of garbage per day (Figure 9.3). In 1960, the amount was substantially less—2.7 pounds per day. According to the federal Office of Solid Waste, total garbage generation will increase 5 percent from the 207 million tons of 1993 to 218 million tons in 2000. New York City generates 14,000 tons of garbage every day, enough to fill a convoy of 1400 trucks that stretches for six miles. For the entire United States, the figures are 500,000 tons each day and trucks stretching for 180 miles.

What is to be done with the ever-growing mountains of trash throughout the United States? Federal, state, and local governments spend billions each year on waste disposal; it accounts for the majority of environmental expenditures. Large cities commonly pay disposal fees of $100 per ton to haulage contractors. There is a lot of trash and a lot of money in it.

In pre-industrial, rural America, trash-disposal problems were few. Communities were small and

(A)

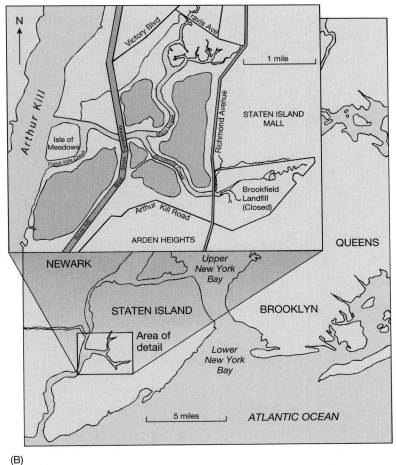

(B)

FIGURE 9.2 Mountain-building in the Big Apple.

Fresh Kills landfill now rises 150 feet above Staten Island, of which it occupies 8 percent. The landfill is projected to rise to about 400 feet elevation, making it the highest topographic feature along the seaboard from Boston to Florida. *Photo: Stephen Ferry/Gamma-Liaison, Inc.*

widely scattered and almost all refuse was biodegradable organic material. Under the then-prevailing philosophy of "waste not, want not," all non-food products were used and reused until they were so worn or decayed that they dispersed easily into the landscape. However, communities grew into cities. A growing

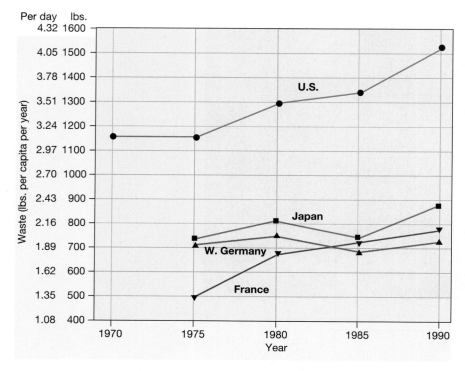

FIGURE 9.3 Wasteful us.

The United States is twice as wasteful as the other major industrialized countries. Shown is municipal solid waste in pounds per person per year. *Source: Characterization of Municipal Solid Waste in the United States: 1992 Update, Franklin Associates, Ltd., and U.S. Environmental Protection Agency Report 530-R092-019, 1992*

industry manufactured huge quantities of products that last a very long time. Consequently, the waste-disposal problem grew rapidly. Early in the twentieth century, disease due to unsanitary piles of garbage was common, and a stroll through a community was best made with severe nasal congestion. As early as 1865, an estimated 10,000 hogs roamed the streets of New York City, gorging on garbage.

Today's solid waste is basically paper, metal, glass, plastic, food waste, and yard waste. The proportions of each are shown in Figure 9.4. Note that 70 percent of the waste consists of manufactured products: plastic, paper,

FIGURE 9.4 How we fill America's garbage can.

Most people think of garbage as food waste, but that forms less than 7 percent of the total by weight. Paper products and yard waste (grass clippings, brush, tree trimmings) comprise over half the total.

Other 3.7%
Ferrous metals 6.3%
Wood 6.3%
Food wastes 6.7%

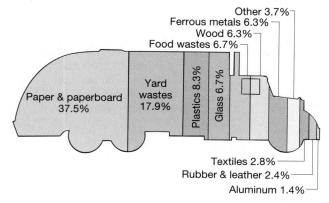

Paper & paperboard 37.5%
Yard wastes 17.9%
Plastics 8.3%
Glass 6.7%

Textiles 2.8%
Rubber & leather 2.4%
Aluminum 1.4%

metals, glass, and other materials that do not occur naturally (Figures 9.5 and 9.6).

A modern problem is household hazardous waste. Despite local restrictions on disposal, landfills are certain to contain old paint cans (paint contains pigments that are compounds of heavy metals like arsenic, cadmium, and mercury), household cleaners and disinfectants (containing mercury and organic solvents), batteries and electrical parts from cars and television sets (containing heavy metals like cadmium, lead, mercury, and the poisonous chemicals known as PCBs), and many other common disposable products that contain potentially harmful substances (Figure 9.7). About one percent by weight of all garbage coming directly from households can be regarded as hazardous, according to the EPA's definitions.

Sanitary Landfills

Over the decades, open dumps have littered the landscape, urban and rural (Figure 9.8). Scavenging dogs, cats, rats, crows, and gulls partied in rubbish dumps, joined by hordes of flies whose maggots riddled the garbage, mosquitoes whose larvae incubated happily in rainwater puddles, and other undesirables. As sources of disease, these open dumps had few equals. They also were receptacles for hazardous household chemicals that oozed into shallow groundwater. **Methane gas** was released into the air as microorganisms decomposed the garbage, and ever-present smoldering fires polluted the

(A)

(B)

FIGURE 9.5 Family portraits.

Families in developing and developed countries differ dramatically in their consumption habits. The Material World project created a "global family portrait" by photographing typical families with their possessions around the world. Shown here are families from Texas and India. Note that the American family not only has many more possessions than the Indian family, but the American family's goods contain few natural materials. The possessions of the Indians are mostly natural, easily degradable materials. *Photo: Peter Menzel and Peter Glinter/Material World.*

air with acrid smoke. The dumps were hazardous to children, who found them attractive as play yards. Older youth found them ideal sites for target practice with slingshots and low-powered arms. "The city dump" hurt nearby property values, although dumps in some locations actually created new real estate (Figure 9.9).

For both health and aesthetic reasons, the U.S. Congress passed a law prohibiting these open refuse dumps—the Resource Conservation and Recovery Act of 1976. Amazingly, this was only twenty years ago. The title of the law is misleading because it did nothing to promote recycling, but it did require that existing

dumps be converted to **sanitary landfills**. In 1996, these landfills received more than 60 percent of solid waste generated in the United States (20 percent is incinerated and 15 percent is recycled). However, the law requiring sanitary landfills is still widely ignored in rural areas, where open dumps persist (Figure 9.10). There are now 75,000 on-site industrial landfills, 5800 municipal landfills, and perhaps 40,000 closed or abandoned municipal landfills.

Sanitary landfills differ drastically from open dumps. The trash is placed in a hole lined with chemically resistant plastic, compacted by heavy equipment

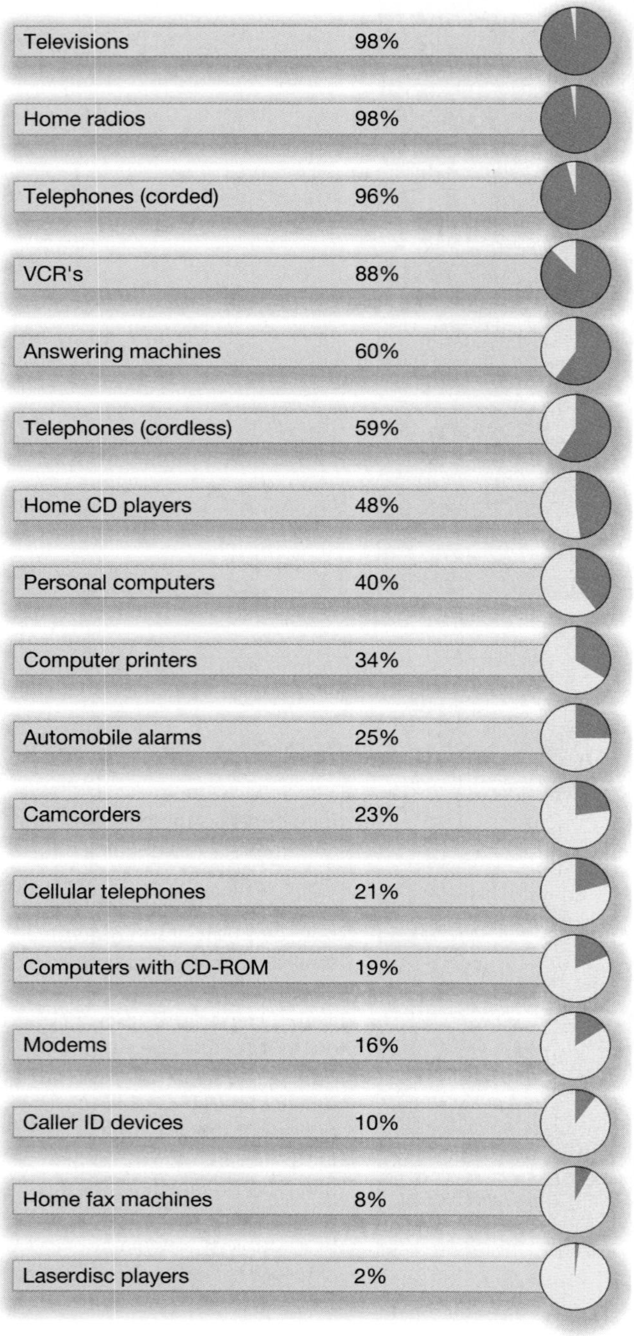

Televisions	98%	
Home radios	98%	
Telephones (corded)	96%	
VCR's	88%	
Answering machines	60%	
Telephones (cordless)	59%	
Home CD players	48%	
Personal computers	40%	
Computer printers	34%	
Automobile alarms	25%	
Camcorders	23%	
Cellular telephones	21%	
Computers with CD-ROM	19%	
Modems	16%	
Caller ID devices	10%	
Home fax machines	8%	
Laserdisc players	2%	

FIGURE 9.6   An electronic society.

The percentage of American households with each type of consumer electronic product in 1995. Nearly 100 percent of each item is composed of artificial materials. *Source: The Electronic Industries Association, as printed in* The New York Times, *April 21, 1996.*

such as bulldozers, and covered each day with a layer of dirt. This process reduces accessibility to rats and other vermin, lessens the risk of fire, and decreases odor. However, part of the reason for the decrease in vermin and odors is the increasing percentage of nonedible garbage in most modern landfills—now 93 percent. Proof of the nonedible nature of nearly all modern garbage is the lack of flies at landfill sites, as a visit will show.

If someone hired you to build and operate a sanitary landfill, here is what you would do to comply with present law (Figure 9.11):

- Locate it distant from population concentrations (this may be impossible around a large city).
- Select a site that has an impermeable bedrock, such as shale, for the landfill's base (a sinkhole in limestone does not qualify!).
- Carefully lay a heavy, chemical-resistant plastic or rubber liner and clay layer atop the shale to prevent leakage of fluids from the landfill down into the water table, where it might contaminate an aquifer (Figure 9.12).
- Install collecting pipes above and below the liner to collect fluids that percolate through the waste pile like water through coffee grounds.
- Each day, after the day's refuse has been placed in the landfill, compact it and cover with a layer of impermeable, clay-rich sediment (Figure 9.13).
- When the landfill finally is full, cover the last day's deposit with the same type of heavy liner used at the base of the fill, to prevent entry of rainwater and snowmelt.
- Install vertical pipes to collect the methane gas (CH_4) that is produced within the refuse as bacteria do their work. The methane may be vented into the air or collected and used as a fuel (it is the chief component of natural gas), depending on how much is produced (Figure 9.14).
- Drill monitoring wells near the landfill to detect contaminants that will escape.

If these procedures are followed, groundwater under the landfill will be protected from contamination, as will local surface waters. Rainwater will have difficulty percolating through the clay-rich sediment layers during the lifetime of the landfill and will be unable to penetrate the heavy plastic overlay when the landfill is full and closed. Following closure, the landfill surface can be landscaped and used for a baseball field, playground, or park area. Of course, even with these procedures many opportunities for trouble exist in a landfill.

Landfill Problems

The first U.S. sanitary landfill opened in 1904. Most sanitary landfills in operation today opened decades

FIGURE 9.7 Hazardous waste from our homes.

Even "empties" contain residual chemicals that are harmful, especially when randomly mixed in a landfill. *Photo: Aaron Haupt/Photo Researchers, Inc.*

ago, when environmental concerns were not focused as they are today. Thus, the average landfill is older, and does not meet current legal standards.

Leachate.   The most common problem is that the waste pile is permeable. This allows harmful substances to dissolve in rainwater as it percolates through the pile, just as water passing through coffee grounds dissolves chemicals in the coffee. This creates **leachate,** which is water that contains ions from the trash. These ions may be both toxic and nontoxic. If the landfill does not have a plastic liner at its base, the leachate will transport the ions into groundwater. If this groundwater feeds a stream, the contamination will affect it too.

An older landfill may have no plastic liner. Or, the liner may be punctured by a sharp object at the base of the landfill. Perhaps the bedrock is not impermeable shale, but a more permeable sandstone or limestone that lets leachate pass through. Or, if it is a shale, it may be fractured, making it more permeable.

FIGURE 9.8   The way it was.

Landfilling a century ago, now illegal in the U.S. and Canada. *Photo: Drs. Nick and C.H. Eyles.*

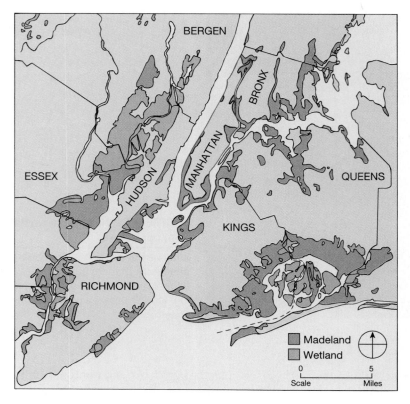

BERGEN

BRONX

MANHATTAN

HUDSON

ESSEX

QUEENS

KINGS

RICHMOND

■ Madeland
■ Wetland

0   5
Scale   Miles

FIGURE 9.9   Growing New York with trash.

New York City and suburbs are a bigger place today, expanded with "madeland" created by landfilling with solid waste. Much of the new land is former wetlands. *After Waste Management, Regional Plan Association, New York, 1968.*

FIGURE 9.10   An open dump in a wooded area.

Note the paper, plastic containers, and large metal objects. Monitoring the thousands of these dumps that exist nationwide is impossible. Do you know where your community's landfill is located? *Photo: Simon Fraser/Science Photo Library/Photo Researchers, Inc.*

Perhaps the daily spreading of clay sediment over the trash was incomplete.

Experts in landfill design believe that no landfill can be guaranteed as permanently secure. Eventually, all pipes clog, preventing the withdrawal of leachate. All buried pumps fail, sooner or later. In all but the most arid conditions, refuse becomes wet no matter now well the landfill is designed. Pockets of moisture are always present, partly due to leaks and partly because of liquids or wet materials discarded in the dump. Experience has shown that rats, badgers, and other burrowing creatures chew holes in the plastic barrier at the top of the landfill. Over the years, precipitation inevitably seeps in and, after mixing with the landfill's corrosive contents, the leachate dissolves the bottom liner and leaks outward and downward into groundwater.

As we saw earlier, ordinary household waste is far from harmless, to say nothing of more obvious hazardous waste from chemical plants and refineries. Every American household sends an average of 1.5 quarts a year of hazardous liquid to landfills—and there are 97 million households in the United States.

Because of these certainties, many designers of landfills advocate using water to speed decomposition of the refuse. They suggest adding perforated horizontal and vertical pipes within each layer of garbage accumulation that is separated by plastic liners. This **wet landfill design** allows the landfill manager to saturate the rubbish with the rank fluids that gather at the base

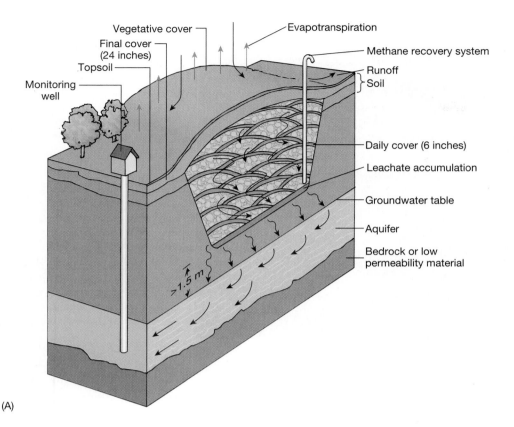

(A)

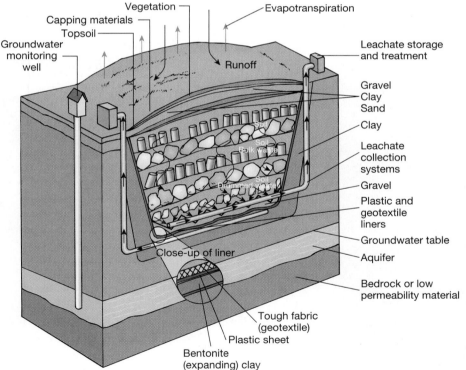

(B)

FIGURE 9.11 Two types of land-fills.

State-of-the-art landfills minimize environmental problems that plague older landfills. (A) The natural attenuation landfill uses soil processes to lessen the impact of (attenuate) contaminants in the leachate before they reach groundwater. (B) The containment-type landfill collects the leachate and pumps it out for storage and treatment. It can be used for more hazardous wastes or when soil conditions are unsuitable for the natural attenuation landfill.

of the pile, which will keep microorganisms flourishing and speed up chemical decay.

In "wet" dumps, this speedier decay means that pipes and pumps have to operate for only five years rather than for decades, as in a "dry" landfill. Consequently, the most contaminated water should never leak into the soil and groundwater. The leachate is circulated rapidly in the early years of landfill use, so

FIGURE 9.12   Every landfill should have a silver lining.

Note the plastic liner at the base. Hopefully, it will remain undamaged for a long, long time. *Photo: Larry Lefever/Grant Heilman Photography, Inc.*

when the bottom liner eventually breaks down and leachate pipes and pumps fail, the bulk of the potentially dangerous contaminants have already been removed. In a four-year field test of the wet landfill concept in Sonoma County, California, the final leachate contained 2700 parts per million (ppm) dissolved substances. In an identical dry landfill after four years, the leachate contained 30,000 ppm. In addition, the wet landfill system produced more than eleven times as much methane gas, a pollution-free source of fuel.

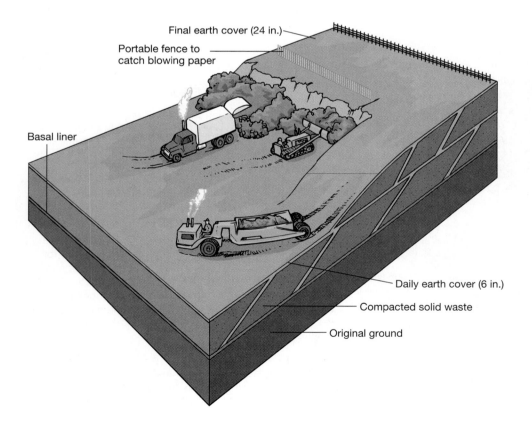

Final earth cover (24 in.)

Portable fence to catch blowing paper

Basal liner

Daily earth cover (6 in.)

Compacted solid waste

Original ground

FIGURE 9.13   Daily burial.

At a modern landfill, each day new earth cover is bulldozed over the day's delivery of compacted solid waste.

FIGURE 9.14 Bacteria at work.

A sign of the landfill beneath: a vent pipe releases methane gas into the atmosphere. The methane is generated as anaerobic bacteria slowly digest their dinner. *Photo: Willie L. Hill, Jr./The Image Works.*

Landfill Subsidence. It is difficult to build atop a finished landfill because installation of underground utilities might puncture the upper (and perhaps lower) plastic liner and allow water to penetrate the waste pile. Also, the refuse in the landfill compacts and settles with time, perhaps not uniformly, leading to settlement cracks in structures built on the surface. Depending on how crushable the landfill contents are and on the weight of overlying garbage, the settlement rate may be several inches or even feet during the first few years.

Gassy Landfills. In the innards of landfills, organic materials decompose away from air, so no oxygen is present. Consequently, oxygenless gases are produced: methane (CH_4) and hydrogen sulfide (H_2S). If air is able to penetrate parts of the landfill, carbon dioxide (CO_2) will also be produced. At the Fresh Kills landfill in New York City, methane and carbon dioxide are produced in approximately equal amounts.

These and other gases may escape into the air, increasing the risk of cancer in people who breathe them over a long period. A study of cancer occurrence in people living near North America's third largest landfill, near Montreal in Quebec, found a significantly higher than expected cancer rate for several types of the disease.

In a sealed landfill, these gases are unable to escape and can accumulate to explosive levels. Methane produced in 82 U.S. landfills is collected and used as a fuel to generate electricity. At Fresh Kills, enough methane is recovered to satisfy the annual energy needs of 12,000 Staten Island households. Some 2,000 to 3,000 other large landfills have the potential for large-scale methane recovery. In small landfills, too little methane is produced to be recovered economically, so the gas either is burned off or vented into the air through perforated pipes inserted into the landfill (Figure 9.14).

Even if the methane concentration is too small for economic recovery, it still can threaten homes above a landfill. In 1991, potentially explosive methane concentrations were discovered in thirty-three of forty-four homes in a subdivision in Savannah, Georgia. Of these, seventeen were directly above an abandoned landfill. The gas entered the homes through defects in basement flooring and was trapped by the excellent insulation in the walls of the houses. Methane is explosive at concentrations of more than 5 percent in room air. Lawsuits were filed by homeowners, and the city purchased most of the homes at prevailing value. Some homeowners decided to remain because the city offered to place meters and vents in contaminated parts of the structure to preclude the possibility of an explosion.

Other industrialized nations have major garbage disposal problems, too. The small island nation of Japan generates 1300 tons of garbage daily. Disposing of it on land is not feasible, so much of their solid waste has been used to create artificial islands in Tokyo Bay. They are built of thousands of tons of household waste, sandwiched between layers of soil washed down from the uplands of this mountainous country. The islands are used for both residential neighborhoods and recreational facilities.

However, an environmental problem recently surfaced on these artificial islands. Methane from anaerobic (oxygenless) decomposition of garbage is wafting up through vents in the ground to the extent that it is not advisable to smoke on some golf courses! So much methane is wafting up from the depths of the islands that golfers are in danger of blowing themselves up, and "no smoking" signs have been placed near some holes.

A Japanese newspaper noted, "Never has the need been greater to review our practice of mass production, mass consumption and mass discarding." An ecology and economics professor at a Japanese university commented, "In short, we need less production, less consumption, [and] a higher tax on imported resources and the discharging of wastes." Might this strategy apply equally to American society? If so, is there any practical way to accomplish it?

Biodegradability

A **biodegradable** substance is one that becomes "biologically degraded," meaning that it is decomposed by living things like bacteria and fungi. Generally, the term implies that biodegradation will proceed quickly, over months or a few years. But will a product touted as "biodegradable" decay in a landfill? Advertising often is deceitful. The product may indeed be biodegradable—but over many thousands of years. A study of biodegradable plastics released by the pro-environment organization Greenpeace quotes a Florida newspaper's interview with a candid Mobil Chemical Company spokesman:

> Degradability is just a marketing tool. We're talking out of both sides of our mouths because we want to sell [plastic trash] bags. I don't think the average consumer even knows what degradability means. Customers don't care if it solves the solid waste problem. It makes them feel good.

Manufacturers define biodegradability as the spontaneous breakdown of a product when it is left outdoors, in contact with sunlight, rain, and oxygen. The problem, of course, is that these conditions are not present within a landfill. When properly designed, a landfill prevents sunlight, rainwater, and air from penetrating the waste pile. Under these conditions, decomposition is very slow. A hot dog or a newspaper may last for many decades (Figure 9.15), as inferred from recent studies of old landfills. Many decades-old newspapers excavated from landfills by **garbologists** are as readable today as when they were buried. Today's landfill is tomorrow's time capsule.

The Dirty Diaper Dilemma. When your author's children were born in the 1960s, all diapers were made of cloth and designed to be reused for many years. The daily laundry was a nightmare of soggy diapers; people with twins did laundry in shifts. Those who could afford it hired diaper services to pick up and launder the soiled ones and supply a clean batch.

Then disposable diapers were developed, revolutionizing baby maintenance. Currently, about 85 percent of U.S. baby bottoms are wrapped in disposable plastic-and-paper diapers (Figure 9.16). EPA estimates that parents discard more than 18 billion disposable diapers each year, enough to stretch back and forth to the moon at least seven times, a distance of 3.5 million miles. Most environmentalists, even some who are parents, are opposed to disposable diapers because they can resist decomposition in landfills for decades, according to excavations by garbologists. It may take hundreds of years for a disposable diaper to decompose in a landfill, but we can't be sure because these family helpers did not come into widespread use until the 1970s. Reusable cotton diapers decompose in about six months. According to the Solid Waste Composting Council, disposable diapers take up about two percent of our landfill space. They add about $400 million annually to the cost of waste disposal.

Another problem with disposable diapers in landfills is that they contain disease. Polio and AIDS are among the many viruses and bacteria that researchers have discovered in landfilled diapers. This exposes sanitation workers to these diseases.

Cloth diapers cost about 4 cents each for parents to wash themselves or about 17 cents per diaper if they use a laundry service. Disposable diapers run about 22 cents per diaper. Apparently, most of today's American parents would rather quintuple their cost rather than wash reusable diapers by hand.

The city of Toronto, Canada, has begun to recycle disposable "nappies" to ease the pressure on landfill space. The metropolitan authority now collects 100,000 nappies each month from its 200 child care centers and twelve drive-through recycling depots. A local company recycles the wood pulp, plastic, and water-absorbent gel from the nappies. About 90 percent of the recycled material is wood pulp with long fibers, much sought after by the paper industry. Toronto hopes to increase its recycling effort to 100 million nappies a year, which would cut its landfill costs by $400,000 annually.

Where the Rubber Meets the End of the Road. Another serious problem of non-biodegradable waste is discarded tires from cars, buses, trucks, motorcycles, and

FIGURE 9.15 Heat and serve?

These hot dogs lay buried for ten years in a sanitary landfill. The idea that a landfill is a large compost pile in which things biodegrade fairly rapidly is a myth. Decomposition in modern landfills is slow because the garbage is tightly packed and exposed to little moisture and essentially no sunlight or air. *Photo: Riley N. Kinman/University of Cincinnati/Cincinnati Department of Civil & Environmental Engineering.*

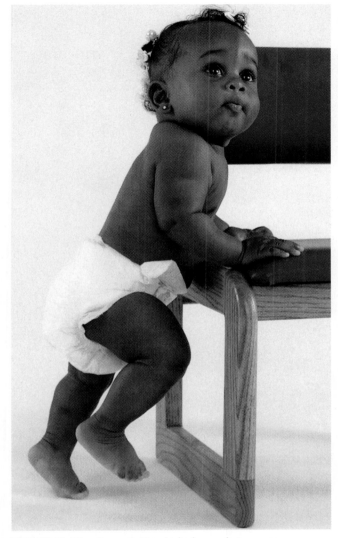

FIGURE 9.16 A revolution in baby maintenance.

It may take hundreds of years for disposable diapers to decompose in landfills, although we can't be sure because these indispensable parenting aids were not used widely until the 1970s. *Photo: Yoav Levy/Phototake NYC.*

bicycles (Figure 9.17). Landfills generally won't accept them because they trap methane gas, fill with water and become mosquito breeders, and burn uncontrollably for years when they catch fire. Rubber tires on fire pour into the air a dense black smoke, loaded with particulates, an assortment of polluting organic chemicals, and sulfur compounds that contribute to acid rain. They also discharge zinc, iron, tars, and soluble organic compounds into local streams, poisoning the wildlife.

There are now 2.5 billion waste tires in the U.S., and another 280 million are thrown away every year—one for each man, woman, and child in the U.S. Unfortunately, tires are among the least biodegradable products produced by modern civilization. Once the rubber used in car tires has been vulcanized (treated with heat and sulfur to make it harder and more elastic), it becomes immune to attack from bacteria that would otherwise break it down. As a result, dumped tires enjoy a long life span on the landscape, unless they catch fire.

What can be done with this ever-growing rubber mountain? Tires make an excellent fuel, when their polluting fumes are controlled. They have 25 percent more heating value per pound than the coal typically used in power plants, and can be cheaper. Electric utilities in the United States and Canada are starting to use old tires as a supplemental fuel in coal-fired plants. Illinois Power Company, one of the major users of tires as fuel, estimates savings of $670,000 and significantly reduced sulfur dioxide emission (a major cause of acid rain; see Chapter 17) by burning annually 7.5 million tires instead of coal.

Of the 280 million tires we discard each year, 60 percent are simply dumped. The other 40 percent are reused, retreaded, burned as fuel, or shredded for surfacing roads. The use of old tires in road resurfacing has been pushed by the U.S. Congress. In 1991, Congress passed legislation requiring states to use this recycled rubber in at least 20 percent of their asphalt roadway repaving jobs by 1997.

When Landfills Become Landfulls

The ultimate problem with landfills is that they become landfulls: they fill up. During the decade from 1978 to 1991, the number of landfills in operation plummeted from 20,000 to 5812. It is likely that half the landfills currently in operation will be full in just a few years. By 2000, EPA estimates that 25 or more states will have no more landfill space, including very populous states like New Jersey, Pennsylvania, Massachusetts, and Florida (Figure 9.18).

Philadelphia, the country's sixth most populous metropolitan area, no longer has an active landfill. The City of Brotherly Love ships its trash many hundreds of miles to Maryland, Virginia, and even Ohio. This drove up the trash bill for Philadelphians from $20 million in 1981 to $128 million in 1989, and it has continued to increase. Philadelphia paid $20/ton in 1980, $90/ton in 1988, and $140/ton in 1990 to dispose of its garbage. Neighboring New Jersey has one of the worst disposal problems in the U.S., shipping its solid waste as far away as New Mexico.

The chief health officer of Washington, D.C. warned that "appropriate places for garbage are becoming scarcer year by year. . . . Already the inhabitants in proximity to the public dumps are beginning to

(A)

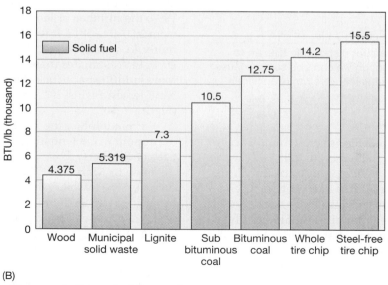

(B)

FIGURE 9.17 Where the rubber meets the end of the road.

(A) A few of the 2.5 billion waste tires in the U.S. This tire collection facility at Westly, California, contains six million tires, the fuel for a power plant that provides electricity for 3500 homes. *Photo: Jose Azel/Aurora & Quanta Productions.* (B) Comparison of the heat energy (BTUs) obtained from different types of fuel. Rubber tires outperform coal.

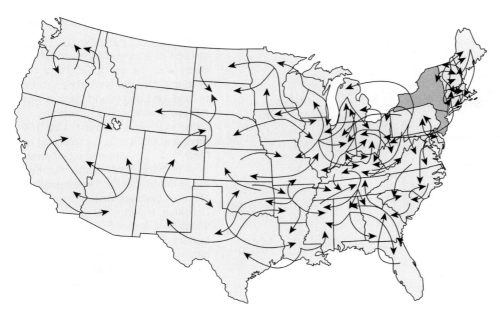

FIGURE 9.18 Throwing trash over the fence.

Interstate traffic in garbage is intensive and complex. For lack of space, this simplified map does not show trash shipping routes from New York and New Jersey, the two biggest garbage-exporting states. New York's garbage is trucked as far away as New Mexico. *Source: National Solid Waste Management Association.*

complain." He said this in 1889! The situation has only worsened in the 100 years since his remarks.

New York City is approaching desperation, continually trying to talk other states into accepting its waste. In 1987, the garbage barge *Mobro* sailed from New York to Honduras and back, with many intermediate stops—a trip of fifty-five days—trying to unload its 3200-ton cargo (Figure 9.19). No one would accept it; six states and three foreign countries declined the offer. The journey finally ended when the trash was incinerated and the ashes were buried in a New York landfill.

The *Mobro* story illustrates the **NIMBY phenomenon** (Not In My Back Yard) that accompanies all attempts to dispose of unwanted substances or to locate industries perceived as hazardous, such as nuclear power plants. In the last few years, NIMBY has evolved into NIMFYE (Not In My Front Yard Either) and BANANA (Build Absolutely Nothing Anywhere Near Anything), and other acronyms (Figure 9.20).

How can we slow the rate at which landfills are filling and closing? Possibilities include generating less waste to begin with, recycling, and incineration.

Hazardous Solid Waste

As noted in Chapter 8, 90 percent of hazardous waste is in liquid form. The other 10 percent is solid and cannot be disposed of by purification or by disposal down deep wells. Common examples of solid hazardous waste are automotive and flashlight batteries, plywood, waxes, some types of air fresheners, many carpets, and old paint brushes. The most common approach to disposing of hazardous solid waste is by incinerating it at very high temperatures in a mass-burn facility. This can reduce waste volume by 90 percent.

The 170 burn plants now operating in the United States range from capacities of a few hundred tons to 4,000 tons of trash a day. Most of these plants are designed to generate electricity by allowing the burning trash to heat water in a boiler to produce steam that drives a turbine to generate electricity. However, garbage is an inefficient fuel. Burning rubbish never generates as much electrical energy as the energy saved by recycling these same materials (Table 9.1) or, better yet, by reducing the amount of rubbish in the waste stream headed for the incinerator. (Of course, even inefficient electrical power generation from garbage is better than getting nothing from it at all.)

Also, incineration can create air pollution. Despite claims of "newer, cleaner" technologies, all incinerators produce large numbers of substances harmful to human health and the environment. These poisons issue from the smokestacks and are part of the ash that remains after burning. The worst smokestack offenders are acid-forming gases that generate acid rain (Chapter 17) and toxic heavy metals such as arsenic, beryllium, cadmium, chromium, lead, mercury, and selenium—elements familiar to us from Chapter 8. They can cause nervous system disorders and cancers. Dangerous organic chemicals are also commonly emitted—pollutants known as *dioxins* and *furans*.

Ideally, all potentially harmful substances can be captured before departing the smokestacks, so that the smoke exiting the burn facility is as clean as a baby's breath. But, as noted by a branch director of the National Institute for Environmental Health Sciences, "If an incinerator is designed well and run efficiently, it should be no problem. The problem is they're not designed that well, not run that efficiently, and they present a problem."

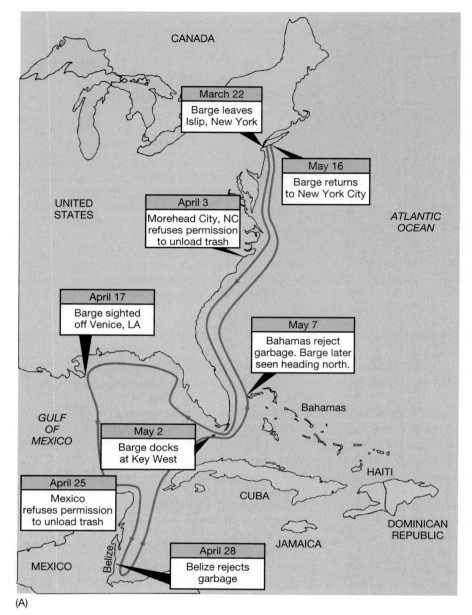

(A)

FIGURE 9.19 Next time, try recycling.

In 1987, the barge *Mobro* sought a place to dump about 3200 tons of garbage from the New York suburb of Islip, on Long Island. *Mobro* was refused permission to unload its cargo everywhere it tried. After 164 days and 6000 miles, it returned to New York City, where it was barred from docking. Finally, after simmering in the harbor for three months, *Mobro's* garbage was incinerated in Brooklyn, leaving 400 tons of ash, which was shipped back to Islip for burial in a local landfill. The publicity catalyzed Islip to develop a recycling program, and, by 1989, the town was recycling 35 percent of its solid waste. This has saved the community $2 million a year and extended the life of its landfill. *Photo: Dennis Capolongo/Black Star.*

CANADA

UNITED STATES

ATLANTIC OCEAN

March 22
Barge leaves Islip, New York

May 16
Barge returns to New York City

April 3
Morehead City, NC refuses permission to unload trash

April 17
Barge sighted off Venice, LA

May 7
Bahamas reject garbage. Barge later seen heading north.

GULF OF MEXICO

May 2
Barge docks at Key West

Bahamas

April 25
Mexico refuses permission to unload trash

April 28
Belize rejects garbage

Belize

CUBA

HAITI

DOMINICAN REPUBLIC

JAMAICA

MEXICO

(B)

The ABCs of Waste Disposal

NIMBY...Not In My Back Yard
NIMFYE...Not In My Front Yard Either
PIITBY...Put It In Their Back Yard
NIMEY...Not In My Election Year
NIMTOO...Not In My Term Of Office
LULU...Locally Unavailable Land Use
NOPE...Not On Planet Earth

*Environmental Protection Agency

FIGURE 9.20 Where should it go?

This poster's grim humor highlights the difficulty of finding places for new landfills.

Even assuming the ideal controls on stack emissions, a large amount of ash is produced. About 5 million tons are generated each year, with the largest cities producing the most ash, just as they produce the most garbage. For example, Philadelphia produces 500 tons of incinerator ash a day, almost 200,000 tons a year.

Incinerator ash takes up far less space, but with its concentrated toxicity it is even more difficult to get rid of than unburned trash. The wayward garbage barges of today may well turn into the unwanted ash barges of tomorrow as incineration grows more popular for reducing the volume of solid waste.

Recycling

Recycling is as old as human society, from hand-me-down clothing to using manure to nourish new vegetables. During World War II, when household items were in short supply, recycling became second nature. Old clothing became part of bed quilts; people reused containers, Daylight Savings Time was started to reduce electricity consumed in lighting, and people straightened and reused old nails. In wartime, recycling was a patriotic duty. It was recognized that recycling saves energy and natural resources.

Since that far-away time more than fifty years ago, recycling has had its ups and downs. Today, recycling is "in" again (Figure 9.21), and about 20 percent of our states require recycling of some products. Many states have "bottle laws" requiring deposits on beverage containers to encourage their return. Many states and communities have successful voluntary recycling programs.

The benefits of recycling are unarguable:

- Recycling one ton of steel prevents the production of 200 pounds of air pollutants, 100 pounds of water pollutants, almost 3 tons of mining waste (Chapter 13), and the use of about 25 tons of water.
- Recycling one ton of aluminum eliminates the need for 4 tons of bauxite (aluminum ore) and almost a ton of petroleum products to provide the energy for mining.
- Recycling a ton of paper saves about seventeen trees, which live to absorb 250 pounds of global warming carbon dioxide gas annually (Chapter 17).

More than half the metals used in the United States each year do not come directly from a mine but from recycling operations (35–40 percent of our steel and aluminum are recycled). Recycled paper accounts for about 25 percent of our annual consumption; recycled plastic for about 2 percent. Considering all types of refuse, about 19 percent is recycled. The percentage of municipal waste going to sanitary landfills dropped from 83 percent in 1985 to 62 percent in 1993, in part because of recycling (incineration was also a factor). Recycling is a big saver of both energy (Table 9.1) and landfill space.

At present, recycling's biggest problem is its cost. A large front-end investment is needed to acquire extra

TABLE 9.1 Environmental Benefits of Recycling

Environmental Benefit	Aluminum (Percent)	Steel (Percent)	Paper (Percent)	Glass (Percent)
Reduction of:				
Energy Use	90–97	47–74	23–74	4–32
Air Pollution	95	85	74	20
Water Pollution	97	76	35	—
Mining Wastes	—	97	—	80
Water Use	—	40	58	50

Source: Robert Cowles Letcher and Mary T. Sheil, "Source Separation and Citizens Recycling," in *The Solid Waste Handbook*, ed. William D. Robinson (New York: John Wiley and Sons, 1986).

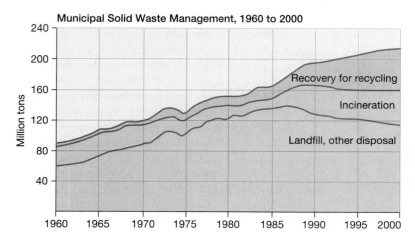

Municipal Solid Waste Management, 1960 to 2000

FIGURE 9.21   Recycling is "in" again.

Landfilling peaked in the later 1980s and now is in decline. Recycling programs, despite their problems, are catching on. Incineration also is keeping some waste out of landfills. *After* U.S. Environmental Protection Agency Journal, *Winter 1995.*

trucks, sorting equipment, operators, and management. Together, the costs of pickup and processing are considerably greater than the current value of recycled material (Table 9.2). The largest private garbage hauler and landfill operator has calculated that it averages $200 per ton to pick up and sort the recyclables that most communities include in their curbside programs—glass, aluminum, steel cans, newspaper, and plastic—but the company receives only half that amount for these materials. The cities must make up the difference, plus a reasonable profit for the operator. However, as landfill space grows more expensive and the cost of environmentally acceptable incinerators climbs, the cost of recycling may become very competitive.

Some recycling casualties have resulted. Philadelphia discovered that its municipal recycling program was too expensive and discontinued it in 1993. San Jose, California, reports it costs $28 per ton to deposit waste in a landfill but $147 per ton to recycle.

Sometimes the cost–benefits balance is closer. Atlantic City, New Jersey, earned $2.45 million from recycling during the first six months of 1995. But collection cost was $1.6 million, sorting cost was $1.1 million, and interest payments on the recycling facility were $325,000. Consequently, recycling actually cost the county $575,000. However, recycling is increasing the life of their landfill by 20 percent, so is it worth the loss of half a million dollars? The answer to such questions is not simple.

Calculating the Environmental Benefits of Recycling.   Table 9.3 shows ways that recycling benefits us all. Reuse seems better than throwing away—but is this always true? Sometimes an apparent environmental benefit turns out to be not so beneficial after all.

A good example is using a ceramic mug to replace those white polystyrene-foam coffee cups used by everyone. One American chemist concludes that it takes up to seventy times more energy to make a reusable ceramic or glass cup than to make the polystyrene cup. So, you would have to drink from the reusable cup seventy times before there was any energy savings compared to drinking from disposables. Not bad, as many of us do use a ceramic cup at least seventy times.

But if we figure in the energy required to wash the reusable ceramic cup in hot water and detergent, the gap widens. If you washed the cup in a dishwasher, you would need to use it a thousand times to make it as environmentally friendly as using the seventy polystyrene cups. Some dishwashers use more energy to wash a ceramic cup than it takes to make a disposable one. But, then, no one uses a dishwasher to wash a single cup.

What if you wash it by hand instead? What about the energy to heat the water and the amount of water used in the washing or rinsing? And what is the average number of times a ceramic cup is used before it is broken? As you can see, the *real* cost of anything is not so easy to determine.

A Dutch Study of Ceramic Versus Styrofoam.   The Dutch Environment Ministry studied the relative "greenness" (environmental friendliness) of ceramic and styrofoam cups. Their report follows the life history of a ceramic coffee mug from extraction and processing of its raw materials, through production and

TABLE 9.2	1996 Prices for Recycled Materials
Solid Waste Component	**Market Price**
Shredded automobiles	$135/ton
Steel cans	$95/ton
Aluminum cans	$50/ton
Clear glass bottles	$40/ton
Newspaper	$24/ton
Car batteries	$15/ton
Plastic bottles	$10/ton

TABLE 9.3 Some Facts about Recycling

Aluminum Cans

Aluminum cans may be recycled indefinitely.

American consumers and industry throw away enough aluminum to rebuild our entire commercial air fleet every three months.

Recycling aluminum scrap saves 95 percent of the energy that would have been required to make new aluminum from ore.

Glass

Each person in the United States uses about 85 pounds of glass each year.

About 75 percent of the glass used in the United States goes into packaging.

Every glass bottle that is recycled can save enough energy to light a 100-watt bulb for four hours.

By recycling one ton of glass, we save the energy equivalent of nine gallons of fuel oil.

About 30 percent of today's average glass soft drink container is made of recycled glass.

Plastic

Americans throw away 2.5 million plastic bottles every hour.

Plastic is virtually immortal. Had the Pilgrims been able to enjoy six-packs of Mayflower beer, the plastic rings holding the cans together would still be around today.

Plastics comprise about 20 percent of the volume in American landfills.

1,050 recycled milk jugs can be made into a six-foot park bench.

The production of virgin plastic employs five of the six worst chemicals that EPA lists as "most hazardous waste."

Paper

Manufacturing recycled paper uses up to 64 percent less energy than manufacturing virgin paper, reduces air pollution by 74 percent, and water pollution by 35 percent.

The paper industry is the largest single user of fuel oil in the United States.

Every day, U.S. businesses generate enough paper to encircle the earth 20 times, and 70 percent of office trash is recyclable waste paper.

Every ton of paper that is recycled saves approximately 17 trees and enough energy to heat the average home for six months.

Paper comprises about 40 percent of the volume in American landfills.

Scrap Metal

Americans throw out enough steel and iron to continually supply all of the nation's automakers.

Each year, Americans abandon 3 million cars.

Recycling ferrous metals (iron and steel) saves 74 percent of the energy used to make them from iron ore and coal and reduces mining waste by 96 percent.

Recycling iron and steel reduces air pollution by 86 percent, uses 40 percent less water, and produces 76 percent less water pollution.

Making steel from scrap saves 90 percent of the virgin, nonrenewable materials used in making steel from ore.

Miscellaneous Recyclable Products

Americans dump waste oil equivalent to at least 25 *Exxon Valdez* oil spills each year, and all this waste oil is recyclable.

Every American family produces approximately 15 pounds of household hazardous waste each year, most of which is disposed of improperly down drains or in landfills.

Although about 80 percent of automobile batteries are recycled in the United States, the remaining 20 percent, containing approximately 330 million pounds of lead, ends up in landfills. Household batteries account for over 50 percent of the mercury and cadmium found in our trash.

About 250 million tires are thrown out every year in the United States but many landfills refuse to accept them. As a result, they accumulate in mountainous tire graveyards, posing a smoky fire hazard to surrounding communities should they be ignited.

Food (80 percent) and yard waste (20 percent) are major components of American landfills. Composting is a far better way to dispose of this material than is burial in a landfill.

Source: Bebi Kimball, 1992, *Recycling in America,* Santa Barbara, California: ABC-CLIO, Inc., pp. 35–65.

use, to its final resting place in a landfill. It takes into account the raw materials, landfill space, energy used for processing, transport, and cleaning, and air pollution and water pollution.

Washing a reusable mug just once in a dishwasher, says the report, has a greater negative impact on the water supply than the entire life cycle of a disposable cup. The main reason is the *surfactants* in the detergents—those chemical compounds that cause grease to float off surfaces. Surfactants release toxic gases into the air and are a major cause of water pollution. Thus, whether it is greener to drink from a reusable mug than from a disposable cup depends on two things: how many times the mug is used and how often it is washed with detergent.

The report concludes that if you wash the ceramic mug every time you use it, you need to use the mug 1800 times before it has less negative environmental impact than a disposable cup. But if you refill the mug at least once between washings, you need to use it only 114 times before it does less damage. (Also, it turns out that disposable polystyrene cups actually are greener than those made of paper.)

It's not that easy being green. Some things that seem to have an obvious environmental benefit at first glance turn out, upon scientific analysis, to be much more iffy. Amidst all the hype and misinformation from vested interests on both sides—environmental and commercial—how are you to learn the truth? The most you can do is to read and view credible sources, and keep an open mind.

Recycling and the Stock Market. In 1995, the Chicago Board of Trade established an exchange for trading recyclable materials, such as secondhand plastic and glass. It enables recycling companies to find buyers for their product. The exchange is funded by EPA. Because more manufacturers are using recycled materials, there is a growing need to match these users to suppliers. At present, there is a glut of some recycled materials and a shortage of others. The new market, it is hoped, will match supply to demand through bidding, as has been done for more than a century with stocks and bonds. The Chicago Board of Trade also has developed a system for grading recyclable materials by quality, so buyers will know what they are getting.

400,000 Problem Waste Sites

Thousands of toxic waste dumps and landfills dot the landscape, and many of them store waste unsafely. Storage barrels often are simply placed in open lots, exposed to rain, freezing, and vandalism. Some landfills are poorly constructed. At many, waste is leaking

into local water supplies. Epidemiological studies have correlated living near a hazardous waste site with increased risk of lung, bladder, and gastrointestinal cancers and birth defects. Approximately 41 million U.S. citizens—about 16 percent of the population—live within four miles of a toxic waste site.

The federal government estimates that the nation has more than 400,000 problem waste sites: (a) hazardous waste sites with leaking chemical storage tanks and drums, both above and below ground, (b) pesticide dumps, and (c) piles of mining waste. This figure excludes the 7233 toxic waste sites at military bases and nuclear weapons facilities, some of which no doubt are as hazardous as the others.

Superfund's 35,000 Sites

Clearly, something must be done to prevent wholesale contamination and almost certain pollution of the nation's water supply with toxic heavy metals and synthetic organic chemicals. In response to public pressure, the U.S. Congress in 1980 passed the Comprehensive Environmental Response, Compensation, and Liability Act (CERCLA), popularly known as **Superfund.**

Financed jointly by federal and state governments, and taxes on chemical and petrochemical industries, the legislation created a fund to pay for identifying dangerous sites, protecting groundwater near the site, remediation of contaminated groundwater, and site cleanup. Funding has increased dramatically over five-year increments:

$1.6 billion for 1980–1985
$8.5 billion for 1986–1990
$9.1 billion for 1991–1995
$6.1 billion for 1996

Superfund is a large part of total public spending on pollution control, which rose linearly from $16.6 billion in 1972 to $102 billion in 1992.

Of the 35,000 sites in the Superfund inventory, the worst (those that seem to present the most immediate and severe threat) are placed on a **National Priorities List** (NPL). This qualifies them for immediate attention (Figure 9.22). As of 1996, the NPL listed 1296 sites in the 50 states, plus nine in Puerto Rico and one on Guam. About 100 more sites are added each year, far faster than cleanup is removing older sites from the list.

As of 1992, with Superfund a dozen years old, only 283 hazardous waste sites had been cleaned up enough to be deleted from the National Priorities List, despite the expenditure of more than $11 billion. Why is cleanup so slow and so costly?

Much of the explanation lies in our legal system. Most companies charged by the EPA as PRPs (potentially responsible parties) either deny it or sue every-

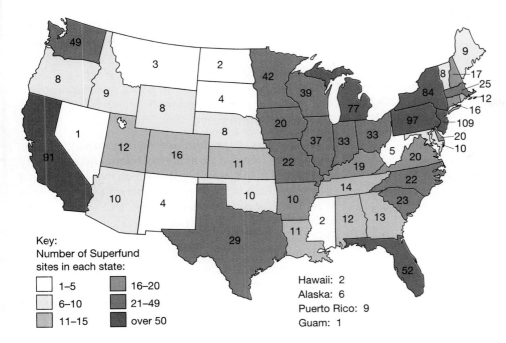

FIGURE 9.22 Superfund sites on the National Priorities List in 1992.

Since 1980, the EPA has identified more than 1200 of these sites. The total number of sites in the Superfund inventory exceeds 35,000.

Key:
Number of Superfund sites in each state:

- ☐ 1–5
- 6–10
- 11–15
- 16–20
- 21–49
- over 50

Hawaii: 2
Alaska: 6
Puerto Rico: 9
Guam: 1

one involved with the site. One independent study found that from 1986 to 1989, insurers spent $1.3 billion on Superfund litigation and cleanup, with $1.2 billion going to lawyers and only $0.1 billion used for cleanup. More recent estimates say that 75 percent is swallowed up by legal fees. Companies readily acknowledge that it is worth spending millions on lawyers to put off spending hundreds of millions on cleanup. (Environmental Law is looking more attractive as a major.)

Environmental concerns have become significant in the balance sheets of companies large and small. Many companies have added environmental specialists. However, some industries, such as mining of metal ores, nearly have been wiped out by environmental concerns and problems.

Given population increase, growth of the American economy, and increasing public concern over environmental problems, Superfund is likely to grow over the next few decades. Pollution control and remediation may take a larger share of the U.S. budget than social security, military expenditures, welfare, or any other governmental program. Environmental cleanup is essentially a bottomless pit into which more and more of our money will pour. (You might not approve of this use of tax revenues. If not, can you suggest a workable alternative?)

Sewage and Septic Systems

Figure 9.23 shows the source of the liquid waste that drains from your home. In cities, this wastewater is piped to a sewage treatment plant, where it is purified

FIGURE 9.23 Down the drain.

Although toilet-flushing produces most of the effluent in the average household, toilet waste is biodegradable. Wastewater from the kitchen or laundry is much more likely to contain toxic or non-biodegradable pollutants.

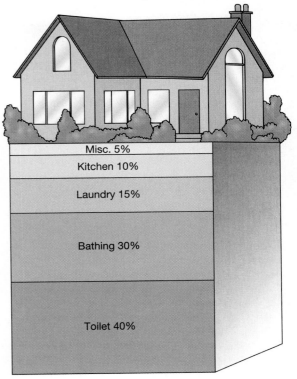

Misc. 5%
Kitchen 10%
Laundry 15%
Bathing 30%
Toilet 40%

before being returned to a river or lake. In small towns and rural homes, the wastewater usually is piped a few feet to a septic tank, typically buried in the yard.

We are very fortunate that domestic sewage is biodegradable and easy to purify—so long as the wastewater does not contain toxic chemicals. The biggest problem with sewage is the risk of overloading the treatment facility, due to increasing population and storm runoff. Such overloading is not uncommon, and fish often are the victims (Figure 9.24).

An example of such overloading occurred in 1993 in Milwaukee, when storm runoff from city streets "drowned" the city's sewage treatment plant, resulting in pollution of the city's drinking water. Parts of the city experienced intestinal disease from bacteria.

About 75 percent of the American public is served by sewer systems. Most of the remaining 25 percent

FIGURE 9.24 Fish killed by a sharp decrease in oxygen.

In the River Seine near Paris, untreated sewage was washed into the river by heavy rains. Bacteria eagerly went to work, consuming both the sewage and oxygen in the water, leaving little oxygen for the fish. *Photo: P. Vauthey/Sygma.*

have septic systems that service single homes (sometimes a few together) in rural areas. The percentage of homes serviced only by outhouses (outdoor toilets built over a pit in the ground) is now quite small.

Sewage Systems

Municipal **sewage system** treatment normally consists of two steps, termed primary treatment and secondary treatment. During primary treatment, raw sewage that enters the plant is screened and put into settling tanks to remove coarse particles, grease, and scum. In secondary treatment, the sewage is **aerated** by trickling it through a bed of rock or flowing it into an aeration basin (much as a fish tank is aerated). The basin contains bacteria and other microorganisms that decompose the organic material in the water. After several hours, the particles and microorganisms are allowed to settle out, forming **sewage sludge,** a slimy mixture of bacteria-laden solids. The water then may be chlorinated to disinfect it, much as swimming pool water is chlorinated. Most wastewater treatment facilities in the United States perform both primary and secondary treatment of raw sewage this way.

Sewage sludge forms about 10 percent of all municipal waste and often is used for fertilizer. Unfortunately, much of this sludge is contaminated with toxic heavy metals and must be disposed of as hazardous waste rather than as a nutrient-rich fertilizer.

A third step, called tertiary treatment, is seldom performed by municipal facilities because of cost. Tertiary treatment removes nitrogen and phosphorous that remain after the primary and secondary treatments are completed. Sewage that is released to surface waters without tertiary treatment may result in eutrophication of nearby waters (especially lakes).

Eutrophication. Oxygen depletion in water is called **eutrophication.** It results when excess nitrogen and phosphorous are present in a lake or slow-moving stream. Such an excess can occur when fertilizer runs off croplands into streams, or from sewage. The excess stimulates uncontrolled growth of algae, which consume the oxygen (Figure 9.25).

Using a lake as an example, the algal growth rapidly blankets the lake surface. This diminishes the penetration of sunlight into the water. Without adequate sunlight for plant photosynthesis, life in the lake waters below the surface is lessened. As the algae die, they fall to the lake bottom and decompose. This decomposing matter consumes oxygen, depleting the oxygen dissolved in the lake water.

The oxygen cannot be replaced because the thick covering of algae at the lake surface inhibits the wind

(A)

(B)

FIGURE 9.25 Early stage of eutrophication.

(A) On this river, eutrophication results from untreated sewage and the river's slow velocity, which cannot flush out excess nutrients fast enough.
Photo: D. Newman/Visuals Unlimited.
(B) Toxic red algae bloom in nearshore waters off the east coast of Italy. The algae feed on pollutants pouring into the Adriatic Sea from the Po River.
Photo: Geospace/Science Photo Library/Photo Researchers, Inc.

from stirring the water and replenishing the oxygen supply from the atmospheric storehouse. Without oxygen, life in the lake below the surface is extinguished. Eutrophication can be reversed by rerouting the harmful runoff and temporarily adding an algicide to the lake water.

Living things demand oxygen, both when they are alive and when they are dead. Animals need oxygen to breathe, and all dead organisms consume oxygen as their tissues decompose. In a body of water, this collective need for oxygen is called **biochemical oxygen demand (BOD).** When BOD is excessive, eutrophication (oxygen depletion) of the water body occurs. The more organic matter there is to be decomposed, the greater the amount of oxygen is needed—and the greater is the BOD.

For example, raw sewage contains generous amounts of organic matter, so its complete decomposition requires plentiful oxygen. The BOD for raw sewage ranges between 40 and 150 milligrams of oxygen needed per quart of water. Water for human consumption should contain very little organic matter and, therefore, should have a very low BOD. The EPA specifies less than 0.5 milligrams of oxygen needed per quart of water.

Sewage Sludge. Sewage treatment produces an enormous amount of sludge, about half the volume of raw sewage that enters the plant. Large cities process ten-millions to hundred-millions of gallons of sludge daily. Because it is a rich source of nitrogen and phosphorous (without tertiary treatment), it commonly is used as fertilizer. Unfortunately, much urban sewage also contains heavy metals, so it cannot be used to fertilize cropland. Some sludge is so contaminated with toxic materials after secondary treatment that it must be further treated to remove heavy metals, or else buried in a landfill. About 20 percent of the sludge from America's sewage treatment plants ends up in landfills.

U.S. sewage sludge was dumped into the ocean through 1991, but federal law made this illegal. European countries have agreed to stop by 1998. Because of the massive amounts of sludge produced in urban areas and the filling of available landfills, many cities are seeking alternatives for handling it.

Septic Systems

Septic systems are used in rural areas to treat raw sewage. They serve 25 percent of the U.S. population (Figure 9.26). Primary treatment is accomplished in a septic tank, which must be emptied of grit and sludge every 2–3 years. Outflow from the tank flows through a drain field, where it becomes aerated and filtered by the sandy sediment. At the same time, organic material in the water is decomposed by microorganisms and oxygen in the soil air above the water table (secondary treatment). You can see that a septic tank is a scaled-down, low-tech version of a municipal sewage treatment system.

Ideally, the water table should be at least several feet below the bottom of the septic tank. To be safe, water wells should be located higher than the drain field of a septic system so that septic effluent cannot flow into the wells.

A properly maintained septic system may function indefinitely. However, organic matter that passes through the tank may enter the drain field and underlying sediment faster than it can be decomposed. If this happens, the sediment pores become clogged and the partially treated sewage is forced upward to the surface where it smells, contaminates surface waters, and is a general health hazard.

Properly siting a septic tank requires an understanding of the permeability and clay content of the surrounding sediment and the depth to the water table. More than 25 million septic systems operate in the U.S., discharging about 1 trillion gallons of wastewater annually into soils and groundwater. It is sobering to realize that more than 50 percent of all drinking water is groundwater that is easily contaminated by septic-tank effluent.

Case Study: Tracing Sewage

Water from a sewage treatment facility that is discharged above ground and percolates into the soil and rock can be traced by its distinctive chemistry, because

FIGURE 9.26 Septic system for a private home.

(A) The sewage enters the septic tank, where decomposing bacteria work. Fat rises to the surface as scum, and particulate matter settles to the bottom as sludge. The water that leaves the tank is clear, but contains dissolved mineral nutrients. Water and nutrients are disposed of by permitting them to leach slowly into the soil. (B) Relation of a water well (left) to the effluent from a septic drain field. Will the well become contaminated?

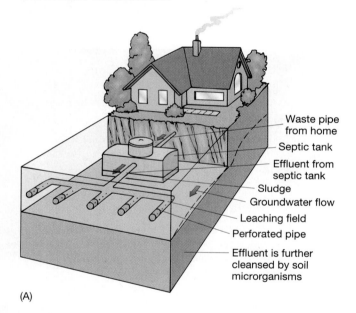

(A)

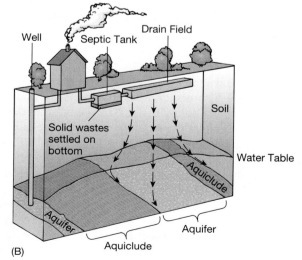

(B)

sewage effluent is rich in nitrogen, phosphorous, and boron. The ease of tracing is illustrated by a study of liquid movements from fuel dumps and secondary sewage treatment at Otis Air Force Base on Cape Cod, Massachusetts (Figure 9.27).

Surface disposal onto sand beds has continued since 1936 and resulted in ten plumes of highly contaminated groundwater. The biggest plume is 2500–3500 feet wide, 75 feet thick, and more than 2 miles long. It extends in the same direction as the regional groundwater flow and is overlain by 20–50 feet of groundwater that recharges the aquifer from precipitation. The bottom of the plume generally coincides with the contact between the permeable sand and gravel and underlying finer-grained, less permeable sediments. The plume should reach Nantucket Sound in a few decades.

More frightening than a few chemicals from sewage are the multiple plumes of solvents, jet fuel, and other complex chemicals that currently threaten four towns (Figure 9.27). The plumes, leached from fuel dumps and other sites on the base, are moving southward at about a foot per day. They already have forced the shutdown of a well that supplied 25 percent of the municipal water for the town of Falmouth and are now about 600 feet from another public well in Bourne. Two other towns, Mashpee and Sandwich, are also threatened. In all, 67,000 residents in the four towns are involved.

The Defense Department estimates it would cost $250 million over twenty years to clean the water by pumping it to the surface, treating it, and pumping it back down into the aquifer. As an alternative, they are experimenting with a new and untested method. They are inserting iron filings into the aquifer. They hope that the iron dust will change the chlorinated hydrocarbons in the plumes and convert them into harmless gases. Results in an above-ground test in New Jersey in 1995 were promising, and the endangered residents on the Cape Cod peninsula hope for results as least as good in their area.

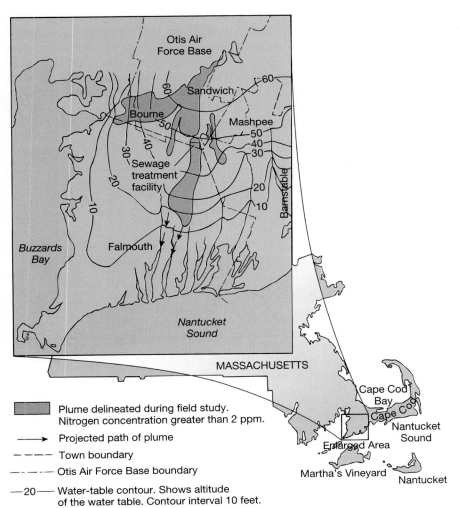

FIGURE 9.27 An underground menace.

Projected path of groundwater plume contaminated by land disposal of treated sewage at Otis Air Force Base, Cape Cod, Massachusetts. *Source: U.S. Geological Survey.*

Plume delineated during field study. Nitrogen concentration greater than 2 ppm.

Projected path of plume

Town boundary

Otis Air Force Base boundary

—20— Water-table contour. Shows altitude of the water table. Contour interval 10 feet. Datum is sea level.

Summary

Municipal solid waste forms only 5 percent of all solid waste but is a major disposal problem for America's cities. Most Americans live in urban areas, are extremely wasteful compared to people in other countries, and are in danger of drowning in their own garbage. Some accumulations form major topographic features, rising hundreds of feet above the surrounding countryside. Some of this waste is hazardous and poses a continuing threat to local water supplies.

The main methods of disposing of these mountains of waste are sanitary landfills and incineration. At present, about 80 percent of our garbage is landfilled, 20 percent incinerated. Incineration will no doubt increase in importance as landfill space quickly disappears. (Incineration has been adopted on a much larger scale in other countries; Sweden burns 50 percent of household waste; Switzerland, 75 percent.) Smokestack emissions must be controlled during incineration to prevent toxic gases from contaminating the atmosphere. Also, the 10 percent ash remaining after burning is a problem because it is enriched in heavy metals that were dispersed through the original waste pile.

Recycling is the best way to reduce municipal solid waste. It saves significant energy that would be needed to manufacture a product from virgin materials. Recycling also generates much less air and water pollution. Unfortunately, the cost of collecting, sorting, and reprocessing recyclables often is greater than the money saved. Because of our economic system, energy savings and cleanliness of our air and water take a back seat to short-term cost-cutting. However, as the cost of landfilling continues to skyrocket, recycling will become less expensive by default. Another 5–10 years should be enough to turn the tide in favor of recycling.

A particularly serious problem are the 35,000 waste disposal sites classed by the federal government as dangerous because they are now polluting, or soon will pollute, our water with toxics or radiation. The money provided to deal with this crisis has risen rapidly from $1.6 billion in 1980–1985 to $6.1 billion for 1996 alone, an increase of more than 700 percent in only fifteen years. No doubt it will continue to increase as we approach a national disaster. When you can't breathe or drink safely, little else matters.

Sewage disposal is another area of concern. In many urban areas, the plants are old, too small to meet the needs of rapidly growing urban populations, and sometimes inundated by floodwaters. Inadequate sewage treatment can cause outbreaks of disease and is a major source of death in undeveloped countries. In recent years, there have been outbreaks of sewage-borne infection in the United States as well but they are still infrequent enough that public concern is still unfocused. This will change as more and more sewage plants become overloaded and unable to provide the needed purification treatment.

Key Terms

aeration
biochemical oxygen demand (BOD)
biodegradable
eutrophication
Fresh Kills landfill
garbologist
hazardous solid waste

leachate
methane gas
sanitary landfill
municipal solid waste
National Priorities List
NIMBY phenomenon
recycling

septic system
sewage sludge
sewage system
Superfund
wet landfill design

Stop and Think

1. What are the major sources of solid waste? Why has one of the volumetrically minor sources assumed such great importance in modern civilization?

2. As an average American, you generate 4.4 pounds of garbage each day, more than twice as much as the average Japanese, German, and French person. Are we wasteful? Why? Is it a habit that can be changed? How?

3. What is a sanitary landfill? List the requirements for a properly designed sanitary landfill.

4. In what ways is a sanitary landfill superior to an open garbage dump?

5. Where is (are) the sanitary landfill(s) for your community located? Are they nearby, or is the community's waste transported elsewhere and dumped in someone else's back yard? Ask the city manager's office or the trash-hauling contractor.

6. Is any of your community's solid waste incinerated? What percentage? How was this percentage established? Get the name of your community's solid waste disposal company from the city manager's office and phone them.

7. What is leachate? Why is it sometimes dangerous to groundwater aquifers?

8. What are the special problems involved in the disposal of tires from cars, trucks, and buses?

9. What are the benefits of incineration as a way of disposing of hazardous solid waste? What are the drawbacks?

10. What percentage of your community's solid waste is recycled? What types of items do you separate from each other before putting out the garbage?

11. What kinds of benefits does society receive from recycling wastes? Why are some kinds of wastes easier to recycle than others?

12. What is Superfund? How is it financed? How successful has it been in cleaning up the worst cases of pollution?

13. The map shown as Figure 9.22 indicates that five states stand out as leaders in the number of Superfund sites—a dubious distinction. The states are New Jersey, Pennsylvania, California, New York, and Michigan. What factors might these five states share that have resulted in this leadership role?

14. What are the advantages and disadvantages of using sewage sludge as an agricultural fertilizer?

15. What are the geologic requirements for the proper siting of a septic tank? If you are a rural homeowner, how might you know whether your septic tank is functioning properly? Improperly?

16. Name three ways that groundwater can become polluted. Explain why it is so difficult or even impossible to cleanse polluted groundwater.

References and Suggested Readings

A Guilt-Free Guide to Garbage. 1994. *Consumer Reports.* February, p. 91–114.

Aubrey, D. G. and M. S. Connor. 1993. Boston Harbor: Fallout over the outfall. *Oceanus,* v. 36, no. 1, p. 61–70.

Black, Harvey. 1995. Rethinking recycling. *Environmental Health Perspectives,* v. 103, p. 1006–1009.

Eyles, N., J. I. Boyce, and J. W. Hibbert. 1992. The geology of garbage in southern Ontario. *Geoscience Canada,* v. 19, p. 50–62.

Goldberg, M. S., and three others. 1995. Incidence of cancer among persons living near a municipal solid waste landfill site in Montreal, Quebec. *Archives of Environmental Health,* v. 50, p. 416–424.

Grove, Noel. Recycling. 1994. *National Geographic,* July, p. 92–115.

Hasan, S. E. 1996. *Geology and Hazardous Waste Management.* New York: Prentice-Hall, 387 pp.

Lave, L. B., C. Hendrickson, and F. C. McMichael. 1994. Recycling decisions and green design. *Environmental Science and Technology,* v. 28, p. 19A–24A.

Lottermoser, B. G. and G. Morteani. 1993. Sewage sludges: toxic substances, fertilizers, or secondary metal resources. *Episodes,* v. 16, p. 329–333.

Mueller, D. K. and Helsel, D. R. 1996. Nutrients in the nation's waters. Too much of a good thing? *U.S. Geologcial Survey Circular* 1136, 24 pp.

O'Leary, P. R., P. W. Walsh, and R. K. Ham. 1988. Managing solid waste. *Scientific American,* December, p. 36–42.

Pellerin, Cheryl. 1994. Alternatives to incineration: There's more than one way to remediate. *Environmental Health Perspectives,* v. 102, no. 10, p. 840–845.

Rathje, W. L. 1991. Once and future landfills. *National Geographic,* May, p. 116–134.

Rathje, W. L. and C. Murphy. 1992. *Rubbish: The Archeology of Garbage.* New York: Harper Collins.

"Dead bottlenose dolphins have washed up on U.S. beaches so laced with chemicals the federal government would consider them toxic waste hazards. They were found to be contaminated by PCBs, a chemical used since the early 1930s in the manufacture of electrical equipment, paints, and plastics."

World Resources Institute, 1993

Shorelines, Erosion, Wetlands, and Pollution

Earth's polar ice caps are melting very slowly, increasing the volume of ocean water. This is slowly raising sea level worldwide, about two-thirds of an inch since you were born, causing a slow landward migration of shorelines, beaches, and their associated near-shore environments of bays and wetlands. Why are the polar ice caps melting? Are our activities causing global warming, intensifying sea level rise? Can sea level rise be halted, or at least slowed, to protect people, property, and scenic vistas?

Storms can severely erode beaches. Are all shorelines equally susceptible? Have attempts to stop beach erosion been successful? Who should pay for them? Can a state prohibit property owners from building on their own beachfront land? Should states provide insurance against property loss along shorelines that are judged vulnerable? Who should decide?

The World Ocean

"How inappropriate to call this planet Earth, when clearly it is Ocean," noted writer Arthur C. Clarke. As he suggests, Earth from space appears to be mostly ocean. Its salty waters cover about 71 percent of Earth's surface. The ocean's depth varies, averaging about 2.25 miles (12,000 feet). The volume of seawater is so great that, were it all piled atop the continental United States, it would be 106 miles high.

Because the ocean dominates Earth's surface, we would expect it to be a major influence on all living species—particularly the 50 percent that live in the sea. In addition to the sea's obvious effect on all life, it strongly influences climate, shoreline shape, coastal erosion, and events like hurricanes and tsunami (earthquake-generated waves, erroneously called "tidal waves"). In this chapter, we will look at the ocean and features of its basins and coast.

The Shifting Shoreline

The most direct contact humans have with the ocean is at the **shoreline,** that ever-shifting boundary of land, sea, and air. It is a quite remarkable boundary of solid, liquid, and gas. At the shoreline, water-meets-rock-meets-air. This busy intersection features erosion, moving sediment, turbulent water, wind, and many life forms covered with hard shells, feathers, skin, and swimsuits. The United States has more than 88,600 miles of shoreline, including that of the five Great Lakes.

Oceanographers distinguish between the shoreline and the coast. The shoreline is that shifting line where land and water meet. The **coast** is a strip or zone that is affected by ocean processes; it includes river mouths (estuaries), bays, beaches, and sand dunes near

◄
Using the ocean as a giant sewer.
Photo: Simon Fraser/Science Photo Library/Photo Researchers, Inc.

257

the beach. A coast varies in width from hundreds to thousands of feet.

Why are shorelines located where we find them? Why does their location shift over time? Is it possible for humans to move a shoreline? How safe is it to build beachfront homes along a shore? The answers can be discovered through observation and technology.

Anyone who has traveled the United States has observed that the eastern and western parts of the country are very different. Very generally, the eastern 60 percent is quite flat (excepting the Appalachian Mountains, Ouachita Mountains in Oklahoma and Arkansas, and some hilly areas). In contrast, the western 40 percent is an unrelieved series of north–south mountain ranges. Their topographic relief is almost twice as great as that of the eastern mountains.

Generally, this same difference exists offshore of the east and west coasts. Off much of the East Coast, the sea bottom slopes very gently, averaging only 9 feet/mile (less than 0.1°) all the way out to about 50 miles offshore (Figure 10.1, East Coast). Rising from the shallow ocean bottom near shore are elongate ridges of sand called **barrier islands**. Familiar examples are Assateague Island (Ocean City) and Chincoteague Island in Maryland/Virginia, North Carolina's Outer Banks, and Florida's East Coast.

Off the West Coast, the seafloor is also an extension of the land onshore—a series of ridges and deep basins (Figure 10.1, West Coast). There is no broad, flat surface such as occurs off New York, New Jersey, Virginia, and southward.

This difference in land characteristics between the east and west coasts is reflected in the relative severity of their erosion problems. Shoreline retreat is severe along long stretches of the East Coast, but **coastal erosion** is much less of a problem in the west (Figure 10.2). On the East Coast, sand along this year's beach is gone next year. The ocean seems to be creeping landward each year, much to the distress of the owners of beachfront property and summer homes. Those who live along the California coast are more likely to maintain their elevated view of the Pacific year after year. (Some retreat of the cliffs does occur as they collapse due to undermining by wave activity.)

What is the oceanographic explanation for this difference? It lies in the ceaseless interaction among wind, the shallow ocean currents created by wind, and water depth.

The Nearshore Zone

The **nearshore zone** is underwater, a strip bounded by the edge of land (shoreline) and a water depth of a few tens of feet. Thus, it extends seaward only a few thousand feet from the beach. It is the place where low

FIGURE 10.1   Profiles of the U.S. East and West Coasts.

Moving out under the ocean water, each margin of our continent is basically an extension of its "topographic style" on land. At left is a typical West Coast profile, with ridges and basins, just like on land, largely the result of faulting. At right is a typical East Coast profile—a fairly smooth, gradual descent to the deep ocean floor.

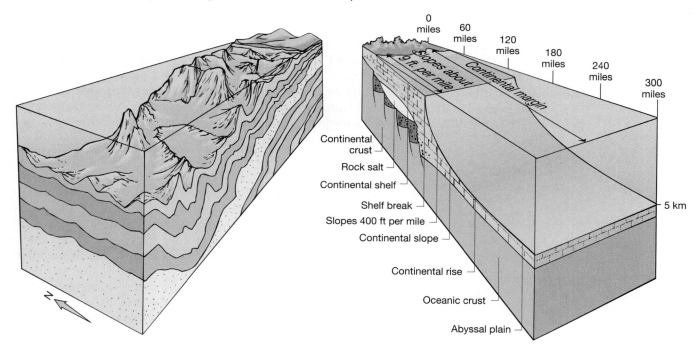

FIGURE 10.2 Coastal erosion.

All 28 coastal states (Atlantic, Pacific, and Great Lakes) continually experience coastal erosion, but to different degrees. (The Great Lakes have significant shorelines, and thus are considered "coastal" states.) About 25 percent of the U.S. coastline is significantly affected by erosion.

incoming waves change shape and grow taller, beginning to scrape bottom, moving sediment along the shallow seafloor. It is a zone of rapid change, only some of it easily observed.

The waves may move sediment shoreward, enlarging a beach, or they may transport sediment offshore, eroding and narrowing a beach. Typically, these waves also generate currents that move parallel to shore and transport sediment, another cause of beach erosion. The incoming waves also create rip currents that can sweep people seaward to their death.

The nearshore zone is where hurricane damage and coastal flooding are most destructive. Here also, seismic sea waves, called tsunami, wreak havoc with lives and structures.

Waves and Coastal Currents

Most of the visible result of the interplay between sea and land results from ocean waves and currents. When wind blows over a water surface, friction between the air and water allows the wind to drag the water, making it move, transferring energy to the water. It begins when a breeze reaches around 3 miles per hour, enough to drag ripples in the water surface. As more wind energy is transferred to the water, and speed reaches around 20 miles per hour, ripples grow into **ocean waves.** The distance between wave crests, called the **wavelength,** may be tens of feet. Wavelength depends on wind strength, the consistency of wind direction, the distance over which the wind blows (called the **fetch**), and the shape of the shore bottom.

In deep water, the water molecules within a wave move in a large circular pattern (Figure 10.3). However, just as there is friction between the air (wind) and the water, friction also exists between the water and the seafloor. In the deep ocean, this friction is too slight to have an effect. But as a wave enters shallow water near shore, friction increases, and the wave's shape becomes distorted. Air-to-water friction creates the circular orbits in water that we call waves, and the water-to-seafloor friction tries to damp them out, distorting their shape.

As waves enter shallow water—in other words, as they head toward shore—friction between the water and the seafloor slows the lower part of the water mass. As this happens, the upper part still rushes forward.

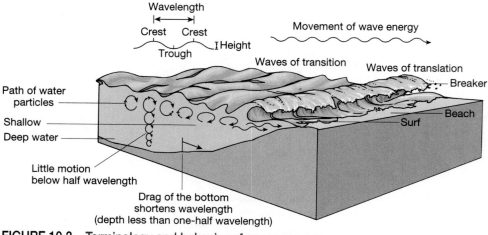

Wavelength

Crest   Crest

Trough

Height

Movement of wave energy

Waves of transition

Waves of translation

Breaker

Path of water particles

Shallow

Deep water

Beach

Surf

Little motion below half wavelength

Drag of the bottom shortens wavelength (depth less than one-half wavelength)

FIGURE 10.3 Terminology and behavior of ocean waves.

In deep water, surface winds cause water molecules to move in circular orbits. Their diameters decrease with greater depth; wave motion becomes negligible at a depth of half the wavelength. As waves move into shallow water—water shallower than half the wavelength—orbits of the water molecules become compressed. The moving water scrapes the seafloor, causing erosion and sand movement. The energy has to go somewhere, so wave height increases until the energy is too weak to overcome gravity, and the waves "break" or topple (breakers), collapsing to form surf. Surf is very turbulent water with great erosive power.

This shoreward rush of water is accompanied by a large increase in wave height, as shown in Figure 10.3. The energy of the onrushing wave becomes concentrated, increasing as the square of the wave height. This means that high incoming waves have great energy and erosive capability.

More than 6000 waves per day break on America's open-coast beaches, an average of one every 15 seconds. As a result, great quantities of sand are continually moved by the water, especially during stormy winter months when winds are strong, waves are high, and beach erosion grows most intense.

During winter, strong storms with high winds generate powerful waves. Beach erosion intensifies. In warm climates like Hawaii's, surfers are ecstatic about the great waves, but owners of beachfront property are not, as they watch their sand wash away beneath the surfboards. Sand is removed far faster than it can be replenished. However, during summer, storms are infrequent, with smaller waves. Along most shorelines, sand eroded away during winter is replaced during summer. This contrast is shown for a California beach in Figure 10.4.

The nearshore zone, where waves grow so tall in shallow water that they break into **breakers,** is the end of the ride for surfers. The water from a broken wave rolls up the beach face, stops for an instant, then flows back into the ocean, carrying sediment with it. The energy in this backflow is strong enough to sweep small children from their feet.

The wind blowing from offshore moves waves shoreward. If a shoreline were perfectly straight, and if the wind were blowing exactly perpendicular to the shore, all of the energy of the wave would be expended in crashing directly onto the beach. But such a simple relationship between the shoreline and wave direction is rare. Typically, a wave front approaches the shore at an angle other than head-on, almost everywhere along the beach.

The interaction between shoreface and waves gives rise to two major types of currents along beaches: longshore currents and rip currents (Figure 10.5).

Longshore Currents

As waves hit the shore and wash up onto the beach at an angle, sand grains on the beach are rolled landward. As the wave retreats back toward the sea, the sand grains are rolled oceanward. Because waves strike the shore at an angle, the sand grains migrate in a zigzag pattern along the beach (Figure 10.5A, at top). Thus, sand is transported downcurrent along the beach surface.

Unless the migrating sand is continually replaced by other sand from up the beach, the beach will erode landward, and shallow water will exist where the beach used to be. Immediately oceanward of the shoreline, the waves lift grains off the seafloor and the **longshore current** carries them downcurrent. This movement of sand parallel to the shoreline is called the *longshore drift.* It transports bountiful sediment and is a major cause of beach erosion, literally hauling beach sand from one beach to the next.

FIGURE 10.4 Changing with the seasons.

The same beach in winter (A) and summer (B). What causes this seasonal difference? *Photos: John Shelton.*

Rip Currents

Rip currents are strong, narrow currents that start at the shoreline and flow directly out to sea through the breaker zone (Figure 10.5B). They are simply returning water seaward that was brought ashore by waves.

Because of their high velocities, up to at least 6 miles per hour, they have the energy to carry considerable sediment away from the beach—including people. They are a quick ride for experienced surfers, an expressway out into the high breakers, but for inexperienced swimmers they can be fatal.

FIGURE 10.5 Typical shoreline situation.

(A) Waves approach the beach at an angle, setting up longshore currents. Sand grains on the beach move in a zigzag motion over 6000 times a day, rolling up on the beach with each incoming wave, and rolling back out as the wave recedes. This creates a net movement of sand *along* the shore, in the direction of the longshore current. This is how sand continually migrates along a coast. (B) Underwater rip currents also develop. They are the water from incoming waves, returning to the sea in narrow channels. Rip currents are marked by discolored patches of turbid water that extend seaward of the surf zone. Rip currents often are called "rip tides," even though they have nothing to do with tides. *Photo: D. A. Rahm, courtesy Rahm Memorial Collection, Western Washington University.*

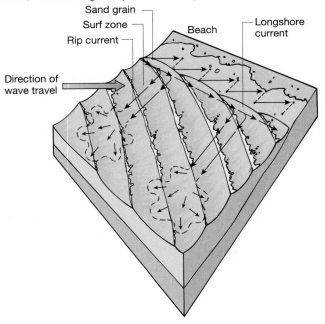

(A)

(B)

Rip currents do not occur everywhere along a beach front. Along many stretches of beach, water brought ashore by waves returns in sheets, not channels. Rip currents occur in certain places: above low places on the seafloor near a beach, places where opposing longshore currents meet head-on, and along points of land, jetties, and piers that deflect longshore currents seaward. Fortunately, rip currents are easily spotted because they make the water turbid with sediment (Figure 10.5B).

A swimmer caught in a rip current should remember not to struggle by trying to swim shoreward (the current is too strong to overcome), but to swim *across* the current, parallel to shore, to exit it quickly. Rip currents are narrow and easily escaped. They also dissipate quickly beyond the breaker zone, so another escape is to ride the current a few tens of feet until it weakens, as surfers do.

Tides and Tidal Currents

Waves, longshore currents, and rip currents all reflect the movement of Earth's winds. But tides, or tidal currents, have a completely different cause: the gravitational attraction of the moon and sun.

All objects attract each other with a force that depends on their mass and the distance between them. This force, called gravity, is not noticeable to us when objects are small. But when at least one of the objects is large—for example, Earth—the effect of gravity is very noticeable: it keeps your feet on the ground. When both objects are large, like Earth and the moon, or Earth and the sun, the combined gravitational force is huge. It is so strong that it distorts Earth's crust slightly on the side that is facing the moon or sun at the moment. It distorts the ocean even more, because it is fluid, allowing a broad bulge to develop in the ocean on the side facing the moon (Figure 10.6A). It also distorts the atmosphere even more dramatically, although we cannot see it.

This **tidal bulge** in the ocean, in combination with Earth's daily rotation, creates the great, broad, slow waves we call **tides**. The moon is far smaller than the sun, but it is only some 238,000 miles distant, so it has a greater effect on the ocean. The sun is huge, but is 400 times more distant (around 93 million miles), so its effect is weaker than the moon's (Figure 10.6B).

As the world turns, the ocean's tidal bulge stays facing the moon. Eventually, the bulge encounters the coast of a continent, so the water rises up the beach: we say "the tide is coming in." As Earth's rotation slowly moves the shoreline away from directly facing the moon, gravitational attraction decreases on the sea at that place, and the bulge subsides: we say "the tide is going out." Because every point on Earth faces the moon once every 24 hours, every point passes through the bulge, and experiences a daily tide. Figure 10.7 shows how dramatic a difference the tides can make.

Complicating this scenario, on the opposite side of Earth from the moon, a smaller tidal bulge also occurs due to the sun's gravity. Further, the strength of a tide depends on the position of Earth, the moon, and the sun at the moment, as shown in Figure 10.6B,C. Suffice it to say here that most shores experience two tides daily, one greater than the other.

A low-lying area affected by the advance and retreat of the tides is termed a **tidal flat**. The tidal flat can be narrow or very extensive (Figure 10.7A), depending on the elevation of the shoreline. Tidal currents on an open beach are not believed to have a significant erosive effect, because their velocities are too low to erode sand. However, tidal currents that enter and leave the shore area through a narrow inlet are greatly accelerated and do have large erosive capacity.

FIGURE 10.6 How the sun, moon, and Earth control daily tides.

(A) The moon's gravity raises a bulge in the ocean. Earth rotates, but the bulge always faces the moon; continental land rotates into the ocean bulge, raising water level in a lunar tide. (B) The sun complicates this situation, its gravity causing a lesser solar tide. Further, the position of each body controls the tides. Here, during *spring tides* (which are not seasonal; they are just high tides that "spring" forth), the three bodies are aligned so that the sun's and moon's gravity exert a combined pull, causing greater tides.
(C) During *neap tides* (which are very low), the sun's gravity and moon's gravity counter one another, reducing the tides.

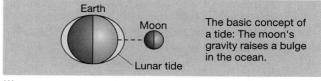

(A)

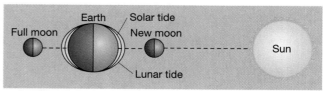

(B) Spring tide

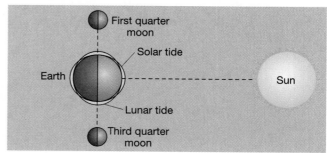

(C) Neap tide

(A)

(B)

FIGURE 10.7 Mont St. Michel on the Normandy coast of France.

(A) At low tide, extensive tidal flats are exposed. *Photo: Bachmann/Photo Researchers, Inc.*
(B) At high tide, the medieval abbey is surrounded by water; it was inaccessible by vehicle until the causeway was built a century ago. *Photo: Porterfield/Chic Kering/Photo Researchers, Inc.*

Some tidal currents in narrow inlets can attain 15 miles per hour velocity and can create a local "wall" of water several feet high. Along parts of British Columbia's coast, a diesel-powered fishing boat cannot make headway against strong tidal currents that flow between closely spaced islands. Fishermen must wait until the slack tide reverses direction before proceeding.

Ocean Currents

Ocean currents strongly affect the climate near a coast. Warm currents warm the coast; for example, the people in Iceland, near the Arctic Circle, enjoy a comparatively moderate climate due to free heat from the Gulf Stream. Cold currents chill the coast; for example, California's coast is kept from overheating by the cool California Current. But what causes these currents?

Currents in the upper few hundred feet of the ocean are driven by winds that blow steadily over them. The pattern of these reliable worldwide winds is shown in Figure 10.8A. The ocean currents arise from friction between these prevailing winds and the water surface, just as waves do. But surface currents also are affected by the presence of continents, which the currents run into, and Earth's rotation, which puts its own spin on current flows.

The interaction among these three controls—winds, continents, and spin—are complex, but create the circulation pattern of currents shown in Figure 10.8B.

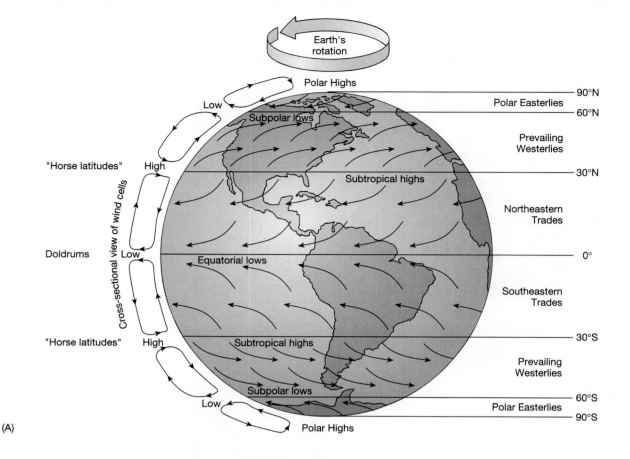

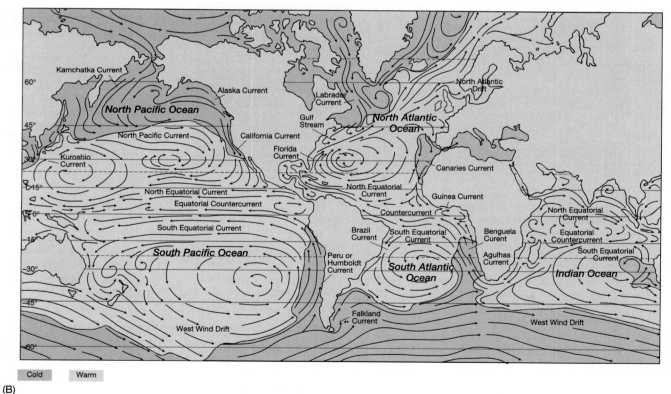

FIGURE 10.8 Earth's winds.

(A) Major wind directions at Earth's surface are essentially permanent, 24 hours a day. The wind directions result from two combined movements: Earth's steady eastward rotation on its axis, combined with the continual circulation of the atmosphere (shown at left). Names like "westerlies" indicate the direction that the wind blows *from*. Other names are sailors' terms for certain latitudes. (B) Winds drive ocean currents, literally by dragging the water at its surface through friction. Compare the pattern of winds in A with these currents. The patterns are distorted greatly by the presence of continents.

Note that surface currents rotate clockwise in the Northern Hemisphere but counterclockwise in the Southern Hemisphere, due to Earth's rotation. The temperature of a current depends on where it originates—near the hot equator or the frigid poles.

Ocean currents strongly affect the climate near a coast. We can see this by comparing sea surface temperatures along the coasts of Virginia and California. You don't need a thermometer to tell the difference, just dive in. Polar Bear Club members enjoy California's summer surf, while others joyously splash in the warm waters off Virginia. The difference is that California's coast is bathed by the cool, southbound California Current, a descendant of the chilly North Pacific Current, but Virginia's coast is caressed by the northbound Florida Current, descended from the warm North Equatorial Current. You can trace these currents in Figure 10.8B.

The Florida Current that warms Virginia's coast continues northeastward, becoming the **Gulf Stream.** (Ocean currents are like streets that change their names as you travel them.) Figure 10.9 shows the Gulf Stream.

In Figure 10.8B, note how the Gulf Stream extends its warm waters across the North Atlantic to northern Europe. As noted, this makes Iceland habitable. But it helps the United Kingdom, too. The United Kingdom and Labrador are at about the same latitude, 55 degrees

north. But Labrador lacks warmth from the Gulf Stream, and so is barely inhabited, whereas 58 million people thrive in the United Kingdom.

However, a penalty accompanies the Gulf Stream. As the warm air above it meets the chilled air at 55°N, atmospheric moisture condenses, and near-perpetual fog results. London is a foggy legend, giving rise to the clothing brand, London Fog.

Upwelling Currents

In Figure 10.8A, you can see that prevailing trade winds blow westward, and they blow surface waters away from shore along the west coasts of North America, South America, and Africa (Figure 10.8B). To replace this water, deep water upwells from the seafloor along these coasts. This water originates in polar regions as meltwater from ever-present ice. It is very cold, barely above freezing, and flows on the seafloor. There it becomes nutrient-rich by dissolving abundant phosphorus and nitrates from decomposing organisms on the seafloor. Continual **upwelling** brings these nutrients to the surface, where they nourish tiny plants, which feed extraordinary numbers of fish. Consequently, the Peruvian coast is noted for its fishery.

FIGURE 10.9 The Gulf Stream.

A river-like current of warm water from the Gulf of Mexico that flows northward through cooler water along the East Coast, the Gulf Stream gives New York and Boston milder winters than they otherwise would have. This satellite image reveals the Gulf Stream by showing differences in water temperature. False colors are used to show colder Atlantic Ocean water (blue, green) and warmer water of the Gulf Stream (yellow, red, brown). The core rings are cut-off meander loops, similar to those created by the Mississippi River (Figure 7.7B).

Photo: Dr. R. Legeckis/Science Photo Library/Photo Researchers, Inc.

Should any part of this elaborate chain break down, the Peruvian fishing industry would be in trouble. This is exactly what happens during an El Niño event.

El Niño

The El Niño effect, so widely reported in the media over the past decade, is an excellent example of the linkages among the hydrosphere, atmosphere, and biosphere. In Spanish, El Niño means "the boy child," specifically, the Christ Child. The term originally was used by Peruvian fishermen to identify a seasonal warm ocean current that appears around Christmastime. Fish become less abundant because upwelling of the deep, nutrient-rich water is prevented, meaning a smaller fish catch. In some years, the warm current remains until May or June, and the term El Niño has come to refer to these exceptional long-lasting warm intervals. Historically, El Niños occur irregularly, every 3 to 8 years.

The **El Niño** phenomenon is basically a current reversal, caused by slack winds. For an explanation, see Figure 10.10.

During the past 40 years, nine El Niños have affected water temperature, not only along the South American coast but in a wide band extending thousands of miles across the Pacific to Indonesia and northern Australia. Stronger El Niños have raised sea surface temperatures about 10 degrees Fahrenheit, decreasing the fish catch by about 25 percent. Sea birds, which also rely on catching fish, have abandoned their young and scattered, seeking food. Fur seal and sea lion pups starve.

During the severe 1982–1983 El Niño, the anchovy catch decreased 50 percent. Anchovies are ground for use as cattle and poultry food. Without them, soybeans were substituted in feed, sharply increasing chicken and beef prices. Fewer anchovies also reduced the marine bird population, reducing the volume of their phosphate-rich guano (droppings), which is harvested for fertilizer by coastal residents.

The El Niño also increased rainfall and flooding in coastal villages along the Andes Mountains. During 1982–1983, 100 inches of rain fell in Ecuador and northern Peru, changing a coastal desert into a lake-dotted grassland. Populations of toads, birds, shrimp, and mosquitoes soared—along with mosquito-borne malaria cases. The current reversal also caused an unexpected harvest of warm-water scallops to wash ashore in Ecuador.

What causes an El Niño? The Trade Winds diminish, but why? We do not know yet. But the weather effects of an El Niño are worldwide, so scientists from many countries are studying the phenomenon. The answers may not come quickly due to the complex interactions between atmospheric circulation and ocean currents. El Niño is a good example of the need for col-

FIGURE 10.10 How El Niño works.

(A) In normal years, Trade Winds blow westward, generating the Equatorial Current, allowing upwelling of nutrient-rich water to support the Peruvian fishing industry. (B) In El Niño years, the situation reverses.

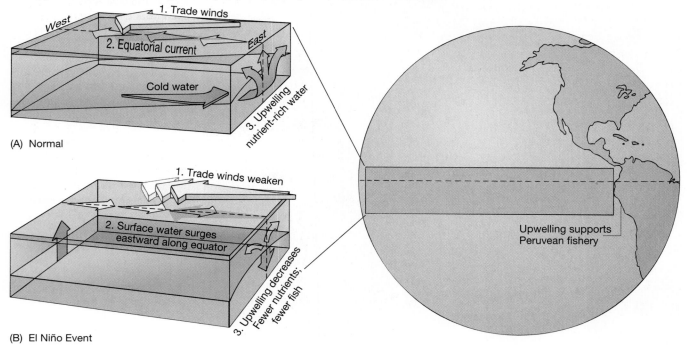

laboration among scientists of different disciplines and countries to solve worldwide environmental problems.

Hurricanes

Thanks to TV, everyone in the United States has viewed the incredible damage caused by the great tropical storms we call **hurricanes**—shorelines shifted by severe beach erosion, buildings flattened, trees uprooted, small boats dumped inland, and homeless people (Figure 10.11). Other parts of the world have them too, although they may be called *typhoons* or *cyclones*. Why and how does a hurricane form? Where are they most likely to occur? How can we minimize the damage?

Meteorologists are still learning exactly how hurricanes form. But we certainly know where they form and the general paths they follow. Hurricanes form over tropical ocean waters where winds are light, the humidity is high in a thick layer, and the sea-surface temperature is warm, typically 79°F or higher, over a wide area. This provides what a hurricane needs most: plentiful moisture and heat energy.

These requirements restrict the area where hurricanes can start, because surface ocean water north or south of 30° latitude is too cool. But the tropical North Atlantic and North Pacific have the right conditions in summer and early fall (Figure 10.12). The official hurricane season is June through November. Prevailing winds in these low latitudes blow westward, so that's the direction the storms move.

The word hurricane comes from a Carib Indian word, *urican*, meaning "big wind." (Now you have a new word to describe certain people on campus.) To be classed as a hurricane, a storm's winds must reach at least 74 miles per hour. The greatest wind speed recorded in these storms is about 200 mph. Hurricanes are rated in Categories 1 through 5, based on wind

FIGURE 10.11   After the storm.

The aftermath of Hurricane Gilbert in 1988 in Kingston, Jamaica. *Photo: Bleibtreu/Sygma.*

speed. Table 10.1 shows the Saffir-Simpson Hurricane Damage-Potential Scale.

To get an idea of the energy level in a hurricane, just 1 percent of the energy in a typical storm, if harnessed, could meet the energy needs of the entire United States for a full year! The force of the winds that surround the eye of a hurricane have energy equivalent to a nuclear bomb exploding every 10 seconds. And, as we know

FIGURE 10.12   Hurricane conditions.

Typical paths of tropical cyclones related to sea temperature. (Composite of summertime temperatures in each hemisphere.)

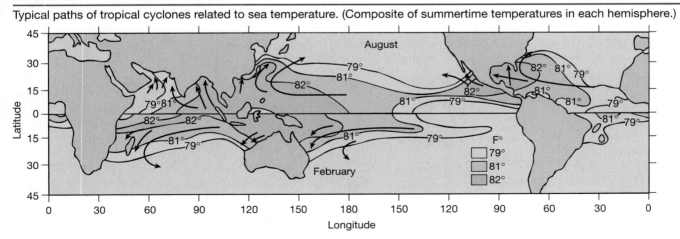

TABLE 10.1 Categories of Hurricane Intensity

Saffir-Simpson Damage-Potential Scale.

Category	Wind Speed (mph)	Storm Surge (ft)	Damage
1	74–95	4–5	Minimal: Damage mainly to trees, shubbery, and unanchored mobile homes. Example: Allison, 1995; Erin, 1995.
2	96–110	6–8	Moderate: some trees blown down; major damage to exposed mobile homes. Example: Roxanne, 1995.
3	111–130	9–12	Extreme: foliage removed from trees; large trees blown down; mobile homes destroyed; some structural damage to small buildings. Example: Marilyn, 1995.
4	131–155	13–18	Extreme: all signs blown down; extensive damage to roofs, windows, and doors; complete destruction of mobile homes; flooding inland as far as 6 miles; major damage to lower floors of structures near shore. Example: Hugo, 1989; Andrew, 1992.
5	greater than 155	greater than 18	Catastrophic: severe damage to windows and doors; extensive damage to roofs of homes and industrial buildings; small buildings overturned and blown away; major damage to lower floors of all structures less than 15 feet above sea level within 1500 feet of shore. Example: Gilbert, 1988.

from TV weathercasts, hurricanes have a lifespan of a week or more as they migrate from their central Atlantic birthplace toward North America.

When a hurricane makes landfall, it can kill and destroy in three ways: wind, rainfall, and storm surge. In serious hurricanes, wind speeds exceed 100 mph and commonly achieve twice that speed. This creates sufficient pressure to pry up roofs, exposing the inside of buildings to inundation by the almost unimaginable intensity of hurricane rainfall. One day's rainfall from a moderate hurricane equals the average yearly discharge of the Colorado River at its point of greatest flow. Talk about getting dumped on! The deluge typically causes severe landslides in hilly areas as the water saturates the already-soggy ground, increasing damage to structures (Figure 10.13).

As fearsome as the wind and rain are, the storm surge is most deadly. The storm surge is a rise in sea level along shore, caused by the drop in atmospheric pressure inside the hurricane, which allows water from surrounding higher-pressure areas to push in, creating a bulge. As this happens, the wind piles up water against the shore. The height of a surge can exceed 40 feet. Table 10.1 shows storm surge heights that can be expected from different categories of hurricanes.

We measure hurricane destruction in lives lost and dollar cost. In terms of deaths, the world record-holder is a typhoon that came ashore in 1737 at the mouth of the Hooghly River on the Bay of Bengal in what now is Bangladesh, killing about 300,000 people. The worst U.S. killer hurricane occurred in 1900 when a ferocious storm struck Galveston on the Gulf Coast of Texas. In those pre-radio, pre-TV days, the storm arrived without warning, killing an estimated 6000 people.

Today, thanks largely to weather satellites that track a hurricane from a baby "tropical wave" to a full-blown storm to its final dissipation, we have plenty of warning. Even though the U.S. Atlantic and Gulf Coast population has increased from 29 million in 1960 to 39 million in 1990, few lives are lost, even in the worst storms.

FIGURE 10.13 Hurricane Andrew, 1992.

The town of Homestead in South Florida was decimated by Hurricane Andrew. Even though the storm had only one major landfall, Andrew was one of the most expensive natural disasters in U.S. history. *Photo: Al Diaz/Miami Herald Publishing Co.*

But financial loss is another story. As deaths go down, dollars go up (Figure 10.14). We continue to expand in hurricane-prone areas, and build costlier structures. The costliest hurricanes to date have been Gilbert (1988), Hugo (1989), and Andrew (1992).

Hurricane forecasts are issued from the National Hurricane Center at Coral Gables, Florida (which was temporarily shut down by Hurricane Andrew in 1992). The Center issues hurricane watches and hurricane warnings. A **hurricane watch** is a caution to part of the coast that a 50 percent chance of being hit exists for the next 36 hours. It alerts residents for possible evacuation. A **hurricane warning** means that a part of the coast is in imminent danger and immediate action is required to protect life and property.

The forces that control the path of a hurricane are not well understood, so the point of landfall cannot be predicted precisely. Some storms follow erratic paths and make surprising turns, taking forecasters by surprise. Numerous times, a storm heading directly for

FIGURE 10.14 Deaths decrease, costs increase.

Changes since 1900 in the number of deaths and cost of damage in U.S. hurricanes. Cost is in billions of 1996 dollars.

Deaths			Damage	
1980 1989	161			
1970 1979	226			
1960 1969	570	$19.1	1980 1990
1950 1959	750	$21.6	1970 1980
1940 1949	220	$21.0	1960 1970
1930 1939	1,050	$13.3	1950 1960
1920 1929	2,130	$5.0	1940 1950
1910 1919	1,050	$5.1	1930 1940
1900 1909	8,100	$1.9	1920 1930

land has abruptly veered, sparing one area, but possibly striking another (Figure 10.15). Recent studies show that the difference between the actual landfall point and the location forecast 24 hours earlier averages 119 miles.

The naming of hurricanes originated in a 1941 novel called *Storm*. During World War II, naming storms for women became commonplace among military meteorologists. The practice was made official by the National Weather Service in 1953, and men's names were added in 1979. The names for each year's hurricanes are chosen by the World Meteorological Organization. Names for the next four years are shown in Table 10.2.

Eroding the Shore

The overall effect of waves and the currents they spawn is to straighten irregularities in coastlines (Figure 10.16). They do so by focusing their erosive energy on headlands and prominences and depositing the debris in recessed areas such as bays. Sediment not so deposited either is transported downcurrent or carried offshore into deeper water, perhaps never to return. Much sediment carried offshore by currents finds its way to the deep ocean floor via channels from the shore that extend to great depth. Waves also cause the gradual retreat of the shoreline.

Tsunami (Earthquake-Generated Waves)

In 1883, the eruption of Krakatau volcano in Indonesia generated a wave 40 feet high, which killed 36,000 people. In 1755, an undersea earthquake generated a similar giant wave, which rushed ashore in Lisbon, Portugal, killing 25,000–30,000. In 1896, a Japanese earthquake generated a great wave that killed 26,000. What are these waves? What causes them?

The Pacific Ocean basin is noted for earthquakes. When they happen underwater, they generate a large, low wave. The Japanese islands have been struck by countless such waves over millions of years. Because larger ones can cause great damage, especially when

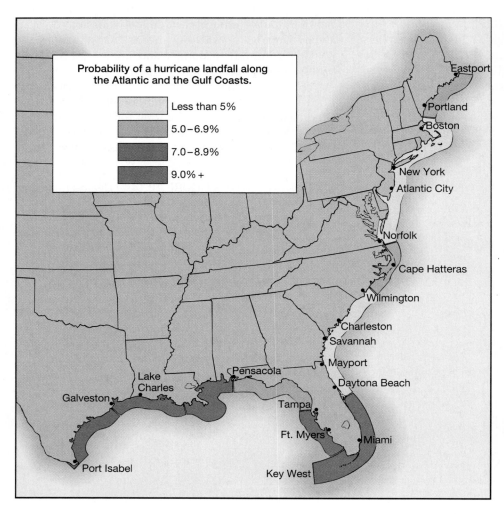

FIGURE 10.15 Where will it land?

Probability of a hurricane landfall along the Atlantic and Gulf Coasts.

TABLE 10.2 Hurricane Names for 1997, 1998, 1999, and 2000

1997	1998	1999	2000
Ana	Alex	Arlene	Alberto
Bill	Bonnie	Bret	Beryl
Claudette	Charley	Cindy	Chris
Danny	Danielle	Dennis	Debby
Erika	Earl	Emily	Ernesto
Fabian	Frances	Floyd	Florence
Grace	Georges	Gert	Gordon
Henri	Hermine	Harvey	Helene
Isabell	Ivan	Irene	Isaac
Juan	Jeanne	Jose	Joyce
Kate	Karl	Katrina	Keith
Larry	Lisa	Lenny	Leslie
Mindy	Mitch	Maria	Michael
Nicholas	Nicole	Nate	Nadine
Odette	Otto	Ophelia	Oscar
Peter	Paula	Philippe	Patty
Rose	Richard	Rita	Rafael
Sam	Shary	Stan	Sandy
Teresa	Tomas	Tammy	Tony
Victor	Virginie	Vince	Valerie
Wanda	Walter	Wilma	William

they enter a harbor, the Japanese call them "harbor waves," or **tsunami** (soo-NAHM-ee).

The media usually call tsunami "tidal waves," which is incorrect and misleading, for they have nothing to do with the tides. A better English-language term is *seismic sea wave*. Most tsunami result from sudden fractures on the ocean floor, where one side of the fracture suddenly drops relative to the other side. This causes a usually harmless earthquake at sea, but it also generates a far more ominous wave. Almost immediately, water from the surrounding area rushes in to fill the depression, generating a very long, low, fast-moving wave of great force.

The average water depth in the Pacific Ocean basin is 18,000 feet, and the velocity of the initial wave of a tsunami originating at this depth would be 758 feet per second, or 520 miles per hour. Tsunami originating at greater depths could generate waves with velocities of 600–700 mph. This is the speed of a jet liner. If you took off from Honolulu a few minutes before a tsunami was generated there, the sea wave might travel immediately beneath you for the entire 5-hour flight to Los Angeles. However, the plane might not be able to land because of the destruction that would occur if the tsunami reached an airport built on flat coastal land. Figure 10.17 shows the Tsunami Warning System for the Pacific basin, centered on Hawaii, which is a perfect sitting duck. Table 10.3 shows the tsunami risk for U.S. Pacific states.

Tsunami travel rapidly outward from their point of origin. In the open ocean, they are exceedingly difficult to observe. Unlike wind-whipped waves, tsunami have extraordinary wavelengths (100–150 miles), but wave heights of only a few feet. Commonly, a tsunami will pass unnoticed beneath ships at sea. The largest tsunami have wavelengths of 600 miles and open-ocean heights of 10–15 feet.

On entering shallow water, however, wave height increases dramatically, just as is shown for smaller wind-driven waves in Figure 10.3. A 16-foot wave moving at 400 miles per hour in the deep water of the open ocean can be refracted into a devastating wave nearly 100 feet high, hitting a coastline at over 30 mph. In 1868, a monstrous tsunami picked up a U.S. warship anchored in a Peruvian port, carried it over the port city, and deposited it 1300 feet inland!

TABLE 10.3 The Danger of Tsunami

State	Total State Population	Cities Susceptible to Tsunami	Total Population of Susceptible Cities	Population Endangered by a 50-foot Tsunami	Population Endangered by a 100-foot Tsunami
Washington	4,866,692	102	1,510,000	96,000	202,500
Oregon	2,842,321	60	95,000	31,500	55,000
California	29,760,021	152[a]	8,700,000	589,500	1,100,000
Hawaii	1,108,229	123	750,000	131,000	314,500
Alaska	500,043	52	170,000	47,000	73,000
Total	39,127,306	489	11,225,000	767,500	1,745,000

[a]Not including urban areas on San Francisco Bay, because they are not considered vulnerable. Updated from 1970 NOAA estimates.

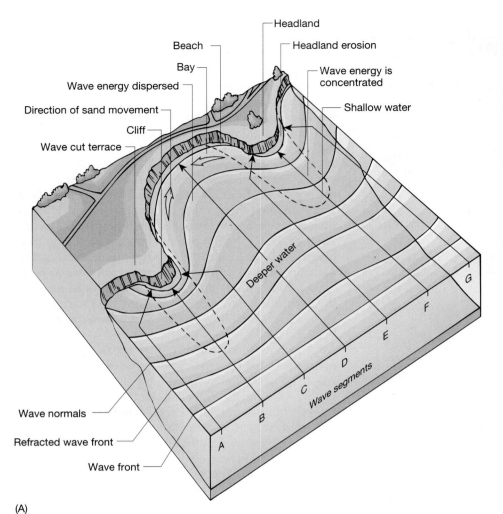

- Headland
- Beach
- Headland erosion
- Bay
- Wave energy is concentrated
- Wave energy dispersed
- Direction of sand movement
- Shallow water
- Cliff
- Wave cut terrace

Deeper water

- Wave normals
- Refracted wave front
- Wave front

Wave segments

A B C D E F G

(A)

FIGURE 10.16 Wave refraction (bending) around land.

(A) Wave refraction (bending) along an irregular shoreline. Refraction focuses wave energy on headlands, but disperses it across bays. As wave fronts enter shallower water, they are refracted progressively so that they become more parallel to the bottom contours. (B) Sea gull's view of wave refraction around headlands and islands is clearly shown in this aerial photograph. Energy concentrated on the headlands has eroded some of them to offshore islands. *Photo: Dr. W. K. Hamblin.*

(B)

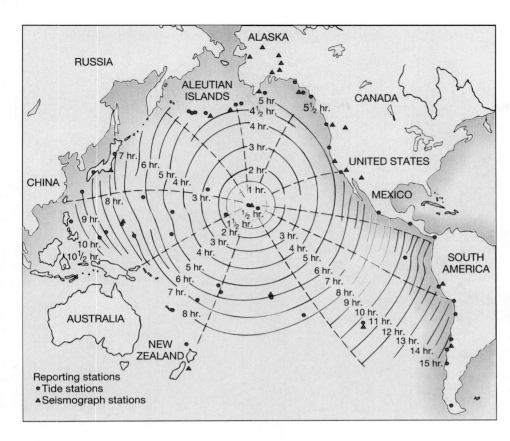

FIGURE 10.17   The Tsunami Warning System for the Pacific basin.

Circles show tsunami travel times to Hawaii. For example, a tsunami generated by an earthquake off the coast of Chile would take about 15 hours to reach Hawaii. *Source: NOAA.*

A 100-foot wave hitting a shoreline has enormous destructive power, much like an earthquake. Figure 10.18 shows typical damage from a large tsunami. Because of this, a *tsunami intensity scale* has been developed to describe the scale of the damage produced (Table 10.4). The frequency of occurrence is shown for the Pacific Ocean, becaue that is where most tsunami occur. Fortunately, only one or two destructive tsunami are generated worldwide each year. Of these, only about one every ten years is catastrophic. This frequency is on the same order as the frequency of occurrence of destructive earthquakes in California.

Should structures along the coast be constructed with tsunami protection in mind? Should breakwaters be constructed offshore to dissipate some of the wave energy? Where should the money come from? Should it be taken from other social programs, or come from higher taxes?

Sea Level Change

The sea is not level, as a quick glance at its surface reveals. Its level constantly shifts inches and feet with tides and waves. It even varies with the seafloor topography beneath it—slightly higher over undersea mountains, and a few feet lower above undersea canyons—as is revealed by satellite measurements. When we speak of **sea level,** we mean the average elevation of the land–sea contact worldwide.

When scientists talk about sea level change, they are referring to its elevation at some earlier time or later time. Is the average position of the land-sea contact higher or lower than 50 or 100 years ago relative to some arbitrary elevation? For example, is a given island any higher or lower above sea level than it was 100 years ago? Has the water level risen or fallen on rocks along the seashore, compared to century-old photos of the rocks?

Peat is an early stage of coal. It forms from incomplete decomposition of trees, and trees grow only on the land surface. So, if we drill a hole at the present shoreline and discover a peat deposit two feet below the surface, it is clear that sea level has risen at least two feet since the peat formed. A part of Earth that once was above sea level now is submerged. The sea has moved inland enough to cover the peat with two feet of sediment. But has the sea risen, or the land sunk? The answer is mostly that sea is rising, but in some places land is sinking, so the answer depends on where you are.

Migrating Shorelines

Figure 10.19 shows the present U.S. shoreline. It also shows the shoreline as it was during the most recent Ice

FIGURE 10.18   Tsunami aftermath.

The Japanese named the big waves that roared without warning into their harbors *tsunami*, or "harbor waves." Here is what a 1993 tsunami, triggered by an undersea earthquake, did to a small town on the island of Okushiri, Japan. *Photo: National Geophysical Data Center/NOAA.*

Age, about 18,000 years ago, when a lot of ocean water was frozen into continental ice sheets. Clearly, sea level is rising along the U.S. East and West Coasts. The shorelines are migrating landward.

What determines the position of a shoreline through time? For at least the last several hundred thousand years, the major cause of sea-level change has been the growth and decay of continental ice sheets, such as those that still blanket Antarctica and Greenland. At the maximum extent of the last glacial advance 18,000 years ago, sea level was about 400 feet lower than at present. As a result, the shoreline of the gently sloping U.S. East Coast was well out to sea, some 100 miles east of New York City (Figure 10.19). Since then, sea level has risen more or less continuously to its present position.

Should Earth's large ice sheets melt completely, sea level would rise about another 150 feet. About 15 percent of the present land area of the lower forty-eight states would disappear, as you can see in Figure 10.19. The *rate* of sea-level rise seems to have slowed during the past few thousand years—Figure 10.20A. But it still

is increasing; over the past century, sea level rose 3.4 inches—Figure 10.20B. This rate is expected to increase, due to global warming and melting of glacial ice. Several investigators have predicted a future sea level rise of 6–20 inches by 2050, and 12–43 inches by 2100.

But Ice Ages are not the only factor in changing sea level. The shoreline can move inland, or move seaward, or remain stationary, in response to several processes. One cause of sea-level rise is the growth of a system of undersea mountains, as you shall see in Chapter 11. Another cause is the thermal expansion of seawater. Water expands as its temperature rises above 40°F because the higher energy in the water makes its molecules more active and spread farther apart. Between 1880 and 1985, sea-level rise due to expansion of water from global warming is estimated to have been 1 to 2 inches—a very significant portion of the total rise of 3.4 inches.

Subsiding Land

In the cases cited so far, the ocean water is doing the moving, not the land. But the shoreline also will move

TABLE 10.4 Scale of Tsunami Intensity

Intensity	Visual Height (feet)	Description of Tsunami	Frequency in Pacific Ocean
4	53	Disastrous. Partial or complete destruction of manmade structures for some distance from the shore. Flooding of coasts to great depths. Big ships severly damaged. Trees uprooted or broken by the waves. Many casualities.	1 in 10 years
3	26	Very large. General flooding of the shore to some depth. Heavy structures near the sea damaged. Light structures destroyed. Severe scouring of cultivated land and littering of the coast with floating objects, fish, and other sea animals. With the exception of big ships, all vessels carried inland or out to sea. Large bores in estuaries. Harbor works damaged. People drowned, waves accompanied by strong roar.	1 in 3 years
2	13	Large. Flooding of the shore to some depth. Light scouring on paved ground. Embankment and dikes damaged. Light structures near the coast damaged. Solid structures on the coast lightly damaged. Big sailing vessels and small ships swept inland or carried out to sea. Coasts littered with floating debris.	1 per year
1	6.6	Rather large. Generally noticed. Flooding of gently sloping coasts. Light sailing vessels carried away on shore. Slight damage to light structures situated near the coast. In estuaries, reversal of river flow for some distance upstream.	1 per 8 months
0	3.3	Slight. Waves noticed by those living along the shore and familiar with the sea. On very flat shores, waves generally noticed.	1 per 4 months
−1	1.6	Very slight. Wave so weak so to be perceptible only on tide gauge records.	

After Soloviev, 1978, in *The Assessment and Mitigation of Earthquake Risk—Natural Hazards,* copyright © UNESCO, 1978. Used by permission of UNESCO.

landward if the land surface subsides. **Ground subsidence** occurs for at least two reasons: depression beneath tremendous loads of ice or sediment, and movements within Earth itself.

Where sedimentation is rapid and great in volume, as in the deltas of the Mississippi, Nile, and Niger Rivers, compaction is rapid. This lowers the land surface and the shoreline moves landward. At the Mississippi River's mouth, the shoreline is migrating landward five times faster than the average rate for the Gulf of Mexico, ten times faster than the world average, and more than fifteen times faster than the U.S. East Coast. The river's

FIGURE 10.19 Shoreline changes.

Three U.S. shorelines: 18,000 years ago, present shoreline, and future location if all existing glacial ice melted. The real estate of Massachusetts, Connecticut, Rhode Island, New Jersey, Delaware, Florida, and Louisiana would be seafloor, as would our nation's capital. Other coastal states would lose only some of their representation in the U.S. Congress, which would be holding session underwater.

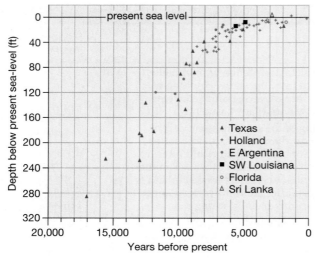

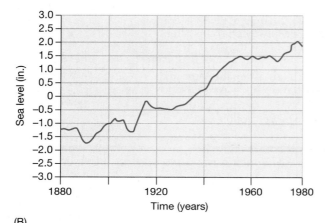

FIGURE 10.20 Sea-level change.

(A) The long-term trend: Sea-level trend over the past 18,000 years, deduced from carbon-14 dating of peat deposits and shells of shallow-water sea creatures. Note that the rate of rise has slowed. (B) The recent trend: Sea-level change, generally a rise, from 1880 to 1980. Worldwide sea level has risen about 3.4 inches during the past century.

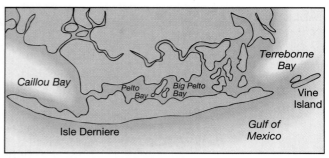

1853

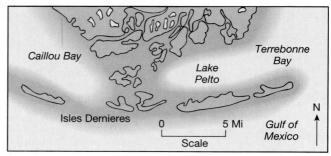

1978

FIGURE 10.21 The case of the disappearing delta.

The effect of 125 years of landward movement of the shoreline in the Mississippi Delta region. Isle Derniere is now submerged into plural islets—Isles Dernieres—and not much remains as it was in 1853. The marshlands (wetlands) also have changed for the wetter, seriously affecting the ecology of the area. Most of the landward movement of the shoreline has resulted from sediment compaction and regional subsidence of about 3 feet per century.

lower delta is sinking at a rate of 0.5 inch per year because the enormous weight of the ever-increasing delta sediment is depressing Earth's crust beneath it. This is shown in Figure 10.21.

Human activities also can depress the land surface, resulting in landward movement of the beach. People are responsible for an estimated one-third of the shoreward movement of the sea in the Mississippi delta region, mostly because of the extraction of groundwater and petroleum from the subsurface. The rock loses its internal support from water and oil, and partially collapses. Other ways in which humans intervene in natural processes to cause landward movement of the shoreline include deforestation, which increases runoff and erosion, and the draining of wetlands for construction and agriculture, which increases erosion.

Rate of Change

Despite the fact that the world ocean is a single interconnected body of water, it is clear that the landward movement of the shoreline is not uniform over Earth. Some land areas are being uplifted by the movement of subcrustal material deep within Earth, as we will see in Chapter 11. The rate of this crustal rise is much less than the rate of sea-level rise from glacial melting, so although the shoreline continues to move landward, the rate of movement is less than the world average.

Another reason for a slowed landward movement of the shoreline in some areas is that those areas that had the thickest accumulations of ice during glacial times are rebounding. When the ice was present, Earth's crust was depressed by the weight. With the ice gone, the crust is springing back at breakneck geologic speed, which is hundred thousands of years. The more ice that was removed, the greater is the rebound. In eastern Canada and Scandinavia, the rebound rate is faster than the increase in seawater volume from glacial melting. So, in these areas, the shoreline is actually moving seaward.

Worldwide, however, seaward movement of the shoreline is localized. The worldwide trend is landward movement, with attendant destruction of shoreline structures, increased flooding in lowland areas, and salinization of the groundwater supply of coastal cities.

Can the shoreline's position in 2005 or 2050 be predicted, so that humans can accommodate to the inevitable, or perhaps intervene to prevent the change? Answers to such questions have enormous social, ecological, and financial implications for us and our children and grandchildren. Virtually all of you reading this book will still be alive in 2005, and many still will be around in 2050.

If you live in a coastal state, is the shoreline moving landward, staying stationary, or moving seaward in your area? What is the explanation for whichever of the three possibilities is occurring? Is there anything that your state government can or should do about it?

Sea Level and Small Islands

The tiny country of Tuvalu in the South Pacific is an example of the life-or-death effect that a small sea-level rise can have on communities built near sea level. Tuvalu's nine little islands occupy less than 10 square miles about 2500 miles southwest of Hawaii. The country averages only 6 feet above sea level. Occasionally, a typhoon (hurricane) inundates nearly the entire country. Tuvalu's 10,000 residents are justifiably nervous about any sea-level rise, no matter how small.

In an attempt to preserve his island country, Tuvalu's Prime Minister has appealed to Japan to send dirt: a transpacific dirt lift. It seems that each year, Tokyo has to dispose of about 30 million tons of excess dirt, produced by construction. This is enough dirt to elevate the entire nation of Tuvalu about 2 feet, giving the country new stature. Tokyo spends tens of millions of dollars each year getting rid of the unwanted soil, and it might actually save money by sending it to Tuvalu. Still, shipping 30 million tons of dirt 3700 miles from Tokyo to Tuvalu is not dirt cheap, and it is unclear that Tuvaluans can pay the cost.

Coastal Erosion

Some scientists project a sea-level rise of 13 inches in the next 100 years in response to global temperature increase. (Global warming is discussed in Chapter 17.) Storms are likely to grow stronger because of the additional heat available to power them. In low areas such as the Mississippi Delta and Bangladesh, little sea-level rise will be needed to cause a dramatic landward movement of the shoreline (Figure 10.19). Located near coastlines are half the world's people, nine of Earth's ten largest cities, and thirty-three of the largest fifty cities, so it is clear that many people, businesses, highways, and entire cities will be moving. This will not happen overnight, but over many decades.

The most reasonable long-term response to this gradual rise is to plan future development farther landward, and accept the fact that any structure now within a few feet of sea level eventually will be abandoned.

About 40 percent of the American population is clustered along the Atlantic and Pacific coasts, and by 2010, less than fifteen years from now, the coastal population is expected to reach more than 100 million. Many of you reading this book will be among these coastal-dwellers. Short-term problems you will see include flooding and coastal erosion. Can anything be done? What methods have been tried so far to restrain the sea and reduce erosion? How successful have they been?

Pacific Coastline

The western United States is a mountainous area from Denver to the Pacific shore, extending for some distance beneath the sea (Figure 10.1, West Coast). A 2-foot sea-level rise over the next century would affect the West Coast differently from the East Coast and Gulf of Mexico (Figure 10.19). In California, the low-lying San Francisco Bay area, extending into the northern part of the San Joaquin Valley, would flood gradually. As noted in Chapter 6, the irrigated San Joaquin Valley is currently the source of a very large percentage of this country's fruits and vegetables. But few edible crops other than rice grow under water, particularly salt water. So, California's agriculture would change dramatically. Low-lying areas around Seattle would have similar problems. Perhaps ocean barriers could be erected at the mouth of San Francisco Bay, as the Dutch have been doing for centuries along their coast. See the section called "Case Study: Rethinking Holland" later in this chapter.

Californians also would experience greater erosion of the cliffs that border the ocean and faster loss of beach width. Beaches have been narrowing for years because the sand being washed into the deep ocean is not being replenished by local rivers, as it once was. The once-abundant supply of sand from coastal mountains has diminished greatly, because the rivers that supplied it have been dammed for irrigation and flood control. Debris from cliff erosion supplies some beach sand, but not enough.

In turn, the narrower beaches cannot slow the waves and protect cliffs from wave impact as they once did. Thus, cliff erosion and retreat has accelerated. Faster cliff retreat will impact existing homes and new

FIGURE 10.22 Great beachfront units, ideal for student housing.

Note that the lower layer of sandstone is more resistant than the upper one, so during collapse you might land on rock instead of in the ocean. Buy a sea-cliff home in southern California at your own risk. *Photo: James R. McCullagh/Visuals Unlimited.*

home construction on the scenic bluffs, as well as highways. Figure 10.22 shows this reality.

Atlantic and Gulf Coastlines

The problems faced by the states and communities along the Atlantic and Gulf coasts are different from the Pacific coast, reflecting the topography:

1. The land is flat and low, very near sea level for many miles inland from the shoreline (Figure 10.19 shows how much would be underwater if the ice sheets melted completely). Consequently, sea level is rising fast enough to be significant, even during the 75-year lifetime of the average person.

2. Erosion rates are very high, shifting some shorelines by feet each year. See how far the shoreline nearest you shifts in Table 10.5.

3. Too many people appreciate the scenic vistas and recreational activities near the beach. This volume of users stresses beaches, increases groundwater demand, and makes the whole shore area more critical economically.

4. Under our legal system, it is very difficult for state governments to restrict the uses to which private property may be put.

Barrier Islands. The Atlantic and Gulf coastlines are characterized by barrier islands, elongate ridges of sand and gravel that parallel the coastline not far offshore (Figure 10.23). Barrier islands protect the coast from storms by taking the brunt of their wave energy. The islands also make the seafloor shallow, which

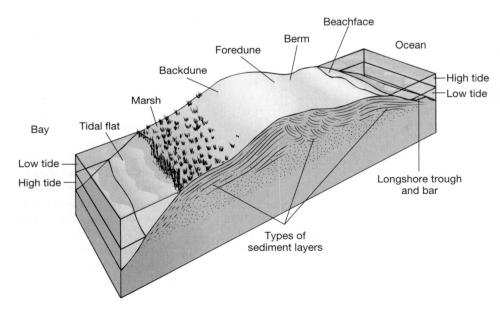

FIGURE 10.23 Cross section of a typical barrier island.

Composed of vulnerable loose sand, the width of the sand pile is a few thousand feet and its height a few tens of feet. Waves and currents act upon the sand like an army of children with shovels and buckets.

TABLE 10.5 Shoreline Erosion Rates

Region	Average rate of shoreline change (in/yr)*
Atlantic Coast	−0.8
Maine	−0.4
New Hampshire	−0.5
Massachusetts	−0.9
Rhode Island	−0.5
New York	−0.5
New Jersey	0.1
Delaware	0.1
Maryland	−1.5
Virginia	−4.2
North Carolina	−0.6
South Carolina	−2.0
Georgia	0.7
Florida	−0.1
Gulf of Mexico	−1.8
Florida	−0.4
Alabama	−0.4
Mississippi	−0.6
Louisiana	−4.2
Texas	−1.2
Pacific Coast	0.0
California	−0.1
Washington	0.5
Alaska	−2.4

*Negative values denote erosion, positive values, accretion.

Source: Adapted from S. K. May, R. Dolan, and B. P. Hayden, EOS 54 (1983): 551–53.

slows approaching waves. These islands are easily moved during large storms, so they are unsuitable for supporting permanent buildings, although this fact has not slowed developers one bit. It is inevitable that buildings on barrier islands will be destroyed; it is only a matter of time (Figure 10.24).

A large storm is not an unusual event and should not be considered an "act of God" in the legal sense (an unpredictable catastrophic event). Severe storms and hurricanes are *normal*—think of the hurricanes you hear about every fall. But island property often is owned by a developer, who wants to erect condominiums, stores, beachfront cabanas, and the other amenities many Americans have come to expect. The barrier islands along the Atlantic and Gulf coast contain 1.6 million acres of oceanfront real estate, worth hundreds of millions of dollars. Developers seek ways to keep the sand from moving, in pursuit of short-term profit. After a devastating storm, they admit that development was a bad idea in the first place, and then start building anew (Figure 10.25). As sea level continues to rise over the years, it will become more difficult to restrain Father Nature's natural impulses.

Protecting Shorelines

Coastal engineers have two basic approaches to stabilizing shorelines: building structures to reduce erosion and replenishing sand that is eroded away.

Building Structures to Reduce Erosion. Breakwaters, seawalls, jetties, and groins all discourage erosion where desired, but increase erosion elsewhere. These structures are made of cement, concrete, or large rocks, and are anchored to the land or shallow seafloor. They are designed to decrease or deflect the energy of the waves as they approach the beach. These structures are shown in Figure 10.26:

- A **breakwater** is built offshore, usually parallel to the beach front. It offers protection against offshore transport of sediment and large incoming waves.
- A **seawall** is similar to a breakwater, but it is located right at the shore. (You can see part of a seawall in Figure 10.27.)
- A **groin** is oriented perpendicular to the shore and offers protection against longshore sediment transport (You can see several groins in Figure 10.28).
- **Jetties** are built in pairs on either side of an inlet to keep it from being closed by migrating sand.

If sea level rose 3 feet, the estimated cost of maintaining the present U.S. coastline using such engineering structures would be at least $500 billion dollars per year.

Sand Replenishment. Coastal engineers also try to stabilize and maintain beaches with sand replenishment—adding new sand to rebuild beaches that have eroded to positions near seawalls or buildings. Sand may be pumped to the beach from places where large accumulations exist, such as inlets, tidal delta shoals, or the continental shelf some distance offshore from the beach (Figure 10.29). In some cases, sand is trucked from inland quarries, but this can be quite costly. In fact, any method of beach replenishment is very costly, in the millions of dollars (Table 10.6). Worse, the replenishment is quickly removed by frequent storms.

The longest-lasting replenishments along the Atlantic coast are Florida beaches south of Cape Canaveral, which average a replenishment life of nine years. As one travels north, however, storm frequency and wave energies increase and durability diminishes: from Georgia to Delaware, replenishments last only two to five years; for New Jersey, it is less than three years. Ocean City, New Jersey, holds the record for destruction: in 1982, storms removed a $5.2 million beach in only two

(A)

(B)

FIGURE 10.24 Before and after. (A) Before Hurricane Andrew struck the Louisiana coast in 1992, Raccoon Island was 3 miles long. (B) After the hurricane, the island lost half its length due to backwash. *Photos: Louisiana State University/Agricultural & Mechanical College.*

and a half months. It could cost $2 billion each decade to replenish New Jersey's entire developed shoreline, and perhaps $200 million per decade for South Carolina's developed shore.

From both a geologic and environmental engineering viewpoint, it is unrealistic to stabilize beaches long enough to justify the large cost. Four states already have opted—sensibly—to retreat from the shoreline—New Jersey, the Carolinas, and Maine. They have made illegal the construction of seawalls and breakwaters.

Retreat from the shoreline, however, is not without its cost. Many states receive bountiful revenue from their coastal resorts via taxes on sales and gasoline paid by out-of-state vacationers. For example, vacationers from landlocked Pennsylvania and Ohio spend lots of money along the coast of New Jersey, Delaware, Maryland, and Virginia. It can be argued that spending government dollars to preserve beaches is a good investment, for it makes money for coastal states. However, most federal funds for such projects come from taxes collected in all fifty states, and most of these states do not benefit from the enhanced tourism.

What do you believe is the best response to the shoreline preservation problem?

Wetlands

Wetlands are known variously as swamps, marshes, bogs, and fens. They are common near any interface between land and water, either in an inland swamp, or along a coastline. Some wetlands are covered or saturated with water throughout the year; others are partly or completely dry for weeks or months. The only reliable indication that an area is a wetland may be the rich organic composition of the soil or the presence of plants adapted to saturated (waterlogged) soils and flooding, like cattails, bulrushes, or weeping willows.

Wetlands are notoriously difficult to define, as legislators have discovered in recent years; even farm hog wallows have been loosely called wetlands. Although salt marshes and mangrove swamps ribbon portions of the Atlantic and Gulf coasts, more than 97 percent of the nation's wetlands are freshwater systems. Overall, wetlands cover 6–9 percent of the lower forty-eight states (Figure 10.30). Seventy-five percent of these wetlands are privately owned.

Most of the nation's coastal wetlands are in Louisiana (along the Mississippi River and its delta) and in Alaska (at the mouths of the Yukon and Kuskokwim Rivers). The Great Lakes and the southeastern seaboard are fringed by other important wet-

(A)

(B)

FIGURE 10.26 Engineering structures for reducing beach erosion and retreat.

In all cases, sand accumulation in one place is balanced by sand removal (erosion) in an adjacent place. Moving water has both an erosive capability and a sediment-transporting capacity; neither can be canceled by artificial structures.

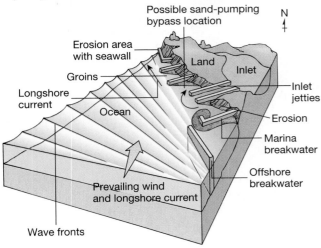

lands, such as Florida's Everglades. Coastal wetlands can be dominated by saltwater, as along the Gulf coast of Louisiana. Or they can contain a complex and changing mixture of saltwater and fresh water, like the estuaries of Chesapeake Bay, Galveston Bay, and San Francisco Bay.

Value of Wetlands

Wetlands are among the most productive ecosystems on Earth. Two-thirds of the species of Atlantic fish and shellfish that humans consume depend on wetlands for some part of their life cycle. Of some 800 species of protected migratory birds, more than 400 rely on wetlands for feeding, breeding, and resting. Wetlands also purify water by filtering and settling sediments and pollutants. During floods, wetlands protect lives and property by slowing the flow of water. During coastal storms, they buffer shores against damage and erosion. Wetlands are sites for much of our hunting, canoeing, and fishing.

Despite their value and relative scarcity, wetlands worldwide are sacrificed increasingly for agricultural expansion and urban sprawl. Today, half the

FIGURE 10.27 A matter of time.

Desperation to protect this hotel near Fort Lauderdale, Florida, is evident. The building is sited on soft sand, and neither the sand-bags nor the attached seawall will save it when a major storm occurs. The owners should not let their insurance lapse. *Photo: P. Moser/American Institute of Professional Geologists.*

FIGURE 10.28 A series of groins at Ship Bottom, New Jersey.

The longshore current is moving from the top to the bottom of the photo, as revealed by the sand accumulations on the upper side of each groin. Each groin slows the current, allowing sand to drop out of the water, depositing a sand wedge. On the lee (sheltered) side of each groin, the water has little sand left to drop, so a sandless void occurs. *Photo: John S. Shelton.*

FIGURE 10.29 Sand replenishment in progress.

The sand is dredged offshore and spread over the beach, eventually to be washed away again by longshore drift. *Photo: American Meteorological Society.*

world's wetlands have been drained. France has lost 67 percent; Italy, 66 percent; Greece 63 percent; The Netherlands, 55 percent. The U.S. has lost more than half of its original 221 million acres of wetlands. California has the dubious distinction of having lost a greater percentage of its wetlands than any other area on Earth, 91 percent over the past 200 years. Ohio has lost 90 percent. Major surviving wetlands are those isolated from advancing civilization in Brazil, Sudan, and Siberia.

Case Study: Ruining Florida's Everglades

"Draining the swamp" has been a familiar activity in Florida since the 1960s, but only recently has the ecological value of these wetlands been realized. After decades of drainage to create dry areas and altered water flow, observations reveal that water pollution now is rampant and that the famed Everglades are dying. This is potentially disastrous for all of south Florida. Wetlands that once replenished underground

TABLE 10.6	Selected Beach Replenishment Projects on the U.S. East Coast				
Beach	State	Year	Volume of sand (yd³)		Cost
Great South Beach	NY	1962	993,500		$ 844,100
Jones Beach	NY	1927–1961	>40,000,000		?
Sea Gurt	NJ	1966	425,211		$ 552,774
Long Beach Island	NJ	1979	1,000,000		$ 4,600,000
Avalon	NJ	1987	1,300,000		$24,000,000
Ocean City	MD	1963	1,050,000		$ 1,517,600
Atlantic Beach	NC	1986	3,600,000		$ 4,750,000
Myrtle Beach	SC	1986–1987	850,000		$ 4,500,000
Tybee Island	GA	1976	2,300,000		$ 3,600,000
Cape Canaveral Beach	FL	1975	2,715,000		$ 1,050,000
Pompano Beach	FL	1970	1,076,000		$ 1,873,437
Hollywood-Hallandale	FL	1979	1,980,000		$ 7,743,376
Miami Beach	FL	1979–1982	12,000,000		$55,000,000
Key Biscayne	FL	1987	360,000		$ 2,600,000

Source: Adapted from O. H. Pilkey, Jr., and T. D. Clayton, *Journal of Coastal Research* 5 (1988): 147–59.

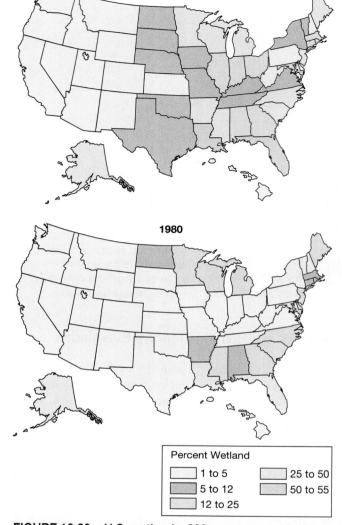

1780

1980

Percent Wetland

	1 to 5		25 to 50
	5 to 12		50 to 55
	12 to 25		

FIGURE 10.30   U.S. wetlands, 200 years ago and today.

In colonial times, U.S. wetlands occupied a total area estimated as the size of Texas and Oklahoma combined. Today, the total area is the size of Montana, a decline of about 55 percent.

aquifers no longer do so, and cities served by these aquifers soon will face severe water shortages. Figure 10.31 tells the story.

Prior to development, wetlands held rainfall for months, continually recharging the aquifer that supplies the growing cities of Florida's southeastern coast. This has now changed. Water that formerly spilled over the banks of Lake Okeechobee (Figure 10.31B) in the wet season and moved lazily through the swamp now is channeled swiftly through 2500 miles of canals and levees, stored in parks called "water conservation areas," and partitioned by countless water-control structures (Figure 10.31C). To prevent flooding, "extra water" is diverted eastward and westward to the Atlantic Ocean

and Gulf of Mexico. This has created an oxygen deficiency in parts of the wetland and threatens the fish population in Florida Bay.

The diversion of water westward away from the Atlantic coast is reducing the water pressure in the South Florida aquifer and encouraging Atlantic saltwater to intrude into marshes and water wells. Everglades wildlife has been damaged by gross disturbance of the natural water system during the past thirty years.

Attempts to reverse the ecological damage in the Everglades have barely begun. Over the next fifteen to twenty years, the U.S. Army Corps of Engineers and other federal and state agencies plan to replumb the entire Florida Everglades ecosystem, at a cost currently estimated at $2 billion. The goal of the restoration is to transform the engineered swampland, now riddled with canals and levees, into a natural wetland that floods and drains in rhythm with rainfall—like it used to. The planners hope that the remaining plants and animals will thrive as a result. But, because the ecological complexity of this 35,000-square-mile area is poorly understood, it is uncertain whether the restoration attempt will succeed.

Case Study: Rethinking Holland

The Dutch, renowned for their invention and development of methods for turning seafloor into farmable land, are rethinking the wisdom of their approach. They began turning coastal salt marshes (one type of wetland) into dry land about 1000 years ago (Figure 10.32A). The oldest written records of dike-building date from the first half of the eleventh century. Systematic embankment to protect inhabited areas from storm surges started in the twelfth century. Around that time, the Dutch also started to reclaim the vast lowlands and swamp forests of the country's interior.

The proliferation of windmill technology in the sixteenth century accelerated land reclamation. Shallow lakes were pumped dry, a feat not possible using only gravity drainage. In the nineteenth century, steam, diesel, and then electric engines powered the pumps. A large part of the territory of modern Holland has been created by removing its former cover of water.

Peat bogs and other wetlands were converted for agriculture. Ten thousands of square miles of The Netherlands have been made "dry if not high" by reclamation. Today, the hard work of the Dutch has added significant useful real estate to their country (Figure 10.32B).

However, since the 1960s, opposition to wetlands loss has caused abandonment of several reclamation projects. Such works likely are a thing of the past. In fact, land once taken from the sea is being

(A)

FIGURE 10.31 Florida's Everglades.

(A) In this false-color image from *Landsat 5,* urban development covers the eastern coastline. Grassy wetlands (green) sweep south from rectangular sugar cane fields at the top of the photo (with red spots of subtropical forest). Dense mangrove trees (red) thrive along the southwestern coast. Cypress swamps (blue) are to the west of the grasslands. Lake Okeechobee is just visible at the top of the photo. Widely spaced crisscrossing lines are water diversion canals. *Photo: Terranova International/Photo Researchers, Inc.*

(B, C) Arrows show the drainage in the Everglades in 1871 and today. Water withdrawal for agriculture and cities, plus canals, have drastically disoriented the drainage, disrupting the delicate Everglades ecosystem.

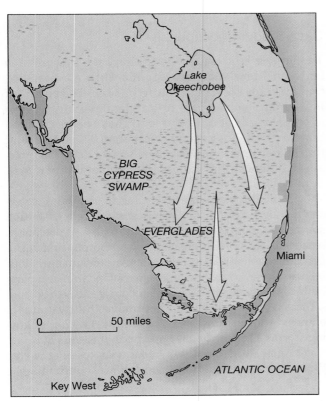

(B) 1871

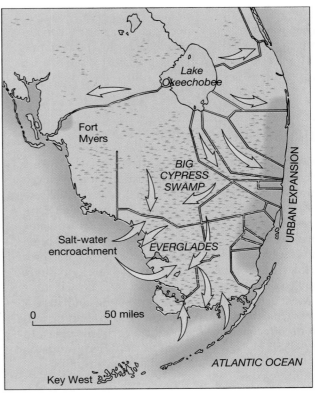

(C) LATE 20TH CENTURY

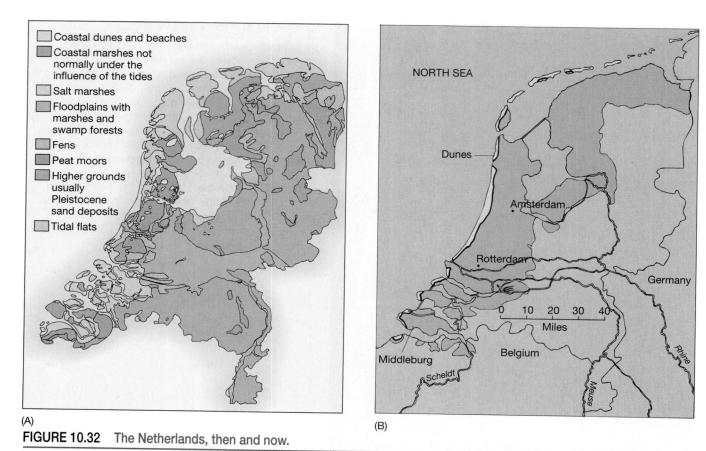

FIGURE 10.32 The Netherlands, then and now.

(A) Then: Paleographic reconstruction of The Netherlands in 1000–1200 A.D. *Source: Ambio, v. 21, 1992, p. 287.* (B) Now: The shaded area is all the land that is lower than 3 feet below mean sea level. It would be flooded if left unprotected. The Dutch have been reclaiming land from the sea for hundreds of years, building dikes to wall off the water.

changed into wetlands once again for nature conservation. In a small, populous, coastal lowland nation like The Netherlands, this is a major reversal of a centuries-old policy.

Estuaries and Pollution

An **estuary** is the lower end of a river valley where a river enters the ocean. Estuaries basically are river valleys drowned during the past 18,000 years by the sea-level rise that accompanied glacial melting. In an estuary, fresh water in the river mixes with salty seawater, creating a zone of intermediate saltiness that often extends several miles. About 850 estuaries exist in North America, from Long Island Sound (New York) to Puget Sound (Washington).

Most ocean pollution comes from the land and air via rivers that empty into the sea. As you might expect, pollution effects are most clearly seen in estuaries. In Figure 10.33, note the sources of pollution: industry, cities, suburbs, construction, and agriculture.

Case Study: Chesapeake Bay

The largest U.S. estuary is Chesapeake Bay, an area between Maryland and Virginia with a watershed that includes six states (Figure 10.34A). The bay itself is 195 miles long and from 4 to 30 miles wide, with an average depth of only 21 feet. It has been estimated that 10 percent of the bay is less than 3 feet deep and that 20 percent of it is less than 6.5 feet deep, an overall shallowness that increases its vulnerability to modern pollution.

More than one-quarter of Chesapeake Bay's watershed is agricultural, so this is a significant pollution source. Although farm acreage shrank by 30 percent in recent decades, the tonnage of phosphorous and nitrogen from commercial fertilizers and animal manure doubled and even tripled per acre of cropland in many parts of the bay's watershed. As a result, large areas of soil are saturated with nutrients. They move into waterways either by trickling underground or in surface runoff during storms. Ultimately, they end up in the estuary.

Agriculture accounts for seven to twenty-five times the nitrogen runoff from urban/suburban areas

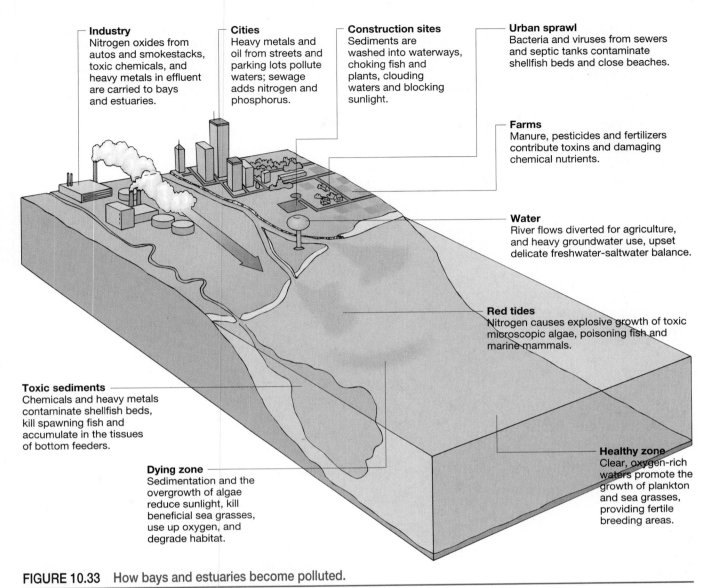

Industry
Nitrogen oxides from autos and smokestacks, toxic chemicals, and heavy metals in effluent are carried to bays and estuaries.

Cities
Heavy metals and oil from streets and parking lots pollute waters; sewage adds nitrogen and phosphorus.

Construction sites
Sediments are washed into waterways, choking fish and plants, clouding waters and blocking sunlight.

Urban sprawl
Bacteria and viruses from sewers and septic tanks contaminate shellfish beds and close beaches.

Farms
Manure, pesticides and fertilizers contribute toxins and damaging chemical nutrients.

Water
River flows diverted for agriculture, and heavy groundwater use, upset delicate freshwater-saltwater balance.

Red tides
Nitrogen causes explosive growth of toxic microscopic algae, poisoning fish and marine mammals.

Toxic sediments
Chemicals and heavy metals contaminate shellfish beds, kill spawning fish and accumulate in the tissues of bottom feeders.

Dying zone
Sedimentation and the overgrowth of algae reduce sunlight, kill beneficial sea grasses, use up oxygen, and degrade habitat.

Healthy zone
Clear, oxygen-rich waters promote the growth of plankton and sea grasses, providing fertile breeding areas.

FIGURE 10.33 How bays and estuaries become polluted.

Factories, farms, and residential areas all contribute to coastal contamination.

and ten to fifteen times the phosphorous. The result of these excess nutrients is quite predictable: large areas of the bay suffer from eutrophication (oxygen deficiency) because of excess growth of algae. Much of this excess nutrition comes from soil erosion in Pennsylvania via river transport. Once again, this demonstrates the interaction between the lithosphere and hydrosphere, and the mobility of pollution.

Pesticides are another land-based contaminant that end up in the estuary. In Maryland, which surrounds much of the Chesapeake Bay, corn is the state's largest agricultural crop, so much of the pesticide in the bay may come from corn fields.

Stormwater channeled through overloaded sewage plants carries sewage bacteria into the bay. (Recall the similar case in Milwaukee in 1993, mentioned in Chapter

8.) Interestingly, the worst violator of environmental pollution laws around the estuary is the federal government's own military bases. Currently, EPA is prohibited by law from taking action against other federal agencies. States have found it difficult to pursue federal polluters. However, the use of negative publicity in local and national media often can initiate cleanup efforts.

The least understood but potentially most damaging pollutants in Chesapeake Bay are the mind-boggling array of toxic chemicals discharged into the water by the region's industry. Toxic chemicals, including heavy metals (arsenic, cadmium, mercury) and organic compounds (DDT, PCBs), have been found throughout the bay's waters and in its bottom sediments.

Not all of this enters the bay directly from polluter's drain pipes. Some enters via the air, having come

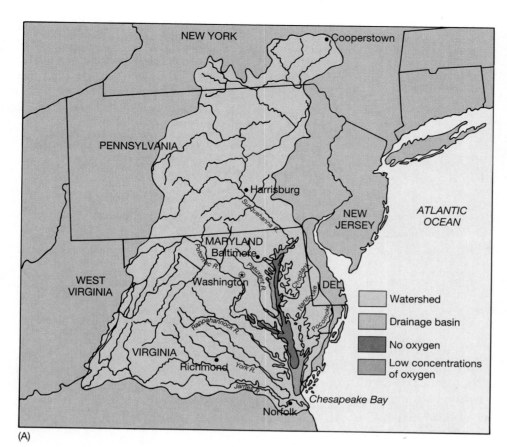

FIGURE 10.34 The Chesapeake Bay, largest estuary in the United States.

(A) Drainage basin, or watershed. (B) Both the watershed and "airshed" of Chesapeake Bay. An airshed is like a watershed, but encompasses the broader area from which winds can carry contaminants into the watershed.

(A)

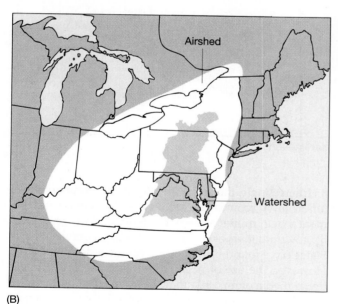

(B)

ago, and even these are dangerous to eat because of pollution. Fish and shellfish (including the bay's famous blue crabs) have been found to contain a wide variety of toxic materials in Maryland's and Virginia's rivers and bays during the past twenty years. Maryland has warned people not to eat catfish and carp (scavenger fish that suck up anything they find on the bottom) from the Baltimore area. A small amount of sediment from Baltimore Harbor is polluted enough to kill fish in the laboratory!

Consumers also have been advised to halt or limit consumption of fish from about 440 miles of rivers in the watershed above Baltimore, in Pennsylvania, New York, and Maryland. Mercury is the culprit for 193 stream miles, PCBs for 44 miles, dioxin (from paper mills) for 90 miles, and kepone (a pesticide) for 113 miles. All of the 440 miles of streams empty into the Chesapeake Bay estuary.

Estuaries typically include a shallow bay fringed with tidal wetlands, and these areas are probably the most productive areas of organic matter on the planet (Figure 10.35). This high productivity is magnified along Chesapeake Bay, because it has an intricate shoreline that totals more than 8000 miles in length, including its islands (Figure 10.34A). No other bay on Earth has so much shoreline in proportion to its length, only 195 miles. Life concentrates in the warm, nutrient-

from more distant points in the bay's "airshed." An airshed is like a watershed, but encompasses the broader area from which winds can carry contaminants into the watershed. Figure 10.34B shows the airshed.

Fishermen in the bay now harvest only 1 percent of the oysters they pulled from the waters a century

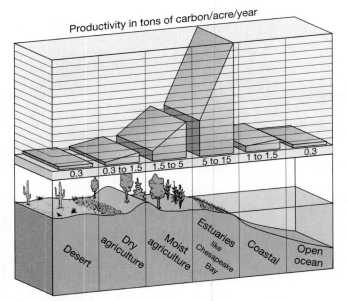

Productivity in tons of carbon/acre/year

0.3 0.3 to 1.5 1.5 to 5 5 to 15 1 to 1.5 0.3

Desert | Dry agriculture | Moist agriculture | Estuaries like Chesapeake Bay | Coastal | Open ocean

FIGURE 10.35 Productivity of the land-water edge.

Productivity refers to the growth or "production" of plants. The biological productivity of estuaries, measured as tons of carbon, is among the highest of any Earth environment.

rich, protective marsh and tidal flats where land and water meet.

From a commercial perspective, the place where land and water meet is known as *waterfront real estate*. It is every bit as attractive to people as it is to crabs, ducks, and rockfish. Nearly two-thirds of all Americans live in the 50-mile-wide ribbon of land that borders the Atlantic, Pacific, Gulf of Mexico, and Great Lakes.

In the Chesapeake Bay region, power plants seek the edge of the bay and its tributary rivers for cooling water; sewage plants locate there for a ready place to dilute treated waste; marinas want expansion to accommodate a growing demand for boat access. In Maryland, a survey in the early 1980s showed that nearly 20 percent of all development activity in the state was occurring within approximately a thousand feet of the edge of the bay and its tidal rivers. In response, Maryland enacted Critical Area Laws that sharply restrict development within 1000 feet of the bay and its tributary tidal streams.

The competition for use of the edge of the sea is great. In this conflict, Chesapeake Bay's long, productive, and embattled shorelines are second to none.

Ocean Pollution

In the United States, lobsters containing up to 20 times the allowable limit of PCBs [a highly toxic organic chemical] have been caught off the Massachusetts coast. Fish with tumors from unknown sources are also

being caught with greater frequency along the eastern seaboard of the United States, as are fish in Florida with high levels of mercury.

Worldwatch, 1989

Until the last few decades, we got away with using the ocean as a giant open sewer (see chapter opening photo). Its vastness seemed able to absorb endless waste, either dispersing it to harmless levels, or storing it for eternity on the seafloor, out of harm's way. But this assumption now is recognized as untrue, because even the ocean has limits. Directly or indirectly, every country now pollutes the sea, largely through river estuaries, as shown in Figure 10.33. Other pollutants come from ships at sea, intentionally or not: solid waste, oil spills, and radioactive waste.

Can any of these pollutants be controlled? How will such controls affect American industry? If other nations do not control their ocean dumping, why should we handicap American industry by controlling dumping here? Can we balance profitability, pollution, and human health?

Case Study: Polluting the Mediterranean Sea

The Mediterranean Sea demonstrates what can happen when ocean pollution continues unchecked for long periods. The Mediterranean is particularly vulnerable to pollution. It is an enclosed sea, and it has plentiful polluters: heavy oil tanker traffic between the Middle East and Europe; heavy industry along rivers that drain into the sea; some 130 million residents living along its 30,000-mile coastline; and a population that nearly doubles annually each summer with 100 million tourists. Observe the drainage into the Mediterranean from all sides in Figure 10.36.

How much *untreated* sewage pours into the Mediterranean? More than 500 million tons per year, or 80 percent of all sewage generated by surrounding countries! Coastal pollution is so bad that dozens of Italian beaches and 10 percent of French beaches fail the cleanliness standards established in 1975 by the European Union (Figure 10.37). And the European standards are much more lenient than those of the EPA in the United States.

For example, the American standard for coliform bacteria (from fecal matter) is 400 bacteria per 100 milliliters (20 teaspoons) of water. Italian, French, and Spanish standards permit 1000 bacteria per 100 milliliters. Among Mediterranean nations, only Israel has adopted the more rigorous American standard and is able to maintain clean beaches. According to Zeev Fish (yes, that's his name), the chief beach inspector for the Israeli Health Ministry, most Israeli beaches in May, 1995 had bacterial counts of less than 100 per 100 milliliters of water.

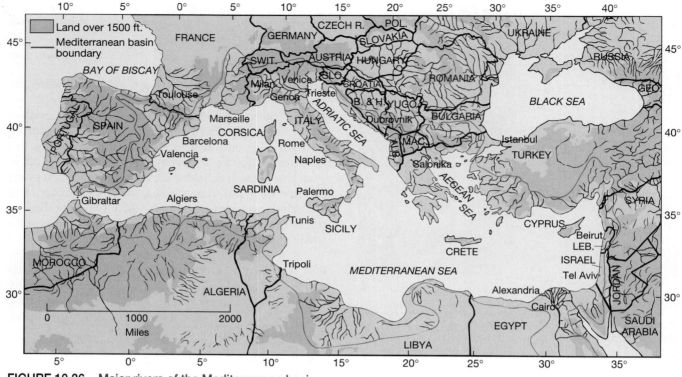

FIGURE 10.36 Major rivers of the Mediterranean basin.

Sewage is only part of the Mediterranean's pollution problem. Each year, human activities on land contribute 130,000 tons of mineral oil, 70,000 tons of detergent, 110,000 tons of mercury, 4200 tons of lead, and 4000 tons of phosphates. Phosphates and nitrates from fertilizer and sewage have resulted in explosive eutrophication in calm, warm Mediterranean waters. Plastic trash, which takes hundreds of years to disintegrate, continues to accumulate. Up to 33 million gallons of crude oil are dumped from ships, it is estimated, often because harbors lack the facilities to collect waste oil or to clean storage tanks.

Three-quarters of the land-based pollution comes from France, Spain, and Italy. Thus, mercury contamination of fish caught off the Mediterranean coast of these countries commonly exceeds even their own liberal standards. The pollution has caused the unchecked growth of weeds that are killing sea-bottom plants where hundreds of fish species spawn and feed. Clearly, what was a pollution problem in the Mediterranean Sea now is a pollution crisis.

Can anything be done to reverse the ominous progression toward the death of this important body of water upon which millions depend for livelihood and recreation? In 1975, governments of the region agreed to a Mediterranean Action Plan (MAP) to clean up the sea and protect its ecosystems. But, by 1985, the plan was faltering, so the nations set "ten priority objectives to be achieved by 1995." These included sewage-treatment facilities for all cities with a population over 100,000, new efforts to cut industrial pollution, protection of endangered marine species, and establishment of fifty new marine and coastal conservation sites.

However, by the early 1990s, MAP was near collapse, with major participants failing to fund the projects. As of 1996, the continued functioning of MAP is in question. Many countries bordering the Mediterranean Sea apparently do not give high priority to the cleanup of their common sewer.

Open-Ocean Dumping

An estimated 6 million tons of garbage is dumped in the open ocean each year from passenger, recreational, military, and commercial ships. The United States is responsible for one-third of this input. One study found quite a variety of items washed ashore on coastal beaches in 1994, and it is an interesting reflection of modern civilization (Table 10.7). Items found included a bag of undelivered mail, two kitchen sinks, a Rolex watch, and a Pontiac Fiero. In past years, even human body parts have washed onto beaches.

Cruise ships account for up to 15 percent of the 7 million tons of waste that is illegally dumped at sea each

FIGURE 10.37 Beach or garbage dump?

Polluted Mediterranean Sea shore. *Photo: Gianni Tortoli/Photo Researchers, Inc.*

TABLE 10.7 Types of Litter Found on U.S. Beaches

The Top Twelve		
	Item	*Total number*
1	Cigarette butts	1,283,718
2	Plastic pieces	354,689
3	Foam plastic pieces	301,793
4	Plastic food bags	259,143
5	Paper pieces	244,468
6	Glass pieces	242,256
7	Plastic lids	215,822
8	Glass bottles	195,503
9	Metal drink cans	182,878
10	Plastic straws	179,986
11	Plastic bottles	140,971
12	Metal bottle tops	125,826

Source: Center for Marine Conservation, 1995

year, according to the Washington-based Center for Marine Conservation. Cruise ships are allowed to dump some garbage at sea, but materials such as glass and metal first must be crushed into one-inch cubes by the trash-disposal system aboard the ship. In 1993, the U.S. government began to prosecute illegal dumping that occurs within 175 miles of our shoreline. Within 10.4 miles of shore, no dumping of any kind is permitted, including any plastic, even garbage bags. The record fine so far was paid by Princess Cruise Lines: $500,000.

An estimated one million birds and 100,000 turtles and marine mammals such as whales and dolphins die each year from eating plastic bags and pellets, or are choked to death by plastic six-pack rings. The plastic is not biodegradable and does not disappear. As noted by the Center for Marine Conservation, "If Columbus had flipped a six-pack ring off the side of the *Santa Maria* during his voyage to America 400 years ago, it could float to shore today."

Crude Oil

Approximately 500 million gallons of petroleum hydrocarbons enter the ocean each year. More than half comes from tankers (including large **oil spills**). Most of the rest is from municipal and industrial waste and runoff. Only 10 percent of the total is from natural sources such as oil seeps in the seafloor.

The bulk of the petroleum occurs on the ocean surface as oil slicks, mostly in the Atlantic Ocean. This is the area of heaviest tanker traffic, traffic that consists of Middle Eastern oil being shipped to western Europe and North America. Transported by currents, slicks may travel very long distances and still be visible (Figure 10.38A).

Our experience with large oil spills varies with the setting. Figure 10.38C shows the locations of visible oil slicks in 1985, a typical year. In the open ocean, oil slicks are dispersed gradually by waves and currents. The oil is partially decomposed by sunlight and atmospheric oxygen. On exposed rocky coasts with much wave action, most of the oil disappears within a year. On quieter beaches, the oil persists for two or three years, soaks downward into pore spaces in the sand, and becomes mixed with the sand by gentle

(A)

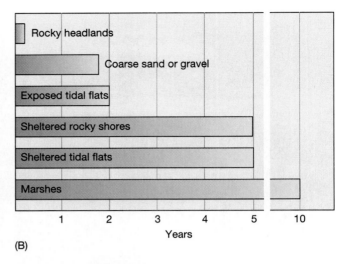

(B)

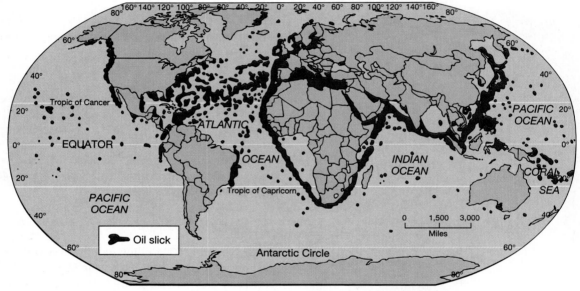

(C)

FIGURE 10.38 Consequences of oil-tanker accidents.

(A) The *Sea Empress,* aground along the shore of Wales in 1996, pours 65,000 tons of crude oil into the sea. This is double the quantity of oil spilled by the tanker *Exxon Valdez* in Alaska in 1989. Seabirds, fishing, beaches, and tourism all suffered great losses. The scope of this disaster was truly international: the ship was owned by a Norwegian, registered in Cyprus, managed from Scotland, chartered by the French, crewed by Russians, flew a Liberian flag, carried American cargo, and damaged the environment of Wales, according to *Newsweek. Photo: AP/Wide World Photos.* (B) This chart shows how long oil persists in various natural environments. *Source: 1987 Oil Spill Conference Proceedings, API Publication 4452, p. 525. Reprinted courtesy of the American Petroleum Institute.* (C) Map shows the location of visible oil slicks in the world's ocean in 1985, a typical year.

wave action. Figure 10.38B shows how long oil persists in various settings.

The most serious nearshore damage from oil occurs in wetlands, where the marsh vegetation as well as the fish and bird populations may be wiped out, although both the vegetation and wildlife return within a few years. Before they return, however, local industries that depend on these animals for their existence may be in desperate straits. Tidal and intertidal ecosystems may suffer the most long-term damage.

There are several ways to deal with the pollution risk from oceanic oil tankers. The best way is to use alternative sources of energy (see Chapters 15 and 16). However, our petroleum-based economy will not change quickly. More realistic approaches are to decrease the chance of an oil spill by requiring double-hulled tankers and to improve cleanup methods.

Many techniques have been tried to scrub oil from beaches and the nearshore. These include booms to control the spread of the spill (Figure 10.39) and vacuum

FIGURE 10.39 Oil boom.

Spill response training required by the Oil Pollution Act of 1990 was tested in 1994 off the coast of San Juan, Puerto Rico, when the barge *Morris Berman* ran aground on a reef. A potential 1.5 million gallon spill was reduced to a 300,000 gallon spill, and containment booms limited the spread of oil around the barge. The barge was subsequently sunk so that the cold bottom waters would solidify any remaining oil. Cleanup cost was estimated at more than $30 million. Photo shows containment boom keeping the oil (top two-thirds of photo) from spreading. *Photo courtesy of Crowley Maritime Corporation.*

cleaners and adsorbents to remove the oil. Burning, sinking, and dispersing spills have been attempted, as has the innoculation of the spill with oil-eating microorganisms. Each method's success depends on wave intensity offshore, the local climate, and the degree of isolation of the area where the spill is located.

In extreme cases of beach pollution, people have been employed to clean gravel stone-by-stone along miles of beach. Such was the case at Prince William Sound, Alaska, following the grounding and cracking of the single-hulled tanker *Exxon Valdez* in 1989. It spilled more than 180 million gallons of crude oil. This highlights the need to develop new, automated methods for dealing with the tanker spills that are inevitable in our petroleum-based world economy.

A novel method of oil cleanup has emerged from a British university. Researchers there noticed that adding starch to crude oil and applying a high-voltage electric current made the oily mixture become solid.

In practice, when oil is spilled on the sea, a helicopter drops starch into the slick and then lowers a metal cage into the mix. The cage is connected to a high-voltage power supply in the helicopter. In less than a second, the oil solidifies and the helicopter can carry off the cage with the solid oil inside to a nearby barge. The pilot stops the current and the oil liquifies, falling into the barge.

Radioactive Waste

Of greater long-term significance to ocean pollution is the dumping of low-level radioactive waste. Radioactive waste is classed as high-level or low-level. High-level waste radiates intensely, and refers to the fuel rods in nuclear reactors. Low-level waste is far less dangerous, but is radioactive nonetheless. **Radioactive dumping** involves sinking materials contaminated with low-level radioactive elements, like rubber gloves and containers, in the ocean. Until 1993, it was permissible to dump barrels of low-level waste into the deep sea.

Between 1971 and 1982, the United Kingdom, Belgium, The Netherlands, and Switzerland dumped low-level nuclear-industry waste in the Atlantic, 1300 miles southwest of England. Almost 82,000 tons were dumped into water 14,600 feet deep (2.75 miles), on the assumption that it would never pose a danger to humans at such a remote site. More than 90 percent of the radioactivity in the waste came from the United Kingdom, which opposed a voluntary ban in 1983. They stopped dumping only because British seamen refused to handle the cargo. Figure 10.40 shows waste dumped by Britain.

The dump sites are monitored by the International Atomic Energy Agency in Monaco. Recent water sampling has revealed that plutonium is leaking from the drums at the seafloor. Both sediment and sea life at the site are being contaminated and, of course, both are mobile and spreading the contamination. The concentrations of plutonium in seawater at the dump site are three to seven times higher than elsewhere. Other radioactive isotopes in the water have similar enrichments. The leakage problem is not yet considered dangerous but clearly should be continually monitored. Eventually, it will spread via circulating ocean currents and through the food chain.

Unfortunately, it appears that monitoring soon will end. European nations that pay the cost of monitoring believe that the leakage is insignificant and want to use the money for other purposes. The International Atomic Energy Agency and the German Fisheries Institute disagree and hope to organize another cruise to take water samples in 1996.

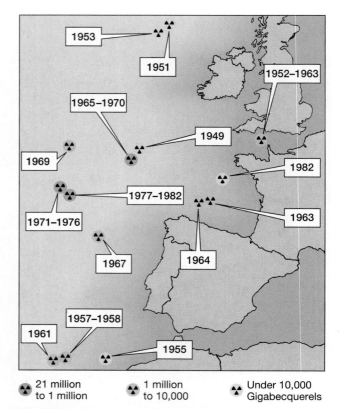

FIGURE 10.40   Gone but not forgotten.

Radioactive waste dumped by the United Kingdom between 1949 and 1982, in gigabecquerels. A gigabecquerel is one billion becquerels. A becquerel is one radioactive decay event per second, so the becquerel is a measure of radioactive intensity. The unit is named for French physicist Henri Becquerel, who discovered radioactivity.

Summary

The oceans cover 71 percent of Earth's surface and thus have a powerful effect on environments everywhere. They strongly influence the weather, the shape of shorelines, and how close to the shore it is safe to live. Sea level is rising worldwide and is unlikely to stop in the foreseeable future. Change in sea level cannot be prevented by our actions, but we may increase the change by removing petroleum and groundwater near coastlines, causing the land to subside. We have no choice but to accommodate our lives to the inevitable rise of the sea. Coastal flooding will increase gradually during the coming decades. If we are prudent, we will stop building within a mile or so of the present shoreline, but those who develop real estate probably will not be prudent.

Ocean waves are caused by, and directed by, the winds that blow across them in deep water, and by the shape of the seafloor in shallow water near the coast. Waves and the surface currents they spawn erode the shore. Coastal storms and hurricanes drive large amounts of ocean water into coastal areas, causing billions in damage annually. On the Pacific Coast, seismic sea waves from undersea earthquakes threaten low-lying areas.

The zone where land, ocean, and atmosphere meet is one of the most productive areas for organic matter on Earth and deserves special care and protection. Unfortunately, it is also much admired by humans for scenery and recreation. When the ecological needs of the environment and the material comforts of people are in conflict, ecology loses. Wetlands are being lost and nearshore areas increasingly polluted. These events are obvious in estuaries but are present along nonestuarine coastlines as well. Preserving the assets of estuarine settings will be very difficult, particularly for estuaries that lie in more than one state.

Key Terms

barrier island
breakers
breakwater
coast
coastal erosion
El Niño
estuary
fetch
groin
ground subsidence
Gulf Stream
hurricanes

hurricane warning
hurricane watch
jetty
longshore current
nearshore zone
ocean currents
ocean waves
oil spills
radioactive dumping
rip current
sand replenishment
sea level

seawall
shoreline
tides
tidal bulge
tidal flat
tidal current
tsunami
upwelling
wavelength
wetlands

Stop and Think!

1. What factors determine the location of the shoreline?

2. The oceans are all connected, yet the world's shorelines do not move uniformly. Give reasons for this.

3. Explain the behavior of waves as they approach the shoreline. Why do they always work to straighten an irregular shoreline?

4. Why do the eastern and western coasts of the United States respond so differently to the attack of ocean waves?

5. Explain how a longshore current originates and describe how it affects the shoreline.

6. What causes tidal currents? Why are they so predictable that tidal current tables can be printed telling, to the minute, when the tide will rise and fall at a particular point along a coast?

7. Explain how ocean currents affect temperatures along coasts, such as in the North Atlantic Ocean at high latitudes.

8. What is an El Niño? How does it affect human societies?

9. What meteorologic conditions produce hurricanes? Describe what happens in the coastal zone as a hurricane approaches the shoreline.

10. What is the difference between a *hurricane watch* and a *hurricane warning*?

11. How do tsunami originate? Why are they concentrated in the Pacific Ocean rather than in the Atlantic?

12. The major causes of shoreline migration are beyond human control. How do you think coastal residents should respond to this fact? Who should bear the cost of an inappropriate response?

13. List things that people do, both at the surface and in the subsurface, that affect the position of the shoreline.

14. What are wetlands? Why are they so important to humans and why are they disappearing?

15. What is an estuary? Why are they commonly some of the most polluted areas along a coastline?

References and Suggested Readings

Bird, E. C. F. 1993. *Submerging Coasts.* New York: John Wiley & Sons, 184 pp.

Cobb, C. E. 1993. Bangladesh: When the water comes. *National Geographic,* v. 183, p. 118–134.

Davidson, Keay. 1995. El Niño strikes again. *Earth,* June, 24–33.

Doerffer, J. W. 1992. *Oil Spill Response in the Marine Environment.* New York: Pergamon Press, 391 pp.

Dudley, W. C. and M. Lee. 1988. *Tsunami.* Honolulu: University of Hawaii Press, 152 pp.

Hodgson, Bryan. 1990. Alaska's big spill: Can the wilderness heal? *National Geographic,* January, p. 5–43.

Horton, Tom. 1993. Hanging in the balance: Chesapeake Bay. *National Geographic,* v. 183, p. 2–35.

Kusler, J. A., W. J. Mitsch, and J. S. Larson. 1994. Wetlands. *Scientific American,* January, p. 64B–70.

List, J. H. et al. 1994. *Louisiana Barrier Island Erosion Study— Atlas of Sea-Floor Changes from 1878 to 1989.* U.S. Geological Survey Miscellaneous Investigations Series I-2150B, 81 pp.

Pilkey, O. H., Jr. and W. J. Neal (eds.). 1978 onward. *Living with the Shore* series. Durham, North Carolina: Duke University Press. (A series of paperback books for lay people dealing with shoreline development and environmental problems along various segments of the nation's coastline.)

Pilkey, O., Sr., W. D. Pilkey, O. Pilkey, Jr., and W. J. Neal. 1983. *Coastal Design: A Guide for Builders, Planners, and Home Owners.* New York: Van Nostrand Reinhold Company, 224 pp.

Platt, R. H., T. Bently, and H. C. Miller. 1992. The failings of U.S. coastal erosion policy. *Environment,* v. 35, July, p. 7–10.

Williams, Jack. 1992. *The Weather Book.* New York: Random House, 212 p.

Williams, S. J., K. Dodd, and K. K. Gohn. 1990. Coasts in crisis. *U.S. Geological Survey Circular* 1075, 32 pp.

Wolff, W. J. 1992. The end of a tradition: 100 years of embankment and reclamation of wetlands in the Netherlands. *Ambio,* v. 21, p. 287–291.

Young, Patrick. 1996. The "new science" of wetland restoration. *Environmental Science and Technology,* v. 30, no. 7, p. 292A–296A.

"I don't know why people worry so much about earthquakes. They are such a small hazard compared to things like traffic."
Geologist Charles F. Richter

Earthquakes and Plate Tectonics

11

Dr. Richter, quoted at left, died in 1985 at age eighty-five from neither an earthquake nor a traffic accident. He was one of his generation's foremost students of earthquakes. He viewed earthquakes and traffic statistically: you are far more likely to have a traffic accident than to experience an earthquake. However, a key difference is that we generally believe—correctly or not—that we have some control over traffic safety. We can drive carefully and alertly and soberly, sharply decreasing our odds of a traffic accident.

But earthquakes are a different story. With our present technology, their occurrence is only generally predictable—some percentage chance over a period of months or years—and their power to destroy is beyond our control. Earthquakes are probably the scariest natural phenomenon that people encounter. (Lightning, tornadoes, hurricanes, and volcanoes certainly are frightening, but they typically give warning signals and therefore can be predicted.)

Most major U.S. earthquakes occur in California and Alaska. However, two of the largest on record occurred in Missouri and South Carolina. Outside the United States, Japan and the Middle East are prime quake areas. This distribution clearly is not random. Is there a discernable pattern to worldwide earthquake occurrence? If so, why? Is any place on Earth immune to earthquakes? Exactly how dangerous is an earthquake? Is the amount of destruction from an earthquake related to the sediment or rock on which buildings are constructed? Can earthquakes be predicted?

Where Do Earthquakes Occur?

Everyone knows that California is earthquake country. More than 500 earthquakes strong enough to cause damage have occurred in the Golden State or near its borders during this century. This is about one every seventy days, on the average. Although tremors are reported from the majority of states, the West Coast is the clear leader. Why? Do other countries also find that their earthquakes are concentrated in one region and are rare in others? Are there any countries that never experience devastating major earthquakes?

Earthquake locations are quite restricted. Almost all can be traced along three belts (Figure 11.1):

◄
Damage from the 1995 earthquake in Kobe, Japan.
Photo: Shinya Inui/Friday/Sygma.

FIGURE 11.1 World distribution of earthquakes.

Earthquake epicenters shown with Earth's crustal plate boundaries. Is there a connection? *Source: American Institute of Professional Geologists.*

1. 80 percent of the world's earthquake energy is released around the Pacific rim, from Chile northward through Central America, the western United States, Canada, Alaska, Japan, Indonesia, and New Zealand.

2. 18 percent of the energy is released along an east–west zone from the Alps in southern Europe through the Middle East and into the Himalaya Mountains of northern India.

3. 2 percent of the energy is released along thin but well-defined zones running through the central parts of the world oceans (for example, the line of quakes that bisects the Atlantic Ocean).

Obviously, earthquake distribution is not random. Scandinavia has almost no significant tremors, whereas Indonesia seems to have nothing but earthquakes. Why?

Earthquakes at Fault

When investigators visit an area where a large earthquake has struck, what do they find? In an urban area, they find the expected damage to people, buildings, automobiles, highways, and utilities (including downed electric, phone, and TV cables; broken water, gas, and sewer lines). Figure 11.2 illustrates earthquake damage in a city.

In unpopulated regions, investigators may spot where Earth's crust has moved a few inches or feet on opposite sides of a crack; the crack may extend for miles (Figure 11.3). The crack is the surface trace of a **fault.** A fault is a crack in rocks along which movement has occurred.

Earth's crust is its "outer shell," about 20–40 miles thick. Faults form because rocks in the upper miles of the crust are brittle and the crust always is under **stress.** The stress may result from movements of molten rock tens of miles beneath the surface, or from forces related to Earth's rotation on its axis, or from other causes. From time to time, the accumulation of stress becomes too great for the rocks to withstand. They break and move to relieve the stress. This break is a fault (Figure 11.4).

Once a fault has formed and the stress is relieved, the stress rebuilds. But it cannot be released continually because of friction between the rock surfaces on the two sides of the fault. This friction prevents smooth, continuous movement. Stress has to build up until enough energy is stored to overcome the friction, and allow the rocks to move abruptly. Hence, movement along faults is jerky. It also is unpredictable, given the state of knowledge in the final years of the twentieth century.

An **earthquake** is the sudden release of pent-up energy that has accumulated along a fault surface. The

FIGURE 11.2 Urban earthquake damage.

Numerous cars were crushed as this apartment building collapsed in Canoga Park, California, 3 miles west of the Northridge earthquake's epicenter in 1994. The large first-story openings weakened this building so that it broke easily during the side-to-side motion generated by the earthquake waves. This was true for most of the 40-some apartment buildings that collapsed during this magnitude-6.7 earthquake. *Photo courtesy P. W. Weigand.*

spot underground where the break occurs is called the **focus.** The point on the surface directly above the focus is called the **epicenter** (Figure 11.5). Because existing breaks in rocks (faults) have less resistance to stress than unbroken rocks, succeeding crustal movements to relieve stress tend to occur along established faults. This is the reason fault movements (and earthquakes) occur repeatedly along the same broken rock surface, like the San Andreas fault in California (Figure 11.6), which has been moving in fits and starts for many millions of years. It is important to note that the movement which causes a tremor occurs only along a small section of the fault at any one time, not along its entire length.

Although movement on a large fault eventually may total many miles—perhaps a few thousand, as is the case with the San Andreas—this total displacement is the sum of a very large number of abrupt little slips. The amount of displacement that occurs along a fault during a single earthquake varies widely. Most fault displacements are measured in inches, but displacements of tens of feet occur occasionally. An example is Yakutat Bay, Alaska, where a stretch of the shore suddenly was lifted up to 40 feet above sea level in 1899. Imagine standing on the upthrown side of this fault when that happened: a reverse bungee jump! In general, the greater the magnitude of the earthquake, the greater will be the displacement along the fault, although this correlation is far from perfect.

Figure 11.7 shows the three main types of faults—normal, reverse, and strike-slip. Please refer to this diagram for the following discussion.

FIGURE 11.3 California has its faults.

The crack running horizontally across the middle is the San Andreas Fault. The low sun angle enhances features, including the offset streams, linear ridges and valleys, and linear scarps (cliffs) associated with recent faulting. The far side of the fault has moved to the right faster than the stream channels could adjust, explaining the jogs. This aerial photograph looks west–southwestward into San Luis Obispo County, near Fellows, California. *Photo: R. D. Borchert/U.S. Geological Survey.*

Normal faults are caused by tensional stresses (stretching forces) that pull apart Earth's crust. Figure 11.7 shows how normal faults happen, and Figure 11.8 offers a real-world example. Note in the diagram that the arrows on either side of the fault surface show only relative movement; they do not indicate whether the left side rose or the right side dropped. Usually, we can't tell; we see only the result.

Normal faults are very common in the Basin and Range Province of the western United States and are the cause of the repeated up-and-down topography of this enormous region (Figure 11.9). The region is dominated by upthrown and downthrown blocks bounded by normal faults.

These faults are called "normal," but there is nothing "abnormal" about the other two types of faults. This

FIGURE 11.4 Under stress.

Stress buildup to the point of rupture and movement along a fault. *After U.S. Geological Survey Circular 1045, 1989.*

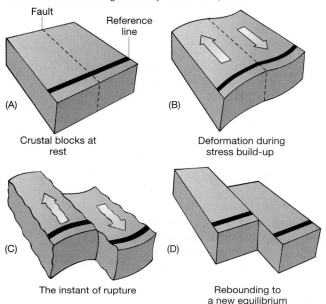

Fault

Reference line

(A) Crustal blocks at rest

(B) Deformation during stress build-up

(C) The instant of rupture

(D) Rebounding to a new equilibrium

FIGURE 11.5 ' Focus and epicenter of an earthquake.

The focus of a quake is the point underground where movement begins along a fault. The epicenter is the point on Earth's surface directly above the focus.

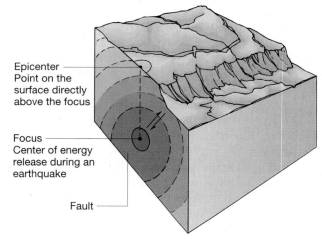

Epicenter
Point on the surface directly above the focus

Focus
Center of energy release during an earthquake

Fault

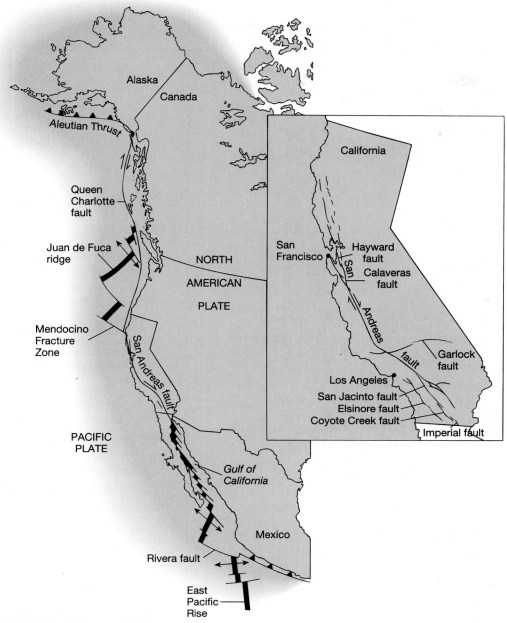

FIGURE 11.6 A famous fault.

The San Andreas fault system in context, at the junction of the Pacific plate and North American plate. The inset shows a few of the many strike-slip faults that are part of the fault system carving up the western part of the state. *After E. J. Tarbuck and F. K. Lutgens, 1994,* Earth Science, *7th ed., New York: Macmillan, p. 287.*

odd name arose because this type of fault is most common in British coal mines, where the types of faults were named in the 1800s.

Reverse faults result from the opposite movement: compression or pushing-together (Figure 11.7). They actually shorten Earth's crust a little, in contrast to normal faults, which expand the crust a bit. As with normal faults, the arrows along the surface of the reverse fault indicate only relative movement, not

which side truly moved and which held still. Very low-angle reverse faults are called **thrust faults.** Mountain ranges such as the Grand Tetons and the Appalachians are the result of thrust faults.

Strike-slip faults are those in which movement of the two sides is side-by-side, as they horizontally slide past one another (Figure 11.7). This type of fault is most common along the boundaries of the large plates that make up Earth's crust.

Block diagram	Name of fault	Definition
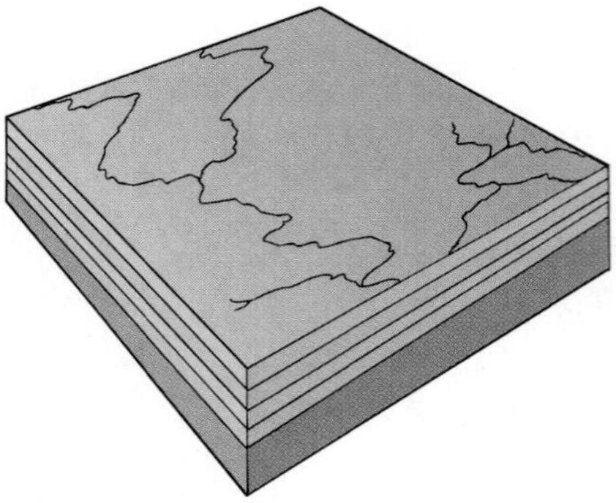		Reference block before faulting.
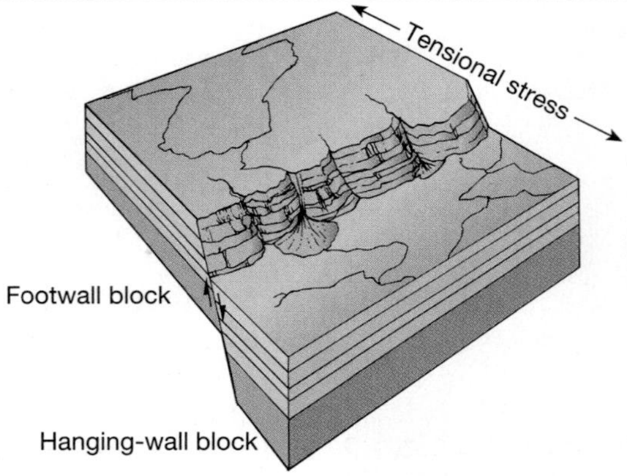 Tensional stress Footwall block Hanging-wall block	Normal fault	A "pull-apart" fault, movement vertical, generally steeply inclined, along which the hanging-wall block has dropped downward.
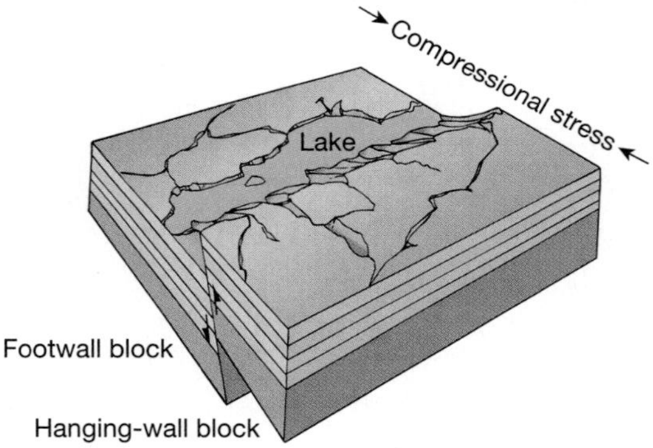 Compressional stress Lake Footwall block Hanging-wall block	Reverse fault	A "push-together" fault, movement vertical, generally steeply inclined, along which the hanging-wall block has pushed upward.
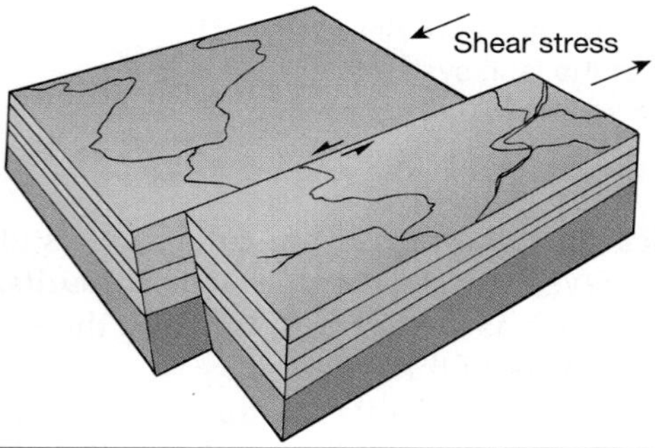 Shear stress	Strike-slip fault	A "shearing" fault, movement horizontal. Notice that horizontal strata show no vertical displacement. The fault plane is vertical so there is no hanging wall or footwall.

FIGURE 11.7 The three principal types of faults.

Note the type of movement in each: pull-apart, shove-together, or slide-past. The direction of movement along a fault usually stays the same during successive movements and does not reverse.

FIGURE 11.8 Normal fault in rocks only 10,000 years old.

In this example, near the Dead Sea, Israel, note the stretching (drag) of the dark layer on the downdropped block, a result of friction during fault movement.
Photo: A. Agnon.

Typically, strike-slip faults are hundreds or thousands of miles long. The San Andreas fault is a long strike-slip fault. It has repeatedly moved in the same direction: the west side has moved northward relative to the east side (see Figure 11.6). Strike-slip faults are prominent all around the Pacific Ocean basin.

Sometimes, investigators may not see any surface expression of a fault, because not all earthquake-producing faults intersect the surface. Some are "blind faults," hidden deeper underground. This is why you hear geologists say that some earthquakes occur along an "unknown" fault. A good example was the 1994 earthquake at Northridge, California, which occurred along a previously unidentified blind fault, surprising everyone. The common feature along all faults, at the surface or deeply buried, is that rocks move and break along either side of the crack.

Earthquake Magnitude and Intensity

The formation of a new fault or renewed movement along an existing fault causes vibrations called *earthquake waves* or **seismic waves.** The size of an earthquake wave is indicated by its height (amplitude) and wavelength, just like the water waves described in Chapter 10. The size of seismic waves reflects the amount of energy released to generate them, called **magnitude.** The damage the waves cause reflects their **intensity.** Sporadic earthquake recording began in the 1890s, which allowed the magnitude of tremors to be measured.

Richter Magnitude (Energy Released)

The well-known **Richter scale,** which describes an earthquake's magnitude, was developed by Dr. Charles F.

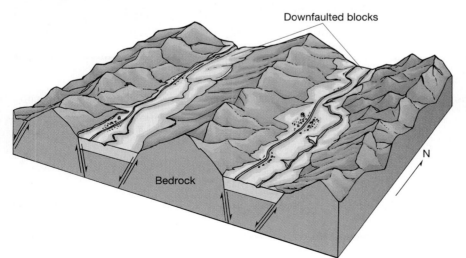

Downfaulted blocks

Bedrock

N

FIGURE 11.9 Common topography of the Basin and Range Province.

This cross-section could be through either Nevada or Utah. It shows the repeated sequence of normal faults that trend north–south, causing the alternating basin-and-range topography.

Richter, mentioned at the beginning of this chapter. The Richter scale is not linear, but logarithmic: each unit of higher magnitude indicates a seismic wave *ten times* greater than the magnitude below. This means that a magnitude-5 quake has ten times the amplitude (ground motion) of a magnitude-4 quake, and a magnitude-6 quake has ten times the amplitude of a magnitude-5 quake, and so on. Therefore, a magnitude-6 quake has an amplitude 100 times that of a magnitude-4 quake ($10 \times 10 = 100$). And a magnitude-7 quake has 1000 times the amplitude of a magnitude-4 ($10 \times 10 \times 10 = 1000$).

Thus, if you hear news of a magnitude-4 quake, it probably means minor damage, but a magnitude-6 quake is 100 times stronger, with serious damage and injuries, and a magnitude-8 quake is 10,000 times stronger, meaning a catastrophe.

This gets a bit more complicated. Whereas the *amplitude* (ground motion) of earthquake waves rises ten times for each Richter unit, the *actual energy released* by an earthquake rises about 28 times for each unit. Thus, a magnitude-6 earthquake releases nearly 800 times the energy of a magnitude-4 quake (28×28), which explains the dramatic increase in damage, and a magnitude-8 whopper releases about 600,000 times the energy ($28 \times 28 \times 28 \times 28$). To give you perspective on these amounts of energy, a 60-watt light bulb can burn you mildly, but a 60 million-watt bulb (600,000 times the energy) would make you one crispy critter.

Is there an upper limit to the energy released by an earthquake? Evidently so, for the largest recorded so far is magnitude-8.9 (Chile in 1960). Apparently, rocks cannot store any more elastic energy than this without fracturing or faulting, which releases the stored elastic energy. In summary, each Richter step represents nearly twenty-eight times the energy, and ten times the ground motion.

Table 11.1 lists the magnitudes and relative energy releases of some North American earthquakes of the past 100 years. Table 11.2 shows that small earthquakes are extremely common; hundreds of thousands occur each year as Earth's crustal rocks constantly relieve the **strain** (deformation caused by stress) from movements deep within Earth and from weight changes at the surface. Most earthquake magnitudes are small and are incremental fault displacements. Displacements of a

TABLE 11.1 Major North American Earthquakes of the Last Hundred Years

Location and Date	Magnitude	Relative Amounts of Energy Released
New Madrid, Missouri, 1811–1812[a]	8.7 (est.)	1070
Anchorage, Alaska, 1964	8.4	400
San Francisco, 1906	8.3 (est.)	270
Mexico City, 1985	8.1	144
Tehachapi, California, 1952	7.7	38
Southern California, 1992	7.5	20
Landers, California, 1992	7.4	14
Kobe, Japan, 1995	7.2	7.5
Loma Prieta, California, 1989	7.1	5.3
Northridge, California, 1994	6.8	2.0
Coalinga, California, 1983	6.7	1.4
San Fernando Valley, California, 1971[b]	6.6	1.0

[a]New Madrid earthquake included because of its notoriety.
[b]The San Fernando Valley quake is given an energy release value of 1 for comparison. At magnitude-6.6, it was a serious quake.

TABLE 11.2 Frequency of Earthquakes of Various Magnitudes on the Richter Scale and the Amount of Energy Released

Note: **The relation between energy released and earthquake magnitude is log energy = 1.44 magnitude + 5.24. In this table, magnitude-2 = 1 unit of energy released (actually 126 million joules).**

Description	Magnitude	Number Per Year	Magnitude	Relative Amount of Energy Released	Ratio
Great earthquake	over 8	1 to 2	8	440 million	28
Major earthquake	7–7.9	18	7	16 million	28
Destructive earthquake	6–6.9	120	6	580,000	26
Damaging earthquake	5–5.9	800	5	22,000	29
Minor earthquake	4–4.9	6,200	4	770	28
Smallest usually felt	3–3.9	49,000	3	28	28
Detected but not felt	2–2.9	300,000	2	1	—

Source: Data from B. Gutenberg in Earth, *2nd ed. by Frank Press and Ray Siever, 1978, W.H. Freeman and Company.*

few inches are much more common than displacements of several feet.

Most quakes with magnitudes less than 2 or 3 are not sensed by people, unless they occur directly underfoot. Thus, more than 98 percent of all earthquakes go undetected by people and are sensed only by instruments. Such an instrument is a **seismometer,** which can detect the smallest earth "seisms" or tremors—including the rumble from a passing truck. Waves from even a tiny earthquake in Indonesia or Antarctica travel thousands of miles to seismometers worldwide. (If that seems amazing, consider that American Indians used to put their ears to the ground to hear buffalo running miles away, or to railroad rails to hear a distant "iron horse" approaching. They used their ears as seismometers.)

Movements sensed by a seismometer are recorded on a paper chart or computer disk. The paper chart is called a **seismogram,** and this is what they show on TV when a major quake strikes. A seismogram is shown in Figure 11.10.

FIGURE 11.10 How a seismometer works.

(A) Seismometer recording horizontal motion. (B) Seismometer recording vertical motion. The "suspended masses" with the pens hold still during an earthquake, but everything else jiggles. Thus, the recording paper on the turning drum is moved against the motionless pen, making a recording. (C) Four types of seismic waves. Each has a different amplitude.

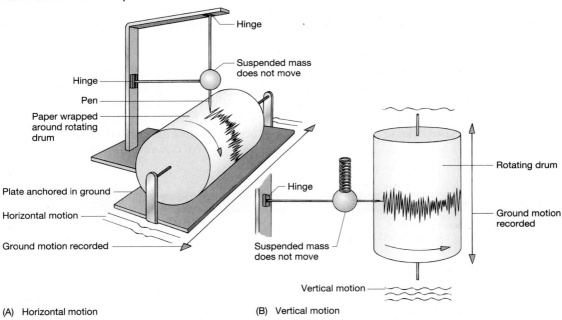

(A) Horizontal motion   (B) Vertical motion

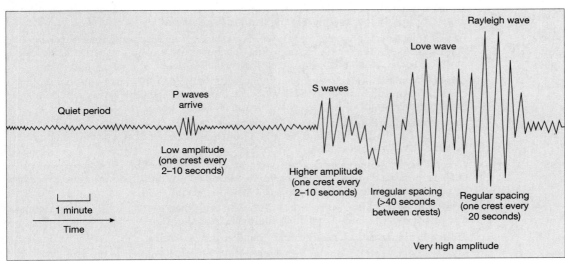

(C)

The seismometer is a simple instrument that makes use of the principle of **inertia**. This principle says that a body at rest or in motion tends to *remain* at rest or in motion. (If you have trouble getting up for your eight-o'clock class, just explain to the instructor that you are a victim of inertia.) The seismometer is anchored to bedrock, and therefore moves as Earth does during a quake. Within the seismometer is a heavy mass that is suspended by a spring or is hanging like a pendulum from the frame of the instrument. The mass has a pen in it (Figure 11.10A, B).

Now, here is where inertia comes in. When the ground moves, the inertia of the heavy mass keeps it *motionless*, while Earth and the seismometer *move in time with the seismic waves*. The mass with its pen hold perfectly still, but the recording paper it is pressed against jiggles from the quake, forcing the pen to make a wiggly line that follows the movement of Earth. This is how a seismogram is made, shown in Figure 11.10C. The stronger the ground motion, the larger the wiggles.

An extensive network of these earthquake recording devices (seismometers) exists worldwide, so that every earthquake (and nuclear bomb explosion) is recorded at many places. Within minutes after a tremor, seismometers around the world record the ground motion.

Mercalli Intensity (Damage Level)

Richter scale magnitudes are important, and are stated in newscasts. But earthquakes also can be classified by the level of damage they produce. Italian seismologist Guiseppe Mercalli devised such a damage scale, the Mercalli Intensity scale, shown in Table 11.3. We would anticipate a high degree of correlation between Richter magnitudes and Mercalli intensities in populated areas, and we are not disappointed. The correlation is useful for estimating the energy released from historic quakes that occurred before modern seismometers were developed.

Locating an Earthquake

Most of us live in well-populated areas, which gives us the impression that Earth is thoroughly covered with people. But in fact most of Earth's surface is either sparsely populated or completely vacant. (Proof: how many people live in the ocean, Antarctica, within the Arctic Circle, or in the Sahara Desert?) Consequently, most earthquakes really do occur "in the middle of nowhere," mostly beneath the ocean, and go unreported because they have no significance to the public. If a quake occurs in an unpopulated area, with no one to report it, how do we determine where it happened?

TABLE 11.3 Modified Mercalli Intensity Scale and Approximate Corresponding Magnitude

Intensity	Magnitude	Description
I	<3.4	Not felt.
II	3.5–4.2	Felt by persons at rest on upper floors.
III		Felt indoors—hanging objects swing. Vibration like passing of light trucks.
IV	4.3–4.8	Vibration like passing of heavy trucks. Standing automobiles rock. Windows, dishes, and doors rattle; wooden walls or frames may creak.
V	4.9–5.4	Felt outdoors. Sleepers wakened. Liquids disturbed, some spilled; small objects may be moved or upset; doors swing; shutters and pictures move.
VI		Felt by all; many frightened. People walk unsteadily; windows and dishes broken; objects knocked off shelves, pictures off walls. Furniture moved or overturned; weak plaster cracked. Small bells ring. Trees and bushes shaken.
VII	5.5–6.1	Difficult to stand. Furniture broken. Damage to weak materials, such as adobe; some cracking of ordinary masonry. Fall of plaster, loose bricks, and tile. Waves on ponds; water muddy; small slides along sand or gravel banks. Large bells ring.
VIII		Automobiles hard to steer on moving roadway. Damage to and partial collapse of ordinary masonry. Fall of chimneys, towers. Frame houses moved on foundations if not bolted down. Changes in flow of springs and wells.
IX	6.2–6.9	General panic. Frame structures shifted off foundations if not bolted down; frames cracked. Serious damage even to partially reinforced masonry. Underground pipes broken; reservoirs damaged. Conspicuous cracks in ground.
X	7.0–7.3	Most masonry and frame structures destroyed with their foundations. Serious damage to dams and dikes; large landslides. Railroads bent slightly.
XI	7.4–7.9	Railroads bent greatly. Underground pipelines out of service.
XII	>8.0	Damage nearly total. Large rock masses shifted; objects thrown into the air.

The precise location of earthquakes is determined by **seismologists,** scientists who study the origin, distribution, and frequency of Earth tremors. They have found that earthquakes generate four basic types of waves, and that each travels at its own speed through different types of rock. Earth's crust (its outermost 40 miles or so) is composed almost entirely of granite and basalt. By subjecting these two types of rocks to simulated earthquakes, the speed of each wave type through them was determined, and is shown in Table 11.4.

The four types of earthquake waves not only travel at different speeds, but shake the ground in different motions.

- Two wave types travel *through* Earth: **P-waves** (compressional waves) and **S-waves** (shear waves).
- Two wave types travel *on the surface:* **L-waves** (Love waves) and **Rayleigh waves.**

The properties of these waves are shown in Table 11.4. Their different effects are illustrated with a fence in Figure 11.11.

The seismograph at a recording station records incoming waves on a seismogram. It records all four types of waves (the P, S, L, and Rayleigh waves). But seismologists focus their attention on the P and S waves, because the large velocity difference between them is the key to discovering where a quake occurred. Follow any of the three traces in Figure 11.12A and you can see that the P-wave arrives first, followed by the S-wave. The farther the recording station is from the epicenter, the *greater is the time between arrivals* of the two wave types.

To find the origin of a quake, a seismologist draws a circle around the seismic station, with the circle's size (radius) corresponding to the distance from the epicenter (Figure 11.12B). This distance is determined by knowing how fast each type of wave travels. When this is done from at least three seismic stations, the point at which all three circles intersect must be the epicenter of the quake. Generally speaking, this usually is where damage is greatest. Many quakes are not at the surface, but occur several miles underground, at a point called the focus of the earthquake (Figure 11.5).

Seismometers operate continuously so that all earthquakes with magnitudes greater than 1 are detected, anywhere on Earth. As noted in Table 11.2, this amounts to several hundred thousand earthquakes per year. (Those with intensities less than 1 require special methods to detect.)

Earthquakes and Plate Tectonics

Look back at Figure 11.1 and the striking distribution of earthquakes. Why do earthquakes concentrate where they do? On land, they seem to follow mountain ranges, particularly volcanic mountain ranges. For example, look at western South America—earthquakes seem to coincide with the Andes Mountains, which are very volcanic.

But what about the strings of earthquakes that run down the middle of oceans? Look at the Atlantic, for example. The string of quakes divides the Atlantic down the middle, between South America and Africa. What is going on here?

The mystery could have been solved centuries ago, had there been a way to drain all the water from the oceans, and fly around Earth for a good look at the bare seafloor, as in Figure 11.13. This amazing view of

TABLE 11.4 Properties of Earthquake Waves

Velocities are approximate because they are faster at higher pressures (greater depths) in the crust.

Type of Wave	Wave Velocity (miles per hour)		Other Characteristics
	In Granite	In Basalt	
Compressional	13,600	15,000	P-waves: vibration is in same direction that energy travels, like waves you can see moving through a Slinky spring toy. (Gently stretch a coiled spring on a table and strike one end sharply. Pulses of energy move horizontally as the spring alternately compresses and expands.)
Shear (transverse)	9150	9400	S-waves: vibration is perpendicular to the direction that energy travels: tie one end of a rope to a post, hold the other end in your hand, and wave the rope up-and-down.
Love	8700	8900	L-waves: like shear waves, except they travel along interfaces or surfaces rather than through objects, as P-waves and S-waves do.
Rayleigh	7800	8000	Surface waves in which particle motion is an elliptical orbit, similar to the motion of water molecules in an ocean wave.

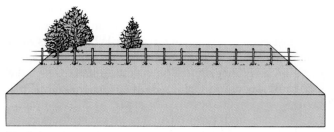

(A) Before the quake.

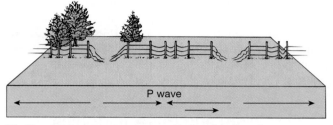

(B) P waves compress and expand in the direction of wave movement.

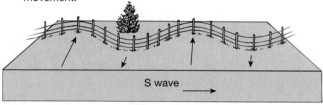

(C) S waves move back and forth at right angles to the direction of wave movement

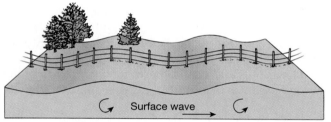

(D) Surface waves move in a circular path at the surface. The motion diminishes with depth, like that produced by surface waves in the ocean.

FIGURE 11.11 Each type of seismic wave produces a distinctive movement.

Note what happens to the fence in each case.

seafloor worldwide reveals mountains, volcanoes, and deep trenches. Until this century, the seafloor was assumed to be pretty flat; no one knew it to be so mountainous. It is thanks to modern depth-sounding equipment that the topography of the seafloor now is revealed.

In the depth-sounding technique, sound waves are sent downward from a ship, echo off the seafloor, return to the ship, and are recorded (Figure 11.14). The time required for the sound waves to travel down and back is determined. Sound waves that return quickly are being reflected from mountaintops. Sound waves that take longer are reflecting from the seafloor, or from deep trenches. The method is called *sonar* and the

recordings of these waves are called **sonograms.**

Figure 11.13 was created based on information gathered by many ships making continuous sonogram recordings in the world ocean. It answers the question about the line of earthquakes that runs down the middle of the Atlantic Ocean in Figure 11.1: they coincide with a high underwater mountain chain. In addition, the surrounding Atlantic Ocean floor on both sides of the mountain chain is sliced into elongate strips that run perpendicular to the ridge. These strips are bounded by faults, the most common source of earthquakes. The north–south slices of oceanic plate bounded by faults are best seen on land in Iceland, a huge island that is part of the Mid-Atlantic Ridge (Figure 11.15). [Note: Despite having so many earthquakes, the Atlantic has few tsunami because Mid-Atlantic Ridge quakes nearly always have low magnitudes compared to Pacific quakes.]

The seafloor mountains are volcanoes. Proof of this exists at the surface, where some of them are tall enough to rise above the water, forming the country of Iceland and the Canary Islands. These islands are simply piles of lava that managed to build above sea level from the ocean floor. The Pacific Ocean floor has a similar pattern in the volcanic Japanese Islands, Hawaiian Islands, and other Pacific islands your parents or grandparents may have visited in World War II: Guam, Tarawa, the Philippines, Solomons, and many others.

The linear mountain chains and trenches mark the boundaries of large slabs of Earth's crust known as **plates.** There are about a dozen major plates and numerous smaller ones (Figure 11.16). The remarkable thing is that all of them continually move with respect to each other. This **plate movement** is very, very slow—at most, inches per year. The plates are roughly 60 miles thick. In places, they grind past one another. In other places, like the seafloor, they are spreading apart. Elsewhere, they push together until the edge of one is forced beneath the other. In all cases, earthquakes result, and sometimes volcanoes. The mystery of the earthquake pattern in Figure 11.1 is solved.

Convection in the Mantle

Now, what makes the plates move? They are believed to be powered by slow, powerful convection currents in the partially plastic material immediately below them. This material moves by **convection,** a concept easily understood by any coffee drinker. Add creamer to hot coffee, and what do you see? The creamer swirls through the hot liquid, falling toward the bottom of the cup, and rising again. The creamer is just a visible tracer that reveals the dynamic little world of hot coffee. The coffee is hotter at the base of the cup than at the upper surface (where it is cooled by air and escaping steam). The hotter liquid at the bottom is less dense, so it bobs toward the surface like a cork. The cooler coffee at the top, being denser, falls

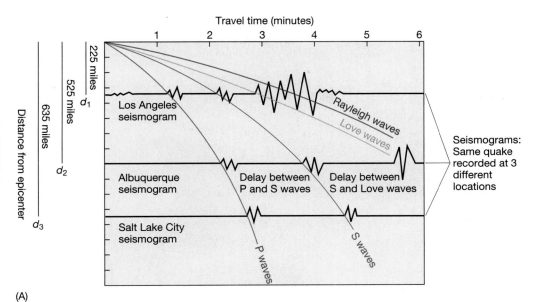

(A)

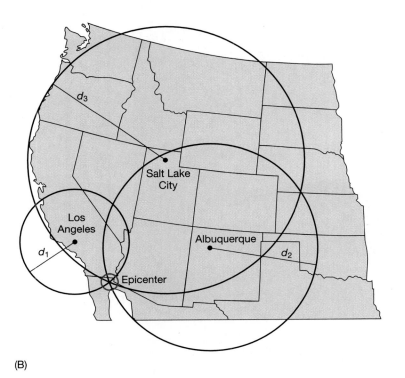

(B)

FIGURE 11.12 Travel-time curves for the four types of seismic waves.

(A) All three seismograms (the horizontal lines with the squiggles) show the same earthquake, but recorded at three different seismic stations in the western U.S. The four curves are drawn to connect the arrival time for each of the four types of seismic waves. For example, the P-wave curve connects the start of the P-wave squiggle on all three seismograms. Notice how the beginning time changes with distance from the earthquake focus. By matching the arrivals of the waves from each seismogram on the travel-time curves, the distance of each seismic station from the epicenter can be calculated. (B) Locating an earthquake epicenter requires data from three seismographs, like the ones in Part A. This process is called *triangulation,* and is the same method used to find a lost ship or aircraft that is transmitting a distress signal. A circle is drawn around each seismic station, with its radius equal to the distance from the seismic station to the epicenter. These distances correspond to $d1$, $d2$, and $d3$ at bottom in Part A. The intersection of the three circles marks the epicenter.

toward the bottom. This movement, convection, continues as long as the coffee is hot. The continual rise of hot coffee pushes the coffee at the surface aside and the cooler coffee on top sinks, continuing the cycle.

So it is within Earth, we think. The temperature deep within Earth is much greater (several thousand degrees) than at the surface. As a result, the material at depth is less dense and slowly rises toward the surface, perhaps at inches per year (Figure 11.17). It reaches the base of the solid crust, is deflected sideways—cooling and growing denser as it goes—and then returns back

toward Earth's hot interior. This zone of convecting material is termed the **mantle** and extends between the base of the crust and Earth's core. The time to complete one convection cycle—from bottom to top and back to the bottom again—is a few seconds in your coffee cup, but millions of years in Earth's mantle.

We think that continual convection in the upper portion of the mantle, a zone called the **asthenosphere,** causes the plates to move. Movement is at different rates and in different directions, just as your hot coffee swirls in different directions. The asthenospheric material is

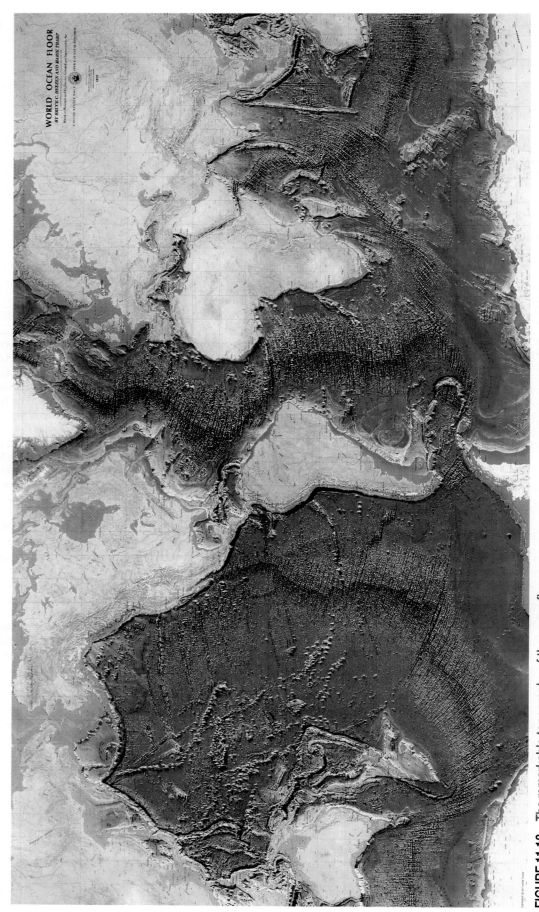

FIGURE 11.13 The remarkable topography of the ocean floor.

It is a vast area—71 percent of Earth's surface—featuring mountains, valleys, and faults, just like the continental surface. Note the 40,000-mile-long mid-ocean mountain ridge, which snakes through the Atlantic, Pacific, and Indian Oceans; it is especially clear between the Americas and Africa. *Source: World Ocean Floor map by Bruce C. Heezen and Marie Tharp, 1977. Copyright 1977 by Marie Tharp. Reproduced by permission of Marie Tharp, 1 Washington Ave., South Nyack, NY 10960.*

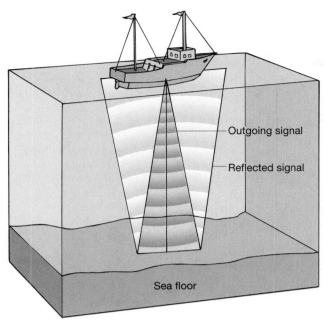

FIGURE 11.14   Sonar to map sea depths.

This sonar system works by generating sound vibrations, transmitting them to the seafloor, and recording the echoes that return. Measuring the timing of the echoes allows determination of water depth. Sound travels through water at 5000 feet per second. Therefore, water depth = 1/2(5,000 ft/sec x echo travel time). *After F. K. Lutgens and E. J. Tarbuck*, Essentials of Geology, *Merrill, 1989, p. 220.*

very thick, more than a billion times more viscous than water. Therefore, it convects very slowly—and as a result, the plates move ponderously and, until the 1960s, when modern instruments permitted measurement,

undetectably. Rates range between about 0.4 and 7 inches per year. As plates lumber along, they scrape each other on their sides or boundaries, causing the tremors we call "shallow earthquakes."

The most prominent **plate boundary** in North America is marked by the San Andreas fault. It is just east of Los Angeles and trends northward through the San Francisco peninsula (Figure 11.6). Movement along the fault is the strike-slip type, shown at the bottom of Figure 11.7. The Pacific side continually moves northward relative to the continental side, at an average 1.9 inches a year. Los Angeles eventually will become a sister city at the western edge of San Francisco.

The journey northward by Angelinos and southward by San Franciscans will not be graceful, but sporadic and jerky, as we have seen, much to the distress of those on both sides of the fault. How soon will they become sister cities? About 11 million years, long enough for everyone to get used to the idea. Will airlines decrease fares between the two cities as the travel distance shortens?

Plate Boundary Earthquakes

About 80 percent of Earth's quakes occur along the contacts between plates that form the crust. In 1995, there were fifteen earthquakes with magnitudes of 7 or greater. To North Americans, earthquakes along the West Coast are most significant, occurring at the boundary between the Pacific and North American plates. In California, an earthquake of magnitude 6–7 occurs about once every three years and a magnitude 7–8 tremor about once every 100 to 150 years.

FIGURE 11.15   Visible plate tectonics.

The Mid-Atlantic Ridge, a diverging plate boundary, surfaces above sea level in Iceland. The cracklike valley indicates that plates are being pulled apart. *Photo: Charles Preitner/Visuals Unlimited.*

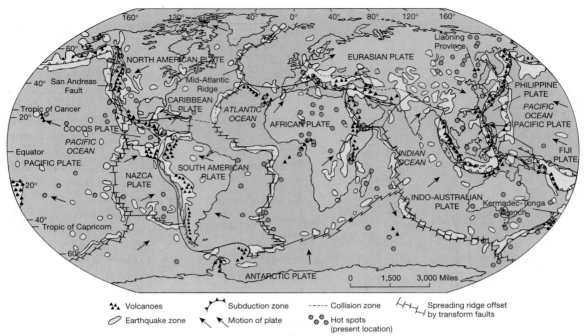

FIGURE 11.16   Presently active plate boundaries on Earth's surface.

Source: Robert W. Christopherson, 1994, Geosystems, 2nd ed., Prentice Hall.

Let us look at three major quakes of the past decade, and the boundaries along which they occurred. Two of them shook California along part of the San Andreas fault system, where the Pacific Plate is grinding against the North American Plate. The third quake occurred in Japan, where the Pacific Plate and Philippine Plate are plunging slowly beneath the Eurasian Plate (Figure 11.18).

Case Study: Loma Prieta, California Earthquake

California's San Andreas fault is not just one long crack in Earth's crust but a complex *fault system*, composed of numerous faults in the area. The entire San Francisco Bay area is sandwiched between two active major strike-slip faults (Figure 11.19). Worse, the peninsula on which the city rests is underlain largely by mud and alluvium (Figure 11.19, blowout). This is a recipe for disaster and the reason for continual concern about the "big one," the feared major quake that is certain to strike the area sometime, and is very likely to do so during the next few decades. Earthquake specialists believe that the likelihood of a magnitude-7 or larger earthquake in the Bay area by the year 2020 is two chances out of three.

A taste of what will happen arrived in 1989, as millions gathered to watch a televised World Series game from Candlestick Park in San Francisco. Shortly before the game was to begin, a Richter magnitude 7.1 quake struck beneath Loma Prieta, California. Three

seismograms of the quake are shown in Figure 11.20A. The quake lasted only 15 seconds but caused $6 billion in damage and killed 61 people. But, within 6 months, the state received 400 requests for $6000 burial grants from people claiming their dead relatives perished because of the quake.

The epicenter was 60 miles southeast of San Francisco along the San Andreas Fault. Despite the distance, most of the damage was not in Loma Prieta but in the Marina District at the tip of the San Francisco Peninsula, adjacent to the Bay (Figure 11.19). Why did the major damage occur so far from Loma Prieta?

The answer lies in the character of the materials underlying this part of California. Near the epicenter, hard rock underlies the soil. But the Marina District is underlain by soft mud, covered with artificial fill. The fill is of two types: sand (either from land nearby or from the Bay) and *hydraulic fill* (a mixture of 70 percent sand and 30 percent mud, pumped from the Bay). This fill has been accumulating since early in this century.

Almost all the damage on the peninsula occurred in areas underlain by the hydraulic fill, because the earthquake waves made it succumb to liquefaction (review: Figure 5.25). The liquefaction, followed by dramatic settlement, broke foundations, streets, sidewalks, and underground utilities. Damage to the Marina District's water system cut off water critically needed for fire-fighting after the earthquake (Figure 11.21). The identical problem occurred in the aftermath of the famous 1906 San Francisco quake, when

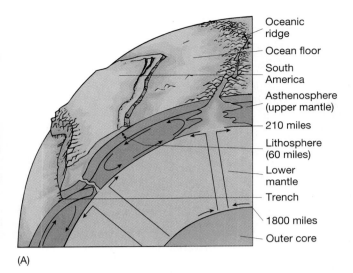

(A)

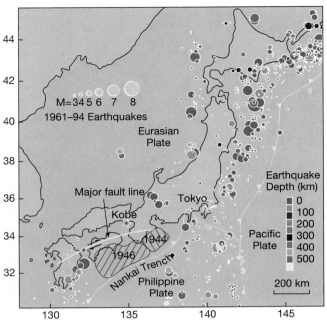

FIGURE 11.18   Earthquake activity in Japan, 1961–1994.

Also shown is the location of the 1995 Kobe earthquake. The outlined areas are the projected rupture areas of the largest historical earthquakes to shake Kobe, which were subduction-zone earthquakes in 1944 and 1946. *Figure prepared by Grant A. Marshall, U.S. Geological Survey.*

(B)

FIGURE 11.17   Earth's interior and its convection cells.

(A) The mantle has a break between its upper and lower parts, so each part convects separately. The convection of the upper mantle causes rupture of the crust, the generation of basaltic magma, and the formation of volcanic mountain chains (*Oceanic ridge*, top center). The opposite plate boundary is the oceanic trench, the line along which oceanic crust is carried back (subducted—one plate sliding under another) into the mantle (*trench*, at left). *Source: W. K. Hamblin, 1992, The Earth's Dynamic Systems, 6th ed., New York: Macmillan, p. 451.* (B) Convection in a Hawaiian crater shows plate motion in miniature. As molten lava rises by convection, older, cooler lava (darker) is spread apart, moving away from the rising material, sinking elsewhere in the crater. At center, a transform fault appears. Front of picture is about 150 feet wide. *Photo: Dr. W. K. Hamblin.*

ruptured water mains prevented fire fighters from extinguishing blazes caused by broken gas pipelines and overturned oil lamps. As in the 1989 quake, damage was greater in the Marina District in 1906 than in nearby areas.

Adding insult to injury, **aftershocks** normally occur for days or weeks following a quake. A typical aftershock pattern is shown in Figure 11.20B.

Case Study: Northridge, California Earthquake

College students certainly have no immunity to earthquakes, as some discovered at 4:30 a.m. on January 17, 1994. A 6.7-magnitude earthquake and hundreds of smaller aftershocks rocked the northwestern edge of suburban Los Angeles, killing sixty-one people. Damage at California State University—Northridge was extensive. Many water conduits ruptured, including two that deliver water to northern Los Angeles. This cut off water to large areas of the city, for up to ten days in some areas. In places, contaminated water had to be sterilized by boiling even after water service was restored. Damage from shaking and an explosion and fire at a critical electrical transformer facility cut power, which was not restored to some areas for five days. A 22-inch gas pipeline ruptured and ignited, causing a fire that consumed five houses.

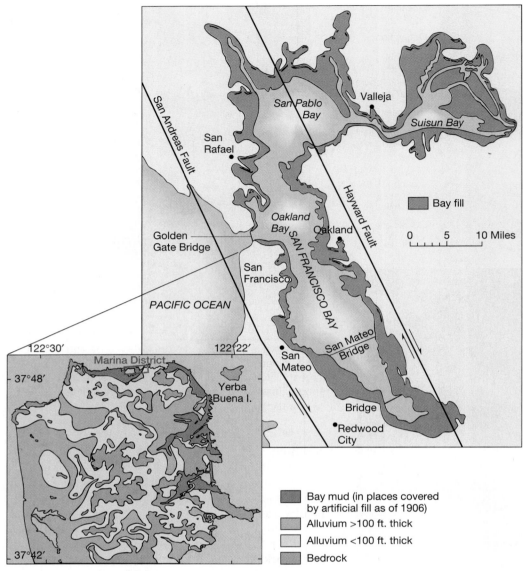

FIGURE 11.19 The unstable San Francisco Bay area.

During historic times, one-third of San Francisco Bay has been filled, either by sediment from streams or with artificial fill. Buildings atop this fill are at risk from amplified earthquake waves. Note how the Bay is flanked by two major faults. The inset shows a very general geologic map of the abundance of unstable sediment on which the city is built. Note at top the Marina District, site of major damage during the 1989 earthquake. The District is underlain by bay mud and thin alluvium. *Source: U.S. Geological Survey.*

In Los Angeles County, 86,000 structures were damaged; eleven freeway overpasses collapsed (one 30 miles from the quake epicenter); eighteen parking structures tumbled; fourteen shopping malls were severely damaged; highway disruption was extensive; and estimated losses totaled $30 billion, making this the most expensive earthquake on record (Figure 11.2).

California State University at Northridge suffered plenty of damage. The main library sustained structural injuries, and all fifty-three major buildings

on campus experienced damage that may exceed $300 million. Some science classes still are meeting in temporary trailers.

Maximum Mercalli intensity values of IX were assigned to three areas in the San Fernando Valley by the U.S. Geological Survey. Subsequent measurements revealed that the 32-mile distance between Pasadena and Palos Verdes had decreased by about half an inch—not much, but think of the energy released when it happened! The nearby Santa Susana Mountains were

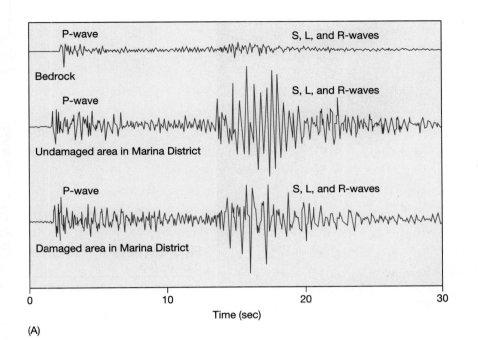

(A)

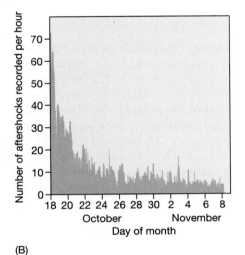

(B)

Loma Prieta aftershocks		
Magnitude	Number	Effect
5	2	Damaging
4	20	Strong
3	65	Perceptible
2	384	Not felt
1	1,855	Not felt
<1	2,434	Not felt
Total	4,760	

FIGURE 11.20 The Loma Prieta earthquake of 1989.

(A) Three seismograms of the Loma Prieta quake. The difference in amplitude of the earthquake waves in the three records reflect differences in the type of sediment and rock in each area. The Marina District is only 60 miles from the epicenter of this quake, which is too close for the S-waves, Love waves, and Rayleigh waves to be separated on the seismogram. (See Figure 11.12A for further explanation of these three wave types.) *Source: N. K. Coch and A. Ludman, 1991,* Physical Geology, *New York: Macmillan, p. 525.*

(B) Record of aftershocks following the Loma Prieta quake. Large quakes always are followed by smaller tremors over the next few days or weeks.

elevated 15 inches and moved more than 8 inches to the northwest along the fault surface.

Beyond the destruction were some medical effects of the earthquake. A study of data from seventy-two Los Angeles-area hospitals revealed 35 percent more heart attacks in the week following the tremor than in the previous week (201 versus 149). A second study examined Los Angeles-area heart-attack deaths in 1991, 1992, 1993, and 1994 during the week before and week after January 17, the date of the quake. The average of about fifteen deaths per day was sharply exceeded on the day of the earthquake, when the toll hit fifty-one. It appears that the stress of experiencing the quake pushed heart attack-prone people over the edge. The study authors suggest that, in the aftermath of a quake, physicians prescribe extra medication for people with heart disease.

A remarkable indirect health effect of the earthquake was reported by the USGS and the Centers for Disease Control. In the two months following the quake, 166 cases of "Valley fever" (the medical name is "acute Coccidioidomycosis") were reported in Ventura County, just west and downwind of Northridge. This was a dramatic, threefold increase over the fifty-three cases reported in the entire previous year. Valley fever is contracted by inhaling airborne fungus spores that live in topsoil. Apparently, numerous landslides triggered by the quake in the area's unstable sedimentary rocks generated enormous dust clouds, filling the air with the spores. Valley fever cases were most plentiful

FIGURE 11.21   Earthquake-caused fire.

Earthquakes rupture gas and electric lines, often initiating fires. This San Francisco blaze followed the Loma Prieta quake.

Photo: M. Williamson/Sygma.

directly downwind from the most concentrated landslides. Little change in incidence of the disease was reported in other parts of the Valley.

Despite the level of destruction, seismologists agree that the Northridge earthquake of 1994 was not the feared "big one"—a quake of magnitude 7 or 8—that seismologists expect to hit southern California within the next few decades. The $30 billion estimated damage from the Northridge quake may look like pocket change when we see the bill for the "big one." More importantly, imagine the loss of life, predicted to be in the thousands. Many Californians are living on shaky ground.

Case Study: Kobe, Japan Earthquake

On January 17, 1995—one year to the day following the Northridge quake—a 6.9-magnitude earthquake rocked Kobe (KOE-bee), a Japanese city the size of Philadelphia. The quake killed 6308 people, damaged or destroyed 152,000 buildings, caused about $100 billion damage, and incinerated the equivalent of seventy U.S. city blocks. Most of the movement was traced to a complex system of small strike-slip faults that run under the city, rather than to a single large fault, as is most commonly the case. Horizontal displacement overall was about 5 feet. Kobe's casualties were ninety times greater than at Northridge because the 6.7-magnitude Northridge quake directed 80 percent of its energy into the sparsely populated Santa Susana mountains rather than toward downtown Los Angeles.

Of 223 major structures in Kobe erected before 1971, 36 percent collapsed or were unsafe to enter (Figure 11.22). But none of the buildings that were completed after a 1981 building code had collapsed, and only 6 percent were unsafe (Figure 11.23). This demonstrates Japan's intensive effort to develop earthquake-resistant structures, more than any other country. Yet, despite this success at reducing earthquake damage, there were serious failures. The elevated track for the famed 150 mile-per-hour "Bullet Train" to Tokyo had been designed to withstand a magnitude-8 quake, about thirty-nine times the energy released by the Kobe quake—yet, it collapsed in nine places.

The earthquake's focus was about 12 miles southwest of downtown Kobe. Despite this distance, about 300,000 people were left homeless, about 20 percent of the population (for perspective, that is the population of Colorado Springs). Some 100,000 buildings were destroyed, but only 3 percent of Kobe's homeowners had earthquake insurance. The rigid granite beneath Kobe University protected most of its buildings, as expected, but widespread liquefaction of fill occurred in the area of reclaimed land in the bay offshore of Kobe. Liquefaction caused subsidence of up to 8 feet in places, and large volumes of mud were ejected. Lateral spreading of the soil wrecked 13 miles of wharfs around the artificial islands that form the Kobe port, through which 30 percent of Japan's shipping passes.

Governmental response to the disaster was inadequate and disappointing, according to those on the scene. Over 300 fires, a dozen of which were major conflagrations, occurred within minutes of the tremor. But emergency response was hindered by failure of the

FIGURE 11.22   Expressways toppled in Kobe.

Supporting columns, built before design codes were upgraded, proved no match for the 1995 earthquake. *Photo: N. Hosaka/Gamma-Liaison, Inc.*

water system and disruption of traffic. Of Kobe's water system, 70 percent was inoperable due to numerous breaks caused by widespread ground failure. Soldiers and fire fighters were paralyzed in miles-long traffic. Residents battled block-long blazes themselves using buckets of sewer water. The city had only one-third of the food and water needed for survivors.

Japan spent more than $100 million in 1995 trying to predict earthquakes, much more than is spent in the U.S. For 1996, the Japanese have budgeted $150 million. Many Japanese earthquake experts believe this money could be better spent, perhaps to improve coordination among disaster-relief agencies.

In characteristic Japanese fashion, cabinet ministers responded to this poor performance by pledging a month's salary, about $10,000 each, for earthquake relief as atonement for their lack of foresight in plan-

ning to cope with the disaster. Do you believe your federal representatives and senators will make a similar gesture when the "big one" strikes Los Angeles or San Francisco?

Earthquakes in Plate Interiors

As noted, most of the world's quakes occur along contacts between the plates that form the crust. However, large and potentially damaging earthquakes also occur in the interior of plates, far from their margins (Figure 11.24).

In the eastern United States, the most famous event, and perhaps the largest in historic times, occurred near the town of New Madrid in southeastern Missouri in 1811–1812. It was a series of quakes over a three-month

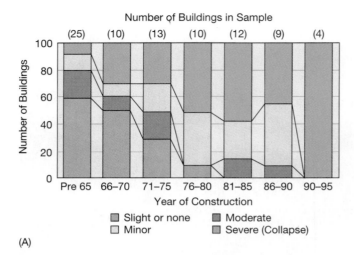

Number of Buildings in Sample

(A)

FIGURE 11.23   The value of reinforced-concrete construction during an earthquake.

(A) Correlation between damage level and construction year for reinforced-concrete buildings. (B) Collapsed sixth story in the eight-story Kobe City Hall Annex, a reinforced-concrete building built in the 1960s. Behind it is the 16-story New City Hall, a 1980s steel-frame building that was not damaged and remained functional after the earthquake. *Photo: Peter Monk/Applied Technology Council.*

(B)

period, with more than 1800 felt by residents, and many others too small to note. The magnitude of the largest tremors has been estimated at 8.7, an event expected in this intraplate region once every few thousand years.

Modern analysis by seismologists indicates that the number and magnitude of New Madrid tremors during the three-month period equaled the tremors in southern California during the forty-one years from 1932 through 1972. Mercalli intensity at New Madrid is estimated to have been XII at the epicenter. Property damage was low because few lived there at the time. If

the New Madrid earthquake were to be repeated today, it would be a full-scale disaster, for the cities of Memphis, Nashville, and St. Louis are not far distant.

Earthquakes between plates occur as they grind past or beneath one another, or spread apart. But within plates, quakes result from reactivation of old movement along ancient and usually buried fault systems. The New Madrid tremors occurred along a weak area in the crust parallel to the Mississippi River.

Earthquakes in plate interiors affect much larger areas than similar quakes at plate boundaries because the

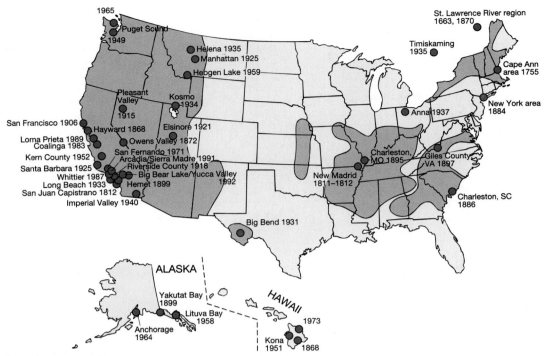

FIGURE 11.24   Earthquakes in the North American plate interior.

Historic earthquakes in the United States and Canada that have caused notable damage. Shaded zones show high-risk areas for earthquake damage. All or part of thirty-nine states—home to 70 million people (27 percent of the U.S. population)—are at risk.

crust within plates is less fragmented. Hence, seismic energy is transmitted more efficiently. For example, the New Madrid earthquake affected an area 2 million square miles in extent, in contrast to California earthquakes, which affect only thousands or ten thousands of square miles.

Manmade Earthquakes

Earthquakes are such powerful and awesome features of the natural world that it is hard to imagine that they could be created by humans. However, people have created thousands of **manmade earthquakes,** although they rarely have exceeded magnitude-5. None in the United States has resulted in death or major damage.

Earthquakes at the Rocky Mountain Arsenal

One of the best-documented examples of human-caused earthquakes is an earthquake swarm near Denver during 1962–1965. Since 1942, chemical-warfare products had been manufactured at the Rocky Mountain Arsenal, about ten miles northeast of Denver. A byproduct of this operation, wastewater, was evaporated from surface reservoirs until 1961.

But the wastewater was found to be percolating into the soil and rock, contaminating the local groundwater supply and endangering crops. So, a well was drilled to dispose of the wastewater deeply enough that it would no longer contaminate groundwater. Such a well is called an *injection well,* because the water is "injected" into the ground. The well bored through sedimentary rock into the metamorphic-rock basement, to a depth exceeding 2 miles. Each month, up to 9 million gallons was pumped down the well, totaling 150 million gallons, enough to fill 156 Olympic-sized swimming pools.

The result was that 710 earthquakes of magnitude 0.7 to 4.3 were recorded, all with epicenters in the Denver area (Figure 11.25). The only tremor previously noted there was in 1892. Clearly, a strong correlation existed between the fluid injection and earthquake frequency. The reason was simple: the injected fluid made slippage easier along existing faults, triggering fault movement. Despite the federal government's repeated denials of responsibility for the quakes, the disposal well was eventually shut down.

However, the increased seismicity caused by the injected fluid has persisted. This suggests a long-term or even permanent change in the rocks beneath the area. This is another example of a harmful interaction between humans and the natural environment.

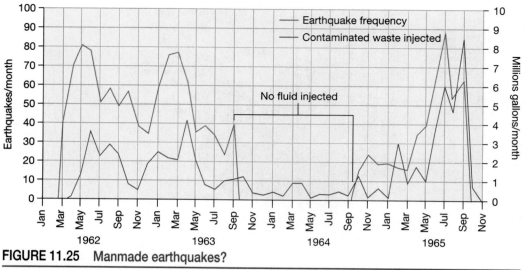

FIGURE 11.25 Manmade earthquakes?

Quakes recorded over four years in the Denver area clearly correlate with the monthly volume of contaminated wastewater injected into the Arsenal well. *Source: Evans,* Mountain Geologist 27, 1966, p. 27, *Rocky Mountain Association of Geologists, Denver.*

Earthquakes at Water Reservoirs

When large dams are built on rivers, the weight of water impounded behind the dam can be great enough to affect the rocks beneath it and induce earthquakes. Most reservoirs do not cause quakes. But more than 100 have, creating thousands of quakes. The most damaging reported so far was a 1967 tremor in India, magnitude-6.5, which killed 177 people and injured 2200.

There are two likely explanations for reservoir-induced earthquakes, and both require unusual geologic or hydrologic conditions.

The geologic condition is illustrated by Lake Mead on the Colorado River between Nevada and Arizona. The lake is artificial, created by the building of Hoover Dam. When the dam was built in the 1930s, the area was thought to be free from earthquakes, although little information was available. Filling of the lake began in 1935 and the first shocks were felt in 1936. The largest—magnitude-5.0—was recorded in 1939, when the lake was filled to 80 percent of capacity. At least 10,000 small quakes have occurred in the area since dam construction, and nearly 10 percent of them have been felt by local residents. The periods of highest seismic activity correlate with months of highest water levels. Apparently, the weight of the added water was sufficient to reactivate existing faults that had been inactive for at least as long as people had lived in the region.

The hydrologic condition relates to water pressure in porous rock. Where the pores connect the reservoir to deeper rock layers, so that water exists continuously from the reservoir throughout the pores in rocks below, water pressure within the rock increases with the weight of the added overlying water. This pressure may reduce friction, causing strain (fault movement) along fractures or weak areas.

Earthquakes at Geothermal Power Stations

Near plate boundaries, magma often is near the surface. The giveaway is the active volcanoes and steam columns that rise from the ground (Figure 11.26). The steam is groundwater that has been vaporized by heat from the magma, a few hundred or thousand feet below. As we will see in Chapter 15, this steam is harnessed for use as a commercial power source in many countries. The United States is the leading producer of this energy source, called **geothermal power.** The steam is used in electrical power plants to turn turbines to generate electricity, just like the steam boiled from water with heat from conventional fuels such as coal.

The major geothermal power site is The Geysers, about ninety miles north of San Francisco. Using this free steam, power has been generated for about thirty years. However, some side effects exist. The removal of so much fluid from underground has caused micro-earthquakes and ground settling, with a subsidence bowl measuring 1.2 square miles in area and 15 feet deep.

Seismic activity nearly doubled from preproduction (1962–1963) to peak production (1975–1977), with regional earthquakes of magnitude greater than 2 increasing from twenty-five per year to forty-seven. These low-magnitude quakes caused no major problems. But an effort to halt subsidence and maintain production by injecting steam condensate into the rock could increase seismic activity by inducing slippage along the many small faults that traverse the area.

FIGURE 11.26 A clue that magma is near.

Steam rises from ground vents in The Geysers geothermal field in California. *Photo: Pacific Gas & Electric Company.*

Intentional Earthquakes

The unplanned generation of earthquakes at the Rocky Mountain Arsenal led to an idea. Where serious earthquakes will occur sooner or later, as along the San Andreas fault, could deliberate triggering of small, harmless quakes relieve stress, thus defusing the potentially big ones? To test this hypothesis, experiments were conducted at Rangely, Colorado, in the early 1970s. Water was deliberately injected into subsurface rocks to increase the pore pressure, promoting slippage and small earthquakes on nearby faults. Water also was pumped out of fracture zones to decrease fluid pressure, to stop the quakes by in-

creasing friction along existing fractures. The project successfully demonstrated that earthquake frequency can be controlled by manipulating the water pressure in rock pores (Figure 11.27).

However, this is not a proven technology. If the method were tried along the San Andreas fault, for example, it could trigger excessive results if more strain energy is released than was thought to be present. Consequently, no one has had the courage to attempt the technique in heavily populated areas. Obtaining a public consensus and agreeing on who would be liable for deaths and damage is probably impossible. Even minor resulting quakes could cause damage. The experiment could be very costly!

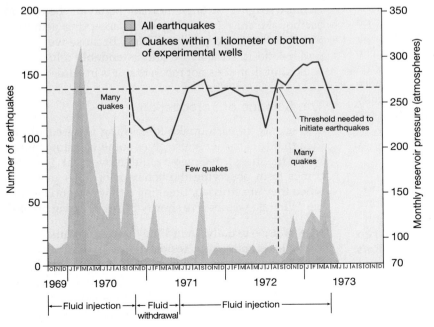

FIGURE 11.27 Earthquake control at Rangely oil field, Colorado.

This is the first case in which earthquakes were intentionally started and stopped by controlling water pressure in the pores of underlying rock. A specific "threshold pressure" was needed to trigger earthquakes. *Source: U.S. Geological Survey.*

Earthquake Forecasting: Can We?

". . . there is a 67 percent chance for at least one earthquake of magnitude 7 or larger in the San Francisco Bay Area between 1990 and 2020. Such an earthquake could strike at any time, including today."

U.S. Geological Survey, 1990

Earthquake forecasting has been a goal since ancient times. The invention of the seismograph shortly before 1900 permitted study of earthquake waves, and, from them, scientists deduced what must be going on within Earth to generate them. The recognition of tectonic plates and their movements in 1967 was another essential step in understanding earthquake origin. However, despite intensive studies over the past three decades, practical earthquake forecasting remains an elusive goal. At best, forecasts are as broad as the one quoted above.

An effective earthquake forecast needs four essential elements:

1. A usefully short time span during which the event is predicted to occur.
2. A usefully small area in which the predicted quake will occur.
3. A fairly accurate prediction of magnitude.
4. A statement of the odds that such an earthquake might occur randomly, without reference to any special evidence.

When we examine these criteria, questions immediately appear. How narrowly must the time period be identified? (Saying "sometime in the next twenty years" will not spur people to action.) How small an area must be specified? (Saying "somewhere along the San Andreas fault" will not make people move.) How much leeway is allowed in the estimate of expected magnitude? (If it is too broad, people will take the forecast as a half-baked guess and ignore it.) How can trustworthy odds be calculated for an event that occurs so infrequently that it has happened only once or twice during the past few hundred years? Such odds really can't be calculated meaningfully. It is demanding too much from too little data.

Can Animals Sense Impending Quakes?

Technology may not be the only key to earthquake prediction. Animals may sense the onset of a quake before humans, their more "advanced" cousins. Shortly before an earthquake, snakes have been observed to awaken from hibernation and slither from their dens. Rats gather in fearless groups, unconcerned about the presence of people. Small pigs have bitten off and eaten their tails before an earthquake. Dogs bark incessantly; horses and cattle run in circles and refuse to enter their corrals. Shrimp crawl onto dry land. Ants pick up their eggs and migrate en masse. Fish jump above the surface of water, and rabbits hop aimlessly about.

Such **animal forecasting** has been reported shortly prior to large tremors in many countries. It was reported in China late in 1974, and a swarm of small earthquakes occurred at December's end. In January, thousands of Chinese reported more unusual animal behavior, especially by larger animals, in an area that would be the epicenter of a major quake (magnitude-7.3) a month later. Based on these animal reports, the government warned residents of the city of Haicheng not to sleep in their homes. As a result, only 1328 people perished, rather than the ten thousands who would have died in collapsing buildings.

We all know that some animals are better-adapted than humans to detecting light, sound, odor, touch, and temperature variations at a very low level. We trust watchdogs to hear intruders and smell smoke too subtle for us to notice. (Gevy, your author's German Shepherd, can smell a teaspoon of ordinary table salt in a gallon of water.) Migrating birds and sea turtles apparently navigate using small variations in Earth's magnetic field. Perhaps some animals can sense very minor Earth changes that occur shortly before earthquakes, such as the sound of microfracturing (Figure 11.28). Should Californians pay more attention to the animals in their zoos?

Curiously, unusual animal behavior was observed prior to a 1971 San Fernando earthquake, but no action was taken. After all, would you leave the area for several months because local dogs barked incessantly or because rats in the poorer sections of town acted funny? Would reports of strange behavior among farm animals near your city incite a mass exodus from town? Acknowledging strange animal behavior before an earthquake may be analogous to accepting acupuncture, another concept from China. Because we do not understand it, we may be understandably reluctant to use it. But that does not mean that it is invalid.

The Science of Forecasting

The science of earthquake forecasting is based on the idea that rock under stress gives some warning signals before the stress is relieved by fault movement. This idea comes from observing the behavior of other materials when they are stressed. Breaking a twig is a good example. What do we observe shortly before the twig breaks?

- *Noise*—usually inaudible, but sometimes audible—comes from inside the twig, as some fibers begin to separate.
- *Sound velocities decrease* through the twig as fibers separate, because sound travels slower through air in the crack than through a solid twig.

FIGURE 11.28   More than a pretty face.

Gevy grows nervous when faults slip in the Middle East. But will anyone take her behavior as a warning of an impending quake?

- *Volume of the twig increases* as spaces (cracks) are opened.
- *Electrical properties* of the twig change because electrical resistance is greater through air than through a wet twig. If we imagine water entering the cracks in the twig, electrical resistance would decrease because electricity flows much better through water than through air.

We have just described the **dilatancy model** of how materials dilate (expand) as strain accumulates on a microscopic scale. Most efforts at earthquake forecasting are based on this model. What is expected when the dilatancy model is applied to rocks is listed here. These are called **precursor phenomena:**

1. *Seismic wave velocities should slow* as cracks and pores develop in the rock, then speed up again as fluid enters the cracks. When a quake occurs, the cracks become closed and wave velocities increase toward normal.

2. *The ground surface should be uplifted or tilted* as the rock expands because of the empty spaces (cracks) that develop in it. When the quake occurs, the cracks are closed and the uplift or tilt becomes minimal.

3. *Electrical resistance should drop* as water from the immediately surrounding area enters and fills the cracks. It then rises again as the quake occurs. Salty water is less resistive than rock.

4. *Radon gas should concentrate* in water wells near the developing fault. Radon is a radioactive gas formed as uranium decays, and a small amount of uranium exists in most rocks. As cracks appear, the radon escapes from the rock and moves into well water near the fault. When the quake occurs, the cracks close and radon in the water decreases (Figure 11.29).

5. *Many small earthquakes* result from development of the small cracks. They peak when the "big one" finally happens.

Each of these five changes can be measured using modern equipment. The measurements can be useful indicators of impending large quakes. However, there is a problem: the changes that produce these precursor phenomena occur in such a small volume of rock that they are difficult to monitor unless the instruments are located at the critical place along the fault. Of course, we don't know where that will be, so we don't know where to put the instruments. Further, not all earthquakes show the same pattern of precursory events.

It is possible, using rather sophisticated equipment, to measure directly the amount of **elastic deformation** in the rock along a fault as load increases. Elastic deformation is like stretching a rubber band and letting it return to its original shape. The greater the elastic deformation, the greater likelihood that the rock (or rubber band) will not elastically return to its original shape, but will break. When coupled with laboratory measurement of rock strength, it is possible to anticipate when a rock might rupture. However, in the real world of Earth's crust, stress is seldom uniform or continuous, so it is impossible to predict the exact time or even date of faulting. No reliable predictive tool has yet been found.

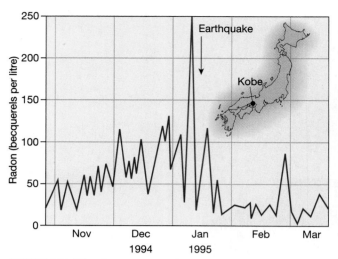

FIGURE 11.29   A precursor phenomenon.

Radon gas concentration in water wells in Kobe, Japan near the time of the 1995 earthquake.

The Japanese on Earthquake Forecasting

As noted, Japan spends an enormous amount of money each year trying to forecast earthquakes. This seems reasonable, for the country lies along the edge of two major plate boundaries (Figure 11.18). Imagine how we would feel if the entire United States were located along the San Andreas fault system.

Around the clock, government technicians monitor seismic activity, crustal tilt, the buildup of strain (rock deformation) in the crust, and groundwater levels. The government also funds dozens of studies of animal behavior and electromagnetic waves in hopes of finding reliable precursors to quakes. What do Japanese scientists and the Japanese public think of spending $100 million in 1995 and even more in 1996 on quake research?

Despite the flowing cash for research, the Japanese public is skeptical. According to a 1995 survey by the Prime Minister's office, about half the public believe that accurate prediction of every earthquake is impossible. Eighty-seven percent believe that even very large quakes are not amenable to reliable prediction. (Presumably, "very large" means quakes with magnitudes of 7 or more.)

Scientists are even more skeptical. A university seismologist reports that, "At this moment, the majority of seismologists in Japan believe earthquake forecasting is either impossible or very difficult."

Controversy over the ability of seismologists to accurately forecast earthquakes led to the resignation of the head of Japan's earthquake prediction program, Dr. Kiyoo Mogi, in March, 1996. Dr. Mogi had served in this program since 1979 and had been chairman since 1991. He advocated warnings ranked by color: black

alert for a certain quake, white for a possibility, and gray for a strong possibility. But the Japanese government refused to use his suggestions and was unwilling to notify the public unless forecasters could guarantee that an earthquake would occur. Critics of the government's stand say that the government has lulled the public into a false sense of security by suggesting to them that earthquakes can be predicted with a high degree of certainty. Who is right? Should the government sound an alarm only when scientists can prove a quake will occur? Will the Japanese public be unnecessarily frightened by regularly issued probability statements? What is the government's responsibility in issuing warnings of impending earthquakes?

When is it reasonable to conclude that a problem cannot be solved, and that we should redirect our efforts to more manageable tasks? Should earthquake forecasting research be stopped? Have we tried long enough? Ask yourself the same questions about flood prediction, poverty, the causes of war, future economic trends, or the possibility of intelligent life in other galaxies. The list is endless. How should you or society decide when "enough is enough"? At what point should you or society say, "I've had it with throwing money down that rathole?"

Fossil Paleoquakes

A **fossil** is any evidence of the past existence of something. Usually, we think of fossils in terms of organisms, like a dinosaur skeleton or an ancient plant in shale rock. But the term can be used for nonliving things as well, like evidence of ancient meteorite impacts or earthquakes. We can find evidence in sediment that earthquakes happened within the past few thousand years. These are called **paleoquakes** (*paleo* means ancient). From this evidence, we can determine the approximate years when the paleoquakes occurred. This may allow us to spot a pattern in how often the quakes occur, so we can predict the next occurrence.

This approach to earthquake prediction was tried along the San Andreas Fault at Pallett Creek, about thirty miles northeast of Los Angeles. The area where the fault and creek intersect is poorly drained. It becomes a swamp during the rainy season, and abundant vegetation develops. Researchers dug trenches across the fault to expose a cross-section of the marsh sediments (Figure 11.30A). The sediments are layers of silt, fine sand, and organic matter from dead plants. Such mucky sediments are prime candidates for the liquefaction caused by earthquake waves.

Liquefaction associated with earthquakes has occurred in many areas. Based on studies of such areas, researchers expected to find vertical columns of sand that cut through the silt, sand, and organic layers.

(A)

(B)

FIGURE 11.30 Evidence of earthquakes at Pallett Creek, California.

(A) Swamp and marsh sediments at Pallett Creek, cut by faults. Each fault represents one or more earthquakes. Radiocarbon dates indicate that sediment age ranges from A.D. 200 at the base to 1910 at the top. *Photo: Kerry Sieh, CalTech.*
(B) Recurrence pattern of earthquakes at Pallett Creek, California. The last two dates are based on historical records. Earlier dates are accurate within ±50 to 100 years. Based on these data, when would you forecast the next quake to occur?

These columns, called *sand pipes*, occur during liquefaction. They develop when the weight of overlying sediment forces water and sand that are a few feet below the surface to rise through weak areas in the overlying sediment. The organic material was dated using the radiocarbon-dating method. By combining the evidence from the sand pipes with radiocarbon dates for the various marsh layers, dates when earthquakes occurred could be determined. (Of course, these dates are approximate, accurate within perhaps ±50–100 years or so.

Evidence was found for at least nine paleoearthquakes at Pallett Creek, extending back more than 1400 years (Figure 11.30B). The two most recent quakes, in 1745 and 1857, also were documented from historical records, which validated the radiocarbon dates. Study the figure and you can see that the average time between earthquakes here was 164 years (1857 – 545 ÷ 8 increments between quakes). However, the actual time between quakes ranged from 275 years to 55 years, a 5:1 ratio. Clearly, this variation is too large to be useful for disaster planning.

It has now been about 140 years since the last quake recorded at Pallett Creek. Do you believe it is now safe or unsafe to live in the area? Is there any year in which you believe it would be unsafe to the point where you would relocate if you lived there? These are the practical decisions that each person must make based on data such as these. How helpful are such data? Do you think most people, even if informed, would simply not worry about it and proceed to build a house in the area?

The Parkfield Prediction

Parkfield, California (population thirty-nine) sits on the San Andreas fault between Los Angeles and San Francisco (Figure 11.31). It is the self-styled "Earthquake Capital of the World" where Earth allegedly "Moves for You." It is the site of the only official earthquake prediction in the United States. Six earthquakes have been reported at Parkfield since 1857, with recurrence intervals of 24 years, 20 years, 21 years, 12 years, and 32 years, the most recent in 1966. In 1985, the USGS announced a 95 percent probability for a magnitude-6 earthquake between 1987 and 1992. The forecast quake did not occur, much to the distress of earthquake scientists at USGS. Whether the thirty-nine residents of Parkfield were similarly distressed is unlikely.

As this is written, we are several years and $20 million in maintenance cost beyond the predicted date. The Parkfield area has proved notable for its *in*activity. When the largest recent quake occurred, a magnitude-4.5 tremor late in 1992, the USGS immediately announced that a magnitude-6 event might follow within seventy-two hours. Fire engines, hospitals, and emergency facilities stood at the ready, residents stashed bottled water, and helicopters from four or five television stations hovered overhead. But nothing happened. A magnitude of 5.0 earthquake occurred late in 1994 but still no magnitude 6. We still have much to learn about predicting the specific time and place of an earthquake.

Yet, the urgent need for earthquake forecasting is demonstrated every time television shows us the devastation from a quake somewhere in the world. Research continues in the United States, Japan, and other quake-prone countries for keys to reliable forecasting.

In lieu of useful predictions, we could try common sense. If a gasoline refinery is in flames, you move away from it. If lightning is known to strike a hilltop with great frequency, you find another place to build your house. If the ground beneath your feet has an active fault with a history of nasty earthquakes, you . . . build public buildings, homes, and even a nuclear power plant on it. In California, these very structures

---

FIGURE 11.31   Where will the "big one" happen?

The probability (in percent) of major earthquakes occurring along the San Andreas fault during the 30-year interval between 1988 and 2018. Estimated earthquake magnitudes are given above each column. This diagram originally was drawn in 1987, before the 1989 Loma Prieta (San Francisco) event. Note the 90 percent probability of motion along the Parkfield segment. The inset shows the alarming number of schools, hospitals, and other public buildings that are uncomfortably close to the Hayward Fault. *Source: U.S. Geological Survey.*

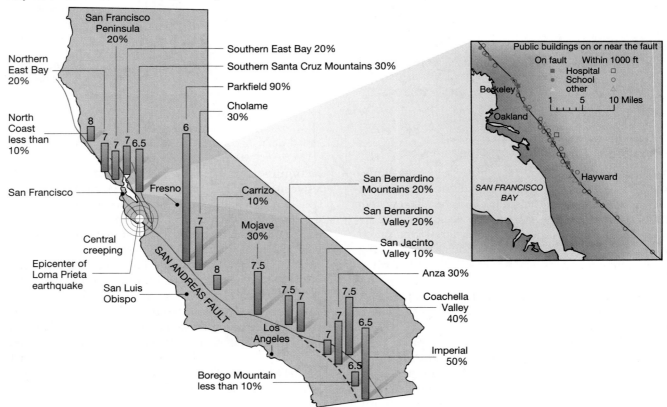

FIGURE 11.32   No respect.

Statue of noted geologist and educator Louis Agassiz, following the 1906 San Francisco earthquake. One observer noted, "I have read Agassiz in the abstract. Now I see him in the concrete." *Photo: Ernst Mayr Library.*

are built with seeming disregard for known faults and recent history (Figure 11.31). The Diablo Canyon nuclear power plant operates today near San Luis Obispo, about halfway between San Francisco and Los Angeles. As you read these words, a structure probably is being built directly atop a major fault in California. We should use our knowledge to avoid siting critical facilities and living quarters in harm's way.

Earthquakes do not distinguish between everyday citizens and noted scientists. One famous scientist was treated with exceptional disrespect by the 1906 San Francisco earthquake (Figure 11.32).

Protecting Yourself from Earthquakes

If you really live in fear of earthquakes, move to Michigan, Wisconsin, Iowa, or Florida. These four states have not experienced a large quake for at least several million years. They are far from tectonic plate boundaries. If you are willing to accept earthquake risk, the most dangerous area is along the west coast of North America, particularly California, where the active plate margin runs the length of the state.

Building Construction and Insurance

"Earthquakes don't kill people, buildings do." This is why, during an earthquake, you are safest outside and away from anything that can fall. Earthquakes cause building collapse for obvious reasons, just as a stack of coins will topple if you jar it hard enough. But collapse also may result from settling of the ground, or from liquefaction of underlying sediment and rock (Figure 11.33).

Equally disastrous is "collateral damage": the rupturing of gas pipelines, water mains, and sewer systems. The gas ignites fires, which cannot be extinguished without water; in San Francisco's 1906 quake, most property damage resulted from uncontrollable fires that swept the city. And most of the human damage was from epidemics caused by untreated sewage and contaminated water.

Because of California's vulnerable location and history of frequent earthquakes, the state restricts the location of new buildings. Where a surface fault is precisely mapped, no new building may be constructed within 48.75 ft of the fault, and no building other than a single-family home may be constructed within 123.5 ft of the fault. Where a surface fault is known to exist, but is only approximately located, no new building may be constructed within 97.5 ft of the fault, and no building other than a single-family home may be constructed within 172.25 ft of the fault. The state's laws also require quake-resistant building construction. It adds only 20 percent to the cost of new construction.

Some criteria for seismic design and construction have been included in the California Uniform Building Code since 1973, updated in 1988 to reflect new information and technology. Nearly all *new* buildings in the state comply with this or a similar code.

California's Uniform Building Code requires greater strength for essential facilities and for building on soft sediment (Figure 11.34). The Code emphasizes life safety while allowing some structural damage and loss of function. (This is the survival principle used in auto racing, wherein a driver is surrounded by a steel "roll cage" that will protect him or her even if the car is

FIGURE 11.33 Some good fruits of planning.

Anticipating quakes, the designers of this bridge on the Harbor Expressway in Kobe, Japan, sank the support columns completely through the potential liquefaction layer and into more stable rock. When the quake hit in 1995, the area between the columns buckled, but the columns and the expressway remained in place. *Photo: Earthquake Geology and Geophysics.*

demolished.) Some building owners want protection for their structures greater than Code minimum, and may insist on higher design standards, with the accompanying higher cost. The down side is that most buildings in California, as in all other states, were constructed before the 1973 code. It costs twice as much to retrofit an existing building as it does to build quake-resistance into new construction.

California homeowners can buy extra insurance to cover quake damage for about $2–$3 per $1000 of coverage. Usually, 10 percent of the cost of any damage must be deducted before the insurance kicks in. Hence, earthquake insurance for a relatively inexpensive house (by California standards) worth $400,000 costs an additional $800–1200 per year, and the homeowner must pay the first $40,000 of any repairs. In the San Francisco Bay area, 30–40 percent of homeowners have purchased earthquake insurance; for all of California, about 25 percent have done so.

Personal Preparation

Earthquakes usually last less than a minute, and may subside within seconds. The safest refuge during an earthquake is an open area outside. If you are in a building, leave it promptly, because aftershocks may collapse the weakened structure. Never use an elevator, even if it still works, for it can jam between damaged floors and become a deathtrap. If you cannot get out, shield yourself under the strongest thing available—a table, metal desk, a doorway arch, or a closet. Keep away from windows, which shatter. If you live or attend college in earthquake country, Table 11.5 provides useful advice.

Official Forecasts and Public Response

When the predicted Parkfield earthquake failed to occur, there were mixed feelings. Earthquake specialists were dismayed that their prediction was wrong; area residents were relieved. But Parkfield is rural and few people would have been affected anyway.

But what about an earthquake forecast for Los Angeles or San Francisco? A magnitude-8.2 quake on the San Andreas fault near San Francisco similar to the 1906 tremor would generate damage intensities of IX or X in parts of the city. An estimated 10,000–12,000 people would die if the quake occurred during business hours, or 3000–5000 if it happened after hours. Injuries also would depend on time of day, but might reach 40,000, with 20,000 homeless (Figure 11.35). Property damage would exceed $10 billion, as would collateral economic costs like lost earnings. Of the damage, 40 percent would occur in the center of San Francisco.

(A)

FIGURE 11.34 Examples of construction according to the California Uniform Building Code. Nearly all new buildings in the state comply with this or a similar code (1973, modified 1988). Older buildings, however, often do not comply. (A) Diagonal steel braces strengthen a concrete frame building that houses U.S. Geological Survey employees. *Photo: U.S. Geological Survey.* (B) Metal bands brace a gas water heater. The flexible gas pipe reduces the risk of releasing flammable gas.

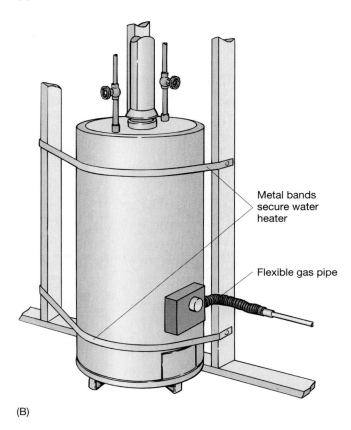

Metal bands secure water heater

Flexible gas pipe

(B)

It can be argued that the effort and money devoted to forecasting earthquakes is wasted. Too many uncertainties exist about the actual stress in Earth's crust, the existence of hidden faults, the difficulty of determining frictional resistance along short sections of known faults, and the impossibility of quickly evacuating millions of people. Thus, it seems reasonable to consider halting efforts at forecasting. The money might better be spent on water pollution, soil erosion, or social problems.

To illustrate the present uselessness of earthquake forecasts, suppose that seismologists forecast a magnitude-7.5 quake in July or August "in the vicinity of Oakland," about 10 miles east of San Francisco. Such a forecast is realistic, given our knowledge of the Hayward fault (Figures 11.19, 11.31). Such a tremor might cause 4500 deaths, 135,000 injuries, and $40 billion in damage.

But how useful is this information to the general public? Will everyone within twenty miles leave their homes and visit relatives in Phoenix or Hermosilla for two months? Will Sam's Wholesale, Price Mart, or Safeway food stores unstock their shelves and give employees two months "earthquake leave" with pay? Will employees who heed the warning be laid off if they leave town? Will children be excused from school attendance? Will insurance companies refuse to renew policies due to expire on buildings during May and June? And what if the forecast turns out to be "another Parkfield" and the quake does not occur—or occurs in September, after everyone returns? See Figure 11.36 for other possible consequences of an earthquake prediction.

FIGURE 11.35 Path of destruction?
Aerial view looking north from the southern edge of San Francisco. The area shown lies across strands of the San Andreas Fault (black lines). The earthquake of 1906 caused as much as 8 feet of horizontal displacement along fault strands like these; much less movement would wreak havoc here because of urban growth during the past ninety years. *Photo: R. E. Wallace/U.S. Geological Survey.*

TABLE 11.5	What to Do if You Live in Earthquake Country

Before a Quake Hits

At home

Have a portable radio, flashlight, first-aid kit, fresh batteries, and a fire extinguisher in your home. Know where they are stored. A cellular phone can be valuable too, because some cellular service may survive smaller quakes.

Take a first-aid class; everyone needs to do this anyway.

Being able to shut off utilities is important. Know the location of your circuit-breaker box, gas shutoff valve, and water shutoff valve. (Some need a special wrench; keep it nearby.) Know how to turn them off.

Don't keep heavy objects on high shelves.

Secure heavy appliances to the floor. Anchor cupboards and bookcases to the wall.

Devise a plan for reuniting your family after an earthquake in the event anyone is separated.

At school or work

Ask what plans the college or your employer has in event of a quake. Note what objects in your classroom, lab, or work area might fall. Learn the emergency plan, if one exists. Volunteer for emergency responsibilities.

If a Quake Hits . . .

Stay calm; panic is a killer.

Many injuries occur as people enter or leave buildings. If indoors, try to get outside, unless you are in a quake-resistant building.

If indoors, get under a sturdy frame, such as a door frame or metal table.

Don't use candles, matches, lighters, or any open flame (you might ignite a gas leak).

Stay away from windows, and face away from them.

Evacuate when told to do so. Use stairs, never elevators.

If outside, stay away from buildings. Keep away from overhead electric wires or anything that might fall (chimneys, parapets and cornices on buildings).

If you are in a moving car, stop. Remain inside until the shaking is over. Stay away from overpasses and bridges.

Expect aftershocks; they often cause additional damage.

Source: Bruce A. Bolt, 1993, *Earthquakes and Geological Discovery*, New York: W. H. Freeman and Company.

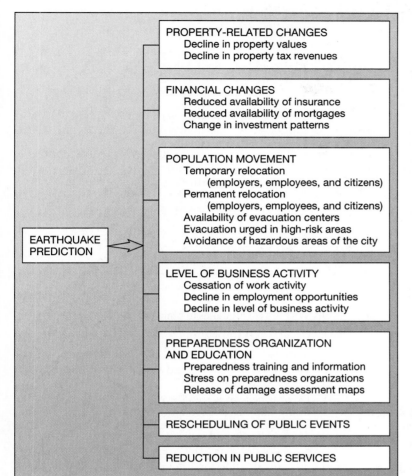

EARTHQUAKE
PREDICTION

PROPERTY-RELATED CHANGES
    Decline in property values
    Decline in property tax revenues

FINANCIAL CHANGES
    Reduced availability of insurance
    Reduced availability of mortgages
    Change in investment patterns

POPULATION MOVEMENT
    Temporary relocation
        (employers, employees, and citizens)
    Permanent relocation
        (employers, employees, and citizens)
    Availability of evacuation centers
    Evacuation urged in high-risk areas
    Avoidance of hazardous areas of the city

LEVEL OF BUSINESS ACTIVITY
    Cessation of work activity
    Decline in employment opportunities
    Decline in level of business activity

PREPAREDNESS ORGANIZATION
AND EDUCATION
    Preparedness training and information
    Stress on preparedness organizations
    Release of damage assessment maps

RESCHEDULING OF PUBLIC EVENTS

REDUCTION IN PUBLIC SERVICES

FIGURE 11.36   Consequences of earthquake prediction.

Socioeconomic impacts and adjustment to an earthquake prediction. *Source: Robert W. Christopherson, 1994,* Geosystems, *2nd ed.,* Prentice Hall, p. 371.

Summary

Earthquakes are movements along ruptures in Earth's crust, termed faults. Movements along faults cause earth tremors because of the release of stored stress as blocks slip past each other. The magnitude of an earthquake is a measure of the amount of energy released along this fracture, as contrasted to the intensity of the quake, which is a measure of the amount of damage produced to human artifacts. Earthquakes can be caused by human activities, but these almost always have magnitudes less than about 4, although a rare few as high as 6 have been measured.

Four types of waves are produced by fault movements: P-, S-, L-, and Rayleigh waves. Each of these waves has a known velocity through crustal rocks. By studying the temporal separation between their arrival times at a recording station, the waves' point of origin can be determined.

Earthquake forecasting is exceedingly difficult and is far from being achieved. Current approaches involve observing animal behavior immediately before quakes, measurements of strain in rocks near known faults, tilting of the crust, occurrence of radon gas in water wells, changes in electromagnetic waves in the crust, and the onset of subaudible noise near faults. Each of these possible predictors has worked in some cases but not in others. Why a predictor works in one area but not others is unknown. At present, our ability to forecast the time, place, and magnitude of future earthquakes is poor, despite billions of dollars in research worldwide. It is uncertain whether the required accuracy ever will be achieved.

Key Terms

aftershock
animal forecasting
asthenosphere
dilatancy model
earthquake
earthquake forecasting
elastic deformation
epicenter
fault
focus
fossil
geothermal power
inertia

intensity
L-wave
lithosphere
magnitude
manmade earthquakes
mantle
mantle convection
normal fault
P-wave
paleoquakes
plates
plate boundaries
plate movement

precursor phenomena
Rayleigh wave
reverse fault
Richter scale
S-wave
seismic waves
seismogram
seismometer
seismologist
sonogram
strain stress
strike-slip fault
thrust fault

Stop and Think!

1. Where do most of the world's earthquakes occur? What causes them?

2. What is the relationship between a fault and an earthquake? What is elastic deformation and how does it relate to earthquake damage?

3. What are the types of faults and what stress orientations cause them?

4. What is the Richter magnitude of an earthquake? How does it differ from the Mercalli intensity of an earthquake? Why is the correlation between the measures imperfect?

5. Explain how a seismometer operates.

6. What are the four types of seismic waves? How do they differ?

7. Why were the effects of the 1989 Loma Prieta earthquake most severe 60 miles away in San Francisco, rather than at the epicenter in Loma Prieta?

8. Building collapses during the Kobe earthquake occurred for the same reason as they did at the Marina District in San Francisco. Explain. What lesson do you learn from this?

9. Why are earthquakes in plate interiors so much less common than quakes at plate margins? Why do earthquakes in plate interiors affect a wider area than those at plate margins?

10. What lessons were learned from the earthquakes at the Rocky Mountain Arsenal?

11. Do you believe we will ever be able to put to use the information gained from the earthquake experiments at Rangely, Colorado? Explain your viewpoint.

12. What four characteristics must an earthquake forecast have to be useful?

13. Can you think of a way to use the apparent high sensitivity of animals to predict earthquakes?

14. Explain the dilatancy model that describes rock behavior before an earthquake. What are the difficulties in applying it to quake forecasting?

15. Explain how "collateral damage" (damage to a city's infrastructure) affects a city's ability to deal with the many problems caused by earthquakes.

References and Suggested Readings

Bolt, B. A. 1993. *Earthquakes and Geological Discovery*. New York: W.H. Freeman and Company, 229 pp.

Buskirk, R. E., C. Frohlich, and G. V. Latham. 1981. Unusual animal behavior before earthquakes: A review of possible sensory mechanisms. *Reviews of Geophysics and Space Physics*, v. 19, p. 247–270.

Davidson, Keay. 1995. Kobe in California. *Earth*, June, p. 20–21, 70–71.

Haas, J. E. and D. S. Mileti. 1977. *Socioeconomic Impact of Earthquake Prediction on Government, Business, and Community*. Boulder, Colorado: University of Colorado, Institute of Behavioral Sciences, 40 pp.

Johnston, A. C. and E. S. Schweig. 1996. The enigma of the New Madrid earthquakes of 1811–1812. *Annual Review of Earth and Planetary Sciences*, v. 24, p. 339–384.

Levy, Matthys and Mario Salvadori. 1994. *Why the Earth Quakes: The Story of Earthquakes and Volcanoes*. New York: Norton.

O'Rourke, T. D. (ed.). 1992. The Loma Prieta, California, Earthquake of October 17, 1989—Marina District. *U.S. Geological Survey Professional Paper* 1551-F, 215 pp.

Page, R. A., D. M. Boore, R. C. Bucknam, and W. R. Thatcher. 1992. Goals, opportunities, and priorities for the USGS Earthquake Hazards Reduction Program. *U.S. Geological Survey Circular* 1079, 60 pp.

Robinson, Andrew. 1993. *Earth Shock: Hurricanes, Volcanoes, Earthquakes, Tornados and Other Forces of Nature*. New York: Thames and Hudson, 304 pp.

Stover, C. W. and J. L. Coffman. 1993. Seismicity of the United States 1568–1989 (revised). *U.S. Geological Survey Professional Paper* 1527, 418 pp.

Svenson, A. G. 1984. *Earthquakes, Earth Scientists, and Seismic-Safety Planning in California.* Lanham, Maryland: University Press of America, 130 pp.

Wakefield, J. and D. Pendick. 1995. Earthquake reality check. *Earth,* August, p. 20–21, 60–61.

Weigand, P. W. 1994. The January 17, 1994, Northridge (California) earthquake: A personal experience. *Journal of Geological Education,* v. 42, p. 501–506

Wuethrich, Bernice. 1995. Cascadia countdown. *Earth,* October, p. 24–31.

"A blast of burning sand pours out in whirling clouds. Conspiring in their power, the rushing vapours carry up mountain blocks, black ash, and dazzling fire."
Lucilius, Jr., A.D. 50

"More than 60,000 people fled walls of boiling mud . . . from Pinatubo volcano in the northern Philippines . . . Steaming rivers of mud and ash . . . reaching 7 feet high swamped hamlets in 11 towns . . ."
International Herald-Tribune, 1995

Volcanoes and Eruptions

12

Volcanoes make pretty scenery. Volcanic rock forms excellent agricultural soils. Volcanoes were the original source of much of our atmosphere. That's the upside of these fissures and mountains that periodically erupt material from Earth's interior. But an active volcano can be frightening, spewing molten lava and poisonous gases and triggering flows of hot mud. In this chapter, we will learn the where, what, why, and how of volcanoes.

How many volcanoes are in your state? If there are some, why? And if there are none, why? Do volcanoes occur randomly over Earth, or do they develop in certain places? Why? Are all volcanoes dangerous? How predictable are eruptions? Are scientists capable of warning citizens of an impending eruption? Can your employer require you to work near a potentially dangerous volcano? Why are volcanic soils so agriculturally productive? Do volcanic eruptions cause air pollution and affect our climate?

Our Heated Planet

Earth's surface is heated almost entirely from the outside, by solar energy. But Earth's interior is heated too, by very different means. As noted, rocks are much hotter a few thousand feet below Earth's surface than topside. At depths of a few ten thousands of feet, temperatures are so high that steel drill pipe behaves like cooked spaghetti and can be bent 30 degrees or more to tap into deposits of oil located off to the side. This increase in temperature with depth, called the geothermal gradient, exists because of two heat sources below the surface: (1) heat trapped in Earth's interior since the time of its formation about 4.5 billion years ago, and (2) heat generated by decay of radioactive elements (like uranium) in Earth's crust.

Because of these sources of heat, temperatures 20 miles down are hot enough to melt rocks, forming magma. If conduits to the surface are available, the magma slowly rises to the surface. If it breaks through the surface, its name is changed to lava. The amount of lava that reaches the land surface during an average year is about half a cubic mile, equal to a cube 4000 feet on a side. A much larger amount erupts onto the seafloor, perhaps ten times this much.

◀

Mount Pinatubo erupts in the Philippines, 1991.
Photo: James J. Mori / U.S. Geological Survey.

The lava may flow gently and form a large plain. Or it may build a **volcano,** a structure built of lava and exploded fragments of igneous rock. Scientists who study volcanoes are called **volcanologists.** It is fortunate for living creatures that the rocks of Earth's crust are good insulators (poor conductors of heat). Were this not the case, we all would be wearing insulated shoes. Cars would sport tires that would not melt or explode from the heat.

How good an insulator (how poor a heat conductor) is rock? Earth has a radius of about 4000 miles, but the crust has an average thickness around 20 miles, only 0.5 percent of the radius. If Earth were a 400-page textbook, the crust would be only two pages thick. Yet, these two pages are sufficient insulation to keep your fingers from getting burned as they hold the book.

Where Do Volcanoes Occur?

Most volcanoes occur along tectonic plate boundaries. A few occur over "hot spots" in Earth's mantle. Let us look at both settings.

Tectonic Plate Boundaries

Study Figure 12.1, and you can see that most active volcanoes are clustered along tectonic plate boundaries, the same boundaries where most large earthquakes occur. Is there a connection? There certainly is. Note especially how the pattern of volcanoes follows the edge of the Pacific plate, forming a rough circle around the Pacific Ocean. Appropriately, this volcanic belt is called the "Ring of Fire."

Figure 12.2 shows a portion of the Ring of Fire, our own Pacific coast. Note that, whereas earthquakes occur very close to tectonic plate boundaries, the volcanoes usually are many miles inland. For example, in Figure 12.2, note the string of inland volcanic peaks: Mount Shasta in northern California, Mount Hood in Oregon, Mount Saint Helens and Mount Rainier in Washington, and on northward through British Columbia. This volcanic belt continues into Canada, Alaska, and westward into the Aleutian Islands and Japan. These volcanoes are relatively young, having built up within the past 2 million years.

Volcanoes older than a few million years now are extinct, unlikely ever to erupt again. These have undergone extensive erosion, reducing the height of their peaks, and thus are not as obvious. An example is Crater Lake in Oregon. This beautiful lake, 6 miles across and North America's deepest, fills an extinct volcano. The volcano's crater collapsed to form a **caldera** (Spanish for *cauldron*) that filled with water (Figure 12.3). The volcano that houses Crater Lake last erupted about 7,000 years ago.

We commonly think of volcanoes as mountains on land. But far more exist in the ocean. Some are visible as

FIGURE 12.1   Volcanoes and tectonic plate boundaries: is there a connection?

This map shows the distribution of active volcanoes (above sea level) in relation to major tectonic plates. Nearly all eruptions are landward of plate boundaries. Two-thirds of the volcanoes are in the Ring of Fire around the Pacific plate. The Hawaiian Islands and Galapagos Islands (off the coast of Ecuador) are located over "hot spots" within major plates.

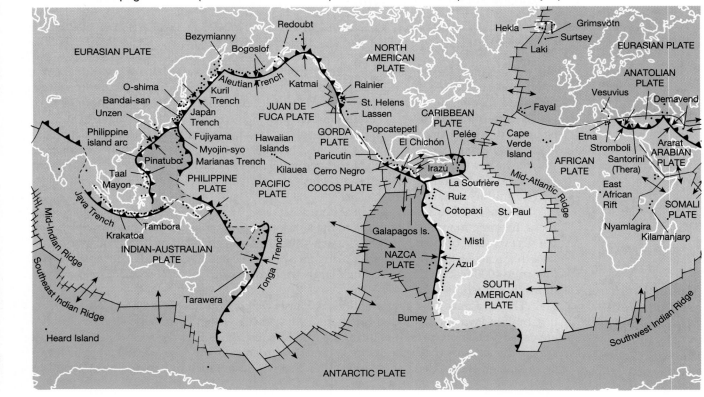

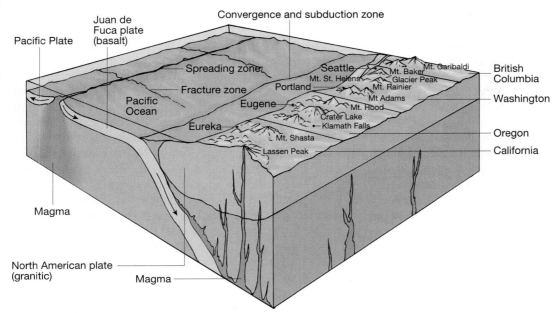

FIGURE 12.2 An American portion of the Ring of Fire.

A section of the convergent plate margin between northern California and British Columbia. The upper few miles of basaltic ocean crust plunges beneath the lighter granitic continental crust. When it has descended deeply enough for melting to begin, magma forms. The magma rises through fractures in the crust and becomes richer in silica as it dissolves granitic crust. At the surface, it appears as lava. Many large volcanoes

FIGURE 12.3 Oregon's Crater Lake is an ancient volcano, now dormant.

(A) The volcano when it was active. (B) Its peak collapsed to form a caldera. (C) The caldera filled with water, forming North America's deepest lake. (D) The volcano's last eruption built Wizard Island. The caldera is about 6 miles in diameter.
Photo: Greg Vaughn/Tom Stack and Associates.

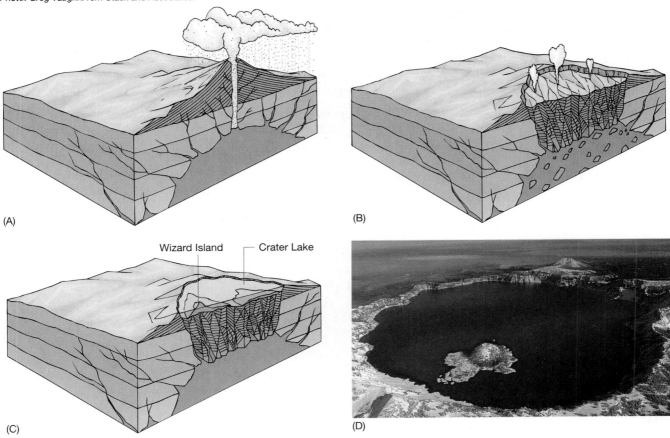

islands, like Hawaii. But most are hidden by the sea, erupting steadily as ships travel overhead, unaware of the molten lava thousands of feet beneath them. Most are located along plate boundaries where the plates are spreading apart. One example is the volcano chain that forms the Mid-Atlantic Ridge, an undersea mountain range. A few volcanoes in this range have built themselves high enough to be above the ocean surface, forming islands. The largest of these is Iceland, a volcanic island the size of Virginia, which marks the Mid-Atlantic Ridge. Undersea volcanoes affect people only when an eruption is large enough to cause a tsunami (Chapter 10).

In summary, most volcanoes and earthquakes occur at plate boundaries, whether in the ocean or on land. You can see this clearly by comparing Figure 12.1 with Figure 11.1 in the last chapter.

Hot Spots

Although most volcanic activity is associated with plate boundaries, there are important exceptions. The most extensive of these is the Hawaiian Islands chain. These islands are composed entirely of lava piles built ten thousands of feet from the deep seafloor over millions of years. They occur in a string in the middle of the Pacific plate (Figure 12.4). Radiometric dating of the lavas (a technique explained in Chapter 16) has revealed that the northwesternmost islands (Nihau, Kauai) are progressively older than islands to the southeast.

One hypothesis for the age pattern is that a stationary **hot spot** exists in the mantle, and the Pacific plate is migrating slowly over it, currently at around 3.4 inches per year. The result is that the area of the plate currently above the hot spot "blisters" into a volcano,

FIGURE 12.4   Island hot spot.

Hawaii is part of a linear volcanic chain of islands called the Emperor Seamounts. The ages of the islands and seamounts in the chain are shown in millions of years. The progressive submergence of the volcanoes resulted from cooling and sinking of the ocean floor as the area moved away from the hotspot. *After Robert W. Christopherson*, Geosystems, 2nd ed., *Prentice Hall, 1994, p. 342.*

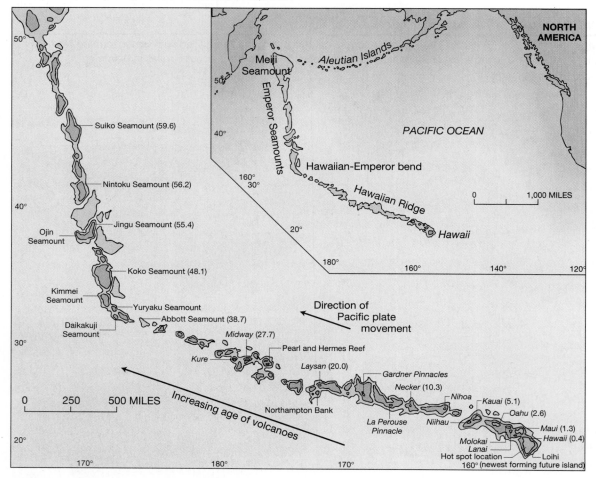

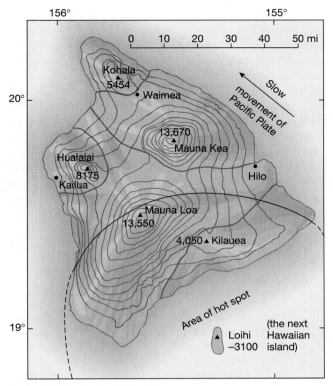

FIGURE 12.5 On the hot spot.

The big island of Hawaii, located directly above a rising current of magma, a hot spot. Five major, overlapping piles of lava rise about 20,000 feet from the ocean floor, all of which have accumulated within the past few hundred thousand years at a rate of 20 feet per thousand years.

which dies out as the plate moves on. Currently, the big island of Hawaii, specifically Kilauea volcano, is over the hot spot (Figure 12.5). Over time, the Pacific plate will move Kilauea off to the northwest, and the next volcano will form.

Actually, it already is forming: an underwater volcano (seamount) called Loihi is growing just 15 miles southeast of the big island of Hawaii. However, Loihi's summit remains about 3100 feet below sea level, so it will be unavailable as an ocean paradise for a while—we'll have to wait a few hundred thousand years.

Hot spots occur beneath continents, too. A large hot spot beneath Wyoming keeps the hot springs and geysers at Yellowstone National Park active.

The Secret Is in the Lava

To scientists, volcanoes have many interesting characteristics, including location, size, shape, and chemistry. But to the public, a volcano's most important characteristic is whether its lava will erupt explosively or flow gently. Explosive eruptions threaten lives over a wide area (for example, the eruptions at Mount Pinatubo and Mount Saint Helens). Flowing eruptions mainly threaten property over a much smaller area (for example, volcanoes in Hawaii and Iceland). What determines how explosive a volcano is? The main factors are (1) how gassy it is and (2) how viscous the lava is, which is controlled by how much silica it contains. Let us look at each factor.

How Gassy Is a Magma?

The most important factor in eruption violence is gassiness. A magma, like any other liquid, dissolves gases within itself. For example, soft drinks have about 0.5 percent carbon dioxide "carbonation" dissolved in them. The typical magma contains about the same percentage of dissolved gases, and 80 percent of these is water vapor. The rest is carbon dioxide, sulfur dioxide, hydrogen chloride, and hydrogen fluoride. When the magma is fairly deep underground, high pressure keeps the gases dissolved in it. But as the magma rises toward the surface, confining pressure is reduced, and the gases are able to escape from solution.

You probably have watched this happen in a clear bottle of carbonated beverage, such as a soft drink, beer, or champagne (depending on your taste and income). Confined in its sealed bottle, the beverage exhibits few bubbles, if any. This is because the bottle is sealed, keeping the liquid under pressure, forcing the carbon dioxide gas to stay dissolved. When the pressure is released by opening the bottle, the carbon dioxide gas abruptly comes out of solution, and gas bubbles quickly appear. If enough gas is released fast enough, it may carry liquid out of the container, erupting like a miniature volcano. In the case of magma underground, whether or not the dissolved gas can escape depends on the magma's viscosity (see the next section), the confining pressure of rocks surrounding the magma, and how fast the magma rises toward the surface.

The explosiveness of an eruption clearly is related to the amount of gassy substances the magma contains. These substances are **volatile,** meaning that they easily change from a compact liquid to a more expansive gas. For example, when confining pressure is removed from magma, the water in it quickly changes from a liquid to water vapor (gas). When this happens, the water molecules separate from one another, going from being packed closely together (as water) to spread far apart (as vapor). Thus, magmas that contain abundant water are more explosive.

The amount of **pyroclastic** material ("fire pieces"; chunks and bits of glassy rock) that can be ejected during an eruption is staggering. It reflects the enormous power of a gaseous eruption.

How Viscous Is a Magma?

Viscosity is the quality of a fluid that describes how easily it flows. Water and gasoline flow easily, so they have low viscosity. Cold syrup and mud flow with more difficulty, so they have high viscosity. In a magma, the main control over viscosity is the amount of silica it contains. The greater the silica content, the more viscous is the magma. Granitic magmas contain 70 percent silica and thus are quite viscous; they are not very fluid and do not flow readily. Basaltic magmas contain only 50 percent silica and thus are less viscous; they are more fluid and flow readily, covering greater distances before solidifying (Figure 12.6). Magma viscosity is important, because it controls how rapidly a magma can flow.

Basaltic magmas form within the ocean basins and, having low viscosity, can rise very rapidly through Earth's crust, typical rates being 10–20 mph. If a basaltic magma could rise without interruption from several miles below the surface, it could reach the surface in a few hours. However, even the most fluid magmas normally rise with many interruptions, and thus may take several million years to rise 30 miles. Frequent stops and starts are typical in magma movement, so rates are hard to predict. When they do reach the surface, they tend to flow or bubble out "gently" because they contain little gas. (They flow "gently" as volcanoes go, but you still don't stand too close to a Hawaiian "lava fountain"—see Figure 12.7.)

Because of its low viscosity, basaltic lavas flow outward from their vent for hundreds or thousands of

(A)

FIGURE 12.6 Magma viscosity.

Basaltic magmas contain only 50 percent silica. This makes them less viscous, so they flow more fluidly, covering greater distances before solidifying. (A) The Columbia Plateau was built around 14 million years ago in this manner. It consists of multiple layers of basaltic lava (B). These flowed 200–300 miles at around 3 mph, some 100 feet high. Volcanoes in Hawaii, Iceland, and the seafloor erupt in similar flows. *Photo: Spence Titley/Peter L. Kresan Photography.*

(B)

FIGURE 12.7 Curtain of fire.

Caused by a fissure eruption along the northeast rift zone of Mauna Loa volcano in Hawaii Volcanoes National Park, "fountains" spray approximately 80 feet high. *Photo: U.S. Geological Survey.*

feet before solidifying. A volcano built of basaltic lava is not a conical peak, but a lower, broad mound. It is shaped like a warrior's shield, and so basaltic volcanoes are called **shield volcanoes.**

Granitic magmas form beneath the continents and are more viscous, so they move much slower than basaltic magmas. As they rise within the crust, more often than not they cool and crystallize completely several miles beneath the surface, forming the rocky cores of mountains like the Rockies.

Because it is easier for runny basaltic (low-silica) lava to reach the surface, the majority of Earth's volcanic eruptions are of such lava. They make fewer headlines because they are less dramatic and seldom kill people. It is the high-silica granitic lava, which has a harder time reaching the surface, that erupts explosively (especially if it is gassy), killing people and making the news.

Volcanic Plumbing

The plumbing system beneath a volcano usually is far more complex than the plumbing in your home. Instead of smooth pipe of uniform diameter, joined at neat right angles, volcanic plumbing is an irregular maze of joints,

faults, cracks, and fissures in the rock that overlies the magma chamber located miles beneath the surface. The rock may be made of different minerals at different depths. The "piping" through which the magma migrates toward the surface twists and turns unpredictably, and is very rough. Unlike your home's plumbing, which has planned exit points (faucets), magma breaks through the surface wherever the volcano's irregular plumbing leads it. The best guide to future eruption openings is the location of past ones.

Volcano Shapes

The shape of a volcano depends on the proportions of fluid lava and pyroclastic debris it contains. As we have seen, lava piles accumulated around a vent or fissure will develop a shield shape, forming a shield volcano. An accumulation of pyroclastic material is much more conical, reflecting the angle of repose of loose sediment (we discussed angle of repose in Chapter 5). This forms a **cinder cone volcano.** Some volcanoes alternate eruptions of lava and ash, forming a **composite cone volcano.** All three types are shown in Figure 12.8.

The Surface of Lava Flows

Very fluid lava cools and solidifies into a distinctive "ropy" shape called **pahoehoe.** More viscous, slower-moving lava cools and solidifies at its surface, but flow continues within and beneath it. This creates a broken surface, forming jagged blocks of lava called **aa.** Both names are Hawaiian. Both types of lava are shown in Figure 12.9.

Hazards of Explosive Eruptions

Undersea volcanoes are not particularly explosive. But volcanoes on the continents near the ocean, like those along the Pacific "Ring of Fire," commonly are explosive because they are rich in silica, compared to oceanic magma. When the high-pressure water vapor finally forces its way through the silica-rich, viscous magma, the gas literally explodes from the magma. The high water content is vaporized by the heat, creating superheated steam. The violence of a steam-enhanced eruption can strain belief.

A classic example is Krakatau, a volcanic island in Indonesia that destroyed itself in 1883. Its explosion, one of the loudest sounds in recorded history, was heard an unbelievable 3000 miles away, in the western Indian Ocean (Figure 12.10). Staggering amounts of ash were propelled as high as 40 miles, causing almost total darkness in the city of Jakarta, 100 miles distant. Red sunsets occurred for several years around the globe.

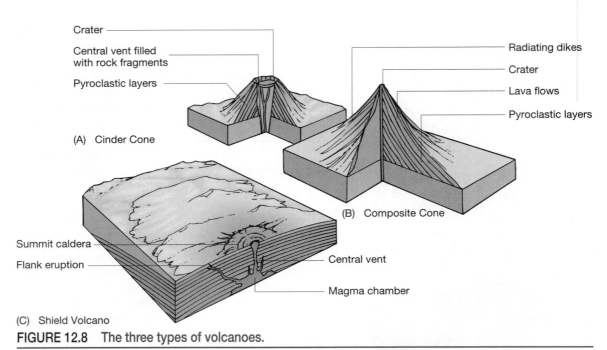

Crater

Central vent filled
with rock fragments

Pyroclastic layers

(A) Cinder Cone

Radiating dikes

Crater

Lava flows

Pyroclastic layers

(B) Composite Cone

Summit caldera

Flank eruption

Central vent

Magma chamber

(C) Shield Volcano

FIGURE 12.8 The three types of volcanoes.

Each has its own distinctive shape, which is controlled by the way it erupts.

FIGURE 12.9 A flow of aa lava advances over a flow of pahoehoe lava in Hawaii.

Photo: J. D. Griggs/USGS Hawaiian Volcano Observatory.

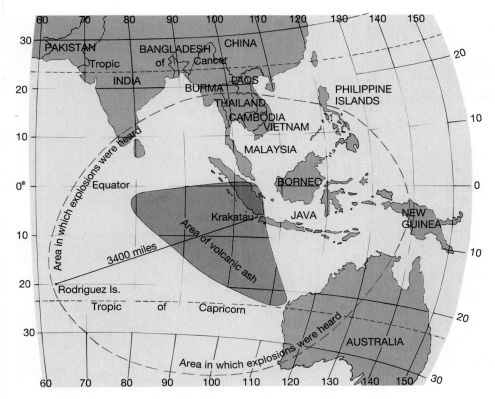

FIGURE 12.10 Krakatau.

When Krakatau volcano erupted violently in 1883, it was not heard around the world, but it tried, as you can see from the distance at which people heard it. Also, note the area covered with tephra (volcanic ash).

Earth's average surface temperature was lowered a few degrees by the ash, which blocked about 13 percent of the sun's radiation from reaching the surface.

Krakatau did not kill many people directly. Instead, a large tsunami generated by the explosion and collapse of the island destroyed 295 coastal towns as far as 50 miles distant and killed over 33,000 people on nearby islands. The tsunami was recorded on tidal gauges as far away as the English Channel.

Tephra.   Material ejected from an explosive volcano may be liquid or solid. But if it is liquid, it does not remain so for long. It quickly chills as it flies through the air and is "frozen" solid when it hits the ground. Such material is composed of glassy fragments (much like window glass) that may shatter explosively in the air or as they hit the ground. All solid material hurled from a volcano is called **tephra** (Figure 12.11A, B). Loosely, tephra is called "ash," because it resembles dark, gritty ash from burned wood, but tephra is really molten glass and rock bits and is not "burned" (Figure 12.11C). The vertical and lateral distances that fragments travel depend on ejection velocity, fragment size, and wind velocity.

Tephra from large eruptions can be dispersed over very extensive areas for several weeks. Figure 12.12 shows how tephra from Mount Saint Helens in Washington State was dispersed by high-altitude winds. The "ash" not only blankets the area around the volcano but

also contaminates the air with very fine suspended particles that enter breathing passages and lungs. Quick, dense coatings of tephra can create remarkable fossils, like the Romans shown in Figure 12.13.

Acid Precipitation.   Gases emitted from the volcanic vent combine with moisture in the air and are transformed into very corrosive acids (sulfur dioxide becomes sulfuric acid, like that used in car batteries; hydrogen fluoride becomes hydrofluoric acid, which attacks glass). These rain (or snow, or fog) over a wide area, falling as **acid precipitation.**

Nuée Ardente (Pyroclastic Flow).   Another product from explosive volcanoes is a lethal mass of hot gases and rock debris that moves downslope like a fluid at speeds that can exceed 60 miles per hour, incinerating everything (and everyone) in its path. It is called a **nuée ardente** (French for "glowing cloud") or a pyroclastic flow. A nuée ardente is shown in a photo in the case study on Mount Pinatubo later in this chapter. The very high velocity is made possible by a cushion of hot, expanding gas that almost eliminates friction between the flowing mass and the ground. These flows can affect areas up to 15 miles from the volcano.

In 1902, a nuée ardente from Mont Pelée in the Caribbean incinerated an entire town of about 30,000 people in moments (Figure 12.14). The lone survivor

The Secret Is in the Lava   **343**

FIGURE 12.11   Tephra.

The common range in size of tephra, from (A) microscopic particles of glass and minerals called volcanic ash to (B) aero-dynamicaly shaped fragments the size of boulders, called bombs. *Photos: (A) P. Schofield/Gamma-Liaison, Inc. (B) Geoscience.*
(C) Tephra (volcanic ash) magnified 140 times. Dark grains are glass with black, hard magnetite; bright grains are plagio-clase feldspar; grains with hint of color are the minerals hornblende and diopside. The hardness of this material makes ash devastating to machinery because it grinds into all moving parts that it contacts. Sample is from the 1980 eruption of Mount Saint Helens. *Photo: American Institute of Professional Geologists.*

was a prisoner in an underground cell. Who says crime doesn't pay?

Lahar (Mudflow).   Yet another common hazard from explosive eruptions is the mudflow, often called by its Indonesian name, **lahar.** These are masses of water-saturated rock debris and mud that move downslope like fast-flowing wet concrete (Figure 12.15A). The debris comes from rock fragments on the volcano's flanks, and the water can come from rain, flash-melted glacial ice and snow, a crater lake, or a reservoir adjacent to the volcano.

Mudflows can be hot or cold, depending on whether they contain hot rock debris. The speed of mudflows depends on their fluidity and the slope of the terrain; they sometimes move 50 miles or more down valley floors at speeds exceeding 20 miles per hour. Volcanic mudflows can reach even greater distances—about 60 miles from the source—than do

nuées ardentes. The chief threat to humans is burial. Structures can be buried or swept away by the vast carrying power of the mudflow (Figure 12.15B).

Hazards of Nonexplosive Eruptions

Non-explosive eruptions usually occur along fissures in the crust. Earth's 40,000-mile-long undersea string of "mid-ocean ridges" is essentially one long, broken fissure that oozes thin, basaltic lava, nonexplosively. On land, as in Hawaii and Iceland, eruption commonly is a comparatively gently flowing and bubbling of lava, which flows downhill along existing valleys. Its speed depends on slope steepness and lava viscosity, but typically it is slower than walking speed (Figure 12.16).

Generally, lava flows destroy any area they cover; property damage is great but loss of life is small or

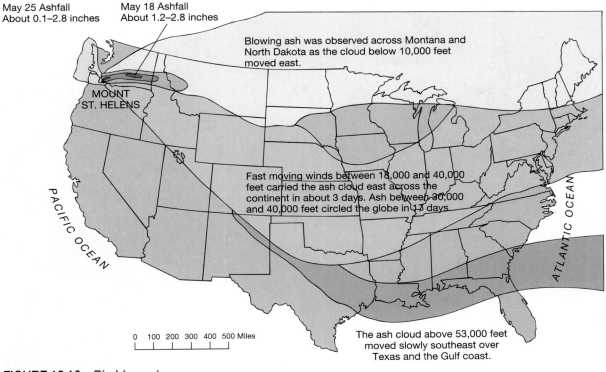

May 25 Ashfall
About 0.1–2.8 inches

May 18 Ashfall
About 1.2–2.8 inches

Blowing ash was observed across Montana and
North Dakota as the cloud below 10,000 feet
moved east.

MOUNT
ST. HELENS

PACIFIC OCEAN

ATLANTIC OCEAN

Fast moving winds between 18,000 and 40,000
feet carried the ash cloud east across the
continent in about 3 days. Ash between 30,000
and 40,000 feet circled the globe in 17 days.

0 100 200 300 400 500 Miles

The ash cloud above 53,000 feet
moved slowly southeast over
Texas and the Gulf coast.

FIGURE 12.12   Big blowout.

Map showing distribution of tephra (ash) from the 1980 eruption of Mount Saint Helens. Some communities were covered by as much as 3 inches of ash. *After U.S. Geological Survey.*

even none. The hazard zone from lava flows normally extends only a few miles because the lava chills and solidifies as it moves, stopping itself. Typically, the surface of the flow hardens in contact with cool air but flow continues below the insulating crust.

Fissure eruptions also occur as spouting fountains that may rise tens of feet above ground level (Figure 12.7). As with explosive eruptions, the key element in the lava fountain is gas content. Usually there is little gas in fissure flows.

FIGURE 12.13   Fossil people.

In Italy, the twin cities of Pompeii and Herculaneum (near Naples) were buried by an eruption of Mount Vesuvius in A.D. 79. At least 2000 people died in Pompeii, many from poisonous volcanic gases. Buried in volcanic ash, their bodies were preserved, actually becoming fossil people. Over the centuries the bodies decayed to dust, leaving a mold in the ash that faithfully reproduced the body shapes. To make the casts in this photo from the ash molds, holes were drilled into the ash and liquid plastic injected, after which the ash was removed. This family apparently suffocated as they slept. *Photo: John Verde/Photo Researchers, Inc.*

The Secret Is in the Lava   **345**

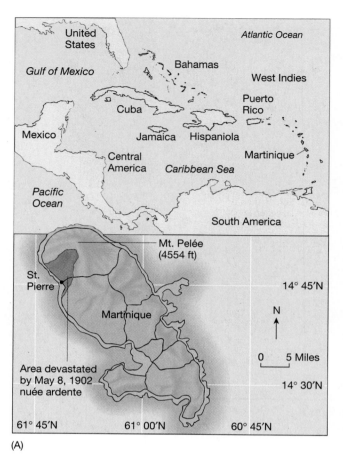

(A)

(B)

FIGURE 12.14   Trouble in paradise.

The beautiful Caribbean island of Martinique lost some 30,000 citizens in minutes, incinerated by a nuée ardente when Mont Pelée exploded in 1902. (A) The location of this island paradise. On the enlarged map, you can see the valleys on the flanks of Mont Pelée that guided the deadly pyroclastics at freight-train speed into the town of St. Pierre, shown in (B). *Photo: Underwood & Underwood/Library of Congress.*

How Many Eruptions Occur Each Year?

How many eruptions occur each year? Has the number changed through time? Figure 12.17 shows eruption frequency for the past 200 years. The top curve (blue) is very misleading, for it gives the impression that volcanic activity is increasing. This is a good example of how inadequate data can mislead the public, the media, and even scientists on occasion. An uninformed or unscrupulous "journalist" might see the upper curve and use it with a 10-second sound bite: "Scientists prove volcanic eruptions are on the rise!" For this period, there does appear to have been an increase (blue curve). But is it meaningful?

It is likely that the apparent increase simply reflects improved reporting of eruptions and better worldwide communication. Note that the number of active volcanoes reported leveled off during the past few decades, to between fifty-five and seventy per year. It is during this period that satellite observation and communication developed.

The interpretation of no change in the number of active volcanoes with time over the past 200 years is supported by the red curve, which tabulates only the largest eruptions, those least likely to have escaped reporting. No significant change through time is present, or expected in the near future.

Eruption Duration

How long can a volcano erupt? The current record-holder is an Italian volcano named Stromboli. It has erupted continuously for 2500 years! At present, fifteen volcanoes have been erupting for at least twenty years and are likely to remain active for some time. All of

(A)

(B)

FIGURE 12.15 Lahar.

(A) A lahar is a mixture of volcanic materials and melting glacial ice. This lahar is flowing down the slopes of Mount Saint Helens during an eruption in 1982. A lahar is basically a volcanic debris flow or mud flow. *Photo: John S. Shelton* (B) One result of a lahar: a mudflow-damaged house along the Toutle River about 25 miles from Mount Saint Helens. Mudflow height is recorded by mud coatings on tree trunks. *Photo: U.S. Department of the Interior Geological Survey.*

FIGURE 12.16 Slower than walking speed.

Glowing orange, hot, viscous lava emerges from East Rift Zone of Kilauea Volcano in Hawaii. The flow moves slowly, so the photographer is in no danger. *Photo: R. T. Holcomb/USGS Hawaiian Volcano Observatory.*

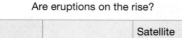
Are eruptions on the rise?

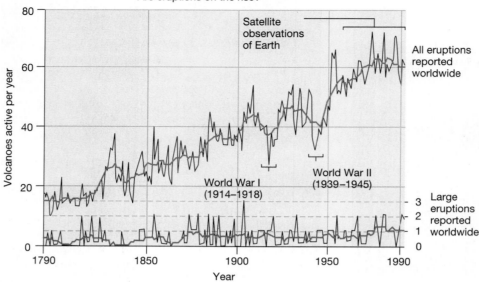

FIGURE 12.17 A lesson in data interpretation.

Volcanic eruptions reported since A.D. 1790. The upper trace shows all volcanoes reported worldwide; the blue line is a 10-year running average. The lower trace shows only large annual eruptions (greater than 0.25 cubic miles of tephra or magma); note the enlarged scale at right. The red line is a 10-year running average. Major declines occurred during the two world wars, when people had greater concerns than reporting volcanic activity.

these are in the Pacific "Ring of Fire," except two (Mount Erebus in Antarctica and Erta-Ale in Ethiopia).

At the other extreme, 10 percent of eruptions end within one day. But most eruptions end within 100 days and few last longer than 1000 days, as you can see in Figure 12.18. Of course, 1000 days—roughly three years—would seem interminable to those living nearby.

Eruption Forecasting: Can We?

Do you, or relatives, or friends live in a volcanic risk area? See the map, Figure 12.19. Will you or someone you know be affected by the next eruption? When will

FIGURE 12.18 How long does it last?

Duration of about 3000 eruptions for which start and stop dates are known. Most eruptions end within 100 days and few last longer than 1000 days.

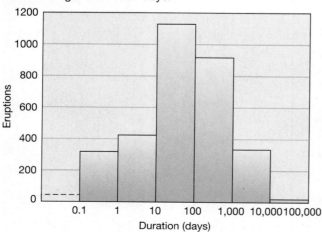

it occur? Will it be explosive? What will be its effects locally, regionally, internationally?

These are a few of the questions asked by the public when the topic of volcanoes comes up. In the last chapter, you saw that seismologists cannot yet predict earthquakes with useful accuracy. But what about **eruption forecasting**? Can scientists predict their specific time and level of eruption? Yes and no.

Forecasting anything is tricky. Consider the daily weather forecast. Your author lived for many years in Norman, Oklahoma, home to a federal weather research center that uses state-of-the-art equipment. You might expect Oklahoma's weather forecasts to be more accurate than most. But analysis revealed that the next day's temperature forecast was correct 82 percent of the time—give or take 4 degrees.

Is that reasonably accurate? An 8-degree band of error for temperatures only a day ahead, and right only 82 percent of the time? (This means that for 18 percent of the year—which is 66 days, more than two months—the temperature forecast was off by more than the ±4-degree error.) Five-day forecasts are even more disappointing; they change almost daily. A forecast made on a Tuesday for the following weekend is likely to be revised three or four times before the week is over. (This is not to criticize meteorologists, who constantly make tough calls from masses of rapidly shifting data.)

If forecasting the weather is difficult, where satellites and other instruments constantly feed measurements made above ground, forecasting volcanic eruptions is much harder, in part because the important variables are hidden. We cannot directly measure the volume of magma, its viscosity, the plumbing system through which it is rising, its velocity, or the strength

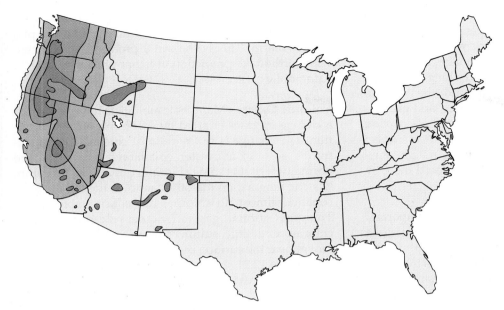

FIGURE 12.19 How dangerous is it?

Map of potential volcanic hazards for the lower 48 states. Severity of risk decreases from red to green to blue. *Source: U.S. Geological Survey.*

of overlying rocks. Meteorologists send helium-filled weather balloons heavenward to measure temperature, pressure, humidity, and wind. But we have no "magma balloon" to drop down a volcano to make measurements! Satellites cannot orbit within the crust to observe conditions. Considering these handicaps, it is a wonder that forecasts of volcanic activity are *ever* correct—but some are. And the track record is better for volcanoes than for earthquakes.

In Chapter 11, we considered the calamity of failing to forecast an earthquake. We also considered the dismaying effect of forecasting one that fails to happen. An example of crying wolf for a volcanic eruption occurred in 1976. A French scientist predicted that Soufrière volcano on the Caribbean island of Guadeloupe would erupt within 24 hours. Hundreds fled, but no eruption occurred. Those who stayed bought land and property from fleeing neighbors at bargain rates, making a healthy profit for their "bravery." Many who heeded the warning were ruined financially. Certainly, the residents of the island will be less likely to respond to future eruption forecasts.

An example of an opposite public reaction—ignoring a warning that turns out to be accurate—is the 1985 eruption of Nevado del Ruiz volcano in Colombia, South America. Scientists predicted the event weeks in advance and notified the government, but people remained in their homes. Heat from the eruption melted large volumes of snow and ice near the volcano's summit, and the resulting mudflows, moving at 100 mph, overwhelmed the town of Amero, burying alive 90 percent of its inhabitants. More than 25,000 people perished.

Forecasting's First Step: Classifying Volcanic Activity

Volcanoes generally are classified as active, dormant, or extinct. An **active volcano** has erupted within recent history. If the volcano has not erupted within recent history, say 5000 years, but is fresh-looking and not significantly eroded, it is considered **dormant,** with the potential to become active again some day. A volcano is **extinct** if it shows significant erosion of its crest and flanks and has no recent eruptive history. Under this classification, there are about 600 historically active volcanoes worldwide, erupting during the past 2000 years.

How accurate are volcano classifications? UNESCO (United Nations Educational, Scientific, and Cultural Organization) estimates that, of approximately 250 volcanoes that erupt every five years, one "extinct" volcano erupts, on average. This error is less than half a percent, so criteria for judging "extinct" volcanoes are very sound.

The definition of "active during historical times" is much more problematic. "Historical times" are those of human record-keeping, but that varies widely, from the past 3000 years in the Middle East, to only 200 years in Hawaii, to none at all for uninhabited volcanic Aleutian Islands in Alaska. Further, the oldest eye-witness account of a volcanic eruption goes back only 3500 years.

What Do Volcanologists Measure to Forecast an Eruption?

The key to forecasting anything—weather, earthquakes, bird migrations, volcanoes—is *continuity of measurement* of whatever is being observed. For weather, it means measuring temperature, pressure, humidity, and wind twenty-four hours a day for long periods. The same is true for volcanoes. For a volcano, scientists must continually monitor seismicity (small earthquakes epicentered in and near the volcano), bulging and tilting of the volcano's slopes, changes in rock around the volcano (changes in the rock's electrical, magnetic, and gravitational properties), and the composition of pre-eruption gases emitted by the volcano. A volcano that is suspect may bristle with seismometers, tiltmeters, magnetometers, gas sensors, and other instruments placed there by scientists who are taking its pulse.

Seismicity. Seismicity probably is the most important measurement. Small earthquakes occur as the magma moves upward, stretching and spreading the plumbing channels through which it moves. This creates the vibrations recorded as earthquakes (a sample seismogram is shown in Figure 12.20). However, this does not make earthquakes an infallible indicator. In a study of seventy-one earthquake swarms and volcanic eruptions, 58 percent showed an increase in earthquake activity before eruption, 38 percent showed an increase without eruption, and, in 4 percent, an eruption occurred with no apparent increase in earthquake activity.

Also, what really predicts an eruption—the number of quakes, or the total energy collectively released by them? And how much of an increase in quakes—or total energy they release—must occur, and at what rate, to reliably forecast an eruption? We don't know, but research continues. Further, even where earthquakes prove to be predictive, the time between increasing quakes and actual eruption varies from hours to months. Consequently, increasing seis-

micity around a volcano warns of possible eruption perhaps six times out of ten, but its timing may cause premature alarm. This is the same problem encountered with earthquake prediction: it is not precise enough to earn public trust.

Volcanic Gases. Many volcanologists have measured a change in gases emitted by a volcano prior to eruption. The change is in the ratio of chlorine gas to sulfurous gas. Groundwater circulating near a volcano normally contains some chlorine, but almost no sulfur. Basaltic magma, on the other hand, is the opposite—it contains almost no chlorine but considerable sulfur. Thus, the "normal" gas from a quiet volcano has mostly chlorine and little sulfur. But, as sulfur-rich magma moves near the surface, more sulfur dioxide gas is emitted. Figure 12.21 shows where a volcano's sulfur dioxide gas has created deposits of sulfur mineral around the vent.

Another example is the 1991 eruption of Mount Pinatubo in the Philippines. In mid-May, the volcano was emitting about 500 tons of sulfur dioxide daily. Two weeks later, the rate rose by a factor of ten to 5000 tons per day. But then it dropped abruptly to only 280 tons daily, which volcanologists interpreted as a blockage, warning of a pressure build-up underground. At about the same time, the foci of associated earthquakes was centering on the summit of the volcano, which had begun to bulge. With all these indicators pointing in the same direction, an eruption clearly was imminent. The main eruption began within a week.

Bulging, Rising, Tilting. As magma approaches the surface, it enlarges its pathways and forces upward the

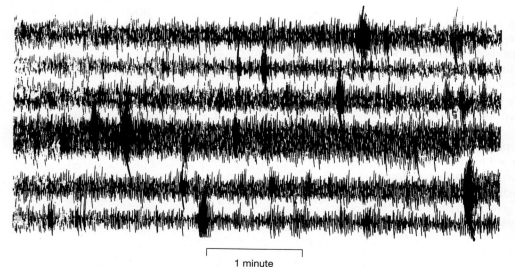

1 minute

FIGURES 12.20 Sample seismogram.

Volcanic tremors recorded on a seismograph during the 1977 eruption of Kilauea volcano. The low-amplitude background wiggles are related to the movement of magma through the plumbing system beneath the volcano, the volcanic equivalent of a freight train lumbering slowly by, continually vibrating the ground. The higher-amplitude vibrations show microearthquakes that occur intermittently as the pressure from moving magma causes movement in the adjoining rocks. Total sample shown is about 28 minutes. *Source: U.S. Geological Survey/James Griggs.*

FIGURE 12.21 Sampling volcanic gases.

A geologist samples gases and measures temperature at Galeras Volcano in Colombia. The temperature is about 500°F, and the gases include up to 5000 tons per day of sulfur dioxide. It is transformed into sulfuric acid droplets in the atmosphere. The yellow mineral around the crater rim is sulfur.

overlying rocks. The result is a raising and tilting of the surface, strong indicators that eruption may be imminent (Figure 12.22). Shortly before an eruption, the change can be quite significant, with areas on a volcano's flanks bulging several feet per day. But, the relation between magma's nearness and how much the surface rises and tilts is not simple. It depends on the viscosity of the magma, its rate of movement, the thickness of overlying rock, and its strength. Once a bulge is detected, either visually or by instruments, it may be minutes, days, weeks, or years before the magma breaks through the ground surface. Again, this makes useful prediction of an eruption very difficult.

Indicators in Combination. Four indicators can be studied together to help predict eruption:

- *Shifts in local gravity.* Magma movement shifts the mass under the volcano, slightly altering local gravity.
- *Shifts in electrical field.* As magma moves, it may alter electrical fields because molten magma is a good electrical conductor, but magma that has cooled and hardened into rock is a poor conductor.
- *Shifts in magnetic field.* Molten magma contains no magnetic minerals, but magma that has chilled and hardened into basalt rock normally contains about 5 percent magnetite, a highly magnetic mineral. Hence, the magnetic field around the volcano may change as the ratio of liquid magma to basalt changes with movement.
- *Ground deformation.* This bulging reflects magma movement below.

Put together, these indicators can help map the movement of magma.

A Prediction that Fizzled

In the 1980s, the USGS studied the Mono Lake area of California, east of San Francisco. This area was chosen because it has been volcanically active within the past 300 years. Geophysical monitoring of the Long Valley caldera began in 1978. Numerous quakes were noted, differing in character from those generated by faults, suggesting they might result from movement of magma. Further, the surface had risen 18 inches in five years, directly above a known magma chamber. The earthquakes' foci migrated nearer the surface, and more recent quakes were less than 2 miles deep.

The geophysicists hypothesized that magma was rising from the top of a chamber about 5 miles down, causing a broad uplift and forming a new finger of injected magma. This caused the USGS to issue an

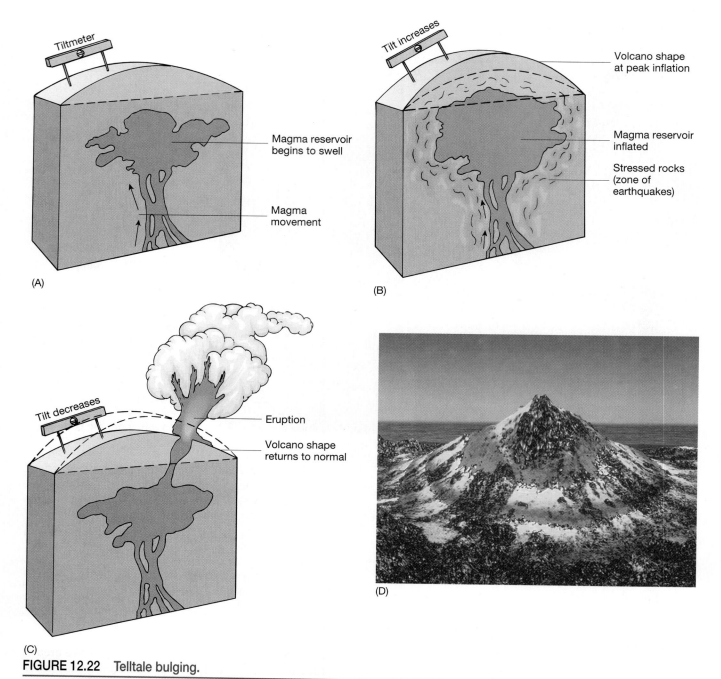

(A)

(B)

(C)

(D)

FIGURE 12.22 Telltale bulging.

(A, B) Pressure from rising magma pushes up the surface of a volcano, warning of impending eruption. (C) Following the eruption, the surface subsides. (D) Using a radar technique from a satellite and computer enhancement, this remarkable false color photo shows amounts of deflation of the summit of Mount Etna following an eruption in 1993. The technique can also be used to monitor uplift and tilting that precedes an eruption. *Photo: D. Massonnet/CNES French Space Agency.*

eruption warning, which showed the possible centers of eruptive activity and likely areal extent and damage potential. The only actual damage in the area was that the warning hurt tourism. In 1983, the uplift slowed and earthquake activity decreased—good for residents, but bad for scientists' credibility. The area now is peaceful, but perhaps some day the magma will continue its upward path, and an eruption will happen.

The Prediction Dilemma

Obviously, no single indicator can predict an eruption accurately. Used in concert, multiple predictors can come close. But timing remains uncertain. At what point should the public be notified of the possible danger? When should a wholesale evacuation be required by local, state, or federal authorities? Should people be permitted to take their chances and ignore the warnings? Should those who stay sign a waiver that they will not sue for damages, should they lose the gamble? These are real-life questions that responsible people need to consider if they live in an area of active volcanoes.

Although eruption forecasts are closer to being useful to the public than earthquake forecasts, neither is there yet. People seek a guarantee that an event will happen on a specific date, plus or minus a few days, and want to know the severity of the event. Such a precise forecast may become possible within the next half century for some eruptions.

Case Study: Mount Saint Helens, Washington, 1980 Eruption

The most devastating North American eruption in recent times was the explosion at Mount Saint Helens in Washington State in 1980 (Figure 12.23). Eruptions began here more than 37,000 years ago. The volcano has destroyed itself in explosive eruptions and rebuilt itself with more lava and tephra many times during this period. Virtually all of the present volcanic mountain has formed since 500 B.C., and most of its upper

FIGURE 12.23 Mount Saint Helens, before and after its explosive eruption in 1980.

(A) Ten years prior to the eruption. *Photo: William E. Ferguson.*
(B) Two months after the eruption. The crater is 6500 feet in diameter. The top of the volcano was lowered from 9675 feet to about 8200 feet (nearly 1500 vertical feet were blown away, a bit more than Earth's tallest building, Chicago's Sears Tower). The floor of the crater lies at about 6070 feet. *Photo: Roger Werth/The Daily News.*

(A)

(B)

part has been created within the past few hundred years. The most recent prior eruption was in 1857. Trenches dug near the volcano disclose multiple pyroclastic deposits (Figure 12.24), some of which may have affected the original American Indian settlers, who knew of the mountain's explosiveness.

Mount Saint Helens' restlessness first appeared on March 20, 1980. A swarm of microearthquakes suggested that magma was rising upward into the volcano. A week later, a small eruption sent clouds of smoke and ash more than 20,000 feet into the atmosphere and formed a new small crater near the mountain summit. Intermittent volcanic activity continued and seismic rumbles renewed, indicating magma movement. In the first week of April, a large bulge developed on the north side of the mountain. By early May, the bulge was rising as fast as *5 feet per day.* Eventually, it measured nearly 0.6 by 1.2 miles and

FIGURE 12.24   A fiery history.

Sequence of tephra deposits from eruptions of Mount Saint Helens prior to 1980, seen in a trench 5 miles southeast of the former summit. Seven units of tephra are labeled; Bi has been dated as about 1800 years old. Units below Bi are older, those above Bi are younger. *Photo: U.S. Geological Survey.*

swelled outward over 300 feet. At this point, geologists and the U.S. Forest Service closed the area to the public. Local inhabitants were evacuated.

On the morning of May 18, two magnitude-5 earthquakes rocked the mountain. They caused the steepened bulge on the mountainside to break loose, sending a huge landslide tearing down the mountainside, partly lubricated by glacial ice from near the summit. The loss of the bulge of rock released confining pressure on the gas-charged magma beneath it, and the magma exploded outward, blowing away a sizeable part of the mountain (Figure 12.23B).

Waves of superheated dust and gas stripped leaves and needles from trees and flattened them like matchsticks, which they might have become had they lived (Figure 12.25). Huge mudflows filled lakes, rivers, and streams, destroying roads, bridges, and life in the lakes. More than 120 houses were partly or completely buried.

In a town 200 miles northeast of Mount Saint Helens, 4-foot-deep ash drifts accumulated. Ash accumulations closed public and commercial facilities across the State of Washington, idling 370,000 workers. Damage estimates to property, timber, agriculture, roads, harbors, and other human works were $2–3 billion.

The force of the blast was much greater than anticipated, resulting in an unexpectedly high death toll of sixty-three. Victims included a geologist near the volcano who was studying its activity when it erupted (his last words over a radio were "This is it!"). The Weyerhauser Timber Company, which was logging in the area, had marked its maps with "red zone" danger symbols, but the red zone was too small. The company paid benefits to surviving employees and their families, and was cleared of negligence by the courts.

The distribution of devastation from the blowout of Mount Saint Helens is shown in Figure 12.26. It reveals several things:

1. The blowout occurred on the north side of the volcano, as this is the side where almost all of the damage occurred and where debris flows are located.
2. Although flattened trees at least 3 miles distant are oriented downblast, closer to the mountain there apparently were swirling motions of the blast materials. Many trees were forced to the ground oriented east–west.
3. Outward from the wide area of felled trees is a scorch zone, in which the heat burned foliage from the trees but the blast was insufficient to fell them. Without leaves and needles, the trees died anyway.
4. Debris avalanches that originated on the north side of the volcano followed existing drainage channels of the Toutle River and flowed westward.
5. Mudflows were confined to the southern side of the mountain. Because mudflows require a great

FIGURE 12.25 Like quills on a porcupine's back.

Large trees lie in alignment, where the eruption of Mount Saint Helens blew them over. Researchers learned that such blow-downs of timber result from the force of heavy, ash-laden clouds that follow the land's contours and flatten any forest they pass across. *Photo: American Institute of Professional Geologists.*

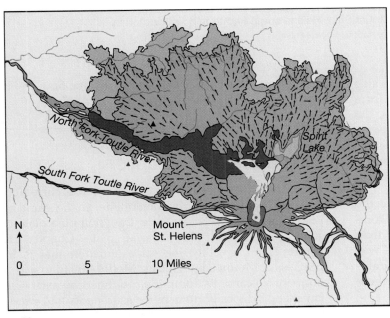

FIGURE 12.26 Effects of the eruption.

Map of Mount Saint Helens region shows effects of the 1980 eruption and lava dome that subsequently formed. The devastated area exceeds 200 square miles. Spirit Lake grew; its old and new shorelines are shown. *Source: U.S. Geological Survey.*

Key:

- ⬜• Lava dome
- ⬜ Pyroclastic flow deposits
- ⬛ Mudflow deposits and scoured areas
- ⬛ Scorch zone
- ⬛ Down-timber; arrows indicate direction of tree blowdown
- ⬛ Debris-avalanche deposits
- ▲ Mountain summits

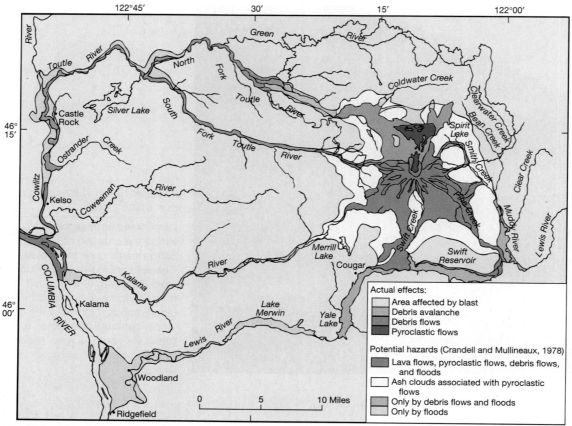

FIGURE 12.27 Forecast vs. Reality.

Actual extent of damage during the 1980 eruption of Mount Saint Helens compared to that anticipated in the hazard-zone map published two years earlier. *Source: U.S. Geological Survey.*

deal of water to form, they occurred on the side where glacial ice was melted by the heat to form water. On the north side of the mountain, the blast instantly vaporized the ice so that water was not available. The mudflows on the south side, like the debris avalanches on the north side, moved downhill along existing stream channels.

In 1978, two USGS volcanologists predicted the damage that an eruption would cause. They had forecast an eruption at any time, probably sooner than later, but could not specify a date. The "sooner" was only two years later. Comparing their predicted damage area with what actually happened reveals where their forecast was in error (Figure 12.27).

They forecast damage based on a normally directed explosion (straight up), about as powerful as previous eruptions in the area. This was reasonable, but they were wrong on both counts. The volcanic blast came from a blowout on the north side of the volcano, not from the summit, and thus was not vertical but directed horizontally, increasing damage on the ground. Further, the blast was much more powerful than expected. Hence,

the damage was greater and the pyroclastic flow, mudflows, and floods occurred largely to the north.

Case Study: Mount Pinatubo, Philippines, 1991 Eruption

In 1991, another violent volcanic eruption occurred on the other side of the Pacific Ring of Fire, on the island of Luzon in the Philippines. Here, the Philippine plate is subducting beneath the Eurasian plate (Chapter 11, Figure 11.16), creating the same situation that is in progress at Mount Saint Helens. Mount Pinatubo was a dormant volcano, its summit surrounded by extensive fan-shaped deposits from pyroclastic flows and mudflows 600 to 8000 years old. On April 2, villagers were startled to see small water-vapor explosions from a line of vents about a mile long near the upper part of the volcano. By next morning, the line had stretched to 3 miles.

The next day, Philippine geologists installed a portable seismograph 5 miles west of Pinatubo and recorded about 200 volcanic earthquakes in its first 24 hours of operation. Evacuation within a 6-mile radius of the summit was recommended, and 5000 people left. More seismographs were installed 6 to 9 miles northwest of Pinatubo. They recorded 40 to 140 earthquakes each day for the next seven weeks, with magnitudes mostly between 0.5 and 2.5, too slight to be felt by people but easily recorded by the instruments.

A concerned Philippine government requested help from the USGS, and a team went to the Philippines on April 22. They installed seven seismic stations to monitor the depth of the earthquake foci and the amplitude of their waves. They distributed a volcanic hazard map, shown in Figure 12.28. It indicated that old pyroclastic flows had reached American-operated Clark Air Base and adjacent heavily populated areas. Alert levels 1 to 5 were established (5 = worst). A Public Broadcasting System-style volcano video was shown to the local populace, improving their response to official warnings.

Between May 13 and May 28, emissions of sulfur dioxide from the volcano's vents increased tenfold, from 500 tons per day to 5000 tons per day. Apparently, magma was rising toward the surface. In early June, the earthquake focus shifted from 3 miles north-northwest of the summit to directly beneath the steam vents, and a lava dome formed near the most vigorous vent.

An explosive eruption appeared imminent. The alert level was raised to 3 (possible eruption within two weeks) on June 5, then to level 4 (possible eruption within 24 hours) on June 7, and then to level 5 (eruption in progress) on June 9. The evacuation radius was increased from 6 miles to 9 miles, and then 12 miles, and 20,000 more people moved into evacuation camps. On June 10, more than 14,000 military personnel, dependents, and retirees evacuated.

On June 12, several explosive eruptions thrust an ash column to at least 12 miles altitude, as the dramatic chapter-opening photo shows. Nuées ardentes (pyroclastic flows) extended about 3 miles from the volcano (Figure 12.29). Sand-size tephra blanketed nearby coastal towns to thicknesses of an inch. Earthquake frequency and intensity increased dramatically, interpreted to mean increasing pressure as gases were released from solution in the magma. On June 15, a continuous, strong eruption occurred and Clark Air Base was evacuated. The base was in total darkness from a heavy, wet ash fall that included golf-ball sized pieces of glassy volcanic rock called **pumice.**

FIGURE 12.29 Nuée ardente.

Nuée ardente races down the slopes of Mount Unzen, Japan, in 1991. The fireman and fire truck are trying to outrun the incinerating ash cloud. *Photo: Yomiuri/AP/Wide World Photos.*

FIGURE 12.28 Volcanic hazard map for Mount Pinatubo.

This map was distributed on May 23—just twenty days before the June 12 eruption. *Source: U.S. Geological Survey.*

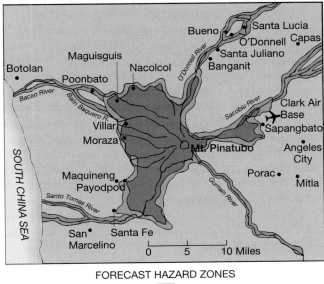

FORECAST HAZARD ZONES
- Mudflow
- Ashfall
- Pyroclastic-flow
- Pyroclastic-flow buffer

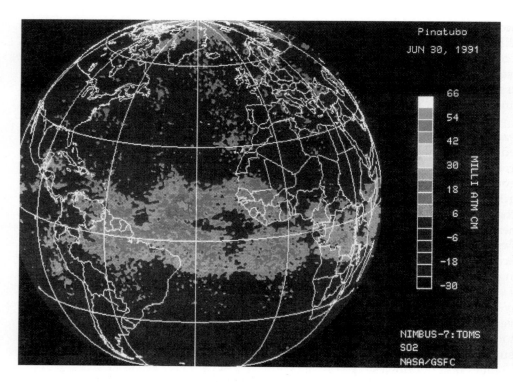

Pinatubo
JUN 30, 1991

66
54
42
30
18
6
-6
-18
-30

MILLI ATM CM

NIMBUS-7:TOMS
SO2
NASA/GSFC

FIGURE 12.30 A volcano's aftermath.

Two weeks after the June 1991 eruption of Mount Pinatubo in the Philippines, a false-color satellite image shows its sulfur dioxide cloud stretching from the eastern Indian Ocean to Central America. Scientists estimate it had circled the globe by July 5, 1991. Such a cloud affects climate change more than ash. Scientists predicted a cooler summer in 1992 worldwide due to the eruption. Pinatubo affected global climate more than any eruption since Krakatau in 1883. Pinatubo's ash in the upper atmosphere also caused red sunsets through 1991 and early 1992. The ash reflected sunlight that otherwise would be converted to heat on Earth. *Source: NASA Headquarters.*

FIGURE 12.31 Mudflow damage.

Abacan River channel in Angeles City near Clark Air Force Base, Philippines. A mudflow has taken out the main bridges; a makeshift bridge allows pedestrians to cross the debris. *Photo: T. J. Casadevall, U.S. Geological Survey.*

Smoke and ash attained 22–25 miles altitude and expanded more than 60 miles in all directions. At least 17 commercial jets inadvertently flew through the drifting ash cloud, all surviving the experience (see later section on airline safety). In about 22 days, particles and gases from the eruption spread worldwide (Figure 12.30). Pyroclastic flows up to 650 feet thick covered an area of roughly 36 square miles and filled deep canyons near the volcano. The worst-case scenario hazard map prepared several weeks earlier closely forecast the extent of the pyroclastic flows. This was a significant forecasting success, based on lessons from Mount Saint Helens.

As if Mount Pinatubo's eruption were not enough, the ash was soaked by heavy rainfall from a hurricane, adding tremendous weight (each gallon of water weighs 8.33 pounds). This collapsed numerous roofs. Rain also remobilized pyroclastic debris, causing debris flows and mudflows that inflicted great damage on all sides of the volcano (Figure 12.31). By June 16, about 200,000 people had fled to makeshift evacuation camps, the largest evacuation ever from a volcano. Casualties included 320 dead, 29 missing, and 279 injured. Most deaths were from collapse of ash-laden roofs, many occurring only because of the coincident hurricane.

Hazard Assessment

In **hazard assessment,** volcanologists estimate whether an eruption will be explosive, how large it will be, and its effects. They must use past eruptions as a guide, assuming that a specific volcano will produce the same events at the same average frequency as in its past. Unfortunately, this often is untrue, and disastrous consequences result. But how else can volcano forecasters proceed? If they did not use the past as a guide to the future, forecasts would have to be pulled from thin air.

Hazards to be expected from a volcano vary with its eruptive style, which varies with tectonic setting and magma composition. For explosive eruptions, hazards include hot ash falls, pyroclastic flows in existing valleys, and the ground-hugging, incinerating clouds called nuées ardentes. Explosive volcanoes have steeper slopes than volcanoes that erupt quietly, so mudflows and avalanches must be anticipated. Clearly, schools and hospitals should not be built where these phenomena have been observed during past eruptions.

Mount Saint Helens is an example of a volcano from which explosive eruptions must be anticipated. Population is light around this volcano, so there are no schools or hospitals nearby. In contrast, consider Mount Pinatubo, another explosive volcano. The entire Philippine Islands consist of active volcanoes, and the islands are small. If the guidelines suggested above were applied to the Philippines, no schools or hospitals could be built anywhere—and the Philippines has 74 million people.

In areas of "Hawaii-type" volcanoes, with fluid, gas-poor magmas, gentle eruptions are the rule. Their chief hazard is lava flows, which generally threaten property, not lives. Structures should not be built in low areas, for the obvious reason that lava flows down existing channels. The higher on a slope a building is located, the safer it will be. The volume of lava cannot be predicted accurately.

So, hazard assessment must consider a volcano's nature, history, and the population around it. UNESCO's criteria for dangerous volcanoes are shown in Table 12.1. The higher the total score in this table, the more hazardous is the volcano.

Using criteria like those in Table 12.1, environmental geologists prepare hazard-zone maps for areas near active volcanoes. In Hawaii, Mount Kilauea has concerned the USGS for many years because of nearly continuous eruptions since at least 1823. Yet, for unknown reasons, the period from 1934–1951 saw no eruptions at all. The USGS prepared a hazard map in the 1980s for the entire island and a more detailed one in 1990 for the area affected by Kilauea on the southeastern side of the island (Figure 12.32).

In 1951, a land developer would have felt safe building on the East Rift Zone (areas 1 and 2 in Figure 12.32). It had seen no eruptions since 1934. But in 1952, the volcano became active again, and over the succeeding thirty-eight years there were thirteen separate East Rift Zone eruptions, two persisting over four years. From the first return of volcanic activity at the lower East Rift Zone to the present, almost 30 percent

TABLE 12.1	Proposed Criteria for Identification of High-Risk Volcanoes	

Hazard Rating   *Score*

1. High silica content of eruptive products (andesite / dacite / rhyolite)
2. Major explosive activity within last 500 yr
3. Major explosive activity within last 5000 yr
4. Pyroclastic flows within last 500 yr
5. Mudflows within last 500 yr
6. Destructive tsunami within last 500 yr
7. Area of destruction within last 5000 yr is > 10 km^2
8. Area of destruction within last 5000 yr is > 100 km^2
9. Occurrence of frequent volcano-seismic swarms
10. Occurrence of significant ground deformation within last 50 yr

Risk Rating

1. Population at risk > 100
2. Population at risk > 1000
3. Population at risk > 10,000
4. Population at risk > 100,000
5. Population at risk > 1 million
6. Historical fatalities
7. Evacuation as a result of historic eruption(s)

         Total Score _____

Note: A score of 1 is assigned for each rating criterion that applies; 0 if the criterion does not apply.
Source: UNESCO

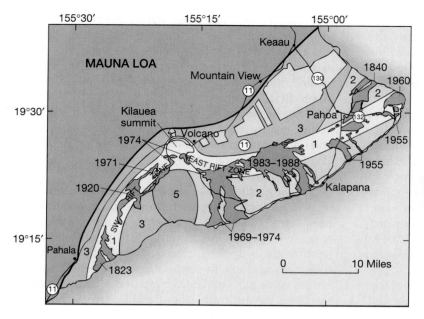

FIGURE 12.32 Hazard map.

Hazard range of Kilauea lava flows, from 1 (high) to 5 (low). Lava flows since 1823 are tinted. Over 90 percent of Kilauea's surface has been lava-covered since people first occupied Hawaii about 1500 years ago. Nearly 30 percent of the surface in Hazard Zone 2 (south of the east rift zone) has been covered since 1955. *Source: U.S. Geological Survey.*

of the land surface between the rift zone and 32 miles of coastline have been covered by lava.

Mauna Loa volcano has had a similar erratic eruption history. Since record-keeping began in 1832, the volcano has erupted thirty-nine times, but irregularly. Over the past forty-five years, eruptions have declined, and real estate value has soared. Table 12.2 shows the probability of a Mauna Loa eruption occurring each day, week, month, and so on.

There is a fundamental irregularity to volcanic processes that makes it impossible to predict volcanic activity with confidence. Hazard-zone maps are valuable for planning, but come with no guarantee.

TABLE 12.2 | **Probability of Eruption of Mauna Loa**

The probability does not change daily. It does change as future eruptions create new average recurrence times.

Time Interval	Probability
1 Day	.00069
1 Week	.0048
1 Month	.020
1 Year	.22
2 Years	.39
3 Years	.53
4 Years	.63
5 Years	.71
6 Years	.78
7 Years	.83
8 Years	.86
9 Years	.89
10 Years	.92
12 Years	.95
19 Years	.99

Case Study: Fighting an Eruption in Iceland

Suppose you live in Iceland, a place of continuous flowing (basaltic) volcanism. You know that the entire country is simply a pile of lava thousands of feet thick with twenty-two active volcanoes. You know that Iceland is part of the Mid-Atlantic Ridge, a crack in Earth's crust that will not heal for millions of years, so eruptions will continue "forever," as far as you are concerned. Yet, atop this assured volcano, you build a house where you believe it to be fairly safe, based on past eruptions.

But then one day it happens. A major eruption occurs. It's not a Pinatubo-style detonation, but plentiful lava is flowing your way. What can you do?

This situation faced residents of Heimaey, a tiny island off Iceland's southern coast. Heimaey is Iceland's largest fishing port. In 1973, Helgafell volcano erupted, pouring out basaltic lava and tephra, threatening to bury the island's lone town (Figure 12.33A). Over a few hours, 3 feet of tephra accumulated in parts of the town near the vent. The defense of Heimaey became history's most ambitious attempt to control volcanic activity.

Although some residents evacuated, many stayed, and with help from rescue workers they continually shoveled ash from roofs before heavy accumulations could cause collapse (Figure 12.33B). They covered windows with sheet metal to prevent breakage from the pressure of ash drifts (like snow drifts). But their greatest triumph was to divert and contain a lava flow from reaching the harbor, saving their livelihood. They used bulldozers to divert the lava flows. With large hoses and powerful pumps, they sprayed seawater onto the front and top of the lava to chill it and hasten solidification.

(A)

(B)

(C)

FIGURE 12.33 A hot time in Iceland.

(A) Basaltic lava pours through the fishing port of Heimaey in 1973. Spraying cold seawater on the advancing lava chilled and solidified it, halting the lava's advance. The town uses geothermal heat from the volcano. *Photo: Vulcain-Explorer/Photo Researchers, Inc.* (B) A sloped roof saved this house from being crushed by tephra. *Photo: Krafft-Explorer/Photo Researchers, Inc.* (C) A hose carries cold seawater to quench the hot lava, halting its flow. *Photo: U.S. Geological Survey.*

The chilly water cooled and hardened the lava, forming protective dikes of volcanic rock (Figure 12.33C).

The pumping continued day and night for several months, totaling about six million tons of water, equal to pouring Niagara Falls on the lava for 30 minutes! Measurements showed that the lava hardened fifty to a hundred times faster in sprayed areas than where the lava was just allowed to air-cool. The harbor was saved, as was much of the island's one city. It was well worth the $1.5 million cost.

How Volcanoes Affect Climate

Large volcanic eruptions hurl vast volumes of tephra and assorted gases into the atmosphere. Among the gases is sulfur dioxide, which combines with water vapor in the air to form droplets of sulfuric acid. Most volcanic ash falls from the stratosphere in a few months, so its climatic effects are temporary, but the sulfuric acid droplets remain. The sulfuric acid also shows up in tests for acid in ice cores from Greenland and Antarctica, and correlates in time with historic volcanic eruptions (Figure 12.34). Further, an apparently permanent layer of sulfuric acid droplets from volcanic sources exists at about 15 miles altitude.

So what? Well, the droplets reflect incoming solar radiation (sunlight) so effectively that the presence of this layer cools the lower atmosphere in which we live and the adjacent ground surface. Following a large eruption, a significant worldwide temperature reduction of 0.5–1.0°F occurs for several years. This effect is shown in Figure 12.35. The coldest summers of the past 400 years resulted from volcanic eruptions.

The most famous such episode happened in 1816. It followed the massive eruption of Tambora volcano on Sumbawa Island in Indonesia in April, 1815. This blast sent tephra 20 miles into the air. After five days of quiet, Tambora shot another blast to 38 miles altitude, which caused three days of total darkness over an area 300 miles from the volcano! The Tambora eruption was one of the largest ash-producers in the past 10,000 years. Figure 12.36 shows Tambora's tephra production compared to other major volcanic eruptions.

The summer months in central England were about 2–3°F cooler in 1816 than in 1815. That may not sound like much, but it was the equivalent of moving England many miles closer to the North Pole or dampening the warming effect of the Gulf Stream. In fact, 1816 was a famine year in England, France, and Germany, as crops failed during the extraordinarily cold summer. In eastern Canada, the summers of 1816 and 1817 on Hudson Bay were the coldest of any in modern times. French grape harvests were delayed in 1816. Frost damage to tree rings for those years has been reported from the western United States and South Africa. Average

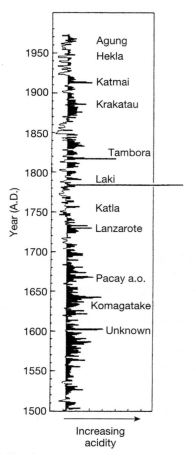

FIGURE 12.34 Ice cores store volcanic evidence.

Acidity profile through an ice core from central Greenland, showing sulfuric acid "spikes" due to acid aerosols from historic eruptions. Several spikes have been related to known eruptions like Krakatau and Tambora; others are more debatable. The "unknown" spike at 1601 may have been due to a major eruption in Peru. *From Hammer et al. (1980). Greenland ice sheet evidence of post-glacial volcanism and its climatic impact.* Nature 288, p. 230–235.

temperatures in the Northern Hemisphere in 1816 were 1–2°F cooler than normal. No wonder that people called 1816 "the year without a summer!" Although the effect was short-lived, the explosion of Tambora was a good lesson to us all. The world about us that seems so stable really is not.

A recent example of the same effect, although much smaller in scale, is the climatic cooling caused by plentiful sulfur dioxide emitted by Mount Pinatubo in 1991. The sulfuric acid droplets it produced caused a temporary reversal of the global warming trend of the past half century (discussed in Chapter 17).

Volcanic eruptions that contain little sulfur dioxide cause no measurable effect on climate. Large amounts of the gas are needed to alter climate, but specifically how much is uncertain. We do know that a few high-sulfur eruptions during the past few centuries have measurably

affected agriculture. Evidence from prehistoric pyroclastic rocks suggests that much larger eruptions have occurred, and will again.

"Volcanic winters" are a real possibility for us, wherein one or more massive eruptions causes such a layer of sulfuric acid droplets that so much solar energy is reflected away from Earth that temperatures plummet, perhaps for years. Such a large sulfurous eruption would drastically affect crop yields and could create food crises, especially where food presently is in short supply. There is no question that such eruptions will recur; the only uncertainty is where and when. As noted by physicist James Lovelock, developer of the Gaia hypothesis (Chapter 1), "Every two or three centuries there is a volcanic eruption so big that it fouls up the atmosphere for a couple of years. According to current estimates [1996], the world's grain reserves are only 45 days."

Benefits from Volcanoes

We have dwelt upon the hazardous aspects of volcanoes. But volcanoes also have important benefits. You may be surprised to learn that one is an increase in real estate. All that lava has to go somewhere, and if it is above sea level, it becomes more land. The 1960 eruption of Kilauea added 0.4 square mile to the state of Hawaii. The emergence of a new volcano off Iceland's southern coast in 1963 added Surtsey Island to that country. The neighboring island of Heimaey grew 20 percent from the 1973 eruptions. Additional land is just below sea level at many places in the world's ocean, in the form of **seamounts**, volcanoes that are building vertically with each eruption, approaching new islandhood. Some need only a few more eruptions to become inhabitable real estate.

In addition to more acreage, lucky countries receive prime rock for agriculture. Volcanic rock normally is broken and porous near the surface and is naturally rich in plant nutrients. Because it weathers rapidly at Earth's surface, soil development on lava and tephra is quick compared to other types of rock. With rapid soil formation and a warm, humid climate, Hawaii's soil is well suited for growing pineapples and coffee.

Other benefits from volcanoes include:

- Pumice, a volcanic rock, is used as an abrasive (it is like crude broken bubbles of fine glass glued together).
- Volcanic cinders are used in construction.
- Fractured volcanic rock forms a useful aquifer in many areas.
- Chemicals such as boric acid, ammonia, and carbon dioxide can be recovered from volcanic gases.
- Of importance in energy-poor areas, such as Iceland, is the steam that pours from numerous

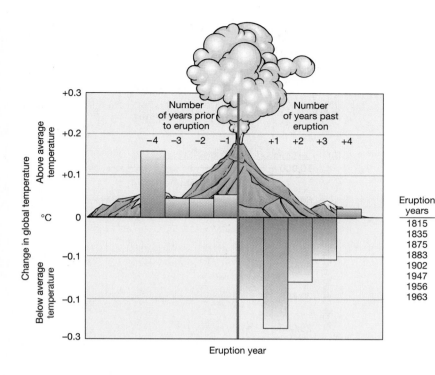

Number of years prior to eruption

−4 −3 −2 −1

Number of years past eruption

+1 +2 +3 +4

Change in global temperature

Above average temperature

Below average temperature

°C

+0.3
+0.2
+0.1
0
−0.1
−0.1
−0.3

Eruption year

Eruption years
1815
1835
1875
1883
1902
1947
1956
1963

FIGURE 12.35 Volcanic cooling.

Temperature change in the Northern Hemisphere in the four years immediately before and after eight large eruptions of the past two centuries.

vents. It is an inexhaustible heat source for homes, schools, and other buildings.

Volcanic Eruptions and Airline Safety

A **volcanic plume** of ash (tephra) can be a serious menace to commercial aircraft. Not only can the tephra clog air intakes; it contains glass and mineral particles with hardnesses about the same as the metals in aircraft engines, causing abrasion.

Some plume heights are shown in Figure 12.37. During the past fifteen years, twenty-three aircraft have reported encountering these eruption plumes. Several aircraft suffered damage from the 1980 eruptions of Mount Saint Helens in Washington State. In 1982, a near-disaster occurred from eruptions of Galunggung Volcano in Java, Indonesia. Volcanic ash entered the engines of two 747s and caused thrust loss in all four engines on both planes. After powerless descents of nearly 25,000 feet (almost 5 miles), the pilots of both aircraft succeeded in restarting their engines and both landed safely at Jakarta Airport.

Between December 1989 and February 1990, five commercial airliners suffered volcanic ash damage from Redoubt Volcano in Alaska. Most serious was a 747's encounter with an ash cloud nearly 200 miles from the volcano. All four engines lost power and the plane dropped about 2.5 miles (from 28,000 feet to 14,000 feet altitude) before the engines were restarted. Nearby mountain peaks were at 10,000 feet. No passengers were injured, but damage to the plane exceeded $80 million.

Although Anchorage airport remained open, most airlines canceled operations for a while. Normal service did not resume until mid-February. Lost revenue to Anchorage International Airport was about $2.6 million.

Unfortunately, plumes are very difficult for aircraft to detect with current technology. In the fifty states, approximately fifty-six volcanoes have historical eruptive activity, and forty-four are in Alaska (80 percent). Because of the hazard, the federal government has established a group of scientists and airline representatives to react to eruptions that affect U.S. air operations. The group must issue a volcanic advisory forecast every 4 hours regarding the status of volcanic clouds, with a 12-hour forecast of ash cloud behavior.

Summary

Volcanoes are located mostly along current plate boundaries, and represent magma welling up from the lower crust and upper mantle. Most volcanoes spew basaltic magma as either lava, pyroclastic material, or both. The most dangerous volcanoes are the ones that are along the margins of continents, because they emit the most pyroclastic material. Their magmas contain plentiful gas, which supplies

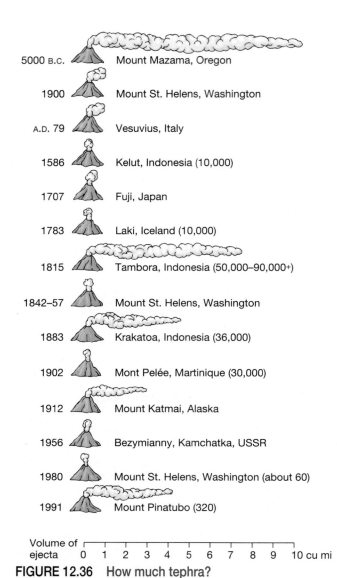

5000 B.C.   Mount Mazama, Oregon

1900   Mount St. Helens, Washington

A.D. 79   Vesuvius, Italy

1586   Kelut, Indonesia (10,000)

1707   Fuji, Japan

1783   Laki, Iceland (10,000)

1815   Tambora, Indonesia (50,000–90,000+)

1842–57   Mount St. Helens, Washington

1883   Krakatoa, Indonesia (36,000)

1902   Mont Pelée, Martinique (30,000)

1912   Mount Katmai, Alaska

1956   Bezymianny, Kamchatka, USSR

1980   Mount St. Helens, Washington (about 60)

1991   Mount Pinatubo (320)

Volume of
ejecta   0 1 2 3 4 5 6 7 8 9 10 cu mi

FIGURE 12.36   How much tephra?

Cubic miles of tephra ejected by historic explosive eruptions. Each cubic mile of these fragments weighs about 350 trillion pounds. Where known, casualties are shown in parentheses. *Source: U.S. Geological Survey.*

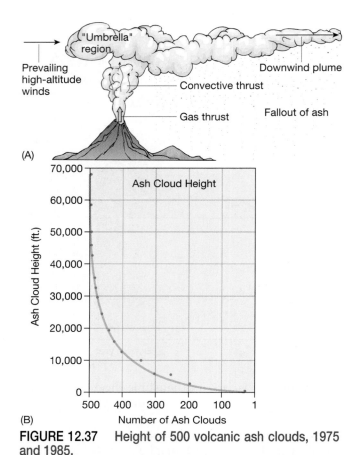

(A)

(B)

FIGURE 12.37   Height of 500 volcanic ash clouds, 1975 and 1985.

Of these 500 volcanic ash clouds, 490 reached 40,000 feet, so it was not possible for a plane to fly above the ash.

the explosive power. Explosive eruptions commonly are followed by incinerating clouds of gas (nuées ardentes) and mudflows (lahars).

The number of eruptions per year has not changed over the past few hundred years, based on historical records. The length of time a volcano can continue to erupt is uncertain but can be several thousand years. For most volcanoes, eruptions are irregular and difficult to predict accurately. Commonly, however, eruptions are preceded by an increased number of microearthquakes, bulging of the surface as the magma rises, and possibly by changes in grav-

itational acceleration and electrical and magnetic properties of the near-surface rocks. The amount of sulfurous gas emitted also generally increases shortly before eruptions.

The USGS and other organizations with environmental concerns prepare maps to show hazardous areas in volcanic, inhabited areas. These maps are helpful guides but do not guarantee safety. Volcanoes are too irregular in behavior to permit forecasting eruptions to the day.

Large eruptions affect climate. In the first few years after large eruptions, average temperatures in affected regions may drop by a few tenths of a degree Centigrade, largely because of sulfuric acid droplets in the lower stratosphere. These droplets stay airborne much longer than ash particles. Presently unpredictable, large eruptions of pyroclastic debris and sulfurous gases are certain to occur from many of the world's active volcanoes, and may cause serious disruptions in agricultural productivity in many countries.

Key Terms

aa
acid precipitation
active volcano
caldera
cinder cone volcano
composite cone volcano
dormant volcano
extinct volcano
eruption forecasting

eruption frequency
hazard assessment
hot spots
lahar
nuée ardente
pahoehoe
plate boundaries
precursor phenomena
pumice

pyroclastic
seamount
shield volcano
tephra
viscosity
volatile
volcanic plume
volcano
volcanologists

Stop and Think!

1. What causes Earth's geothermal gradient? Why has it continued to exist for all 4.5 billion years of Earth's history? Will it always exist?

2. Why do many regions of Earth suffer from both earthquakes and volcanoes? Explain their relationship.

3. What distinctions exist among extinct, dormant, and active volcanoes? How reliable are these distinctions?

4. What characteristics of magma determine how explosive its lava will be?

5. Why are basaltic volcanoes restricted to the ocean basins, whereas more silicic volcanoes occur at continental margins?

6. List and describe the hazards from explosive volcanoes.

7. How do the hazards from nonexplosive volcanoes differ from those of explosive ones?

8. Is it safer to build your house next to a cool stream channel in Hawaii or adjacent to a similar stream near Mount Saint Helens? Explain your reasoning. Is it more important to care about the predominant wind direction when locating a house near Kilauea or when locating a house near Mount Saint Helens? Why?

9. Hawaii and Iceland are islands. What else do they have in common? (Hint: What are they made of, and how did it get there?)

10. Why do small earthquakes commonly precede a volcanic eruption? How would you interpret a swarm of small quakes that was *not* followed by an eruption?

11. List precursors useful for predicting volcanic eruption.

12. Explain why damage predictions from the eruption at Mount Saint Helens missed the mark.

13. What assumptions must be made to construct a hazard zone map? How reliable are these assumptions?

14. How do volcanic eruptions affect global climate?

15. In what ways are lava flows and even explosive eruptions beneficial to humans?

References and Suggested Readings

Casadevall, T. J. (ed). 1994. Volcanic ash and aviation safety: Proceedings of the First International Symposium on Volcanic Ash and Airline Safety. *U.S. Geological Survey Bulletin* 2047, 450 pp.

Chester, David. 1994. *Volcanoes and Society.* New York: Chapman & Hall, 288 pp.

Decker, Robert and Barbara Decker. 1989. *Volcanoes.* New York: W. H. Freeman, 285 pp.

Duncan, D. E. 1996. Volcanoes. *Life,* June, p. 52–62.

Francis, Peter. 1993. *Volcanoes.* Oxford: Clarendon Press, 443 p.

Lipman, P. W. and D. R. Mullineaux (eds.). 1981. The 1980 eruptions of Mount Saint Helens, Washington. *U.S. Geological Survey Professional Paper* 1250, 844 pp.

National Research Council. 1994. *Mount Ranier: Active Cascade Volcano.* Washington, D.C.: National Academy Press, 114 pp.

Pinatubo Volcano Observatory Team. 1991. Lessons from a major eruption: Mt. Pinatubo, Philippines. *EOS,* v. 72, No. 49, Dec. 3, p. 552–555.

Rampino, M. R., S. Self, and R. B. Stothers. 1988. Volcanic winters. *Annual Review of Earth and Planetary Sciences,* v. 16, p. 73–99.

Rhodes, J. M. and J. P. Lockwood (eds.)., 1995. *Mauna Loa Revealed.* Washington: American Geophysical Union monograph 92, 348 p.

Simkin, Tom and Lee Siebert. 1994. *Volcanoes of the World.* Tucson, Arizona: Geoscience Press, 349 p.

Simkin, Tom. 1993. Terrestrial volcanism in space and time. *Annual Review of Earth and Planetary Sciences,* v. 21, p. 427–452.

Tanaka, J. C. 1986. Volcano trial. *Earth Science,* summer, p. 20–24.

Tiedemann, Herbert. 1992. *Earthquakes and Volcanic Disasters: A Handbook on Risk Assessment.* Zurich, Switzerland: Swiss Reinsurance Company, 951 pp.

Wright, T. L. and T. C. P. Pierson. 1992. Living with volcanoes. *U.S. Geological Survey Circular* 1073, 57 pp.

"In 1991, the operation at Colorado's Summitville mine leaked cyanide [using cyanide is a standard technique in gold recovery], killing all aquatic life in a 17-mile stretch of the Alamosa River. The company was fined $100,000 by the state, but the mine leaked cyanide two more times in 1991. Then it went bankrupt, leaving taxpayers with an estimated $20 million cleanup."
Environmental Almanac, 1993

Nonfuel Mineral Resources and the Environment

If the typical American was asked to name products that we harvest out of Earth, likely answers would be oil, natural gas, coal, gold, and gemstones. But that is only a start. When you drink from a glass bottle or metal can, or eat with a metal fork or spoon from a ceramic plate or bowl, you are using materials that came from within Earth. When you ride a car, bus, or aircraft, you are surrounded by materials that once were underground. The computer you use is made of silicon chips, wire, plastic, and glass, all from Earth. Most of the raw materials used to build your home or dormitory were produced by the mining industry (Figure 13.1). Every day, you use things that had their origin within Earth.

This book does not focus on the economic geology of rocks and minerals. But it is important for you to know how these materials are recovered from Earth, for the recovery and processing of ores and non-metallic deposits can create serious environmental problems.

Earth is a treasure trove of useful materials. What are these materials, where in Earth are they located, how do they form, and how are they taken from Earth and processed to become the base for modern civilization?

And on the environmental side, what harm comes from mining and processing mineral resources? Surface mining leaves a huge hole; what should be done with it after the desired rock or mineral is removed? What chemical contamination is associated with metallic mining activities? Should the present owner of a mine be financially liable for environmental damage produced by previous owners? Who should bear the cost of remediating damage caused by companies that now are bankrupt?

◀

In 1850, this California Gold Rush gold prospector swirled his gold pan, separating sand and mud from flecks of gold. Gold occurs almost entirely as the pure yellow metal.

Photo: Seaver Center for Western History Research, Los Angeles County Museum of Natural History.

What Are Mineral Resources?

A **mineral resource** is any mineral that has *value* to people (for whatever reason) and can be extracted from Earth at a *profit*. If it is of no value to people, then it is just a mineral. It is the value that turns it into a resource. Minerals from which we can derive useful metal elements (gold, copper, etc.) are called **ore minerals**, and they exist in ore-bearing rocks. Examples:

- The mineral galena is a lead ore, in the form of lead sulfide (PbS). It is the major source of the lead used in car batteries.
- The mineral hematite is an iron ore, in the form of iron oxide (Fe_2O_3). It is the source of most of America's iron, from which steel is made.
- The mineral ilmenite is a titanium ore, in the form of iron-titanium oxide ($FeTiO_3$). It is our major source of titanium, mixed with iron to make specialty steels.

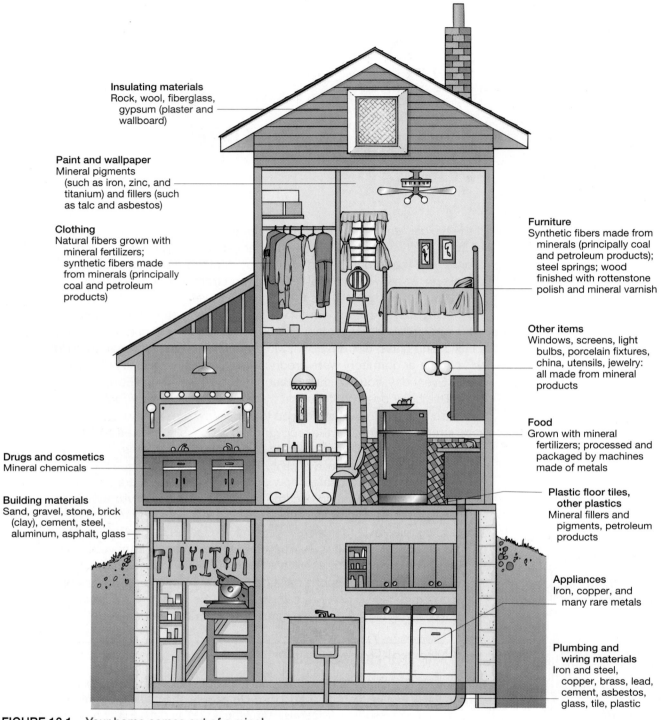

Insulating materials
Rock, wool, fiberglass, gypsum (plaster and wallboard)

Paint and wallpaper
Mineral pigments (such as iron, zinc, and titanium) and fillers (such as talc and asbestos)

Clothing
Natural fibers grown with mineral fertilizers; synthetic fibers made from minerals (principally coal and petroleum products)

Furniture
Synthetic fibers made from minerals (principally coal and petroleum products); steel springs; wood finished with rottenstone polish and mineral varnish

Other items
Windows, screens, light bulbs, porcelain fixtures, china, utensils, jewelry: all made from mineral products

Food
Grown with mineral fertilizers; processed and packaged by machines made of metals

Drugs and cosmetics
Mineral chemicals

Building materials
Sand, gravel, stone, brick (clay), cement, steel, aluminum, asphalt, glass

Plastic floor tiles, other plastics
Mineral fillers and pigments, petroleum products

Appliances
Iron, copper, and many rare metals

Plumbing and wiring materials
Iron and steel, copper, brass, lead, cement, asbestos, glass, tile, plastic

FIGURE 13.1 Your home comes out of a mine!

Many of the materials used to construct your home come from Earth's crust. *Concept from Society of Mining Engineers, Foundation for Public Information and Education.*

And, of course, there are gold ores, silver ores, zinc ores, tin ores, chromium ores, and so on. About 70 percent of the natural elements in Earth's crust are metallic, and there is no particular reason why all of them might not be useful to human civilization.

The metals we use most are those that have useful properties, are relatively abundant, are relatively easy to locate and mine, and, consequently, are relatively inexpensive. For example, copper is almost always the metal used for electrical wires in homes and commercial buildings. Why? It is not the metal that best conducts electricity; gold and silver work better (and are used in some pricey electronic products). Also, because copper wire is a poorer electrical conductor than gold or silver, it gets hotter than gold and silver when carrying electricity. To avoid having hot wires snaking through wooden houses, we simply use larger wire, which runs cooler, although this uses more copper.

Also, copper is much cheaper to bring from a copper mine to your home than is gold or silver. Can you imagine gold power lines gleaming in the sunshine, carrying power to your home? What would the power company charge for the use of such extravagant cables, and for how many minutes would they be safe from thieves?

Although isolated deposits of pure copper metal occur, the element copper is most commonly bound up in an ore called chalcopyrite. Pronounced *KAL-co-PIE-right*, it is a compound of copper, iron, and sulfur ($CuFeS_2$). Chalcopyrite must be mined from the ground, crushed in a mill, and cooked in a smelter to recover the copper. Gold and silver have far less processing cost, for they occur most often as pure metals. But chalcopyrite is far more abundant than isolated deposits of pure gold or silver.

All of these variables are reflected in the relative prices of copper, silver, and gold in the commodities market (see the business section of a newspaper). As this is written, gold is selling for roughly $370 per ounce, silver for roughly $5 per ounce, and copper for about 6 cents per ounce.

Table 13.1 lists metallic minerals and their uses. Each metal shown has its own unique properties (hardness, corrosion resistance, electrical conductivity, melting temperature, and so on). Each also has its own cost, worldwide distribution, mining problems, processing problems, and environmental problems.

To cite another example, contrast mercury and tungsten. Mercury is a free-flowing liquid metal (sometimes called quicksilver) at room temperature, whereas tungsten is a solid, easily drawn into wire. Mercury is used in thermometers and tooth fillings; tungsten is the glowing filament in ordinary incandescent light bulbs. We mine much of our own mercury ore in the United States, but import tungsten from South America and Germany. Mercury is highly toxic and a dangerous polluter; tungsten is not. Mercury costs $0.14 per ounce, and tungsten costs $0.03 per ounce. As you can see, each metal has its own story.

Of equal importance, we also mine deposits of *nonmetallic* minerals. These include asbestos, phosphate rock (used for agricultural fertilizer), sand and gravel for construction, quartz for glassmaking, graphite for lubricants, and so on. Accumulations of these materials are simply called *deposits*, not ores.

TABLE 13.1 Metallic Minerals and Their Uses

Metal	Main Ore (mineral)	Some Important Uses
Aluminum	Bauxite (soil)	aluminum siding, foil
Antimony	Stibnite	alloys, batteries, bullets, bearings
Chromium	Chromite	chrome plating, paint pigments, steel alloys
Cobalt	Cobaltite	paint pigment, alloys, nuclear medicine, varnish
Copper	Chalcopyrite	electrical wiring, pipes, brass, bronze, coins, cooking pots
Gold	(uncombined metal)	jewelry, money, dentistry, alloys, plating
Iron	Hematite	the major ingredient of all types of steel
Lead	Galena	pipes, solder, batteries, pigments, bullets, wheel weights
Manganese	Pyrolusite	steel, alloys, batteries, chemicals
Mercury	Cinnabar	thermometers, dental fillings, medicine
Molybdenum	Molybdenite	high-temperature applications, lamp filaments, rifle barrels, lubricants
Nickel	Pentlandite	alloys, metal plating, jet engines, coins
Platinum	(uncombined metal)	jewelry, electrical equipment, industrial catalyst
Silver	(uncombined metal)	jewelry, tableware, film, mirrors, plating
Tin	Cassiterite	cans and containers, bronze, solder
Titanium	Rutile	paint pigment, aircraft/satellite construction
Tungsten	Scheelite	high-temperature applications, dentistry, light bulb filaments
Uranium	Carnotite	nuclear energy, weapons
Zinc	Sphalerite	brass, metal coatings, batteries

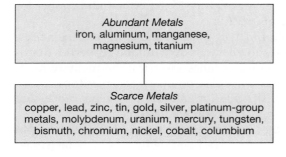

| METALLIC MINERAL RESOURCES | NONMETALLIC MINERAL RESOURCES | FUEL RESOURCES |

Abundant Metals
iron, aluminum, manganese, magnesium, titanium

Scarce Metals
copper, lead, zinc, tin, gold, silver, platinum-group metals, molybdenum, uranium, mercury, tungsten, bismuth, chromium, nickel, cobalt, columbium

Minerals for Industrial and Agricultural Use
phosphates, nitrates, carbonates, sodium chloride, fluorite, sulfur, borax

Construction Materials
sand, gravel, clay, gypsum, building stone, shale and limestone (for cement)

Ceramics and Abrasives
feldspar, quartz, clay, corundum, garnet, pumice, diamond

Fluid Hydrocarbons
petroleum, natural gas

Solids
coal, tar sand, oil shale

Non-Carbonaceous
uranium, geothermal waters

FIGURE 13.2 Three types of mineral resources.

The three broad classes of mineral resources are *metallic, nonmetallic,* and *fuels. From James R. Craig, David L. Vaughan, and Brian J. Skinner,* Resources of the Earth, *Englewood Cliffs, New Jersey: Prentice Hall, 1988.*

Also of equal importance, we extract *fuel resources* from our planet. Figure 13.2 summarizes these three broad classes of mineral resources. We cover metallic and nonmetallic mineral resources in this chapter and fuel resources in Chapter 14, "Energy from Fossil Fuels."

All mineral accumulations, whether metallic, nonmetallic, or fuels, are more valuable when they are larger, when they contain a higher percentage of the desired mineral, when they are near processing plants, and when they are near good transportation facilities.

How Are Mineral Resources Formed?

There are very good reasons why each metal ore and nonmetallic mineral deposit occurs where it does. It comes down to geologic environment. Environments that are igneous, metamorphic, sedimentary, active biologically, and active erosionally all concentrate mineral resources, each in its own way. An overview of these environments and examples of the resources they concentrate is given in Table 13.2.

Study Figure 13.3A and you can see that many metallic minerals are concentrated at and near convergent plate margins. The reason is that temperatures here are unusually high, enabling the metallic elements to dissolve. This segregates them from most other elements. They later precipitate as the solution cools, usually accumulating as sulfides and oxides (review Table 2.3 and related text). For example, Figure 13.3B shows how the copper ore chalcopyrite has become concentrated in mountains that were formed by converging tectonic plates along the western coasts of North and South America.

And, as Figure 13.3C reveals, groups of elements tend to be concentrated together because they are chemically similar. The copper ore vein shown in the photo includes not just copper, but zinc, lead, iron, and sulfur.

As noted, mineral resources can occur in all three types of rocks—igneous, metamorphic, and sedimentary (Table 13.2). We will now look at all three.

Resources from Igneous Rocks

Which minerals occur in which type of rock depends on the chemical makeup of the magma, the local subsurface environment, and the properties of the mineral.

An interesting example is chromium, a valuable metallic element used for plating car trim, for making yellow paint pigment, and for alloying with iron to make stainless steel. The element chromium is abundant in magmas that have very low silica (SiO_2) content. These silica-poor magmas are the ones that crystallize to form basalt at Earth's surface. (They also crystallize deep underground to form basalt's coarser-grained sister rock, gabbro. These two varieties of igneous rock differ only in crystal size; see Chapter 2.)

The major chrome-bearing mineral is chromite (an iron-chromium oxide, $FeCr_2O_4$). Chromite is among the first minerals to crystallize from chromium-bearing magmas. It is dense, with a high specific gravity of 4.6, which is much greater than the specific gravity of the liquid magma, about 3.0 (recall that water is 1.0). Hence, the newly formed chromite crystals settle to the bottom of the magma and accumulate in a layer at the base of the liquid. Due to **crystal settling,** this bottom zone of the magma now is enriched in chromite.

TABLE 13.2 Sources of Mineral Resources

Mineral resources come from the three rock types, from biological activity, and from weathering.

Source	Example
Igneous	
Disseminated (scattered in rock)	Diamonds (South Africa)
Crystal settling in magma (see text)	Chromite (Stillwater, Montana)
Late (during crystallization in magma)	Magnetite (Adirondack Mountains, New York)
Pegmatite (very large crystals; see text)	Beryl and lithium (Black Hills, South Dakota)
Hydrothermal (watery solution around magma)	Copper (Butte, Montana)
Metamorphic	
Contact metamorphism (see text)	Lead and silver (Leadville, Colorado)
Regional metamorphism over many square miles at high temperature and pressure	Asbestos (Quebec, Canada)
Sedimentary	
Evaporite deposit	Potassium (Carlsbad, New Mexico)
Placer (stream deposit)	Gold (Sierra Nevada foothills, California)
Glacial deposit	Sand and gravel (northern Indiana)
Deep-ocean precipitate	Manganese oxide nodules (central and southern Pacific Ocean)
Biological	Phosphorous (Florida)
Weathering	
Residual soil	Bauxite (Arkansas)
Secondary enrichment	Copper (Utah)

Source: Modified from Robert J. Foster, *General Geology*, 4th ed. Columbus, OH: Charles E. Merrill, 1983.

After a few million years, the magma cools until all its minerals have crystallized as igneous rock. Now, enter the geologist: where does she explore for chromium ore? Knowing how it concentrates early at the bottoms of magma chambers, she goes where silica-poor rocks occur, and looks around. If she knows her geology and is lucky, she will find something like the dark-banded rocks shown in Figure 13.4. (Unfortunately, despite the chromium enrichment, most silica-poor magmas do not contain enough chromium for an economic deposit of chromite to occur.)

Another example of a mineral deposit in igneous rocks is the mineral beryl, a beryllium-aluminum silicate, $Be_3Al_2(Si_6O_{18})$. Beryl is of interest for two reasons. First, its dark green varieties are called emerald, a much-desired precious gem. Second, beryl is our source of the metallic element beryllium, which we alloy with copper to harden and strengthen it.

Beryllium is quite rare overall. But in a beryllium-bearing magma, as it cools, nearly all other minerals crystallize first, concentrating beryllium in the remaining liquid. As crystallization nears completion, the remaining watery liquid invades cracks in overlying rocks. As these watery solutions crystallize, they may contain enough beryllium for the mineral beryl to crystallize in veins (the other necessary components, aluminum and silica, are plentiful). If the crystals in the veins are very large, perhaps several feet in length, the vein is called a **pegmatite.** Beryl-containing pegmatites (Figure 13.5) are valuable.

Many other metallic ores are associated with igneous rocks; some notables include:

- The lead ore galena (lead sulfide, PbS), main source of lead for your car battery.
- The zinc ore sphalerite (zinc sulfide, ZnS), chief ore of zinc, alloyed with copper to make brass.
- The mercury ore cinnabar (mercury sulfide, HgS), used in thermometers, mercury-vapor lights, and tooth-filling amalgam. (Mercury is very toxic, but tests show no harmful effect from the little chunks that dentists embed in your mouth.)

Resources from Metamorphic Rocks

The most common metallic ores in metamorphic rocks are formed during **contact metamorphism.** This phenomenon occurs in the unusually high temperatures but very low pressures around a hot magma chamber (Figure 13.5). Loosely speaking, the magma "roasts" adjoining rocks, metamorphosing them, converting some minerals into others (review: metamorphic rocks in Chapter 2).

Metallic ore deposits are particularly common where limestone surrounds the magma chamber, because limestone is more chemically reactive than quartz-rich rocks. The intense magmatic heat decomposes the limestone, releasing carbon dioxide gas, which speeds the transport of elements. Economic deposits of lead sulfide and silver sulfide may form in the remaining limestone; they come from the watery fluids that escape the top of the magma chamber as it cools.

Resources from Sedimentary Rocks

A very wide variety of metallic ores occur in sedimentary rocks. Examples include uranium in New Mexico and Wyoming, lead and zinc in southeastern Missouri,

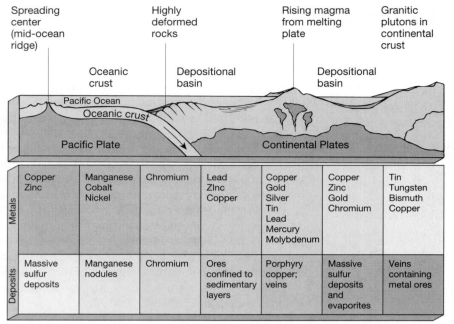

Spreading center (mid-ocean ridge) | Highly deformed rocks | Rising magma from melting plate | Granitic plutons in continental crust

Oceanic crust | Depositional basin | Depositional basin

Pacific Ocean
Oceanic crust
Pacific Plate | Continental Plates

Metals	Copper Zinc	Manganese Cobalt Nickel	Chromium	Lead Zinc Copper	Copper Gold Silver Tin Lead Mercury Molybdenum	Copper Zinc Gold Chromium	Tin Tungsten Bismuth Copper
Deposits	Massive sulfur deposits	Manganese nodules	Chromium	Ores confined to sedimentary layers	Porphyry copper; veins	Massive sulfur deposits and evaporites	Veins containing metal ores

(A)   Location of some mined deposits with respect to plate boundaries

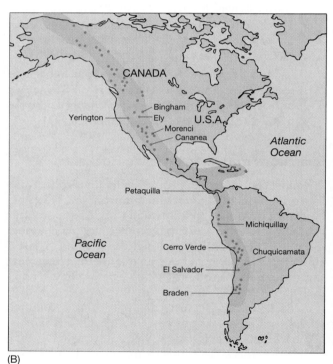

(B)

FIGURE 13.3   Why do metallic mineral deposits occur where they do?

The reason is their *geologic environment.* (A) Spreading plates along a mid-ocean ridge concentrate lighter-weight metals like copper and manganese. Converging plates, generally near the edges of continents, span wide areas and spawn more varied environments, generating a greater variety of metals of all weights. Rising magma that cools to form plutons also concentrates metals. (B) An example of metal concentration is copper, in its most important ore mineral, chalcopyrite. The entire western coast of the Americas has a well-defined belt of chalcopyrite deposits, running parallel to the converging plate margin. (C) This rich vein of copper ore cuts through igneous rock. The ore includes three valuable minerals: chalcopyrite (copper ore, $CuFeS_2$), sphalerite (zinc ore, ZnS), and galena (lead ore, PbS). *Photo: B. J. Skinner and S. C. Porter, 1992,* The Dynamic Earth, 2nd ed.

(C)

FIGURE 13.4   Dark bands of chromium ore (chromite) in South Africa.

Each chromite band was the base of a former magma chamber that existed during settling of the chromite crystals to the bottom. There are multiple bands because there were multiple magma chambers: each dark band and the lighter band above it represents a fresh infusion of magma. *Photo: William E. Ferguson.*

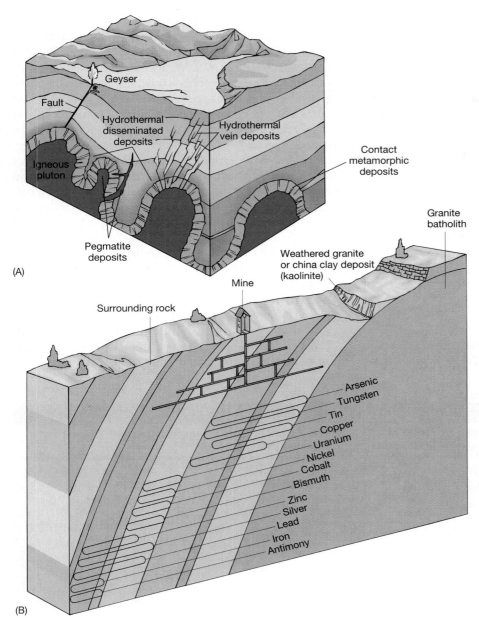

(A)

(B)

FIGURE 13.5 Underground igneous formations.

(A) Magma that rose millions of years ago but never reached the surface. It cooled to crystallize large plutons of igneous rock. As the crystallization was nearing completion, residual hydrothermal fluids (hot water) invaded cracks above the magma and deposited valuable minerals in veins (pegmatites). Geologists seek such veins, for they may contain beryl and other valuable minerals. (B) Simplified cross section at Dartmoor, England, shows concentric zoning of metallic ores produced by decreasing temperature, outward from the granite as it intruded. After D. Elson, 1992, *Earth*, New York: Macmillan, 1992, p.65.

gold in California and Colorado, titanium in Florida, copper in Germany, tin in Indonesia, and aluminum (**bauxite**) in humid tropical areas of Brazil, central Africa, and India. There are many more; these are the more outstanding examples.

Each of these occurrences of metallic ore has its own reason for forming where it did. Sometimes the reason is chemical, as when copper-bearing waters move through sedimentary rocks and encounter new chemical conditions that force the copper to leave the solution (it *precipitates*). Or, sometimes the reason is mechanical, as when a stream current curves around a bend and slows, allowing heavier grains (like gold) to drop out and concentrate on the stream bottom.

Uranium ore in sedimentary rocks illustrates how changing chemical conditions can form ore deposits. Uranium concentrates late in the crystallization of magma, so subsurface water may become enriched in uranium. If a uranium-enriched subsurface water infiltrates a sandstone containing lots of burned wood chips (such as residue from a forest fire), a chemical reaction causes the uranium to precipitate as a yellow mineral, carnotite. Extensive carnotite deposits formed in this way occur in western sandstones that are about 180 million years old.

In the 1950s, when nuclear weapons and nuclear power both were high government priorities, the federal government supported uranium exploration and many deposits were located. However, aversion to nuclear weapons has grown, and enthusiasm for nuclear power has waned (see Chapter 16). It seems unlikely that uranium exploration will boom again soon, if ever.

Sedimentary rocks also are our source of non-metallic resources, including building stone, sand and gravel, clay deposits, phosphate rock, and sulfur. Like metals, each of these resources has distinctive characteristics that affect its usefulness. For example, sand and gravel deposits must be within a short trucking distance of where they are to be used, or they are essentially worthless. Transportation cost for the large volumes needed in construction is very high for these otherwise cheap materials. The same is true of building stone. Gold and silver, on the other hand, are valuable enough to justify long transport.

How Abundant Is Each Mineral Resource?

Remember that a mineral is a *resource* only if it can be extracted from the ground at a profit. Most metals we use occur in surprisingly small percentages in Earth's crust (Table 13.3). Looking at the table, the top two metals, aluminum and iron, total about 14 percent of the weight of Earth's crust. The remaining metals obviously are much scarcer, many forming only a tiny percentage of the crust.

Consider the copper used in wiring buildings and homes. If we could grind up all of Earth's crust and evenly mix it, and then scoop up a 10,000-pound sample, copper would constitute only 55 pounds (0.0055 percent). This tiny percentage is too small to justify the cost of extracting the copper, so the trick is to locate natural concentrations of the most common copper ore (chalcopyrite). For copper mining to be profitable today, the chalcopyrite deposit must be rich enough to contain at least 0.5 percent copper. This is a ninety-times concentration over the average abundance in the crust (0.0055 percent average abundance × 90 times enrichment = about 0.5 percent copper concentration; see Table 13.3).

To mine tungsten, used in steel products and light bulbs, the ore must occur in a natural enrichment that is 2000 times greater than the average concentration in Earth's crust. Lead, used in automobile batteries, must be naturally concentrated 2400 times over its crustal

TABLE 13.3 It's All Economics

The average abundance of metal elements in Earth's crust, multiplied times the natural concentration needed to create ore worth mining, equals the minimum concentration necessary for profitable mining.

Metal	Average Abundance of Metal in Earth's Crust (weight % of crust)	X	Natural Concentration Needed to Form an Ore Worth Mining	=	Minimum Concentration of Metal Needed for Profitable Mining (weight %)
Aluminum	8.3%		5 times		40%
Iron	5.6		4		25
Titanium	0.57		25		15
Manganese	0.095		260		25
Vanadium	0.0135		35		0.5
Chromium	0.010		4,000		40
Nickel	0.0075		130		1.0
Zinc	0.0070		350		2.5
Copper	0.0055		90		0.5
Cobalt*	0.0025		80		0.2
Lead	0.00125		2,400		3
Uranium	0.00027		40		0.01
Tin	0.00020		250		0.5
Molybdenum*	0.00015		660		0.1
Tungsten	0.00015		2,000		0.3
Mercury	0.000008		12,500		0.1
Silver*	0.000007		700		0.005
Platinum	0.0000005		400		0.0002
Gold*	0.0000005		250		0.0001

*Much is recovered as a byproduct of mining other metals.

average for a deposit to be worth mining. Because of the very high ore concentrations needed for a mining venture to be profitable, you can see that minable ore deposits can be hard to find.

Is Any Country Self-Sufficient in Mineral Resources?

No. No country is self-sufficient in all metallic and non-metallic resources, but some are better endowed than others. Tiny countries clearly are at a disadvantage, for their limited real estate is unlikely to encompass many geologic environments, and therefore many resources. Large countries like the United States, Canada, Russia, and Australia have better odds of having all they need—but they don't.

This is illustrated by a self-sufficiency analysis for the United States. Table 13.4 shows what percentage of metals and other minerals we now import because we lack a profitable source at home. For some items, like salt, iron ore, and lead, we are self-sufficient or nearly so, and should remain so for many years. But it is obvious that many metals and minerals, some critical, must be imported constantly to satisfy our growing demand.

TABLE 13.4 U.S. Reliance on Foreign Supplies of Minerals

Note that our dependence on foreign suppliers continues to increase most minerals. Currently, we import less than half our demand for selenium, silicon, gypsum, zinc, sulfur, salt, iron, and lead.

Mineral	Percent Imported in 1995	Percent Imported in 1992	Major Sources (1990–1994)	Major Uses
Columbium	100%	100%	Brazil, Canada, Germany	Steelmaking, superalloys
Graphite (natural)	100	100	Mexico, Canada, China, Madagascar	Refractories, brake linings, packings
Manganese	100	100	South Africa, France, Brazil, Australia	Steelmaking
Mica (sheet)	100	100	India, Belgium, China, Argentina	Electronic and electrical equipment
Strontium (celestite)	100	100	Mexico	Television picture tubes, pyrotechnics, ferrite magnets
Bauxite and alumina	99	100	Australia, Jamaica, Guinea, Brazil, Guyana	Aluminum production, abrasives, refractories
Asbestos	95	95	Canada, South Africa	Roofing products, friction products
Diamonds (industrial)	95	92	Ireland, Britain, Zaire	Machinery for grinding and cutting
Tungsten	87	85	China, Bolivia, Peru, Germany	Machinery, lamps
Platinum	91	94	South Africa, Russia, Britain	Catalysts, electrical and electronic equipment
Fluorospar	88	87	China, Mexico, South Africa	Hydrofluoric acid production, steelmaking
Tantalum	80	87	Germany, Australia, Canada, Brazil	Electronic components
Tin	84	73	Brazil, Bolivia, China, Indonesia	Cans, electrical, construction
Barite	65	44	China, India, Mexico	Oil and gas well drilling fluids
Cobalt	82	76	Zambia, Zaire, Canada, Norway, Finland	Aerospace alloys, catalysts, paint driers, magnetic alloys
Chromium	78	74	South Africa, Turkey, Zimbabwe, Yugoslavia	Ferroalloys, chemicals, refractories
Potash	74	67	Canada, Israel, former USSR, Germany	Fertilizer
Nickel	61	64	Canada, Norway, Australia, Dominican Rep.	Stainless steel, other alloys
Antimony	60	58	China, Mexico, South Africa, Hong Kong	Flame retardants, batteries
Iodine	62	52	Japan, Chile	Animal feed supplements, catalysts, inks, disinfectants
Cadmium	50	49	Canada, Mexico, Australia, Belgium	Batteries, pigments, plating and coating of metals

Source: Bureau of Mines, U.S. Department of Interior.

From which countries do we import? First, they must be those that have affordable resources of the substance we need. But equally important, the countries should be friendly toward us. Countries with which the United States has disagreements are not anxious to sell to us. Fortunately, our relations with most foreign governments are friendly, and most have stable governments, so we can purchase what we need. Canada, in particular, is a major U.S. supplier.

Note that there is no mention in Table 13.4 of governments with which we have strained relations, such as Cuba, Iran, Iraq, or North Korea. Cuba is particularly noteworthy. Our major suppliers of nickel are Canada, Australia, Norway, and the Dominican Republic, but not Cuba, despite its possession of abundant nickel. Nickel from Cuba would be cheaper to import, but we do not trade with Cuba for political reasons.

Another example: as shown in Table 13.4, we import 100 percent of our graphite, much of it from Mexico and China. But, suppose Mexico resents our treatment of illegal Mexican workers and stops exporting graphite to us. And suppose China resents our trade sanctions and our criticism of their human rights abuse, so they also halt graphite exports. What might we be willing to pay for graphite from more expensive exporters? We use graphite for steelmaking, brake linings, lubricants (mixed with oil) and in ordinary writing pencils (mixed with clay). Could we substitute something for graphite? Substitutes would cost more, which is why we use graphite now. How much are we willing to pay for graphite?

All imports are expensive and are a large drain on every country's economy. This has given rise to armed conflicts throughout history, most recently the 1991 Gulf War, essentially fought to preserve the flow of petroleum to the Western world and Japan. In 1996, the Turkish prime minister said "As the Arab world is not thinking of sharing its oil with others, Turkey is not ready to share water flowing in its territories with others." A possible future conflict over a resource is a disagreement between Turkey and her southern neighbors, Syria and Iraq, over water. The issue is control of water in rivers that arise in Turkey's mountains, but which flow through these other parched countries.

Can We Run Out of Resources?

The problem here is the expression "running out." What does it mean? Look at Table 13.4 and note that we import all of our bauxite, the ore for aluminum. Does this mean that the United States has not one ounce of aluminum in its generous share of Earth's crust? When

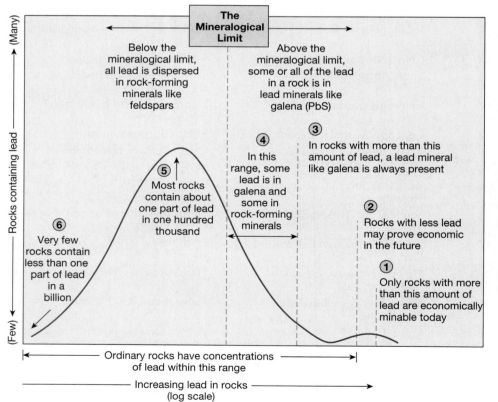

FIGURE 13.6   The mineralogical limit.

This curve shows the concentration of the element lead in rocks (horizontal axis) versus the occurrence of rocks that contain lead (vertical axis). The (1) line shows that rocks with economically minable lead concentrations are few. The (2) line indicates that rocks with potential economic lead are not common, either. Anything to the left of the (2) line probably never will be worth mining. Similarly shaped curves can be drawn for copper, chromium, tungsten, and other metallic elements. *After Solid Earth Sciences and Society, 1993, National Academy Press, p. 151.*

people say "this mineral is in short supply" or "we are running out of this metal," it is not literally true. Such statements mean that the supply is *too small* or *too poorly concentrated to be extracted at a profit using current technology*. **Running out of a resource** is an economic issue, not a geologic one. If a metal is needed and the resource has been diminished by earlier mining, scarcity drives up the price, and deposits previously considered too costly to mine suddenly become a minable resource.

Gold mining is a good example of this principle. Until about 1970, gold sold at a fixed price of $35/ounce, a price enforced by U.S. policy. We controlled the price of gold, much as Saudi Arabia controls the price of crude oil today. But then other countries decided they wanted us to pay for imports with gold rather than with dollars. So our gold reserve diminished, and we no longer could enforce our price of $35/ounce. Gold skyrocketed to about $800/ounce before falling to its present open market price of about $375. When the market price rose from $35 to $800, exploration for gold increased dramatically, and previously ignored deposits abruptly became a minable resource. In a free economy, demand has a major effect on supply and price.

Despite this economic reality "we are running out of _____" someday will become essentially true for our most-used mineral resources. Eventually, some will become so scarce and expensive to mine that, as a practical matter, they will have "run out." Figure 13.6 shows how minable concentrations of lead are distributed in few rocks. On the curve, we are presently moving from the extreme right (easy pickings) toward deposits of lower concentration. At some point, lead will become too expensive to mine; that is point (1) on the graph. Future demand for lead may extend mining to point (2), but mining cost will grow so great that all lead mining will cease. This same pattern is true for other metals.

This is why *recycling* of materials is increasingly important. Figure 13.7 shows the history of recycling in the U.S. during 1900–1990 for five important metals. Metals that are used in their pure form and in large or loose pieces are easiest to recycle: lead plates in automobile batteries, copper in wire and pipe, and aluminum in beverage cans. Consequently, lead and copper already are heavily recycled, as the graphs show, and aluminum is a good candidate for increased recycling.

Mixtures of metals require expensive separation of the desired metal, so the savings are less dramatic, which discourages recycling. For some metals, it remains cheaper to use virgin metal from ore than to scavenge the metal from discarded objects. It all comes down to cost.

For example, the rare metal platinum is used in the exhaust system of every car, in its catalytic converter.

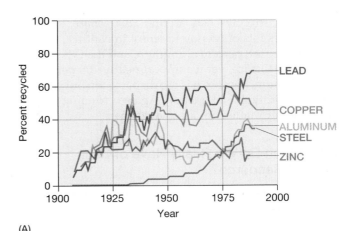

(A)

(B)

FIGURE 13.7   Recycling is on the rise.

(A) Metal recycling in the U.S., 1900–1990, showing percent recycled. *Source: I. K. Wernick, R. Herman, S. Govind, and J. H. Ausubel, "Materialization and Dematerialization: Measures and Trends," in J. H. Ausubel and D. Langford, eds.,* Technology Trajectories and the Human Environment. *Washington, D.C.: National Academy Press, December 1995.*
(B) Who says Coke is not popular? *Photo: Liz Dew/Banson Marketing Limited.*

Platinum currently costs around $400/ounce, so recycling makes lots of economic sense! In contrast, Earth's land surface is 8 percent aluminum, and the U.S. owns a lot of land surface, yet we import nearly 100 percent of our aluminum, because it is too dispersed in our share of crust for profitable extraction. Again, "running out" is an economic matter, not a geologic one.

How Much Do We Really Have of Each Mineral Resource?

In assessing the amount of a mineral resource, we distinguish between reserves and resources. **Reserves** include only the ore or mineral deposit that we are confident exists and that can be extracted profitably with available technology. **Resources** include not only reserves but potential ores that are not profitable to extract at present, but may be in the future as demand, price, and technology change.

Reserves are classified as either measured, indicated, or inferred:

- **Measured ore** has been sampled in some detail, either through limited mining or by drilling from the surface. A high degree of certainty exists in the calculated tonnage and quality.
- **Probable ore** is only a little less certain, based on reasonable estimates from known ore localities.
- **Possible ore** rests on less complete knowledge, but is based on some physical evidence and knowledge of how this ore occurs.

Thus, mineral resources include not only minable ores but deposits that are too low in quality or too difficult to mine at present. A rise in price or advance in technology may make some ores valuable in the future. Resources also include deposits that have yet to be properly investigated, and those whose existence is inferred from the geologic environment. Such resources are, of course, very "iffy."

These careful distinctions, made by geologists and environmental scientists, are not always observed by others. Politicians and businesspeople manipulate numbers to achieve goals. Journalists sometimes fail to do their homework and do not understand fully what they report. When one of these people uses the word "resource," it may be difficult to know precisely what he or she is talking about, and the individual may not know, either.

Mining: Surface and Underground

A geologist locates a good body of ore. Should it be mined at the surface by "open-pit" methods, or should it be mined underground by "deep mining" methods?

It depends on the size of the ore body, its shape, depth, and the grade of ore.

If the ore body is three miles long, 5000 feet thick, and is exposed at the surface, then **surface mining** obviously would be best. On the other hand, if the ore body is intermittent and extends to great depths, then **underground mining** is better. Far less waste rock must be moved to "surgically" mine ore underground than the great volume that must be removed from an open pit.

In the real world, ore bodies seldom are at either of the extremes just described. Thus, complex engineering is needed to determine whether surface or underground mining is cheaper. At times, a combination of surface and underground methods is used. In the United States, more than 85 percent of mineral production is from open-pit mining, so we shall focus upon this method first.

Surface Mining

As an example of a surface mine, we shall consider a copper deposit. But first, a few words about copper and its value to us. Several properties make copper useful. It is very ductile, meaning that it can be drawn into thin wire. Copper also is an excellent conductor of heat and electricity. We take advantage of these properties to make wire used in our computers, washing machines, stereos, and so on.

About 40 percent of U.S. copper use is for residential construction. The average single-family home of 2100 square feet uses 439 pounds of copper:

195 pounds of building wire

151 pounds of plumbing piping, fittings, and valves

24 pounds of plumber's brass (brass is an alloy of copper and zinc)

47 pounds of built-in appliances

12 pounds in builder's hardware

10 pounds in miscellaneous wire and pipe

Nonresidential copper uses include coins, jewelry, and bronze (an alloy of copper and tin). Copper ranks second only to iron as an essential metal in modern civilization.

When early civilizations first mined and used copper more than 4000 years ago, we assume they used pure, uncombined, 100 percent copper that occurs naturally in some areas (Figure 13.8). However, pure copper in nature is uncommon, and most copper occurs combined with other elements in copper-bearing minerals. The most common is chalcopyrite, a copper-iron sulfide ($CuFeS_2$), shown in Figure 13.3C. Early societies

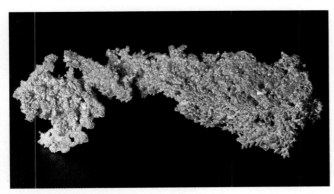

FIGURE 13.8 Pure native copper.

Copper occurs as a pure metal, and in ore. Pure native copper sometimes occurs naturally, as you see it here. (Specimen length is 9 inches.) *Photo: U.S. Geological Survey.*

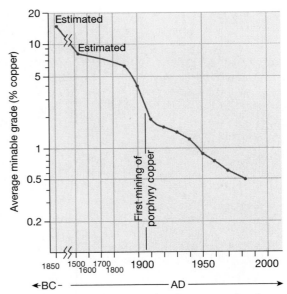

FIGURE 13.9 Decline in grade of copper mined.

Curves drawn for other metals mined from antiquity have a similar shape. The easy pickings went first to primitive low-technology miners, leaving the lower-grade, harder-to-mine ores for modern mining technology to recover. *After Richard Edwards and Keith Atkinson, 1986,* Ore Deposit Geology and Its Influence on Mineral Exploration, *New York: Chapman and Hall, p. 70.*

quickly learned to mine this copper ore and smelt (melt) it to obtain pure copper.

Prior to about 2000 years ago, the quality or *grade* of copper ore being mined needed to be at least 15 percent (15 pounds of copper recoverable from 100 pounds of ore) to obtain useful amounts of the metal. Technology improved, and by A.D. 1500, this grade had been reduced to 9 percent (Figure 13.9). By 1800, it had reduced to about 6–7 percent, and this value continued until early in this century, when technology was developed to mine and recover low-grade copper from granite.

Today, with modern technology and an understanding of chemistry not possessed by earlier people, a grade of only 0.5 percent copper is sufficient for profitable mining. At present, the most important copper ore mineral is chalcopyrite (Figure 13.3B, C), which is only about one-third copper by weight.

Case Study: Copper Mining at Bingham, Utah

The largest excavation on Earth is at Bingham, Utah, about twenty miles southwest of Salt Lake City (Figure 13.10A). Open-pit mining for copper (and gold as a by-product) began here in 1906. Copper reserves then were estimated to exceed 260 million tons, with grades from 0.75 percent to 2.5 percent copper. Approximately 3.3 billion tons of material have been removed from this giant hole, about seven times more than was moved to build the Panama Canal.

Bingham mines great tonnages of low-grade ore and supplies 15 percent of all U.S. copper. To mine enough chalcopyrite to recover the copper at a profit, about 200,000 tons of material (ore plus overlying rocks and soil) are removed from the Bingham pit every day.

Bingham was the first to mine porphyry copper deposits. A **porphyry** is an igneous rock that mixes two different crystal sizes of the major minerals. In the Bingham porphyry copper deposit, the granite "host" rock also

contains small crystals of minerals composed of copper, iron, molybdenum, lead, and zinc. These are thinly dispersed throughout the granite (Figure 13.10B). Such porphyry copper deposits currently supply more than half the world's copper, and our possession of numerous such mines makes us self-sufficient in this essential metal.

At the mine, if you pick up one chunk of rock, its content of desirable minerals may be too slight to recognize. If you choose another, it may contain noticeable yellow chalcopyrite. This variable material must be processed to separate the desired minerals from worthless host rock.

Processing the Ore. Briefly, when ore is processed it is crushed in a mill, treated chemically to separate the desired mineral from the host rock, and smelted to extract the pure metal. This is shown in Figure 13.11.

Ore ranges in size from large chunks down to dust. To chemically process the ore, it must be readily mixable with water, so the first step is to crush it all into small particles in a mill. Next comes chemical separation of "the wheat from the chaff" (the desired mineral from the host rock). Finally, the mineral is cooked in a **smelter,** a structure in which heat separates the pure metal from the sulfur or oxygen or whatever it is bonded within the mineral. This process is basically the same for all metal ores.

(A)

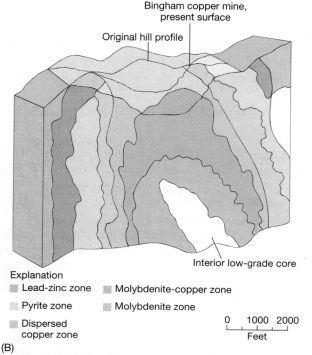

Bingham copper mine, present surface

Original hill profile

Interior low-grade core

Explanation

■ Lead-zinc zone   ■ Molybdenite-copper zone

■ Pyrite zone   ■ Molybdenite zone

■ Dispersed   0   1000 2000
   copper zone   Feet

(B)

FIGURE 13.10   Bingham copper mine near Salt Lake City.

(A) Aerial view reveals this vast open-pit excavation, about 2.5 miles across and 3000 feet deep. Its reserves now are estimated at 1700 million tons, including grades down to 0.71% copper. This is about 6.5 times the reserve originally estimated ninety years ago. This spectacular pit is visited by 200,000 tourists each year. *Photo: Kennecott Corporation.*
(B) Cross-section of granite host rock and sulfide mineral ore zones at Bingham mine. The ore deposit is shaped like a huge blanket, roughly parallel to the ground surface, with extensions that follow fractures to unknown depths. Metals mined here include copper (the ores are chalcopyrite $CuFeS_2$ and bornite Cu_5FeS_4), iron (pyrite, FeS_2), and molybdenum (molybdenite, MoS_2). *Source: U.S. Geological Survey.*

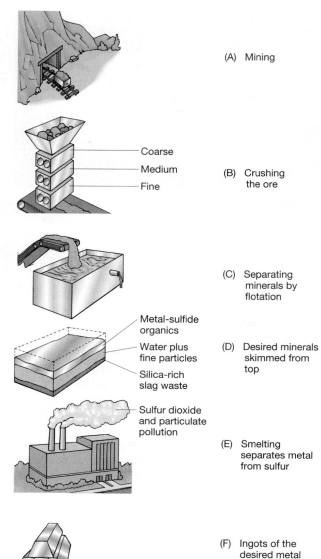

(A) Mining

(B) Crushing the ore

Coarse
Medium
Fine

(C) Separating minerals by flotation

Metal-sulfide organics
Water plus fine particles
Silica-rich slag waste

(D) Desired minerals skimmed from top

Sulfur dioxide and particulate pollution

(E) Smelting separates metal from sulfur

(F) Ingots of the desired metal

FIGURE 13.11   From raw ore in the ground to pure metal for manufacturing.

This diagram shows how a typical metal sulfide ore is processed to recover copper, molybdenum, lead, or whatever metal is involved. (A) Ore is mined and moved to a crushing mill. (B) In the mill, raw ore is crushed to a fine size by rollers. (C) The crushed ore then is mixed with water and organic solvents. The key to separation occurs when the organics affix themselves to the metal sulfides but not to the host rock debris. The low-density organics float, with the desired sulfide minerals adhering to their surface. This is separation by *flotation*. (D) The froth of organics and metal sulfide minerals is skimmed off and sent to the smelter. The residual rock debris (a silicate slag) and water are disposed of, usually drained into a nearby low area. (E) At the smelter, the sulfide ore mineral is roasted with oxygen to remove the sulfur. The sulfur combines with the oxygen, leaving the pure metal behind. (F) The goal: pure metal, ready for manufacture into products. If more than one metal is involved (often the case), the metals are separated from each other, magnetically or electrically.

Some mines have an ore-crushing mill and smelter immediately adjacent. But more commonly, a central mill and smelter facility serves numerous mines that may be a few hundred miles apart. Heavy, bulky ore is transported by railroad, river barge, or truck.

About 10 percent of copper production at Bingham is accomplished by *leaching* of the waste dumps. Leaching also is known as **solution mining** and was discovered accidentally in the 1930s. Most rainwater is slightly acidic. As rainwater percolated through waste from copper mining, it was observed to dissolve some of the residual copper and to redeposit it on cans in a garbage dump near the waste rock dump. This natural recovery method has been refined and now is part of copper recovery at Bingham.

As you might guess from Figure 13.10B, copper is not the only product from Bingham Mine. Bingham also is an important producer of molybdenum, gold, silver, lead, and zinc. For example, each year they produce 2 million ounces of silver (worth $5 an ounce) and a quarter-million ounces of gold (worth about $370 an ounce).

Case Study: Copper Smelting in Ducktown District, Tennessee

The Ducktown District lies in the southeastern corner of Tennessee. Copper smelting began at Ducktown in the mid-1850s. By the early 1870s, one publication described the area around the smelter as:

> . . . denuded of timber . . . large quantities have been destroyed by the fumes from the smelting furnaces, which, charged with sulfurous acid, wither and deaden all vegetation by their poisonous contact.

The chemistry of the problem is simple: smelting copper ore to release pure copper metal also releases a noxious gas, sulfur dioxide (SO_2). This has been known since 1736 but until recently no one cared. Emitted from the smelter's smokestacks, the sulfur dioxide mixes with atmospheric moisture and oxygen and thus is converted into sulfuric acid droplets (acid precipitation is described in Chapter 17). The Ducktown District has plenty of atmospheric moisture, averaging 58 inches of annual precipitation. Its humid, temperate climate is ideal for the area's luxuriant forest and for generating acid precipitation.

An area of 22 miles by 15 miles (180,000 acres) was alleged to be severely affected, with significant damage beyond (Figure 13.12A). In 1896, a mining engineer wrote:

> No civilized community ought to be afflicted with the nuisance resulting from the continuous dissemination of scores of tons of sulfurous acid gas daily through that part of the atmosphere nearest the surface of the ground.

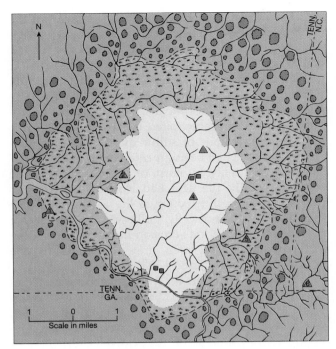

(A)

(B)

FIGURE 13.12 Environmental damage from copper smelting at Ducktown, Tennessee.

(A) The Copper Basin in Polk County, Tennessee. The damage from airborne sulfur dioxide gas is obvious. The central bare zone is the largest completely barren area in the U.S. Note four smelters and the six meteorological stations established to study the local climate. *Source: U.S. Department of Agriculture.* (B) The Ducktown Desert. The denuded soil quickly eroded, forming deep gullies. *Photo: Pat Armstrong/Visuals Unlimited.*

By the 1890s, alleged damage to crops and surrounding forests had generated many lawsuits against the smelter's operators. Anti-smelter forces prevailed, so the smelter operators developed a technology to capture most of the sulfur dioxide gas. They converted the gas to sulfuric acid, which they were able to sell. Thus, instead of natural atmospheric processes converting the gas to acid and spreading it over surrounding vegetation, the process was accomplished within the smelter facility and the acid put to use.

Lightly damaged forests at some distance from the smelter began to recover. Near the smelter, however, recovery was slight, despite decades of replanting programs. These programs failed because of continuing lower-level emissions of sulfur dioxide, and because the land surface had been greatly altered, both chemically and physically, by its prior dosing with sulfuric acid. Chemically, the acid had leached nutrient elements away from the portion of the soil where plants send their roots. Physically, gullies 20 feet deep were created, and this serious erosion hampered reseeding and replanting. The area grew infamous as the Ducktown Desert (Figure 13.12B).

A century later, in the 1970s, the environmental movement began to change U.S. law, and the Ducktown District benefited from the stricter air-quality standards. New revegetation programs are now proving more successful. Reforestation of the "Ducktown Desert" may be completed around 2000. Ironically, this has generated some interest in preserving part of the Ducktown Desert in its denuded, eroded state, for historical purposes! This is not as silly as it sounds. Such preservation can serve as a graphic reminder of the damage that thoughtless development can cause.

Waste and Pollution from Mining and Processing

Surface mining, at Bingham copper mine or any other site, causes serious environmental problems, including tremendous rock waste, smelters that release particulates and acid-generating gas, water pollution, and toxic heavy

metals. The mining industry has a place among the greatest environmental problems in American industry.

Waste Rock

In all surface mining, the surface vegetation and soil are first removed. Then, **overburden** rock covering the ore is removed. This work is done with huge equipment: large bulldozers, trucks large enough to carry a small cottage, power shovels, and dinosaur-like draglines that scoop automobile-size volumes of rock. As the mine develops, waste called **tailings** builds like giant dunes around the operation.

In 1991, about 990 million tons of copper ore were mined in the United States to produce 9 million tons of copper. In other words, because of the copper ore's low grade, 99 percent of the ore rock excavated was waste. (For most ores, the waste is typically 80–90 percent.) As the mine excavation deepens, its width at the surface increases sharply, increasing the proportion of waste (Figure 13.13). As a result, the National Academy of Sciences has predicted that copper mining in the year 2000 will produce three times as much waste as it did in 1978 for each ton of copper produced.

Federal law requires that the mine area be restored as closely as possible to its premining condition when the mine is closed. However, in the interim, waste piles may be undermined by streams at their base. Some landslides of waterlogged material have wiped out entire communities.

Water Pollution

Of greater environmental concern than the waste piles is polluted groundwater. Around the Bingham Mine, at the bottom of the adjacent canyon, a toxic plume of heavy metals leached from the mine spreads over two square miles. But a plume of sulfates (oxidized sulfur from the sulfide minerals) continues to spread for

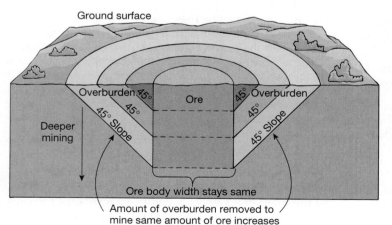

FIGURE 13.13 The effect of increasing the depth of a surface mine.

The amount of overburden that must be removed to mine a given volume of ore increases dramatically with depth—look at the obviously increasing volume shown in different colors. The mine sidewalls cannot exceed a slope of 45°, because to do so would cause collapse of the walls into the mining area at the bottom of the pit.

another fifty to seventy square miles, giving well water a taste you wouldn't want to drink.

To help remediate this water-pollution problem, the mine operator, Kennecott Corporation is building an underground retaining wall along the lower slope of the mountains to catch the groundwater—a Berlin Wall against pollution—at a cost of $200 million. The company also plans to pump and clean the toxic core of the plume in the valley, while piping clear water to homes that have abandoned their wells. Kennecott calls their remediation program the "largest voluntary cleanup" ever by a mining company. Perhaps so, but it is a bit late. The company has known about the contamination for twenty years. The cleanup is in response to a lawsuit filed by the state of Utah.

Solution mining has its problems, too. Just as residual copper is recovered using an acid solution, trace gold is scavenged from waste piles, using a sodium cyanide solution in a method called **cyanide leaching.** The cyanide is poisonous and can leak into the soil and underlying aquifers. However, in 1995, scientists from the U.S. Bureau of Mines discovered a bacterium *(Pseudomonas pseudoalcaligenes)* in processing waters and tailing ponds that decomposes cyanide into harmless compounds. Adding nutrients increases this bacterium's population and speeds cyanide breakdown. Successful at several mines, the technique is being extended to others where cyanide residue poses an environmental threat. Unfortunately, the agency that discovered the bacterium, the U.S. Bureau of Mines, was disbanded by Congress in 1996 as a cost-cutting measure.

EPA's Superfund list includes seventy-seven mining-related sites among its top-priority hazardous-waste cleanups (review: Chapter 9, Superfund's 35,000 Sites). One is the abandoned Iron Mountain Mine near Reddings, California. It leaks acids, copper, cadmium, zinc, arsenic, and lead into the Sacramento River. This river provides drinking water for two-thirds of California's 32,000,000 residents. At the peak of a six-year drought in 1992, water managers had to release water from upstream reservoirs to dilute toxic seepage from the mine. EPA estimates that cleanup cost could reach $35 million.

Acid Mine Drainage. As you have seen, most valuable elements occur combined with other elements, commonly oxygen and sulfur. (In Chapter 2, Table 2.3 shows that ores of important metals commonly are sulfide minerals.) In addition, pyrite (iron sulfide, FeS_2) is a very common companion in metallic ore deposits. Consequently, lots of sulfur is involved in mining. This is unfortunate, because sulfur released to the environment around mines often forms sulfuric acid, causing **acid mine drainage** (Figure 13.14).

FIGURE 13.14 Runoff from a smelter seeping into an aquifer in Arizona.

In the background, the smelter operates in Sonora, Mexico. The water's orange color reflects the presence of pyrite, which has oxidized to hematite to create the color. *Photo: Peter Chartrand/D. Donne Bryant Stock Photography.*

During mining, rock is broken and crushed, exposing fresh rock surfaces. Upon this contact with air and water, sulfur in minerals like pyrite and chalcopyrite is converted to the sulfate ion (SO_4^{-2}). This is the key ion in sulfuric acid (H_2SO_4). The acid drains into streams. Not only is acidity generated, but ore metal is released into streams. Thus, ordinary rainwater that percolates through a mine and its dumps often contains unacceptable levels of lead, copper, or other elements. (Recall "solution mining" at the Bingham Mine.)

We determine how acidic or alkaline water is by measuring the quantity of hydrogen ions in it. This is expressed as numbers from 0 to 14 on an acidity-alkalinity scale called the **pH scale.** For general perspective, 0 is extremely acidic, 2 is lemon juice, 3 is vinegar, 4 is an acidic rain, and 5.6 is normal rain (slightly acidic); 7 is neutral. On the basic side, 8 is a baking soda solution,

12 is strong ammonia, and 14 is extremely basic. (The pH scale is illustrated in Chapter 17, Figure 17.27.)

Immediately adjacent to mines that are rich in sulfide minerals, the pH may be as low as 1–2, which is 100,000 to 1,000,000 times more acidic than normal streamwater. Few plants can survive such acidic waters. The acidity is lethal for many water-dwelling animals as well—fish, snails, and others. (Chapter 17 discusses acid precipitation.)

Pollution from Smelters

Smelters produce great volumes of particulates that spew from the smokestack unless controlled. No other industrial activity produces more particulate matter than smelting. Smelting also produces **slag,** a voluminous residue left after the valuable metals have been removed. Slag looks like dark, frothy volcanic rock—not surprising, for slag cools and hardens from molten material just like a lava cools and hardens from a volcano. Slag is glassy rather than crystalline because it

cools rapidly upon ejection from the smelter (review: Chapter 2, Igneous Rocks).

Typically, slag is enriched in toxic heavy metals that were not recovered from the ore, such as cadmium, lead, or arsenic. These toxic elements make slag disposal a major headache, for they can pollute surrounding soil and groundwater.

Underground Mining

Figure 13.15 is a cross-section of a typical underground mining operation for a metal ore. If you find it complex and bewildering, then you can appreciate why mining requires lots of engineering and major investment. In this example, two separate ore bodies ("A" and "B") are being mined, and they run at steep angles into the ground. Further, these ore bodies are not continuous rich veins; the ore grade varies with depth.

Today's mines employ far fewer people than decades ago because a few powerful, high-tech machines

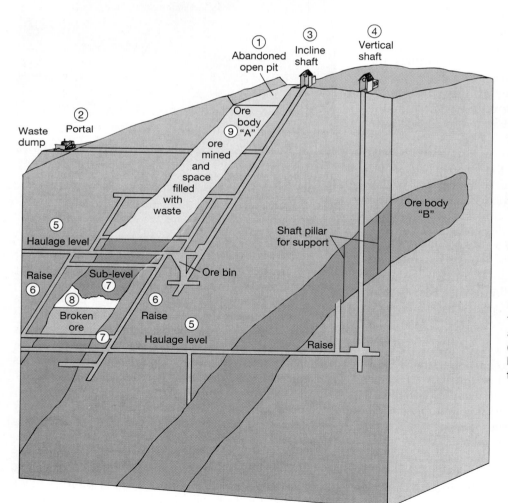

FIGURE 13.15 Cross-section of a complex underground mining operation.

The sequence in which this mine developed: (1) Ore body "A" was first mined from the outcrop as an open pit, now abandoned. (2) A horizontal shaft was excavated into the hillside to continue mining ore body "A." This and other entries are called *portals.* (3) An incline shaft was sunk later to mine ore body "A" still deeper. Mined ore is removed through this shaft. (4) A vertical shaft eventually was sunk to access both ore bodies ("A" and "B") more efficiently. (5,6,7) Ore is mined by excavating two haulage levels (5) and connecting them by raises (6). These then are connected by sub-levels (7). (8) Ore is mined upward from the lower sub-level. (9) Waste from the processing mill is pumped in.

have replaced dozens of workers. This equipment is carried part-by-part into the mine, through the same narrow shafts and haulageways that workers use. The parts are assembled into machinery underground. The ore also moves through the haulageways and shafts to the surface. From there, the ore is transported to the crushing mill and smelter.

Years ago, underground mining for metallic ores was very dangerous. Fatalities and injuries still occur today, but the mines are far safer because of federal mining regulations, better technology, and fewer workers in the mines. Collapses and accidental explosions now are uncommon.

Unfortunately, the extraordinarily deep gold mines of South Africa are a notable exception. Here, the rocks sometimes burst from the walls with explosive force. They do so because excavation exposes rock that has been intensely confined for millions of years, allowing it to expand explosively. Also, this rock is under great pressure from the weight of thick rock overhead (South Africa's deepest mine extends over two miles below the surface). Can you imagine the legal and environmental restrictions that would be imposed by the U.S. government if American workers were subjected to such dangerous conditions?

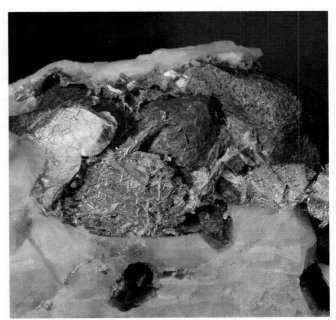

FIGURE 13.16   Gold nugget, still attached to quartz, from South Africa.

Such nuggets fired the 1848 California Gold Rush and other gold rushes in Colorado, Canada, and New Zealand. *Photo: Hartmann/Sachs/Phototake NYC.*

The Allure of Gold

Gold was among the earliest metals to be used because it occurs mostly in pure form, glittering in its host rocks or as loose nuggets and flakes in stream gravels. Gold is aloof, reacting with few other elements, remaining pure and gleaming (Figure 13.16). Art and jewelry are shaped readily in this soft metal. Gold objects more than 5000 years old have been found in Egypt, and archeologists have discovered numerous ancient gold mines in the Middle East.

Throughout history, gold has been the most reliable "money," holding its high value through times of war, inflation, and depression. Because gold is beautiful, easily shaped, and highly valued, most of it goes into jewelry. In the United States, student class rings are gold's biggest use.

But gold's other properties give it other uses. It does not combine with oxygen, sulfur, or other elements, as most other metals do, so gold does not tarnish or corrode. This makes it valuable in computers, communications equipment, spacecraft, and jet engines. Its ability to reflect heat protects astronauts, satellites, and electronic components from damage by X-radiation and cosmic rays in space. Gold used in dental fillings lasts a lifetime, even in the acidic environment of our mouths. Medically, gold treats arthritis and cancer, and it patches damaged blood vessels, nerves, and bones.

Gold occurs in significant amounts in many countries, so no monopoly of the world supply exists, as it does for diamonds (described later in this chapter). Gold is freely bought and sold. Over the past twenty-five years, gold's price has fluctuated, sometimes wildly, between $35 and more than $800 an ounce. It has stabilized around $400 an ounce in recent years, but this price could jump up or down at any time. World events and investors' psychology affect the price, but changes in the actual gold supply also occur. For example, new discoveries can be expected in Russia, which is vast and incompletely explored, now that a free market has replaced centralized planning.

Two other factors can change the gold price. One is the need of third-world countries for food and other commodities, which they can buy with gold. The other is that gold sometimes is "dumped" on the market by countries such as Russia, whose paper currency is not valued everywhere. Everyone values gold.

Thar's Gold in Them Thar Hills!

You probably have heard of the California Gold Rush of 1848, in which thousands of people streamed westward to get rich. What Earth processes placed the gold where people could find it?

In 1848, small chunks of gold were spotted at Sutter's Mill on the south fork of California's American River (Figure 13.17). Word of the find quickly reached

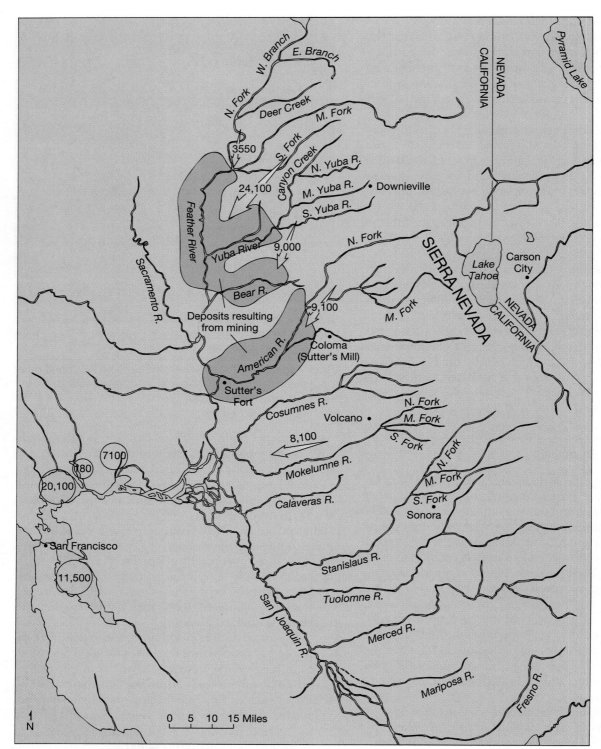

FIGURE 13.17 Gold Rush legacy.

California's Gold Rush brought thousands of prospectors to the Sierra Nevada foothills around 1850. The names of the hastily established gold camps tell the story of those days—including the camp named Rough and Ready. Thirty years of mining took an environmental toll: this map shows areas where millions of cubic feet of rock debris washed downstream from mining operations and were deposited in the bays near San Francisco.

San Francisco, then a sleepy coastal town of 1000 people. The population quickly fell below 100 as most left to seek the "gold in them thar hills." Gold fever lured the wannabe wealthy from the East. They came by train to St. Louis and then rode in covered wagons to Sutter's Fort. They also sailed around South America (there was no Panama Canal until 1914) and north to San Francisco. Both routes took several hazardous months, but this only heightened the allure of the Gold Rush.

The Gold Rush tales are wonderful, and many are true. On the Middle Yuba River, one miner recovered 30 pounds of gold from 4 square feet of ground in less than a month. At the Volcano mining camp, miners sometimes found up to $370 worth in a single panful of gravel (see chapter opening photo). Near Downieville, four men got $80,000 richer from a 60-square-foot claim in six months. Tin Cup Diggings was named for three men whose rule was to fill a tin cup with gold each day before quitting. A 161-pound nugget (141 pounds gold, 20 pounds quartz) was uncovered on the Feather River. A record find was a pure gold nugget weighing 162 pounds. By 1865, $750 million worth had been recovered from the hills and stream beds of the Golden State.

In those days, gold was worth about $20 an ounce, small compared with the $400 an ounce it brings today. (Calculate how much that 162-pound nugget would sell for today!) But then, in 1850, a restaurant meal might cost a dime. More important, there was no income tax (that invention lay sixty-three years in the future.) And the preposterous idea of an inheritance tax (a tax on income that has already been taxed) was sixty-six years away.

The chance for ordinary people to get rich quickly was almost unprecedented, for the gold belonged to its finder, not to some hereditary king or prince who owned everything, as in much of the world at that time. By fortunate timing, California's settlers had declared themselves independent of Mexican rule in 1846, only two years before the gold strike.

Most of the gold found by those lucky folks 150 years ago occurred in placer deposits. What is a placer deposit and how does it form?

Placer Deposits

A **placer deposit** (PLASS-er) is a concentration of free metals (like gold) or heavy minerals (like diamonds) that has been created by stream currents or waves. The concentration can be accomplished by a reasonably strong current in a stream or river, or by waves along the seashore. Common placer deposits include dense materials such as gold, platinum, diamonds, and semiprecious gems. *Density* is the key concept: all placer minerals are denser than the more abundant surrounding minerals, like quartz and feldspars.

Another way of saying this is that placer minerals have a much greater *specific gravity* than the surrounding common minerals. (Recall that specific gravity is a substance's weight compared to an equal volume of water; water is assigned a specific gravity of 1.0.) Generally, placer minerals have higher specific gravities of 3.5 (diamond) to 19 (gold), whereas the more common minerals are lightweights around 2.7 (quartz, feldspar). This difference allows swirling waters and waves to separate (sort) minerals according their specific gravity, simply because a given water movement will carry the lighter ones farther and the denser ones not as far. Table 13.5 shows major placer minerals and their use.

Gold placers are typically rather pure, without other heavy metals. The reason is that gold has a specific gravity of 19, much greater than any other placerable mineral except platinum (17). As moving water sorts minerals of different specific gravities, it will widely separate the extremely dense from the extremely light. Thus, a mineral with a specific gravity of 3–6 normally will not be deposited with a mineral of 19.

Could you find a gold-platinum combination placer? After all, their specific gravities of 19 and 17 are similar. Sorry. Both precious metals occur in igneous rocks, but platinum occurs in silica-poor rocks and gold occurs in silica-rich rocks. A stream would have to drain both types of igneous rock to contain both gold and platinum—an unlikely occurrence.

Most placer minerals are durable and not very reactive chemically. Both traits allow them to resist breakdown. In an outcrop, placer deposits are most noticeable when they are dark-colored, because the dominant minerals in sand deposits and sandstones are light-colored quartz and feldspar.

Placers and Environmental Setting

A rare mineral does not simply wash down from the mountains and form a rich placer deposit. To become really concentrated, sediment containing a rare mineral must undergo repeated sorting. It is somewhat like separating sand of twenty mixed grain sizes into twenty different size piles: to do so, you would have to sift it nineteen times, through screens of different size.

Such repeated reworking of rare mineral grains is common where Earth's crust is unstable, along converging plate margins. For example, the entire Pacific area has experienced plate motion for at least 100 million years. This has caused continual emergence of older gold-bearing zones and continual reworking of their sediment. The result today is gold-rich placers in California, British Columbia, the Yukon, Alaska, Siberia, and New Zealand.

TABLE 13.5 **Major Minerals that Occur Commonly in Placer Deposits**

Name of Mineral	Specific Gravity (Water = 1.0)	Chemical Formula	Important Use
Waste rock	2.7	SiO_2 (quartz) (Na, K, Ca, Al, Si, O)	Undesired waste surrounding of placer
Olivine	3.3	$MgSiO_4$	semiprecious gem
Diamond	3.5	C	precious gem
Topaz	3.5	$Al_2SiO_4(F,OH)_2$	semiprecious gem
Garnet	3.5-4.3	$Fe_3Al_2Si_3O_{12}$	semiprecious gem, abrasives
Corundum	4.0	Al_2O_3	precious gem (ruby, sapphire)
Rutile	4.2	TiO_2	paint pigment
Ilmenite	4.7	$FeTiO_3$	major source of titanium
Zircon	4.7	$ZrSiO_4$	semiprecious gem
Magnetite	5.2	Fe_3O_4	important iron ore
Cassiterite	7.0	SnO_2	major ore of tin
Cinnabar	8.1	HgS	major ore of mercury
Silver	10.5	Ag	tableware, jewelry, film, mirrors, plating
Platinum	17.0	Pt	catalyst in industrial processing, jewelry, electrical equipment
Gold	19.0	Au	basic currency, jewelry, dentistry, alloys, plating

Most placers that we mine formed during the past few ten thousands of years. They are at the surface, in streams and along beaches, and are easily found. However, a few important *paleoplacers* are known, ancient placers buried deeply beneath sediment or rocks. A prominent paleoplacer is in the Witwatersrand region of South Africa, which supplies more than half the world's annual gold production. Here, the placer gold occurs as fine particles in ancient river gravel, in a 2.5-billion-year-old rock sequence. Another important paleoplacer yields uranium ore from Blind River in Ontario.

There are two basic placer environments—streams and beaches—which create stream placers and beach placers. We shall look at stream placers and how they concentrate gold, for example. In a later section in this chapter, we will examine beach placers, and how they concentrate diamonds, for example.

Stream Placers and Gold

Stream placers form when flowing water carries sediment particles of mixed size and specific gravity. If the water slows, sediment begins to drop out, starting with the largest pieces and those of greatest specific gravity. In other words, first to drop out are the larger gravel and the denser gold (specific gravity = 19). Next to drop out are smaller gravel and less-dense silver (specific gravity = 10.5). And so on downward in size and density, to tiny clay-size particles and to lightweight quartz and feldspar (specific gravity = 2.7). This process "sorts" the sediment as water slows, forming placer deposits.

But what makes a stream slow down? It slows where it flows around a bend or obstruction, or empties into a larger water body. These are the places to look for a placer deposit (Figure 13.18A). The Gold Rush prospectors knew this and sought good placers accordingly.

The gentleman in the chapter opening photo demonstrates the simplest way to "work" a placer: scoop some river-bend gravel into a shallow pan and start swirling, which separates gold from lighter minerals. More sophisticated miners shoveled gravel into *sluices*, running water through them to sort gravel and sand from the gold. Later, large dredges were used to process tons of gravel for gold (Figure 13.19).

On a smaller scale, an inexpensive "vacuum cleaner" technique has been developed that can extract worthwhile amounts of gold from placers (Figure 13.20). The diver inspects the bottom for likely deposits and vacuums their contents. At the surface, a companion monitors a floating sluice that separates the gold from the gravel.

Gold Mining and Mercury Pollution

Not all gold occurs as pure nuggets and flakes in placers—remember, those bits and pieces had to come from a host rock somewhere upstream. Gold

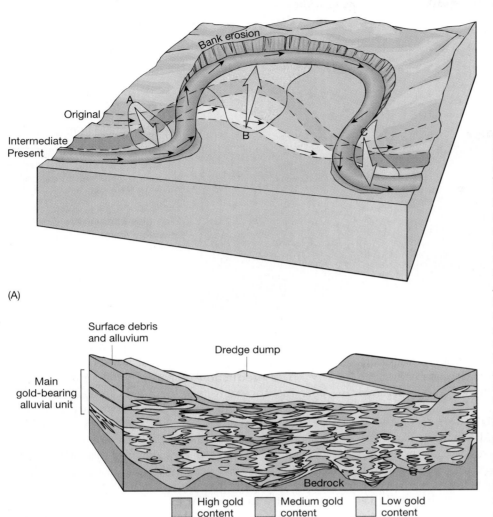

(A)

(B)

FIGURE 13.18 Placer formation in a fast-flowing meandering stream.

Like a serpent very slowly snaking its way downhill, meanders gradually migrate both laterally and downstream. (A) Shown are the stream's original position, an intermediate position, and the present streamcourse. Deposits formed at *A, B,* and *C* (which began on the inside of meanders of the original streamcourse) have grown both downstream and laterally in the direction of the large arrows; it is here that buried placers result. *After Jensen and Bateman,* Economic Mineral Deposits, *copyright © 1979, John Wiley & Sons, Inc.* (B) Cross-section of a placer in a river (in this example, the Kuranakh River near the Arctic Ocean in Russia). Richness of the gold deposit is irregular, both laterally and vertically, due to irregular currents that occurred as sediment was deposited. *After Karatashov,* Economic Geology, *v. 66, Economic Geology Publishing Co., 1971.*

often is associated with other minerals, from which it must be separated. A widely used method is to add the liquid metal element mercury to the ore, because it combines with the gold to form an *amalgam.* (An amalgam is any alloy of mercury with another metal; many dental fillings are a mercury-silver amalgam.) Later, the mercury is separated chemically from the gold. The ability of gold and mercury to combine has been known for a long time and was the most common method used by miners to extract gold from ore during 1860–1890 in northern Nevada, at the famous Comstock Lode.

But using mercury to separate gold has caused **mercury pollution** in streams—in particular, the Carson River basin near Lake Tahoe on the California–Nevada border. About seventy-five ore-processing mills in the basin lost three-fourths of a pound of mercury for every 1000 pounds of ore they processed. This adds up to about 12 million pounds of mercury being dumped into the basin. Today, the area's water and sediment contain 10–100 times the mercury of unaffected areas nearby, among the highest values in North America. There are high mercury concentrations in plants, fish, and insects in the drainage basin. The Lahontan Reservoir, which receives water from the Carson River, has the highest mercury content ever reported from North American surface waters. Consequently, EPA added the Carson River Superfund Site to its National Priorities List in 1990.

The Allure of Diamonds

Diamond is a mineral formed in Earth's mantle, about 100 miles down. Here, the temperature is about 2000°F and the pressure is 50,000 times higher than the atmospheric pressure we live in (15 pounds per square inch at sea level). These conditions force carbon atoms to form the hard crystals we call diamonds. Diamonds are pure

FIGURE 13.19 A gold dredge separating gold from gravel.

The dredge nibbles gravel from the wall of the pond, runs the gravel through itself to extract the gold, and then dumps the waste in the pond behind itself. Old gold-dredging areas are now ugly heaps of dredged gravels, and sometimes the dredges themselves were abandoned when the gold or the money ran out. *Photo: Fred Bruemmer/DRK Photo.*

carbon and are the hardest naturally occurring substance. They reach the surface volcanically, through small **kimberlite pipes.** A typical pipe at the surface is small, so the environmental effects of kimberlite mining are minimal (Figure 13.21).

Why do we value diamonds? The basic answer is that they are beautiful, rare, and the hardest natural substance. But there is more to the story. Diamonds are classified broadly into gem quality and industrial quality. *Gem-quality diamonds,* the beauties used in jewelry, produce 90 percent of the profit in the retail diamond trade. *Industrial-grade diamonds* are ugly ducklings, not pretty but just as hard as their flashier sisters. They are used for cutting, grinding, machining, and in electronics. When drilling a well in hard rock, drillers use drill bits that are tipped with industrial diamonds. Industrial diamonds are 75 percent of all natural diamond production.

The production of natural diamonds long has been insufficient to meet demand. Thus, synthetic industrial diamonds have been manufactured since 1953. These are tiny crystals, seldom exceeding 0.03 inch (less than a millimeter), suited only for industrial use. In 1970, General Electric produced a few gem-quality diamonds weighing up to a carat. But these large synthetic diamonds are exceedingly difficult and costly to make, far exceeding the cost of mining and processing equivalent natural stones. Synthetic diamonds of gem quality cost twice as much as natural ones.

FIGURE 13.20 Vacuuming for gold.

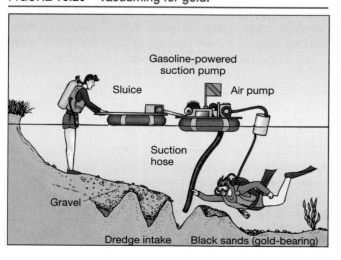

Gasoline-powered suction pump

Sluice

Air pump

Suction hose

Gravel

Dredge intake   Black sands (gold-bearing)

The Diamond Market

In India, a diamond trade flourished 2400 years ago. India was the only significant diamond-producer from ancient times until Brazil joined the trade in the 1700s. Today, however, the major producers of gem-quality diamonds are African: the Republic of South Africa and Namibia. De Beers Consolidated Mines, headquartered in South Africa, holds a virtual monopoly. Russia is the only significant independent producer of gem-quality diamonds, but the Russian government is cooperating with De Beers to keep the diamond market "stable."

"Stable" means that De Beers controls worldwide diamond production and distribution and thus controls world diamond prices. The normal economic law of supply and demand is suspended in the diamond industry because this company has a monopoly over

FIGURE 13.21 The Mirny diamond mine in Siberia.

This mine's shape is that of a kimberlite diamond pipe. *Photo: B. Bisson/Sygma.*

the supply. De Beers sets diamond prices artificially high to maximize its profit.

As an example of the economic power of the De Beers company, consider the Japanese market for diamond engagement rings. In 1967, De Beers commissioned the Chicago-based J. Walter Thompson advertising agency to promote diamond engagement rings in Japan. At the time, fewer than 6 percent of Japanese brides possessed a diamond. By 1972, 27 percent owned diamond rings; by 1978, about 50 percent; by 1981, 60 percent; by 1989, 77 percent. Japanese brides now have the highest diamond acquisition rate on Earth, paying a higher average price than in any other country. De Beers presently is targeting a rapidly emerging affluent population in southeast Asia, particularly China.

Beach Placers and Diamonds

Beach placers are similar to stream placers, but they form in a different environment. Typically, the dense minerals are eroded by streams, which carry them to the coast, where they are deposited on the beach and along the shoreline. Incoming waves and the longshore current continually sort the sediment (review: Figure 10.5). Thus, beach placers constantly are forming on beaches worldwide. Next time you are at a beach, dig a short trench about 6 inches deep perpendicular to the shoreline. You probably will see a dark concentration of magnetite (iron ore) and ilmenite (titanium ore) in thin layers that dip gently seaward. When such placers are thick and abundant, they are worth mining.

Older placers exist where beaches used to be (recall from Chapter 10 that shoreline position can migrate rather quickly). A beach of 1000 years ago now may be a few hundred feet offshore because of sea level rise, its valuable placer minerals concealed beneath the waves. Beach placers are worked on a vast scale around the coasts of Japan, Australia, India, Brazil, Malaysia, and many other countries.

Beach placer miners seldom consider the environmental impact of their workings. The chief effect is increased shoreline erosion. As you saw in Chapter 10, natural shorelines are fed by stream-carried sediment that replenishes sand that is hauled away by longshore currents. Placer dredging at the shoreline disrupts the natural balance of sediment addition and removal, stimulating increased erosion.

Along Africa's southwestern coast, in the country of Namibia, a 200-mile-long gravelly beach placer offers bountiful diamonds, both above and below sea level (Figure 13.22A). The diamonds were weathered from igneous rocks in the country's interior and carried coastward by the Orange River. Longshore drift spread the diamonds southward. Subsequently, some beaches were raised by tectonic forces, whereas others were submerged by rising sea level.

The Namibian diamonds are exceptional in quality and quantity. During transport from their inland source, the weaker or "flawed" diamonds are fractured and reduced to small particles. Stronger, unflawed diamonds survive to enrich the Namibian diamond placer and those who sell its product (Figure 13.22B).

To recover the diamonds, great earth-moving machines excavate the sand down to 60 feet, while powerful pumps keep the ocean from flooding the excavation. In one year, the shoreline area produced more than 2 million carats of diamonds (about 880 pounds; 1 carat = 7/1,000 of an ounce).

Mining Nonmetallic Rock and Minerals

When you think of a "mine," what comes to mind? You probably do not think first of a quarry for stone, or a dredging operation for sand and gravel, or a cement mine, yet these materials are by far the most-mined mineral commodities. Excluding fossil fuels, more than

(A)

(B)

FIGURE 13.22 Diamond mining along the Namibian coast of southwestern Africa.

(A) The diamonds occur as broken fragments in sediment, at and below sea level. This bulge in Namibia's coastline is waste sand, processed during mining. *Photo: Anthony Bannister/Natural History Photographic Agency.* (B) Gem-quality diamonds from the De Beers Oranjemond Mine on the Namibian coast, the richest diamond mine on Earth. *Photo: Peter Johnson/Corbis-Bettmann.*

TABLE 13.6	Production of Rock and Mineral Resources in the U.S., 1992 (Tons)
Nonmetals	*U.S. Production*
Crushed stone	1,155,000,000
Sand and gravel	805,700,000
Phosphate rock	42,770,000
Clay	38,357,000
Salt	36,218,000
Gypsum	14,833,000
Sulfur	9,646,000
Potash (KCl)	1,602,000
Dimension and facing stone	1,086,000
TOTAL	2,105,212,000 = 97.3%
Metals	
Iron ore	51,142,000
Aluminum	3,640,000
Copper	1,565,000
Zinc	473,000
Lead	373,100
Nickel	5,000
Silver	1,638
Gold	291
Platinum group	7
Tin, chromium, cobalt, manganese	0+
TOTAL	57,199,936 = 2.7%

Source: U.S. Bureau of Mines.

TABLE 13.7 Some Nonmetallic Minerals and Rocks and Their Uses

Mineral	Main Source Rock	Important Uses
Apatite	Sedimentary phosphate rocks	agricultural fertilizer, water treatment
Asbestos	Low-grade metamorphic rocks	incombustible fibers, insulation
Calcite	Limestones	building stone, cement, soil conditioning
Clays	Sedimentary rocks	ceramics, paper, brick, drilling "mud" in oil drilling
Feldspar	Igneous rocks	glass and ceramics, paint, plastics
Fluoride	Igneous rocks (hydrothermal)	steel-making, aluminum refining, glass
Graphite	Low-grade metamorphic rocks	pencil "lead," "dry" lubricants, batteries
Gypsum	Evaporite sedimentary rocks	wallboard, plaster, soil conditioning
Halite	Evaporite sedimentary rocks	table salt, water treatment, ice control
Muscovite	Igneous rocks (hydrothermal)	electrical insulator, plasterboard
Quartz	Igneous rocks (hydrothermal)	glass manufacture
Sulfur	Sedimentary rocks	chemicals, fertilizer, sulfuric acid
Talc	Low-grade metamorphic rocks	paints, cosmetics, toiletries

Rock	Important Uses
Granite	building stone, monuments, tombstones
Limestone	building stone, cement, road aggregate, metal smelting, coffee tables
Dolostone	building stone, road aggregate
Sandstone	building stone, source of glass sand (quartz)
Shale	source of clays, lightweight aggregate
Slate	building stone, roofing, blackboards, billiard tables
Marble	building stone, monuments, coffee tables

95 percent of our consumption of mined minerals is nonmetals (Table 13.6). These are unglamorous materials—crushed rock, stream gravel, sand, and limestone (Table 13.7). Construction materials dominate mineral and rock mining in the United States.

Crushed Stone (Aggregate)

At a **quarry**, rock is mined. It then is crushed to form **aggregate** of a size specified by the customer. In the United States, limestone and dolostone account for about 71 percent of the aggregate produced, followed by granite (14 percent) and very fine-grained igneous rock (8 percent). The remaining 7 percent includes sandstone, quartzite, marble, and slate.

The construction industry consumes 90 percent of all crushed stone and 95 percent of natural sand and gravel. Consequently, demand for these products depends more on economic cycles than most products. During economic expansion, production is about 1,200,000,000 tons. But during recessions, construction activity slows, and annual production drops roughly a third to between 800,000,000 and 900,000,000 tons.

Because it is the most fundamental component of construction, aggregate is needed for every building, highway, bridge, and dam that is built. Although the quarrying process is relatively inexpensive, rock weighs an average of about 2.7 times what an equal volume of water weighs (specific gravity = 2.7). It also is very bulky. Transporting crushed rock or sand costs about 10 cents per mile for each ton, so, for example, 100 tons transported 50 miles costs about $500. For this reason, quarries rarely are farther than 50 miles from a city. Generally, they are as close as environmental laws and zoning will allow.

Because of the high transportation cost, quarries tend to be numerous and small, operating near construction activity. About 6000 different companies own 9300 rock quarries in the United States, an average of 186 quarries per state. If you drew a grid of squares over the United States, in which each square measured 18 miles by 18 miles, each would have a rock quarry, on average (although they are unevenly distributed, depending on population and rock availability). A few quarries are quite large (Figure 13.23).

Of the 9300 rock quarries, 99 percent are surface operations, with a few operating underground.

FIGURE 13.23   Reed Quarry, near Paducah, Kentucky.

This is the largest quarry in the United States, producing 10 million tons of limestone a year. It covers 200 acres and is 500 feet deep. *Photo: Charles Beck Studios Incorporated.*

didn't plant the quarry in the middle of Faultless; Faultless grew around them. The owners believe they have been good corporate citizens for many years, supplying an essential commodity at a fair price. Much of the concrete in the county has been made using aggregate from the quarry. Blasted Stone fenced the quarry to exclude youngsters and vandals, scheduled blasting infrequently, announced blasting plans in advance, and beautified the property by planting evergreens around its perimeter. Their trucks travel only on designated streets, at times of minimal residential traffic, and during daytime so no one's sleep is interrupted.

Nevertheless, Blasted Stone's days probably are numbered. They own enough land to keep operating for years, but Faultless residents are trying to get the property rezoned from "industrial" to "residential." The county zoning commission, whose chairwoman lives beside the quarry, is likely to approve this rezoning request, giving the quarry operators twelve months to move out. The community, which has been able to grow at will because of this cheap local source of construction materials, now is out to kill the very industry that made its growth possible. The Faultless citizenry plan to drive out that Blasted Stone Company.

This story is ironic but realistic: many such companies are being zoned out of business in high-growth areas of southern California and the New York City region.

Sand and Gravel

Like crushed stone, the need for construction sand and gravel is cyclic—900 million tons in times of economic expansion, dropping to 600 million to 700 million tons during economic recessions.

Commercial sand and gravel deposits are nearly all either stream deposits located near highland source areas or are deposits formed by glacial meltwater. The debris dropped by glacial ice as it melted was sorted by the meltwater, removing the commercially undesirable silt and clay from the sand and gravel.

Most commercial sand and gravel deposits were deposited by streams within the past million years or so. Older deposits tend to be at least partially cemented into rock (sandstone and conglomerate). Fast-moving streams act as grinding and washing mills, pulverizing soft and weak rock fragments, leaving hard, sound particles in sand and gravel deposits. Only a few miles of transport in a fast-moving stream pulverizes the unsound grains. Hence, even fairly close to the mountain front where the grains originate, the hard quartz content of these deposits is substantially increased. An important benefit of quarrying gravel close to its source (mountains) is that the thickest sediment accumulations are there.

Occasionally, quarrying is below ground, usually due to environmental restrictions. The higher cost of underground workings sometimes is preferred by companies to the environmental constraints on surface mining imposed by spreading urbanization. Residential communities often view a surface quarry as a nuisance and try to limit its activities. Indeed, quarrying brings dynamite blasting, heavy truck traffic on light-duty roadways, limestone dust, and blemishes on the landscape.

Let us consider a hypothetical but typical case. Blasted Stone Company began quarrying limestone decades ago. At that time, the quarry was miles from the town of Faultless, but the town has since sprawled, now surrounding the quarry with expensive homes. Former cornfields have become a country club, complete with artificial lake.

The owners of Blasted Stone are painfully aware of their unpopularity. But what can they do? They

FIGURE 13.24 Layers of stream gravel and sand.

The stream gravel contains conspicuously rounded gravel grains. Crushed stone particles are very angular, a normal result of mechanical crushing in a mill. *Photo: U.S. Geological Survey.*

California has a fortunate juxtaposition of urban construction and thick sand and gravel deposits. For millions of years, coarse sediment has been deposited by westward-flowing streams from the Sierra Nevada and eastward-flowing streams from the Coast Range mountains. (These are the same streams that offered placer gold deposits to Gold Rush prospectors.) Many urban centers are located between these two mountain chains.

Ideally, a commercial deposit contains 60 percent gravel and 40 percent sand. This provides coarse material to crush for a highway base or asphalt paving, and sand in the optimum size and proportion for mixing with cement to make concrete. Deposits that depart from the desired percentages of gravel and sand can be adjusted by screening, washing, and combining of size grades. Thus, several deposits having different proportions may all be commercially valuable (Figure 13.24).

The ideal sand and gravel deposit is one that consists entirely of quartz, because it is hard, strong, and comparatively inert chemically. Particularly undesirable are opal and silica-rich volcanic rocks, because these materials react chemically with the sodium and potassium that is present in some types of Portland cement. When concrete is made by mixing Portland cement with such unde-

sirable aggregate, the result is expansion, cracking, and deterioration of the concrete (Figure 13.25).

Politics, Economics, and Stone

The demand for crushed rock and sand can be separated into three groups: highways and public works, industrial and commercial projects, and housing. The use by highways and public works is easy to determine. It depends largely on the money appropriated by the Federal Highway Administration for distribution to the states to finance their highway construction and maintenance projects. These projects are commonly termed "pork barrel" appropriations. The congressional appropriation for 1996 was about 15 percent larger than for 1995.

The amount of crushed rock and gravel needed for industrial and commercial projects is more dependent on the strength of local economies. The fluctuations are much greater from year to year than the variation in highway funds. To take a shining example, Atlanta's preparations to host the 1996 Summer Olympics stimulated highway construction, commercial projects, and

FIGURE 13.25 "Map cracking" pattern in concrete.

A chemical reaction between the aggregate and the cement within the concrete has caused this "map cracking" pattern to appear. Soon the finished surface will "spall" away (break away in flakes) and complete deterioration will follow. Careful selection of aggregate can prevent this expensive mistake. *From American Institute of Professional Geologists, The Citizens' Guide to Geologic Hazards, 1993, p. 18, photo courtesy E. Nuhfer.*

major drain on our economy. As economic activity slows, the Federal Reserve Board may lower interest rates to stimulate economic development. Thus, the aggregate business depends on both the U.S. and world economy.

Maps Help Locate Resources

Resource maps are among the most valuable tools used by geologists to find mineral resources. A geologist starts with detailed geologic maps of the area, which show what rocks exist at the surface, or immediately beneath the soil. Geologic maps usually cover a few square miles. They often are topographic maps with the geology added, like the example shown in Figure 13.26. From these maps, the geologist locates outcrops of a desired rock (outcrops are places where bare rock is exposed at the surface and not covered by soil).

Next, rock samples are collected from these outcrops and chemically analyzed in a laboratory to determine the quality and minability of the resource present. For example, factors considered for a limestone are its percentage of calcium carbonate (the purer, the better), what impurities are in it (some impurities can ruin limestone for use as aggregate), and how fractured it is (the more it is naturally fractured, the easier it will be to excavate).

All of these characteristics then are noted on more detailed (larger-scale) maps to determine the best places to open a quarry. Permits are acquired and land is purchased from the current owner.

Mapping of river sand and gravel may require greater field work, because stream deposits shift rapidly with changes in flow, particularly flooding. Figure 13.26 shows a sand and gravel map. Following the mapping, sediment samples are measured to determine the different sizes present. Categories are defined by possible use.

housing. Hosting the Olympics contributed prominently to a record output of stone and gravel in Georgia in 1995.

Each new home built in the U.S. uses an average of 200 tons of crushed rock, sand, and gravel. Foundations, concrete blocks, bricks, mortar, and roofing shingles all require sand. But the number of new houses built depends heavily on interest rates charged by the federal government and banks, because most people borrow money to build a house. If interest rates are excessive, people don't build. Interest rates fluctuate with economic factors not only in this country, but in others. For example, if Saudi Arabia allows OPEC to raise oil prices, the American economy becomes depressed, because we import about half the oil we use and paying for it is a

Summary

Mineral deposits are naturally occurring concentrations of useful materials that can be extracted from Earth at a profit. An ore deposit is a body of rock that contains a natural concentration of a mineral from which a needed metal can be extracted at a profit.

Nearly all the metals needed to keep modern industrial civilization functioning exist in exceedingly low concentrations. To be economically exploitable, these rare elements must occur in concentrations several hundred to several thousand times their average amounts in the crust. Some metals are obtained in large amounts as byproducts from the mining and processing of ores of other elements.

Ore deposits can be mined either at the surface or underground. The choice depends on the ore's volume, shape, and variations in quality. The volume of rock

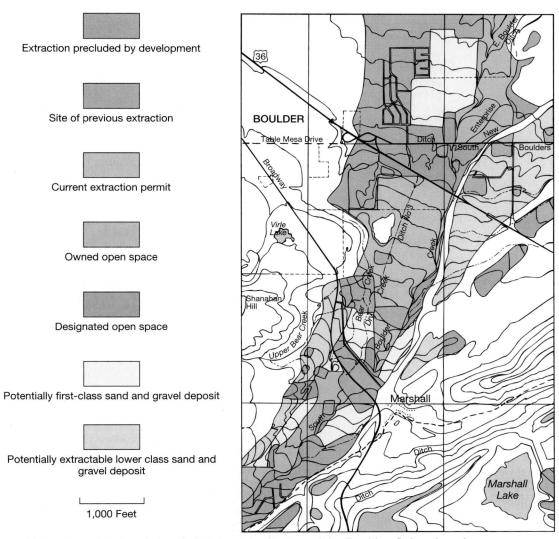

Extraction precluded by development

Site of previous extraction

Current extraction permit

Owned open space

Designated open space

Potentially first-class sand and gravel deposit

Potentially extractable lower class sand and gravel deposit

1,000 Feet

FIGURE 13.26 Sand and gravel planning map for part of the Boulder, Colorado, urban area.

The sand and gravel area defines an ancient stream channel where the sediment was deposited. *Source: U.S. Geological Survey.*

excavated to extract an ore can be enormous, as shown by the copper mine at Bingham, Utah. Federal law requires that a mined area eventually be restored as close as possible to its original condition.

Another major environmental concern is generation of acidic water, most common around metal sulfide ore deposits. Sulfur in the ore mineral changes to sulfuric acid when exposed to oxygen (the atmosphere), and the water can pollute adjacent streams. Metallic mineral deposits also are commonly associated with pyrite; this iron sulfide also releases sulfur upon contact with the atmosphere.

Placer deposits are concentrations of grains that are chemically resistant, durable, and have high density (like gold flecks). Placer deposits form in streams

and beaches by the sorting action of moving water. They are most easily seen when the placered mineral is dark-colored, because nearly all the abundant mineral grains (quartz and feldspars) are light-colored.

The major nonmetallic "mineral" deposits are rocks and sediment: crushed stone, sand, and gravel, used almost entirely by the construction industry. Almost three-quarters of the crushed stone is limestone and dolostone. Commercial sand and gravel deposits are nearly all either stream deposits located near highland source areas or are deposits formed by meltwater from glaciers. Active quarries, particularly those for crushed stone, can be environmentally undesirable and frequently are difficult to develop because of opposition from local residents.

Key Terms

acid mine drainage
aggregate
bauxite
contact metamorphism
cyanide leaching
crystal settling
kimberlite pipes
measured ore
mercury pollution
mineral resources

ore mineral
overburden
pegmatite
pH scale
placer deposit
porphyry
possible ore
probable ore
quarry
reserves

resource maps
resources
running out of a resource
slag
smelter
solution mining
surface mining
tailings
underground mining

Stop and Think!

1. What characteristics classify a mineral concentration as an ore deposit?

2. Give as many reasons as you can why ore deposits are so uncommon and hard to locate.

3. Why do most metallic ore deposits occur near tectonic plate margins?

4. Why do you think the percentage of zinc recycled is so much lower than the percentage of lead, copper, or aluminum (Figure 13.7A)? What would have to happen in manufactured products for zinc recycling to boom?

5. What is the difference between a mineral *reserve* and a mineral *resource*?

6. List variables that determine whether it will be more efficient to use surface or subsurface methods when mining an ore deposit.

7. Describe how an ore mineral such as chalcopyrite is transformed from pieces of a yellow mineral in granite host rock to a piece of copper wire in your home.

8. What types of pollution are associated with metallic mining activities?

9. Why do most metallic ore deposits generate acid mine waters?

10. Describe how a placer deposit forms. Why are metallic minerals more likely to form placer deposits than non-metallic minerals?

11. Gold usually occurs as the pure metal rather than combined with other elements such as sulfur. Yet, we have managed to make gold mining and processing a polluting activity. Describe the environmental problems of gold recovery.

12. Quartz forms only 20 percent of Earth's crust but more than 90 percent of the sand and gravel in quarries. Why?

13. Only food, water, shelter, and clothing have fundamental value to people. Yet, from earliest recorded history, humans have chosen to assign to gold an unassailable monetary value. Can you think of any reasons why supposedly intelligent creatures do this? Answer the same question for diamonds.

14. Where is the nearest stone or gravel quarry to your community? Check the Yellow Pages. Would you object if a quarry opened within a mile of your home? Do you want to increase the cost of housing construction by transporting aggregate long distances? Where is the middle ground that most people could live with?

References and Suggested Readings

Barnes, J. W. 1988. *Ores and Minerals: Introducing Economic Geology.* Philadelphia: Open University Press, 181 pp.

Bates, R. L. and Jackson, J. A. 1982. *Our Modern Stone Age.* Los Altos, California: William Kaufmann, Inc., 132 pp.

Earney, Fillmore C. F. 1990. *Marine Mineral Resources.* New York: Routledge, 387 pp.

Green, T. S. 1981. Diamond diggers in Namibia sift ocean sands for gemstones. *Smithsonian,* v. 12, May, p. 48–57.

Gurney, J. J., Levinson, A. A., and Smith, H. S.. 1991. Marine mining of diamonds off the west coast of Southern Africa. *Gems and Gemology,* v. 27, #4, p. 206–219.

Gustin, M. S., Taylor, G. E., Jr., and Leonard, T. L. 1994. High levels of mercury contamination in multiple media of the Carson River Drainage Basin of Nevada: Implications for risk assessment. *Environmental Health Perspectives,* v. 102, p. 772–778.

Hodges, C. A. 1995. Mineral resources, environmental issues, and land use. *Science,* v. 268, June 2, p. 1305–1312.

Jackson, J. O. 1996. Diamonds: Is their luster fading? *Time,* March 4, p. 44–50.

King, Trude V. V. (ed.). 1995. Environmental considerations of active and abandoned mine lands. *U.S. Geological Survey Bulletin* 2220, 38 pp.

Langer, W. H. and Glanzman, V. M. 1993. Natural aggregate: building America's future. *U.S. Geological Circular* 1110, 39 pp.

Poulin, R., Pakalnis, R. C., and Sinding, K. 1994. Aggregate resources: production and environmental constraints. *Environmental Geology,* v. 23, p. 221–227.

"If the United States wants to save a lot of oil and money and increase national security, there are two simple ways to do it: stop driving Petropigs and stop living in energy sieves."
Amory B. Lovins, Energy Conservationist and Director of Research, Rocky Mountain Institute

Energy from Fossil Fuels
14

About 88 percent of U.S. energy is obtained from fossil fuels—petroleum, natural gas, and coal. Of this 88 percent, petroleum supplies 40 percent. So, why is oil production declining in this country? Can the decline be stopped? Should it be? How does petroleum form? How is its location discovered by geologists? What are the odds of drilling a successful oil well? Given the black, polluting smoke that comes from the tailpipes of old cars and diesel-powered trucks and buses, should we continue to rely on petroleum? Are there other options?

About 60 percent of America's electricity is generated from coal-fired power plants, and the U.S. contains 20 percent of Earth's coal reserves. But coal is a source of serious acid mine drainage when it is mined, and it is an even worse source of acid and radioactive air pollution when burned in a power plant. Is possessing a lot of coal a sufficient reason for using it, despite its drawbacks? Should we seek alternative sources of energy?

Everything We Do Takes Energy

Our earliest ancestors consumed just enough energy to keep alive. When they began to cook with wood, their energy consumption doubled. When they started to cultivate crops, their energy consumption tripled. Today, we consume about 125 times more energy than our early forebears. Although we still use energy to cook and keep warm, most energy today is used to maintain our comfy and convenient lifestyle.

Energy is defined as the capacity to do work, a definition that becomes meaningful when you mow the lawn or climb a steep hill. Energy is expended (work is done) when anything is moved (kinetic energy), burned (chemical energy and heat energy), connected to a battery or power line (electrical energy), and so on.

Types of Energy

All objects—atoms, living cells, rocks, orbiting planets—possess energy in some form. It is an inherent property of all matter. Energy exists in seven basic forms, all of which surround us in daily life, and without which we would not have the energy to live. It is essential to know these, for they are central to understanding this chapter on fossil fuels and the next two on alternative energy and nuclear energy.

1. **Chemical energy** is the energy stored in chemical bonds within materials. Common examples of substances that possess chemical energy are food, batteries, and petroleum. In each case, this chemical energy is simply in storage, awaiting use; it must be

◀
Workers lifting drill pipe from a well on the floor of a drilling platform.

Photo: Randy Taylor/Sygma.

converted to some other form of energy to accomplish work. For example, your muscles convert the chemical energy in food into the kinetic energy of motion. Batteries convert chemical energy into electricity. Coal-burning power plants also convert the chemical energy in coal into electricity. Wood-burning stoves convert the chemical energy in wood into heat energy.

Petroleum, natural gas, and coal are rich in chemical energy that is easily converted to a useful form, simply by adding oxygen (we call this *burning*). These fossil fuels are abundant, easily found, and easily extracted from Earth's rocks, so they form the backbone of the world's industrial energy supply. They are called **fossil fuels** because they are the remains (usually unrecognizable) of ancient plants or animals that lived millions of years ago. Fossil fuels continue to be produced today, but this happens so slowly in the human timeframe that production is at a negligible rate, especially compared to our rate of consumption.

2. **Thermal (heat) energy** is the energy an object possesses due to the local motion of its atoms or molecules. A hot object is hot because its atoms are vibrating at a more rapid rate than atoms in a cold object. When water is heated on a stove, it becomes hot because heat energy is being transferred from the burning gas under the pot to the water in the pot.

The sun is extremely hot because it is busy converting nuclear energy into heat and light energy (it is a giant nuclear reactor). This solar energy radiates outward through space. Upon reaching Earth, it is partly absorbed and partly reflected and reradiated back into space. The part that Earth absorbs is the energy that powers our weather, ocean circulation, and virtually all movement of anything on Earth's surface. Symbolically, Figure 14.1 shows the sun, source of most energy on Earth, with wells that tap another source—petroleum.

A very large amount of heat energy is also present deep within Earth's interior. Part of this heat is left over from the time around 4.5 billion years ago when Earth formed. Part of it results from continuing disintegration of radioactive substances. As noted earlier, at depths only a few tens of miles below the surface, the environment is hot enough to melt rock, about 1300°F. At Earth's center, the temperature is estimated to be 12,000°F. It is indeed fortunate for us that surface rocks are such good insulators.

When molten rock (magma) rises to within a few thousand feet of the surface, some of its heat energy is transferred through overlying rock to groundwater in the rocks. Large magma bodies require ten thousands of years or longer to cool and crystallize. So, for human purposes, this heat is available indefinitely. In places such as Iceland or around the Pacific rim, as in Japan, the magma supply is continually renewed because these countries are located at plate boundaries. Hence,

geothermal energy is inexhaustible. In fact, the ever-present subsurface magma often makes its presence known by rising to the surface as lava (volcanoes were discussed in Chapter 12).

3. **Potential energy** is the energy an object has because of its distance from the center of Earth (which is Earth's center of gravity). Thus, a boulder on a hilltop has greater potential energy than a boulder lying in a valley bottom. Some of the valley boulder's potential energy already was converted into kinetic energy (see below) and heat energy as it rolled down from the hilltop.

Similarly, river water high in the mountains has greater potential energy than water in a lowland valley. This difference in potential energy can be converted into electrical energy wherever water falls, by inserting turbine blades in the falling water, absorbing some of the energy to spin the turbine and generate electricity.

FIGURE 14.1 The sun, source of most energy on Earth.

Oil wells tap another energy source—petroleum. *Photo: Garry D. McMichael/Photo Researchers, Inc.*

We enjoy the beauty of water losing potential energy in a waterfall.

4. **Kinetic energy** is the energy of an object in motion. If you are struck by a falling rock, this energy of motion is transferred unpleasantly to your body. The kinetic energy of water from a showerhead has a more pleasing effect, as does the vibrating motor in a mattress or foot massager. The difference is simply in the distribution of the kinetic energy; the falling rock focuses too much energy in one small area. The kinetic energy of water and wind in motion is the basis of water power (hydroelectric power) and wind power (wind turbines).

5. **Nuclear energy** is the energy released as the nucleus of an atom disintegrates. In popular parlance, this is called "splitting atoms." This process occurs spontaneously in unstable atoms of many elements, such as uranium, carbon, potassium, and others. In a nuclear reactor, the disintegration process is highly controlled, so the nuclear energy is released slowly, mostly as heat energy. The heat can be used to boil water into steam, which turns a turbine to generate electrical power. But if the reaction is uncontrolled and proceeds too rapidly, we have a nuclear explosion.

6. **Electrical energy** is the energy of electrons in organized motion. We set the electrons in motion at a power plant and ship them to where they are needed through conducting wires, like the copper wires in the walls of your home. We seldom use electrical energy for its own sake (what could you do with only a battery or an electrical outlet—sit and look at it?). Instead, we convert electricity into more useful energy forms—heat (cooking and space heating), light (lighting, TV, computer monitor), sound (your stereo), or motion (electric motors).

We tailor the conversion of electrical energy into other forms of energy for our purposes. For lighting, we force electrons to race through tungsten wire inside a light bulb. This causes atoms in the wire to vibrate so rapidly that they release energy as visible light. The thicker the wire, the more electrons it uses and the greater the glow—and the higher the wattage of the bulb. Unfortunately, most of the electrical energy in an ordinary light bulb is converted not into light energy but into heat energy. (Many home fixtures are limited to 60-watt light bulbs to avoid overheating.) Using a different kind of wire, electrical energy can be changed deliberately into heat energy in the elements of an electric stove or electric space heater.

7. **Radiant energy,** also called electromagnetic energy, is energy that moves as an electric-magnetic field through space, air, and other substances. Radiant energy includes radiant heat we feel through the air, plus waves of light, radio, TV, radar, microwaves, ultraviolet, and so on. These all are part of what we call the *electromagnetic spectrum* (explained in Chapter 17, Figure 17.3). (Note that these waves are very different from the mechanical waves in the ocean and earthquakes.) The wavelength of the energy (similar to the wavelength of an ocean wave) can be very short (ultraviolet radiation) or very long (infrared radiation). Civilization's future energy supply will come largely from conversion of radiant energy from the sun (solar energy) directly to heat and to electrical energy.

Interestingly, most of the energy we use today began as radiant energy from the sun. Scientists often refer to fossil fuels as "fossilized sunshine," and this is literally true. Our fossil fuels started out mostly as plants that received their energy from the sun through photosynthesis. The plants converted this radiant energy into chemical energy in their tissues. When the plants died, this chemical energy became warehoused in the form of coal, oil, and natural gas (if geologic conditions were right).

When we recover fossil fuels from Earth's crust, we take this "fossilized sunshine" and release it again, by adding oxygen to oxidize (burn) the fuels, which converts their stored chemical energy to heat energy (plus some light energy in the flames). In a coal-fired power plant, this heat energy is added to water until it turns to steam, which applies its pressure to spin turbines, which rotate generators, producing electrical energy. We may then choose to convert this electrical energy back into thermal (heat) energy (in a toaster, oven, space heater), or into mechanical energy (motors), or into radiant energy (television) and kinetic energy (the TV sound).

Converting Energy from One Form to Another

From the preceding, it is clear that our daily lives involve endless conversions of energy. Figure 14.2A shows a common daily example of the energy conversion involved in cooking an egg. Electrical energy we buy from the power company is converted into heat energy by the stove burner. This heat energy causes chemical changes within the egg, so some of the heat energy of cooking is converted into stored chemical energy. The heat energy also is transformed into mechanical energy when the water molecules move in the boiling water. Figure 14.2B shows some other everyday energy conversions we use.

The problem with energy conversions is that some energy is always lost as heat. The percentage of energy that is converted describes the **energy efficiency** of the system. For example, an automobile engine is only about 25 percent efficient in converting the chemical energy in gasoline into the desired kinetic energy of motion. An ideal conversion is 100 percent, but in reality, conversion efficiency is always less.

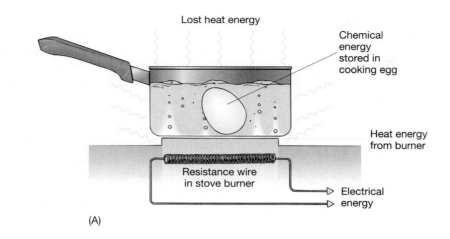

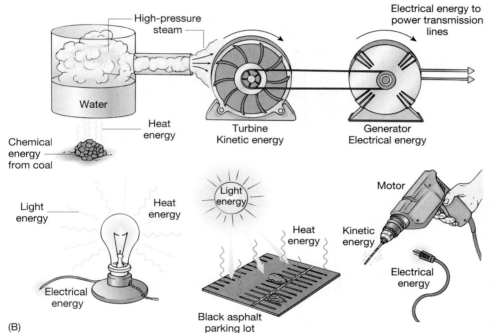

FIGURE 14.2 Energy conversions.

Any kind of energy can be converted into another kind.
(A) Electrical energy is being converted into heat energy, and some of the heat energy is being converted to stored chemical energy in the egg. (B) Other kinds of conversions in everyday life.

Some natural energy conversions are extremely inefficient, photosynthesis being a good example. This is the process by which a plant uses solar energy to manufacture food. This actually is an energy conversion, with the plant converting radiant light energy into stored chemical energy in the form of sugar (plant food). Of the light energy that strikes a plant leaf, 74 percent is dissipated from the leaf surface as heat, 15 percent is transmitted through the leaf and back into the air with no photosynthetic effect, and 10 percent is reflected from the leaf surface. Only the remaining 1 percent is used for food production by the plant, so the efficiency of this conversion is very poor. Plants can get away with this poor efficiency because the energy source is so plentiful; it makes no difference how inefficient the process is.

Many energy conversion processes employed by humans use sources other than the sun, and these sources (like coal, oil, and natural gas) eventually will be exhausted. So, maximum conversion efficiency is very desirable to conserve the energy resource. And better efficiency means lower energy bills. The efficiencies of some common energy conversions are shown in Table 14.1. As the examples suggest, human civilization uses some very inefficient energy transfers; energy waste is rampant.

Normally, the main loss of energy is in the form of waste heat. Figure 14.3 contrasts the waste heat in passive solar heating and the waste heat in nuclear power generation. Clearly, the fewer steps involved in an energy conversion, the more efficient it tends to be. These heat losses can be reduced by using the waste heat to power other processes. This dual use is called **cogeneration** and is done in many coal-fired power plants to increase efficiency.

TABLE 14.1 Efficiencies of Common Energy Conversions

Equipment	Efficiency	Energy-to-Energy Conversion
Incandescent light bulb (with hot glowing wire filament)	5%	electrical to radiant (light)
Solar cell (capturing solar radiation)	10%	radiant (light) to thermal
Nuclear power plant	14%	nuclear to thermal
Gasoline engine	25%	chemical to thermal to kinetic (thermal)
Human body	20–25%	chemical to kinetic (motion)
Aircraft jet engine	35%	chemical to thermal to kinetic (motion)
Diesel engine	37%	chemical to thermal to kinetic (motion)
Coal-fired steam-generating power plant	40%	chemical to thermal
Steam turbine	46%	thermal to mechanical
Electric motor	60–90%	electrical to mechanical
Fuel cell (uses oxygen and hydrogen to produce water)	75%	electrical to chemical
Gas home furnace	85%	chemical to thermal
Electric generator	98%	mechanical to electrical
Hot water heater	99%	electrical to thermal
Electric heating element	100%	electrical to thermal

FIGURE 14.3 Passive solar heating vs. baseboard electric heat from a nuclear power station.

Passive solar heating—the sun shining through a window—is 90 percent efficient. But generating electrical power and transmitting it to an electric heater involves so much activity and so many steps, with heat loss at each point, that the efficiency is only 14 percent! You can see why many now advocate direct use of solar energy in the home, to the extent possible.

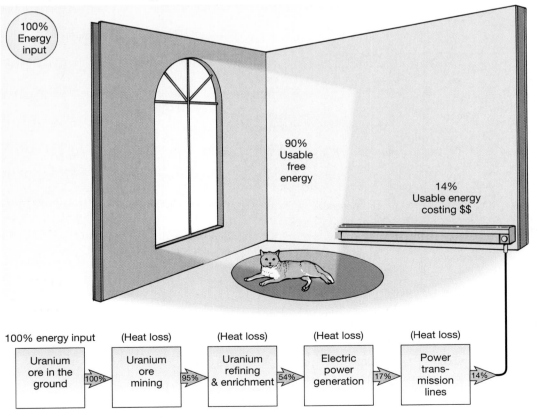

Energy Is Empowerment

Modern industrial economies need enormous amounts of energy to maintain their progression from animal skins and stone clubs to animal skins and nuclear weapons. In fact, the industrial and economic development of a nation roughly corresponds to the energy it consumes. A nation that consumes more energy produces more goods and services. This is reflected by a nation's gross national product (GNP). Figure 14.4 plots GNP against energy consumption, and shows what we might expect: the most highly developed nations gobble the most energy, and the Third World consumes the least. The demand for energy increases as nations develop. Energy demand is projected to increase by more than 50 percent over the next twenty years.

The citizens, however, do not necessarily receive the benefits of the energy their country consumes. That is a political problem, not a scientific one. For example, Libya and Italy use the same amount of energy, but their standards of living are distinctly different. And look at the startling contrast in GNP between equal-energy consumers Poland and Japan.

Although less than one-quarter of the world's population lives in developed countries, these nations use about three-quarters of the world's energy. That means that each person in a developed country uses approximately ten times as much energy as each person in an undeveloped country. (Compare the energy you consume daily—for heating, cooling, cooking, water heating, transportation, lighting, and electricity-powered entertainment—to the energy consumed by most individuals among the billions in Africa, India, or Southeast Asia.)

Figure 14.5 illustrates how we use energy in the United States. About 40 percent is consumed by industry. Another third makes buildings comfortable with heating, air conditioning, lighting, and hot water. The remaining energy we consume is used for transportation, with the automobile being the major consumer, and an inefficient one at that (Table 14.1).

FIGURE 14.4 Correlating gross national product (GNP) with energy consumption.

Generally, the more energy a country consumes, the greater its value of goods and services and the higher its technological development. *Source: Data from 1992* Information Please Environmental Almanac, *World Resources Institute.*

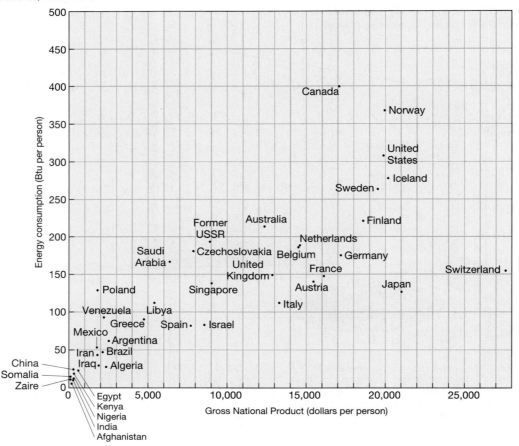

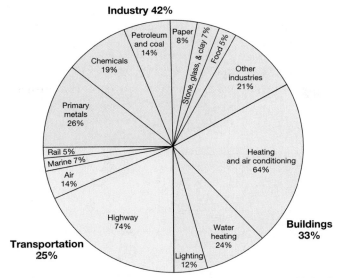

Industry 42%

Paper 8%
Stone, glass, & clay 7%
Food 5%
Petroleum and coal 14%
Chemicals 19%
Other industries 21%
Primary metals 26%

Rail 5%
Marine 7%
Air 14%
Highway 74%

Heating and air conditioning 64%
Water heating 24%
Lighting 12%

Buildings 33%

Transportation 25%

FIGURE 14.5 How we consume energy in the United States.

Petroleum and Natural Gas

U.S. oil production also declined in 1993. For the year as a whole, U.S. oil output averaged less than 7 million barrels daily—its lowest level since 1955. This boosted U.S. oil import dependence to a record figure of 52 percent—higher even than in the late seventies.

Worldwatch Institute, 1994

Our economy literally is fueled by petroleum (40 percent) and natural gas (25 percent), providing the energy needed to maintain our manufacturing, transportation, and lifestyles. Figure 14.6 shows how much we have consumed of various sources of energy over the past

fifty years. It is clear why petroleum and gas companies wield the political power they do. The proportions of use of each energy source are similar for the rest of the industrialized world.

Because petroleum is abundant, easy to transport, and is produced with a well-developed refining technology, it is unlikely to be replaced for at least the next fifty years. This is the reality, despite the fact that petroleum does major damage to the environment when it is extracted, refined, burned, and spilled.

In principle, drilling an oil well is like using an electric hand drill to make a hole in wood. The differences are that the oil well is drilled in rock, is larger in diameter (several inches), and is drilled much deeper. The tip of the drill has diamond chips that chew into the rock. The drill "bit" consists of lengths (joints) of pipe that are continually added as the well deepens. Oil and natural gas wells can be several thousand feet deep, using many dozens of pipe joints.

As a well is drilled, mud is circulated in the hole to keep the drill bit cool, so it will last longer. (The mud consists of water, clay particles, and additives.) As this circulating mud returns to the surface, it carries with it bits of rock chewed from the rock layers through which the drill has cut. A geologist monitors these rock chips for shows of oil.

The discarded drilling mud is a major waste product of well-drilling. The used mud is deposited in a waste pond near the well site, from where it commonly seeps into the soil and groundwater, contaminating them.

If both oil and natural gas are produced from a well (they commonly occur together), further contamination is typical. An oil-and-gas-bearing rock layer may be under considerable pressure, so when it is drilled into, gas and oil may be blown from the hole,

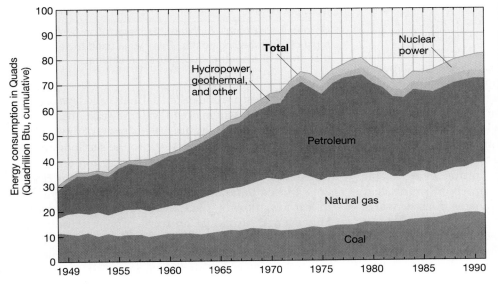

FIGURE 14.6 U.S. energy use, 1949 to 1991.

Clearly, the great majority of our energy comes from fossil fuels. Nuclear and alternative sources are significant but remain small. (Energy consumption is shown in "quads" or "quadrillion Btu." Translation: 1 quadrillion is 1,000,000,000,000,000. A Btu is a British thermal unit, the amount of heat energy needed to raise the temperature of 1 pound of water 1°F. Use of "quads" makes the numbers smaller and easier to use.) *Source: Annual Energy Review 1994, U.S. Energy Information Administration.*

carrying with them hundreds of feet of drill pipe and many thousands of gallons of oil. If this happens, the well must be capped quickly. The blown-out oil seeps into the ground, polluting soil and water.

In the Middle East, it is common to ignite any natural gas from oil wells to get rid of it, because it is relatively worthless. Wasteful though this is, natural gas cannot be economically transported by pipeline unless it is first liquified, an expense that reduces the profitability of gas, compared to the oil. Small amounts of oil mixed with the burning gas create a black, polluting smoke.

When the oil is refined, large volumes of sulfur and other noxious chemicals are produced. These liquid chemicals are often dumped into an adjacent waterway or pumped into disposal wells, typically polluting both surface water and groundwater (review Chapter 8). The much publicized oil spills from tankers that carry the oil from the Middle East, where most of the oil is located, to western Europe, Japan, and North America are another source of environmental damage (Chapter 10).

Finally, oil production from some layers can cause ground subsidence, just as extraction of groundwater does (review Chapter 6). This may cause serious damage to water pipes, sewage pipes, gas pipes, as well as disrupting surface water drainage.

Raw petroleum (crude oil) is separated by distillation into a wide variety of fuels and products, as shown in Figure 14.7. Although not shown in the diagram, about 40 percent of petroleum is used for nonenergy purposes: the manufacture of plastics, chemicals, drugs, and clothing.

Exactly what are petroleum and natural gas, how do they form, how are supplies located, and how long will supplies last?

How Does Petroleum Form?

Crude oil (unrefined petroleum) and natural gas are **hydrocarbons,** chemical compounds composed almost entirely of carbon (82–87 percent) and hydrogen (12–15 percent). (An example is methane, one part carbon to every four parts of hydrogen, or CH_4.) Almost all petroleum includes some sulfur, ranging from under 0.5 percent to 10 percent, but averaging 1.5 percent. Modern refineries remove much of the sulfur, which is used to manufacture sulfuric acid, a widely used industrial chemical. A little nitrogen may be present, averaging 0.1 percent. Released into the air through burning of petroleum products, sulfur and nitrogen are major atmospheric pollutants. Petroleum, both crude and refined into fuels, often contaminates surface water and groundwater.

Most crude oil forms from the decaying organic tissue of microscopic organisms living at or near the ocean surface. Most of these are one-celled plants called

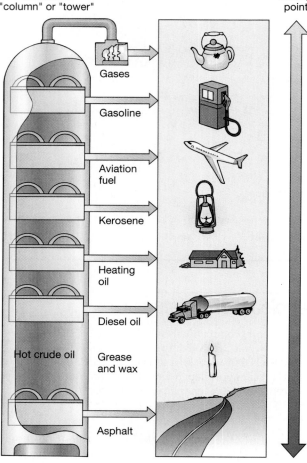

FIGURE 14.7 Sample fuels and other products refined from petroleum.

Crude oil is a complex mixture of different components that have different boiling points. This fact is used to separate them from one another in a distillation column.

diatoms; the rest are one-celled animals called *radiolarians.* Diatoms are present in ocean water in such uncountably large numbers that they are called "grass of the sea."

When they die, they settle to the seafloor and are buried in other sediment (mostly clay minerals, very fine-grained quartz, or calcite). If their organic tissues are exposed to abundant gaseous oxygen, they decay to carbon dioxide plus water, leaving no residue. However, if buried in an oxygen-poor setting, they decay by the less efficient process of **anaerobic decomposition.** This results in a residue of black organic (carbon-rich) material. This "black mud," if compressed long enough beneath additional sediment, becomes a black shale.

As the black organic residue is buried to greater depth, subjecting it to greater temperature and pressure, chemical reactions form the hydrocarbon molecules that are **petroleum** and **natural gas.** Petroleum forms at temperatures between 140°F and 210°F. As burial depth increases with additional sediment, and as temperature consequently rises, the complex and heavy molecules that comprise petroleum break down into simpler gas molecules, chiefly methane (CH_4). This occurs largely at temperatures between 210°F and 285°F. At still higher temperatures, the natural gas is driven off, and no commercially significant hydrocarbons remain.

In sum, preserving dead ocean plants and animals so they can become petroleum requires that gaseous oxygen must be eliminated soon after rapid burial.

How Do Oil and Natural Gas Migrate and Accumulate?

After petroleum forms, it is liberated from the source rocks. The mechanism that releases it is not completely understood, but probably involves fracturing of the black shale source rock and/or movement of the hydrocarbons with the involvement of water. The released petroleum, being of fairly low density (oil floats on water), migrates sideways and upward into porous and permeable sandstones, limestones, or dolostones.

Once in the permeable rock, the hydrocarbons migrate from areas of higher pressure toward areas of lower pressure, which normally means shallower depths, until stopped by some geologic barrier. Three common barriers are shown in Figure 14.8.

How Are Oil and Gas Discovered?

Petroleum and natural gas typically exist many thousands of feet below the surface. The depth of the average U.S. oil well is about 5000 feet; for natural gas, average depth is about 500 feet deeper. However, many wells produce from shallower or greater depths. The world's deepest producing natural gas well produces from depths between about 24,000 and 26,000 feet (nearly 5 miles) in west Texas. Petroleum-bearing rock layers exist in continental crust, which extends beneath the sea. Consequently, oil wells are drilled on land and into the underwater continental shelf (Figure 14.9).

How do geologists decide where to drill? The first step is to identify rocks of suitable type. Petroleum does not form in igneous or metamorphic rocks, so we can eliminate such areas. Petroleum forms in sea-bottom rocks, particularly sandstones and shales, so this is where

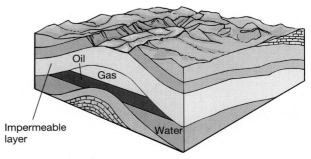

(A) Anticlinal (upfold) trap. The fluids cannot rise farther and are trapped beneath an impermeable layer of shale.

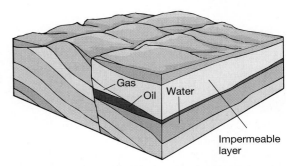

(B) Fault (fracture) trap. Fluid migration updip is stopped because a fault has positioned an impermeable layer in the path of movement.

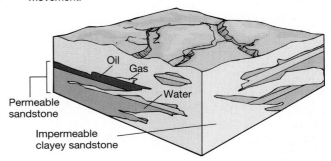

(C) Lithologic change. The permeable sandstone becomes more clayey and less permeable. The change is usually more gradual than shown here.

FIGURE 14.8 Common petroleum traps.

To be recoverable at a profit, oil and natural gas must be concentrated in a "trap." The three most common petroleum traps are anticlinal, fault, and lithologic. In each, the fluids (natural gas, petroleum, water) have migrated through a permeable sandstone toward the ground surface (the updip direction) until stopped by an impermeable barrier. The three fluids always reside in layers according to their densities (specific gravities), in the order of gas on top, then oil, then water.

we look. A geologist seeks thick marine sandstones or carbonate (limestone) rocks that are porous. Such "sands" and "limes," as well-drillers call them, originate in an environment that makes them good candidates for petroleum, and their porosity means that they probably will transmit oil and gas to a well drilled into them.

Petroleum and Natural Gas **409**

FIGURE 14.9 Offshore petroleum production platforms drill into the continental shelf.

These platforms are operated by Marathon Oil Company in Cook Inlet, just south of Anchorage, Alaska. The State of Alaska contains a significant part of America's petroleum reserves. *Photo: Grant Heilman/Grant Heilman Photography, Inc.*

Once a good sandstone or limestone layer is selected, the most useful exploration approach is the **seismic method.** It works like sonar, and is similar to depth-sounding in the ocean. Essentially, a microearthquake is generated (hence the term *seismic*) at the surface by setting off a small dynamite blast. Waves from the blast, similar to sound waves, travel downward several thousand feet into the crust. They are reflected by rock layers back to ground level, where a device (essentially a seismometer) records them. The reflected waves return at different speeds, depending on the density of each rock type. The travel time of the wave through each type

of rock is known, so knowing the time and rate of travel allows the depth to each rock layer to be calculated.

Figure 14.10 is an example of a seismic recording, enhanced by a computer program. What you are looking at is old seafloor that has lithified to become layers of sandstone and shale rock. Later, Earth's tectonic forces elevated the ancient seafloor above sea level and bent and folded the layers. The folded rock layers are evident, as is a salt dome to the right and an upfold on the left.

Note the tall mass under the 750-foot mark. Within the mass, there is no layering, so we interpret it to be a salt dome that has risen from a thick layer of salt (halite, NaCl) far below the maximum depth of this recording. We think of salt as quite solid, but it flows plastically under the great pressure prevailing at these depths. The salt was deposited long ago, when shallow seawater evaporated. Millions of years later, new sediment atop the salt layer became cemented into rock layers. A weak area in the rocks allowed the plastic, lower-density salt to flow upward through them.

Salt is impermeable, so any petroleum that exists in the sandstone layers around the salt dome cannot pass through it. This forms a trap for oil and gas. If they are present, they probably will be in rock strata that have been tilted upward, approximately between 500 and 650 feet and between 800 and 1000 feet. These are places where a geologist might recommend drilling.

An upward fold (an *anticline* to geologists) is centered at about 1350 feet. Permeable beds there might contain petroleum, if they are capped by an impermeable rock layer. This is the same kind of "anticlinal trap" shown in Figure 14.8A. Unfortunately, there is no good way to tell if any of the rock layers contain oil or gas without drilling into them.

In the U.S., the chance of finding a profitable well in a new area is only about one in twenty. In other words, an oil company might drill twenty wells and get just one that makes money; the rest are "dry holes." There is no need to pity the poor oil companies, but it is important to understand how expensive the discovery process is, and how much environmental effect all this drilling has. High drilling cost and low success rate make the search for oil an expensive venture.

Is the U.S. Running Out of Oil?

American oil production has been declining since about 1980 and will continue to decline in the future. The reason is simple: we have used more oil for more years than anyone else. We were the world's first major producer of petroleum and have increasingly exploited this resource for a century. This country provided much of the oil needed (as fuel and lubricants) to fight World War I (1914–1918) and World War II

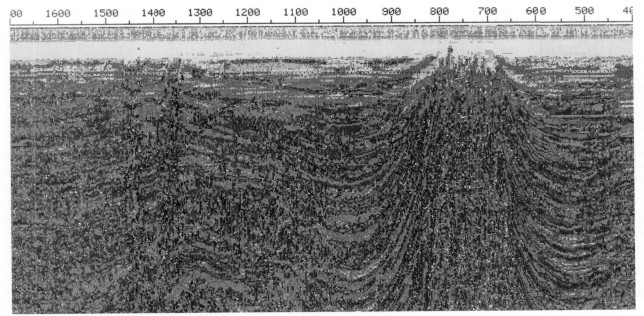

FIGURE 14.10 Computer-colored seismic image from the Middle East.

You can see the pattern of horizontal rock layers. They are interrupted by a dome, probably of salt, centered at 750 feet. To its left is a much gentler upward fold (an anticline), centered at 1350 feet. The image covers about 20 miles horizontally and 6000 feet vertically. *Source: J. Pigott.*

(1939–1945), plus the Korean War (1950–1953) and Vietnam War (1964–1975).

Our supply of this nonrenewable fossil fuel is decreasing, even as our need increases. Domestic production is slowing, and new environmental concerns have made exploration for new reserves even more costly. For example, there are now controls on the storage and disposal of wastewater and drilling mud generated during drilling. Concern about possible effects of oil exploration on wildlife have restricted development of the huge oil field in Alaska. Further, the world price of oil is low, compared to the cost of exploration and production in this country.

Consequently, almost all major American petroleum companies have sharply reduced their scientific staffs in the United States and are concentrating exploration effort overseas. Although substantial oil remains to be discovered, especially in Alaska, the future is not particularly bright for domestic petroleum production.

Since the mid-1980s, the United States has imported nearly half of its petroleum needs. Much of it has come from countries other than those in the Middle East, such as Mexico, Venezuela, and Nigeria. In 1994, U.S. oil imports broke through the halfway mark at 50.4 percent, the first time that our foreign-oil consumption exceeded that of our domestic production. *Any industrial nation with a petroleum-based economy that imports half its oil is on shaky ground regarding oil price and stability of supply.* This great volume of imported oil is a great financial drain on the American economy. Even a

brief interruption in imports leads to massive economic social disruption.

The Arab Oil Embargo of 1973: Taste of the Future

The United States received a small taste of such a disruption in 1973, when we were importing only 36 percent of our needs. At that time, the oil-producing countries in the Middle East disapproved of U.S. support for Israel in the ongoing Arab-Israeli territorial dispute. Consequently, they cut off oil exports to the U.S.; the event was called the "Arab oil embargo."

The prevailing petroleum price of $2.30 for a barrel of oil (42 gallons) leaped to $10.50. By continuing to limit their oil production throughout the 1970s, the oil cartel led by Saudi Arabia kept supplies tight and forced prices even higher. In 1980, the price for a barrel of oil peaked at $41 before beginning a rapid decline to its present $15–25.

U.S. gasoline prices at the pump had been about $0.30 per gallon in 1973. "Gas wars" were common among neighboring gasoline stations, shaving pennies to suppress prices. But during the Arab oil embargo, the price rapidly doubled, then tripled, then quadrupled. To most Americans, this seemed an incredible rise in gasoline prices. Yet, the average price in the United States is still among the lowest in the world; Europeans are accustomed to paying over $3 per gallon (Table 14.2).

TABLE 14.2 Gasoline Pump Prices Worldwide

Prices are based on analysis of fifty international locations in December 1993 and reflect the lowest average price available, including self-serve/full-serve unleaded regular and full-serve leaded regular. Prices averaged 20% higher in 1995.

High-Priced Locations (Over $3 Per Gallon)

Location	Average Pump Price Per Gallon
Tokyo	$4.58
Hong Kong	3.87
Paris	3.58
Milan, Italy	3.54
Amsterdam, Netherlands	3.43
Munich, Germany	3.37
Brussels, Belgium	3.14

Low-Priced Locations (Under $1.50 Per Gallon)

Location	Average Pump Price Per Gallon
Caracas, Venezuela	.21
Riyadh, Saudi Arabia	.33
Quito, Ecuador	.74
Abu Dhabi, United Arab Emirates	.83
United States (national average)	1.11
Bangkok, Thailand	1.23
Mexico City	1.44
Nairobi, Kenya	1.49

When you consider that the average American income far exceeds that in other countries, you see that petroleum-based energy consumes an amazingly small proportion of income in the U.S. Why? Taxes. We have a lower gasoline tax at the pump than almost any other country. The price Americans pay includes an average of about $0.40 in federal and state taxes. Most other industrially developed nations have a $2–3 tax per gallon. Why are their taxes so high and ours so low?

It comes down to the prevailing U.S. philosophy of individual rights and freedoms, versus the "common good" view more prevalent in other countries. Most foreign governments have decided that individual transport (via automobiles and small trucks) is a luxury, and those who demand it should pay the cost it incurs, including road construction and maintenance. Further, gasoline-powered vehicles are serious air polluters, so the cost of cleaning up the atmosphere also should be borne by those who cause the problem: drivers of vehicles.

What do you think of this method of discouraging gasoline consumption (or smoking, or fatty beef consumption, or some other demonstrably harmful activity)? Keep in mind that American life is structured around personal, on-demand mobility in the automobile. In contrast to all other developed countries, we have almost no inexpensive public transportation, except in a few major cities.

A quarter-century has passed since our brief lesson in the risk of depending on others for a major part of the energy that runs our industrial economy. Yet, we remain at the mercy of foreign governments for about half our main energy resource.

As noted, petroleum is a major pollutant and needs to be phased out for that reason alone, without even considering the serious economic cost of importing half our needs. We have no choice but to begin the switch to alternative fuels. This is inevitable. As expensive and disruptive as the changeover will be now, it will only be more expensive and just as disruptive later.

Improving Automobile Efficiency

A bright spot in the U.S. since 1973 has been a sharp improvement in automobile fuel efficiency, from an average of 13 miles traveled per gallon of gasoline burned to the present 22 mpg. The fuel efficiency of individual car models has risen from 15 mpg to 27 mpg. Because 25 percent of the commercial energy used in the United States is consumed in transportation, the increased mpg values are a significant improvement.

Unfortunately, much of this improvement has been nullified by a 50 percent increase in number of miles driven (Figure 14.11) and by a tripling of air miles flown. About 10 percent of the oil consumed on Earth each day is used by American motorists on their way to and from work, 69 percent of them driving alone. In your lifetime, you probably will be part of a dramatic change in this statistic, as electric vehicles become dominant and public mass transit increases.

Although the United States has about 40 percent of the world's cars, the increased use of automobiles is a worldwide problem. The world's auto population rose from 60 million in 1950 to 200 million in 1970, and on to today's 500 million vehicles. It is expected to double again over the next twenty to forty years. In Europe, car traffic is growing at an annual rate of 6 percent. An extreme example of auto proliferation is occurring in the Greek city of Athens, where car ownership grew 2600 percent from 35,000 (1964) to 900,000 (1994) in only thirty years. One result of this increase became evident in the summer of 1988, when about 800 people died from smog-related illness. Smog results when hydrocarbon emissions from automobile exhaust react with nitrogen oxides in the air (see the section in Chapter 17 titled "Ozone"). Athens has now banned cars from the city center.

In Mexico City, considered by many to have the world's dirtiest air, smog levels are routinely triple the limits recommended by the World Health Organization.

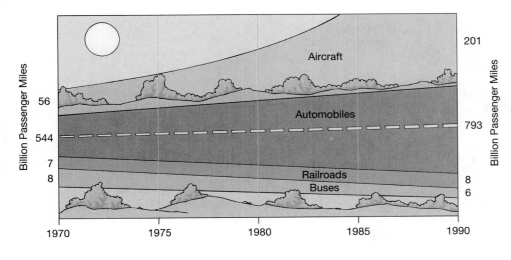

Billion Passenger Miles

56
544
7
8

1970   1975   1980   1985   1990

Aircraft
Automobiles
Railroads
Buses

Billion Passenger Miles

201
793
8
6

FIGURE 14.11 We Americans love our cars!

The automobile is king of travel, its use increasing by nearly 50% since 1970. Air travel grew nearly four-fold, but remains only a fourth of passenger car miles traveled. Mass transit by bus has stayed about the same, as has rail travel. *Source: Ahrens, Meteorology Today, 5th ed., St. Paul, MN, West Pub.*

The city's 36,000 factories account for some of this pollution, but roughly 80 percent of it emanates from the tailpipes of the 3 million cars and trucks (many poorly maintained, with no pollution controls whatsoever) that burn, every day, nearly 5 million gallons of low-quality leaded gasolines. (Mexico City's residents not only get an ample dose of smog; they also get plenty of airborne lead, inhaled and sprinkled upon food and water.)

At present, transportation consumes 63 percent of all oil used in the United States, up from 50 percent at the time of the Arab embargo in 1973. Combustion of gasoline and diesel fuel accounts for about one-third of total U.S. emissions of carbon dioxide.

Electric Cars

The gasoline-burning internal-combustion engine is woefully inefficient—only about 25 percent. Most of the waste occurs as heat when the gasoline is burned. Looking ahead to the inevitable decline of the world petroleum resource, domestic and foreign automakers have devoted much research to electric-powered cars in recent years. General Motors began the sale of electric vehicles in California and Arizona in 1996.

Unlike gasoline-powered vehicles, the **electric car** is 95 percent efficient in converting the stored chemical energy in its batteries into useful motion. But the problem is the batteries. Storage batteries in current electric cars can power a vehicle only for about 120 miles before needing an overnight recharging. The batteries must be replaced about every 25,000 miles. This cost and inconvenience makes today's electric cars about twice as expensive to operate as gasoline-powered cars. Also, electric cars are inconveniently slow, even when batteries are freshly charged. And, don't forget that the electricity needed to recharge electric car batteries comes from—guess where—fossil fuels that are burned in most power plants.

But electric-car technology is improving fairly rapidly. For example, Electric Fuel, Ltd., an Israeli company, has developed a prototype zinc-air battery. Instead of recharging it like a conventional car battery, you simply replace a cartridge in the battery after 200 miles. It can power a car to 70 mph. The manufacturer projects the cost of the battery at under $4000. Before you laugh at that, consider that mass production for 100,000 cars would bring the operating cost down to $0.06 per mile. At current gasoline prices, the cost of operating a conventional gasoline-powered vehicle is $0.11 in Italy, $0.08 in Germany, and $0.06 in the United States. It appears that competitive electric cars are just around the corner.

What about the use of other nonpolluting fuels in cars, such as clean-burning hydrogen or free solar power? Substitutes for gasoline and diesel fuel are being very actively sought, and even soybean oil (Figure 14.12) is being used. This search has been propelled by the EPA and by state regulations in California, the state with the nation's worst air pollution problems. Several years ago, California decreed that by 1998, 2 percent of the cars sold there by major manufacturers must emit *no pollutants at all*. The figure rose to 5 percent for 2001, and to 10 percent for 2003. However, in 1996, the state backed off because a recent study concluded that a new generation of more compact, longer-range batteries would not be ready until at least 2000. Automakers argued that without such batteries consumers would not buy electric cars.

Oil in the Future

You do not hear much in the news about the approaching oil crisis in the United States, simply because it is not a crisis yet and our country's political style is more one of reacting to emergencies than planning for them. But as domestic production declines and the share of imported oil goes up, we will be increasingly at the mercy of other countries, friendly or not, for oil to fuel our lives.

FIGURE 14.12 Bean-powered bus.

This bus runs on a 20/80 blend of soybean oil and diesel fuel. This reduces particulate emission by 11 percent, hydrocarbons by 21 percent, and carbon monoxide by 12 percent. One bushel of soybeans yields enough oil to make 1.5 gallons of fuel, so the vehicle gets 250 miles per acre of soybeans. At least thirty cities coast to coast are using soy diesel, mostly in buses. *Photo: Nebraska Soybean Board.*

For some other countries, prospects are considerably brighter. The Middle Eastern nations, which currently have about two-thirds of world petroleum reserves (Figure 14.13), continue to discover large new oil fields. It costs only about $1 to produce a barrel of oil in Saudi Arabia, versus $8 or more in the United States. Also, some countries of the former Soviet Union and the People's Republic of China are likely to experience major oil and gas development in the near future. Despite the great size and potential petroleum richness of these countries, the increase in their proven reserves has come slowly because of political and technical problems, some of which have now disappeared. As they open their doors to Western know-how, major new discoveries can be anticipated.

FIGURE 14.13 Earth's proven oil reserves (1994): about 1 trillion barrels.

Note who has about 7 percent of the oil, and who has about 60 percent of it. Expect U.S.–Middle East relations to remain difficult for quite some time. *Source: World Almanac.*

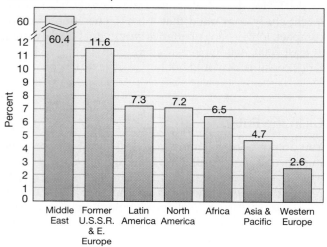

Considering only the likely oil finds, and neglecting political considerations, Earth probably has sufficient petroleum to satisfy the world need through most, if not all, of the twenty-first century. Whether Earth's citizens will want to continue using enormous amounts of this highly polluting fuel is another question. This will be decided on a nation-by-nation basis.

But recall that the major polluting effect of petroleum combustion occurs in the atmosphere. Air is not exclusive to any country; it circulates globally. Pollution is a very exportable commodity, particularly as atmospheric pollution, and a country cannot ban imports of air pollution. A relatively isolated country such as the United States can control its own water pollution and soil pollution, but not its air pollution.

Is the U.S. Running Out of Natural Gas?

The domestic future for natural gas is considerably brighter than for petroleum. Our known reserves are estimated to last another fifty years. Recent studies suggest that undiscovered resources of natural gas are probably much larger than previously thought. Natural gas is far more versatile than either coal or oil, and with a little effort can be used in more than 90 percent of energy applications. Yet, until recently, its use has been largely restricted to household and industrial markets, where it has thrived. In the U.S., natural gas is by far the most popular heating fuel; by the early 1990s, nearly two-thirds of all U.S. single-family homes and apartment buildings had such heating systems (Table 14.3).

Natural gas is a much more desirable fuel than petroleum. It is taken from the ground as nearly pure methane (CH_4), so that sulfur and other contaminants are not released when it burns (except CO_2). Natural

TABLE 14.3 Relative Abundance of Heating Fuels in the United States

Home Heating Source	Number of Households	
	Main Heating Fuel	Secondary Heating Fuel
Natural Gas	51,700,000	2,500,000
Electricity*	21,500,000	16,600,000
Fuel Oil	10,400,000	700,000
Liquid Propane (LP)**	4,400,000	900,000
Wood	3,900,000	20,800,000
Kerosene	1,100,000	4,200,000
None	600,000	—
Other (solar plus coal)	300,000	600,000
Coal	12,900	—
TOTALS	94,512,900	46,300,000

* Electricity, in turn, is generated in power plants having various sources of energy: burning coal, burning gas, burning oil, running water (hydropower), or splitting uranium atoms (nuclear).

** Sometimes called *bottled gas.*

Source: Department of Energy, Energy Information Administration, 1990

gas is cheaper than petroleum for the same amount of energy. Increase in natural gas use has been slow, but within the past few years, more cities have converted their municipal vehicles from gasoline to compressed natural gas, a hopeful sign. The vehicles are refueled at central municipal locations. Very few service stations for the general public provide natural gas, however.

The United States has 5 percent of the world's proven reserves of natural gas, and this nonpolluting fuel is widely available in the industrialized countries of the world. Unfortunately, natural gas is more difficult and expensive to transport unless it is liquified. Hence, it is unrealistic to consider shipping unliquified natural gas from, for example, the huge reserves in the Middle East to the U.S. East Coast.

Natural Gas Hydrates

A **natural gas hydrate** is formed of methane gas and water. It is a solid ice-like material that occurs just beneath the seafloor where water depths exceed 900 to 1500 feet. It also occurs in conjunction with permafrost at high latitudes (see Chapter 5). You may not have heard of natural gas hydrates, but worldwide, the energy that might be obtained from them is estimated at twice that in existing crude oil, natural gas, and coal combined!

The hydrates are, therefore, an important potential future energy resource. Efforts are underway to locate the best exploration areas and to develop methods of producing this resource. Drilling and production problems are severe because the hydrate can decompose explosively when confining pressures upon it are released by drilling into it.

Petroleum Removal and Land Subsidence

Crude oil, natural gas, and groundwater have something important in common: their presence in rock pores commonly provides some of the support for overlying rocks (see Chapter 6). If much of this support is removed from relatively uncompacted rocks near the ground surface, settling of the land can occur. This **land subsidence** has happened in many places where oil and gas have been removed.

A well-studied example is the subsidence caused by withdrawal of crude oil in the harbor area of Long Beach, near Los Angeles. Subsidence first was noted in 1940, and, by 1994, the ground surface had dropped nearly 30 feet in the central area (Figure 14.14). Starting in 1958, an attempt was made to raise the ground surface by pumping up the rocks with water. Large quantities were pumped into the compacted rocks to replace the oil that had been removed, raising the fluid pressure. By 1963, a small rebound was observed, with the ground surface rising back up as much as 15 percent of the initial subsidence in some areas. But the rocks reached a limit beyond which they would not decompress, and the rebound stopped. Most of subsidence was not reversible.

In Japan, coastal land around the port city of Niigata has subsided due to natural gas extraction. The area of sinking covered about 20 miles by 12 miles. The greatest sinking was concentrated along the shoreline on both sides of the city, with some local areas depressed about 5 feet.

Oil-Impregnated Rocks: A Solution?

Oil-impregnated rocks include several rock types that are impregnated with petroleum that ranges from highly viscous to solid. The most common oil-impregnated rock is called **tar sand,** although it is neither tar (a product refined from petroleum) nor sand (again, this is the well-driller's term for a sandstone).

Hydrocarbons in these sediments have a viscosity at least 10,000 times greater than that of normal crude oil. Thus, they cannot be recovered by conventional methods. Basically, the hydrocarbons must be "cooked" from the tar sands. The rock is saturated with high-pressure steam, either after being mined or *in situ* (in place, unmined). The steam lowers the viscosity so that its lighter fraction (lighter petroleum) can be extracted through wells. It then is filtered and refined.

The world's largest surface deposits of tar sand are in northern Alberta and Venezuela. Since 1985, two

FIGURE 14.14   Land subsidence.

Aerial photograph of the coastal area around Long Beach, California. Withdrawal of petroleum from the Wilmington oil field resulted in ground subsidence of 30 feet. *Photo: City of Long Beach, California.*

Alberta processing plants have supplied 12 percent of Canada's petroleum needs. The cost is reportedly $14–15 per barrel, quite competitive with the current worldwide price of $15–25 per barrel for conventional oil.

One reason the tar sand price is lower is that it includes no cost for controlling air-pollution emissions. The Canadian government does not require controls in this desolate area, which is 200 miles north of Edmonton, the largest city in the region. In another location, pollution controls certainly would be mandatory. The tar contains a higher percentage of sulfur than conventional oil, plus some toxic heavy metals. In addition, the waste piles from mining occupy more space than the original deposit, and the large waste disposal ponds continue to grow.

Economically recoverable deposits of oil from the tar sands have the potential to supply Canada's projected need for about thirty-three years at current consumption rates. Because the tar is only supplying 12 percent at present, the recoverable reserve could last several hundred years.

Smaller tar sand deposits exist in the United States, almost entirely in Utah. Since we are running short on conventional domestic oil reserves, why don't we solve the problem by recovering petroleum from our tar sands? We could—but only for about three months, at $50–60 per barrel, if all U.S. deposits were developed!

This is a good lesson in the economics of energy sources. The cost of oil production from tar sand is so much higher in the western United States because water needed for processing the sediment at the mining site is scarce in this arid region and because of the environmental controls required anywhere in the U.S. Also, start-up costs for a tar sand processing plant are quite high. The federal government would have to share this cost with any private developer for it to happen. Given the very limited tar sand supplies in the United States and the high cost of production, this U.S. oil source has a limited future.

Oil Shale: A Solution?

Oil shale is another loosely named material. It is not a true shale, nor does it contain oil. It is a sedimentary rock with layers of silt-sized dolomite and organic matter, actually an immature petroleum source rock. Thus, we must supply the heat energy to complete what has not yet occurred in nature. When oil shale is crushed and heated to 900°F, the organic material is converted partly to liquids and gases. Some of the gases condense to liquids upon cooling, and are used to help fuel the conversion facility. The product is crude shale oil, and it is very crude. Even small quantities must be prerefined before it can be processed in conventional refineries.

The United States has immense oil shale deposits in Colorado, Utah, and Wyoming (Figure 14.15). These could yield twice the oil ever produced from conventional sources. Why, then, has this resource not been exploited?

A major difference between oil shale and tar sand is that tar sand contains true petroleum that has formed through natural processes. Nevertheless, the two resources share some problems. An enormous capital investment is required to mine, crush, transport, and heat oil shale to recover oil on a major scale. When all costs are considered, the cost to produce a barrel of oil from oil shale is greater than it can be sold for.

And then there are the environmental problems. Voluminous water is needed to process the rock, but water is scarce in these semiarid oil shale areas. Voluminous waste rock is produced. There are sulfur and toxic metal emissions. Further, the area has no infrastructure for mining: the population is small, so new towns would

(A)

(B)

FIGURE 14.15 Oil shale processing.

Oil shale's cost remains higher than that of conventional oil, and this has canceled major oil shale projects. As competing fuels grow more expensive, oil shale will become more attractive. (A) The Unocal facility for processing oil shale at Parachute Creek, Colorado, designed to produce 10,000 barrels of crude shale oil per day. However, the oil requires further refining to convert it into petroleum fuels. *Photo: UNOCAL.* (B) Sample of oil shale and the oil extracted from it. *Photos: William E. Ferguson; Ken Lucas/Visuals Unlimited.* (C) Oil shale distribution. The darker areas have at least a 10-foot thickness of oil shale and significant yield. The lighter areas are unappraised or of low grade. *Source: U.S. Geological Survey.*

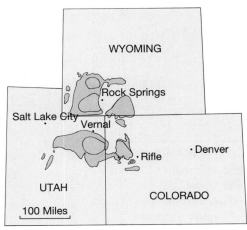

(C)

have to be created, with utilities and services for thousands of migrants needed to operate the processing facility.

As with tar sand in Utah, private business would undertake such a project only with subsidies from the federal government. It is estimated that the price of conventional crude oil must rise to $50 per barrel for oil shale development to be worthwhile (without subsidy). This is about two to three times the current price of a barrel of crude oil.

Coal: The Rock That Burns

Coal ranks third behind oil and gas as a supplier of energy in the United States (Figure 14.6). It accounts for about 23 percent of our current needs. About 60 percent of the nation's electricity is generated in coal-fired plants, and if they all shut down right now, much of the U.S. would be plunged into darkness. It is interesting to realize that 60 percent of our electricity comes from a burning organic rock. This percentage is nearly double that in 1950 (Figure 14.6).

The U.S. has about 20 percent of Earth's coal reserves (map, Figure 14.16), so our domestic supply is good for two more centuries or so. Because coal exceeds 80 percent of our total recoverable fossil fuel reserves (the other 20 percent is oil and gas), coal will be our major energy source for years to come, despite the pollution from burning it and the environmental damage from mining it.

Exactly what is coal? How does it form? How are coal reserves located? How is it mined? How does it generate electricity? What are its environmental problems?

How Does Coal Form?

Coal is a combustible rock formed of the remains of ancient land plants. The greatest concentration of these plants, both today and in the past, is in swamps (like Okefenokee Swamp in Georgia and Florida) and in tropical river deltas (like the Amazon River rain forest in South America). In such environments, coal formed in the past, and continues to form today. Figure 14.17 shows how it happens.

Preserving terrestrial plant debris so it can become coal has the same requirement as preserving plant and animal remains so they can become petroleum: oxygen must be eliminated soon after rapid burial. In contact with oxygen, all living tissue decomposes to carbon dioxide plus water, leaving no residue. Thus, only oxygen-free decomposition leaves the residue needed to form fossil fuels. In a swamp, plant corpses fall into the muck. Oxygen is rapidly removed by decaying matter in the stagnant water.

FIGURE 14.16   Coal-bearing areas of the U.S. and Canada.

Robert W. Christopherson, Geosystems, 2nd ed., Englewood Cliffs, New Jersey: Prentice Hall, 1994.

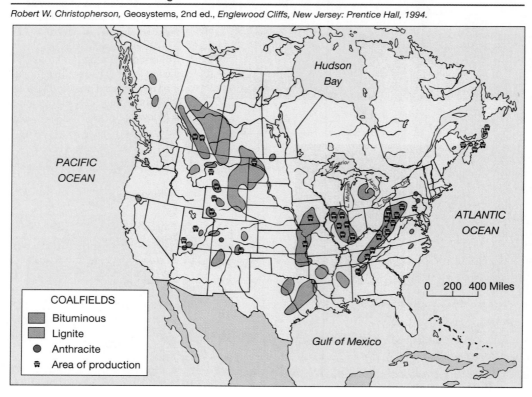

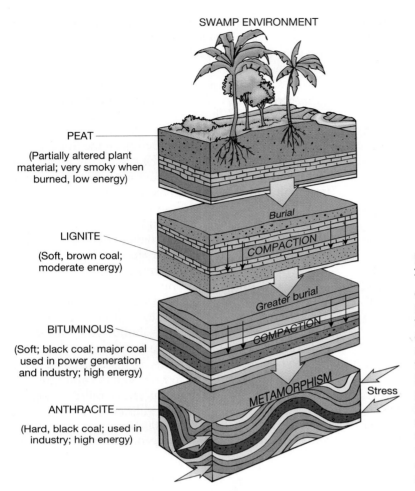

SWAMP ENVIRONMENT

PEAT
(Partially altered plant material; very smoky when burned, low energy)

Burial

COMPACTION

LIGNITE
(Soft, brown coal; moderate energy)

Greater burial

COMPACTION

BITUMINOUS
(Soft; black coal; major coal used in power generation and industry; high energy)

METAMORPHISM   Stress

ANTHRACITE
(Hard, black coal; used in industry; high energy)

FIGURE 14.17 How coal is formed.

The process begins as dead plants in a swamp. If conditions are right, the dead plants become peat. If buried under sediment, and given enough time, the peat becomes lignite coal. If the proper conditions continue, lignite becomes the highly desired bituminous coal. Finally, should temperature and pressure conditions grow sufficiently intense, bituminous coal may become metamorphosed into anthracite ("hard") coal. Note that the percentage of carbon increases with each stage of coal formation. Note also that coal forms from *land* plants, not from ocean seaweed or the diatoms that form petroleum. *Source: E. J. Tarbuck and F. K. Lutgens, 1996,* Foundations of Earth Science, *New York: Prentice Hall, p. 47.*

Living terrestrial plants contain about 50 percent carbon. The remainder is oxygen, hydrogen, nitrogen, sulfur, and other elements in lesser quantities. As the dead plant debris is buried to greater depth by new sediment, the temperature and pressure increase. The plant corpses become tightly compacted. The temperature rises, causing chemical changes in the plant remains. Most significant is the loss of carbon dioxide (CO_2) and methane (CH_4). The major result of this change is that, as the hydrogen and oxygen diminish, the percentage of carbon increases by default, because that is all that remains. Each level of carbon content has significant characteristics, and a name:

- When the amount of carbon is 50–60 percent, the material is called **peat,** a "pre-coal" that still is used as a very low-grade, smoky fuel for home heating in some places.
- At 60–70 percent carbon, it is called **lignite,** a coal that is soft, low-grade, and makes a smoky fuel.
- At 70–90 percent carbon, it is called **bituminous coal.** This is the most-desired and most-used coal; bituminous coal is used worldwide in coal-fired electric plants and in making steel.

- Above 90 percent carbon, it is called **anthracite coal,** a hard coal that burns virtually without smoke.

The amount of carbon in coal is very important because carbon determines how much heat is released upon burning. The greater the carbon content, the more desirable is the coal for heating. Very little heating difference exists between bituminous and anthracite coals. But bituminous is far more common and easier to mine, so it is the coal of choice for coal-fired power plants and steelmaking.

Searching for Coal Deposits

Unlike petroleum and natural gas, which are fluids that migrate after being formed, coal is a solid and normally remains where it forms. Thus, to locate coal deposits, geologists analyze the environment in which ancient rocks formed. They do not seek rocks that once were on the seafloor, for those are more likely to contain petroleum. Instead, they seek rocks that formed near swampy ancient shorelines, because that is where the largest modern swamps are located. The coal fields shown on the map (Figure 14.16) were lush swamps long ago.

What is now the eastern coal fields were great swamps some 300 million years ago. What is now the western coal fields were vast swamps 70 million years ago.

Just as swamps today vary in area, depth, and how much sand and clay sediment wash in, so did they vary in ancient times. The result today is that coal layers vary in area, thickness, and quality. These factors are critical to finding a good-quality layer of coal that is big enough to be worth mining. Of particular importance is the sulfur content. Normally, sulfur is a small percentage, yet it becomes a big problem when coal is burned, for sulfur compounds in the smoke contribute to acid rain.

A common saying among geologists is that *the present is the key to the past,* meaning that we can look at processes in motion today to see what happened in the past. The search for new coal reserves is an excellent illustration of this, for geologists look at how present swamps form to find potentially coal-rich ancient swamps. It is the reverse of how a historian might express the same thing—past events suggest what is likely to happen in the future.

Mining Coal

Coal deposits can be located at any depth from the surface to thousands of feet underground, depending on the geologic history of the area. These deposits are called **coal beds** or **coal seams.** As a very rough rule, coal beds less than 200 feet deep are surface-mined, and deeper seams usually are underground-mined.

Surface Mining

Surface mines also are called **open-pit mines** or **strip mines.** How deep they go depends on the thickness of overlying rock, thickness of the coal seam, its quality, the distance from the mine to a railroad or river barge terminal, and the distance the coal must travel to its ultimate user. Each of these factors is an important financial consideration in planning a mine. If any factor is too costly, the coal may never be mined. A modern surface mine is shown in Figure 14.18.

The first step in surface mining is to remove vegetation and soil, stockpiling the soil for reclamation after mining is complete. Next, tens of feet of overburden (overlying rock) are blasted with explosives and removed with power shovels, bulldozers, and draglines (Figure 14.19). This waste rock is set aside in *spoil piles.* When the surface of the coal is cleared of overburden, coal is mined with front loaders and bulldozers, loaded onto trucks, and taken to a railroad hopper or river barge terminal for shipment. This mining method can recover close to 100 percent of the coal. As coal is removed, reclamation proceeds right behind it,

FIGURE 14.18 Aerial photo of a modern surface mine.

A modern surface mine harvests low-sulfur coal (center) from the Appalachian Mountains in West Virginia. The coal will be burned to generate electricity in a power plant. Note the tremendous volume of light-colored rock (overburden) that has been removed to get at the coal. Federal and state laws require the mining company to "reclaim" this land by replacing the rock and planting the surface with grass and trees. *Photo: Melvin Grubb/Grubb Photo Service.*

FIGURE 14.19   Giant dragline strip-mining coal in West Virginia.

The dragline does not mine the coal; instead it moves overburden (waste rock) so that bulldozers, front loaders, and trucks can remove the coal. In some locations, a dragline must remove many cubic yards of waste rock to mine a single ton of coal. This raises the cost, sometimes enough to close an operation. *Photo: Melvin Grubb/Grubb Photo Service.*

replacing rock and stockpiled soil according to federal and state laws.

In the United States, the total coal mined has grown steadily for decades. And the percentage of coal mined by surface stripping (as opposed to underground mining) has increased: 23 percent in 1950, to 29 percent in 1960, to 36 percent in 1970, to 60 percent in 1980 and 1990.

Unfortunately, surface mining can increase sharply the erosion rate and the sediment load of nearby streams. The sediment may fill stream channels and increase the frequency of flooding. Before tough mining laws required thorough reclamation of mined land, mining companies simply walked away from the mess when the coal was gone (Figure 14.20A). But today, such problems are remedied by concurrent land reclamation, required by federal and state laws. The disturbed land must be restored as closely as possible to its premining condition (Figure 14.20B). Smoothing artificially chopped-up topography and replanting vegetation are costly but necessary to prevent major environmental damage. Restoration is particularly difficult in the semi-arid areas of the West where much of the nation's coal reserves are located.

Underground Mining

In contrast to the near-100 percent coal recovery from a surface mine, underground mining typically recovers about 40 percent of the coal in a seam. Mining underground generally is feasible down to a depth of a thousand feet or so, depending on the thickness, quality, and distance factors noted for surface-mined coal. In mountainous terrain, an underground mine is started by digging into the side of a mountain where the coal crops out. In flatter topography, the mine is opened by digging a vertical shaft down to where the coal occurs. Horizontal shafts then branch from the vertical to create access tunnels.

In underground mining, removing rock and coal creates "caverns" and tunnels into which the overburden can collapse. Obviously, this is dangerous to mine workers. Thus, holding up the "roof" of the mine is important. In traditional *room-and-pillar mining*, coal is removed entirely from "rooms," but is left in "pillars" to support the roof. These pillars may be removed as the miners vacate the mine, allowing the roof to collapse. This may cause the ground surface above to subside, cracking highways and buildings. A typical room-and-pillar mine is shown in Figure 14.21.

In *longwall mining*, several million tons of coal can be produced from a single mine in a year, using computer-controlled equipment and fewer workers. Typically, mining is accomplished by a spinning cutter that travels along a 600-foot "long wall" of coal. In a single pass taking just a few minutes, it chews a 3-foot thickness of coal from the wall, dropping the cut coal onto a conveyor that transports it out of the mine. The spinning cutter then is ready for its next swath.

(A)

(B)

FIGURE 14.20 Effects of strip mining.

Restoration of new surface-mined areas has been required for a couple of decades in the United States, but many older mined-out areas have been abandoned (orphaned) and not reclaimed. (A) Aftermath of strip mining before the law required reclamation. *Photo: Anson Eaglin/USDA.* (B) Modern reclamation. With the land returned to its approximate original contour and vegetation planted to hold the soil in place, it is hard to tell that this once was a surface coal mine. *Photo: Sam C. Pierson, Jr./Photo Researchers, Inc.*

FIGURE 14.21 Typical room-and-pillar coal mine.

The mining machine is in a "room," and the machine, operated by the miner at the left, chews into a "pillar" of coal at the right. *Photo: Melvin Grubb/Grubb Photo Service.*

Dozens of powerful shields support the roof over the cutter. With each pass of the spinning cutter, these roof shields roll forward, and the unsupported roof tumbles in behind them. This method is safer, faster, cheaper, and recovers more coal than any other underground method. Unfortunately, it also causes more surface subsidence. Figure 14.22 shows a typical longwall mining operation.

Life in a Mine.　What's it like in an underground coal mine? In some mines, the coal seam is thick enough that you can stand up. But many mines extract coal from thinner seams, some under 4 feet high, so you work lying down, squatting, on your knees, or duck-walking. A modern underground coal mine is pitch black with all the lights out, but white with lights on, because the walls and roof are sprayed with "rock dust" (ground limestone) that resembles flour. Rock dust suppresses highly explosive fine coal dust in the air, reducing the risk of explosion. Many mine roofs bristle with the heads of "mine roof bolts," driven several feet into overlying rock layers to prevent roof falls.

Air is fresh, cool, and damp in a mine. It is circulated by powerful fans, so effective that diesel vehicles can discharge their exhaust directly in the mine. Multimillion-dollar machines that run on high-voltage electricity (generated at a distant power plant by the coal they mine) claw out the coal and convey it on belts to the surface. Miners operate the cutting machines with remote controls and computers (they haven't used picks and shovels in years). And when they return to the brilliant daylight and head for the bath house, they shower off only a fraction of the coal dust that covered miners decades ago.

Underground Mining Problems.　Underground coal mining has its own set of problems. In areas of old and abandoned coal mines, collapse of mine roofs is common. It may be deliberate as described above, or accidental as the mine roof and support pillars deteriorate with age, but the damage is the same to the surface (Figure 14.23). In some areas of Pennsylvania and West Virginia, where underground coal mining has continued for generations, highways and homes often crack from subsidence.

Another serious problem in some abandoned mines is mine fires (Figure 14.24). In the mid-1980s, the U.S. Bureau of Mines estimated that more than 250 uncontrolled mine fires were burning in 17 states. Such fires can release noxious sulfurous fumes and increase surface-structure collapse. Some Pennsylvania mine fires have been burning for twenty-five years, traveling several miles during that time. Repeated attempts to extinguish the fires have failed.

Mine Safety

Earlier this century, coal mining was extremely hazardous. Each year, hundreds of miners were crushed

FIGURE 14.22 Typical longwall coal mine.

Note the thick steel shields that support the roof while the miner operates the cutter on the mine wall. *Photo: Melvin Grubb/Grubb Photo Service.*

FIGURE 14.23   Land subsi-dence.

This view near Sheridan, Wyoming, shows the character-istic rectangular pattern of sink-holes above an abandoned coal mine. The mine operated from the early 1900s through the 1940s. If you live in a coal-mining area, is your home insured against mine subsi-dence? *Photo: C. R. Dunrud, U.S. Geological Survey.*

by falling mine roofs. Mine roofs still fall, but far less often, due to better support with roof bolts, chain-link fencing fastened to the ceiling, timbers, and the steel shields used in longwall mining.

Another hazard is methane gas, which escapes from coal as it is mined. It is fatal to animals and people, and it can combine explosively with oxygen. Methane gas is colorless, odorless, and tasteless, so it is impossible for us to detect with our senses. Miners used to take caged canaries into the mine to detect methane. Not that they smelled it—their tiny bodies simply suc-

cumbed to the gas before it affected larger humans. So when the canary keeled over, miners knew it was time to leave, and not to cause any sparks on the way out.

Today, automatic sensors for methane are used, saving both miners and canaries. Modern mining equip-ment has onboard methane detectors that automatically shut down the equipment if the methane level rises to 1 percent. Coal mine fatalities today are only 5 percent of what they were early in this century (Figure 14.25).

The major remaining danger to the coal miner is *black lung disease*, discussed in Chapter 2. It is caused by

FIGURE 14.24   Underground fire.

The red, lava-like glow reveals the high temperature developed in underground coal fires. This fire follows abandoned mine workings around Sheridan, Wyoming, leaving voids that are certain to cause subsidence of the surface above. Vapors released from such fires are a health hazard. Fires are often started by careless burning of trash near a coal seam; others start by natural processes. Extinguishing such fires is never easy, and some have burned for decades. *Photo: David H. Ellis/Visuals Unlimited.*

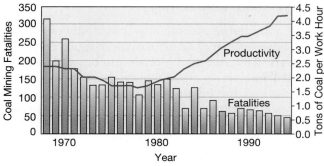

FIGURE 14.25 Mining fatalities are down and production is up.

The drop in fatalities reflects safer mining practices ("a safe mine is a productive mine"), fewer workers in modern automated mines, strict safety regulations, and the shift from underground mining to surface mining. *U.S. Department of Labor.*

inhaling microscopic particles of coal dust over many years. Government estimates suggest that about 4000 U.S. miners die each year from this malady, many after their retirement. Over the past twenty years, however, coal dust has significantly decreased in mines, even though production has increased. This attests to the U.S. mining industry's ability to comply with stringent dust-control regulations. The current generation of coal miners works in far cleaner surroundings than their predecessors, but coal mining can never be like working in a hospital operating room.

All subsurface miners should wear air-filtering masks on the job, although this is not presently required. The effects of inhaling coal dust do not appear for many years, so most miners choose not to wear such equipment. This attitude resembles that of motorists who refuse to wear seatbelts, and for some of the same reasons.

Sulfur Content of Coal

Coal has a problem in common with most metallic mineral deposits: it normally contains several percent of sulfur in the form of pyrite (iron sulfide, FeS_2). When exposed to air, pyrite decomposes (becomes oxidized) and the sulfur generates sulfuric acid, as described in Chapter 13. More than 43 percent of the coal that is most used—bituminous—is too rich in sulfur, containing at least 3 percent of the element (by weight).

Worse, a single coal seam can vary in sulfur content from less than 1 percent to 5 percent over a horizontal distance of a few hundred feet, so the seam may be low-sulfur in one spot but high-sulfur only tens of feet away. Low-sulfur coal and high-sulfur coal often occur in the same coal mine.

Environmentally, pyrite is nothing but trouble. It generates acidic water during and after mining; this pollutes streams, killing plants and fish. The sulfur in pyrite is converted to sulfur dioxide during combustion in coal-fired power plants. The sulfur dioxide gas is released from smokestacks of the power plant, is converted to sulfuric acid by atmospheric oxygen, and falls as acid precipitation (see Chapter 17).

Existing environmental regulations have sharply suppressed the level of sulfurous emissions from coal-burning power plants. Even more stringent rules will be imposed after the year 2000. Clean coal technologies are being developed that will not contaminate the atmosphere with sulfur oxides and will significantly reduce nitrogen oxide. But these technologies are costly and now account for about 40 percent of the cost of building a power plant.

Environmental cleanliness is a cost of doing business that generally has not been included in price calculations by American (and world) business enterprises. That is changing rapidly in response to increasing control at all levels of government. Consumer prices for all commodities will increase as our environmental responsibilities are increasingly recognized.

Most concern about the pollutant effect of burning coal has focused upon release of sulfur gases from smokestacks. Recently, however, another concern has surfaced. A typical coal contains 1.3 ppm (parts per million) of uranium and 3.2 ppm of thorium. Both are radioactive elements. As U.S. power plants burn coal, they release about 2000 tons of these elements into the environment each year. This release from coal combustion far exceeds the annual U.S. consumption of nuclear fuels.

The amount of radioactive material released each year from coal combustion would probably provoke an enormous public outcry if it were released from nuclear facilities. Consequently, coal ash should be treated as low-level radioactive waste! (Radioactive materials are discussed in Chapter 16.)

Energy Conservation

The industrialized countries, led by the United States, consume ten times more energy per person than the less developed countries. Two-thirds of this energy is supplied by petroleum (40 percent) and coal (23 percent), and the use of coal continues to grow worldwide.

Mainland China is the clearest example of rapidly increasing coal use. This great country, with its more than 1 billion people, is hurtling into industrialization with little thought for the environment. An unusually large proportion of China's rocks are continental rather than marine, meaning that many swampy areas existed in which coal could develop. Sure enough, China has many large coal deposits.

China's energy capital is Taiyuan, a city 240 miles southwest of Beijing. Within a 300-mile radius of the city, hundred thousands of miners are tearing into underground coal seams thicker than ocean liners. Nationwide, more than 5 million Chinese are engaged in the largest coal-extraction exercise in human history. In 1994, China's mines produced 1.2 billion tons of coal, more than any other country, and the numbers are expected to triple to 3.1 billion tons by 2020.

China, like other newly industrializing countries, views the diversion of scarce funds toward environmental protection as a luxury of wealthier nations. It has no intention of slowing development of its coal resource. After all, coal is 76 percent of China's commercial energy supply. Heavy reliance on coal, along with wasteful, inefficient energy use, will make China the largest single producer of carbon dioxide, the major greenhouse gas, by 2025. It will pass the United States as the largest emitter of waste gases into the atmosphere. The burning of coal already is causing a serious air-pollution problem in China. Extrapolating from U.S. data, a whopping 900,000 Chinese may be dying each year as a result of pollution-related lung disease.

But China views any attempt to restrict its use of coal as a Western attempt to limit its emergence as a great power. Quoting from the Environmental Protection Committee of the National People's Congress: "Developed countries discharge more carbon dioxide than developing countries on a per capita basis, and the United States discharges ten times more than China on a per capita basis." Another Chinese development expert says: "Two hundred years after the Industrial Revolution, the world economy has greatly advanced and the developed countries are the main beneficiary. About 80 percent of the world's pollution is caused by developed countries and they should be responsible for these problems."

This means, say Chinese officials, that the developed countries should help pay for cleaner coal-burning technologies in the Third World, as well as helping to finance expensive hydroelectric projects, nuclear power stations, and such alternative energy sources as wind and solar power. China is trying to get money for these things from the World Bank of the United Nations, a bank financed to a great extent by American money—in other words, your tax dollars. Do you view this as America's responsibility to help China develop, or is it a crude form of blackmail?

Serious environmental pollution and damage from fossil fuels has been in progress for a century. Equally serious attempts to reduce the problem have been developing only since the 1970s. Fundamentally, there are two approaches: (1) decrease the use of fossil fuels by increasing the efficiency of their use, or (2) decrease the use of fossil fuels by replacing them with less-polluting alternative sources. In this chapter, we consider ways to improve efficiency. The next two chapters look at alternative and nuclear fuels.

Energy Need: Degree-days

Before considering ways to improve energy efficiency, let us consider energy need. Because so much energy is used to heat buildings for human comfort, and because this demand varies with climate, heating engineers have devised the **heating degree-day.** In the United States, we generally start to heat our buildings when the temperature drops below 65 degrees.

Thus, if the average temperature falls to 64 degrees, heating engineers count it as 1 heating degree-day. If the average temperature falls to 60 degrees, they count it as 5 heating degree-days. If it falls to 60 degrees for a week, they count it as 5 degrees a day times 7 days, or 35 heating degree-days. For any location, they add up the heating degree-days for a time period, such as a month or year. For example, Chicago totals 6400 heating degree-days for the year, whereas Miami has only 200 (Figure 14.26A). The more heating degree-days for a particular period, the greater the heating requirement and energy use. The number of heating degree-days in an area tells a builder how much insulation to use and what size of heating equipment to install.

The complementary energy use unit for cooling (air conditioning) is the **cooling degree-day.** If people start turning on the A.C. when the temperature warms to 65 degrees in the summer, then a day on which the temperature reaches 70 degrees adds 5 cooling degree-days. Just as the number of heating degree-days in an area tells a builder how to size heating equipment, the number of cooling degree-days enables intelligent planning for air conditioning.

Important for energy conservation, heating and cooling degree-days also give power companies a way to predict the energy demand for any month, season, or a whole year. Of course, getting the numbers right depends on having good temperature records over a long period, so the average heating days and frequency of extreme variations can be determined.

Improving Fossil Fuel Efficiency

Technology already has reduced energy consumption and will continue to do so. Examples:

- *Gas furnaces*—condensing furnaces reabsorb much of the heat from their own exhaust gases, thereby reducing by 28 percent the fuel demanded by conventional gas furnaces.
- *Indoor environmental control systems* monitor indoor/outdoor temperature, sunlight, and the location of people. From these inputs, the systems provide light, heat, and air conditioning as needed, typically saving 10–20 percent energy.
- *Lighting improvements*—using fluorescent lamps, reflectors, and minimal daytime lighting can cut energy consumption for lighting by 75 percent or

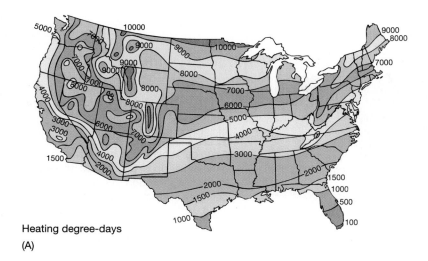

Heating degree-days
(A)

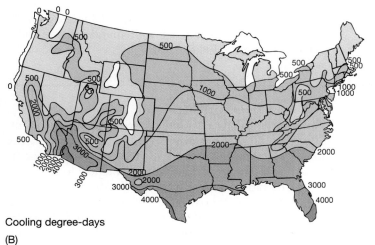

Cooling degree-days
(B)

FIGURE 14.26 Average annual degree-days.

(A) Average annual heating degree-days.
(B) Average annual cooling cooling degree-days. The assumption is that typical Americans turn on the heat when the average temperature drops below 65°, or the air conditioning when the temperature rises to 65°F. (In Great Britain, where people are cooler, a value of 60° is used.) *Source: Lutgens and Tarbuck,* Atmosphere, 6th ed., *Englewood Cliffs, New Jersey: Prentice Hall, p. 67.*

more. Compact fluorescents are 54 percent efficient, whereas a standard incandescent bulb is only 15 percent efficient and a halogen bulb only 17 percent.

- *Household appliances* (water heaters, cooking ranges, refrigerators, air conditioners) have improved markedly, yet the potential for further savings is large. The electricity used by U.S. refrigerators was roughly halved between 1972 and 1992 due to better insulation and more efficient motors. Further improvements could reduce refrigerator energy use another 30–50 percent by 2000.

- *Advanced building materials* can sharply reduce heat loss through windows, doors, walls, and ceilings (Figure 14.27). "Superinsulating" a home can reduce heating energy by 70 percent, important because 85–90 percent of home heating is done using fossil fuels (Table 14.3). For some homes in Sweden, the energy saved has been nearly 90 percent.

- *Industrial cogeneration*—industry is increasing the use of cogeneration, the combined production of

heat and electricity. Only a third of the energy from steam produced in a boiler in a conventional coal-fired power plant is converted to electricity; the other two-thirds goes to waste. But in a cogeneration plant, much of the energy remaining in the "used" steam becomes a heat source for other industrial processes.

But perhaps the greatest potential for fossil-fuel savings is in transportation. Cars and light trucks consume more than one-third of this nation's oil. During the quarter-century since the Arab oil embargo, efficiency improvements have included use of lightweight fiberglass instead of metal, radial tires that reduce rolling resistance, and the redesign of car exteriors to decrease aerodynamic drag. Newer cars now average 27 miles per gallon; prototypes attain efficiencies of 75 mpg or better.

The present efficiencies largely were mandated by federal legislation: automakers were forced by law to improve energy efficiency. Future improvements probably will have to be obtained the same way, because improving efficiency takes a lot of money, which raises

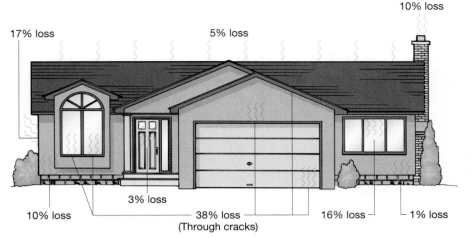

17% loss

5% loss

10% loss

10% loss

3% loss

38% loss
(Through cracks)

16% loss

1% loss

FIGURE 14.27 Energy sieve.

A typical ranch-style home has severe heat leaks: up to 85 percent of heat is lost rapidly through cracks, up the chimney, windows, and poorly insulated walls. Added together, these leaks are equivalent to having a window-size hole in a wall!

prices, can lower profits, and can affect employment. Expect to see two approaches. One will be general mandates that require overall mileage improvements for auto manufacturers. The other will be much higher gasoline taxes, which will force us all to demand more efficient vehicles from automakers.

Summary

Energy is a fundamental property of all materials. Human societies exist because they are able to obtain some of the energy contained in natural materials and use it. The most widely used energy sources are petroleum and coal, which are among the worst polluters of the natural environment.

Petroleum and natural gas are formed over millions of years from the organic tissues of microscopic marine organisms. These fossil fuels are produced, released from their source rocks over time, and migrate slowly through adjacent permeable rocks, mostly sandstones and carbonate rocks. When movement is halted by some sort of physical barrier, the fluids halt, waiting to be discovered.

Exploration for oil and gas is accomplished by indirect means and is a very inexact science. Ninety percent of exploratory drill holes find no useful deposits. The price of oil is determined by a combination of political considerations by the producing nations and the tax policies of the consuming nations.

Natural gas is an abundant fossil fuel but is underused in the United States at present. Use is certain to increase during the next few decades because of the relatively nonpolluting nature and relative abundance of this energy source. Natural gas is transported through a complex web of underground pipelines. Its chief disadvantage is that it is not practical to transport it long distances, unless it first is liquified.

Coal use in this country has increased consistently over the past few decades because the United States has a very large reserve. Unfortunately, coal contains abundant sulfur and causes serious pollution problems if the sulfur is not removed. In the U.S. today, most coal is used to generate electricity in power plants.

Coal deposits are easier to locate than petroleum and natural gas, but mining coal has more environmental problems. Although deaths due to mine collapse and explosions now are below fifty per year, problems continue with coal dust inhalation by miners. Many have decreased lung function as a result.

There exist many ways to use coal and oil more efficiently and additional ways are being discovered as technology develops. Most of these increased efficiencies are initially costly to the consumer but the cost is recovered fairly rapidly during the life of the product.

Key Terms

anaerobic decomposition
anthracite coal
bituminous coal
chemical energy
coal
coal beds (coal seams)
cogeneration

degree-days (heating/cooling)
electrical energy
electric cars
energy
energy efficiency
fossil fuels
geothermal energy

hydrocarbons
kinetic energy
land subsidence
lignite
natural gas
natural gas hydrate
nuclear energy

oil migration
oil shale
open-pit mines
peat

petroleum
potential energy
radiant energy
seismic method

strip mines
surface mines
tar sand
thermal energy

Stop and Think!

1. What is energy? Cite the different types and the differences among them.
2. What is a fossil fuel? Which ones are important as energy sources?
3. Why is it not possible to get 100 percent efficiency in energy conversions?
4. What are the major consumers of energy in the United States? Why is the relative importance of these uses different in less developed countries?
5. Explain how petroleum production can pollute the environment.
6. Describe how your lifestyle would change if all petroleum production ceased tomorrow.
7. Why is decomposition in the absence of oxygen necessary to form oil and coal?
8. What are the three common trapping mechanisms for oil and natural gas?
9. Explain how seismic exploration for oil and gas works.
10. Why is domestic oil production in decline and unlikely to revive?
11. Do you believe the states and federal government should greatly increase the tax burden placed on gasoline consumption? Explain your viewpoint.
12. Public transportation by train and bus is highly developed in almost all developed and industrialized nations except the United States. It never pays for itself and requires a continuing government subsidy. Do you believe the American government should subsidize public transportation? Explain your viewpoint.
13. The federal government subsidizes many things, such as farming (price supports), business activities (tax write-offs for many things), and home ownership (a deduction for mortgages). Should oil exploration be subsidized? Explain your viewpoint.
14. What are the advantages of using natural gas instead of oil?
15. What is tar sand and why is it not a good prospect as a supplementary energy source in the U.S.?
16. What is oil shale and why is it not in wide use as a source of energy in this country at present?
17. Describe the stages that plant material must undergo to become coal.
18. What hazards face coal miners? How have these dangers changed over the past hundred years?
19. What are heating degree-days and cooling degree-days? Why are they useful concepts?
20. Describe some ways that fossil fuel efficiency can be improved.

References and Suggested Readings

Bleviss, D. L. and Walzer, P. 1990. Energy for motor vehicles. *Scientific American*, September, p. 102–109.

Corcoran, Elizabeth. 1991. Cleaning up coal. *Scientific American*, May, p. 106–116.

Dillon, W. P., and seven others. 1995. Resource and climate implications of natural gas hydrates. *U.S. Geological Survey Circular* 1108, p. 68–70.

Fulkerson, W., Judkins, R. R., and Sanghvi, M. K. 1990. Energy from fossil fuels. *Scientific American,* September, p. 129–135.

Gibbons, J. H., Blair, P. D., and Gwin, H. L. 1989. Strategies for energy use. *Scientific American*, September, p. 136–143.

Hubbard, H. M. 1991. The real cost of energy. *Scientific American*, April, p. 36–42.

Lee, F. T. and Abel, J. F., Jr. 1983. Subsidence from underground mining: Environmental analysis and planning considerations. *U.S. Geological Survey Circular* 876, 28 pp.

McCabe, P. J., and three others. 1993. The future of energy gases. *U.S. Geological Survey Circular* 1115, 58 pp.

Smith, J. W. 1980. Oil shale resources of the United States. *Mineral and Energy Resources*, v. 23, no. 6, 20 pp.

Taylor, O. J. (ed.). 1987. Oil shale, water resources, and valuable minerals of the Piceance Basin, Colorado: The challenge and choices of development. *U.S. Geological Survey Professional Paper* 1310, 140 pp.

Turney, J. E. 1985. Subsidence above inactive coal mines: Information for the homeowner. *Colorado Geological Survey Special Publication* 26, 32 pp.

Wilson, Alex and Morrill, John. 1996. *Consumer Guide to Home Energy Savings, 5th ed.* Washington, D.C.: American Council for an Energy-Efficient Economy, 274 pp.

Xu, L., Bhaskar, R., and Vazirnegad, A. 1993. Statistical analysis of U.S. coal mine dust exposure in continuous mining sections: 1971–1990. *Canadian Mining and Metallurgical Bulletin*, v. 86, p. 39–45.

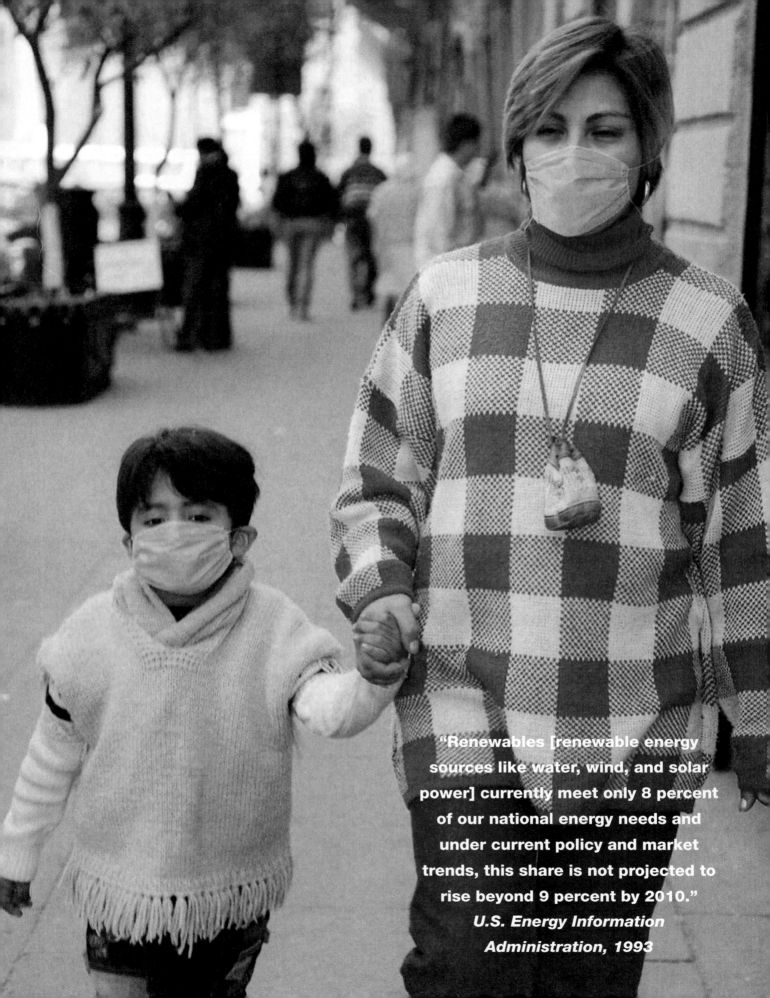

"Renewables [renewable energy sources like water, wind, and solar power] currently meet only 8 percent of our national energy needs and under current policy and market trends, this share is not projected to rise beyond 9 percent by 2010."
U.S. Energy Information Administration, 1993

Energy from Alternative Sources

15

In the preceding chapter, we saw how dependent we are upon fossil fuels, particularly oil and coal. Unfortunately, these fuels cause major environmental problems, including atmospheric pollution, which we shall examine in Chapter 17. Further, they are **nonrenewable resources,** meaning that they cannot be replaced within a useful timeframe for humans (fossil fuels take millions of years to develop).

Consequently, scientists and governments are actively searching for alternative energy sources that are nonpolluting and are renewable. **Renewable resources** are ones that always will be available as long as the sun shines (solar energy, moving water, wind, and biomass) or as long as Earth has a hot interior (geothermal). We expect the sun to shine and Earth's insides to remain hot for many millions of years to come.

Recall that all substances contain energy in some form. The trick is for humans to be clever enough to invent ways of tapping into a nonpolluting energy supply at an acceptable cost. At present, the **alternative energy sources** being actively developed are grouped into five technologies: moving water, wind, biomass, solar, and geothermal. The energy obtained from these sources is expected to total 9 percent of U.S. needs by the year 2010. It will continue to rise during the early decades of the twenty-first century as nonrenewable oil and natural gas resources dwindle and grow more expensive.

Electricity is so technologically developed and so easily delivered that it is the most used type of energy in the U.S. Therefore, it is in electrical generation that alternative energy sources will have their greatest impact in your lifetime.

Examine Figure 15.1 and you can see that most alternative energy sources already are cheaper than nuclear energy for generating electricity. Some alternative sources are growing competitive with the low-cost leaders, coal and natural gas. This is particularly true if we include the environmental clean-up cost for coal and petroleum. As fossil fuels gradually become more expensive, the alternative sources will grow in use—moving water, wind, biomass, solar energy, and geothermal. A good guess is that, within the twenty-first century, alternative energy will supply most of our energy needs worldwide.

How do these alternative energy sources work? Why do they seem to be taking so long to develop? What are the advantages and disadvantages of each? Are some of them practical in some regions but not in others? Do any of them pollute the air or water?

◀
Mother and son walk down a Mexico City street. Air contamination levels have reached high levels, making breathing difficult, especially for asthma sufferers.
Photo: Jose Luis Magana/AP/Wide World Photos.

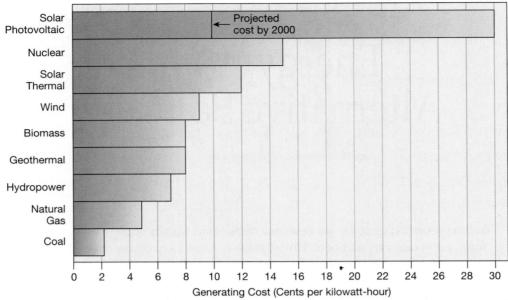

FIGURE 15.1 Generating cost of electricity per kilowatt-hour (maximum and minimum) using different technologies in 1994.

The cost of electricity from solar cells is plummeting and is expected to be about 10¢/kilowatt-hour by 2000, making it more competitive. This will leave the nuclear option as the most expensive.

Kinetic Energy from Moving Water

Harnessing moving water to do work is a technology as old as humanity. Moving water is literally energy in motion; recall that this is called kinetic energy. The challenge is to capture this energy and put it to work for us.

Anyone who has ridden a canoe or kayak down a stream or drifted along a seashore has used the kinetic energy of water currents for transportation. The energy of water currents also can generate hydroelectric power wherever they flow fast enough (or can be dammed and released to create rapid flow).

Plentiful kinetic energy also resides in the ocean tides. Recall that the tides are the dependable rising and falling of sea level that occurs daily along seashores. This change in sea level can be harnessed to generate electrical power in some areas.

Ocean waves are another source of kinetic energy in water. Waves are less dependable, but there certainly are enough of them. Is there some way to capture their energy?

We now will look at each of these energy sources.

Hydroelectric Power

Water power has been harnessed in all cultures for thousands of years. Use of large waterwheels to capture the energy in river water first appeared in western Europe in the 1600s. During the 1700s and 1800s, large waterwheels provided the energy to power grain mills, sawmills, and other machinery in the United States.

Today, we use the kinetic energy in moving water mainly to generate **hydroelectric power,** which provides 15 percent of all electricity produced in the United States. Great kinetic energy resides in rapidly moving floodwaters and high waterfalls. But floodwaters are too erratic to use as a dependable energy source, and high waterfalls are uncommon (how many Niagara Falls are there?). A way to produce dependable water flow is to dam rivers, impounding large reservoirs of water that can be released as needed to spin the turbines of hydroelectric plants (Figure 15.2). The first hydroelectric plant went online in 1882 in Wisconsin. Today, federally funded hydroelectric dams dot our country's rivers.

Best-known in the eastern United States is the 34-dam Tennessee Valley Authority or TVA (review Chapter 7, Flood Control). Many hydroelectric dams exist in the West and Southwest, such as the network of dams along the Colorado River (an example is Hoover Dam, Figure 15.3). Of about 50,000 dams in the United States, few are suitable for electric power generation. Most favorable sites for large hydroelectric dams already are in use, and public sentiment has turned against further dam construction (see Chapter 7).

Numerous existing dam sites have the potential for small-scale electrical power generation. Consequently,

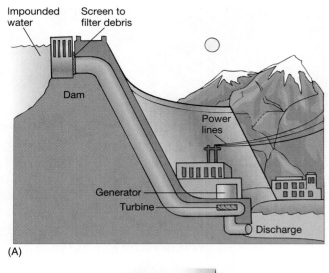

(A)

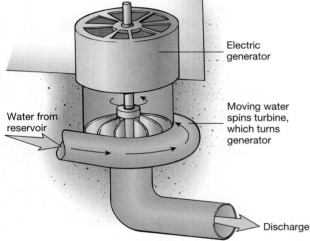

Electric generator

Water from reservoir

Moving water spins turbine, which turns generator

Discharge

(B)

FIGURE 15.2 Converting the kinetic energy in moving water to electricity.

A cutaway of a typical reservoir and hydroelectric plant is shown at top, with detail of power generation below. *From D. Kash and others, Energy Alternatives, Report to the President's Council on Environmental Quality, 1975.*

FIGURE 15.3 Hoover Dam on the Colorado River, 1990.

This hydroelectric dam has impounded Lake Mead. *Photo: Lowell Georgia/Photo Researchers, Inc.*

small hydropower systems are becoming more common for individual homes, farms, and small industry. In mountainous areas, where streams have high gradients and the water flows reliably throughout the year, water flow may be sufficient for small-scale electrical power generation without building dams.

Some hydroelectric plants have added water storage reservoirs at higher altitudes than their main reservoir. This enables them to generate extra power to meet peak demand. An example is the Robert Moses hydroelectric facility five miles from Niagara Falls, operated by the New York Power Authority. When electric demand is low, some of the plant's own electricity is used to pump water from a lowland lake or reservoir to a higher elevation, such as a mountaintop reservoir. This gives the water great potential energy.

When needed, water in the upper reservoir is released, providing kinetic energy. On its downward trip back into the lower reservoir, the water flows through turbines, spinning them to generate additional electricity. It's a good way to add generating capacity to cover peak demand, but it adds pumping cost. Much of this extra cost might disappear if solar-generated power could be used to pump water to the elevated reservoir.

Hydropower can yield plenty of electricity, and hydroelectric plants have low operating and maintenance costs. Environmentally, they can be designed to produce absolutely no air pollution: no carbon dioxide, no sulfur dioxide, no nitrogen oxides, no particulates.

Kinetic Energy from Moving Water **433**

Further, a hydro plant has two to ten times the lifespan of a coal-fired or nuclear power plant.

Damnation. Despite the advantages of hydro power just cited, there are drawbacks to hydro dams:

- Reservoirs behind the dams eventually fill with mud. The reservoir must be dredged.
- Large stretches of river behind the dam become permanently flooded. This affects property ownership and land value, and drowns farms, forests, and highways. In areas where the flooded area contains lush vegetation, as in humid subtropical and tropical regions, the drowned and rotting plants release enormous amounts of methane and carbon dioxide. These potent greenhouse gases are a significant cause of global warming (Chapter 17).
- A dam disrupts the natural pattern of erosion and sedimentation, both above and below it.
- Creating a large lake obviously causes major ecological changes to an area. For example, different plants and aquatic animals move in to the environment of a quiet lake, and organisms that thrive in shallow flowing streams depart. Fish that once traveled freely up and down the stream are blocked by the dam, although fish ladders have been built around some dams (Figure 15.4).

To avoid these problems, an alternative to conventional dams went online in 1990 on the Mississippi River near Natchez. It is a "run-of-river" hydro power station that uses the natural river flow, taking advantage of an existing small drop in elevation (Figure 15.5). This avoids building a major dam, creating a large impoundment, disrupting navigation, disturbing the ecology and the natural erosion/sedimentation pattern, and creating flooded land upstream.

Tidal Power

The harnessing of ocean tides as an energy source for processing grain started in the eleventh century in England, France, and Spain. Water wheels 20 feet in diameter were installed in 1580 under the arches of the London Bridge. They used tidal energy to pump water, and were still pumping some of the city's water supply in 1824. During the 1600s and 1700s, much of Boston's flour was ground from grain at a tidal-powered mill. Water wheels, of course, cannot serve the large energy needs of today's civilization. More sophisticated methods are required to capture the energy in the tides.

Most of Earth's coastlines experience two high tides and two low tides in approximately 24 hours. Along most open coastlines, the difference in height of the ocean water between high and low tides (tidal range) is less than 13 feet. But in confined areas, such as

FIGURE 15.4 Fish gotta swim.

Fish ladders like this one let fish bypass dams as they migrate upriver to spawn. *Photo: Alan Pitcairn/Grant Heilman Photography, Inc.*

narrow estuaries or bays, the tidal range is amplified, and can exceed 20 feet (Figure 15.6). In such places, the water rushes in and out at high velocities. The faster the water, the more kinetic energy in the tide, and the greater the potential that it can be harnessed to generate electricity. This is called **tidal power**.

Several points on the world's coastlines have tidal ranges sufficient for tidal power generation (Figure 15.7). However, only two large tidal energy facilities are operating today—one in France (La Rance) and the other in Canada. The Canadian tidal power plant is in Nova Scotia's Bay of Fundy, where the world's highest tides attain a 50-foot difference between low and high tide. In the United States, tidal energy is being harnessed only at a small plant at Passamaquoddy Bay in Maine, where the tidal range approaches 48 feet.

FIGURE 15.5 Non-dam of the future?

This 192-megawatt "run-of-river" hydro power station may become a model for future non-dam power facilities. It is near Natchez on the lower Mississippi River. *Photo: Sidney A. Murray/Louisiana Hydroelectric.*

The La Rance tidal power station illustrates the potential of this energy source. Completed in 1968, the station straddles an estuary on France's northern coast, where the tidal range attains 45 feet. A 2500-foot-long dam was built across the Rance River (Figure 15.8). Turbines are located beneath the dam. The kinetic energy of inflowing water at high tide and the out-flowing water at low tide turns the turbines, generating electricity.

Why are there so few tidal power plants around the world? Few locations have the unusually high tidal range required. Also, building a long dam is extremely expensive, and such a facility is vulnerable to storm damage and corrosion by salty seawater. However, construction cost can be recouped quickly because a tidal power station requires no fuel.

Ocean Wave Power

Wave energy might also be harnessed as a power source. This requires a steady pattern of waves from a single direction, which does not occur everywhere. In a wave-power plant, a hollow concrete box is sunk into a gully off the coast to catch waves. As each new wave enters the chamber, the rising water forces air in the chamber through a turbine, making it spin, which drives a generator to produce electricity. When the wave recedes, it draws air back into the chamber, and the air movement again spins the turbine.

OSPREY (Ocean Swell Powered Renewable Energy), the world's first commercial wave-powered electrical generator, began operation off the Scottish coast in 1995 (Figure 15.9). It not only generated electricity from waves but included a wind turbine as well, taking advantage of the windy location. It produced 2 megawatts of electricity from the waves and another 1.5 megawatts from the wind, and sent the power to shore through a submarine cable.

Its first use was to power desalinization plants at a cost of $0.10 per kilowatt-hour. This is double the cost of other forms of electricity (see Figure 15.1), but an eventual cost of $0.06 was expected. In some parts of the eastern United States, electricity today costs $0.16 per kilowatt-hour, so this wave-power design would be competitive. As with most renewable energy sources, the flow of electricity from *OSPREY* would be intermittent. Calm July seas generate little electricity, but averaged over the year, *OSPREY* was to produce the same energy as a continually operating 600-kilowatt generator.

Note that this project is described in the past tense. *OSPREY*'s stabilizing tanks leaked in a storm, sinking the structure after only a month. A replacement, *OSPREY II*, is planned. Future plans envision a bigger wave-power generator producing 8 megawatts of power.

A Japanese company plans an *OSPREY*-type wave power generator. The *Mighty Whale* not only will turn wave power into electricity, but reduce the size of waves and calm the sea, creating a safe swimming area. *Mighty Whale* should convert 16–20 percent of wave energy into electricity at a cost three times that of conventional power plants, a reasonable start for a new pollution-free technology.

(A)

(B)

Thermal Energy from Groundwater and Seawater

In addition to the kinetic energy that water has when it is moving, there is heat energy in water. This is obvious; if water lacked heat energy, it would be frozen solid. Even frozen solid, it still has heat energy, just not enough to make it liquid. But how can we recover heat energy from water? Let us examine two ways: heat from hot groundwater (geothermal energy) and heat from ocean water.

Hot Groundwater (Geothermal Energy)

In some areas, groundwater is heated by magma not far beneath the surface. This happens in volcanic areas like Hawaii, Iceland, New Zealand, the Philippines, and Japan. It also occurs in all western states and in Arkansas, Georgia, West Virginia, and Virginia, where hot springs exist. The hot springs reveal where magma lies within a few thousand feet of the surface, heating the groundwater. This is called geothermal water, and it can be tapped as a geothermal energy source.

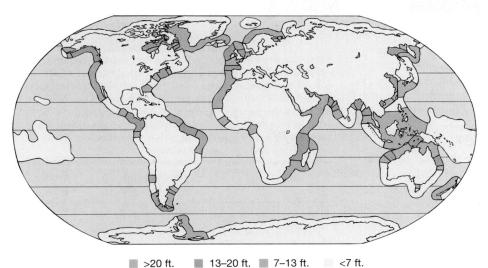

FIGURE 15.7 Potential sites for tidal power generation.

The twenty-one sites shown have the necessary tidal range for generating electricity. *From Thurman, Essentials of Oceanography, 5th ed., Prentice Hall, 1996.*

>20 ft.   13–20 ft.   7–13 ft.   <7 ft.

FIGURE 15.8 Earth's first tidal-power electrical-generating facility.

This tidal power plant at Saint-Malo on the English Channel coast uses flood and ebb tides to generate electricity. *Photo: Science VU/API/Visuals Unlimited.*

FIGURE 15.9 *OSPREY* (Ocean Swell Powered Renewable Energy).

The 65-foot-high structure, which began operation in 1995, sat on the seafloor in 45 feet of water about 1100 feet off the stormy coast of northern Scotland. It survived only one month before sinking. *Photo: Applied Research & Technology Ltd.*

There are two ways to use geothermal energy. The hot water can be piped directly into buildings for heating, or it can be used to generate electricity. At present, about two-thirds of the geothermal energy that is harnessed is used for direct heating, and the other third for power generation. As of 1991, the countries using geothermal energy numbered thirty-nine. Not surprisingly, two-thirds of these are along active plate boundaries. (See Table 15.1)

Along the Mid-Atlantic ridge in Iceland, geothermal energy is almost the only source of energy available. Iceland is a volcanic basalt island, and basalt rock normally does not contain oil, gas, coal, uranium, tar sand, or oil shale—the fuels obtained from sedimentary rocks. Consequently, Iceland's quarter-million people harness geothermal sources to obtain their energy.

Currently, about twenty countries are extracting energy from geothermal sites and producing about 6000 megawatts of electricity, the same amount now generated by wind power (discussed later in this chapter). They are supplying enough heat for more than 2 million homes in a cold climate and enough electricity for over 1.5 million homes. The United States accounts for 41 percent of the electricity generated worldwide from geothermal sources. However, this is less than half of one percent of the energy consumed in this country.

The biggest developers of geothermally generated electricity are the U.S., the Philippines, Mexico, and Indonesia. Earth's largest geothermal power-generation facility is at The Geysers in northern California (Figure 15.10). The Geysers has been an important regional energy source since the early 1970s. Operated by Pacific Gas and Electric Company, the field has twenty-eight plants generating about 2000 megawatts of electricity a year, enough to meet the electrical needs of a city the size of San Francisco.

A related source of geothermal power is the possibility of mining heat from hot rock, rather than hot water. Two holes could be drilled to a 15,000–20,000-foot depth directly above a magma chamber, where the temperature at the bottom of the holes would be about 900°F. The rock at the bottom of the holes would be fractured by explosives so that the two holes would be connected by rock fractures. Then, cool water would be pumped down one hole, would circulate through the hot, fractured rock and absorb heat, and would flow into the second well, where it would be pumped to the surface.

Geothermal Problems. According to the National Academy of Sciences, the potentially recoverable geothermal energy could meet U.S. energy needs at current levels for 600 to 700 years. However, most sites for geothermal development are in the West, where water is scant. Nothing ever is where you want it: the best

TABLE 15.1 Worldwide Installed Geothermal Capacity as of 1991

Some nations use geothermal energy mostly to generate electricity (Mexico, Philippines, United States) while others use it mostly as a source of heat (China, Hungary, Japan). MW indicates megawatts of power. One MW supplies the energy for a city of 400 homes in the U.S. but 1000 homes in England (Americans keep their buildings warmer than the English and also use more electrical appliances and air conditioning).

Country	Non-electric MW thermal	Electric MW
Japan	3321	214.6 + 55*
China	2143	20.9
Hungary	1276	—
former USSR	1133	11 + 50*
Iceland	774	44.6
USA	463	2770
France	337	4.2
Italy	329	545 + 160*
Bulgaria	293	—
New Zealand	258	293 + 2*
Romania	251	—
Turkey	246	20.6
Indonesia	142.3	—
former Yugoslavia	113	—
Czechoslovakia	105	—
Belgium	93	—
Tunisia	90	—
Kenya	45	—
Ethiopia	38	—
Mexico	28	700 + 50*
Switzerland	23	—
Greece	18	2
Algeria	13	—
Colombia	12	—
Australia	11	—
Taiwan	11	—
Guatemala	10	—
Poland	9	—
Germany	8	—
Austria	4	—
Canada	2	—
Thailand	2	0.3
UK	2	—
Denmark	1	—
Philippines	1	894 + 110*
Azores	—	3
Costa Rica	—	55*
El Salvador	—	95 + 20*
Nicaragua	—	35

*Under construction.

FIGURE 15.10 The Geysers.

A field of natural steaming vents 90 miles north of San Francisco, The Geysers is the world's largest geothermal power-generating development, now nearly forty years old. Its wells, almost 2 miles deep, produce nearly pure steam. It is piped to turbines, rotating them to operate electrical generators. But are The Geysers running out of steam?

Photo: Science VU/Visuals Unlimited.

heat sources are in the West, but the best water sources are in the East. At present, one small plant is operating in New Mexico, generating electricity from hot subsur-

face rocks. The economic advantage of this project has not yet been demonstrated.

Geothermal waters typically are very saline, several times the salinity of seawater, because hot water is a better solvent than cold seawater. The subsurface water can be sulfurous as well. Sulfur compounds are highly corrosive to turbines and pipes and can pollute local water supplies if released untreated. Hot brines released into streams can be ecologically disastrous, killing plants and fish.

The geothermal resource in any area is not permanent. For example, there are signs that The Geysers is running out of steam. Steam is being withdrawn faster than groundwater can be replenished by natural inflow, and so water pressure is declining.

Another problem with withdrawing geothermal water is land subsidence (see Chapter 6). At the Wairaki power-generating site in New Zealand, geothermal wells have been drilled 2000–4000 feet deep. Withdrawing hot water has created a subsidence bowl 1.2 square miles in area with a maximum depth of 15 feet. At The Geysers geothermal field, 5-inch vertical drops have developed during the past thirty years. Just like water pumped from an aquifer or oil pumped from rocks, the geothermal water fills pores in the rocks and thus has supplied some of their support. As the support is removed, the ground subsides.

Pumping Heat. Few Americans live where magma is near enough to the surface to be useful. Thus, we might assume that we cannot use geothermal energy. However, a "cooler" form of geothermal energy is available to most of us, not from magma, but from the everyday heat in the ground. Ground greater than around 10 feet deep is unaffected by seasonal changes, and maintains a constant temperature year-around. The difference in temperature between the ground surface and deeper ground can be used for heating or cooling with a heat pump.

A **heat pump** moves heat from one place to another. Its operation is similar to an air conditioner or refrigerator. Pipes are buried deep enough to be in the constant-temperature zone, and a water/antifreeze solution circulates through them, absorbing heat from the ground. The warmed solution then is pumped into a building, where the heat is released to warm its interior. In summer, when it is warmer in the building than in the ground, the process is reversed. Thus, a heat pump provides heat during the winter and cooling during the summer. Its effectiveness depends on the temperature difference between the building's interior and subsurface and on the system's efficiency.

Oceanic Thermal Energy Conversion (OTEC)

Heat pumps can operate in the ocean as well as on land, wherever a significant temperature difference exists

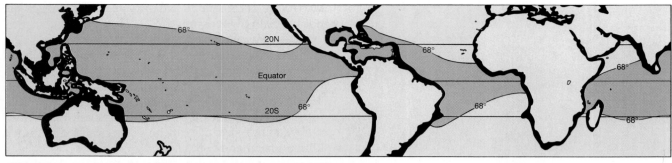

FIGURE 15.11   Oceanic thermal energy conversion (OTEC).

Black areas are too shallow (less than 3300 feet deep) to provide the temperature difference needed for OTEC. Blue areas lack the necessary temperature difference because of too little solar heating of the surface waters. Only the yellow areas have average monthly temperature differences of 40°F or greater between the ocean surface and depths of about 3300 feet, making them suitable for OTEC operation.

between warmer water at the ocean surface and colder water in the deep ocean. Because of solar heating, surface water in the lower latitudes that straddle the equator has temperatures of 75–80°F. But deeper down, at depths as shallow as 2000 feet, the water may be only 40°, and at the ocean floor it may be only 30°.* The process of extracting energy using this temperature difference is called **oceanic thermal energy conversion (OTEC)**. Usually, the thermal energy is used to generate electricity.

In an OTEC plant, warm surface water is pumped in to heat liquid ammonia. The liquid ammonia has a very low boiling point (–28.6°F), so it readily vaporizes. This ammonia vapor then drives a turbine to generate electricity. The vapor then is cooled by cold water pumped from depths of 2000 feet or deeper. Cooling the ammonia condenses it back into a liquid, so it is ready for the next cycle. This same process is used in traditional steam power plants, using water instead of ammonia. The only atmospheric contaminant produced by this process is carbon dioxide exhaust from whatever fuel is used to operate the ammonia compressor.

The electrical power thus generated could be carried landward by cable. Or, it could used on-site to desalinate seawater, to extract minerals from the seawater, or even to split water into hydrogen and oxygen so the hydrogen gas could be pumped ashore for use as a fuel.

Several technical problems have slowed OTEC's development. It is less efficient than other power-generating methods, because a third of the energy produced is needed to pump the deep cold water to the surface. Construction cost is two to three times that for comparable coal-fired power plants. The ocean environment is hostile and expensive: routine storms, hurricanes, saltwater corrosion of machinery, and fouling from algae and barnacles. The temperature difference

between surface water and deep water must be substantial, which limits OTEC sites to the area shown in Figure 15.11.

An OTEC power plant could be stationary or built on a barge to travel where needed. A prototype in Hawaii is desalinating water, cooling buildings, and pumping up nutrient-rich deep water for vegetables and aquaculture, all while generating 210 kilowatts of power. Large OTEC plants could be built off the Florida coast or anywhere enough temperature difference exists within 100 miles of the shore. Other countries actively studying OTEC include Japan, Sweden, and Holland; all of these countries lack significant fossil fuels.

Energy from the Wind

For thousands of years, people have harnessed the energy in the wind. Mostly this has been done on the scale of single units, such as a sailing vessel, a windmill to grind grain or pump water from a well, or a kite. The harnessing of wind energy to generate mass power, such as electricity for thousands of homes, was not seriously considered until the 1970s when the U.S. government initiated development of large-scale wind-power systems.

How much energy is in the wind? Considering the damage caused by hurricane-force winds, there must be a lot. In fact, the power available from wind increases with the cube of the wind speed. For example, if a 10 mph wind increases by only 50% to 15 mph, the available energy is $1.5 \times 1.5 \times 1.5$, or 3.4 times the energy. (Try calculating how much more energy is in 180-mph hurricane winds than is in a 20-mph wind.)

The U.S. Department of Energy has financed experimental **wind farms** in mountain passes known to have strong, steady winds. The most striking is at Altamont Pass east of San Francisco, which now sports 7500 wind turbines (Figure 15.12). Another cluster of 800 turbines is in Solano County in northern California, and

*Normal seawater, because of its of dissolved salts, freezes at about 28° rather than 32°. (Review Chapter 8, "Salt on Roadways.")

FIGURE 15.12   Wind farm at Altamont Pass, California.

Independent companies operate 7500 wind-powered turbines that sell electricity to Pacific Gas & Electric. *Photo: Russell D. Curtis/Photo Researchers, Inc.*

another group is in Tehachapi Pass east of Los Angeles. Currently, about 17,000 of these machines operate in clusters in the state. A plan exists to develop wind farms capable of supplying 8 percent of the state's electricity by the year 2000.

As of 1996, however, wind turbines produced only 1.3 percent of California's electrical demand. The cost of the electricity is 5 cents per kilowatt hour, the same as

electricity generated by conventional sources such as coal, but without the environmental problems of carbon dioxide and sulfur. California currently produces one-third of the world's wind-generated electricity.

Figure 15.13 shows U.S. wind resources, on land and offshore. Wind power could supply up to 25 percent of U.S. electricity by the year 2050, according to wind-power experts. Major wind-power projects are

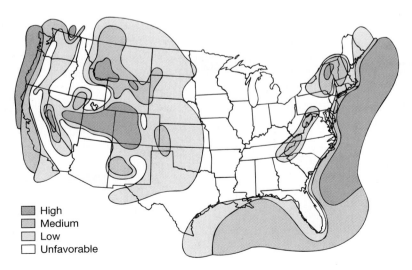

High
Medium
Low
Unfavorable

FIGURE 15.13   The average annual U.S. wind resource.

California currently produces one-third of the world's wind-generated electricity. Interestingly, the city of Chicago, long nicknamed "The Windy City," lacks the prevailing winds necessary to provide useful energy.

now being built or planned in several states. What wind resources are available in your area?

It appears that wind turbines could meet 20 percent of current U.S. power demand if installed on only 0.6 percent of the land area of the lower 48 states (this is a total area of 135 miles by 135 miles, or 18,000 square miles). Just three states (the Dakotas and Texas) could supply the entire electrical demand. Minnesota's energy commission has calculated that a single wind farm on one of the state's southwestern ridges could supply almost all of the state's electricity needs! Wind power is certain to become an important contributor to satisfying America's growing energy demand.

Until the 1990s, ideas for tapping wind energy centered around sites on land. However, winds offshore tend to be stronger and steadier than winds on land. For example, in New England, wind power available about 60 miles offshore is three or four times greater than that available on land (Figure 15.13). To exploit this energy advantage, Danish electric companies built the world's first offshore wind farm in 1991, a mile off the coast in water only a few yards deep. Its present energy cost is 50 percent greater than land-based wind energy because of construction and transmission expense, but the cost should decrease as technology improves.

The dominance of the United States in wind-generated electricity has ended. World wind-generating capacity is increasing dramatically:

10 megawatts in 1980

1020 megawatts in 1985

1930 megawatts in 1990

4700 megawatts in 1995

5800 megawatts in 1996

More than 25,000 wind turbines now are connected to national electric power grids. Europe now boasts half the world's wind-generating capacity, with Germany, Spain, and Greece leading the way, but many other countries have begun to exploit their wind resource.

Energy from Biomass

Biomass ("biological mass") is organic plant matter. In Chapter 14 we noted that it is "fossilized sunshine," produced by conversion of solar energy to chemical energy through the process of plant photosynthesis (review Chapter 14, energy conversions). Biomass includes forests, grass, straw, cut wood, waste paper, agricultural plant residue like corn husks and cobs, manure, and gases and liquids obtained from these sources. **Biomass energy** is energy obtained by burning (which is simply oxidation at a rapid pace). Most

biomass is burned for heat rather than for generating electricity.

In the U.S., we are accustomed to "turning up the thermostat" and enjoying heat from gas, oil, or electric heaters. To us, biomass in the form of firewood is something we enjoy on a cold winter's night for fun. But biomass in the form of **fuelwood** is the primary heating source—often the *only* source—for half the world's population and for most people in Third World countries (Figure 15.14). Biomass constitutes only 4–5 percent of energy consumed in fossil fuel-rich North America.

If biomass is already dead, burning it simply speeds up the natural decay of dead plants to carbon dioxide and water, which is environmentally benign. Such use is not the same as cutting live trees for firewood, which is harmful to the environment. In many Third World countries, live trees are being harvested faster than they can be regrown. This causes widespread

FIGURE 15.14 Global consumption of energy.

Numbers at ends of bars are the energy equivalent in million tons of oil. Note that Earth's developed countries (color), which consume two-thirds of all energy, use the majority of each fuel type except one: biomass. Underdeveloped nations rely overwhelmingly on biomass as their energy source, because it is "free," widespread, does not require mining or drilling or high technology, and is renewable. However, it also has low energy content compared to the energy concentrated in fossil fuels, uranium, and moving water.

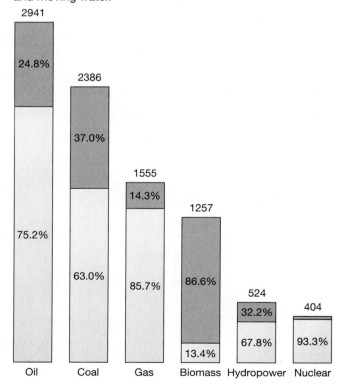

deforestation. As trees are harvested, their soil-holding roots are lost, sharply increasing soil erosion. Killing live vegetation increases erosion and slows the rate of removal of the greenhouse gas carbon dioxide from the atmosphere.

In the United States, wood stoves have become popular again. About 5 million homes rely entirely on wood for heating and another 20 million burn wood for partial heating. Many sawmills and woodworking companies now are burning wood waste, and some sugar refineries are burning sugarcane waste to supply all or most of their electrical power.

Small, wood-burning power plants in heavily forested areas such as New England can produce electricity at about the same price as burning coal. In fact, wood-fired power plants now produce 23 percent of the electricity used in Maine. The National Wood Energy Association estimates that burning wood waste from paper and lumber mills, agriculture, urban land clearing, and tree trimming in the United States could provide electricity equal to that from 200 large conventional fossil-fuel power plants.

Much wood in this country is used in fireplaces by individual homeowners, either to generate primary heat or to create a romantic setting on cold winter days. Unfortunately, heating a house from a fireplace may be so inefficient that it results in a net energy *loss.* The hot air rising up the chimney causes a low-pressure region in the room. Higher-pressure cold outside air thus pushes its way into the home through the cracks that exist in virtually all houses, sometimes creating uncomfortable drafts.

To solve this problem, some fireplaces and stoves now are built with an air duct that draws cold air directly from the outside and into the fire. It may cost money and waste energy to create a romantic atmosphere by a fireplace, but often this is not a difficult cost-benefit analysis.

Although less desirable as a fuel for romantic evenings at home, cow manure can serve as a fuel source for heating and cooking. The Mesquite Lake Resource Recovery Project in southern California burns cow manure in special furnaces to generate electricity for thousands of homes. The manure is too salty and contaminated with weed seeds to be used as fertilizer, so its use as an energy source helps solve the problem of its disposal.

Biogas

Plants, organic waste, sewage, and other biomass solids can be treated with bacteria to release **biogas,** a mixture of 60 percent methane and 40 percent carbon dioxide. The process is called **biodigestion,** and it uses anaerobic bacteria in a sealed tank. Digesters are usually brick or metal tanks inside which fermenting organic materials give off flammable gases such as methane. A simple biodigester like those on family farms in China produces 70–140 cubic feet of gas per day. For perspective, 35 cubic feet will cook about three meals for a family of five or run a 1 horsepower engine for two hours. The solid residue from the digester is used as a nutrient-rich fertilizer.

In the United States, perhaps the most highly developed biogas facility is located on a large farm near Gettysburg, Pennsylvania. Manure from 2000 cows in a barn drops through a grated floor and slides into anaerobic digesters. The biogas produced drives generators that provide enough electricity for the entire dairy, with the excess sold to the local utility company. Heat from the generators is used to warm the digesters and stimulate faster decomposition. The solid residue from the digesters is used as fertilizer. The farmer estimates that he nets about $100,000 per year on his energy system.

Another source of biogas is landfills. (Review "Gassy Landfills" in Chapter 9.)

India and China have almost 8 million small biodigester facilities in operation (Figure 15.15). There is a good reason for this: India has scant fossil fuels and China's huge reserves are in early development.

FIGURE 15.15 Methane generator.

Methane gas produced from the fermentation of animal and human waste is used for cooking and heating. *Photo: Sylvan Wittwer/Visuals Unlimited.*

Bioliquids

Solid biomass can also be converted into liquid fuels called **bioliquids.** Crops such as sugarcane, sugar beets, sorghum, and corn can be fermented and distilled to produce ethyl alcohol (ethanol), which can be burned in automobiles as a substitute for gasoline. (One bushel of corn yields 2.4 gallons of alcohol.) Engine modifications costing only a few hundred dollars are required. Gasoline can also be mixed with 10–23 percent ethanol to make *gasohol,* sold as super-unleaded or ethanol-enriched gasoline.

The only country that has gone forward in a major way with alcohol-based automobile fuel is Brazil. Since 1983, three-quarters of the cars sold in Brazil run on pure ethanol. Alcohol-fueled cars are, however, difficult to start in cold weather. Americans who live in the northern tier of states are unlikely to adopt ethanol-fueled cars.

Biomass Pros and Cons

Burning biomass is like burning coal (which is fossilized biomass): it produces smoke and carbon dioxide. So is biomass a polluter? It depends. As long as the rate of burning does not exceed the rate of regrowth, there is no net increase in atmospheric carbon dioxide. But if destruction exceeds replenishment, atmospheric carbon dioxide will increase and, perhaps even worse, soil erosion will increase. This is already occurring in some areas (Figure 15.16).

Deforestation is a major environmental threat to the entire planet. Deforestation increases soil erosion (Figures 3.25, 4.9, and 4.11) and increases flooding risk. It also contributes to *desertification* (expansion of desert areas), which is occurring now in northern Africa's Sahel region on the Sahara Desert's southern fringe. Most deforestation does not result from harvesting firewood but from clearing land for agriculture. Unfortunately, felled trees typically are burned just to dispose of them, so their energy is wasted.

Crop residue (like wheat stubble and corn stalks) and animal manure can be collected and burned without adding significant sulfur or nitrogen to the air. The State of Hawaii burns residue from sugarcane to supply 10 percent of its electricity; the island of Kaui gets 58 percent from this source, and the island of Hawaii gets 33 percent. Brazil also derives 10 percent of its

FIGURE 15.16 The vanishing rain forest.

Deforestation is so rapid that major rain forests shown may disappear shortly after 2000. *From B. J. Nebel and R. T. Wright, 1993, Environmental Science, 4th ed., Englewood Cliffs, New Jersey: Prentice Hall, p. 430.*

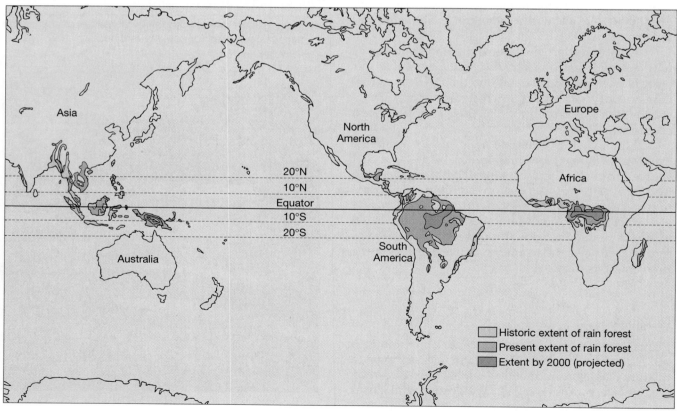

electricity from sugarcane residue and plans an increase to 35 percent by 2000.

A drawback to generating electricity this way is that the residue must be close to the power plants to avoid costly transportation of bulky crop residue. Another is that using agricultural residue for fuel rather than plowing it back into the soil can steal away the soil nutrients, diminishing its fertility. U.S. studies have shown that only 35 percent of residues can be safely removed without damaging the soil.

Burning crop residue for energy suggests the idea of growing plants specifically as a fuel source. The United States produces large surpluses of food grains and other crops, so we could cut back on this surplus production and instead use the land to grow "fuel crops." Such use raises moral questions about our relations with underdeveloped countries, which rely on our surplus export. Should we grow less grain so we can increase our use of "petropig" cars, while most of the world's people go hungry? What is our obligation to those less fortunate than we?

Energy from the Sun

Have you ever started a fire by focusing the sun's rays onto a newspaper with a magnifying lens? This is just one way to harness energy from the sun—cost-free and pollution-free. Solar energy has been exploited for thousands of years. According to legend, the Greek scientist Archimedes saved his home 2500 years ago by using solar energy. Faced with an attack by a foreign navy, he had soldiers assemble in the shape of a parabola (like a satellite dish), turn their shields toward the sun, and focus the sun's rays on the cloth sails of the invading navy, igniting them and leaving the ships helpless before Archimedes' forces. (Duplicating that feat would make an interesting science project.)

At the 1878 World's Fair, an inventor displayed a solar collector that generated enough power to drive a printing press. In 1891, a solar water heater was being sold in California. By the early 1900s, farmers in the Southwest were focusing the sun's rays on a boiler to produce steam that powered an irrigation pump. In 1912 in Egypt, a parabolic solar collector boiled water to make steam that powered a 100-horsepower engine. However, for the next fifty years, solar energy development languished as cheap, plentiful fossil fuels dominated world industry. Pollution and the environment were not concerns in those times.

The sun radiates an enormous amount of energy, but only about one two-billionth of it strikes Earth's atmosphere. Of that, about half reaches Earth's surface. If Earth were flat and perpendicular to the sun, all points on Earth's surface would receive identical amounts of energy. But our round Earth receives solar energy very unevenly, because it arrives at different angles (see Figure 6.7A). Consequently, the average solar radiation is greatest at the equator (where it arrives closest to perpendicular) and diminishes toward each pole. The equator receives about 2.5 times as much solar radiation as the polar regions.

Much of the most intense solar radiation falls on Earth's desert areas. Most are far from population centers, so little of this energy is put to use (Figure 15.17A). In the United States, the solar energy input is greatest in the Southwest (Figure 15.17B). This suggests the cities of Las Vegas, Phoenix, Tucson, Albuquerque, and El Paso are prime sites for solar power. Lowering the cost of electricity for garish casino signs in Las Vegas might encourage the casino management to improve the odds for players.

Today, solar cookers are widely used by Third World households and backpackers. Houses are built to gather solar energy for free heating of the interior space and water. Chances are you own a light-powered calculator or other device whose solar cell converts light energy into electricity. These examples illustrate the two basic approaches to using solar energy: directly for heating, and indirectly to generate electricity. We will now look at each.

Solar Heating

The two fundamental types of solar heating are *active* and *passive*.

Passive Solar Heating. As everyone is aware, a room or a car interior grows warmer when the sun shines through the windows. In fact, a closed-up car sitting in direct sunlight can become a thermal deathtrap for a child, an elderly person, or a pet. This occurs because the glass allows solar radiation in the visible wavelengths to pass through, where it strikes the interior and is converted to longer-wavelength heat (infrared) radiation. The window glass, which is so transparent to sunlight, partially inhibits the passage of outgoing heat radiation. Thus, more radiation enters than leaves, and temperature rises. This is known as the *greenhouse effect*, described further in Chapter 17. It is an example of **passive solar heating,** which simply captures solar radiation and stores it.

In the United States and Canada, solar radiation comes from the southern part of the sky, as you can see in Figure 6.7A. Hence, the simplest passive heating system in a home is having big windows in the wall that faces south. This technique was employed in Greece 2500 years ago. South-facing windows, coupled with better insulation and more airtight construction,

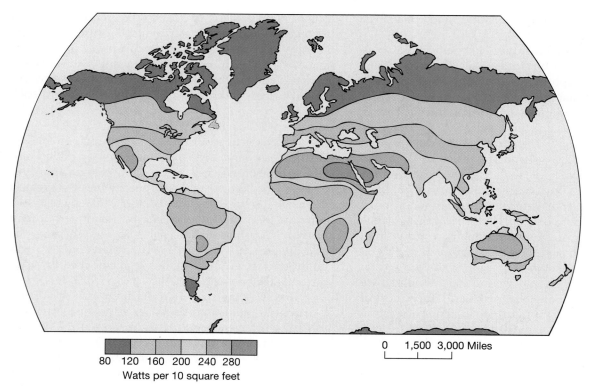

(A)

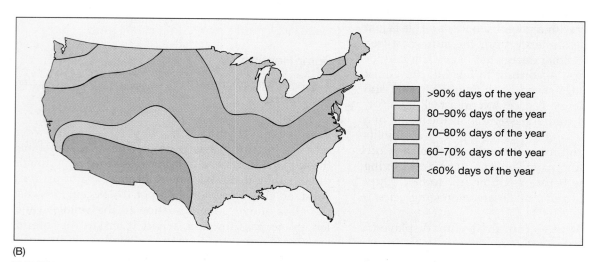

(B)

FIGURE 15.17 Solar energy received on Earth.

(A) Solar energy reaching Earth's surface varies widely, clearly varying with latitude. It is maximum between 5° and 30°, both north and south latitude, where Earth's major deserts lie. The red and orange areas, which receive the greatest intensity of sunlight, are most suitable for solar power plants. However, these desert areas are distant from population centers, where the energy is most needed. *Robert W. Christopherson, 1994, Geosystems, 2nd ed., Prentice Hall, p. 97.* (B) Usable solar energy received by the lower forty-eight states. The maximum is received in the desert Southwest. *From National Wildlife Federation, 1978.*

can significantly lower heating cost, which is more than half of residential energy use.

When building a new house, proper orientation of windows, well-insulated construction, and regular maintenance of openings (weatherstripping and caulk-

ing doors and windows) are inexpensive ways to save energy cost. But orientation and construction are, of course, fixed in an existing home. Only 2–3 percent of existing homes are abandoned each year in the United States. Thus, it will take many decades before most homes are built to take advantage of simple passive solar heating.

In addition, land developers generally do not rank home orientation of great importance. Of far greater concern to them are the shape of a lot, its slope, drainage, location of utility lines, and view from the living room and dining room, because these factors sell houses. Only continued pressure by knowledgeable, environmentally conscious buyers will force builders to give home orientation the top priority it deserves.

Active Solar Heating. **Active solar heating** systems use motors or pumps to circulate air, water, or other fluids. The fluids transport heat from **solar collectors** to a storage unit where the heat is stored until it is needed (Figure 15.18). The most common collection device is an insulated box that faces south, toward the sun (in the Northern Hemisphere). Inside is a flat, black solar panel. (Why black? If you have stepped barefoot onto blacktop under a blazing sun, you know why! Black absorbs the shorter light wavelengths of the sun's radiation, converting them to heat.)

Heat from the solar panel heats the air around it, which transfers heat to a heat exchanger. It raises the temperature of water in the heat exchanger to at least 140°F, recommended for household hot water. Pumps circulate the heated water to an underground storage tank. Another set of pipes and a pump circulate the heated water to radiators in the home. Such active solar heating is especially effective for water heating; about 8 percent of the solar energy consumed in the United States is used to heat water.

Hence, active solar heating in a large percentage of American homes and small buildings could decrease the consumption of natural gas for heating. This would free up a large amount of natural gas for fueling vehicles. This in turn would decrease the nation's need to import large amounts of crude oil for refining into gasoline. (This is a good example of why every country needs a carefully planned energy policy.)

The most intensive development of solar heating has been in Israel. This is no accident: Israel has no significant fossil fuel resources, and for political reasons does little trading with its energy-rich Arab neighbors. The result of this deprivation can be seen on almost every Israeli rooftop: nearly a million solar collectors, which in 1994 provided hot water for 83 percent of Israel's homes, including your author's (Figure 15.19).

In addition, Israeli law requires all residential buildings up to six stories high to use solar energy for

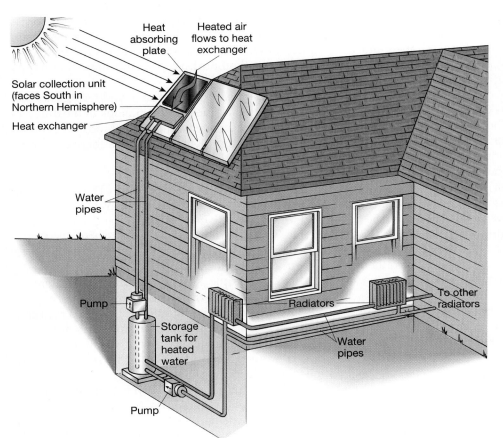

Solar collection unit (faces South in Northern Hemisphere)

Heat absorbing plate

Heated air flows to heat exchanger

Heat exchanger

Water pipes

Pump

Storage tank for heated water

Pump

Radiators

To other radiators

Water pipes

FIGURE 15.18   Solar power at work.

This active solar system uses a solar collector (panel) to warm water for heating a building. In this example, the sun's light energy strikes a dark surface, which absorbs the light energy and converts it into heat energy. This heats the adjacent air, which warms the water in a heat exchanger. This warmed water is pumped to an underground storage tank, where it is available to provide heat to the building on demand.

FIGURE 15.19 Proof that solar power works.

Solar panels collect heat for a 50-gallon water tank on the author's home, and neighbors' homes, in Jerusalem. An apparatus for active solar water heating costs about $2600 to provide hot water for an average family of four. *Photo: Harvey Blatt.*

water heating, a move that has reduced the national electric bill by 3 percent. Today, Israel has one of the highest concentrations of scientists and engineers dedicated to solar research of any country.

How does the conversion from electric or gas water-heating to a solar system affect the average homeowner? An apparatus for active solar water heating costs about $2600 to provide hot water for an average family of four. It consists of two 4′ × 8′ solar panels and an 80-gallon water-storage tank, plus the necessary piping and pumps. Each solar panel provides enough heat for 40 gallons of water. Upsizing to three panels and a 120-gallon tank costs about $3600. Depending on local utility prices and the amount of hot water used by a family, it takes ten to fifteen years of savings on utility bills to recover the cost of the solar unit.

Solar space heating of a home's interior is more expensive. At present, at least ten solar panels are needed to heat an average house in the central United States. Because each panel presently costs about $750, conversion to solar space heating is not feasible for most Americans. However, it is quite cost-effective to construct new homes with solar water heaters rather than with traditional gas or electric systems, as is done in Israel.

The widespread use of active solar water-heating systems would benefit everyone by reducing our consumption of and dependence upon fossil fuels.

Electricity from Solar Energy

Use of solar energy to generate electricity on a small scale is now everywhere. The sun powers calculators, battery chargers, water-well pumps, portable communication equipment, and remote instruments that monitor weather, stream levels, and such. And how about a solar-powered camel? See Figure 15.20.

But solar energy is much less widespread in the generation of mass electrical power for utilities. This is a problem of geography, technology, and economics. The geography problem is easily understood; a solar power-generating facility in the Arizona desert would do well, but the Midnight Sun Solar Electric Company at the South Pole would have few happy customers.

To understand the economic and technical problems of using solar energy, we will look at the different approaches in use. The two main methods for generating electricity from solar power are thermal electric generation and solar cells.

Thermal Electric Power Generation. The basic idea in a thermal electric system is to concentrate solar energy, use it to boil water to generate steam, and use the steam pressure to spin turbines, generating electricity. Figure 15.21 shows two approaches, the "power tower" water-boiler in A and the heated-oil pipe-collector in B.

In A, multiple mirrors reflect sunlight onto a boiler to concentrate the solar energy there. The mirrors are computer-guided to reflect the sun's rays constantly onto the boiler as the sun migrates across the sky each day. This tremendous concentration of energy boils the water, generating steam, which operates a turbine to generate electricity.

In B, 2000 mirrors are curved to focus the sunlight just a few feet away, onto pipes. (It's the same idea as a satellite dish focusing the weak signal from a satellite onto a small antenna mounted at its center.) The pipes

FIGURE 15.20   A solar-powered camel?

Actually, the solar panel mounted on this camel in the Somali Desert powers the small refrigerator on the side of the camel! The refrigerator contains vaccines that must be kept cold until used by nomadic tribes who live in the region. *Photo courtesy Siemens Solar Industries.*

are filled with a synthetic oil or molten salt, which becomes heated to 730°F (if oil) or 1050°F (if salt). The hot fluid is pumped to a boiler, where the heat boils the water, and the resulting steam pressure spins a turbine, which turns a generator to generate electricity. About one-quarter of the solar energy that hits the mirrors can be converted to electricity; the rest is lost to inefficiencies in the conversion process.

The largest installation for thermal electric power occupies 2000 acres and is located in the Mojave Desert of southern California, where eight "solar farms" are operated by Southern California Edison Company (Figure 15.21B). Each farm consists of a power plant surrounded by hundreds of solar collectors and requires about one year to build. It is estimated that solar-thermal plants occupying less than 1 percent of the area of the Mojave Desert could supply the electricity needs of Los Angeles.

A disadvantage of solar-thermal plants is that they lose their energy input at night or on cloudy days. This problem is circumvented in various ways. In the case of the power tower and the synthetic-oil plants, a backup system is needed to generate electricity "when the sun doesn't shine." In the case of a plant that uses molten salt, the salt stays hot for two to twelve hours, so the plant operates twenty-four hours a day.

Solar Cell Power Generation.   **Solar cells,** technically known as **photovoltaic cells,** convert light energy directly into electricity (Figure 15.22). Most solar cells are thin wafers composed of purified silicon crystals (straight Si, not silica, SiO_2). To the silicon, trace amounts of the elements gallium and cadmium are added. When these wafers receive light energy, they emit electrons, which flow as an electric current. The energy from a single solar cell is small, about that of a flashlight battery. Thus, many cells must be wired together as a **solar panel** to provide 30–100 watts of electric power. Groups of panels are then wired together to increase the wattage output.

Current solar cells have a conversion efficiency (light energy to electrical energy) between 15 and 20 percent. Experimental solar cells have reached 25 percent efficiency in the laboratory. This suggests that the future for solar-cell electricity may be quite bright.

The cost of generating electricity from solar cells is five times that from coal-fired power plants, but it is dropping dramatically. An important development is the

(A)

(B)

fabrication of cells from lower-grade silicon, reducing silicon cost from $75 to $1 per pound. This lowers the cost of electricity by about 80 percent. When this new type of cell is mass produced within the next few years, the cost of solar-cell electricity will become comparable to that from coal-fired plants, and without the pollutants.

Enron, the nation's largest natural gas company, is building a solar-cell facility in the southern Nevada desert to generate 100 megawatts of electricity. This plant will be more than twelve times larger than any

other solar-cell solar plant and will produce enough electricity to power a city of 100,000 people. The plant is expected to deliver electricity at a cost of only 5.5 cents per kilowatt-hour.

Arrays of solar panels require lots of space. The acreage required depends not only on the electrical demand but on the efficiency of the panels. Panels with a 20 percent efficiency require only half the space of panels with a 10 percent efficiency. Assuming a 20 percent efficiency, 5790 square miles of solar cells could supply

FIGURE 15.22 Solar cell.

This photovoltaic cell converts light energy into electrical energy. *Photo: Bernard J. Nebel.*

all of America's electrical demand. (For perspective, that would be a plot 76 miles by 76 miles, or three states of Delaware put together.)

Is this an unreasonably large space to dedicate to collecting energy from a permanent source that is cost-free and pollution-free? The U.S. military occupies about five times more land for its bases and nuclear testing. Nearly forty times more land is left unplanted each year to keep crops from being overproduced. Coal-based energy exploits considerably more than 5790 square miles when we include the land that is mined. About 1500 square miles of U.S. land is used just to raise Christmas trees each year.

Further, that 5790 square miles will shrink as efficiencies improve. Research is proceeding so rapidly that estimates of cost and efficiency of silicon-based solar panels and the space they require soon could be obsolete. In 1994, an organic polymer film that generates electricity was developed that could provide electricity 99 percent cheaper than the lowest-cost power now available. The film can be rolled like wallpaper for easy transport, but when laid flat in sunlight, it converts solar energy into electrical energy.

Solar panels can be placed on roofs of homes to generate supplemental power, or even primary power for homes remote from electric power lines. The newer solar panels are the same size and shape as roof shingles and can be used to cover a roof instead of normal shingles. But use of solar power seems to be growing more rapidly in countries other than the United States. Between 1981 and 1990, the U.S. share of the worldwide solar-cell market fell from 75 percent to 32 percent, while Japan's grew from 15 percent to 32 percent. Is this yet another area in which the Japanese are show-

ing more foresight than Americans? What do you think the relative degrees of energy independence of the two countries is likely to be in 50 to 100 years when the supply of fossil fuels becomes unaffordable?

Since 1980, world-wide shipments of photovoltaic cells have increased from 6.5 megawatts to 69 megawatts, increasing 15 percent between 1993 and 1994 alone. In 1995, the world's largest producer of solar cells expanded its production capacity by nearly 50 percent. If this trend continues, as seems likely, photovoltaics could soon become one of the world's largest industries, and a commonplace energy source.

The Best Energy Mix

A wide range of energy sources is available to power a modern industrial society. Two fundamental choices are possible: (1) continue to develop and use the nonrenewable, highly polluting sources such as crude oil and coal, or (2) quickly develop relatively non-polluting, renewable sources, such as wind and solar power. Because of the time required to develop and implement alternative ways of doing things, and the politics of funding, and the dominance of powerful lobbies for fossil fuels (the oil companies and coal companies), the major energy sources for at least the next few decades will be the same as today's: fossil fuels. But there is no doubt that renewable energy sources are the wave of the future.

No single energy source is best at all times and in all locations. Solar power obviously lacks the potential in Alaska that it has in Alabama. Geothermal energy is not a viable option in Kansas. Biomass will not fly in arid western Texas and southern New Mexico. Probably no single alternative energy source can be used exclusively in any location. What is needed is an energy-generation mix, depending on location and the season of the year. In North Dakota, a good mix might be wind power, supplemented by solar power during the less windy but sunnier summer months. In Nova Scotia, a good mix might be tidal power from the Bay of Fundy plus wind power. In Florida, the Sunshine State, solar energy might be supplemented by power from an OTEC facility. States and municipalities need to consider these long-range options as soon as possible.

Social Cost of Power Generation

Most statements concerning the relative cost of different energy sources do not consider their "social cost." This includes land disruption and environmental pollution of the air, soil, and water. For example, the cost of electric power generated by coal does not include the cost of damage and remediation from land subsidence or of mine fires that continue to smolder. The cost of exploiting subsurface petroleum or geothermal power does not include the cost of remediating the damage caused by subsequent land subsidence.

If federal laws require energy companies to pay these "hidden costs" over the next few decades, the real cost of fossil fuels is certain to increase relative to renewable sources. Many industries already are factoring into the price of their products the cost of remediating the environmental damage they cause. Soon, most industries will have to do so.

Fossil Fuels Versus Renewable Energy Sources: The Public's View

A 1994 poll indicated that more than three-quarters of Americans want something done about U.S. dependence on foreign oil. Among the options, renewable energy had the most support (42 percent), with energy-efficient technologies and energy conservation a solid second (22 percent). Natural gas had only slight backing. Fossil fuels and nuclear power each showed a negative net score. Cutting federal funding for nuclear power and fossil fuel programs was favored by 73 percent as a way to reduce the Department of Energy's budget.

The survey also asked whether those questioned would be willing to pay more for electricity generated from renewable sources. Three out of four said yes. Americans have consistently shown support for renewable energy. Nevertheless, the federal budget for renewable energy technologies was cut 29 percent from 1995 to 1996. How did your federal representatives vote on this issue? Why?

Summary

Fossil fuels are limited in supply and are major pollutants. They must be replaced by renewable and less polluting sources of energy over the coming decades. The sooner society accomplishes this the better. Promising technologies are those that obtain energy from water, wind, biomass conversion, and solar radiation.

Energy can be obtained from water in several ways. Hydroelectric power means the conversion of the kinetic energy of moving stream water to mechanical energy, a process that usually requires the construction of dams across large streams. The number of suitable streams is not large enough for hydropower to be the major source of energy for American society. A deficiency of suitable locations also limits the harnessing of the energy of ocean tides. Ocean waves might be usefully harnessed, but the technology to do this efficiently is not yet developed.

The potential for geothermal power is greatest in geographic areas where magma is within about 10,000 feet of the surface and the plumbing system for the groundwater in the rocks above the magma is adequate to support extensive water removal. However, this commonly causes land subsidence above the removal area. Another possible source of thermal energy is the temperature difference at different depths in the ocean (OTEC). Efficient technology for this is not presently available.

Wind power has a large potential as an energy source and is currently in a phase of rapid development, particularly in California. Wind farms are also sprouting in the American Midwest and experimental wind farms are being erected in some offshore areas where wind speeds are higher than on the adjacent land surface.

Direct burning of biomass for heat is a much more common practice in impoverished nations than in developed ones. In some countries, forests are being decimated at an alarming rate, greatly increasing the rate of erosion and loss of topsoil needed to increase crop yields. The people may be warmed temporarily but may starve eventually. Biomass can be converted into gaseous and liquid fuels and this can be locally important.

The most important alternative energy source is solar radiation. Technological progress to date has reduced the cost of electricity generation from solar cells to near that of fossil fuels. Within a few years, solar electricity will cost less than electricity generated by nuclear reactors (Chapter 16) and cost about the same as electricity generated from other fuels, both fossil and alternative. When the non-polluting nature of solar power is factored in, there seems good reason to begin large-scale conversion to solar energy as a source of electric power.

Key Terms

active solar heating
alternative energy sources
biodigestion
biogas

bioliquids
biomass
biomass energy
fuelwood

greenhouse effect
heat pump
hydroelectric power
nonrenewable resources

oceanic thermal energy conversion
OTEC
OSPREY
passive solar heating

photovoltaic cells
renewable resources
solar cells
solar collectors

solar panel
tidal power
wind farms

Stop and Think!

1. In what ways can the power of moving water be harnessed to generate electrical power?
2. What are the drawbacks of large hydroelectric dams?
3. Describe the operation of a plant that generates electricity from the ocean tides, and where it might be located.
4. Describe the relationship between plate boundaries and geothermal energy.
5. What problems are associated with geothermal energy?
6. Describe how a heat pump is used for both heating and cooling.
7. Explain how oceanic thermal energy conversion (OTEC) works.
8. Which areas in the United States are most suitable for harnessing the wind as a source of electrical power?
9. Compare the environmental effect of using dead plants versus live plants for biomass heating and electric power generation.
10. What is the difference between biogas and bioliquids?
11. Explain the principle of passive solar heating. How does it differ from active solar heating?
12. Explain the two methods of thermal electric power generation.
13. What are photovoltaic cells? How efficient are they at present?
14. What do you believe would be the best energy mix (electrical power generation methods) for the area where you live? Explain your reasons.

References and Suggested Readings

Anonymous. 1991. *America's Energy Choices.* Cambridge, Massachusetts: The Union of Concerned Scientists, 124 pp.

Anonymous. 1991. *Green Energy: Biomass Fuels and the Environment.* New York: United Nations Publications, 54 pp.

Bowen, R. 1989. *Geothermal Resources, 2nd ed.,* New York: Elsevier, 485 pp.

Brower, Michael. 1992. *Cool Energy: Renewable Solutions to Environmental Problems.* Cambridge, Massachusetts: The MIT Press, 219 pp.

Davis, G. R. 1990. Energy for planet Earth. *Scientific American,* March, p. 55–74.

Duffield, J. H., J. H. Sass, and M. L. Sorey. 1994. Tapping the earth's natural heat. *U.S. Geological Survey Circular* 1125, 63 pp.

Elliott, D. L., C. G. Holladay, W. R. Barchet, H. P. Foote, and W. F. Sandusky. 1987. *Wind Energy Resource Atlas of the United States.* Golden, Colorado: Solar Energy Research Institute , 210 pp.

Gray, C. L., Jr. and J. A. Alson. 1989. The case for methanol. *Scientific American,* November, p. 108–114.

Greenberg, D. A. 1987. Modeling tidal power. *Scientific American,* November, p. 128–131.

Kammen, D. M. 1995. Cookstoves for the developing world. *Scientific American,* July, 72–75.

Kozloff, K. L. and R. C. Dower. 1993. *A New Power Base: Renewable Energy Policies for the Nineties and Beyond.* Washington, D.C.: World Resources Institute, 196 pp.

Mclarty, L. and Reed, M. L. 1992. The U.S. geothermal industry: three decades of growth. *Energy Sources,* v. 14, p. 443–455.

Moretti, P. M. 1986. Modern windmills. *Scientific American,* June, p. 110–118.

Pearce, Fred. 1996. Trouble bubbles for hydropower. *New Scientist,* May 4, p. 28–31.

Penney, T. R. and Desikan B. 1987. Power from the sea. *Scientific American,* January, p. 86–92.

Weinberg, C. J. and R. H. Williams. 1990. Energy from the sun. *Scientific American,* September, p. 146–155.

"Nuclear fission energy is safe only if a number of critical devices work as they should, if a number of people in key positions follow all their instructions, if there is no sabotage, no hijacking of the transport, if no reactor fuel processing plant or repository anywhere in the world is situated in a region of riots or guerrilla activity, and no revolution or war—even a "conventional" one—takes place in these regions. No acts of God can be permitted."

Hannes Alfven (Nobel Laureate, Physics)

Energy from Nuclear Power
16

Core meltdown! Radioactive contamination! Leakage from nuclear power plants! Radiation sickness! Genetic damage! Leukemia! Radon in homes! Scary stuff, with a serious basis: radioactivity is dangerous, whether inside a nuclear power plant, at a nuclear waste site, or in your dwelling.

You are being continuously bombarded by radiation right now, no matter where you are (although the radiation probably is very light). A radiation counter placed in your dorm room, classroom, or home would indicate continual, random radioactivity (radiation-activity). Most of it is in low doses, which probably are harmless. But at higher doses, this same radiation is indisputably hazardous.

Radiation comes from natural sources, including both space and Earth. Earth's rocks and sediments contain elements that *always* are radioactive, like potassium, uranium, and radon gas, and elements that *sometimes* are radioactive, like carbon. Radiation also comes from artificial sources, including hospitals (nuclear medicine) and nuclear power plants.

In the preceding two chapters, we looked at fossil fuels, solar energy, and other sources for generating electricity. In this chapter, we focus on another alternative for generating electricity: nuclear energy. From its beginnings in the 1940s, nuclear energy has been very controversial. Exactly what is radiation? Why is it a hazard? How much radiation will sicken or kill you, and how long will it take? Do nuclear power plants release radiation? (They can.) Can nuclear power plants explode like a nuclear bomb? (No, but they can have nasty steam explosions that scatter radioactive materials.)

How and where should radioactive waste be isolated? (We've been arguing about that for half a century.) Can nuclear energy be a safe, clean source of electrical power? (Read again the quotation that opens this chapter.) Can we trust our "nuclear watchdogs"? (They include scientists, activist groups, power companies, the military, and the Nuclear Regulatory Commission.)

We also examine the radon gas problem. Where does radon come from? What's the radon level in your classrooms, dorm room, or home? Should you test your living space for radon contamination?

◀

Reindeer herd in Scandinavia contaminated by the nuclear accident at Chernobyl, Ukraine.

Photo: Karen Kasmauski/Matrix International.

What Is Radioactivity?

The 90 naturally occurring elements, plus more than a dozen created in nuclear reactors, all consist of atoms. Each of these elements has some atoms that are stable, and others that are unstable. This chapter is about the unstable ones, including some you have heard about, like uranium, plutonium, and radon. Unstable atoms seek to stabilize themselves by expelling tiny particles and bursts of energy, collectively called **radioactivity.** This process is called **radioactive decay** because the atom loses particles and energy, and readjusts internally. Figure 16.1 illustrates the process of radioactive decay.

Uranium is a notable example. Unstable uranium atoms keep disintegrating (decaying) until they eventually become atoms of a stable metal, lead. This does not happen all at once: individual atoms disintegrate at random, in steps. Depending on the element, the process of attaining stability can take seconds, days, years, or even billions of years. It is this radiation energy that we put to work for us, using it to heat water in a power plant, making steam to generate electricity.

As stated, every element has some atoms that are stable and some that are unstable, or radioactive. These different forms of an element are called isotopes. Isotopes of an element always have the same number of protons in their nucleus (that is what gives an element its identity), but they vary in the number of neutrons. An example is the oxygen in your lungs right now. Oxygen has six isotopes. All oxygen atoms have eight protons in their nucleus, but their neutron count varies from six to eleven. Three of oxygen's isotopes are stable, and three are unstable (radioactive). The nitrogen in your lungs is similar, also with six isotopes (two stable, four radioactive). Fortunately, nearly 100 percent of the oxygen and nitrogen in your lungs consists of their stable isotopes. The unstable, radioactive isotopes of these gases are rare and short-lived, so the oxygen and nitrogen in your lungs is radioactive only at ultra-low levels.

The carbon that makes up every cell in your body can include six isotopes (two stable, four unstable), so

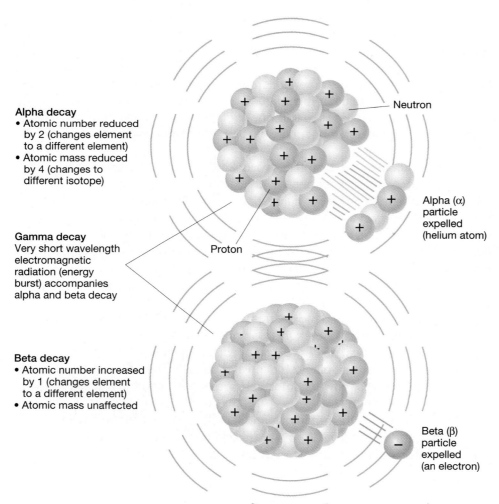

Alpha decay
- Atomic number reduced by 2 (changes element to a different element)
- Atomic mass reduced by 4 (changes to different isotope)

Gamma decay
Very short wavelength electromagnetic radiation (energy burst) accompanies alpha and beta decay

Beta decay
- Atomic number increased by 1 (changes element to a different element)
- Atomic mass unaffected

Neutron

Proton

Alpha (α) particle expelled (helium atom)

Beta (β) particle expelled (an electron)

FIGURE 16.1 How do unstable radioactive atoms seek stability?

They *decay* as shown, by expelling particles or bursts of energy, which we collectively call *radioactivity.* Illustrated are three ways in which an atom can decay. An *alpha particle* is emitted from an atom's nucleus. The particle is two protons and two neutrons, and its departure changes the name of the element. A *beta particle* leaves an atom's nucleus. It is an electron, and its departure also changes the name of the element (because the electron is formed by the decay of a neutron in the nucleus into a proton and an electron). A *gamma ray* leaves the nucleus. It is not a particle, but a high-energy burst like a radio wave or X ray.

your entire body is radioactive, again at low levels. The great danger in radioactivity lies in isotopes of other elements, like uranium and plutonium, which radiate far more strongly, as you shall see shortly.

Discovering the Invisible

Radioactivity is invisible, and its effects generally are subtle, so it was discovered only a century ago. Radioactivity's discovery is a fascinating example of how serendipity (dumb luck) can play an unexpected role in scientific research. In 1895, German physicist Wilhelm Roentgen (pronounced RENT-gen) discovered X rays. X rays are similar to light waves, but their wavelength is too short to be seen by our eyes. They also have much greater penetrating power than light.

Roentgen quickly grasped the practical importance of his discovery: ordinary light cannot pass through the human body very well, but X rays can. X rays have enough energy to penetrate flesh, although not bones. So, putting an X ray generator in front of your chest and photographic film on your back will produce on the film a clear image of your broken ribs. Soon X rays were being used in medicine, to peer inside the body.

Other scientists sensed that Roentgen's discovery was something big, and they experimented to learn more. French scientist Henri Becquerel (BECK-uh-rell) knew that some substances **fluoresce** (glow or give off light) when stimulated by exposure to sunlight. He wondered if these fluorescent substances also might give off X rays, in addition to the light. A way to find out was to place a fluorescent substance in sunlight, and place a photographic plate beneath it. If the sunlight made the substance generate X rays, their tracks would show up on the plate.

Becquerel chose a fluorescent mineral called pitchblende, a uranium sulfate. In bright sunlight, he sprinkled a layer of the mineral on top of a photographic plate wrapped in black paper to keep it from being exposed by the sunlight. If the mineral radiated X rays, they would pass through the black paper and expose the photographic plate. When he developed the plate, sure enough, it showed streaks that were X rays, or so he thought.

The weather changed and Becquerel had to interrupt his experiments, for clouds and rain dimmed the sunlight. He put his materials in a desk drawer. A few days later he developed the plate, expecting to see faint evidence of the X rays. But he was amazed to see a strong image of the pitchblende sample on the plate! Becquerel realized that the mineral was producing an intense spontaneous radiation all by itself, independent of sunlight. But was it X rays, or something else?

A young student, Marie Curie, joined Becquerel to do work toward her doctorate. In those days, this was extraordinary; physics was "a guy thing." Becquerel asked Curie to work on the X ray-like phenomenon he had discovered, which she named radioactivity. As Marie and her husband, Pierre, explored radioactivity, they discovered new radioactive elements, named radium (because it radiated) and polonium (in honor of her native Poland). The result of her doctoral research was described as the single greatest contribution made by a doctoral thesis in the history of science (Figure 16.2).

The danger of radioactivity was not fully recognized at first. Early researchers like the Curies handled radioactive substances freely, their hands growing red and raw from radiation burns. At least 330 early workers succumbed to cancer from handling radioactive substances, including Marie and Irène, her daughter

FIGURE 16.2 Marie Curie at work in her laboratory.

In 1903, she and her husband Pierre, with Henri Becquerel, jointly received the Nobel Prize in physics. In 1911, Marie won the Nobel Prize in chemistry. Her daughter Irène also studied radioactivity and received a Nobel Prize in 1935.
Photo: Science VU-NLM/Visuals Unlimited.

and fellow-researcher. Both died of leukemia caused by the radiation they had spent their lives studying. (Pierre Curie did not live long enough to die from radiation sickness. He was crushed under the wheels of a horse-drawn wagon in 1906 at age 47.

Because the radiation–cancer link was not understood, radioactive substances were used in industry without proper protection. During World War I (1914–1918), fluorescent instrument dials became important for night use. The dials were painted with luminescent (glow-in-the-dark) radium compounds. At the U.S. Radium Corporation in New Jersey, "dial painters" used their lips to shape a point on their paint brushes, ingesting tiny amounts of radioactive radium. By 1931, eighteen cancer deaths of dial painters had been attributed to ingesting the radioactive material. The hazard to living organisms of the nuclear decay process was discovered the hard way.

Types of Radiation

Researchers identified three distinct types of radiation, named simply A, B, and C using the Greek alphabet: alpha (α) radiation, beta (β) radiation, and gamma (γ) radiation. They differ in penetrating ability, which means that they pose different hazards.

Alpha (α) particles are tiny particles of matter, consisting of two protons and two neutrons, actually atoms of helium gas (Figure 16.1). Alpha decay occurs mostly in large, complex atoms, those that have high atomic numbers starting with number 83, the metal bismuth (see the Periodic Table of the Elements in Chapter 2, Table 2.1). Major alpha radiators include uranium and plutonium (used in power plants and nuclear weapons). Alpha particles are heavy (atomic mass of 4) and thus have very low penetrating power. They can travel in air only two to three inches from their source before they are stopped by colliding with air molecules. They cannot even pass through a sheet of paper (Figure 16.3)!

If uranium and plutonium emitted only alpha particles as they disintegrated, you could safely carry chunks of them in your jeans. (However, they also emit more powerful beta and gamma radiation.) In relatively dense human tissue, such as your skin, alpha particles can travel only one-thousandth as far as in air (0.002–0.003 inches). Thus, if they are to cause skin damage, they must be generated directly on your skin or within your body. The low penetrating power of alpha rays does not make them harmless. They dissipate their energy by breaking chemical bonds in your cell tissue, setting the stage for cancer. Your daily exposure to alpha radiation comes from small amounts of uranium and radon gas that pervade our environment.

Beta (β) particles are fast-moving electrons emitted from a disintegrating atom. They are 100 times more penetrating than alphas, able to pass through about 0.1–0.2 inches of human skin. They form when a neutron (no charge) in the atomic nucleus spontaneously breaks up into a proton (positive charge) and

FIGURE 16.3 Radiation varies in energy and in its ability to penetrate.

Alpha particles are low-energy and can be stopped by a sheet of paper. Beta particles are higher-energy and more penetrating. Gamma rays have very high energy and can penetrate metals, depending on the metal and its thickness. *After United Nations, 1985.*

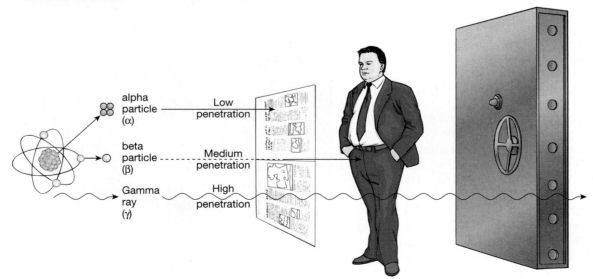

an electron (the beta particle). The proton remains in the atomic nucleus, adding to the atomic number and thus creating a new element. (An element is defined by the number of protons it has; see Chapter 2.)

Isotopes of virtually every element expel beta particles, from the lightest element (hydrogen) to the heaviest elements generated in nuclear reactors. Consequently, beta radiation is common in your daily environment. For example, the element potassium is found in many foods you eat and is common in many rocks, such as feldspar—and potassium has seven isotopes (three stable, four radioactive). You cannot escape radiation, no matter where you live or what you eat.

When unstable atoms that emit alpha or beta radiation enter your body through eating, drinking, or breathing, they may do serious internal damage. The damage will be localized in the part of the body where these atoms become concentrated. Internal human tissue is not as dense as your outer skin and is less able to resist penetration by alpha and beta particles. Obviously, you should not ingest or inhale alpha or beta emitters—but how are you going to avoid them? As noted, isotopes of virtually every element expel beta particles.

Gamma (γ) rays are not particles but energy waves, similar to light and X rays. They have very short wavelength and high energy (Figure 16.4). Gamma rays are emitted from most radioactive isotopes as they decay. They have very high penetrating power (10,000 times that of alphas) because they have such short wavelengths. A thick shield of lead or concrete is required to stop high-energy, penetrating gamma rays.

Natural Radiation Sources

Despite all the talk about nuclear power plants, waste dumps, and nuclear warfare, our greatest dose of radiation by far comes from natural sources, as shown in Figure 16.5. Throughout Earth's history, terrestrial radiation has risen from radioactive materials in the crust and cosmic radiation has arrived from space. You are irradiated from the outside when you stand on a rock or sediment that contains uranium, or from the inside when you eat a food that contains the element potassium.

Which foods contain potassium? Apples, bananas, chicken, raisins, peanuts, potatoes, tomatoes, and dietetic salt substitutes all are high in potassium. A small percentage—0.01 percent—of all potassium is the radioactive isotope potassium-40. In fact, most of the irradiation your body receives is internal. If you were a dead laboratory animal, the amount of radioactive potassium and carbon present in your body would require you to be disposed of as low-level nuclear waste!

Although everyone receives natural radiation, some people receive much more than others. The level of naturally occurring radiation is referred to as the **background level**. It varies with the height above sea level (more cosmic radiation at higher altitudes), the rocks you live on (granite contains more uranium than other rock

FIGURE 16.5 Where does radiation come from?

Most is natural, coming from Earth itself (terrestrial) or from space (cosmic). Human sources are mostly medical, with very little from nuclear power plants or nuclear weapons (so far). The numbers shown are average annual doses in rems. *After United Nations, 1985.*

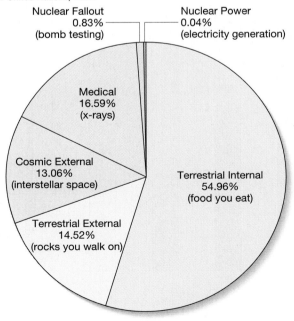

Nuclear Fallout 0.83% (bomb testing)

Nuclear Power 0.04% (electricity generation)

Medical 16.59% (x-rays)

Cosmic External 13.06% (interstellar space)

Terrestrial Internal 54.96% (food you eat)

Terrestrial External 14.52% (rocks you walk on)

FIGURE 16.4 Shorter-wavelength waves have greater energy.

The radio and TV waves we use daily have long wavelengths and are not very penetrating. Visible light has greater energy, but its penetrating power is low; light is easily blocked by most objects. X rays are used to study your skeleton because their shorter wavelength and greater energy allow them to penetrate your flesh. Gamma and cosmic rays are very high in energy, and dangerously penetrating.

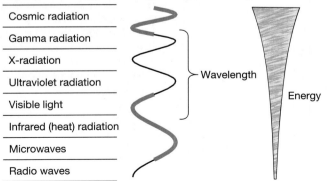

Cosmic radiation

Gamma radiation

X-radiation

Ultraviolet radiation

Visible light

Infrared (heat) radiation

Microwaves

Radio waves

Wavelength

Energy

types), whether your house is made of brick (higher level) or wood (lower level), whether you burn coal for heating (radioactivity concentrates in the ash), and whether you use natural phosphate fertilizer (it contains radioactive elements). There is no escape from natural radiation.

Get a Half-Life!

Each radioactive element disintegrates or decays differently. Some give off alpha particles, some beta particles, some gamma rays, and some a combination. Some decay from an unstable isotope to a stable one in a single step. Others decay through a sequence that produces many intermediate unstable elements before changing into a stable isotope. Isotopes have different lifespans because they decay at such different rates. Scientists measure the time it takes for an unstable isotope to decay as its **half-life,** the time required for half of the starting amount to disappear.

Some half-lives are tiny fractions of a second; others are thousands of years, and still others are many millions (uranium-235) or even billions of years (uranium-238). The idea of half-life is shown in Figure 16.6. We use the example of plutonium, because its long half-life is partly what makes it an environmental problem. A

curve like the one for plutonium could be drawn for any radioactive element; the only difference would be the time involved. For example, the brief half-life of radon gas is only 3.8 days, so most of it is gone after 10 half-lives, or about thirty-eight days. But uranium-238, with its half-life of 4,470,000,000 years, would be mostly gone after 10 half-lives, or 44,700,000,000 years, which is three or four times longer than the Universe has existed!

There is no way to tell when an individual atom of an unstable element will decay, for it occurs at random. But over time, a specific percentage of atoms will decay. It is analogous to flipping a coin. You cannot predict which flip will come up heads and which tails, but if you make a very large number of flips, half will be heads and half tails.

Consider uranium-238, which has a half-life of almost 4.5 billion years (Figure 16.7). The first step in its breakdown occurs when it expels an alpha particle (two protons and two neutrons). This loss changes it from the element uranium (92 protons) to the element thorium (90 protons), with a mass loss from 238 down to 234. Look at the half-life column and you can see that this is by far the slowest step in the long chain of decay events, requiring almost 5 billion years.

Thorium-234, like uranium-238, is unstable. Half of it decays to protactinium-234, but in only 24.1 days.

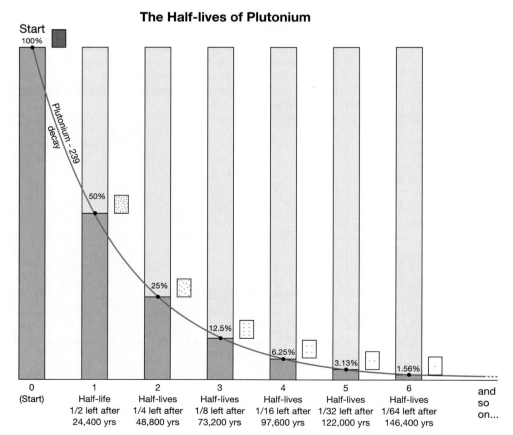

The Half-lives of Plutonium

| 0 (Start) | 1 Half-life 1/2 left after 24,400 yrs | 2 Half-lives 1/4 left after 48,800 yrs | 3 Half-lives 1/8 left after 73,200 yrs | 4 Half-lives 1/16 left after 97,600 yrs | 5 Half-lives 1/32 left after 122,000 yrs | 6 Half-lives 1/64 left after 146,400 yrs | and so on... |

Start 100%
50%
25%
12.5%
6.25%
3.13%
1.56%

Plutonium - 239 decay

FIGURE 16.6 Radioactive decay of plutonium-239.

Its half-life is 24,400 years, meaning that only half of it will be disintegrated into other elements after 24,400 years. This demonstrates why nuclear waste disposal is such a problem: 24,400 years is far longer than any civilization has existed. Of all the plutonium that now exists, in weapons and in nuclear power plant fuel rods, half will remain in the year A.D. 26,400! It takes ten half-lives to diminish any radioactive isotope to around 0.1 percent of its original size; in the case of plutonium, that is 10 × 24,400 = 244,000 years! Our nuclear waste probably will outlive any country now existing, and has a good chance of outliving our species.

Atomic Number (No. of protons)	Atomic Weight (Protons and neutrons)	Element and major radiation	Half-life			
			Seconds	Minutes	Days	Years
92	238	Uranium (α, γ)				4,470,000,000
90	234	Thorium (β, γ)			24.1	
91	234	Protactinium (β, γ)		1.17		
92	234	Uranium (α, γ)				245,000
90	230	Thorium (α, γ)				8,000
88	226	Radium (α, γ)				1,600
86	222	Radon (gas) (α, γ)			3.823	
84	218	Polonium (α, β)		3.05		
82	214	Lead (β, γ)		26.8		
83	214	Bismuth (β, γ)		19.7		
84	214	Polonium (α)	0.000164			
82	210	Lead (β, γ)				22.3
83	210	Bismuth (α, β)			5.01	
84	210	Polonium (α, γ)			138.4	
82	206	Lead	Stable after all those years			

FIGURE 16.7 A nuclear decay chain for uranium-238.

It begins with unstable uranium-238, which has the extremely long half-life shown. Note that it begins with ninety-two protons and an atomic weight of 238. At random, this unstable atom will expel an alpha particle, dropping the proton count to ninety, turning itself into the element thorium. The new thorium atom, during its brief half-life of about 3½ weeks, expels beta particles, changing itself into the element protactinium. And so on, and on, until we arrive at stable lead. Note halfway down the chain that radon gas is generated. Eventually, all the uranium-238 will decay to stable lead-206.

The protactinium is so unstable that half of it is gone in only 1.17 minutes. This long series of steps continues until, after nearly 5 billion years, polonium-210 decays to lead-206, a stable isotope at long last. Note that some decay steps show the atoms expelling alpha particles, some beta particles, and most gamma rays.

What Is the Biological Impact of Radiation?

A century ago, radiation was believed to be healthful. In 1899, a natatorium (spa) opened in Idaho Springs, Colorado (Figure 16.8). Tourists and locals believed the mineral water from local springs to be healthful, and consumed it heartily. Their belief was based on the water's content of significant radium, a radioactive element. In the 1920s, Idaho Springs was touted as a sanitarium for polio called "The Radium Hot Springs." Advertising stated "Come and drink the water. It is FREE. The radium water, taken internally, radiates outward, as if the sun were shining out from inside." Celebrities frequented Radium Hot Springs and the adjoining hotel.

Given the scientific and medical knowledge at the time, no cancers or other harmful effects of drinking the radium-laced water were recognized. But each atom of radium consumed in that water was—and still

FIGURE 16.8 The Idaho Springs Natatorium, near Denver.

In the early 1900s, radium-laced water was consumed here—intentionally. The business should have been named "Radiation Hazards Are Us."

Photo: Denver Public Library Western History Collection.

is—in the process of giving off six alpha particles, five beta particles, and a bunch of gamma rays during its decay to stable lead. This story illustrates why responsible scientists are so cautious in giving opinions on little-understood phenomena. Sometimes we don't know what we don't know.

You don't have to drink radium water to get your daily dose. Every plant and animal on Earth is continuously exposed to natural radiation from the sun, rock, and soil. The amount that reaches us from these sources is considered normal, because we have adapted to it over many thousands of years. We assume it to be harmless because we are unaware of disease that can be traced to natural radiation levels. However, many studies have demonstrated the harmful effects of radiation. It may be that natural radiation is one more component of our "normal" environment that limits the life span of all organisms.

The Hazards of Ionizing Radiation

Alpha, beta, and gamma radiation is destructive to living cells because it dislodges electrons from atoms in the cells. An atom with one or more electrons removed suddenly becomes a positively charged ion. Consequently, we call such destructive radiation **ionizing radiation.** The positively charged ions behave different chemically, and therein lies their hazard. The new ions cause chemical bonds to break in large molecules that carry out life processes. If radiation is acute, the bone marrow that produces red blood cells is destroyed, reducing the number of red blood cells, causing anemia.

Tissues *inside* a human body are easily damaged. Hence, if radioactive dust particles are inhaled, delicate tissues are directly exposed to radiation, and lung damage results. The radioactive materials actually become incorporated into the lung tissue and reside there for a considerable time, continually emitting damaging radiation as they decay.

The major cell damage results from the alpha particles. They are not very penetrating, but they have a large mass (recall that they are composed of two protons and two neutrons) and they interact exceptionally well with lung tissue. They penetrate the cells and either kill them outright or affect the DNA, causing mutations that result in cancer.

Radiation-induced genetic damage also is of great concern. Such damage may not become apparent for many years after exposure. As we have learned more about the effects of ionizing radiation, the dosage once considered safe has been lowered steadily. For example, the U.S. Atomic Energy Commission has dropped the maximum permissible concentration of some radioactive isotopes to *less than one ten-thousandth* of levels considered safe in the early 1950s. Small amounts of radiation cause only minor damage to DNA and are repairable by the body. But, above some threshold yet to be determined, radiation damage to DNA becomes irreparable.

Radiation received by people is expressed in rads and rems. **Rad** (Radiation Absorbed Dose) measures the overall ionizing radiation absorbed by tissue. But radiation varies in penetration, so different damage can result from the same number of rads. As a more practical unit, **rem** (Roentgen Equivalent Man) is the amount of ionizing radiation having the same biological effect as one unit of X rays. To measure the small doses common in our environment, we use the *millirad* and *millirem* (one-thousandth of a rad or rem).

Your daily dose of radiation comes from cosmic rays and from decaying radioactive elements in rocks and soil. It enters your body directly through your skin (particularly high-energy gamma rays), and indirectly in your food, water, and air. This is illustrated in Figure 16.9.

The annual radiation dose to which we all are treated is almost 0.3 rem, or 300 millirem (Figure 16.5). The greater the rem dosage, the greater is the degree of biological damage. Table 16.1 shows the effects of increasing radiation exposure.

In high doses, radiation can cause enough damage to prevent cell division. This is the principle used to halt the growth of cancer cells in *nuclear medicine*. The radiation from the isotope cobalt-60 is focused on a cancerous tumor to destroy its ability to reproduce by cell division.

Alpha, beta, and gamma rays all are energetic enough to cause such disturbances in your body's atoms. For relative biological effectiveness, alpha particles are assigned a value of 10, and beta and gamma

rays are given values of 1. This may seem backwards, for gamma rays are the most penetrating. But an alpha particle is so large (it is a helium atom) that it leaves a wider trail of damage along its path.

Still, gamma rays are particularly dangerous because of their high penetrating power. An indication of this risk comes from the similar but less energetic X rays used by doctors and dentists. You may have observed an X ray technician at a hospital or your dentist's office wearing a heavy apron that contains the metal lead. Lead is so dense that it absorbs alpha, beta, and some gamma radiation. The doses used in medical and dental X rays are small and believed to be safe, but the technician does this work continuously and must be protected from cumulative effects of the radiation. When undergoing an X ray, you may be protected with a lead apron, too. This shields you from any radiation hazard, and shields the medical facility from a lawsuit.

Radiation Sickness

If the whole body is exposed to uncontrolled high levels of radiation, as in a nuclear weapon explosion or

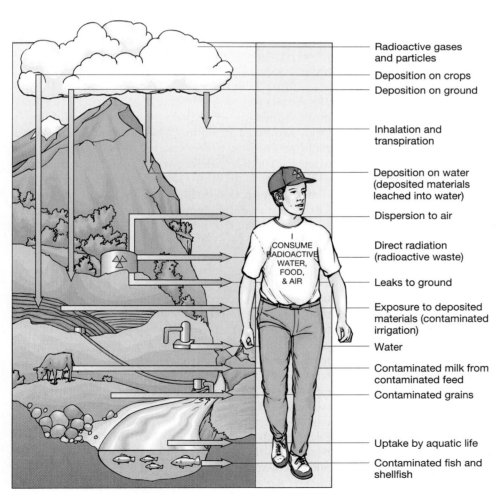

FIGURE 16.9 Radioactive pathways into your body.

Radiation permeates our environment, air, food, and water, so it permeates us, too.

Radioactive gases and particles
Deposition on crops
Deposition on ground

Inhalation and transpiration

Deposition on water (deposited materials leached into water)

Dispersion to air

Direct radiation (radioactive waste)

Leaks to ground

Exposure to deposited materials (contaminated irrigation)
Water

Contaminated milk from contaminated feed
Contaminated grains

Uptake by aquatic life

Contaminated fish and shellfish

TABLE 16.1 The Relationship Between Dosage of Ionizing (High-Energy) Radiation and Biological Damage to Humans

Other, longer-term effects of high doses of radiation include eye cataracts, cancers of various types, genetic birth defects in children, and sterility in adults. The relations between dosage and the time required for these problems to develop is unclear.

Dose (rems)	Effects
0.1–0.5	None detected. This is the total of natural background radiation plus a medical and dental X ray.
0.5–25	Dosage near 25 rem may reduce the white blood cell count.
25–100	Nausea and vomiting for about half of those exposed. Some blood changes. Not fatal.
100–200	Nausea, vomiting, fatigue, low white blood cell count (increased susceptibility to infection). Long-term risk of cancer increases.
200–500	Low white blood cell count. Blotchy skin in 4–6 weeks. Lethal for 50 percent of those exposed, especially without treatment. Bone marrow and spleen (blood-forming organs) damaged.
500–1000	White blood cells destroyed. Chance of death 80–100 percent.
1000–5000	Bowel malfunction, diarrhea, fever. Blood chemistry upset. Damage to central nervous system. Death in 1–14 days.

severe radiation leak from a nuclear power plant, a generalized blockage of cell division occurs. This prevents the body's normal replacement and repair of blood, skin, and other tissues. The result is **radiation sickness,** which may result in death if the radiation dose is great enough.

In cases of extreme exposure, radiation sickness and death may be quite rapid. The most dreadful example in history is the nuclear weapon attacks on the cities of Hiroshima and Nagasaki, Japan, in 1945 (Figure 16.10). These attacks, which exploded just two "atomic bombs," as they were called, ended World War II very quickly. This was the world's first, and so far only, use of nuclear weaponry. The two bombs killed about 150,000 people outright and immersed 200,000 more in intense ionizing radiation; they died within five years.

In cases of lesser exposure, noticeable biological injury may not occur for months or even years. In 1986, the Chernobyl nuclear power plant in Ukraine experienced a steam explosion. The biological effects of radiation scattered by the explosion are becoming clear only now, as discussed later in this chapter.

Uranium: Our Source of Nuclear Energy

The source of our nuclear energy, uranium, is spread widely but thinly through Earth's crust. It constitutes only 0.00016 percent of the average crustal rock. In pure form, it is a heavy metal, but it is too chemically active to occur as a metal in nature, like gold, silver, or platinum. Instead, it always is combined with other elements to form uranium minerals, such as uraninite (common name pitchblende) and carnotite.

Most U.S. uranium ore is concentrated in the sandstones of New Mexico and Wyoming. Millions of years ago, uranium-bearing groundwater passed through these sandstones and left its load of uranium atoms in the form of carnotite (Figure 16.11). This often happened where petrified trees now exist, because carbon in the wood removed oxygen from the water, forcing the uranium to precipitate.

During the 1950s and 1960s, the U.S. and the former Soviet Union were in a heated arms race. Each built thousands of nuclear weapons, demanding a large supply of uranium. Also, electric utilities were building nuclear power plants, which required uranium for fuel. A "uranium rush" ensued, reminiscent of the 1850s Gold Rush. Miners probed petrified logs with radiation detectors. Some logs contained small fortunes in uranium, but most of the uranium lay in the sandstones beneath the logs.

Why Uranium?

Uranium is the densest natural element, even denser than lead (the military sometimes uses uranium bullets for greater impact). Uranium is heavy because its chubby nucleus contains the most protons and neutrons of any natural element. Every uranium atom contains ninety-two protons, plus many neutrons, depending on which of uranium's ten unstable isotopes we consider. Of greatest interest are uranium-238, which has 146 neutrons, and uranium-235, which has 143. This insignificant-sounding difference—just three neutrons per atom—made Earth a far more dangerous habitat, once we learned how to exploit the difference to build nuclear weapons and generate nuclear power.

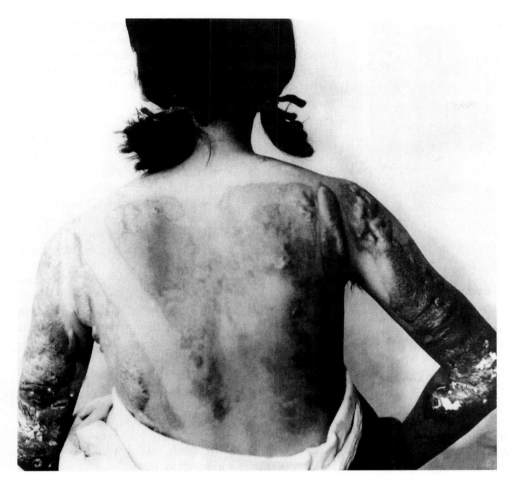

FIGURE 16.10 Aftermath of the bomb.

Victim of severe radiation burns from the nuclear bomb exploded over Hiroshima, Japan, in 1945.
Photo: AP/Wide World Photos.

As with so many of our inventions, our nuclear technology had its roots in warfare. In the 1930s, scientists speculated that atoms could be split to create a superbomb. During World War II, the United States poured a fortune into the top-secret Manhattan Project in a frantic, round-the-clock effort to build a nuclear bomb before the Germans or Japanese could. We succeeded, as proven by the nuclear blasts over Japan in 1945 (Germany gave up before we had the bomb ready). After the war, an "atoms for peace" program promoted use of nuclear power.

Research focused on uranium. All isotopes of uranium are unstable and have been decaying since the time of their creation. As you saw earlier, uranium-238 has a half-life of 4,470,000,000 years, roughly Earth's present age, so half of it is still around; it currently forms 99.3 percent of naturally occurring uranium. Uranium-235 has a shorter half-life of 713 million years, so it has gone through about six half-lives, and much of it is gone. U-235 now constitutes only 0.7 percent of natural uranium. (Put simply; if we took a random sample of 141 uranium atoms, 140 would be U-238, and only 1 would be U-235.)

It is the scarcer U-235 that we need for bombs and nuclear fuel, because it has a unique property among the natural elements: U-235 atoms can be made to split. When they do so, they release the amazing amount of energy that we exploit in power plants and weapons.

Nuclear Fission (Splitting Atoms)

We are so accustomed to "burning" (rapidly oxidizing) fossil fuels and firewood to release their chemical energy that it is hard to conceive of any other way to release energy. But a uranium atom does not release energy through burning. A uranium atom releases its energy when it is physically "split." This is accomplished by hurling a neutron into the unstable nucleus. You might think of it as setting off a mouse trap.

Most Earth processes speed up or slow down with changing temperature and pressure, because they involve only the outer electron shell of atoms. But unstable atoms decay from the inside, as their nuclei disintegrate. They do this at a constant rate, ignoring temperature or pressure. The only factor in radioactive decay is time. For uranium-235, we can "warp" time by bombarding its atoms with neutrons, stimulating them to disintegrate faster.

FIGURE 16.11 What uranium miners mine.

This yellow mineral is *carnotite*, a radioactive complex of uranium oxide, vanadium, potassium, and water, plus traces of radium, a decay product of uranium. It is among the most common uranium ore minerals. *Photo: G. Tomsich/Tomsich Rome/Photo Researchers, Inc.*

When an atom disintegrates, it is called **nuclear fission,** because the atom's nucleus fissions or "splits." This is what happens in a nuclear power plant or nuclear bomb; the basic difference is the *rate* at which fission occurs. In a power plant, fission is carefully controlled at a slow rate, to produce steady heat. In a bomb, countless atoms are forced to fission almost simultaneously, releasing an incredible burst of energy (picture the infamous "mushroom cloud" of nuclear explosions).

Just how much energy is locked in the nucleus of a uranium atom? The fissioning of one little ounce of U-235 produces the same heat as the burning of 400 barrels of oil or 85 tons of high-grade coal! Such wonderful efficiency allows bombs and power plants to be more compact.

In addition to releasing heat, a split U-235 nucleus also leaves debris in the form of simpler isotopes such as strontium-90, barium-141, and krypton-92. All of these are highly radioactive and dangerous.

A fissioning U-235 atom also ejects two or three neutrons from its nucleus. These neutrons may hit the nuclei of adjacent U-235 atoms, causing them to fission. Each atom that fissions in turn ejects two or three neutrons, causing its neighboring atoms to fission. This creates a continuous **chain reaction** (Figure 16.12). When a chain reaction gets rolling, neutrons are hurtling all over the place, hitting and missing U-235 nuclei. Each hit causes fission and releases a burst of heat. *The key to controlling a nuclear chain reaction is to control the neutrons.*

As the chain reaction proceeds, the energy released as heat is formidable. In a weapon, the heat is used to annihilate. In a nuclear power plant, the heat boils water to generate steam, which turns a turbine to generate electricity. At a power plant, the fission is contained in a **nuclear reactor.**

Inside a Nuclear Reactor

To make the chain reaction more efficient, the proportion of U-235 in ore is enriched several times from its natural 0.7 percent to around 3 percent. This enriched uranium is fabricated into pellets (Figure 16.13). The pellets are packed into long, thin **fuel rods.** They are placed in a water bath, which moderates the rate of fission and cools the rods (Figure 16.14).

The fuel rods are mounted vertically. They are interspersed with other vertical **control rods,** which contain neutron-absorbing material. To start the chain reaction, the control rods are lifted, allowing neutrons from each fuel rod to strike the other fuel rods, bombarding each other with neutrons, stimulating the U-235 in the rods to fission. Once the chain reaction is in motion, the control rods are partially lowered back among the fuel rods until the reaction slows to the desired level. The control rods are mounted vertically for safety; in an emergency, such as loss of control or an earthquake, the control rods will deadfall among the fuel rods, halting the reaction.

A further safety feature is the use of water surrounding the fuel rods in U.S. reactors. It moderates the speed of neutrons to achieve an optimum reaction, but if it abruptly leaks away, the reaction becomes inefficient and slows down.

So, a nuclear reactor consists of fuel rods, control rods, and water as a moderator-coolant. All is housed in a **containment building** that prevents leakage of radioactive material into the environment, should disaster strike. (The Russian-built reactors at Chernobyl and other eastern European sites are *not* housed in containment buildings, part of the reason that radioactive material from the 1986 steam explosion now is scattered all over Europe.)

Obviously, the rate of fission is very critical. If anything goes wrong, and the reaction proceeds too fast, the reactor will overheat. Can this cause a nuclear explosion in a power plant? No! For a nuclear explosion to occur, a concentration of high-grade U-235 must be present in a small space. But uranium fuel is low-grade, not enriched as much as bomb-grade material, and it is packaged in pellets and rods to prevent too much of it from being packed to snugly. These are safeguards designed into the nuclear power generation process.

However, the excessive heat from a runaway reaction can cause a **steam explosion.** A steam explo-

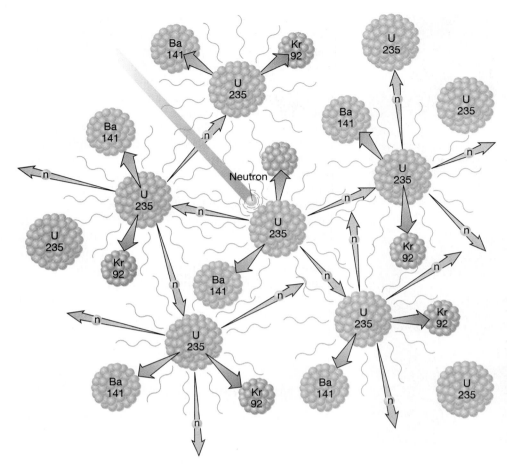

sion is very local, occurring in a single building. But it still is very dangerous, for it can scatter radioactive material outside the plant. This is what happened at Chernobyl in Ukraine in 1986, releasing radioactive debris into the stratosphere, scattering it over thousands of square miles.

Recall that reactor fuel is only about 3 percent fissioning U-235. The other 97 percent is the nonfissioning uranium isotope, U-238. In a reactor, this U-238 is unavoidably bombarded with neutrons as its sister isotope fissions. The U-238 absorbs some of these neutrons, grows heavier, and changes to **transuranic elements.** Transuranic means "beyond uranium" in weight. Examples are neptunium, americium, and plutonium. All transuranic elements are artificial, created in reactors and bomb explosions. (The sole exception is traces of transuranic elements at an African fossil-reactor site, described later.) U-235 is the only natural fissionable isotope that can produce a chain reaction. However, one transuranic element can fission: plutonium.

Breeding Fuel and Bomb Material

Like U-235, plutonium-239 can sustain a chain reaction. When first created as part of nuclear bomb

FIGURE 16.13 Uranium dioxide pellets, held in a gloved hand.

These contain about 3 percent U-235, enriched from the 0.72 percent in natural uranium. The pellets are placed in assemblies called fuel rods. The core of a reactor contains about 40,000 long, thin fuel rods, each packed with the pellets. Each pellet contains the energy potential of one ton of coal. *Photo: Westinghouse Energy Center.*

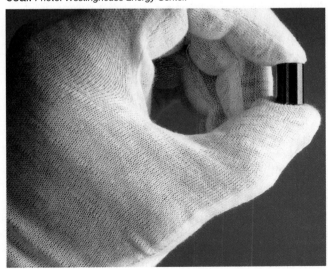

FIGURE 16.14 Technicians ready the reactor core housing to receive fuel rods.

They will be surrounded by water for cooling and to moderate the speed of neutrons to obtain the most efficient nuclear chain reaction. *Photo: Erich Hartmann/Magnum Photos, Inc.*

research during World War II, plutonium was named for the planet Pluto, but the name proved more fitting because Pluto was the Roman god of the underworld. As described by one scientist:

> If ever there was an element that deserved a name associated with hell, it is plutonium. This is not only because of its use in atomic bombs—which certainly would amply qualify it—but also because of its fiendishly toxic qualities, even in small amounts.

A **breeder reactor** "breeds" its own fuel, plutonium-239, from the uranium-238 present in a reactor. Each atom of U-238 that absorbs a neutron yields an atom of plutonium. At present, no commercial breeder reactors exist in the United States. One reason is that breeder reactors are more dangerous than U-235 reactors. The heat released is too intense to be handled by water cooling systems, so they use liquid sodium, which has a much greater heat capacity than water. But this requires more elaborate plumbing, increasing the risk of accident.

Less scientific reasons also discourage breeder reactors in the United States. Plutonium has replaced uranium-235 as the key ingredient in nuclear weapons, and breeder reactors generate plutonium. The fear is that a rogue government (such as Libya, Iran, Iraq, North Korea), terrorist organization, or ultra-nationalist group might steal plutonium and build a crude nuclear bomb, using it to blackmail other countries. Making a nuclear bomb requires about 10 pounds of plutonium. Smugglers are already at work. A few have been caught in Europe. The technology of building a nuclear bomb is now widely known; the main challenge in bomb-building is to obtain the nuclear material. Doing so does not require breaking into a nuclear facility. Simply hijacking some plutonium during transport is easier.

For example, Japan transports uranium for its breeder reactors by ship from a processing plant in France. These transport ships are sitting ducks for plutonium thieves. Many countries object to this transport for fear of either a nuclear accident (less likely) or an open-sea hijacking (more likely). We live in a very dangerous world.

Fossil Nuclear Reactors in Africa

U-235 is the only naturally occurring radioactive isotope that can maintain a chain reaction. As noted, it is rare—only one of every 141 uranium atoms. If it were more common, we might encounter spontaneous nuclear reactions in our rocks. This raises an interesting question: could a uranium deposit become sufficiently enriched in U-235 for a natural nuclear reactor to develop? Until twenty-five years ago, scientists said no, because (a) the ore must contain at least 10 percent uranium, a very rare concentration, and (b) U-235 must be enriched several times (from its normal 0.7 percent to at least 3 percent).

But in 1972, several extinct nuclear reactors were discovered in rocks at the Oklo uranium mine in the Republic of Gabon in West Africa. Here, the ore is exceptionally rich (50–70 percent uranium). The first reactor was discovered when the uranium's isotopes were analyzed. The U-235 isotope was not at the normal 0.7 percent expected, but as low as 0.3 percent. Where had it gone? Such a depletion could be explained only by nuclear fission in a natural chain reaction, accelerating the decay of U-235.

Further study of the Oklo rocks revealed traces of transuranic elements that modern reactors generate. This confirmed that a natural chain reaction had occurred. Eventually, seventeen fossil natural reactors were found in and near the Oklo mine. Each was about 10 feet square and not more than 3 feet thick.

Modern reactors use U-235 that has been artificially enriched to 3 percent. Knowing that the half-life of U-235 is 713 million years, it was possible to back-calculate when the reactors had the necessary 3 percent-enriched fuel available: 1.8 billion years ago. The Oklo reactors operated for about 500,000 years at temperatures between 320 and 680°F. During their lifetime, they generated 17,800 megawatts of energy and consumed close to 6.5 tons of U-235. The Oklo shales have been a repository for spent reactor fuel for the past 1.8 billion years.

Uranium Miners, Smokers, and the Radiation Risk

Uranium miners are surrounded by uranium ore. Does their occupation place them at special risk from radiation? This question is not as easy to answer as you might think. Uranium ore usually is not very concentrated, so its radiation is not intense. However, we know that even light radiation damage is cumulative. A miner with twelve years of experience, for example, has received a pretty good cumulative dose of radiation. How do we assess the risk?

We know that uranium miners who smoke have more lung cancer than nonsmoking miners. Of course, this is true of smokers and nonsmokers in general. But does the miner's exposure to radiation from uranium increase the number of lung cancer deaths to a degree large enough to be detectable? The answer is yes.

However, the increased lung cancer risk depends on the concentration of uranium in the specific ore being mined, and on how many years an individual miner has worked. Further, when uranium mining was booming in the 1950s, mines were much smaller "dog holes" than large modern mines, and ventilation was much poorer. How much worse for the miners' lungs was it to work in a "dog hole" rather than in a well-ventilated mine?

The *qualitative* answer is easy: certainly it is better for a miner's health to have been mining only one year rather than ten; better to have mined poorly concentrated uranium than a financial bonanza; better to have worked in a well-ventilated mine. But difficulty arises when we try to *quantify* these dangers:

- If a uranium miner smokes, are the harmful effects of cigarette smoking and radiation exposure simply additive? Or do they multiply because the miner's lungs have been weakened by smoking, allowing radiation to do its damage more effectively?
- Is a ten-year smoker who mines uranium for five years at greater risk than a five-year smoker who mines for ten years? If the risk is different, how much different?
- How much does the risk for a "dog hole" miner who recovers low-grade ore differ from that of a

worker who spends the same time in a well-ventilated mine recovering higher-grade ore?

Reliable numerical answers to such questions are difficult. If this sounds like a boring academic argument, let's make it practical: suppose you work for Cadaver Life Insurance Company and need to decide how much to charge uranium miners for health insurance or life insurance. Where do you get meaningful numbers on which to base your rates? Suppose you are a staffer for a congresswoman, assigned to "come up with good numbers" to determine how your boss should vote on a "Uranium Miners Health and Safety Act"? How do you sort it out?

How Safe Are Nuclear Power Plants?

How safe are nuclear power plants? For a direct answer, read again the quotation from Hannes Alfven that opens this chapter. Now, here is an explanation of why he and others feel as they do about nuclear power.

During normal operation of a nuclear power plant, the fission products remain within the fuel rods. No discharges of radioactive gases or other materials are permitted. Careful measurements have shown that public exposure to radiation from the normal operation of a nuclear power plant is about 1 percent of the natural radiation a person gets from cosmic rays, rocks and soil, voluntary X rays for medical or dental purposes, or other normal activities.

Public fear of nuclear power plants stems not from normal operations but from the accidents that are inevitable whenever human beings and technology interact. Accidents such as oil spills, chemical explosions, or release of sewage into the public water supply are things that can be cleaned up quickly, or, at worst, within a few years. Nuclear accidents, however, have the potential to cause harm for generations or even thousands of years.

Most frightening is the fact that accidents at nuclear power plants are not uncommon. Between 1971 and 1986, in fourteen countries, 152 accidents were documented, of varying severity. The best-publicized nuclear accidents have been at Three Mile Island in Pennsylvania (1979) and Chernobyl in Ukraine (1986). Let us look at each.

Case Study: Three Mile Island, Pennsylvania

Pennsylvania's capital city of Harrisburg sits astride the Susquehanna River, which flows southward about 75 miles into the Chesapeake Bay. Just below Harrisburg, several islands sit in the river, an ideal site for a nuclear power plant: plenty of cooling water and near to major

FIGURE 16.15 The infamous nuclear power plant at Three Mile Island.

Located near Harrisburg, Pennsylvania, this facility became a symbol of the dangers of nuclear power. The two low, round containment buildings house the reactors, isolating them from the environment. The four tall towers are cooling towers that condense used steam back to liquid water for reuse. *Photo: Breck P. Kent/Animals Animals/Earth Scenes.*

cities that demand electricity. Thus, the Three Mile Island Nuclear Power Station came to be (Figure 16.15). The station had two reactors.

All went well until March 28, 1979. At 4 o'clock in the morning, two water pumps failed in one of the reactors. The emergency cooling system activated automatically, but an operator turned it off manually—twice—because warning lights on the console malfunctioned and did not confirm the failure. The pump failure caused a loss of coolant and partial exposure of the reactor core. More than 70 percent of the core was damaged and about 50 percent of it melted and fell to the bottom of the reactor. Coolant water started draining from the core assembly into the containment building, carrying highly radioactive material with it. Some radioactive water vapor was released into the outside air, causing evacuation of 144,000 people from the area.

Fortunately, there was no explosion, and no significant environmental damage from the emitted gas. Despite hundreds of lawsuits, a study conducted within a 10-mile radius around the plant a decade after the accident found no increase in cancer rate among inhabitants. But it was a near-miss: investigators found that if a stuck valve had stayed open for another 30–60 minutes, the reactor core would have melted completely. The consequences are uncertain, but the possibility of a major steam explosion existed. It would have propelled radioactive material over the area.

The partially melted reactor core and containment building were so contaminated that the building could not be entered for two years. Since then, cleanup

has been underway at a cost likely to exceed $3 billion. (The facility originally cost only $700 million to build.)

The materials used in cleanup will themselves become contaminated and require costly disposal as low-level radioactive waste. The contaminated cleanup materials are estimated to include:

    1,000,000 paper coveralls
    1,000,000 plastic coveralls
    20,000 cloth coveralls
    100,000 raincoats
    100,000 plastic bottles
    100,000 pairs of rubber boots
    1,000,000 pairs of rubber gloves
    100,000 surgical caps
    1,000,000 feet of plastic sheeting

Once the partial cleanup is completed, the reactor building will become a mausoleum, sealed in concrete, a perpetual reminder of nuclear power's risks. The other reactor at Three Mile Island is still operating.

An important result of the Three Mile Island event has been a considerable improvement in safety devices and stronger regulations for all U.S. nuclear plants. This improved safety margin comes at a high price, however. The Seabrook, New Hampshire, reactor obtained its operating license from the Nuclear Regulatory Commission in 1990—eleven years later than originally planned, due to delays to improve safety and to contest lawsuits. The plant cost $6.45 billion, twelve times the original estimate and ten times the cost of the facility at Three Mile Island.

The problem of cost may override even safety concerns as a reason for halting construction of nuclear power plants in the United States. New nuclear power plants produce electricity at an average of 15¢ per kilowatt-hour, which is likely to increase as further safeguards are added and lawsuits increase. As noted in Chapter 15, the cost of solar power generation is expected to be only 10¢ per kilowatt-hour five years from now, and solar power does not carry with it the danger of radioactive contamination.

The Three Mile Island event hardened the public resistance to nuclear power that had been building (Figure 16.16). Solar energy looks better all the time for generating electricity.

Case Study: Chernobyl, Ukraine

The former Soviet Union was host to the worst accident in the history of nuclear power generation at 1:23 A.M. on April 26, 1986. At Chernobyl, in what is now the country of Ukraine, a reactor experienced both a steam explosion and a core meltdown (Figure 16.17). The roof blew off the reactor building and enormous volumes of radioac-

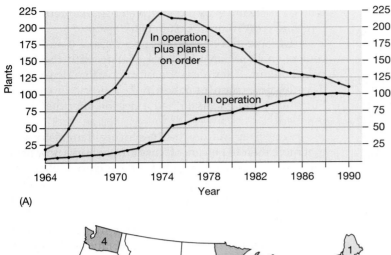

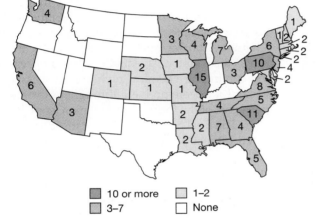

FIGURE 16.16 Nuclear power generation in the U.S.

(A) The rise and fall of U.S. nuclear power. New-plant orders peaked in the 1970s, but many utilities have canceled them. The most recently ordered plant order that still is active was ordered in 1974. Nevertheless, the number of plants in service increased steadily, because plants under construction were completed. (B) States having nuclear power plants either operating or on order. *Source: U.S. Department of Energy.*

tive isotopes were hurled into the atmosphere. Radiation measured within 16 miles of the plant reached about 3000 times higher than normal background levels (using the Savannah River nuclear reactor in the southeastern United States as a reference). More than 375,000 people were evacuated (Figure 16.18).

It got worse as winds spread the fallout over Europe within days (Figure 16.19A). Some western nuclear experts suggest that 15 tons of radioactive fuel may have been pulverized and injected into the atmosphere. This involved far more than the 3 percent U-235 and the 97 percent U-238 that exists in nuclear fuel rods when they are new. This fuel was used, so the fissioning U-235 atoms had produced their deadly potpourri of nasty isotopes, including strontium-90 (a bone cancer agent), iodine-131 (thyroid cancer), and transuranics like plutonium-239. What normally becomes high-level radioactive waste in a reactor's spent-fuel storage tank was instead sprinkled upon the citizens of Ukraine, Belarus, and western Europe. Radioactive particles eventually were deposited across the whole Northern Hemisphere.

Chernobyl's Biological Time Bomb. The scattered radiation is a biological time bomb. Before the explo-

sion, only two thyroid cancers per year were typical in children in Belarus; by 1994, 82 cases per year were reported (Figure 16.19B). One expert says that up to 40 percent of the children who were under a year old when exposed to the highest radiation levels could develop thyroid cancer as adults. (American children exposed to excessive X rays in the 1930s developed thyroid cancer up to forty years later.) Also increasing in children are tumors of the brain, bones, and kidneys. Cancers of the skin, breast, and lungs are predicted; these did not appear in Japan until twenty years after the nuclear bombings in 1945. Overall, illness is 30 percent above normal in contaminated areas of Ukraine, probably due to damaged immune systems.

Among adults, the World Health Organization reported about 200,000 people suffering from fear, stress, tension, heart and circulatory problems, stomach trouble, ulcers, and depression from the blast. The suicide rate quadrupled in the year after the nuclear accident.

A former science advisor to Russian President Boris Yeltsin says that health problems will plague Chernobyl's victims "practically forever." This may be truer than he thinks: scientists report inherited genetic damage in children born in 1995 to parents who were

FIGURE 16.17 The burning Chernobyl reactor, three days after the 1986 steam explosion.

The worst nuclear-power accident so far, the explosion killed at least 33 people outright and placed millions in Ukraine, Belarus, and western Europe at heightened cancer risk. As of 1996, the plant's other reactors—which are of an older design, are poorly shielded, and have no containment building—still were operating, perhaps awaiting the next mistake. *Photo: Shone/Gamma-Liaison, Inc.*

exposed to the fallout. The damage is to DNA in both sperm and eggs. Such mutations become part of the genetic code and are passed down through generations.

"I'm not worried. Soviet radiation is the best in the world," observed one contaminated Russian worker, who obviously was not a rocket scientist. Some of this "quality radiation" was deposited on farmland hundreds of miles from Chernobyl. A decade later, grazing lands in Scotland and northern Scandinavia remain too radioactive to raise cattle and reindeer for human consumption. The Minister of Agriculture in Ukraine, a grain-growing "breadbasket" country, stated in 1996, "We know this land should not be farmed, but if we don't . . . we will simply starve." How would Americans

react if a steam explosion in Nebraska contaminated large areas of America's grain belt?

Ukraine and Belarus still spend more than 10 percent of their national budgets dealing with the Chernobyl fallout. After a decade of radioactive decay, they have 200 times more radiation in their affected areas than Hiroshima and Nagasaki had a decade after they were bombed.

A 1995 report from the Russian health ministry says that more than 10 percent of Russian newborns now have serious birth defects, 50 percent have chronic illnesses, and the percentages are growing. Congenital deformities in newborns are increasing faster than any other health problem in Russia today. Is Chernobyl to be blamed for this, too? No one is certain, but there may be a broader cause: decades of nuclear irresponsibility by the former Soviet government. This is demonstrated not only at Chernobyl but at nuclear waste dumps and test sites across the country. An equally devastating cause may be the continuing chemical pollution of Russia's soil and streams, a severe problem throughout the eastern European part of the former Soviet Union (see Chapter 8).

The Russian and Ukrainian governments have listed almost 600,000 potential victims; some 300,000 citizens per year are being treated for radiation exposure. The eventual death toll from the Chernobyl disaster will be in the millions, over decades. Can victims outside the former Soviet Union recover financial damages? Can victims sue a government that no longer exists (U.S.S.R.)? Can they sue a country that did not exist at the time of the accident (Ukraine)?

Could We Have a "Chernobyl" in the U.S.? It is important to know that *the Russian reactors at Chernobyl are of a design not used in the United States.* Their reactors moderate the flow of neutrons with graphite, not water. (Graphite is the black material in an ordinary "lead" pencil.) Large energy surges are possible with graphite, but not with the water we use. Our water reactors contain more fail-safe backup systems to prevent core overheating. Also, Western reactors are operated in containment buildings of thick concrete to withstand steam explosions.

As of 1996, fifteen of the dangerous Chernobyl-type reactors still operate in Russia, Ukraine, and Lithuania. Eleven improved reactors—which are still not safe enough to be licensed in the West—are operating in Russia, Bulgaria, Slovakia, and Armenia. One Armenian reactor is near a city of 1.2 million and is built in an earthquake zone. Armenians realized this and closed the plant, but had to reopen it in response to harsh winters and only two hours of electricity per day. One citizen noted, "I was active in fighting to close the plant, but we had to reopen it or we would have frozen. No other nation would have put up with what we have."

FIGURE 16.18 A soldier guards a barrier outside a town evacuated following the Chernobyl event.

He is working in a contaminated area. Will he live a full, healthy life? Will his children have birth defects from the radiation? If he dies slowly from radiation-induced cancer, will his wife care for him and be the breadwinner too? The impact of a mistake like Chernobyl affects millions of lives for decades. *Photo: Igor Kostin/Imago/Sygma.*

So, a Chernobyl-style disaster is quite unlikely to occur in the United States. In general, the risk is less in technologically advanced Western countries than in eastern Europe and the poor nations of the former Soviet Union, which have inherited the relatively unsafe reactors of the U.S.S.R. without the money for their maintenance.

Rethinking Nuclear Power After Chernobyl. The Chernobyl event affected world opinion far more than did the Three Mile Island incident. Japan, Great Britain, and France, once committed to nuclear energy, now are rethinking the risk. With about 430 nuclear power plants operating worldwide today, some in unstable countries, the risk is real. If the chance of a major accident at a nuclear power plant is 1 in 1 million each year, and 500 plants are in operation, then the risk of an accident somewhere each year is 1 in 2000—not the most comfortable odds. In fact, accidents have been as frequent as they are in any industrial facility, although not catastrophic.

France depends heavily on nuclear power. In 1996, on the tenth anniversary of the Chernobyl disaster, they decided to prepare citizens for the possibility of a nuclear accident by distributing iodine pills to 400,000 people who live within 3 miles of a nuclear power plant. Across the country, firefighters delivered bottles of iodine pills, door-to-door. Austria and Switzerland already have iodine distribution programs.

Why iodine? Iodine concentrates in the thyroid gland. Radioactive iodine-131 was released by the Chernobyl explosion and is blamed for the thyroid cancer epidemic among children in Belarus, Ukraine, and Russia. Swallowing iodine a few hours before radiation exposure can saturate the thyroid with stable iodine, so the radioactive iodine cannot be absorbed.

Unfortunately, the iodine-pill program is clearly inadequate to protect the public. Those within three miles of an exploding reactor probably would not have time to take the pills before being dowsed by radiation. And restricting the iodine pills to those living very near France's nineteen nuclear power plants "shows the government still hasn't learned the lessons of Chernobyl— because we were contaminated 2000 kilometers (1200 miles) away," according to an independent monitoring group sponsored by Greenpeace. The movement of radioactive clouds is unpredictable. A steam explosion or core meltdown can spread radiation over hundreds of thousands of square miles and endanger a sizeable part of Earth's inhabitants. Thus, the group believes that everyone in France should have immediate access to iodine pills.

Some countries are deeply committed to nuclear power. France and Lithuania generate more than three-fourths of their electricity in nuclear power plants. Belgium, Sweden, and Slovakia obtain about half their power this way. None of these countries produces much oil and gas, so nuclear power is their only present choice (Figure 16.20). But, considering all the problems and cost of nuclear energy, solar energy seems a far better choice for generating electricity.

What Do We Do with Old Nuclear Power Plants?

The nuclear industry is continually upgrading the quality of nuclear reactors and improving reactor safety. More recent American reactors are safer and

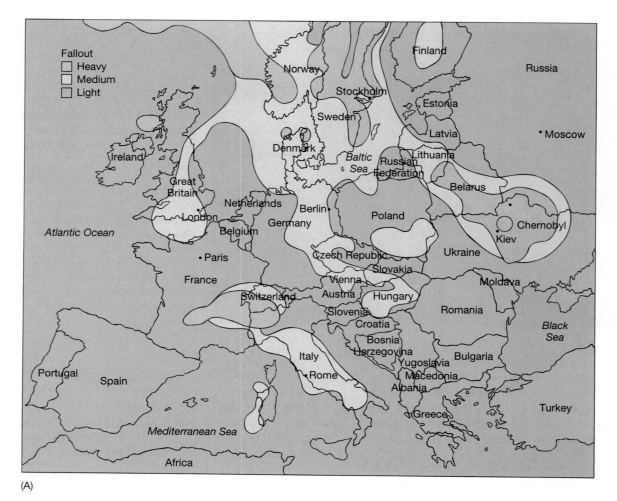

(A)

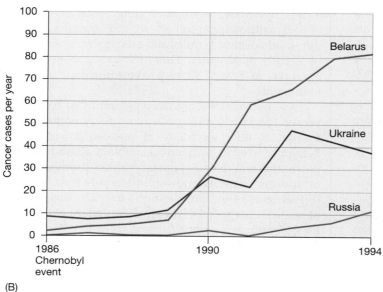

(B)

FIGURE 16.19 Radioactive fallout from Chernobyl.

(A) Fallout was highly variable due to wind patterns and weather systems moving across Europe. *After P. H. Raven, L. R. Berg, and G. B. Johnson, 1993,* Environment, *Fort Worth, Texas: Saunders College Publishing, p. 218.*
(B) Statistics tell the story: cancer rose where radiation fell from the air. *Source: World Health Organization.*

more reliable than older ones. But what do we do with older, less safe, less efficient, and worn-out reactors? The reactor core and surrounding parts are highly radioactive and will remain so forever, measured on a human time scale. The metal parts and the concrete used in the construction have probably become brittle from long exposure to radioactivity. Clearly, we cannot simply abandon the plant to decay of its own accord.

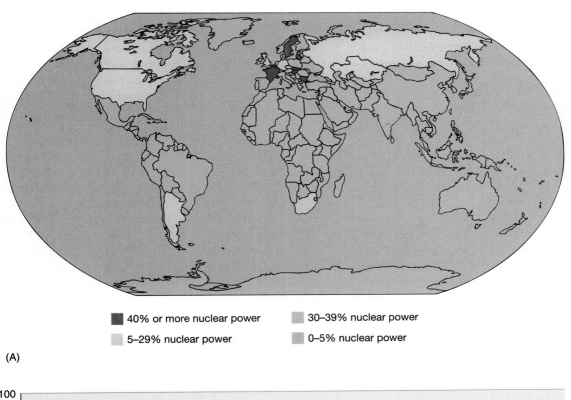

■ 40% or more nuclear power   30–39% nuclear power

5–29% nuclear power   0–5% nuclear power

(A)

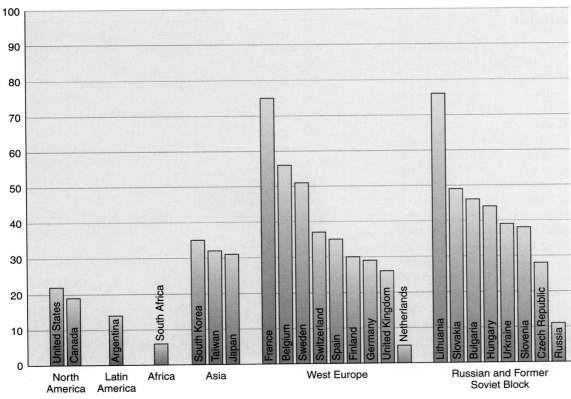

(B)

FIGURE 16.20   Global spread of nuclear power (not nuclear weapons).

The geographic spread of nuclear power reflects each country's fossil fuel resource, technology, history, and international power politics. Note that physically vast countries like the U.S., Canada, and Russia generate smaller percentages of their electricity by atomic means, because they have large fossil fuel resources. China is an exception, with slow nuclear development.

Nor can we attack it with a wrecking ball or dynamite, as we would an old office building.

Several methods for dismantling nuclear power plants have been proposed, but all are expensive—perhaps $5 billion per reactor. No one knows the true cost, or even whether complete dismantlement is technically feasible. The plant must be shut down, the reactor emptied of fuel, and all pipes drained. Then all radioactive material, both solids and liquids, must be sent to a permanent waste storage site, which does not exist yet. Great care would be required during all phases to avoid spills, escape of material, or accidents during transportation.

Worldwide, about 250 commercial reactors are scheduled to be retired and evacuated during the next fifteen years, seventy of them in the United States. The cost will be large, the work very dangerous, and not a single storage site for highly radioactive waste has yet been established. It will be a new experience for us: no full-sized reactor anywhere in the world has yet been fully decommissioned.

Can We Ever Dispose of Radioactive Waste?

Radioactive waste has been accumulating in the United States for about fifty years, with concern about it growing faster than the waste itself. For safety, radioactive waste is classified as either low-level waste or high-level waste, depending on its origin.

Low-level radioactive waste is 99 percent of all radioactive waste. It is likely to remain significantly radioactive for 300–500 years. Some of it is essentially harmless. Low-level waste is not reactor-related, but comes from hospitals (cancer treatment), university laboratories (research), industrial laboratories (research), manufacturing (measurement), and military facilities (nuclear weapons manufacturing and research). The range of contaminated objects is large: fabric (gloves, laboratory clothing), metal (tools), plastic (dishes, containers, bags), glass (bottles, syringes), paper (filter paper, towels), wooden objects, and animal remains used in radiation experiments. This is one-third of the total volume produced.

High-level radioactive waste is produced by nuclear power plants, research reactors, and military reactors. Our nuclear Navy has 20 percent more reactors than the entire commercial nuclear power generating industry. The most abundant high-level waste is spent fuel rods. It is dangerous to humans for thousands to millions of years. Waste from civilian and military nuclear power plants provides 94 percent of the radioactivity from all nuclear waste, and two-thirds of the volume.

How is the period of danger determined? Two factors are involved: (1) the half-lives of the isotopes involved and (2) the volume of the waste material. The important radioactive isotopes for disposal purposes are shown in Table 16.2; you can see that the range of half-lives is very large, from about 7 hours to 4.5 billion years. How many half-lives must elapse before each isotope has decayed sufficiently so the combination of its amount and type of radiation are low enough to pose no danger? *The rule of thumb is ten half-lives of the longest-lived, dominant radioisotopes.* A period of ten half-lives decreases the radioactivity to 1/1000 of its initial value.

High-level radioactive waste consists of spent fuel rods and other materials from functioning or decommissioned nuclear plants. After three to four years in a reactor, the concentration of fissionable U-235 in a fuel rod becomes too low to keep the chain reaction going. Each year, about one-third of the spent fuel rods in the reactor are replaced and, because the rods are highly radioactive,

TABLE 16.2 **Radioisotopes Important in Waste Disposal**

Isotope	Half-Life
Protactinium-234	6.8 hr
Radon-222	3.82 day
Bismuth-210	5.0 day
Xenon-133	5.25 day
Polonium-210	138.4 day
Cobalt-60	5.27 yr
Krypton-85	10.7 yr
Tritium (hydrogen-3)	12.33 yr
Lead-210	22.3 yr
Strontium-90	28.8 yr
Cesium-137	30.17 yr
Americium-241	433 yr
Radium-226	1630 yr
Radiocarbon (carbon-14)	5730 yr
Plutonium-240	6570 yr
Americium-243	7370 yr
Curium-245	8500 yr
Plutonium-239	24,000 yr
Thorium-230	80,000 yr
Tin-126	ca. 10^5 yr
Technetium-99	214,000 yr
Uranium-234	245,000 yr
Neptunium-237	2,140,000 yr
Cesium-135	3,000,000 yr
Iodine-129	16,000,000 yr
Uranium-235	704,000,000 yr
Uranium-238	4,470,000,000 yr

they are stored temporarily in large concrete water pools at the plant site until a suitable long-term storage site can be built.

For how long must the rods be safely stored? Plutonium's half-life is 24,400 years, so fuel rods containing plutonium must be isolated for about 10 half-lives, or 244,000 years! How many civilizations have lasted even 1 percent of that time? For that matter, how many have lasted more than a few hundred years? If the plutonium and other long-lived isotopes were removed from the rods, the remaining waste would need isolation only for 10,000 years or so. But whether the radioactively dangerous period is 10,000 years or 244,000 years, the question remains: where are we going to put this stuff?

Low-Level Waste Repositories

Of sixteen repositories that once received low-level radioactive waste, all but two are closed, in South Carolina and Washington State (Figure 16.21). Little action has resulted from the Low-Level Radioactive Waste Policy Act of 1980, which requires states to build regional disposal facilities. Although most states have banded together to form the "compacts" shown, not a single new site has opened for disposal.

The Nuclear Regulatory Commission, the federal government agency charged with overseeing reposi-

tory sites, has received no applications for a low-level radioactive waste disposal site. This is NIMBY (Not In My Back Yard) at work, and it has raised disposal cost for low-level waste to $400 per cubic foot. The volume of low-level waste is increasing at a rate of 2 million cubic feet per year, so that each year an additional cost of $800 million is incurred.

Storage of low-level waste is done above and below ground, in metal drums. However, metal drums corrode over time, plus stress is placed upon them by the continuous radiation from the material in the drums. Many drums at some sites have leaked radioactive liquids into the soil and groundwater, a situation that has plagued storage at industrial chemical plants as well (Chapter 8). Numerous examples of leakage, presumably long-term, have been documented. Unless the local hydrologic environment is well understood, the rusting drums can be a source of contamination for many decades.

Current scientific opinion favors storage of low-level radioactive waste in shallow excavations above the groundwater table in arid or semiarid areas, such as those in the western and southwestern United States. At such disposal sites, the water table is hundreds of feet beneath the excavation and there is little danger that water would infiltrate the excavation from either above or below. Excavation of the storage rooms

FIGURE 16.21 States that have joined in low-level radioactive waste compacts.

Despite these agreements, not a single new site has opened for disposal.

Source: Nuclear Regulatory Commission.

★ Active disposal site.
• Disposal site under license review.
○ Ward Valley, CA. (being considered)

tens of feet deep would be easy, and the rooms could be monitored easily for a few centuries until the waste ceased to be hazardous and could be landfilled.

High-Level Waste Repositories

Worldwide, about 430 nuclear power plants operate in thirty-five countries, with more coming online each year (Figure 16.21). For every 100 pounds of uranium used, 99 pounds remain as waste. The U.S. Department of Energy estimates that by 2000 the highly radioactive spent fuel and reprocessing waste generated by power plants would fill a football field to a height of 254 feet. It grows a bit higher each day. Compared to the volumes of other forms of hazardous waste, this is tiny. But it is so radioactive, dangerous, and undesirable that we have wrangled for years over where to put it.

At present, more than 20,000 spent fuel rods are stored "temporarily" in cooling pools at power plants. This storage is nearly full and was never designed for such long-term use. The dangers are illustrated by the "temporary" high-level waste storage facility at Hanford, Washington, where 560 square miles of desert enclose the silent hulks of reactors and processing plants that once produced plutonium for nuclear weapons (Figure 16.22).

Hanford has been called "the dirtiest place on Earth." It has 1377 individual waste sites, large and small—trenches, tanks, ponds, sand-covered pits, and underground storage areas. Together they contain about 15 billion cubic feet of hazardous materials. Beneath the sand at Hanford, 2000 pounds of plutonium may be buried—but because of poor record-keeping over the past forty years, no one is sure!

The magnitude of the problem at Hanford is almost beyond comprehension. High-level radioactive waste totals *ten millions of gallons*. More than 300,000 gallons of it has leaked from storage tanks into the soil and groundwater over the past twenty years. Its destination is the Columbia River, which flows through the facility and on to Portland, Oregon, and the Pacific. The U.S. Department of Energy (DOE) and Westinghouse Hanford Company, manager of the facility, are unable to determine what is in 55 million gallons of highly radioactive waste stored in 177 underground tanks. The volume of contaminated soil is about a cubic mile.

The U.S. Senate reports that cleanup at Hanford is foundering in a regulatory morass. Officials lack a clear, realistic vision for the cleanup, the schedule is unrealistic, the public is being misled, and the $7.5 billion spent thus far (of a projected total of more than $50 billion) is totally out of line with what is being achieved. Engineers who have studied the site describe it as "the largest civil works project in history, it has no end." Hanford is only one of many government sites with formidable pollution problems.

A broader study of government high-level waste storage facilities reveals that up to 29 tons of highly radioactive plutonium could leak into the environment. Plutonium has been found in leaking and corroded packages, cracking plastic bottles, old decaying buildings and in pipes, ventilation vents, equipment, and machinery.

DOE says that repairing the environmental damage from the federal nuclear weapons program could cost up to $1 trillion and take until 2070. Even then, some sites could not be restored to pristine "green fields." Hopelessly contaminated areas like Hanford would be fenced off to protect the public. They did not address how the polluted and radioactive groundwater would be fenced off; obviously, it cannot be.

It is far too late to say "we never should have gone nuclear" or "waste disposal should have been in place long before nuclear plants were built." In the

FIGURE 16.22 Hanford Nuclear Weapons Facility on the Hanford Reservation in southeastern Washington State.

Hanford, now managed by the Department of Energy, is a former plutonium production site from which radioactive material has escaped. An engineering firm reported to the U.S. Senate that the Hanford cleanup is "the largest civil works project in history; it has no end." *Photo: Calvin Larsen/Photo Researchers, Inc.*

rush to develop nuclear power and the typically American optimism that technology can solve any problem, Pandora's box was opened decades ago. Now a dangerous problem has been created. Welcome to the wonderful world of nuclear energy!

Criteria for a High-Level Radioactive Waste Storage Site. What criteria must be satisfied for an acceptable site?

1. The site must be impenetrable to water that could transport radioactive material from the site. This means the site must be underground. It must be high, so the local water table can never submerge the canisters.

2. Surrounding rock must be unfractured to prevent leakage. This means granite, volcanic rock, or some varieties of metamorphic rock. (Sedimentary rock is unsuitable because sandstones are porous, limestones are soluble, shales often are fractured, and rock salt is plastic, flowing easily under very small stresses.)

3. The site must be earthquake-free for ten-thousands of years to come. (They could collapse the facility and faults can become pathways for groundwater to migrate into the storage area.)

4. The site must be free from volcanic activity.

5. The site must have good access to roads and rail lines, for waste will be shipped from all over the country.

6. The chosen site must be acceptable to those living nearby. Given the seeming eternity required for any controversial issue to crawl through the American judicial system, site selection might be delayed indefinitely wherever strong opposition exists (NIMBY in action again).

You can see that a perfect site is unlikely to exist in the United States. Requiring the water table to be a few hundred feet below ground level limits the possibilities to the semiarid West. But, faults, earthquakes, and volcanic activity are common in the West. With ten-thousands of years involved, and radioactivity discovered only in 1896, and plate tectonic theory accepted only since the 1960s, forecasts of any kind are dubious. To illustrate: 15,000 years ago, much of the western U.S. was rainy, had high water tables, and contained many lakes, the largest of which was 1000 feet deep and was as large as Lake Michigan (Figure 16.23). Less than 10,000 years ago, volcanoes were erupting in what is now central France; the English Channel did not exist 7000 years ago; and much of the Sahara Desert was fertile just 5000 years ago.

As stated in 1990 by a former President of the Geological Society of America, "No scientist or engineer can give an absolute guarantee that radioactive waste will not someday leak in dangerous quantities from even the best of repositories." Nevertheless, we must do our best with existing technology and scientific understanding.

Don't Hold Your Breath . . . The nuclear waste problem will not go away. So, in 1982, Congress passed the Nuclear Waste Policy Act. It requires the federal government to choose a suitable storage site for high-level radioactive waste. After five years of scientific study, Congress directed the Department of Energy to focus upon Yucca Mountain, 100 miles northwest of Las Vegas, Nevada, as the potential site (Figure 16.24A). Is Yucca Mountain a good choice? Figure 16.24B gives some clues; apply what you have learned in this course about volcanoes, earthquakes, and faults.

Feasibility studies are investigating whether the mountain's volcanic rock can secure up to 77,000 tons of spent fuel rods and similar waste products for at least 10,000 years. The studies are due in 2001. The first delivery of spent fuel rods is scheduled for 2010. Chances of this timetable being met are about those of a snowball's in hell: in 1975, the United States had planned operational storage by 1985, which was moved to 1989, then 1998, then 2003, then 2010, and now 2015 (as of December, 1995). In the meantime, the nuclear industry

FIGURE 16.23   The way it was.

The character of Earth's landscape can change dramatically in a brief period of time (when viewed from a geologic, not human, perspective). For example, this map shows how a mere 15,000 years ago the western U.S. was rainy and contained many huge lakes. A site that is suitable for storing radioactive waste today might become a terrible location before the long-lived radioactive material has completely decayed.

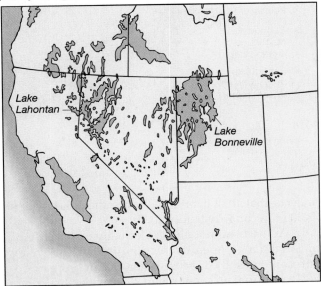

(A)

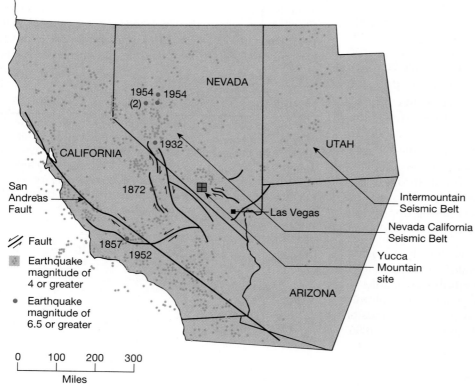

(B)

FIGURE 16.24 A good place to store radioactive waste?

Yucca Mountain in Nevada is the site selected for permanent deep-burial of high-level radioactive waste. The mountain is composed of volcanic tuff. However, faulting and seismicity surround the site. The small dots are earthquakes of magnitude 4 or greater, and the large dots are quakes of 6.5 or greater. *Photo: Nevada Department of Energy.*

projects that it will produce more than the facility's 77,000-ton capacity before 2020.

And we are not alone: Germany expected in the mid-eighties to open its deep burial facility by 1998, but now cites 2008. Most other countries plan deep geologic burial no sooner than 2020, with a few aiming

even later. No nation has yet established a site for storing its high-level nuclear waste, so "temporary" storage sites continue to grow.

And, there is the little issue of money. So far, American taxpayers have spent $1.3 billion on feasibility studies, and these studies will total around

$6 billion. The facility itself is projected to cost $30 billion. Historically, cost estimates on these mammoth projects have gone the same direction as estimated completion dates.

Why Is Radon Gas a Health Hazard?

Domestic radon exposure is estimated to result in 10,000–15,000 excess lung cancer deaths per year in North America and is, therefore, considered to kill more people than any other single environmental hazard.

Geoscience Canada, 1993

Let us briefly review Figure 16.7 (the U-238 decay chain). As U-238 decays step-by-step, thirteen of the isotopes that form are solids, but one is not. Radon-222 is a gas. **Radon** is rarified, clear, colorless, odorless, tasteless, and mixed with far more plentiful gases, so it is undetectable by human senses. It is aloof, not becoming bound up with other elements. People would ignore radon, were it not for one other trait: it is radioactive.

Unlike its thirteen solid sisters, radon gas can drift away from the uranium-bearing rock to roam the mine atmosphere. It enters the miner's lungs and joins the inhaled radioactive dust to further damage tissues. Radon has a half-life of only four days, during which it converts to radioactive metals (Figure 16.7). One has a half-life of several months (polonium-210) and another a half-life of twenty-two years (lead-210). These radioactive metals remain radioactive in the lungs a long time.

You probably are not a uranium miner, so you probably have not inhaled uranium dust. But you remain exposed to radon nevertheless. Radon gas is everywhere, mostly at insignificantly low levels. But it has been measured in thousands of homes, and, in some, the concentration is greater than the EPA deems acceptable. Where does this radon come from, how does it get into homes, how do we determine the level of danger to the inhabitants, and what can or should be done about it?

We know that radon is formed from the decay of uranium. Therefore, our starting point is to determine where uranium occurs. This means identifying the kinds of rocks that contain sufficient uranium to emit radon gas in significant amounts. (Note that uranium often is abundant enough to emit hazardous levels of radon, yet it seldom is concentrated enough to mine economically.) Uranium concentrations are greatest in granites, in rocks metamorphosed from granite, and in soils and sandstones derived from such igneous and metamorphic rocks. The map in Figure 16.25 gives a general indication of radon gas levels that may exist, based on rock type.

Radon in the Home

If a dwelling is built on a uranium-rich rock, the potential exists for radon to leak into the basement or lower floor of the house. Gas molecules can move through very tiny spaces and can enter the house through many pathways (Figure 16.26).

Let us pause here to define the unit by which this radioactivity is measured. The *curie* is a very large unit, much too cumbersome for measuring radon, so we use the **picocurie,** a trillionth of a curie (*pico* means trillionth). A picocurie is the decay of 2.2 atoms of radon in a liter of air each minute (1 liter roughly equals 1 quart).

The average American home contains between 1.0 and 1.5 picocuries of radon in each liter of air. This

FIGURE 16.25 How's the air where you live?

This is a radon *potential* map, showing likely radon gas concentrations on the basis of the general type of rocks at the surface. You *cannot* pinpoint your home on this map and say "Every liter of air I inhale contains so many picocuries of radiation from radon gas." The map is only a very general guide. The only way to determine the true radon content in your dwelling is through careful testing.

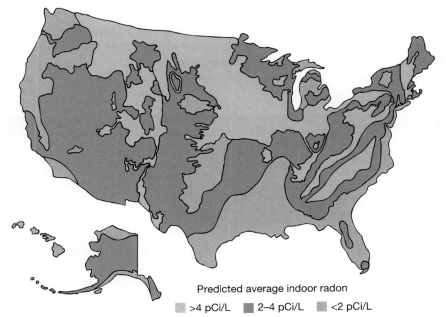

Predicted average indoor radon

■ >4 pCi/L   ■ 2–4 pCi/L   ■ <2 pCi/L

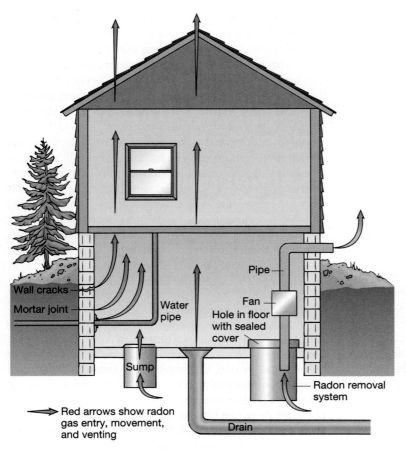

FIGURE 16.26 How radon slips into homes.

The heavy gas seeps in through the foundation and drains, and is circulated as air stirs through the house. A radon removal system is basically an air pump that exhales basement air directly outside.

Red arrows show radon gas entry, movement, and venting

amount comprises more than half of the total radiation dose you receive each year as an average U.S. resident. From this, it is clear that radon exposure is the single greatest radiation source for the typical American.

How Much Radon Is Safe?

Radiation experts estimate that the level of radon in the average American home increases your chance of contracting lung cancer by about 0.3–0.5 percent. In homes with radon levels above 8 picocuries/liter, the lung cancer risk increases by about 2 percent.

The EPA's Radon Division has designated the "safe" level of radon exposure for Americans to be up to 4 picocuries/liter (Table 16.3). Fortunately, only a small percentage of U.S. homes have radon levels greater than this—about 7 percent of homes. But how did EPA establish this "safe" level? They say it is the lowest level achievable at a reasonable price—$500–$2500 per household—with current technology. In plain English, the EPA number is simply an estimate by a government bureaucrat of what

TABLE 16.3	Radon Risk Evaluation	
Annual Radon Level	If 100 People Were Exposed to this Level—	This Risk of Dying From Lung Cancer Compares to:
100 pCi/1	About 35 might die from radon	2,000 chest X rays a year
40 pCi/1	About 17 might die from radon	2 packs of cigarettes a day
20 pCi/1	About 9 might die from radon	1 pack of cigarettes a day
10 pCi/1	About 5 might die from radon	500 chest X rays a year
4 pCi/1	About 2 might die from radon	1/2 pack of cigarettes a day
2 pCi/1	About 17 might die from radon	100 chest X rays a year

Risks shown are for the general population, as well as smokers or nonsmokers. Children may not be at a higher risk.
Source: Environmental Protection Agency.

the average American can afford to pay. However, as EPA notes, "evidence suggests that radon is a national health problem and we'd like this action level to be even lower."

In Finland and Canada, the maximum acceptable radon level has been set more than five times higher, at 22 picocuries/liter. Where did they get that number? Both countries contain exceptionally large areas of granite, a rock relatively rich in uranium. Thus, the radon gas level generally would be higher than in the U.S. A maximum level of 4 picocuries/liter probably is unattainable by most residents of these countries. Recognizing this, and not wanting to alarm the public, the Finnish and Canadian governments set higher limits.

These are examples of how science often takes a back seat to economics and politics. Many African and South American nations contain more granite and other uranium-rich rocks than the world average. But these countries have more pressing concerns than exposure of their citizens to radon, such as food, shelter, clothing, and warfare.

Up to 11,000 Americans might contract lung cancer due to radon exposure in their homes, it is estimated. This is about the same number who die from a fire or serious falls at home each year. The lung cancer threat from a radon level of 8 picocuries/liter is about the same as your risk of dying in an automobile accident.

Despite the apparent minimal impact of radon exposure on nearly all Americans, EPA's position is that radon is a major cause of lung cancer and should be tackled immediately. "If the EPA just sat on the data and did nothing for a decade while waiting for more studies, a lot of people would be exposed when the risk could have been prevented," says their Radon Division head. EPA normally targets hazards that present a risk to the general population of more than 1 in 100,000. For radon exposure, the risk is apparently 4 in 100,000 (10,920 lung cancer cases among 250 million people).

The source of federal money for such an aggressive approach to a possible environmental concern is not clear. But in the case of radon, federal funds are not an issue, because abatement is the responsibility of the homeowner. If you own a dwelling, you must decide whether you wish to spend perhaps $1000 to purify your home, if the level of radon exposure exceeds 4 picocuries/liter of air.

If You Don't Like the Radon Here, Move! Clouding the picture is a study published in 1995, which raises a significant point not considered by the EPA in deciding who is at risk from radon exposure. EPA's estimates assume that you always have, and always will, live in the same "risky" residence, or that all of your residences over a lifetime will expose you to the same average level of radon. This is a poor assumption, because the average American moves ten to eleven times over a lifetime. When this fact is coupled with EPA's estimate that only 5 percent of American homes are at risk, almost all risk vanishes! There is only one chance in twenty of moving to another high-risk house each time you move.

Consequently, EPA's risk estimate for excessive radon exposure has little bearing on actual risk, and greatly overstates it. Remediation of high-radon homes usually has a very small and probably insignificant health benefit for those who have the remediation done.

Testing for Radon

How much radon is in your airspace? There are two kinds of indoor radon measurements, short-term and long-term. A short-term test kit is placed for up to a week in a frequently used part of the house. The charcoal in the kit adsorbs radon gas molecules in the same proportion in which they exist in the air. Laboratory technicians then measure the radon disintegration products in the charcoal and estimate how much radon is in your airspace. This short-term test costs about $25, including the test kit, laboratory analysis, and results.

The long-term test involves placing a detector in your home for three to twelve months. The detector records on film the tracks made by alpha-particles as the radon decays. The long-term test is conducted only if the short-term test indicates a level of radon higher than 4 picocuries/liter of air, and if the homeowner wants a more accurate determination. A laboratory technician counts the alpha tracks on the film to estimate the radon content in your home.

If the long-term test indicates a level exceeding 4 picocuries/liter, EPA recommends reducing the level. This might involve repairing cracks in basement flooring, sealing loose-fitting pipes that enter the house where the floor contacts the ground, or installing a fan to exhaust radon-bearing air from the house (Figure 16.26).

As consumers, we all are justly suspicious of health-related gadgets. Are these radon detectors (a) highly accurate, (b) pretty good, (c) shaky and not to be trusted much, or (d) frauds to make money for radon-detector companies? The answer is: (b) pretty good.

Summary

The words "nuclear" and "radiation" evoke fearful images. Part of the reason is our phobia of the unknown, and part reflects reality—two nuclear bombs dropped on Japan, the nuclear weapons race of the 1950s and 1960s, and nuclear accidents at Three Mile Island and Chernobyl. And part results from the indisputably dangerous properties of radiation; even the smallest amount may be harmful.

Radiation is natural and results from the decay of atomic nuclei. It is of three types. Alpha particles are helium nuclei, composed of two protons and two neutrons. Beta particles are fast-moving electrons. Gamma rays, the most dangerous type, are rays of very short wavelength and high penetrating power. Alpha and beta radiation normally are not considered dangerous unless the particles are inside your body, as when radon-laden dust is inhaled.

Nuclear power plants rely on uranium-235 as a fuel source. As the uranium disintegrates, the heat energy released is used to power steam turbines and generate electricity. When combined with U-235, U-238 in a reactor will produce plutonium-239, perhaps the most poisonous and dangerous gift of the atomic age. Plutonium also is the substance of choice for nuclear explosives.

A principal problem with nuclear power plants is their high risk of accident from equipment failures and human error. The frequency of equipment failures may decrease with time but human error is eternal. A second, and no less important, problem is what to do with highly radioactive waste like spent fuel rods, which remain extremely dangerous for many thousands of years. No disposal method has yet been agreed upon. No permanent storage facility exists anywhere. Current storage is "temporarily" at nuclear plants.

Key Terms

alpha particles	fuel rods	rad
background level	gamma rays	radioactive decay
beta particles	half-life	radioactivity
breeder reactor	high-level radioactive waste	radiation sickness
chain reaction	ionizing radiation	radon
containment building	low-level radioactive waste	rem
control rods	nuclear fission	steam explosion
core meltdown	nuclear reactor	transuranic elements
fluorescence	picocuries	

Stop and Think!

1. Is it possible to completely avoid exposure to ionizing radiation? Explain.
2. For whom was the picocurie named? What killed the namesake?
3. Why is the background level of radiation higher for some people than others?
4. Explain how a decay chain works. What happens to the atomic number and the atomic weight as we move down the chain? What is ejected at each stage?
5. Why does the percentage of U-235 need to be enriched before fission can occur?
6. What salary would you demand to work for a year in a uranium mine? Why?
7. Explain why a natural nuclear reactor was possible prior to 1.8 billion years ago, but not today.
8. Explain the purpose of control rods and water in a nuclear reactor.
9. Why are there no commercial breeder reactors operating in the United States?
10. What salary would you demand to work for a year in a nuclear power plant? Why?
11. Suppose Chernobyl were in Nebraska. How far would the contamination have traveled across the United States? Which way?
12. What is the difference between low-level and high-level radioactive waste? Why do they pose different disposal problems?
13. List some criteria for safe disposal of high-level radioactive waste.
14. What is the origin of radon? Why is it thought to be dangerous?
15. Is the EPA's action level for radon exposure reasonable? Why?

References and Suggested Readings

Ahearne, J. F. 1993. The future of nuclear power. *American Scientist*, no. 81, p. 24–35.

Aldreson, Laura. 1994. A creeping suspicion about radon. *Environmental Health Perspectives*, v. 102, p. 826–830.

Cobb, C. E., Jr. 1989. Living with radiation. *National Geographic*, April, p. 403–437.

Cole, L. A. 1993. *Element of Risk: The Politics of Radon.* Washington, D.C.: American Association for the Advancement of Science Press, 246 pp.

Fishlock, David. 1994. The dirtiest place on earth. *New Scientist*, February 19, p. 34–37.

Freemantle, Michael. 1996. Ten years after Chernobyl consequences are still emerging. *Chemical and Engineering News*, April 29, p. 18–28

Grossman, Dan and Seth Shulman. 1994. Verdict at Yucca Mountain. *Earth*, v. 3, p. 54–63.

Hafele, Wolf. 1990. Energy from nuclear power. *Scientific American*, September, p. 136–144.

Levi, B. E. 1992. Hanford seeks short- and long-term solutions to its legacy of waste. *Physics Today*, March, p. 17–21.

Pooley, Eric. 1996. Nuclear warriors. *Time*, March 4, p. 47–54.

Rhodes, Richard. 1986. *The Making of the Atomic Bomb*. New York: Simon & Schuster, 886 pp.

Rhodes, Richard. 1995. *Dark Sun: The Making of the Hydrogen Bomb*. New York: Simon & Schuster, 731 pp.

Whipple, C. G. 1996. Can nuclear waste be stored safely at Yucca Mountain? *Scientific American*, June, p. 56–64.

Young, J. P. and R. S. Yalow. 1995. *Radiation and Public Perception*. Washington, D.C.: American Chemical Society, 346 pp.

Zorpette, Glenn. 1996. Hanford's nuclear wasteland. *Scientific American*, May, p. 72–81.

"This most excellent canopy, the air."
William Shakespeare

Our Air and Its Quality

Decades ago, only professional meteorologists or climatologists cared about—or even knew about—global warming, air pollution, the ozone layer, the greenhouse effect, climate change, and acid rain. During your lifetime, the quality and purity of our atmosphere has become a serious concern. What has brought these once-obscure topics to our daily news? To what extent are these problems controllable?

To answer these questions, you first need to understand what the atmosphere is and how it works. Then we can consider the changes that are happening to it, and learn which of them result from human activities like driving, generating electricity, or using hair spray.

A basic problem we will address is global warming. There is no question that Earth's average surface temperature has increased since your great-grandparents were young. But how large has the increase been? Why has it occurred? Is this good or bad? Are we causing this warming, or is it natural?

A second basic problem we will examine is atmospheric pollution. It has many sources. Natural sources include Earth's 600-plus active volcanoes and pollen from billions of plants. Artificial sources include blowing dust from construction sites and plowed fields, industrial smokestacks, vehicle exhaust, burning trash, smoking (of any substance), some leaking air conditioners, spray cans, and noise. How serious a problem is each? Which ones can be controlled, and at what cost?

What Is in the Atmosphere?

The air in your next breath is composed almost entirely of two **atmospheric gases,** nitrogen and oxygen. Less than 1 percent of it is other gases, like argon and carbon dioxide (Figure 17.1). Over the past few million years, the proportions of these gases have stayed about the same. However, as shown, a few of the scarcer ones vary. These low-percentage, variable-percentage gases are the troublemakers: water vapor, ozone, carbon dioxide, carbon monoxide, sulfur oxide, nitrogen oxide, and even a little radon. Without these variable gases, this chapter might be only four pages long and this book would be cheaper at the bookstore.

The *proportion* of gases in the air changes very little with elevation above the ground—several miles up, it's still roughly 80/20 nitrogen/oxygen. The major changes as one rises through the atmosphere, as in a hot-air balloon, are *density* (the air quickly gets "thinner," or fewer molecules of it per bucketful), *pressure* (it decreases for the same reason), and *temperature* (the air gets much cooler, fast).

◄
Typical expressway traffic for Bangkok, Thailand.
Photo: Will & Deni McIntyre/Photo Researchers, Inc.

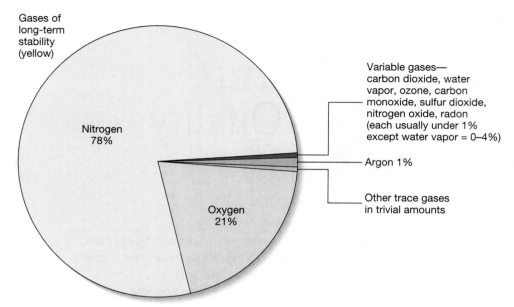

Gases of long-term stability (yellow)

Nitrogen 78%

Oxygen 21%

Variable gases—carbon dioxide, water vapor, ozone, carbon monoxide, sulfur dioxide, nitrogen oxide, radon (each usually under 1% except water vapor = 0–4%)

Argon 1%

Other trace gases in trivial amounts

FIGURE 17.1   What you will inhale with your next breath.

This pie chart shows the gases that make up Earth's atmosphere near the surface. Note that most of the atmosphere is stable in composition. It is the small group of "variable gases" that cause most of the problems. (All percentages are rounded, so they don't add to exactly 100%.)

Let's take our balloon to the seashore (sea level) and go up. At 1 mile height (the equivalent elevation of Denver, the "Mile-High City"), there is only 80 percent of the air at sea level. At 2 miles height, there is less than 70 percent, which is why breathing becomes more difficult high in the Rockies. At 5 miles height, nearing the elevation of Mt. Everest, the oxygen in each breath is only about 35 percent of the amount at sea level. This effect is simulated by the tinted air in Figure 17.2.

Our bodies react to the diminishing oxygen with **altitude sickness,** as shown in Figure 17.2. Headaches develop, some water accumulates in our lungs, brain

function becomes impaired, and eyesight weakens in dim light. Our symptoms will leave and we will recover when we return to lower altitude. (Should we foolishly continue upward, we will pass out in the thin air and freeze in the bitter cold. This happened to a few early researchers who rose too high in hot-air balloons, stayed there too long, and died.)

The highest permanent human settlement is in Tibet, at an elevation of about 16,730 feet. Some dwellings are inhabited briefly at elevations near 20,000 feet. No humans have adapted permanently to altitudes higher than 17,000 feet.

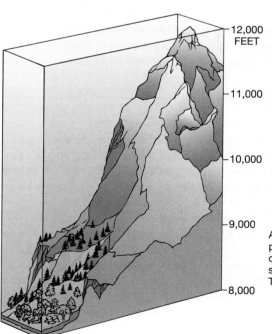

12,000 FEET

11,000

10,000

9,000

8,000

Chronic mountain sickness affects those intolerant to high altitude or cannot acclimate — fatigue and chest pain, increase in red blood cell count, and sometimes, heart failure. The cure: go back down.

Brain problems can occur at 9,000 feet but are more common above 10,000 — mental confusion, hallucinations, drunkenlike walking.

Lung problems commonly occur above 9,000 feet — shortness of breath, severe cough, blood-tinged mucus, headache, lethargy and mild fever.

Acute mountain sickness affects 15 to 17 percent of people who climb to 8,000 feet or higher too rapidly — headache, fatigue, shortness of breath, disturbed sleep, nausea. The cure: go back down.

FIGURE 17.2   What less oxygen does to us.

If you travel to a higher elevation, the "thinner" air can become a problem, or even dangerous. Most of us are adapted to air at lower elevations.

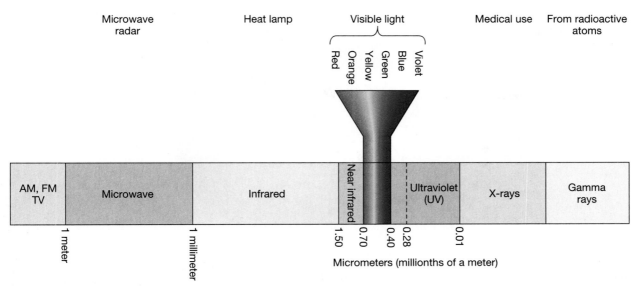

FIGURE 17.3   Electromagnetic spectrum of the sun's radiation.

This range of wavelengths has bombarded Earth for billions of years. The greatest portion of the energy is concentrated in the wavelengths of visible light and invisible near-infrared. This explains why major life processes are adapted to these wavelengths, like vision in animals and photosynthesis in plants.

How Solar Energy and the Atmosphere Interact

The sun is a huge nuclear reactor. It doesn't split uranium atoms, like the human-made nuclear reactors we looked at in the preceding chapter, but instead fuses hydrogen atoms together. Regardless, the sun radiates enormous energy into space. This energy is in the form of electromagnetic waves.

These waves have many different wavelengths, important because different wavelengths have different characteristics. Some are visible (light); some feel warm (infrared); some sunburn us (ultraviolet, or UV); some encourage cancer in cells (recall the X rays and gamma rays from the preceding chapter); and some play havoc with our radio and TV communications (low frequencies). If we spread out the sun's radiation for study, like a rainbow or a prism, we call it the **electromagnetic spectrum** (Figure 17.3). Most of the sun's radiation (81 percent) occurs in the visible and near-infrared wavelengths. (**Infrared** means "below red light" in wavelength. *Near-infrared* means wavelengths near those of red light; *far-infrared* means wavelengths farther from those of red light.)

Fortunately, not all of this radiation reaches Earth's surface, for some of it is dangerous, as indicated. Much of this energy is absorbed by our atmosphere. This absorption is shown for ultraviolet radiation in Figure 17.4. (**Ultraviolet** means "higher than violet" in the spectrum.) But the absorption is not a simple process, like a sponge absorbing all the water it touches. Each gas in the air behaves *selectively* in absorbing different wavelengths.

For example, let us contrast carbon dioxide (CO_2) and oxygen, which absorb energy very differently. The headline is: "CO_2 does not absorb much energy, but oxygen does." Why? Please refer to Figure 17.5. Part A shows the intensity of incoming solar radiation, with an obvious peak in the visible light part of the spectrum. As solar energy arrives, it filters through the atmosphere. When it encounters CO_2 molecules, the far-infrared wavelengths are absorbed, because CO_2 molecules have the right banding to do so. (Note the CO_2 graph in B.) In fact, CO_2 absorbs all wavelengths longer than about 20 micrometers.* But very little solar radiation occurs in such long wavelengths (part A), so carbon dioxide doesn't have much solar energy to absorb. The ultraviolet and visible light energy streams freely past the CO_2 molecules on its way to Earth's surface.

In contrast, molecules of oxygen (O_2) and ozone (O_3) are bonded differently from CO_2, and are able to absorb all wavelengths at the ultraviolet (UV) end of the spectrum (Figure 17.5B). This shields us from dangerous UV radiation, and explains why scientists are so concerned about the ozone layer. If ozone in the atmosphere decreases, as when an ozone "hole" develops, more ultraviolet radiation reaches Earth's surface. In

*A word about *micrometers*—this book uses English measurements, but when discussing solar radiation, we use metric measurement because it is standard practice and the numbers are easier to work with. A micrometer is a very tiny unit, approximately 1/25,000 inch. It is abbreviated μm.

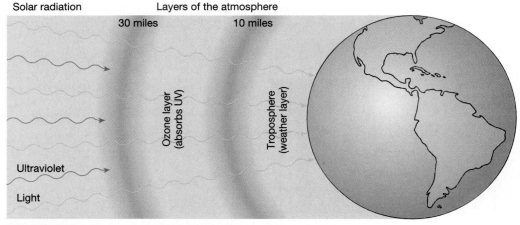

Solar radiation   Layers of the atmosphere
          30 miles   10 miles

Ultraviolet

Light

Ozone layer (absorbs UV)

Troposphere (weather layer)

FIGURE 17.4   Why we need the atmosphere.

Without the atmosphere, we would be immersed in unfiltered solar radiation, including danger-ous ultraviolet (UV) and gamma rays. The atmosphere's oxygen is more than our breath of life. A form of it, ozone, is our overhead shield, absorbing dangerous UV.

FIGURE 17.5   The atmosphere and incoming solar radiation.

(A) Most of the sun's energy is radiated in the wavelengths of visible light and near-infrared (heat), as the peak indi-cates. (B) How Earth's atmosphere filters (absorbs) the incoming solar radiation. The top five graphs show the wavelengths absorbed by each gas. The sixth graph shows the composite absorption of the entire atmosphere. (C) Outgoing re-radiation of longer-wavelength heat energy (far-infrared) from Earth.

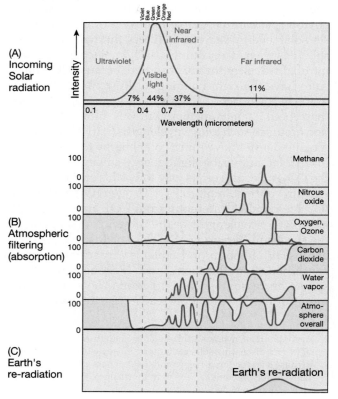

(A) Incoming Solar radiation

(B) Atmospheric filtering (absorption)

(C) Earth's re-radiation

fact, 7 percent more ultraviolet radiation now reaches Earth's surface than did so ten years ago. Increased ultraviolet radiation is harmful to the reproductive cycle of living organisms and increases the risk of skin cancer and vision disorders such as cataracts.

Figure 17.5B has six graphs showing the absorp-tion of different gases in the atmosphere. You can see that no atmospheric gas absorbs much energy in the visible and near-infrared wavelengths. This is the range in which the sun does most of its radiating, so the bulk of these wavelengths reach Earth's surface unimpeded. As you might expect, this is the range sensed by ani-mals (including us), simply because we have evolved over time to fit our environment, adapting to the avail-able radiation. We sense the visible wavelengths with our eyes and the near-infrared (heat) wavelengths with our skin.

Overall, the atmosphere absorbs or reflects back into space about half of the solar radiation that tries to enter. The remaining half reaches Earth's surface, where a third of what remains is reflected back into space. (Think of the light reflecting from snow, ice, and the vast ocean surface.) What is left, about 35 percent of the energy originally received at the top of the atmosphere, is absorbed by Earth's surface, warming it.

Earth's warmed surface then acts like any heated surface: it radiates its heat energy into the air. In effect, Earth receives light energy, *translates it into heat energy*, and re-radiates it back into the atmosphere as longer far-infrared wavelengths. This is shown in Figure 17.5C. The energy re-radiated by Earth is tiny com-pared to that radiated by the sun—about 1/300,000 as much (measured at the 10-micrometer wavelength). Earth's re-radiation is important to understand, for it is the key to the greenhouse effect and to global warming.

Carbon Dioxide and the Greenhouse Effect

As Earth's re-radiation tries to pass through the atmosphere and back out into space, it runs into a brick wall of sorts. Remember the carbon dioxide that absorbed so little of the sun's incoming light and near-infrared wavelengths? About half of the heat energy that is re-radiated from Earth is in longer far-infrared wavelengths, right in the range where carbon dioxide molecules absorb best. The result: this re-radiated heat energy (far-infrared wavelengths) is trapped (absorbed) by carbon dioxide in what we call the **greenhouse effect,** first explained in 1847.

It is so-named because the glass in a greenhouse allows 90 percent of the incoming short-wavelength solar radiation (light) to pass through to the interior, but allows passage in the opposite direction of almost none of the outgoing heat radiation (infrared) emitted by surfaces and plants in the greenhouse. This phenomenon is shown in Figure 17.6. (To experience the greenhouse effect first-hand, sit in a car in the July sun with all the windows rolled up tight. And remember the experience the next time you see a pet in a car on a sunny day!)

Heat retention by the atmosphere due to the greenhouse effect helps maintain Earth's overall warm climate. (Aside from the poles, Earth's climate is quite warm, which allows life to flourish on our planet as it can nowhere else.) If the carbon dioxide in the atmosphere increases, it retains more of the heat being re-radiated by Earth's surface. This can raise the average atmospheric temperature near the surface. Conversely, reducing the CO_2 in the atmosphere can cool temperatures worldwide.

Other Greenhouse Gases

Carbon dioxide is not the only greenhouse gas. The others include water vapor, methane, chlorofluorocarbons (CFCs), and nitrous oxide. The degree to which each contributes to the greenhouse effect is shown in Figure 17.7. Each of these gases is a major absorber of far-infrared radiation, the kind re-radiated by Earth. Water vapor has the greatest effect by far, about 90 percent, and is the most variable day-to-day (think of how the humidity changes with the weather). Together, these gases keep Earth's lower atmosphere about 63 degrees warmer than it would be otherwise.

FIGURE 17.6 The greenhouse effect.

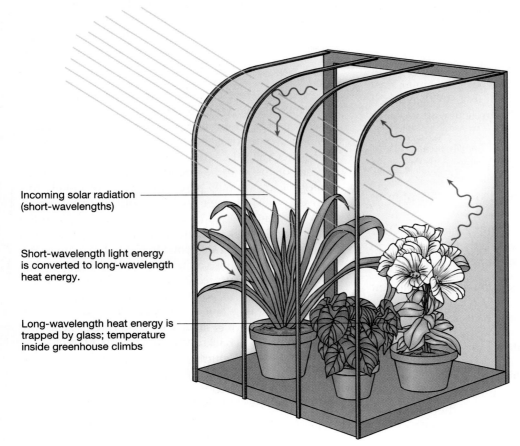

Incoming solar radiation (short-wavelengths)

Short-wavelength light energy is converted to long-wavelength heat energy.

Long-wavelength heat energy is trapped by glass; temperature inside greenhouse climbs

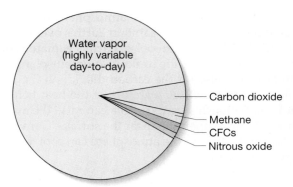

FIGURE 17.7   Gases that contribute to the greenhouse effect.

Water vapor is overwhelmingly dominant, but its presence varies greatly day-to-day with humidity. This pie diagram gives you a sense of relative importance of the different greenhouse gases. Note that CO_2 is the most significant contributor, after water vapor, accounting for about half of the effect. Natural methane gas and human-made CFCs also are significant contributors.

The major concern right now is that our industrial and farming activities are causing these gases to increase, leading to an overall **global warming.** Data that indicate their increase come from two sources: direct measurements made during the past few decades and measurements made on ancient air bubbles trapped in glacial ice (Figure 17.8). Both sources tell the same story:

- Carbon dioxide in the atmosphere remained constant at 280 parts per million (ppm) from at least 10,000 years ago until about A.D. 1850, when it began a steady rise to the present value of 358 ppm (0.036 percent).
- Methane was constant at 0.7 ppm until about 1850, when it began a steady rise to the present value, more than doubled at 1.7 ppm.
- Nitrous oxide has increased from 0.275 ppm to 0.311 during this period.
- CFCs first were formulated in the late 1920s. Over the subsequent seventy years, they have risen to become 0.1 ppm in the atmosphere.

The reason carbon dioxide is such a significant greenhouse gas is not its potency. CO_2 is a comparative "wimp," about sixty times less heat-absorbent than methane (the addition to the atmosphere of a single methane molecule has the absorptive effect of sixty molecules of CO_2). And CO_2 is 4000 times less heat-absorbent than some CFCs. What CO_2 lacks in potency, it makes up for in volume. Many tons of it are pumped into the atmosphere by world industry, the burning of tropical rain forests, and the rotting of plants everywhere.

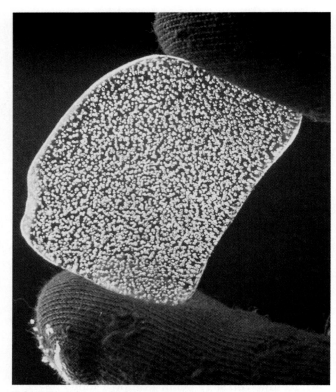

FIGURE 17.8   Bubbles of air trapped in glacial ice.

These bubbles contain microsamples of the air as it was whenever the ice formed, hundreds or thousands of years ago. Measuring gas in these bubbles reveals how much CO_2 existed at that time. *Photo: CSIRO/Science Photo Library/Photo Researchers, Inc.*

Prior to the 1990s, the twenty-four major developed nations were responsible for more than half of greenhouse gases emitted. But this has shifted as less-developed countries have increased their industrial base. As of 1992, these developing nations produced 52 percent of the world's energy-related carbon dioxide emissions, up from 43 percent in 1970. Very likely this increase will continue into the twenty-first century.

In countries like the United States, we have the wealth, legal system, technology, and management skill to create environmental controls and enforce them. But countries with millions of impoverished citizens are in a poor position to enforce environmental controls while maintaining a stable government. Their first priority is food and clothing, not environmental protection. The burning of petroleum, coal, and wood is the fastest way to accomplish this. The result is bountiful CO_2 molecules warming the atmosphere.

Sulfate Aerosols and Global Cooling

Some aerosols cause a natural counterbalance to the heating effect of the greenhouse gases because they reflect solar radiation back into space. An **aerosol** is a microscopic particle suspended in air, like dust or a liquid droplet, not more than a few micrometers in size. **Sulfur dioxide** gas forms an aerosol that has considerable reflective ability. Sulfur dioxide is emitted by explosive volcanic eruptions and burning fossil fuels. The gas reacts with oxygen and is converted to microscopic sulfate droplets (aerosols). These aerosols circle the globe, carried by high-altitude winds.

Aerosol particles less than 2 micrometers in size are weak absorbers of solar wavelengths, but they are efficient *scatterers*. Thus, the net effect of these aerosols is to make the atmosphere more of a mirror to incoming solar radiation, reflecting it back into space. As noted in Chapter 12, a permanent layer of sulfate aerosols exists about 15 miles above Earth. Explosive volcanic eruptions have been a permanent feature of Earth's surface since the time of its formation about 4.5 billion years ago.

The effect of explosive volcanism on aerosol production, and the aerosol's effect on Earth's surface temperature, was clearly demonstrated by the eruption of Mt. Pinatubo in the Philippines in 1991 (Chapter 12). The cloud of sulfate aerosols, gases, and water vapor belched skyward during the eruption circled the globe in twenty-two days, carried by prevailing winds at 12–16 miles altitude. For a year after the eruption, temperatures in the lower atmosphere where we all live were 1 or 2F° cooler than had been measured in recent years. A complete reversal of greenhouse warming was produced—temporarily. But as the volcanic aerosols from Mt. Pinatubo dissipated, greenhouse warming resumed.

Based on both recent volcanic history and inferences from ancient eruptions, the 1991 eruption of Mt. Pinatubo was not exceptional in the volume of sulfate aerosols it generated. It was simply a brief spike in the normal background level of aerosols in the air. That background level, however, has increased significantly during the past 150 years due to human activities, particularly the burning of fossil fuels (Chapter 14).

To reduce acid precipitation and other pollution effects, there is an ongoing attempt to decrease the sulfur emitted by smokestacks. To the extent this succeeds, Earth's average temperature will increase, because fewer sulfur emissions will mean fewer sulfate aerosols and less reflectivity of the atmosphere, allowing more solar radiation to reach Earth's surface. The decrease in lung-damaging aerosols and acid precipitation is good, but other effects are debatable. It is ironic that the phasing out of fossil fuel use in favor of non-polluting energy from water, wind, or solar power will significantly increase global warming!

Burning of biomass, such as forests and plant stubble, has increased steadily worldwide for more than a century. Aerosols from this biomass burning include fine black soot, which is extremely effective at blocking sunlight. One ounce of airborne soot aerosols spread over an area of 300 square feet in the atmosphere can decrease the solar radiation reaching the surface by perhaps half. The biomass that is burned deliberately spews about twelve times more soot into the air than is produced by fires from lightning strikes or other natural causes.

Global Warming: Is It, or Not?

A change in temperature might have serious consequences. Global warming could cause a rise in sea level that would flood coastal lowlands, an increase in weather extremes, and damage to forests and croplands.

New York Times, 1995

There is no dispute among scientists that greenhouse gases are increasing their presence in the atmosphere. It follows that the average world temperature should increase, and it appears to have done so. Several studies indicate an overall rise of about 0.9F° during the past hundred years (Figure 17.9, red and orange curves).

Global Warming: Most Scientists Are Convinced

Meteorologists and climatologists who study global change have predicted an increase of 4F° by 2100, based on current understanding of Earth's atmosphere. But, considering that CO_2 has increased about 25 percent, that methane has more than doubled, and that large volumes of other greenhouse gases have been added during the past 150 years, it is surprising how *small* the temperature increase has been. What this demonstrates is our inadequate understanding of the atmosphere and how it interacts with lithosphere, hydrosphere, and biosphere.

Our poor grasp of the atmosphere is not for lack of effort. It simply reflects the great complexity of Earth's systems. Three examples illustrate what scientists are up against.

Example 1: Clouding the issue. An increase in greenhouse gases raises temperature, which increases evaporation from the oceans that cover 71 percent of Earth. The addition of this water vapor, which absorbs strongly in the infrared range (Figure 17.5B), should further enhance the greenhouse effect. But water vapor is what clouds are made of, and clouds at present reflect about 30 percent of the incoming solar radiation, plus absorb another 15 percent of it. Thus, more water vapor creates more reflective clouds, which actually may *decrease* the solar radiation that reaches Earth's surface. Current opinion is that the net effect of cloudiness is to *counter* the effect of increasing carbon dioxide.

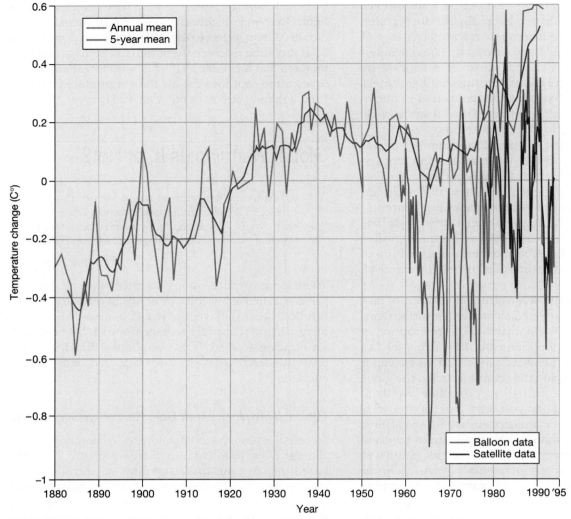

FIGURE 17.9 Average global temperatures—increasing, decreasing, or can't tell?

The upper curves (red and orange) show an overall increase of about 0.9° from 1880 to 1992, quite significant. For these upper curves, each point is an average of thousands of ground-based measurements, and most recording stations lie in or near heat islands centered upon cities. The lower curves (blue and purple) show no detectable trend, either warming or cooling. *Upper curves from NASA; lower curves from* Science, *1995, v. 267, p. 612.*

Example 2: Sinking CO₂ in the ocean. The preindustrial atmosphere contained about 0.028 percent carbon dioxide. Industrial processes may have caused the increase to the present 0.036 percent over the past 150 years. But this increase is less than we might expect. The reason is that carbon dioxide is fairly soluble in water, and thus much of it now is dissolved in the vast ocean, removed from the air. Thus, we call the ocean a *sink* for storing carbon dioxide. The increased carbon dioxide also is moderated by its interaction with existing limestone deposits.

Example 3: Adding heat may not raise the temperature. Water has a very high capacity to store heat, and the world ocean is huge. Therefore, the world ocean can absorb a vast amount of heat *without its temperature rising.* As greenhouse gases increase the heat retained, the ocean absorbs a lot of it, so world temperatures do not rise as much as we would expect.

Because of our limited understanding of interactions such as these, some scientists believe that the effect of increasing greenhouse gases will be far less than predicted in some doomsday scenarios. But again, we just don't know.

An Alternate View: What Global Warming?

When we look at Figure 17.9 (upper curves), a continuing rise in average world temperature seems evident.

But is it possible that these data are biased, and that global warming is *not* happening? A minority of scientists think so. What is their evidence?

Their evidence is data from weather balloons and satellites (Figure 17.9, blue and purple curves). The balloon data start only in 1958 and the satellite data did not become available until 1979, so both records are quite short compared to the ground-based data that started in 1880 (red and orange curves). But the point is that the ground-based data and airborne data sharply disagree. Neither set of airborne data shows an upward trend in temperature. How can this be? The balloon and satellite data were adjusted for any effect of clouds between them and the ground, So if the scientists crunched their numbers correctly, all three sets of data should look the same.

Those who believe the satellite data to be correct say they have the only truly global measurements. The Earthbound data are suspect because they come mostly from urban heat islands that give the appearance of warming at ground level. Also, there are few measurements from the oceans that cover 71 percent of Earth's surface. Because of the high specific heat of water, changes in air temperature near the ocean's surface are almost certainly less than over the land, if any are present at all.

So, the data in the red and orange curves do not show the true picture of Earth's temperature changes. Advocates of the ground-based data respond that we live on the land, not the sea. Even if the temperature is warming only on the land because of expanding urban heat islands, that's where most of us live, so the temperature for most of us *is* rising.

Has Earth's Surface Temperature Always Varied?

From the media, you might get the impression that before people and their fuel-burning industries came along Earth's average surface temperature was constant. Hardly. During the last few hundred thousand years, much of North America has been covered with thick ice as glaciers grew and advanced southward. At other times, the continent has been largely ice-free, as it is now. No significant fossil fuel use occurred until about 200 years ago, so the causes of these earlier air-temperature swings (of perhaps 10 degrees) must have been natural. Possible agents were variations in solar radiation, Earth's orbit, or the amount of greenhouse gases in the atmosphere. Natural variations of several degrees have occurred over only tens of years without human intervention.

So, if worldwide temperatures now are rising, as most scientists believe, what is the cause? How much of the increase is natural and beyond our control, and how much is from polluting we can control? Look again at Figure 17.9. From 1940 to 1965, there was concern about global *cooling* rather than global warming! We generously pumped greenhouse gases into the air during this twenty-five–year period, yet the climate cooled. Does this mean that the greenhouse gases we produce are insignificant as a cause of Earth's warming or cooling?

These questions are hotly debated among meteorologists, environmental scientists, and others interested in global change. The social and political consequences will affect you for the rest of your life. Should more of your tax money be spent on research in this area? Should industry use technologies that reduce greenhouse gases, and raise their prices to pay for it? Will the Antarctic ice sheet melt, inundating large seaports? Are the media capable of reporting this issue intelligently and responsibly?

Global Warming: Hype City

Unfortunately, the global warming debate is driven more by hype and political agenda than by scientific information. Witness this infomercial run a few years ago by the Sierra Club (italics added):

1. America's heartland *might* have to live with temperatures over 90° for almost one-third of each year . . .
2. Chicago *could* expect two months of 90° temperatures each year...
3. *If* the weather in Dallas goes over 100° for two and a half months a year, imagine what *could* happen in Phoenix...
4. America's heartland *could* have trouble growing the corn, wheat, and oats that we need . . . The same is true for California and Florida . . .
5. Drinking water *could* become a problem.

None of these statements is patently untrue, but each is clearly designed to inflame public passion for legislative action. The ifs, mights, and coulds are there, but are likely to be ignored by the casual listener.

Further, these ads reference specific locations, such as America's heartland, despite the impossibility of predicting the climate for any region using current technology. Forecasting climate trends for specific cities is even farther in the future. Also, extreme high temperatures normally bear little relation to *average* temperatures, which are the only ones that researchers of global change attempt to predict.

Now, consider the following statements, just as accurate as those listed above but based on current understanding of the likelihood of global change (Michaels, 1993):

1. The growing season in America's heartland *could* lengthen by up to three months.

2. Chicago currently has more than seven weeks of below-freezing temperatures. In the next century, this *could* be down to three.

3. If there are no more below-zero readings in Chicago, imagine what *could* happen in International Falls, Minnesota.

4. Over 800 laboratory experiments demonstrate that carbon dioxide enhances plant growth. Fruits and nuts *could* proliferate in California.

5. All our climate predictions say that rainfall *should* increase globally. We may have more water available to feed a growing population.

The Sierra Club ads ran as free PSAs (public service announcements), following supporting messages by Meryl Streep, William Shatner, and other Hollywood luminaries. Would a television station run the contrasting five predictions as a public service? What is the definition of a public service announcement, and who decides which messages should be heard? Don't believe everything you hear.

Who Wins and Who Loses in Global Warming?

The gradual average temperature increase of 0.9F° worldwide has not been catastrophic for humans, our fellow creatures, or plants, as far as we can determine. Our parents and grandparents had no problems traceable to the change. But what about a gradual change of 2F° or 4F°?

It happens that Mother or Father Nature already performed this experiment, and not long ago. It occurred within historic times, between the years A.D. 1400 and 1850, a period known as the "Little Ice Age." This was global *cooling,* not warming, but it indicates the scope of change.

In eastern Europe, where your author's ancestors lived at the time, average temperatures were 2–3F° lower than today, and alpine glaciers expanded. The water in the mouth of the Hudson River at New York City froze occasionally, so that one could walk across the ice from Staten Island to Brooklyn or Manhattan. American colonists endured winters more bitter than we do today. However, in Miami and Houston, it probably was cooler, less humid, and more pleasant in August than is now the case. (Equally low temperatures occurred about 1000 B.C. and 6000 B.C., but records of the consequences are poor from that long ago).

Gradual temperature change of a few degrees, either up or down, is to be feared only if we assume that the world's present climate is the best of all possible worlds. But it is "best" only because we are adjusted to it at the moment. Gradual change might require decades, or a century, or longer, allowing time to adjust. (An exception would be a volcano with the punch of Krakatau or worse, which could pump so many aerosols into the atmosphere that world temperatures could drop a few degrees within months, causing tremendous disruption in agriculture and availability of heating fuel.)

Winners in Global Warming

Who might emerge as winners if a significant temperature increase occurs gradually? Earth's present climate is harshly frigid toward the poles and at high mountain elevations. These regions probably would experience nicer living conditions—milder winters and more rainfall. Canadians, Norwegians, Swedes, Finns, Russians, Alaskans, and those dwelling high in the mountains might appreciate more warmth. Perhaps they were pleased that 1995 was the warmest year wordwide since record-keeping began in 1856. The world average temperature now is 58.7°F.

It is generally agreed among those who construct computer models of future climates that global rainfall will increase as Earth warms. More heat evaporates more seawater, which forms more clouds and more rain. One study predicts a 3–11 percent increase in average global precipitation. However, it would be unevenly spread over Earth's surface. It also would fall unevenly within individual countries like the U.S., as shown in Figure 17.10A. Proof is that, over the past century, as average global temperature increased 0.9F°, precipitation has changed variously at different latitudes—look at the patterns in Figure 17.10B.

Apparently, the tropical rainforests near the equator are now receiving 11 percent less rain than they did in 1900, a decrease that occurred almost entirely during the 1960s. There is no obvious explanation. The low-latitude deserts along 30° North have seen an increase in rainfall of perhaps 1–2 percent, while the cold-to-frigid regions at high latitudes are now receiving about 4 percent more precipitation than in 1900. (Data for the Southern Hemisphere are too sparse to break down into latitude bands.)

These precipitation data are the best currently available, but we must cautiously interpret them. To illustrate why, consider that two or three thousand years ago the Middle East was considerably wetter. This is revealed by sand-piercing radar that sees well-developed stream channel patterns a few feet beneath the drifting sand of the Sahara Desert (Figure 17.11). Stream channels of similar age also exist beneath desert sands in Kuwait and Saudi Arabia. What made the climate change so drastically? Modern tampering with the atmosphere by injecting greenhouse gases began long after the Middle East and North Africa dried up. We simply don't know why the climate changed here.

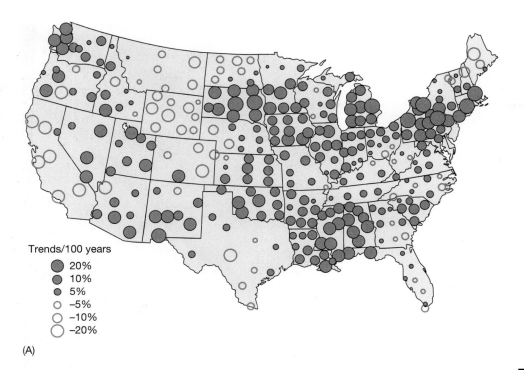

(A)

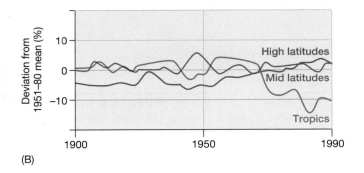

(B)

FIGURE 17.10 How rainfall patterns change over time.

(A) Precipitation trends reveal uneven distribution. *Source: Consequences, Spring 1995, v.1, no. 1, p. 5.* (B) During this century, as average global temperature increased 0.9F°, precipitation has changed variously at different latitudes. *Source: Thomas R. Karl, Richard W. Knight, David R. Easterling, and Robert G. Quayle,* Trends in U.S. Climate during the Twentieth Century.

In summary, investigators agree that our present understanding of small-scale climate change is rather primitive. Forecasts of future changes should improve over the next few decades as our understanding improves.

The amount of carbon dioxide in the atmosphere is increasing, and an increase in carbon dioxide accelerates photosynthesis and plant growth. There is evidence that this has already begun. One recent study predicts that a doubling of atmospheric CO_2 would result in an enormous rice surplus in Japan. In fact, all food grains might benefit from higher atmospheric CO_2 and more rainfall. Who would not want more photosynthesis and crop growth, and a general shrinking of drought-prone regions? One meteorologic expert says that "increasing the carbon dioxide content of the atmosphere is the most cost-effective action man could take to improve planet Earth as a cradle of life." Global warming may make many of Earth's peoples winners.

Losers in Global Warming

The losers from the temperature rise are coastal cities. We already know what will happen: as temperature rises, glaciers and ice sheets melt faster to fill the ocean basins deeper, the warmer ocean waters expand, sea level rises, coastal erosion increases, and people are forced to move inland (or, perhaps more accurately, to the new coastline—review Chapter 10). Natural sea-level rise has been in progress for the past 18,000 years—and it has proceeded *in the absence of civilization's greenhouse gases.* The added gases only make the inevitable slightly more rapid.

Sea-level rise will affect a large part of the world's population, because so many people live along coastlines, near sea level. The important question becomes: How fast will temperatures increase and sea level rise? Will there be time for a gradual, orderly retreat from the rising waters? During the past hundred years, world

(A)
(B)

FIGURE 17.11 Ghost rivers.

The sands of the Sahara desert conceal ancient river systems, now made visible by *Landsat's* penetrating radar. It can peer through the sand to depths of 30–90 feet. The exposed area on the right covers about 30 miles by 180 miles. The big valley equals the present Nile River valley in width. Clearly, it formed when the region had a far moister climate. *Photos: (left) NASA/Johnson Space Center and (right) Jet Propulsion Laboratory/U.S. Geological Survey.*

average sea level has risen 3.4 inches, not a notable problem (see Figure 10.20). But the future is murky. Leading climate scientists differ widely on how fast it will happen. In no case will waters rise like the tides, but the fate of homes within a mile or two of a present seashore over the next 50 to 100 years is more problematic.

It makes good sense to discourage new construction within a mile or so of the present coastline, slowly phasing out existing settlement. Existing industrial plants and home owners should be given tax incentives to move inland. This seems a wiser approach than the present subsidies to rebuild storm-damaged structures in continually threatened nearshore areas along low-lying coasts.

The global warming issue provides cultural anthropologists with a case study of how we protect established dogma and reject information that does not fit, or that we don't like. We need systematic, thorough studies of worldwide impact and adaptation to climate change. From these we could identify places of greatest vulnerability and plan accordingly.

Ozone

Ozone is in the news. You hear of ozone alerts and of thinning of the ozone layer high overhead. What is ozone? How does it form? How is it destroyed? What are its hazards?

Most of the oxygen molecules you inhale are the garden variety that have two atoms bound together (O_2). But sometimes three oxygen molecules become bound together to form **ozone** (O_3). It has a pungent scent and is very active chemically, attacking your mucus membranes. At ground level, ozone normally exists in harmless concentrations of only 0.2–0.4 ppm (0.00002–0.00004 percent). But it can become more concentrated as a result of car-exhaust pollution in cities, creating breathing and lung problems. Ozone alerts are now common in some larger cities.

However, most of the atmosphere's ozone does not exist near the ground. About 90 percent of it is concentrated in the famous **ozone layer,** between about ten and twenty-two miles altitude. Here its concentration attains about six times that at sea level (Figure 17.12). The ozone layer is crucial to life on Earth because it shields us from most of the sun's ultraviolet radiation. (Recall ozone's filtering effect on incoming solar radiation shown in Figure 17.5B.)

If the ozone layer did not exist, intense UV radiation would cause more skin cancer, damaged immune systems, eye cataracts, genetic mutations in our offspring, and poorer farm crops. Increased UV also disturbs the reproduction and growth of plankton in the ocean. These floating, one-celled plants are the base of the entire ocean food chain, so a large decline in plankton would shake sea life at its roots. A healthy ozone layer is essential to healthy life on Earth.

Ozone is formed in the stratosphere from regular oxygen (O_2) in the air. When struck by the sun's rays, some O_2 molecules split apart into single atoms of oxygen (just plain O_1, written O). These atoms are very unstable in this state and immediately bond with remaining O_2 molecules to form ozone, O_3. These ozone molecules are unstable too; they begin to break up by combining with nitrogen, hydrogen, and chlorine in the atmosphere. Prior to human technology, the ozone concentration in the upper atmosphere remained pretty stable, naturally balanced between its continual creation and destruction.

The Troubled Ozone Layer

In 1985, however, satellite measurements revealed that the protective ozone layer was thinner above Antarctica

FIGURE 17.12 Where ozone occurs in the atmosphere.

The high-altitude concentration of ozone is the ozone layer that protects us from harmful UV radiation. The lesser concentration of ozone near the surface can become a health hazard in cities.

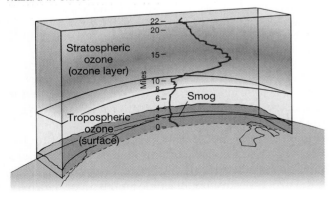

than elsewhere around Earth. This alarmed scientists, for they knew the potential consequences of increased UV penetration. They monitored the thinning. It has grown worse, and now is called the "ozone hole" (Figure 17.13A).

Ozone in the atmosphere is measured in Dobson units. A **Dobson unit** is one part per billion of ozone in the air. In 1955, above Antarctica in the springtime, ozone measured 320 Dobson units. By 1995, it had declined to only 90 (Figure 17.13B). If ozone keeps declining at the same rate, virtually none will remain by 2005.

The thinned area now is larger than the continent of Antarctica, and extends over southern South America. Further, a decrease in ozone concentration is occurring around the globe. An Arctic (North Polar) thinning is becoming nearly as severe as the Antarctic thinning. Ozone levels during winter 1995 were as much as 40 percent below pre-CFC levels (that is, pre-1920s), and a general thinning is spreading worldwide (Figure 17.14). Human skin cancer rates are rising, probably in response to the thinning.

What has caused this problem in the ozone layer?

Ozone and CFCs

In the late 1920s, chemists invented chemicals that combined chlorine-fluorine-carbon atoms into **chlorofluorocarbons (CFCs).** Because of their unusual chemical and physical properties, they have many uses: as coolants in refrigerators and air conditioners (under DuPont's tradename, Freon), as cleaning agents for electronic components, as aerosol propellants in many spray cans, and in styrofoam drinking cups and insulation.

CFCs are large, heavy molecules. Nevertheless, the churning atmosphere easily lofts them up into the ozone layer. The chlorine, fluorine, and carbon atoms in CFC molecules are arranged in an extremely stable way. As a result, CFCs survive in the atmosphere for several decades. Some types can last several centuries.

When the chemists trotted CFCs out of the laboratory, they never dreamed that their creation would destroy ozone. But it works like this: CFC molecules are quite stable at Earth's surface. However, once they rise into the ozone layer, they are broken apart by solar radiation. This releases chlorine atoms, which proceed to break up ozone. Figure 17.15 shows this process.

Each chlorine atom can destroy 100,000 ozone molecules, so a little chlorine (a little CFC) goes a long way toward destruction of the protective ozone layer. And there are now five times more chlorine atoms in the upper atmosphere than before CFCs were invented.

Some deny the effectiveness of CFCs as destroyers of the ozone shield. They point out that many volcanic eruptions spew hundreds of times more chlorine into the

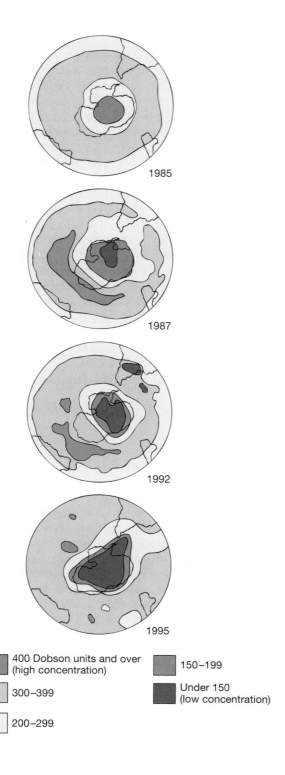

1985

1987

1992

1995

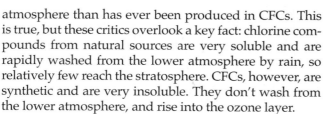

■ (dark)	400 Dobson units and over (high concentration)	■ (medium)	150–199
■ (light)	300–399	■ (darkest)	Under 150 (low concentration)
□	200–299		

(A)

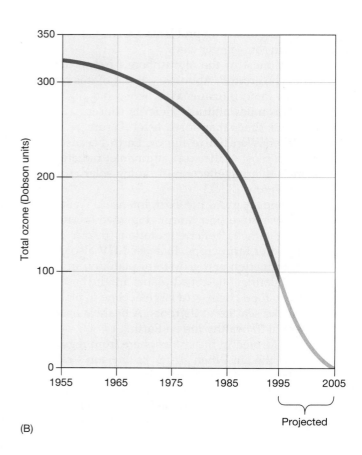

(B)

FIGURE 17.13 Thinning of the ozone layer over Antarctica.

(A) First noted in 1985, the thinning of the ozone layer over Antarctica has progressed dramatically. Notice that the area of lowest ozone concentration is nearly centered over the South Pole. The thinned area covers about 8.5 million square miles, slightly larger than the combined areas of the U.S., Canada, and Mexico. *From* Our Planet, *v. 7, no. 5, 1996, p. 19.* (B) If the thinning continues at present rate, the ozone layer will virtually disappear over Antarctica by 2005, leaving the continent wide open to UV radiation. *Data from* New Scientist, *September 10, 1994, p. 11.*

atmosphere than has ever been produced in CFCs. This is true, but these critics overlook a key fact: chlorine compounds from natural sources are very soluble and are rapidly washed from the lower atmosphere by rain, so relatively few reach the stratosphere. CFCs, however, are synthetic and are very insoluble. They don't wash from the lower atmosphere, and rise into the ozone layer.

The chlorine atoms in CFCs are responsible for 75–80 percent of the ozone depletion in the Antarctic. The remaining 20–25 percent results from bromine, an element similar to chlorine. Bromine atoms are 100 times less common than chlorine, but are about 50 times more powerful in destroying stratospheric ozone. Half of the ozone-destroying bromine atoms occur naturally

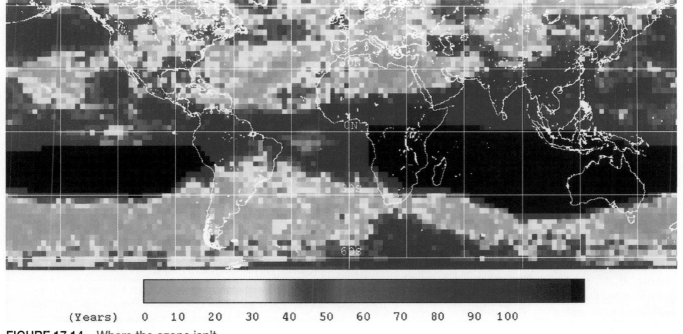

(Years) 0 10 20 30 40 50 60 70 80 90 100

FIGURE 17.14 Where the ozone isn't.

The lighter patches are thinning areas of ozone. These areas are at greater risk from UV radiation, which now can pass through the atmosphere more easily. *Scripps Institute of Oceanography, image courtesy of SeaSpace Corporation.*

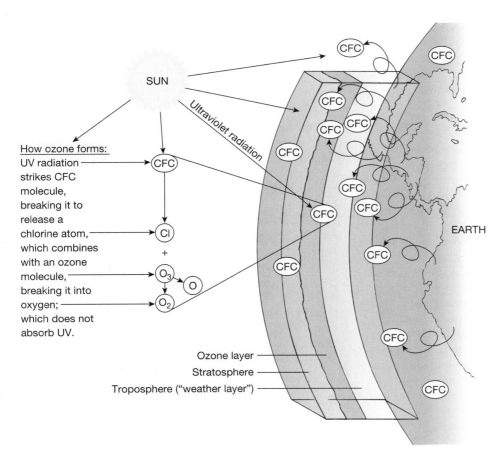

FIGURE 17.15 A dangerous partnership.

How human-made CFCs and ultraviolet radiation from the sun work together to break up the protective ozone layer.

SUN

Ultraviolet radiation

How ozone forms:
UV radiation strikes CFC molecule, breaking it to release a chlorine atom, which combines with an ozone molecule, breaking it into oxygen; which does not absorb UV.

CFC

Cl

+

O₃   O

O₂

EARTH

Ozone layer
Stratosphere
Troposphere ("weather layer")

and the other half are released into the environment by human activities—spreading of pesticides and biomass burning. Unlike CFCs, bromine is short-lived, and will disappear shortly after the U.S. stops using it in 2001.

Is There a Solution to the Ozone Problem?

The ozone thinning is most severe over Antarctica because of the extreme cold. The ozone-destroying reactions seem to occur on the abundant ice crystals in the frigid Antarctic air. During the Antarctic spring (in October down there), conditions conspire against ozone. The reappearance of sunlight after months of darkness increases chlorine production in the stratosphere. The extreme cold forms more ice crystals that allow the chlorine to destroy the gas. This is when the seasonal thinning known as the ozone "hole" appears.

The decrease has been so alarming to many nations that an accord was reached almost instantly (considering the normal continental-drift pace of international negotiations) to ban the production of CFCs and related chemicals by the year 2000. CFC production peaked in 1988 at 1260 tons and has declined yearly, to below 300 tons in 1994. Chlorine abundance in the troposphere seems to have peaked in 1994 and begun a slow decline in 1995. However, stable CFCs will not decrease significantly for some years (Figure 17.16).

Meteorologists estimate that the ozone thinning over the Antarctic will persist until about 2050. Commercial alternatives have already been developed to replace CFCs, but they are more costly. If the projected savings in medical expenses materialize, however, the elimination of CFCs may be quite cost-effective.

Ozone Depletion, Cancer, and You

All of this has been pretty abstract, so let's make it real and apply it to you. You are at a sunny beach, scantily clothed. Hopefully, you have covered your body with a high-rating sunblock. But suppose you did not. Ultraviolet radiation that the ozone layer failed to catch comes beaming down upon your healthy young skin.

Figure 17.17 shows how UV attacks you. UV penetrates to different depths, depending on its wavelength. Particularly worrisome are the wavelengths of about 315, 366, and 440 micrometers, to which your skin is particularly transparent. Looking back to Figure 17.5B, note that ozone and oxygen are the only absorbers of solar radiation at these short UV wavelengths. Hence, depletion of stratospheric ozone lets more of this ionizing radiation reach your skin, and more passes through your protective outer layers of skin. As we saw in the last chapter, excessive exposure to ionizing radiation can spell cancer.

Like most organisms exposed to sunlight, we have evolved defense mechanisms against damaging

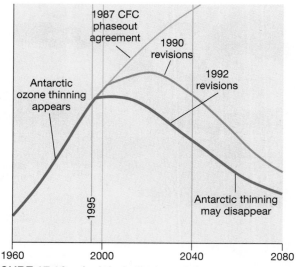

FIGURE 17.16   A global effort.

World leaders seldom agree on anything, yet they perceived the urgent need to halt CFC production. In 1987, they approved the Montreal Protocol, an agreement to phase out CFCs (top curve). In 1990, they tightened it, based on new scientific evidence (middle curve). In 1992, they agreed to phase out CFCs even faster (lower curve). You can see the effect these decisions had on the projected improvement in the ozone layer over the South Pole. *Data from World Meteorological Organization.*

UV radiation. Our skin responds to increased exposure by thickening its epidermis and by developing pigmentation that protects the more vulnerable and deeper cells. This is the reason all people "tan" when sunning for hours at the beach. Radiation that passes through the darker pigmentation and causes molecular cell damage is repaired or replaced.

Skin cancer historically has been commoner for Caucasians at low latitudes, where solar radiation is more intense (Figure 17.18). After a sunny day at the beach, a typical exposed cell of your epidermis has between 100,000 and 1,000,000 damaged sites in its DNA. Despite your body's formidable DNA repair system, some damage may persist and result in the uncontrolled cell growth we call cancer.

The thinner the ozone layer, the more UV radiation you receive, and the greater your likelihood of skin cancer. For each 1 percent loss of ozone concentration, skin cancer is expected to increase about 2 percent. If you lived near the South Pole, where ozone concentrations have decreased 70 percent in recent years (from 350 Dobson units to 100), your skin cancer risk would increase by 140 percent.

Eye Damage

Your eyes lack a protective covering of skin, but they are less exposed, shielded by your eyebrows and eyelashes.

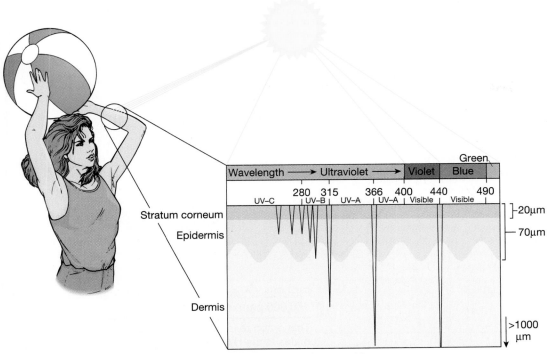

FIGURE 17.17 What really happens when you frolic in the sunshine.

Without a good sunblocker, the UV penetrates your skin's layers, causing ionization in the cells, with potential for developing skin cancer. Enjoy your next trip to the beach!

Long-term eye damage from UV is difficult to prevent because you feel no pain. By the time you become aware of the effect, it is usually too late. Researchers estimate a 0.5 percent increase in cataract incidence for every 1 percent decrease in ozone concentration. Hence, based on the estimated 6–7 percent reduction in ozone at midlatitudes during summertime, cataracts might increase up to 3 percent early in the next century.

Protective Measures

How can you protect yourself from the increased UV? Don't stay home with the shades drawn! Common sense and technology come to your rescue. The partial loss of our ozone shield does not mean that death rays will strike when you venture outdoors. Excessive UV exposure always has been carcinogenic, and the ozone thinning only adds to the risk. Even without the thinning, you have a one-in-six chance of developing skin cancer in your lifetime. The loss of up to 10 percent of ozone since the 1980s could raise the risk at least 15 percent, but no one is certain. A significant increase in eye cataracts, which now afflict one in ten of us, could also occur.

Doctors advise:

- Minimize your time in direct sunlight between 10 A.M. and 3 P.M.

- Wear sunglasses outdoors in bright sunlight. Use sunglasses treated to absorb UV rays. Ordinary sunglasses can be worse than none at all, because your pupils dilate under dark lenses, making it easier for UV to damage the retina, the screen at the back of your eye.

- When in the sun for extended periods, wear fabrics that have a tight weave so radiation cannot pass through. Cover your head with a broad-brimmed hat so the delicate rims of your ears are shaded.

- In summer, when wearing shorts and T-shirts, use a broad-spectrum sunscreen with a sun protection factor (SPF) of at least 15. A lotion with an SPF of 15 screens 94 percent of the UVB radiation.

The National Weather Service publishes a daily UV index. The index covers fifty-eight U.S. cities and is based on the effects of UV radiation on skin types that burn easily. Levels of UV exposure are:

0–2 minimal
3–4 low
5–6 moderate
7–9 high
10+ very high

For example, Atlanta's UV index from July 19 through August 4, 1995, never fell below 5, and peaked

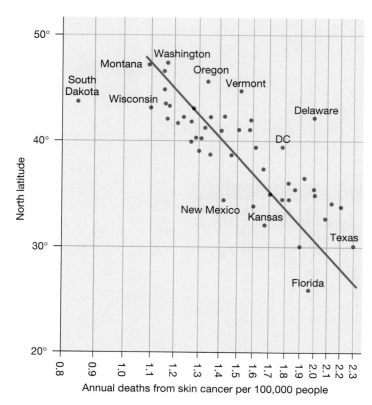

FIGURE 17.18   Annual deaths from skin cancer per 100,000 people.

This graph reflects skin cancer deaths among white males in the U.S. during 1960. *From J. D. Coyle, R. R. Hill, and D. R. Roberts, eds., 1982,* Light, Chemical Change and Life, *Milton Keynes: Open University Press.*

at 9. Think of the UV exposure to participants during the 1996 Olympic games!

Air Quality and Pollution

Residents [near Houston, Texas] say air contaminants, such as trichloroethane, have been responsible for personal tragedies. Among them has been a rash of birth defects: in one four-month period, 11 deformed children were born; other children suffered serious heart and reproductive-organ problems. Most of the citizens have fled their homes.

Time, 1993

When you can't breathe, nothing else matters.

American Lung Association

Air pollution has been around as long as human technology. The Greeks and Romans smelted ores to recover lead and copper, polluting the air with smoke from their fires and fumes from the smelters. But big-league air pollution awaited the Industrial Revolution of the mid-1800s. The changeover from firewood to hotter-burning coal for home-heating and industry assured bountiful air pollution. Burning coal pumped smoke, soot, and sulfur gases into the air.

Ideally, the air we breathe should contain nothing but naturally occurring gases, those that were here before human technology began to puff its stuff into our air-space. Unfortunately, no such clean air exists anywhere

on Earth today. The World Health Organization estimates that 70 percent of urban dwellers breathe air that is unhealthy at least some of the time; another 6 percent breathe marginal air. Mexico City has the worst air on Earth; machines dispense oxygen for those who need it. In Britain, canisters of pure oxygen are sold as an "aid to healthier living, countering the effects of smog and pollution." In Beijing, you can spend $6 to breathe pure oxygen in an oxybar.

The problem is less severe in the United States, because we are doing something about it. The Clean Air Act of 1970 (amended in 1977 and 1990) has improved air quality nationwide. However, of our 255 million people, about 100 million live where the Act's air-quality standards have not been attained.

Certainly some places have cleaner air than others (Figure 17.19A). Antarctic air is cleaner than what people breathe in Los Angeles, Houston, New York City, and New Orleans (Figure 17.19B). For that matter, Antarctic air is cleaner than the air at Arizona's Grand Canyon, where you can't see across the canyon for about 100 days each year because of pollution. Restricting our view to large cities worldwide, the air is generally bad. In smoggy Los Angeles, the coroner can tell how long you have lived there by the condition of your lungs! Researchers estimate annual deaths worldwide from air pollution at between 50,000 and 100,000.

What pollutants are in our air, and how do they get there? The contaminants that are monitored regularly

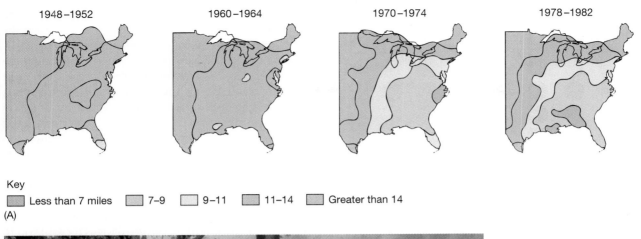

1948–1952   1960–1964   1970–1974   1978–1982

Key

◾ Less than 7 miles   ◾ 7–9   ◾ 9–11   ◾ 11–14   ◾ Greater than 14

(A)

(B)

FIGURE 17.19 Air quality in the eastern United States, based on summertime visibility.

(A) Southern Louisiana "clearly" has some of the worst air in the United States. *Source: Reidel Publishing Company.* (B) These women express an all-too-common sentiment in areas with poor air quality. *Photo: Sam Kittner/Sam Kittner Photography.*

by urban air-quality authorities are sulfur dioxide (SO_2), particulate matter, nitrogen dioxide (NO_2 or more generally NO_x), carbon monoxide (CO), ozone (O_3), and lead (Pb). Based on the most recent data for twenty major world cities, none can pass muster for all six categories of contaminant (Figure 17.20). Where does all this stuff come from? Let us look at all six.

Sulfur Dioxide (SO_2)

Sulfur dioxide (SO_2) is created by burning fossil fuels (coal and oil) and by smelting sulfide ores to recover metals (review Chapter 13, smelting). On average, every 100 pounds of coal contains about 2.5 pounds of sulfur, so burning coal in electric power plants produces 80–85 percent of annual U.S. sulfur dioxide emissions. Most of the other 15–20 percent comes from petroleum refining and smelting of sulfide ores.

Once sulfur dioxide leaves the smokestack at a power plant or smelter, it dissolves in water vapor to form sulfuric acid. When the water vapor coalesces into raindrops, snowflakes, or hail, acid precipitation occurs. This process is illustrated in Figure 17.21. (Acid precipitation is discussed later in the chapter.)

Sulfur dioxide gas has a strong, irritating odor. It corrodes paint and metals and injures (or kills) plants and animals, especially crops such as alfalfa, cotton, and barley. Inhaling it can severely damage your lungs. Children exposed to sulfur dioxide have more respiratory infections. Healthy adults may experience sore throats, coughing, and breathing difficulty when exposed to high concentrations. The constriction of air passages caused by this gas is particularly bad for asthma sufferers.

Emissions of sulfur oxides in smokestack gases can be reduced with **scrubbers,** which use chemical

	Sulfur dioxide	Particulates	Lead	Carbon monoxide	Nitrogen dioxide	Ozone
Bangkok	◔	●	◔	◔	◔	◔
Beijing	●	●	◔	○	◔	●
Bombay	◔	●	◔	◔	◔	○
Buenos Aires	○	◑	◔	◔	◔	◔
Cairo	○	●	●	◑	○	◔
Calcutta	◔	●	◔	◔	◔	○
Delhi	◔	●	◔	◔	◔	◔
Jakarta	◔	●	◑	◑	◔	◑
Karachi	◔	●	●	○	○	○
London	◔	◑	◔	◑	◔	◔
Los Angeles	◔	◑	◔	◑	◑	●
Manila	◔	●	◔	○	○	○
Mexico City	●	●	◑	●	◑	●
Moscow	○	◑	◑	◑	◑	○
New York	◔	◔	◔	◑	◔	◑
Rio de Janeiro	◑	◑	◔	◔	○	◔
São Paulo	◔	◑	◔	◑	◑	●
Seoul	●	●	◔	○	◔	◔
Shanghai	◑	●	○	○	○	○
Tokyo	◔	◔	○	◔	◔	●

● Serious pollution (WHO guidelines exceeded by more than a factor of two)

◑ Moderate to heavy pollution (WHO short-term guidelines exceeded regularly at certain locations)

◔ Low pollution (WHO short-term guidelines exceeded occasionally)

○ Insufficient data

FIGURE 17.20 Overview of air quality in 20 large cities.
This is a subjective assessment of monitoring data and emissions inventories, based on guidelines from the WHO (World Health Organization).

reactions to remove sulfur dioxide from combustion gases. Most are of the *lime-limestone wet scrubber* type and are used in several hundred U.S. coal-fired power plants. The scrubber converts the sulfur oxides into a sludge of calcium-sulfur compounds, which is easily disposed of. Also, The Clean Air Act is forcing many power plants to burn coal that contains less sulfur.

Instead of burning coal with 2 or 3 percent sulfur, some coal with 1 percent sulfur now is burned.

Despite these measures, total sulfur dioxide emissions into the atmosphere have barely declined over the last several decades. Yet, more urban areas now meet air-quality standards for this contaminant. How can this be? The slight decline was accomplished by building taller smokestacks so that the gases are inserted higher in the atmosphere, thereby decreasing ground level concentrations. In addition, older power plants in urban areas (where most air-quality measurements are made) were shut down and new ones were built in rural areas.

In other words, the sulfur oxides were simply moved somewhere else, not a happy solution in the long run. Air still circulates worldwide. The solution to pollution is not dilution or moving away. The solution to pollution is abatement, or to cut back.

Particulate Matter

Particulate matter consists of solid and liquid particles (aerosols) suspended in the atmosphere. There are thousands of different kinds, but the common ones include soil, soot, lead, asbestos, and substances that result from chemical reactions in the air (Figure 17.22).

The smaller the particle, the more likely it is to slip past your natural filters (nose hairs, mucous membranes) and pose a threat to your respiratory tract. Particles smaller than 2.5 micrometers get by your body's defenses and cause the most lung damage. Some particulates, such as diesel soot and wood smoke, also carry toxic compounds like benzene and dioxin, increasing the cancer risk. As discussed in Chapter 2 (Figure 2.29), the size of the particle determines how far into your respiratory tract it can travel.

A 1996 study by the Natural Resources Defense Council says that particulates cause 64,000 deaths each year in the U.S. This is about twice the number who die in automobile accidents and more than three times the number who die from homicides. In the most polluted cities, lives are shortened an average of one to two years. The five worst U.S. areas for particulates are in California.

Coal-fired power plants are the largest soot generators. Their emissions can be controlled with an **electrostatic precipitator,** which causes small particles to clump together. Precipitators are now used at nearly 1000 U.S. power plants. The devices are up to 99 percent efficient and use little electric power.

An alternative to the precipitator is the **baghouse,** which is composed of fabric bags that capture ash before it leaves the smokestack of an industrial plant. Air is drawn up through the bags in much the way that a vacuum cleaner operates. The air passes through, but the particulates are captured by the bag. Efficiency is the same as with the electrostatic precipitator. The baghouse

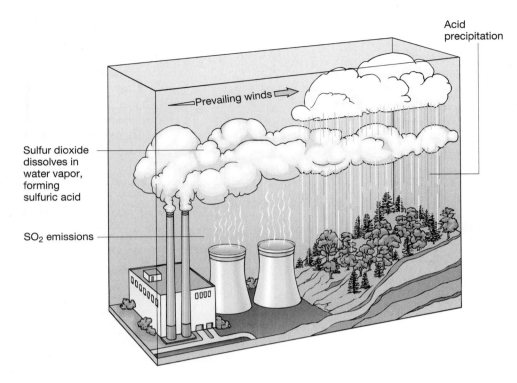

FIGURE 17.21 Formation of acid precipitation.

Sulfur dioxide from power plants and smelters can generate acid precipitation downwind.

Acid precipitation

Prevailing winds

Sulfur dioxide dissolves in water vapor, forming sulfuric acid

SO_2 emissions

also can capture sulfur dioxide. If limestone is injected into the combustion gases, sulfur dioxide bonds to it, forming solid particles that are removed by the baghouse.

Studies in more than a dozen cities demonstrate that the number of airborne particulates correlates with increased mortality, particularly among the elderly and those with cardiovascular disease. Other possible causes were considered and found not applic-

able, so the correlation is seen as cause-and-effect. The particulates likely are bad for younger and healthier people as well, but the effect is probably cumulative and does not show up until late in life.

Nitrogen Dioxide (NO_2)

Nitrogen dioxide (NO_2) and related nitrogen oxides (NO_x) are produced when hydrocarbon fuels are

FIGURE 17.22 You see this woman's hands; what must her lungs look like?

A carbon factory in Romania has spread soot over a 10-mile radius. *Photo: Heidi Bradner/Panos Pictures.*

burned, especially in power plants and motor vehicles. Like sulfur dioxide, nitrogen dioxide generates acidity (nitric acid) when it reacts with oxygen and water vapor in the atmosphere. The harmful effects of the nitrogen gases are the same as those of the sulfur gases.

Modifications of automobile engines and use of the catalytic converter have successfully reduced emissions of hydrocarbons, ozone, and carbon monoxide. But they have not affected NO_x levels, which climbed steadily from 300 to 309 parts per billion during 1977–1990. Procedures have been modified at coal-fired power plants to decrease nitrogen oxide emissions, but increases in coal combustion at electric generating plants and increased vehicle miles have canceled the effectiveness of this change (Figure 17.23).

Carbon Monoxide (CO)

Carbon monoxide (CO) is poisonous because it interferes with the ability of the blood to carry oxygen to your body. Unborn children and those with heart disease are particularly at risk. So are people who live at high elevations—where the air is thinner, a reduced oxygen supply is more critical. Depending on the CO concentration and how long you are exposed, CO poisoning symptoms include headache, nausea, dizziness, coma, and even death. Unfortunately, CO is a colorless, odorless gas, so you cannot detect it with your senses.

In every city studied, hospital admissions for heart failure have increased whenever the CO concentration increases. It is dismaying to note that these CO increases have been below the concentration that initiates federal action.

Earth's number-one source of CO is from burning gasoline in vehicles. Thus, the greatest CO concentrations occur in congested, high-volume traffic. To remove CO from vehicle exhaust, a catalytic converter has been required on American cars since the 1970s. However, vehicles in most other countries lack catalytic converters. For example, Earth's air-pollution capital, Mexico City, has relatively few of its 2.5 million vehicles so-equipped.

A catalytic converter eliminates most carbon *mon*oxide by converting it into less-harmful carbon *di*oxide. But carbon dioxide is a major greenhouse gas. As often happens, one technological solution (the catalytic converter) diminishes one problem (carbon monoxide), but adds to another (global warming).

Ozone (O_3)

As pointed out in the introduction to this chapter, ozone is both friend and foe. In the stratosphere, ozone is vital because it absorbs harmful UV radiation. But at ground level, ozone can become a serious air pollutant. It is corrosive, irritating to mucous membranes, and it is the chief component of urban **smog.**

Most ground-level ozone is a "secondary" air pollutant, so-called because it doesn't come belching directly from smokestacks or car exhaust pipes. Instead, it forms indirectly, in three different ways:

1. *Ozone forms when sunlight stimulates a reaction in car exhaust, between nitrogen oxides (NO_x) and hydrocarbon fumes.* Nitrogen oxides (NO_x) absorb solar energy, splitting into nitric oxide (NO) and single-atom oxygen (O). The single-atom oxygen

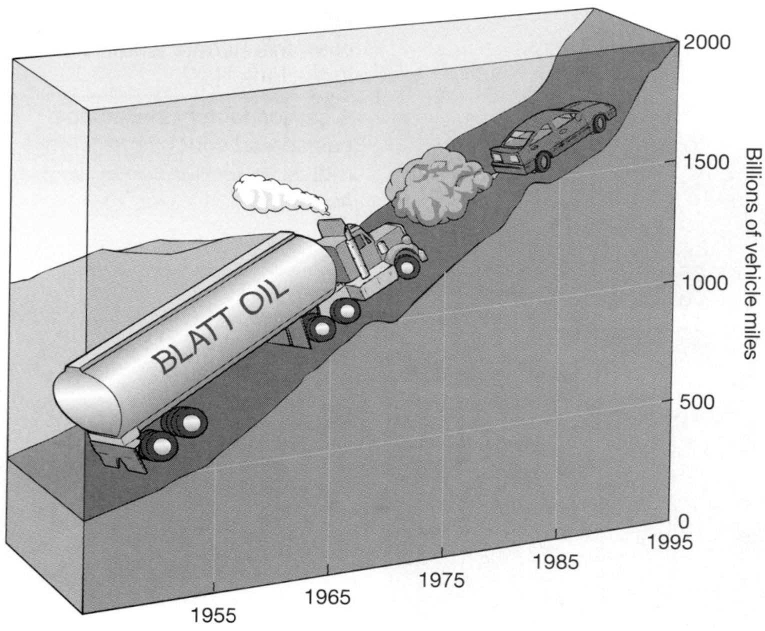

FIGURE 17.23 Why nitrogen oxide levels continue to rise.

We drive more every year. Despite the pollution-control technology added to our cars in the past forty years, automotive air pollution remains a serious nationwide problem. We have reduced emissions per car, but have *quadrupled* the number of miles driven! If we assume an average fuel economy of 25 miles per gallon, the 2000 billion miles traveled in 1995 consumed about 80 billion gallons of gasoline. *Source: U.S. Department of Energy.*

(O) then rapidly combines with normal oxygen gas (O_2) to form ozone (O_3). Because of this, ozone forms where traffic is heavy.

2. *Ozone is produced when methane oxidizes.* Methane levels have increased over the past century, due to increased food production worldwide (methane originates in wetlands, rice fields, and the intestinal tract of cattle). When methane oxidizes, ozone is released.

3. *Ozone is produced when carbon monoxide oxidizes.* Carbon monoxide originates during the combustion of gasoline in motor vehicles (see the previous section, Carbon Monoxide).

Ozone is the main cause of smog-induced eye irritation, impaired lung function, and damage to trees and crops. It also reduces visibility (Figure 17.24). EPA has estimated that roughly half of the American population is routinely exposed to more than 0.12 ppm ozone. This is the maximum deemed acceptable under the Clean Air Act. The Los Angeles basin has by far the worst ozone problem, violating the EPA's standard nearly one-third of the days each year (Figure 17.25).

In the late nineteenth century, the level of ozone was only a tenth of what is now the EPA maximum, about 0.01 ppm. Today, abundances greater than the EPA level often are recorded in western Europe, California, the eastern U.S., and Australia.

Lead (Pb)

The symptoms of **lead poisoning** include damage to blood, nerves, and organs; reduced mental ability; and elevated blood pressure (Chapter 8). The lead in urban

(A)

(B)

FIGURE 17.24 Mexico City on a rare clear day (A) and on a normal day (B).

The main culprit is thousands of automobiles. Their exhaust contains the nitrogen oxides and hydrocarbons necessary to form dense smog. *Photos: Larry Reider/SIPA Press.*

FIGURE 17.25 Areas that violated EPA's ozone standards in September 1994.

From U.S. Environmental Protection Agency's National Air Quality and Emissions Trend Report.

☐ Ozone

air comes mostly from leaded gasoline. What is lead doing in gasoline? In the 1920s, a lead compound was added to automotive fuel to make engines perform better, and, for the next seventy years, lead from car and truck exhausts was part of our daily diet. Burning leaded gasoline emits microscopic lead particles into the atmosphere. They can remain airborne for weeks before settling, circulating around the globe. When inhaled, many of the particles reach the deepest part of the lungs, where they are absorbed with 100 percent efficiency. Over the past twenty years in the United States, lead-based gasoline additives have been phased out, resulting in a 50 percent decrease in the blood level of lead in the average American.

In 1996, leaded gasoline finally became extinct in the United States. By federal law, only unleaded fuel now can be sold. Because the major industrial countries have most of the cars and are phasing out leaded gasoline, the lead added to the world's gasoline dropped 75 percent between 1970 and 1993. But in Third World countries, leaded gasoline still dominates. In Mexico and China, the market share of leaded gas is 70 percent; in Spain and Russia, it is 95 percent; and in India and Nigeria, it is 100 percent. In these countries, leaded gasoline continues to power millions of vehicles, pumping lead into the atmosphere for all of us.

Other sources of lead in the environment include automotive batteries (ever lift one? They have lead plates inside), waste motor oil, and lead smelters (mostly in the West). Another source, particularly dangerous to inner-city children, is old house paint (Chapter 8). Sometimes it contains lead pigment, which children ingest by chewing on painted objects. Modern paints do not use lead pigments.

Non-Automotive Pollution Sources

The combustion of fossil fuels in motor vehicles is the greatest single source of air pollution, simply because cars and trucks pump pollutants into the air around the world every day—188 million vehicles in the U.S. and 500 million worldwide. However, the car is not the only source. According to EPA, 10 percent of all ozone and carbon monoxide in American air can be blamed on millions of other gasoline-powered motors. For example, America's 80,000,000 gasoline-powered lawn mowers emit as much air pollution each year as do 3,500,000 new cars. Here are some other examples, from EPA data:

Driving a car:	Emits the same pollution as:
100 miles	a leaf blower running for 1 hour
200 miles	a chain saw running for 1 hour
500 miles	a farm tractor running for 1 hour
800 miles	a motor boat running for 1 hour

Temperature Inversions

Why does it grow cooler as you climb a mountain or ascend in a hot-air balloon? You are moving into thinner air that has fewer molecules with heat energy. Fewer molecules are available to transfer heat to your skin. So, it is normal for air temperature to drop as you rise in elevation. Temperatures are quite frigid where commercial aircraft fly.

Sometimes, however, a **temperature inversion** develops. In this situation, the normal temperature gradient of the lower atmosphere is turned upside-down. Instead of warmer air being at the surface, colder air is near the ground. Figure 17.26 shows the

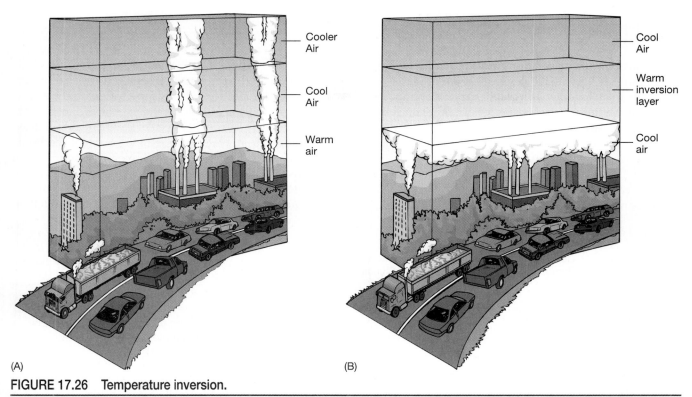

FIGURE 17.26   Temperature inversion.

(A) Normally, air temperatures are warmer at ground level and grow cooler with elevation. Warmer air at ground level is less dense, so it rises, carrying up pollutants with it. (B) In a temperature inversion, air is cooler at ground level, with a layer of warmer air overlying it. The cooler, denser, polluted air at ground level cannot rise. This keeps pollutants at ground level, where they accumulate, trapped beneath the warmer air layer overhead.

result: ground-level air, which normally rises (A), is cold and stays at ground level (B). The air just sits there. If the air is clean, this is not a problem. But if the air is smoggy and polluted, it can spell respiratory problems, pollution alerts, and even disaster.

A brief, harmless temperature inversion occurs on most nights. At night, the ground and the air near it cool faster than the air higher up, so an overnight inversion develops. But the inversion quickly changes the next morning when the sun reheats the ground. On cloudy days, this daily breakup of the inversion cannot occur and the previous day's smog remains. If this situation persists for several days, pollution buildup can become so severe that a smog alert is issued. Those with breathing problems are advised to remain indoors, and everyone is advised to reduce outdoor activities. Most affected are people with heart disease, emphysema, or asthma, as well as joggers and bicyclists.

Temperature inversions occur for another reason. When a high-pressure weather system compresses the upper air, it becomes warmer than the air below, creating an inversion. Inversions caused by stationary high-pressure systems are common in winter and late summer. On the West Coast, they are common throughout the year. For example, Los Angeles has smog and ozone levels that are considered dangerous for more than 100 days each year. Many result from an inversion caused by high atmospheric pressure.

Cities that lie in bowl-shaped valleys have an added problem, for the pollutants cannot even escape sideways. Examples are Los Angeles and Mexico City.

How dangerous can a long-lasting inversion be? Smog trapped by an inversion can mean a quick death for the vulnerable—infants, the elderly, and those with heart or respiratory problems like asthma or emphysema. The first pollution emergency of this type occurred in Donora, Pennsylvania (near Pittsburgh) in 1948. Twenty people died and about 7000 were injured when a temperature inversion trapped sulfur dioxide gas emitted by a zinc smelter in a confining river valley.

Four years later, coal-burning by London homeowners created four days of severe air pollution (sulfur dioxide and particulate matter) that killed 4000. This happened again in London in 1991, when a buildup of traffic fumes created smog during four windless days of temperature inversion. Deaths during this period shot up 10 percent over normal, implicating the smog in the death of 160 people.

Acid Precipitation

We are surrounded by solutions that are acidic and basic. Lemon juice is very acidic; household ammonia is very basic. Mix them together and they neutralize one another. Extremely acidic or basic substances can be hazardous.

We measure acidity on the pH scale. pH means "potential of hydrogen," and refers to how many hydrogen ions exist in a solution. The pH scale is shown in Figure 17.27. Note that 7 is neutral (neither acidic nor basic). Numbers lower than 7 are increasingly acidic; numbers higher than 7 are increasingly basic.

The pH scale is not linear. It is logarithmic, so a pH reading of 6 has ten times more hydrogen ions (is ten times more acidic) than a pH of 7. A pH of 5 has 100 times more hydrogen ions than pH 7. A pH of 4 has 1000 times more hydrogen ions than pH 7. Thus, as numbers get lower, the solution quickly grows much more acidic.

Pure natural rainwater has a pH of 5.6, so it is mildly acidic. (The acidity is caused when carbon dioxide dissolves in water vapor to form weak carbonic acid.) This is not the "acid rain" we talk about. When scientists speak of "acid rain," they mean moisture that is more acidic than 5.6. Figure 17.28 shows examples: pH values of 4 abound in the East. One rain that fell on

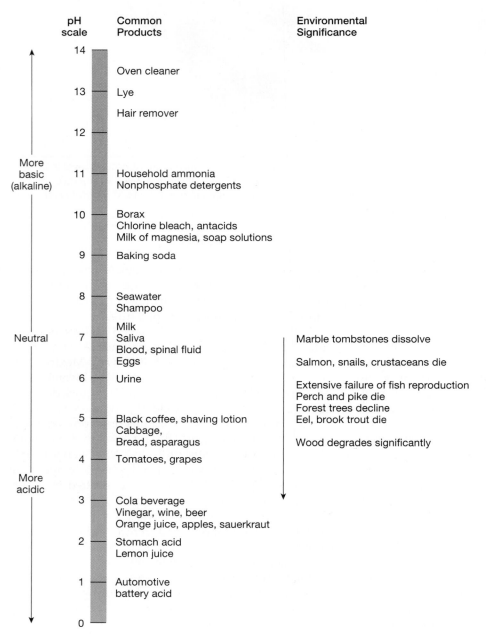

FIGURE 17.27 How we measure acidity: the pH scale.

Numbers below 7 are acidic; numbers above 7 are basic. Shown are the acidities of familiar products and the environmental significance of certain pH levels.

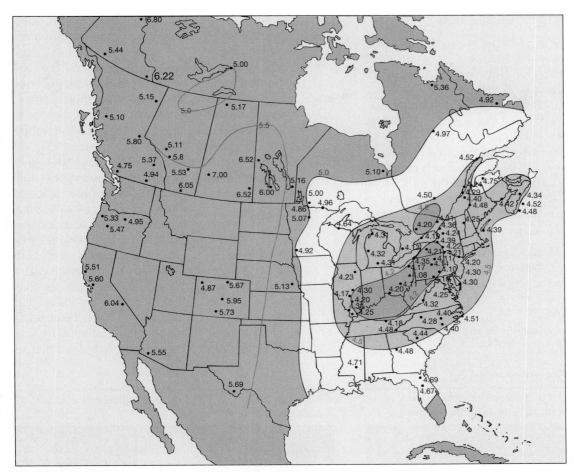

FIGURE 17.28 This map of pH reveals a pattern of air pollution.

Values shown are the annual average pH. Natural acidity of around pH 5–6 generally appears in western North America. But in the East, most values drop to a pH of 4-point-something, which is distinctly acidic and a potential problem. *From Environmental Protection Agency.*

Wheeling, West Virginia, had a measured pH of only 1.5, which is 12,640 times more acidic than unpolluted rain!

Acid precipitation can occur as acid rain, acid snow, acid sleet, and acid hail. Acid fog also occurs. Some acidity is normal, as noted. The real problem is human-caused acid precipitation. Sulfur dioxide from smokestacks combines with water vapor in the atmosphere to produce sulfuric acid. Earth's plants and creatures often cannot tolerate this additional acidity.

As noted, the slight decline in sulfur dioxide emissions from coal-fired power plants has been accomplished in large part by using taller smokestacks. The emissions are thus lofted higher into the atmosphere, to be carried by winds away from the urban areas. The average height of U.S. smokestacks has tripled since the 1950s; the height of the tallest has more than doubled.

The redistribution of sulfur dioxide emissions in this manner is good news for cities but bad news for rural areas. In essence, the city's pollution is being dumped on the countryside. In the case of these sulfur emissions, the dumping occurs as acid precipitation.

The smokestacks that emit the sulfur dioxide are located mostly in the upper Midwest, and the prevailing winds in this area blow toward the northeast. The result is the pattern of acid precipitation shown in Figure 17.28. But acid precipitation is not confined to the Northeast; the acidity of precipitation has increased rapidly during the past twenty years in the southeastern states, too. Pollution does not recognize political borders.

Further west, rainfall acidity also is increasing. Along the West Coast, the main cause is different. It appears to be the oxides of nitrogen released in automobile exhaust, which form nitric acid. Because the nitrogen compounds are emitted at ground level in car exhausts, rather than hundreds or thousands of feet high from smokestacks, the nitric acid is deposited closer to its source. In Los Angeles, acid fog is a more serious problem than acid rain. Acid precipitation is equally bad in industrial areas of China and Europe.

What Are Acid Precipitation's Effects?

At pH 5.5, most freshwater fish decline in health; at 5.0, few survive; and below 4.0, few species of anything remain. A study of lakes in New York's Adirondack Mountains, where the pH is commonly about 4.3, showed that a third of 214 mountain lakes had no fish. In Nova Scotia, the pH of some rivers has fallen so low that salmon cannot live in them.

Three factors determine the impact of acid precipitation on an area: How acidic is it? How much precipitation? What is the chemical composition of the soils, rocks, and water in the area? Considered together, these factors determine how living organisms will be affected. Here are some of the impacts:

- In acidic water, leaves and other organic debris decompose slowly because decomposing bacteria cannot survive in the acidic solution. The plankton and insects die, depriving trout of their food; in turn, the fish starve.
- Acidic solutions dissolve heavy metals, so they become mobilized from the rocks around the lake.

Consequently, many lakes have very high concentrations of highly toxic aluminum, cadmium, zinc, and mercury.
- Acidic water leaches nutrients from the soil faster, reducing their availability to plant roots. Acidic water retards the decomposition of forest litter, slowing the natural recycling of nutrients.
- Acid precipitation injures plants and retards their growth (Figure 17.29A).
- Acid precipitation deteriorates buildings and monuments in urban areas two to three times faster than in rural areas (Figure 17.29B). Cement, concrete, and marble all are composed of calcium carbonate, which is extremely soluble in acid solutions. Remember the midnight ride of Paul Revere? His tombstone doesn't; acid precipitation has erased his memory. Many world-famous buildings are crumbling slowly under the acidic onslaught: the Parthenon in Greece, the Taj Mahal in India, and Mayan ruins in Mexico. Some of Michelangelo's great sculptures in marble are dissolving. In Krakow, Poland, stone monuments are described as

FIGURE 17.29 Acid precipitation damage.

(A) Acid rain devastated this group of pine trees. *Photo: Lange/Mauritius/Phototake NYC.* (B) The effect of air pollution and acid rain is quite evident on this "defaced" statue at the cathedral at Rheims, France. *Photo: William E. Ferguson.*

(A)

(B)

"melting." Acid rain is very hard on modern sculptures, too: it damages automobile finishes and makes the metal bodies rust faster.

Is Air Quality Improving?

Yes, it is improving overall. Public consciousness of environmental quality, government regulations, and technological innovations now in our automobiles, power plants, and industrial plants have improved air quality. Figure 17.30 shows the downward trend for six pollutants over the past twenty years. Lead is the brightest spot, due to phasing out of leaded gasoline. Most other pollutants have dropped by 20–40 percent and are still heading downward. Some of the drop in sulfur dioxide is illusory because measurements are made near ground level and miss the SO_2 vented far above ground by extra-tall smokestacks.

But the basic problem remains—the burning of coal and petroleum products to drive the American industrial machine. As long as we burn fossil fuels, we will have bad-air days and will need the Pollutant Standards Index shown in Figure 17.31.

Until fossil fuel use is reduced significantly, air quality will continue to suffer. Reduction will not come easily, as evidenced by the public's resistance to electric cars that need frequent recharging, expensive solar panels, and nuclear power (which has other environmental problems; see Chapter 16).

However, the switch to alternative energy sources is inevitable, as the world warehouse of fossil fuels dwindles in 50 or 100 years. The sooner we devote more resources and mental energy (which does not burn coal or oil) toward finding suitable replacements for them, the better for human civilization.

Civilization's Microclimates

We have been looking at global atmospheric problems: global warming, the troubled ozone shell that surrounds Earth, air quality, and acid precipitation. Let us now look at atmospheric concerns that are more local: the urban heat island, humidity, wind chill, pollen, mold, cigarette smoke, and noise.

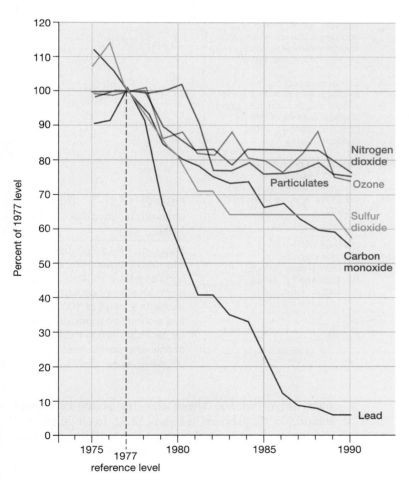

FIGURE 17.30   Air pollution has improved over the past twenty years.

Most dramatic is the reduction in lead, achieved by phasing out leaded gasoline. Sulfur dioxide levels have declined about 40 percent, due partly to increasing the smokestack height at power plants (but paradoxically, this has increased acid rain, with taller stacks pumping SO_2 higher into the atmosphere). *From B. J. Nebel and R. T. Wright, 1993, Environmental Science, 4th ed., Englewood Cliffs, New Jersey: Prentice Hall, p. 346.*

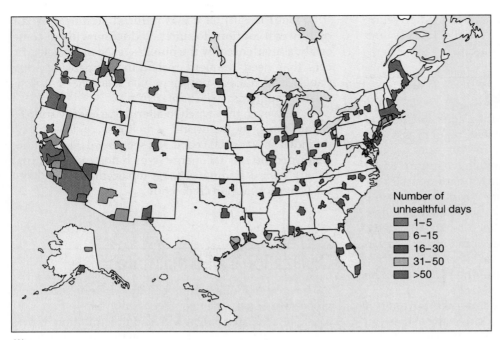

(A)

PSI Value	Description	General Health Effects	Cautions
400–500	Hazardous	Premature death of ill and elderly. Healthy people experience adverse symptoms.	All persons should remain indoors, keeping windows and doors closed. All persons should minimize physical exertion and avoid traffic.
300–399	Hazardous	Premature onset of certain diseases; aggravation of symptoms; decreased exercise tolerance in healthy persons.	Elderly and persons with existing heart or lung disease should stay indoors and avoid physical exertion. General population should avoid outdoor activity.
200–299	Very unhealthful	Aggravation of symptoms, decreased exercise tolerance in persons with heart or lung disease; widespread symptoms among the healthy.	Elderly and persons with existing heart or lung disease should stay in-doors and reduce physical activity.
101–199	Unhealthful	Mild aggravation of symptoms in susceptible persons; irritation symptoms among the healthy.	Persons with existing heart or respiratory ailments should reduce physical exertion and outdoor activity.
51–100	Moderate	None	
0–50	Good	None	

(B)

FIGURE 17.31 Number of unhealthful days by county in 1990.

Based on EPA's Pollutant Standards Index (PSI), a day is considered unhealthful when any one of five pollutants exceeds a value of 100. The five are carbon monoxide, sulfur dioxide, nitrous oxide, ozone, and particulates. *From U.S. Environmental Protection Agency.*

Urban Heat Island

Large cities are home to 75 percent of Americans, and the percentage of city dwellers is increasing worldwide. Urban centers have grown large enough to create their own microclimates. These take the shape of a **heat island** like that shown in Figure 17.32. In cities, natural vegetation is replaced by concrete, asphalt, and glass. Vertical buildings replace horizontal forests and agricultural land. Fossil fuels power most cities (oil, natural

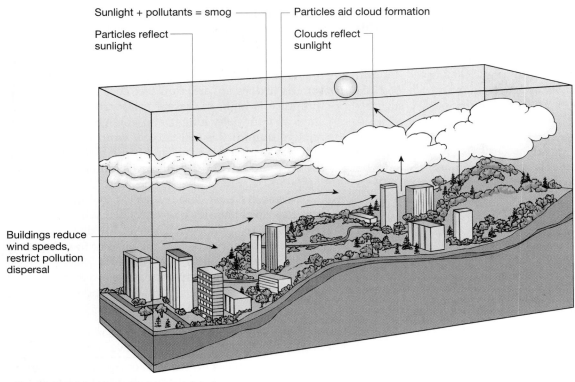

FIGURE 17.32 The urban heat island.

Concrete, brick, and asphalt surfaces absorb solar energy during the day and re-radiate it day and night, warming the urban area, often by several degrees. Dust and smog often concentrate in this warmed environment, creating a dust dome of murky, polluted air.

gas, gasoline, coal), creating air pollution. The result is that urban environments usually are warmer and have more pollution, cloudiness, fog, and noise. Rural environments usually are cooler, windier, and sunnier.

City temperatures are warmer than rural temperatures because the dark surfaces of urban streets and structures absorb up to three times more energy than the lighter-colored surfaces of sandy soil and rural greenery. The principle that darker colors absorb more radiation (because they are less reflective) is the reason you wear lighter or white clothing in the summer and dark colors in the winter. Dark colors absorb more radiation and retain the heat longer.

After sunset, the city becomes a heat source that raises nighttime temperatures in the area (Figure 17.32). Cities indeed are heat islands: not only do they re-radiate the greater energy they receive as heat, but the heat energy's escape at night is slowed by vertical city walls, which inhibit the escape of heat more than open horizontal surfaces of the countryside. Temperatures in a city at night are commonly 10F° warmer than in the countryside. (Confirm this by watching your local late-night weathercast, especially in winter.)

Cities also change the water balance. In rural areas, rainfall does not all run off but is partly retained as soil moisture. Plant roots carry the water upward through the plant and return it to the air by transpiration from leaf surfaces. Cities, in contrast, have only a small percentage of green area, so transpiration is limited. Also, urban areas are paved so that moisture cannot slowly sink into the ground. It runs off very rapidly into storm sewers (review Urbanization and Flooding in Chapter 7).

Evaporation and transpiration require solar energy. Because both are reduced as just described, more solar energy becomes available for heating city surfaces. Just as evaporation of moisture cools your body, evaporation cools all surfaces. Less evaporation in cities means cities are hotter.

Fossil fuels consumed in cities release large amounts of waste heat, adding to a city's thermal reservoir. In addition, microscopic particles from smoke-stacks create air pollution: every major city has an atmospheric *dust dome* over it. These microscopic particles also act as nuclei onto which water vapor condenses to form clouds. Thus, cities have more rainfall, lightning, and thunder than the surrounding countryside. Power outages are more frequent in cities than in rural areas because of the increased lightning.

The soot particles also reflect more solar radiation, so less reaches the ground. As a result, London receives

270 fewer hours of bright sunshine annually than the surrounding rural countryside, according to one study. This has a cooling effect, although it is more than counterbalanced by the temperature-raising effect of dark city surfaces. (Recall that the ash particles erupted by Mt. Pinatubo in 1991 had a similar cooling effect; Chapter 12.)

The roughness and irregularity of city surfaces compared to rural areas increases turbulence as wind blows through the city. The gusty, erratic windflow through the maze of urban canyons is well known to city dwellers. However, gusts are more likely to be faster in open countryside; wind speed generally is slower in cities.

Humidity and Comfort

On weather forecasts, humidity is stated as **relative humidity.** This is the amount of water *actually* in the air compared to the amount the air *could* hold *at that temperature.* In warm weather, when the relative humidity is high, we usually feel uncomfortable. The reason is that the human body cools itself by losing moisture through perspiring. If the humidity is low, sweat evaporates easily, absorbing a great deal of heat from your skin as it does so, and you feel cooler. If the relative humidity is high, the sweat evaporates more slowly, stays on your body longer, and you feel clammy. If the relative humidity is 100 percent, your body moisture will not evaporate at all and your clothing will get soggy. Figure 17.33 shows how comfort varies with temperature and relative humidity.

Over the past century, world average temperature has risen about a degree, and it will continue to rise slowly for some time. Before stabilizing, it may rise another degree or two, perhaps three or four. Not only will it be warmer, it will also be wetter. Rising temperature is accompanied by increasing evaporation from Earth's surface waters: oceans, lakes, and streams. The

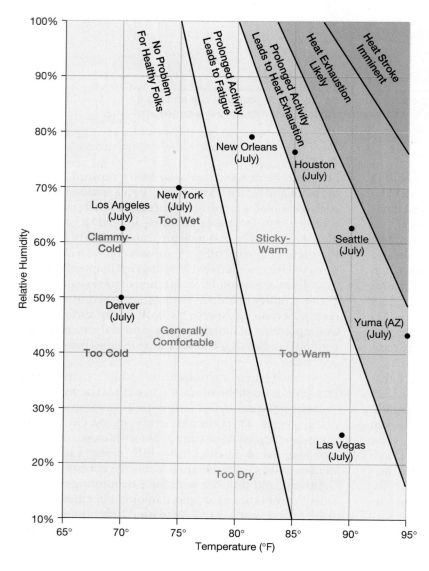

FIGURE 17.33 Human comfort is controlled by temperature and relative humidity.

Extremes of heat and humidity together can overpower the body's ability to regulate its own temperature, creating a health hazard. What are the average temperature and humidity during each season where you live? *From Lennox Industries.*

TABLE 17.1 | Windchill Factor

Wind speed (mph)	Actual Air Temperature in °F											
	50°	40°	30°	20°	10°	0°	–10°	–20°	–30°	–40°	–50°	–60°
	Temperature Effects on Exposed Flesh											
Calm	50°	40°	30°	20°	10°	0°	–10°	–20°	–30°	–40°	–50°	–60°
5	48°	37°	28°	16°	6°	–5°	–15°	–26°	–36°	–47°	–57°	–68°
10	40°	28°	16°	4°	–9°	–21°	–33°	–46°	–58°	–70°	–83°	–95°
15	36°	22°	9°	–5°	–18°	–36°	–45°	–58°	–72°	–85°	–99°	–102°
20	32°	18°	4°	–10°	–25°	–39°	–53°	–67°	–82°	–96°	–110°	–124°
25	30°	16°	0°	–15°	–29°	–44°	–59°	–74°	–83°	–104°	–113°	–133°
30	28°	13°	–2°	–18°	–33°	–48°	–63°	–79°	–94°	–109°	–125°	–140°
35	27°	11°	–4°	–20°	–35°	–49°	–64°	–82°	–98°	–119°	–129°	–145°
40	26°	10°	–6°	–21°	–37°	–53°	–69°	–85°	–102°	–116°	–132°	–148°

greater the water vapor in the air, the greater the humidity. If you like it hot and sticky, you may love global warming.

Windchill

Global warming is also expected to create windier weather, and wind speed is another factor in our comfort. In cold weather, strong winds make it seem colder because the moving air increases heat loss from your body. The wind conducts heat away from the skin. When winds are calm, a body at rest is surrounded by a layer of still air that acts as an insulator, because air is a poor conductor of heat. As wind speed increases, the layer of still air disappears and heat loss from the body increases. In cold weather, the danger of frostbite and hypothermia (freezing to death) is increased. If body temperature drops from its normal 98°F to 79°, it's all over.

Weather reports, particularly in cold weather, typically include a **windchill** factor, referring to the sensation of lowered temperature you feel because the wind is blowing. A table relating apparent temperature to wind speed was first compiled in the 1940s by polar scientists and has since been refined (Table 17.1). In the table, for example, at an air temperature of 30°, a wind of 20 mph will make bare flesh feel as cold as it would in calm conditions at about 4°. The windchill chart is a useful guide to protect yourself during winter. The effect of wind is greater in winter because humidity is generally lower.

Those Other Pollutants

As if sulfur dioxide, particulate matter, nitrogen dioxide, carbon monoxide, ozone, and lead were not enough, we also have pollen, molds, smoldering tobacco, and noise. Here is a look at each.

Pollen, Molds, and Allergies

The bulk of the dust you inhale daily is either particles of soil or particulate matter from industry or vehicles. (Daily, you inhale tiny rubber particles worn from car tires.) But some of the dust is **pollen,** the yellow, powderlike male sex cells of grass, trees, weeds, and flowers (Figure 17.34A).

Many plants produce pollen grains in great numbers; for example, a single sorrel plant may produce 393 million grains, and a single rye grass plant can produce 21 million grains. On windy days, the pollen is widely distributed because of its tiny size, 15 to 100 micrometers (100 micrometers is about the diameter of a hair from your head). The smaller the pollen, the farther it will spread. If you have pollen allergies, there are few places to hide.

Mold reproduces by means of **spores,** which are the same size as pollen and cause similar reactions in people whose immune systems are hypersensitive to them. During the major pollinating and molding months, The Weather Channel and other weather reporting services present data on pollen and mold spore counts (Figure 17.34B).

More than 50 million Americans are afflicted with allergies of some kind, and they account for 9 percent of the visits to physicians' offices. An **allergy** is an abnormal reaction to usually harmless substances. The *allergen* itself is not a threat, but the person's antibodies mistakenly identify it as an enemy. This causes cells in tissues to produce chemicals such as **histamines** that create a range of symptoms. If you are allergic to anything, you already know the symptoms: runny nose, teary eyes, drowsiness, headache, and perhaps a slightly elevated body temperature, the fever part of "hay fever" (so-named because it occurs during the hay-cutting season).

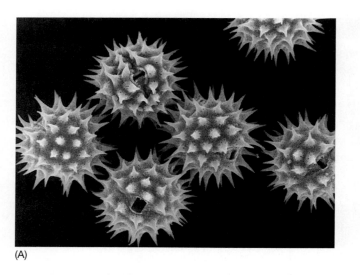

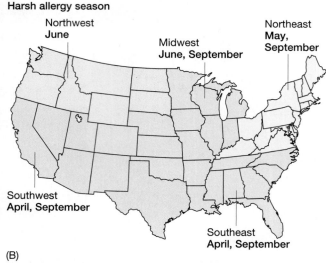

Harsh allergy season

Northwest
June

Midwest
June, September

Northeast
May,
September

Southwest
April, September

Southeast
April, September

(A)

(B)

FIGURE 17.34   The sexual activities of plants are something to sneeze at.

(A) Pollen grains from the Sunflower family of plants. Grains measure about 15 micrometers, so about 1700 pollen grains lined up spike-to-spike would equal 1 inch. *Photo courtesy Samuel Nobel Electron Microscopy Laboratory, University of Oklahoma.* (B) The peak months for allergy sufferers in different regions of the country.

Antihistamines help alleviate the symptoms, but they also increase drowsiness. The best solution is to avoid the irritating agent. Keep windows closed during the day, use air conditioning, and avoid going outside in the afternoon. Grasses pollinate during the middle of the day and wind keeps the pollen in the air until it drops to the ground at night. Pollen clings to hair, so frequent showers and avoidance of furry animals can help.

Smoldering Tobacco: Air Pollution Indoors

> Fifty-four year old Chao Boonchu of Thailand proclaimed himself his country's champion chain smoker. He proudly claimed to have smoked 120 cigarettes, six packs a day, for thirty years with no ill effects. . . . Within a year . . . he collapsed with severe breathing difficulties and was rushed to a Bangkok hospital, suffering with severe heart and lung complaints. After lying in a coma for some days, he died.
>
> The Blunder Book, p. 83–84.

The late Mr. Boonchu is only one of millions who die each year from a totally preventable cause—cigarette smoking. The federal government blames smoking for more than 400,000 deaths in America per year, an estimate doubted only by tobacco company executives. The difference in average life span between smokers and nonsmokers can be simply translated: each cigarette reduces your life span by 10 minutes.

Smoking causes lung cancer, emphysema, and heart disease, and has killed more people than died in all the wars of the twentieth century. If you don't know your history, that's World War I (117,000 dead), World War II (407,000), Korean War (54,000), Vietnam War

(58,000), plus numerous skirmishes. Active smoking causes 120,000 of the 140,000 annual lung cancer deaths in the U.S., it is estimated.

Until the 1960s, these annual deaths were concentrated among males, because it was considered unladylike for women to smoke. But "you've come a long way, baby" as the ads say. In a grim form of equal opportunity, women's death rate from lung cancer is approaching that of men. The rate of lung cancer deaths among women who smoke increased sixfold from the 1960s to the 1980s. People who smoke two or more packs a day have cancer mortality rates much greater than nonsmokers (Figure 17.35A).

Regular smokers obviously increase their risk of an early death. But what of those who are trapped in rooms with smokers? Recent evidence strongly indicates that the health of the second-hand smokers is being compromised. EPA declared second-hand smoke a human carcinogen in 1993. Second-hand smoke killed an estimated 30,000 in 1994.

Sidestream smoke (that given off between puffs) is enriched in partially burned products, compared to the mainstream smoke directly ingested by the smoker. Consequently, sidestream smoke has much higher concentrations of some toxic and carcinogenic substances than mainstream smoke. However, dilution by room air markedly reduces the amount inhaled by second-hand smokers. So, is sidestream smoke safe?

One way to find out is to examine the lung cancer rate of nonsmoking spouses of smokers. They have a 30 percent greater lung cancer risk and suffer more respiratory infections, allergies, and other chronic respiratory diseases. Knowing this, consider workers such as

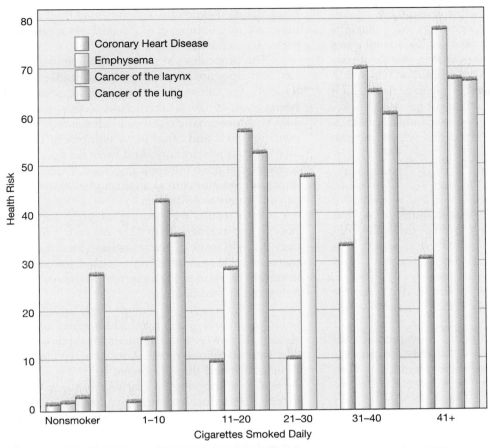

(A)

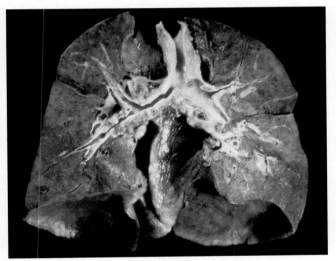

(B)

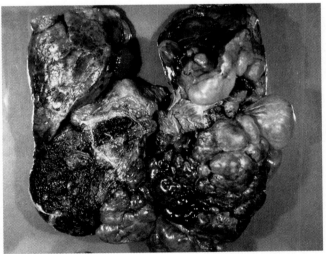

FIGURE 17.35   A bad habit to avoid.

(A) Cigarette smoking is strongly correlated with the development of cancer, emphysema, and heart disease. (B) Normal, healthy human lung (left) and lung from an emphysema cadaver (right). The cadaver's lung looks like it has been cooked on a grill, which is close to what happens when hot cigarette smoke is repeatedly forced into lung tissue. *Photos: O. Auerbach/Visuals Unlimited.*

musicians and bartenders, who earn their living in closed, smoky environments. These people inhale the equivalent of ten or more cigarettes per day (estimated). Second-hand smoking is particularly harmful to infants and young children, pregnant women, the elderly, and people with chronic lung disease.

Among U.S. nonsmokers, 80 percent have detectable nicotine residue in their blood. Over half neither live nor work with smokers! So, is it possible to avoid this air pollution? In the United States, nonsmokers are a majority, and seem to be gaining the upper hand. Nearly all states now restrict smoking in public places to some degree. Smoking is banned on most domestic airline flights and most employers now restrict smoking in their facilities. Of the nation's shopping malls, 70 percent are now smoke-free. Smoking soon may be banned in all public buildings. As for smoker's rights, "Your rights end where my nose begins" (Figure 17.35B).

Turn It Down!

Noisy restaurants, the din of street traffic, annoying alarm clocks, irritating dog-yapping, thunder from storms, the cacophony of car stereos and bands. Our eardrums seem under attack. What is it about sound that can be so irritating?

When strings, reeds, air columns, drum heads, cymbals, speaker cones, bells, and buzzers vibrate rapidly back and forth, they disturb the air. We call these disturbances *sound*. The oscillating movement forces neighboring air molecules to be alternately compressed and rarefied, like pushing and pulling something very rapidly. The vibrations have frequencies: how many times the compressions and rarefactions happen each second. The lowest piano tone (full keyboard) is 27 vibrations per second (hertz); the highest, around 4000.

Adult humans hear sounds from about 20 **Hertz** to 20,000 Hertz, babies up to 40,000 Hertz. We are fortunate we cannot hear the very lowest frequencies, down around a few Hertz. If we could, we might never know quiet, because of the sound of molecules colliding in the air. Some animals can hear higher frequencies than we can. Cats, for example, are better at catching mice than you are because mice utter high-frequency squeaks that are inaudible to you but not to your cat.

Loudness is the sound pressure level that sound waves exert on the nerve cells of the inner ear. Sound pressure level is expressed in *decibels (dB)*. Zero decibels is defined as the bottom threshold of hearing. Humans experience sound levels ranging from 0 to well above 100 decibels (Figure 17.36). Sounds of 100 decibels will produce hearing loss in humans if continued too long. Sounds of 85–100 decibels can produce

hearing loss if exposure lasts for more than a few minutes each day. Sounds of 65–85 decibels are at least disturbing if persistent.

The proportion of people who live in cities is rising, and cities are very noisy places. Traffic roars at 90 decibels, the subway screeches at 95 decibels, and car horns blare at 120 decibels. Add to this, construction sites with pneumatic riveters, jackhammers, screaming concrete saws, and other noisy equipment that generates 100 decibels or more. And there are boom cars and concerts that push the sonic envelope. Even private listening, piped into your skull through headset implants, can add to **noise pollution.**

As a result, the hearing of your generation is poorer, on average, than that of your parents. The nerve cells in some of your ears have been damaged by the continued assault of high-decibel sound. Workers in certain types of noisy factories also have a high incidence of reduced hearing.

A 1995 survey of French high school students found that about one in five had impaired hearing. The abnormalities were more frequent among those who said they listened to loud music. A study at the University of Tennessee found that 61 percent of freshpersons had a detectable hearing loss. A Norwegian study found that rock bands produce a sound pressure of 120–130 decibels when heard from a speaker about 3 feet away, so don't sit that close. Many researchers have found that some—but not all—of those who subject themselves to powerhouse wattages and nuclear bass suffer at least temporary hearing loss.

Musicians and DJs are at greater risk than most listeners. In one study, significant hearing loss was found in one-third of seventy DJs. They were still in their twenties, an age at which hearing disability is expected in less than 1 percent of the population. In contrast, the hearing of people in tribes that live far from our high-decibel civilization remains acute into their seventies.

In an attempt to prevent hearing loss in young people, France passed a law in 1996 limiting the noise levels of personal stereos. The output of all personal steroes sold in France is capped at 100 decibels. They will also have to carry a notice warning that listening at full volume for a prolonged period can damage hearing.

Summary

The Industrial Revolution that began in the mid-1800s certainly has benefited mankind, but we are now realizing its harmful effects on the atmosphere. Carbon dioxide has increased by 25 percent and the amounts of other greenhouse gases have grown. This has caused a small increase in world average temperature and probably in precipitation as well. Sea level has been rising for the last 18,000 years because of glacial melting, but the rate of rise has been increased only slightly by the increase in greenhouse gases. It is

Sound	dB	Effect	Relative Intensity (Loudness)
Carrier Deck Air Raid Siren	140	Painfully Loud	
Jet Takeoff	130		
Thunder (maximum) Auto Horn (3 feet)	120	Maximum Vocal Effort	One Trillion
Rock Band Power Mower Motorcycle	110		
Garbage Truck Subway	100		
Food Blender City Traffic Noisy Office	90	Very Annoying Hearing Damage (8 hours)	One Billion
Lightly Mechanized Factory Alarm Clock	80	Annoying	
Noisy Restaurant Man's Voice (3 feet)	70	Telephone Use Difficult	Hearing Damage Begins
Air Conditioning Unit (20 feet)	60	Intrusive	One Million
Light Traffic (100 feet)	50	Quiet	100,000
Bedroom Quiet Office	40		10,000
Library Soft Whisper (15 feet)	30	Very Quiet	One Thousand
Broadcasting Studio	20		100
Rustle of Leaves in a Gentle Breeze	10	Just Audible (whisper)	10
	0	Hearing Begins	0

FIGURE 17.36   Sound levels in our environment.

Sound pressure level is shown in decibels (dB). Note that the decibel scale is like the pH (acidity) scale: it is logarithmic. For example, going from 10 dB to 20 dB does not double the sound level; it increases it tenfold.

unclear whether the global warming, which probably will continue for some time, will be a net benefit or net loss for plants and animals, including humans.

One of our chemical products, chlorofluorocarbons, appears to have damaged the stratospheric ozone layer that shields Earthlife from harmful UV. The loss of ozone is much more severe over the Antarctic continent than elsewhere. Because CFC production will be banned worldwide in 2000, CFCs gradually will decrease. But full natural restoration of the ozone shield will take 100 years or more.

The air we breathe in the lower atmosphere contains varied contaminants, generated largely by the combustion of coal and oil in power generation, industry, and vehicles. These contaminants include sulfur dioxide and nitrogen oxides that produce acid precipitation, a major environmental hazard to plants. Carbon monoxide, ozone, and airborne lead are other noxious materials produced by the burning of fossil fuels.

Since public consciousness was aroused about thirty years ago, levels of these substances in the atmosphere have decreased, but major problems still exist. The only solution is reduction and eventual elimination of coal and oil as energy sources. This will occur eventually, as supplies of these fuels are exhausted by the year 2100. During the period from 2000 to 2100, major efforts should be made to develop and use alternative fuels.

Other environmental atmospheric hazards include passively inhaled cigarette smoke, plant pollen, and mold spores. Society is advancing rapidly toward a goal

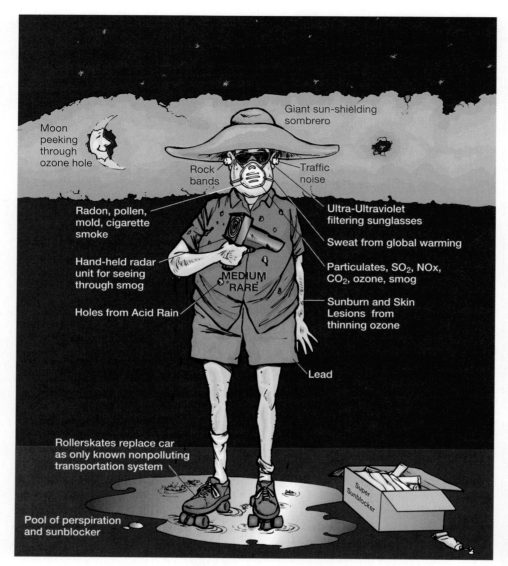

Moon peeking through ozone hole

Giant sun-shielding sombrero

Rock bands

Traffic noise

Radon, pollen, mold, cigarette smoke

Ultra-Ultraviolet filtering sunglasses

Sweat from global warming

Hand-held radar unit for seeing through smog

MEDIUM RARE

Particulates, SO_2, NOx, CO_2, ozone, smog

Sunburn and Skin Lesions from thinning ozone

Holes from Acid Rain

Lead

Rollerskates replace car as only known nonpolluting transportation system

Super Sunblocker

Pool of perspiration and sunblocker

FIGURE 17.37   How bad might it get?

Will it be safe to come out only at night? Will "aerial assault" take on a new meaning?

of removing cigarette smoke from the airways of non-smokers. Pollen and spores remain seasonally trouble-some for many, with no effective cure in sight.

Let us hope that your future environment will not require the equipment shown in Figure 17.37!

Key Terms

aerosol
allergy
altitude sickness
atmospheric gases
baghouse
carbon monoxide
chlorofluorocarbons (CFCs)
Dobson unit
electrostatic precipitator
electromagnetic spectrum
global warming

greenhouse effect
heat island
Hertz
histamines
hype
infrared
lead poisoning
nitrogen dioxide
noise pollution
ozone
ozone layer

particulate matter
pollen
relative humidity
scrubbers
sidestream smoke
smog
sulfur dioxide
temperature inversion
ultraviolet
windchill

Stop and Think!

1. Describe the effects of altitude sickness. What is the simplest cure for it?
2. Explain the meaning of *electromagnetic spectrum.* Describe some of the effects of encountering different wavelengths in the spectrum.
3. Would you rather be exposed to large doses of radiation in the wavelength range of 0.1–0.4 micrometers, 0.4–0.7 micrometers, or 0.7–1.0 micrometers? Explain your reasoning.
4. What is the *greenhouse effect?* Why should people other than professional meteorologists care about it?
5. What are the important greenhouse gases? Which ones have increased during the past 100 years? What has been the source of the increase for each gas?
6. What are aerosols? Are they produced only by humans or has nature also played a part? What might be the effect on climate of large increases in volcanic activity and particulate air pollution?
7. Explain the disagreement over whether global warming is happening or not. Which view seems more credible? Why?
8. What might be the negative effects on human societies if global warming occurs? What might be the positive effects?
9. Discuss the rate at which global warming might occur in relation to people's ability to adapt.
10. How do you believe the media should deal with the "threat" of global warming?
11. Should the federal government or individual states restrict construction along the coastline? If you favor restrictions, spell out what they should be and how you would enforce them.
12. Explain the difference between high ozone concentrations in the stratosphere and at ground level. Why is one good and the other bad?
13. What are CFCs? Explain how they interact with the ozone layer.
14. How does increasing world urbanization increase air pollution?
15. What are temperature inversions? Is it possible to prevent them? If so, how?
16. Explain how acid precipitation forms. How would you stop it if you had the authority? Consider what various interest groups in society would think of your plan.
17. Why have various pollutants decreased over the past twenty years?
18. Explain the cause and possible cures for urban heat islands.
19. Why does high humidity make you feel hotter? Why does a high wind make you feel cooler?
20. How can we decrease the noise level in urban areas?

References and Suggested Readings

American Lung Association. 1992. *Health Effects of Ambient Air Pollution.* New York, 63 pp.

Anonymous. 1995. Secondhand smoke: Is it a hazard? *Consumer Reports,* January, p. 27–33.

Bartecchi, C. E., T. D. Mackenzie, and R. W. Schrier. 1995. The global tobacco epidemic. *Scientific American,* May, p. 26–33.

Bazzaz, F. and E. D. Fajer. 1992. Plant life in a CO_2-rich world. *Scientific American,* January, p. 68–74.

Berk, R. A. and D. Schulman. 1995. Public perceptions of global warming. *Climatic Change,* v. 29, p. 1–33.

Bricker, O. P. 1993. Acid rain. *Annual Review of Earth and Planetary Sciences,* v. 21, p. 151–174.

Brink, Susan and Corinna Wu. 1996. Sun struck. *U.S. News & World Report,* June 24, p. 62–70.

Anonymous. 1996. *Global Warming.* September, p. 38–44.

Gribbin, John and Mary Gribbin. 1996. The greenhouse effect. *New Scientist* (insert), July 6, 4 pp.

Lee, John and Lucy Manning. 1995. Environmental lung disease. *New Scientist* (insert) September 16, 3 pp.

Lents, J. M. and W. J. Kelly. 1993. Clearing the air in Los Angeles. *Scientific American,* October, p. 32–39.

Macdonald, N. J. and J. P. Sobel. 1995. *Changing Weather? Accu-Weather.* State College, Pennsylvania, 28 pp.

Mage, David and others. 1996. Urban air pollution in megacities of the world. *Atmospheric Environment,* v. 30, p. 681–86.

Michaels, Patrick. 1993. *Global Warming: Failed Forecasts and Politicized Science.* St. Louis, Missouri: Center for the Study of American Business, 23 pp.

Moore, Curtis. 1994. *Urban Air Quality.* United States Information Agency, 15 pp.

Royal Swedish Academy of Sciences. 1995. Environmental effects of ozone depletion. *Ambio* (special issue), v. 24, no. 3, May, p. 137–196.

Smith, A. T. 1995. Environmental factors affecting global atmospheric methane concentrations. *Progress in Physical Geography,* v. 19, no. 3, p. 322–35.

Thompson, R. D. 1995. The impact of atmospheric aerosols on global climate: A review. *Progress in Physical Geography,* v. 19, no. 3, p. 336–50.

Warrick, R. A., Barrow, E. M., and Wigley, T. M. L., eds. 1993. *Climate and Sea Level Change:Observations, Projections and Implications.* Cambridge University Press, 424 pp.

World Health Organization. 1992. *Our Planet, Our Health.* Report of the World Health Organization Commission on Health and Environment, Geneva, Switzerland.

Glossary

aa fragmented, blocky-looking surface of hardened lava

accumulator plants plants that selectively take in certain metals

acid mine drainage waters coming from a mine that have low pH

acid precipitation rain more acidic than about 5.6 on the pH scale

active solar heating a system in which machinery circulates a fluid from a solar collector to a storage facility for use as needed

active volcano a volcano that has erupted within the past 10,000 years

adobe a baked calcareous clay and silt deposit used for construction "brick" in the American southwest and in Mexico

aerobic an organism or a chemical process that requires oxygen gas

aerosol a mechanical mixture of a gas and a colloid, either solid or liquid

aggregate rock particles used in the construction industry

A-horizon the uppermost soil horizon; also called topsoil, a mixture of organic material, clay, and quartz

allergy an abnormal reaction to usually benign substances such as pollen, and characterized by runny nose and watery eyes

alluvial referring to a stream, as in alluvial sand deposit

alpha particles nucleii of helium atoms released during radioactive disintegration

alternative energy sources energy sources other than fossil fuels

altitude sickness temporary illness caused by a shortage of oxygen at high elevations

amorphous noncrystalline

amplitude half the height of a wave above the adjacent trough

anaerobic decomposition decomposition in the absence of oxygen

anaerobic an organism that cannot live in the presence of oxygen gas or a process that occurs in the absence of oxygen

angle of repose the maximum angle of slope at which loose, cohesionless material will come to rest, usually about 30 degrees

anion a negatively charged ion

anthracite coal type of coal containing more than 90 percent carbon

aquiclude a body of impermeable rock overlying an aquifer

aquifer a body of rock that is sufficiently permeable to supply useful amounts of water to wells and springs

aquitard a nearly impermeable layer of rock overlying an aquifer

artesian water aquifer water that rises above the aquifer

asbestos a group of silicate minerals composed of thin, flexible fibers.

asbestosis noncancerous lung disease caused by asbestos particles inhaled into the lungs

asthenosphere upper part of the Earth's mantle

atmosphere mixture of gases that surrounds the Earth, held close to the Earth by gravity

atom particle about 10^{-8} inches in diameter and composed of protons, electrons, and, with the lone exception of hydrogen, neutrons

atomic number number of protons in an atom

atomic weight combined weight of protons plus neutrons

avalanche large mass of snow, ice, soil, or rock falling rapidly from heights far above a lowland area

background level radiation received by everyone in an area

baghouse a system of containers designed to stop the release of pollutants from smokestacks

barrier island long, narrow coastal sandy island that is parallel to the shore and above high tide

basalt dark-colored microcrystalline igneous rock

bauxite soil composed largely or entirely of hydrated aluminum compounds

bed load sediment carried at or very near the stream bed, dominantly gravel and sand

bedrock rock that underlies soil or other unconsolidated superficial material

benzene six-sided arrangement of carbon atoms

beta particles electrons emitted from an atom during radioactive disintegration

B-horizon soil horizon below the A-horizon, also called subsoil or zone of accumulation and characterized by enrichment in clay, iron oxide, aluminum hydroxide, and/or calcium

biochemical oxygen demand (BOD) the amount of oxygen required to decompose the organic matter available

biocide also called pesticide; chemicals that destroy unwanted animals or plants

biodegradable substances that are decomposed by the action of air and water at or very near the Earth's surface

biodigestion decomposition of organic matter by bacteria

biogas gas produced by biodigestion

bioliquid liquid produced by biodigestion

biological magnification accumulation of increasing amounts of pollutants at higher levels of organic complexity

biomass energy energy from burning of plants

biomass the amount of living material in an area

biosphere the zone at and near the Earth's surface in which life exists

bituminous coal coal that contains 70–90 percent carbon

black lung disease noncancerous condition of lung damage caused by inhalation of coal dust

bonding linkage between atoms that results from their electron distributions

breakers waves that lose their form to chaotic motion in shallow water

breakwater rock or concrete barrier offshore and parallel to the shoreline

breeder reactor nuclear reactor composed of a core of U-235 and a coating of U-238 that "breeds" plutonium-239

caldera depression at the summit of a volcano that is much wider than the included vent

capillary forces the action by which a fluid is drawn up by surface tension into a channel of very narrow diameter

carbonate mineral whose essential building block is the ionic group CO_3^{-2}

cation a positively charged ion

chain reaction (nuclear) process in which fissioning of one nucleus induces fissioning of others

channelization making a stream channel more geometrically perfect by artificial means

chelation retention of a metallic or nonmetallic cation by two atoms of a single organic molecule

chemical bond force that holds two atoms together

chemical element an atom distinguished by a particular number of protons

chemical energy energy generated by breaking chemical bonds

chemical weathering decomposition of rock at Earth's surface

chert microcrystalline quartz

C-horizon soil horizon below the B-horizon consisting of unconsolidated rock material only slightly affected by weathering

clastic rock sedimentary rock composed of pieces of rock that have been transported, deposited, and cemented

clay minerals microscopic minerals characterized by a sheeted crystal structure and composed largely of aluminum and silicon

cleavage (mineral) breakage of a mineral along certain planes because of weak bonding across the planes

coal combustible rock composed of more than 50% by weight of carbonaceous material formed by alteration of plant remains

cogeneration production of two useful forms of power from the same source, such as both steam and electricity from a coal-fired plant

colloid solid material with particle size smaller than about 10^{-4} inches

compound combination of two or more elements that cannot be separated by mechanical means, distinguishing it from a mixture

computerized drip irrigation method of minimizing water wastage in agriculture

concentration factor (ore) amount of enrichment of a chemical element required to constitute an ore deposit

cone of depression depression in the potentiometric surface of a body of groundwater; forms around a well from which water is being withdrawn

confined aquifer aquifer bounded above and below by relatively impermeable beds

conglomerate clastic rock composed of grains coarser than 2 mm

conservation farming farming that either decreases the amount of plowing or eliminates it altogether

contact metamorphism metamorphism at high temperature but low pressure

containment building building housing a nuclear reactor and designed to prevent the escape of radioactive material when a nuclear accident occurs

contamination presence in water of noticeably larger amounts of a substance than is normally present in the absence of human interference

contour plowing plowing horizontally around a slope so that a plowed row has the same elevation above sea level along its length

control rods neutron-absorbing rods used to moderate the chain reaction in a nuclear reactor

convection circulation of subcrustal material that results from the difference in temperature between bottom and top of a zone in Earth's mantle

cosmic ray radiation emitted from the sun, consisting mostly of protons

covalent bond chemical bond formed by the sharing of electrons in the outer shell of atoms

creep (rock or soil) slow downslope movement of rock or soil under gravitational stresses

crude oil liquid hydrocarbons recovered from the subsurface; also called petroleum

crumb structure aggregated structure of soil particles that results from the sticky nature of soil clays, colloids, and organic matter.

crust (Earth's) outermost shell of Earth, defined by characteristics such as density, seismic velocity, and composition

crystal settling settling of denser minerals in a magma

crystalline solid materials characterized by a regular periodic internal arrangement of their atoms

cyanide leaching scavenging a valuable metal from a waste pile using a solution of sodium cyanide

dam a barrier to water flow

Darcy's Law the mathematical relationship between aquifer discharge, permeability, slope of the water table, and the cross-sectional area of the aquifer

debris flow a moving mass of rock fragments, soil, and mud, more than half the particles being larger than sand size

debris slide rapid movement of coarse sediment downslope without water being present

decay (radiation) spontaneous disintegration of an atom into a new species

decomposition chemical destruction of a mineral or rock

degree day (heating/cooling) one degree Fahrenheit lower or higher than an arbitrary standard for one day

delta an alluvial sediment mass at the mouth of a river forming a fan-shaped deposit, part of which is usually below sea level (or lake level)

desalination removal of dissolved material from salty water

diatom one-celled microscopic plant found in both fresh and salt waters

dilatancy increase in bulk volume during deformation of the material, resulting from the change from a close-packed to an open-packed structure

dip the angle formed by the intersection of a rock layer and a horizontal plane, measured at 90° to the strike

discharge (stream) the volume of water or sediment passing a point per unit of time

dissolved load dissolved material carried by a stream

diurnal tide a tide with only one highstand and one lowstand per day

Dobson unit equals one part per billion of ozone

dolostone sedimentary rock consisting largely or entirely of dolomite

dormant volcano a volcano that is fresh-looking but has not erupted during the last 10,000 years

dowsers people who search for groundwater using a forked stick or metal rod

drainage basin the area from which a stream obtains its water

drainage divide the line on the ground separating one drainage basin from an adjoining basin

Dust Bowl the area in the midwestern United States ravaged by drought in the 1930s

earthflow slow movement of a mass of sediment and water down a gentle slope in response to gravity

earthquake sudden trembling in the Earth caused by the abrupt release of strain

ecology study of the relationship between organisms and their environment

eddy an upward-moving mass of water in a turbulent stream

El Niño warm current along the northwest coast of South America that lasts longer than normal every few years

elastic deformation deformation of a substance that disappears when the deforming forces are removed

electrical energy energy of electrons moving along a wire

electromagnetic spectrum range of wavelengths in energy emitted by the sun

electron shell volume of space in which an electron is most likely to be located

electron negatively charged particle located in the outer part of an atom

electrostatic precipitator a method of removing pollutants before they leave smokestacks of facilities that burn fossil fuels

element an atom with a specific number of protons

energy efficiency the ratio of energy obtained to energy put in during an energy conversion; always less than 1.0

energy capacity to do work

environmental geology application of geology to environmental concerns

Environmental Protection Agency (EPA) government agency in charge of controlling pollution in the U.S.

epicenter (earthquake) point on the Earth's surface directly above the hypocenter (focus)

equilibrium state of minimum energy

erosion removal of rock material from one location to another by a moving medium such as water, wind, or ice

estuary funnel-shaped tidal mouth of a river valley where fresh water and sea water meet

eutrophication process by which a body of water becomes unfit for aerobic life because of uncontrolled algal growth stimulated by an excess of nutrients

evaporite a mineral or a deposit of very soluble minerals formed by evaporation of nearly all the water in which the ions were dissolved

exponential growth growth in which the amount doubles in a fixed unit of time, e.g., 1, 2, 4, 8, 16, 32, . . .

extinct volcano volcano that shows significant erosion on its crest and flanks and has no recent eruptive history

fault break in Earth's crust along which there has been movement parallel to the fault surface

fauna group of animals in an area or environment

fertilizer material added to soil to enhance plant growth

fetch distance over which wind blows over the ocean surface during formation of surface waves

fission (atomic) process in which an atom is stimulated to break apart by bombardment by neutrons

flash flood river flood that occurs suddenly

floatation process used in ore processing in which the ore mineral is caused to float at a liquid surface by the addition of organic chemicals

flood rising body of water that overtops its normal confining banks and spills over onto normally dry land

flood-frequency curve graph showing the number of times per year that a flood of a given magnitude is equalled or exceeded

floodplain zoning restricting construction on river floodplains

floodplain strip of smooth land adjacent to a river channel that was constructed by the existing river when it overflowed its banks (floods)

floodway land around a river that is covered with water during floods of a given frequency, e.g., the 100-year flood

flora group of plants in an area or environment

fluorescent light light produced in a bulb or tube when a substance coating the bulb or tube is hit by a stream of electrons (differentiated from incandescent light)

fluoridation artificial addition of soluble fluoride compounds to drinking water

focus the underground point of rupture of a rock that starts an earthquake

foliation planar arrangement of platy mineral grains in a metamorphic rock

fossil evidence of the past existence of something, usually applied to past life but sometimes used for nonliving things such as evidence of ancient meteor impacts or earthquakes.

fossil fuel applied to hydrocarbons petroleum and natural gas and to coal because they are formed from ancient plant tissues

friction resistance to sliding motion of two surfaces that touch

frost wedging separation of two pieces of rock, or splitting of a single piece, by the expansion of water when it freezes in a crack

fuel rods long tubes containing pellets of enriched uranium

fungicide chemical that kills fungi

fusion (nuclear) combination of the nuclei of two atoms to form a heavier nucleus; accompanied by a large release of energy

gabion wire-enclosed bundle of rocks designed to stop landslides

Gaia hypothesis hypothesis that the Earth and the life on it are a self-regulating network in which a disturbance in one part must result in disturbances in other parts to keep Earth a fit place for life

gamma rays short-wavelength, high-energy radiation

garbologist a scientist who studies garbage

gasohol mixture of gasoline and up to 23% alcohol (ethanol) used in place of 100% gasoline; super unleaded gasoline

gauging stations locations along a stream where discharge is measured

geology study of Earth

geothermal energy energy in naturally heated water that is used to generate power

geothermal gradient the rate of change in temperature with increasing depth in Earth

geyser hot spring that intermittently emits jets of hot water and steam

glacier land-based mass of ice whose thickness is great enough that its lower part is forced to move outward and downslope because of the pressure of overlying ice

glass an amorphous solid

global warming warming of Earth's atmosphere by increases in heat-absorbing gases, possibly caused by human activities

gneiss foliated metamorphic rock characterized by bands of different minerals, with micaceous minerals being a minor part

gradient (stream) angle between a stream channel floor or water surface and the horizontal, measured in the direction of stream flow

granite coarse-grained, light-colored igneous rock that contains quartz

gravel clastic particles greater than 2 mm (about 0.1 inches) in diameter

gravity an invisible force of attraction between objects that results from their mass

greenhouse effect heating of Earth's atmosphere resulting from absorption of Earth's infrared radiation by carbon dioxide and other gases in the atmosphere

groin concrete or rock barrier perpendicular to shore, designed to stop longshore drift of sediment

ground subsidence sinking of the ground as a result of withdrawal of fluid from the subsurface

groundwater subsurface water below the water table

Gulf Stream a stream of warm water in the Atlantic Ocean that moves northward from Florida to England and beyond

gullies V-shaped erosional channels so large they prevent farming

half-life time required for a radioactive substance to lose half of its radioactivity

halide compound formed of fluorine, chlorine, bromine, or iodine and a cation

hardness (mineral) relative resistance to scratching of a mineral

hardness scale (water) the amount of cations with a +2 charge in water; *see also* water hardness

hazard assessment evaluation of the degree of danger

hazardous solid waste solid waste that is either toxic, unstable, flammable, or corrosive

heat capacity the amount of heat a substance can absorb without its own temperature being raised

heat island a limited area that produces more heat than its surroundings

heavy metal metals of high atomic weight that accumulate to toxic levels in living organisms

herbicide chemical that kills plants

high-level radioactive waste waste directly related to nuclear reactors, such as spent fuel rods

histamines chemicals released by your body in response to inhaling substances to which you are sensitive

hot spot volcanic center in the same location for tens of millions of years; commonly has no relation to a plate boundary

humidity amount of water vapor in the atmosphere, usually given as a percent of the amount the air can hold at a given temperature (relative humidity)

humus dark-colored, decomposed, unidentifiable organic matter in soil

hurricane violent storm with winds exceeding 73 mph

hurricane warning notice that a hurricane is expected to hit a certain area very soon

hurricane watch notice that a hurricane landfall may occur in a certain area within 36 hours

hydraulic conductivity measure of the ease with which water flows through a rock or sediment; mathematically related to permeability

hydrocarbon organic compound composed almost entirely of hydrogen and carbon

hydroelectric power power generated by the flow of moving water through a turbine

hydrogen bond weak chemical bond formed between hydrogen and an anion, when hydrogen has its main bond with another atom

hydrogeology the scientific study of surface and subsurface waters

hydrograph a graph that shows the discharge of a stream at a location as a function of time

hydrologic cycle the movement of water from ocean to atmosphere to land, and back to the ocean

hydrostatic head height of a vertical column of water above a given subsurface level

hydrostatic pressure pressure exerted by a given hydrostatic head

hydrothermal waters that have temperatures much higher than is normal for the depth at which they occur

hypocenter the initial rupture point of an earthquake; focus

igneous rock or mineral that solidified from a molten condition

incandescent light light produced when a metal filament is heated by an electric current

incineration burning of waste

inertia tendency to remain in a fixed condition, without change

infrared long wave lengths just beyond those that produce red light

insecticide chemical that kills insects

integrated pest management control of insects on farms without using pesticides

intensity (earthquake) scale of relative damage produced by an earthquake

ion an atom that has become charged by the loss or gain of an electron

ionic bond chemical bond produced by a giving of an electron by one element and its acceptance by a different element

ionizing radiation radiation that causes ions to form

isotope species of a chemical element distinguished from other atoms of that element by having a different number of neutrons

joint one of a series of parallel fractures in a rock body with no movement on opposite sides of the fracture

karst topography type of ground surface formed in soluble rocks such as limestone by underground solution and subsurface drainage

kimberlite pipes fragmented igneous intrusions from the mantle that sometimes contain diamonds

kinetic energy energy of motion

lahar volcanic mudflow

landslide synonym for mass movement

land-use planning investigations aimed at determining the most appropriate use for land

laterite tropical soil consisting largely or entirely of iron or aluminum oxides and hydroxides

latitudinal desert a desert at low latitude whose location is determined by the presence of descending air

lava magma that flows out onto Earth's surface

leachate liquid containing dissolved substances obtained from soil or garbage

leaching removal of elements from a mineral or rock by moving water

levee a natural or artificial embankment along the borders of a stream that protects the surrounding land from inundation at times of high water

lignite brownish-black low-grade coal; brown coal

limestone sedimentary rock composed largely or entirely of calcite

liquefaction transformation of a soil from solid to liquid condition as a result of increased pore pressure

lithification the process by which a loose sediment is turned into a rock

lithosphere outer part of the Earth including the crust and upper mantle

loam soil composed of subequal amounts of sand, silt, and clay

logarithmic referring to numbers whose value increases by exponential amounts; contrasted to arithmetic

longitudinal profile stream gradient in a side view

longshore current ocean current that moves parallel to the shoreline

low-level radioactive waste waste that is not reactor-related, such as gloves used in handling radioactive materials

L-waves earthquake waves that move along surfaces of discontinuity

macronutrients nutrients needed by plants or animals in large amounts

magma molten silicate material formed below Earth's surface

magnitude (earthquake) measure of the strain energy released by an earthquake

mantle zone of Earth's interior below the crust and above the core

mass quantity of matter in an object

mass movement downslope movement of a part of the land surface; may include both soil and underlying bedrock

meander freely developed bend or loop in the course of a stream

mechanical (or physical) weathering producing small pieces of rock from larger pieces without chemical change

medicinal water water containing dissolved substances considered beneficial to health

meltdown partial or complete melting of the reactor core due to loss of coolant

mesothelioma cancer of the lining of the lung and stomach caused by ingestion of asbestos fibers

metabolism chemical and physical processes that occur during the life of an organism, subdivided into anabolism (building up) and catabolism (breaking down)

metallic bond bonds among elements that result from the aggregate action of the outer electrons of all atoms in the mass, rather than only one or two in the outer shell of each atom

metallogenic province   area characterized by particular kinds of mineral deposits

metamorphic   category of rock formed by recrystallization of minerals in the absence of large-scale melting, usually at temperatures between 600°F and 1300°F

micronutrients   nutrients needed by plants or animals in relatively small amounts

milling   physical process of grinding large pieces of ore rock into smaller pieces before sending them on to a smelter

mineral resource   a valuable mineral

mineral   naturally-occurring inorganic crystalline solid with a definite chemical composition

mudflow   viscous mass of mud and water moving rapidly downhill

municipal solid waste   solid garbage from urban areas

National Priorities List   the worst sites on the Superfund list

native element   an element that occurs uncombined with other elements, e.g., gold

natural gas hydrate   solid composed of methane gas and water

natural gas   hydrocarbon that exists as gas at ordinary temperatures and pressures

nematocide   chemicals that kill nematodes (parasitic worms)

neutron   particle without charge located in the nucleus of an atom

NIMBY phenomenon   don't solve a pollution problem at my expense (Not In My BackYard)

noise pollution   excessive noise

nonpoint source   a pollution source that extends over a wide area

nonrenewable resources   resources that are used faster than they are naturally replaced

normal fault   a fault in which the relative movement is downward on the hanging wall side

no-threshold concept   every atom of a potentially harmful substance must be removed

nuclear fission   splitting of an atomic nucleus into two or more parts with a large energy release

nuclear reactor   the device in which nuclear fission occurs

nucleus   central part of an atom where the protons and neutrons are located

oasis   an area in a desert where the water table intersects the ground surface

oceanic thermal energy conversion (OTEC)   generation of electricity by using the temperature difference between the ocean surface and water at depth

oil shale   rock composed of microscopic dolomite crystals and immature hydrocarbons

open-pit mines   surface mines

ore minerals   minerals from which valuable metals are recovered

ore   rock or sediment from which a mineral of economic value can be extracted at a profit

organic matter   solid material whose origin is in living or formerly living organisms

outcrop   bedrock exposed at Earth's surface, or covered only by soil or alluvium

overburden   rock removed to uncover ore-bearing rock

oxidize   to burn by combining with oxygen

ozone   triatomic oxygen, O_3

ozone layer   zone where most of the triatomic oxygen atoms occur, above an elevation of about 10 miles (stratosphere)

pahoehoe   ropy-looking surface of hardened lava

paleoquakes   ancient earthquakes

paradigm   the ruling model under which research results or new information are interpreted

passive solar heating   heating using solar energy without using machinery

peat   an unconsolidated deposit of plant remains in a water-saturated environment that shows some evidence of enrichment in carbon and loss of other elements

pedalfer   one of two broad classes of soil, characterized by an accumulation of iron oxide in the B-horizon

pedocal   one of two broad classes of soil, characterized by an accumulation of calcium carbonate in the B-horizon

pegmatite   igneous rock composed of very large crystals

perched aquifer   unconfined groundwater separated from an underlying larger body of groundwater by an unsaturated zone

periodic table   table of the chemical elements arranged in a way that reflects their electron distributions

permafrost   surficial deposit in cold regions that has remained frozen for at least several years and commonly for thousands of years

permeability   a measure of the ease of flow of a fluid through a rock or sediment; mathematically related to hydraulic conductivity

pesticide   chemical that kills "pests"; also called biocide

petroleum   naturally-occurring complex hydrocarbon liquid from which energy can be extracted; also called crude oil

pH scale   a logarithmic scale used to specify the amount of hydrogen ions in a liquid, to determine acidity or basicity

phaneritic   igneous rock texture with crystals coarse enough to be seen with the unaided eye, or with a 10 power lens

pheromone   substance that is secreted externally by one individual and which elicits a specific response from other individuals of the same species

photovoltaic cell   device that converts solar energy directly into electricity

placer deposit   surficial mineral deposit formed by mechanical concentration of mineral particles by moving water

plastic   synthetic nonmetallic solid formed from organic compounds that can be molded and formed for commercial use

plate (tectonic)   rigid thin segment of Earth's lithosphere bounded by major crustal fractures

plate boundaries   margins of tectonic plates

Pleistocene   period of geologic time from about 2 million years ago until about 10,000 years ago, and characterized by advances and retreats of large continental ice sheets

plutonic   an igneous rock formed at great depth

point source   a localized source of pollution

polar front   the leading edge of a cold air mass that moves from the poles toward the equator

pollen   male sex cells of seed plants

pollution   level of contamination so high that it has a harmful effect on one or more types of living organisms

pore   space between grains filled with air or fluid

pore pressure   pressure transmitted by a fluid that fills the pores of a soil or rock mass

porosity   the volume of space in a sediment or rock not occupied by grains

porphyry   igneous rock with two distinct crystal sizes

potential energy   energy an object has because of its distance from the center of Earth

potentiometric surface   an imaginary surface defined by the level to which water will rise in a well

precipitation   liquid or solid H_2O that falls out of the atmosphere

proton   positively charged particle located in the nucleus of an atom

pumice   light-colored, porous, silica-rich, glassy volcanic rock

P-waves   compressional earthquake waves that travel through Earth

pyroclastic   clastic sediment formed by explosive ejection from a volcanic vent

quarry   place where rock or sediment is removed from Earth and used for construction purposes

quickclay   clay that loses all its shear strength after being disturbed and behaves like a liquid

quicksand   mass of fine rounded sand grains saturated with water which yields easily to pressure of heavy objects on its surface

rad   measure of the amount of ionizing radiation absorbed by a person

radiation   particles or energy emitted by atomic nuclei as they change either by decay or combination with other nuclei; radioactivity

radiation sickness   illness caused by excessive exposure to ionizing radiation

radioactive decay   nuclear disintegration

radioactivity   radiation emitted when atomic nucleii decay

radiolarian   one-celled microscopic animal found only in saline waters

rain forest   tropical jungle

rain shadow   area on the downwind side of a mountain range that receives diminished rainfall

rating curve   the best fit line through a plot of stream discharge versus height of the stream surface above a datum

Rayleigh waves   earthquake waves that move along discontinuity surfaces within Earth

recharge area   the geographic area from which an aquifer receives its water

recycling   reusing the materials from a discarded product

relative humidity   the amount of moisture in the air compared to the amount it is able to hold at a particular temperature

rem   amount of ionizing radiation that has the same effect as a unit of X rays

renewable resources   an energy source that is replenished on a human time scale

reserves   the amount of a valuable material whose location is known and which can be recovered at a profit using existing technology

residual soil   soil formed from the underlying bedrock

resources   rocks and minerals that are valuable

reverse fault   fault in which the relative movement is upward on the hanging wall side

rhyolite   light-colored microcrystalline volcanic rock rich in silica

Richter scale   scale for earthquake magnitudes

rill   small trickling stream of water; first channel formed by water on a slope

rip current   narrow near-surface near-shore ocean current flowing normal to a shoreline through the breaker zone

rock cycle   continual change of Earth materials from igneous to sedimentary and metamorphic and back to igneous

rock   an aggregate of one or more minerals or naturally-occurring amorphous materials such as coal

rockfall   freely falling large piece of rock newly detached from a steep slope

rockslide   rapid movement of large rocks downslope in the absence of water

root wedging   prying apart of rocks by root growth in cracks

runoff   that part of precipitation that appears in surface streams

salinization   precipitation of harmful salts in an agricultural field

salt wedging   prying apart of rocks by crystallization of salt in cracks

saltation   mode of sediment transport in which the grains move in a series of short, intermittent jumps from the ground surface

sand   clastic fragment with diameter between 2 mm and 0.06 mm

sandstone   rock made of sand grains cemented together

sanitary landfill   disposal pit for refuse of various kinds which is designed to pose no danger to humans

schist   strongly foliated metamorphic rock with abundant micaceous minerals oriented parallel to layering

scrubbers   chemicals that remove sulfur dioxide from smokestack emissions

sea level   average elevation of the ocean

seamount   volcanic edifice on the ocean floor that has not yet reached the surface

seawall   steep-faced embankment along the high tide line on a shore that temporarily prevents wave erosion; may be natural but usually is artificial

seawater incursion   movement of seawater into a freshwater aquifer

sediment   fragmental or precipitated mineral matter formed at Earth's surface

sedimentary   pertaining to origin as a sediment

seismic method   method of petroleum exploration in which explosions send waves into Earth to be reflected back to the surface

seismicity   earthquake activity

seismogram   recording of an earthquake made by a seismograph

seismograph   instrument that detects, magnifies, and records earthquakes

seismologist   scientist who uses seismic methods to study Earth

septic system   system of disposing of solid waste by decomposing it in an underground tank

sewage sludge   sediment and organic matter remaining after secondary treatment of raw sewage

sewage system   system of disposing of waste above ground by decomposing it

shale   sedimentary rock formed by the lithification of mud or clay; usually shows prominent orientation of clay flakes as evidenced by easy splitting along closely spaced parallel planes

shear strength   internal resistance of a body to shear stress

shear stress   the component of stress that acts at an angle to the direction of compression of a rock

sheetflow   overland downslope movement of water as a thin continuous film over a relatively smooth flat surface

sheeting   physical weathering process resulting from removal of overlying rock to produce horizontal fractures in unlayered rock

shoreline   where the land meets the sea

sidestream smoke   cigarette smoke released between puffs

silicate   minerals whose essential elements are silicon and oxygen

silicosis   noncancerous lung damage caused by ingestion of large amounts of quartz dust

sinkhole   circular depression in a karst area, typically formed by roof collapse over a subterranean cavity

slag   frothy, glassy rock spewed from a smelter

slump   downslope movement of a coherent slab of muddy sediment along a basal rupture surface

smelter   ore processing facility in which the ore is heated to high temperatures and melted to permit recovery of the ore mineral and the valuable elements it contains

smog   ground-level air pollution caused by the reaction between sunlight and high concentrations of ozone, resulting from hydrocarbon emissions, largely from automobiles

soil creep   slow movement of clayey soil downslope under the influence of gravity

soil crumb   soil particles formed of moist organic and inorganic soil material

soil horizon   layer of soil that is distinguishable from layers above and below

soil profile   vertical section of soil showing soil horizons

soil stability   resistance of a soil to downslope movement

soil   surficial material capable of supporting plant life

solar cell   same as photovoltaic cell

solar panel   wafers of silicon or other material that collects solar energy and converts it to electricity

solifluction   slow viscous downslope movement of waterlogged soil, particularly in regions underlain by frozen ground

solution mining   ore recovery by passing chemical solutions through waste piles

sonogram   recording on graph paper of depth to the ocean floor

spa   a resort that has a mineral spring

spores   reproductive units of mold

spring (water)   location where groundwater flows naturally from a rock or sediment

stage   height of a stream above a datum

stalactite   conical or cylindrical deposit of mineral matter (usually calcite) that hangs from the ceiling of a cave

stalagmite   conical or cylindrical deposit of mineral matter (usually calcite) that projects upward from the floor of a cave

steam explosion   ejection of radioactive water from a nuclear reactor

storage capacity (of an aquifer)   the volume of pore space in an aquifer

strain   change in shape or volume of a body caused by stress

stratosphere   outer layer of the atmosphere, lying above the troposphere

stream discharge   amount of water passing a point per unit of time

stream gradient   decrease in stream elevation per unit of distance

stream load   total of solids and dissolved material carried by a stream

stress   force per unit area acting on a surface

strike-slip fault   fault along which movement is horizontal

strip mines   open-pit or surface mines

subsoil   B-horizon of a soil

sulfide   mineral characterized by linkage of sulfur with a metal

Superfund   pot of money from which the federal government tries to remediate polluted areas

surface mines   strip mines or open-pit mines

suspended load   sediment carried in suspension by moving water, normally silt and clay

swamp   an area in a humid climate where the water table is at the surface

S-waves   earthquake waves that travel along surfaces of discontinuity within Earth

swelling soil   a soil that absorbs water to increase its volume

tailings   portions of washed or milled ore too poor to be treated by smelting

talus   rock fragments at the base of a steep slope or cliff, formed by gravitational action without stream flow

tar sand   sand or sandstone containing viscous petroleum from which the volatile hydrocarbons have escaped

tectonics   large-scale movement and deformation of Earth's crust, such as plate movements

temperature inversion   atmospheric condition in which a layer of warm air lies above a layer of colder air

tephra   pyroclastic sediment

terracing   creating a series of large, flat surfaces on a steep slope to halt erosion

texture   size, shape, and arrangement of mineral grains in a rock

thermal conductivity   rate of heat transfer through a solid or liquid

thermal energy   energy resulting from atomic vibrations within a substance

thermal inversion   situation in which a layer of warmer air comes to exist above a layer of cooler air, blocking updrafts so that polluted air cannot escape upward and accumulates

thermal pollution   heating of a body of water by human activities to the point at which it harms life

thrust fault   a low-angle reverse fault

tidal bulge   rise in level of ocean water resulting from the combined pull of the moon and sun

tidal flat   nearshore area affected by tidal processes

tides   the diurnal advance and retreat of the shoreline resulting from the pull of the moon and sun

tillage   overturning of the soil in preparation for cultivation

topography   configuration of the land surface

topsoil   A-horizon of a soil

toxic   poisonous

trace ions   ions present in very small amounts

transpiration   evaporation of water from plant leaves

transuranic elements   chemical elements with numbers of protons greater than uranium

tributary   stream that supplies water to a larger stream

tropopause   imaginary surface in the atmosphere that separates the troposphere from the stratosphere

troposphere   layer of the atmosphere defined by the presence of thermal convection of air from ground level to about 50,000 feet; lies below the stratosphere

trunk stream   main stream

tsunami   gravitational sea wave caused by sudden disturbance of the ocean floor

turbulence   fluid motion characterized by disturbed and confused lines of stream flow in a channel

ultraviolet   wavelengths slightly shorter than those which produce violet light

unconfined aquifer   aquifer that extends to the ground surface

Uniform Building Code   set of guidelines established to avoid construction problems resulting from soil and slope variations

upwelling   rise of cold water from ocean depths

urbanization   growth of cities

vertisol   soil that contains abundant montmorillonite clay, so that the soil absorbs water and changes volume easily and repeatedly

viscosity   resistance to flow of a fluid; molasses has a greater viscosity than water

volatile   components of a liquid that evaporate easily

volcanic plume   particulate and gaseous material rising from a volcano

volcanologist   scientist who studies volcanoes

water cycle   *see* hydrologic cycle

water hardness   the amount of calcium and magnesium in water

water table   top of the saturated zone in unconfined ground water

wave refraction   bending of ocean water-wave crests because of differences in water depth as the wave trains approach the shore

wavelength   distance between successive wave crests in a series of harmonic waves

weathering   destructive process by which minerals and rocks formed below the surface adjust to chemical and physical conditions at the surface

weed   any undesirable plant, or, as one humorist put it, "a plant out of place."

wet landfill design   landfills designed to keep the solid waste wet so it will decompose faster

wetland   areas that are periodically inundated, such as coastal marshes, tidal swamps, and many low-lying inland areas

wind farms   a large array of windmills used to generate electric power

Index